Analysis, Design and Construction of Foundations

Analysis, Design and Construction of Foundations covers the key concepts in the analysis and design of foundation systems, balancing theory with engineering practice.

The book examines in depth the methods used for the analysis, design and construction of shallow foundations, deep foundations, excavation and lateral support systems, slope stability and stabilization and ground monitoring for proper site management. Some new and innovative foundation construction methods are also introduced. It is illustrated with case studies of failures and defects from actual construction projects.

This second edition is extensively revised and developed to include a new chapter on numerical methods in geotechnical engineering, as well as a large number of new construction drawings, project photos and construction method statements from existing projects to give the book a stronger professional application and connection to engineering practice. It also covers some new advanced theoretical concepts not covered in other texts, making it useful in both the theoretical and practical aspects. It is ideal for senior undergraduates and graduate students, academics and consulting geotechnical engineers.

Analysis, Design and Construction of Foundations

Second Edition

Yung Ming Cheng, Chi Wai Law and Leilei Liu

CRC Press
Taylor & Francis Group
Boca Raton London New York

CRC Press is an imprint of the
Taylor & Francis Group, an **informa** business

Front cover image: Yung Ming Cheng, Chi Wai Law and Leilei Liu

Second edition published 2024
by CRC Press
4 Park Square, Milton Park, Abingdon, Oxon, OX14 4RN

and by CRC Press
2385 NW Executive Center Drive, Suite 320, Boca Raton FL 33431

First edition published by CRC Press 2021

CRC Press is an imprint of Informa UK Limited

British Library Cataloguing-in-Publication Data
A catalogue record for this book is available from the British Library

Library of Congress Cataloging-in-Publication Data
Names: Cheng, Y. M., author. | Law, Chi Wai, author. | Liu, Leilei, author.
Title: Analysis, design and construction of foundations /
Yung Ming Cheng, Chi Wai Law and Leilei Liu.
Description: Second edition. |
Abingdon, Oxon ; Boca Raton, FL : CRC Press, 2024. |
Includes bibliographical references.
Identifiers: LCCN 2023050844 | ISBN 9781032650562 (hardback) |
ISBN 9781032650579 (paperback) | ISBN 9781032650586 (ebook)
Subjects: LCSH: Foundations.
Classification: LCC TA775 .C465 2024 |
DDC 624.1/5–dc23/eng/20240223
LC record available at https://lccn.loc.gov/2023050844

ISBN: 978-1-032-65056-2 (hbk)
ISBN: 978-1-032-65057-9 (pbk)
ISBN: 978-1-032-65058-6 (ebk)

DOI: 10.1201/9781032650586

Typeset in Sabon
by Newgen Publishing UK

Contents

Preface to second edition

After the publication of the first edition, the authors have received much support and many suggestions from engineers. Some minor errors were also found in the first edition of the book. In view of the comments and supplementary information, drawings and photographs from different construction projects and engineering companies, the authors have expanded on the first edition with the following changes:

(1) More photos from different projects are included;
(2) More drawings for different types of construction works are supplemented;
(3) More worked examples on the design and construction method statements from different projects are added'
(4) Some advanced materials are added for advanced readers;
(5) Cheng has delivered a short course recently to engineers in Hong Kong about the bugs, limitations and wrong practice in commonly used engineering computer programs, codes of practice and design practices. This short course is based on many reported problems from engineers in different companies as well as the research works by Cheng. Some of the materials in this short course are included in the second edition of this book.

Engineers will find that the figures and many materials are highly related to real construction projects, which will help them in the design and construction of different foundation works. The aim of this book remains unchanged, to bridge the gap between research and practical engineering.

This edition should be useful to senior undergraduates, post-graduates as well as engineers.

Preface to first edition

Foundation analysis and design is vital to most of the civil engineering construction works, as it affects the safety and serviceability of the construction. Foundation helps to distribute the different types of loads and movements to the ground. In many cases, different types of soil and rock and ground conditions may be encountered during construction, and in response to these constraints, different solutions are available for different conditions and country. The design and construction of foundations are further complicated by the time, cost, availability of construction materials and machines, labourers, environmental factors, social factors and others. Usually, there is also no unique solution for most of the projects, and different engineers may hold different views on the final solution of the project. A very good example is the selection of many smaller diameter bore piles or limited large diameter bore piles. This problem is faced by many engineers in Hong Kong and other countries. In the past, the adoption of many small diameter bore piles may be the more feasible solution, however, the use of limited very large diameter bore piles are currently favoured by some engineers (not all) in Hong Kong. Again, there are no strong reason for selecting limited very large diameter bore piles or many small diameter bore piles, and the final choice will depend heavily on the location, contractor and the availability of plants.

The authors have worked on soft ground tunneling, rock tunneling, immersed tube tunneling, different types of deep excavation and lateral supports systems, shallow and deep foundation, different types of ground treatment works, slope protection works, reinforced and steel works, construction management and safety works in many large scales construction in Hong Kong and other countries before joining the university. After joining the universities, the authors have worked on finite element/difference method, distinct element method, discontinuous deformation analysis and manifold method, meshless method, limit equilibrium/analysis, plasticity theory, innovative ground stabilization method, pile foundation problems, soil constitutive models and others. Based on the various research works, the authors have also developed many computer programs for the analysis and design of foundation works. The authors have noticed the great differences between the views of the academics and engineers towards the analysis and design of foundations as well as other engineering works. Both views are equally important towards the improvements in engineering. It is a pity that while many engineers are not aware of the various advances in both theory and practice in foundation engineering, many researchers are also not aware of the practical limitations of many modern theories and research works. There is a very strong need to bridge the gap between these two directions. In particular, Cheng has participated in the preparation of the new foundation code 2017 and the foundation handbook in

Hong Kong, and the gap between the latest research and practical engineering is clearly reflected during the various meetings about the new foundation code. Besides that, Cheng also viewed that some of engineers have over-emphasized the use of codes. Building/foundation codes are useful in many respects, and the use of codes usually provide reasonable design for engineering application. Nevertheless, there are also many cases where deep understanding of the basic theory, practical application of the theory and the site measured results and experience are also very important towards a better engineering solution. That means, design code provides the guideline for good practice in engineering, but a good understanding of the basic theory is also required for the flexible use and deviation from the design code which may be necessary in some cases. Similar to the design codes, most of the engineers rely heavily on the use of engineering computer programs for the analysis and design of various foundation engineering problems. There is however a major drawback in the current engineering practice in that most of the engineers are not familiar with the basics of the numerical methods, the methods of implementations and the limitations of the numerical methods/programs. In fact, to a certain extent, the methods of implementations and the limitations of the numerical methods are related. In many internal studies using different commercial numerical programs, the authors sometimes found noticeable or even completely different results with different programs or the same program with different default setting for a given problem, and this situation is not *uncommon*. For a problem with unknown solution, how an engineer assess the acceptability of the computer results is a difficult issue that needs serious attention. In several technical meetings in the Hong King Institution of Engineers, the authors have discussed with some engineers about the appreciation of the limitations of the daily-used engineering programs. If two computer programs can produce significantly different results, how an engineer determine the acceptability of the results actually require deeper knowledge about the basics of the numerical methods and implementations. Interestingly, the authors like to ask the students a question "Different answers can be obtained from different commercial programs. Which results should be accepted, and why should those results be accepted?". In general, the authors challenge the students (undergraduate and graduate students) every year for this question, and virtually this question is never answered properly. The problems in the assessment of the numerical results will also be discussed in this book, which is seldom addressed in other books or research papers.

The current book is based on the teaching materials and practical experience of the authors, and the authors hope to bring a balance between theory and practical engineering. This book is not strongly tied with any design code (Euro code, Hong Kong code, China code or others), but the uses of different design codes will also be discussed if possible.

About the authors

Yung Ming Cheng has extensive engineering experience in soft, rock and immersed tunnels (MTR and EHC), steel, reinforced and prestressed concrete bridges design and construction, ground treatment, deep excavation in poor soils with different types of retaining wall types, slopes and piles engineering before joining the Hong Kong Polytechnic University. His research interests include different numerical methods, including: finite element, distinct element and meshless methods, plasticity theory, slope stability, pile foundation, field tests, IoT monitoring in construction. Prof. Cheng has published about 200 papers, 11 books based on the research works, and has served in various engineering committees and government departments as an advisor and as an editor in several international journals. Prof. Cheng was rated in the world's top 2% of scientists by Stanford University in 2020 to 2023, top 59 national ranking in civil engineering discipline in 2022 and top 30 national ranking in geotechnical engineering discipline in China, and has also received several best paper awards from engineering journals.

Chi Wai Law is a structural engineer who has ample structural design experience in high-rise buildings, deep basements and various types of foundations throughout his over 40 years' working experience. In addition to practical design work, he has also carried out research work on structural element and foundation design and published a number of academic papers, design manuals and handbooks. He also has teaching experience in various higher diploma and undergraduate courses of civil engineering.

Leilei Liu is an associate professor in Geological Engineering at the Central South University, China. He received his PhD degree in Geotechnical Engineering from The Hong Kong Polytechnic University in 2018. His research mainly focuses on the geotechnical uncertainty analysis and risk assessment and the associated applications in practical engineering problems. He is currently the principal investigator of several science and technology projects. He has authored and co-authored more than 20 scientific papers in international and national refereed journals.

Acknowledgements

The authors would like to acknowledge support from the National Natural Science Foundation of China (Grant No. 51778313), the Cooperative Innovation Center of Engineering Construction and Safety in Shangdong Blue Economic Zone, Natural Science Foundation of China (No. 41902291).

Chapter 1

Introduction to geotechnical analysis, site investigation and in situ tests

1.1 INTRODUCTION TO GEOTECHNICAL ANALYSIS AND DESIGN

Before introducing the details about site investigation, the authors would like to discuss the importance and special features of geotechnical engineering as compared with other disciplines, such as structural engineering, transportation engineering or others. Some of the special features (but not all) about geotechnical engineering are:

1. Materials are as left by nature, and it is difficult or impossible to control the properties or distributions of materials. The actual constitutive behaviors of geomaterials are complicated, nonlinear, path/stress dependent, and the existence of discontinuity, size effect and mixtures of geomaterials and water (or other fluid) have created further difficulties in providing a versatile constitutive model for their analysis.
2. Most of the problems or structures are actually three-dimensional (3D) in nature, which has created difficulties in carrying out sufficient site investigation works, analysis and design of the construction works. Up to now, most engineers still adopt one- or two-dimensional analysis in geotechnical design, and the use of a high factor of safety is used to cover any deficiency in the analysis. In structural engineering, most of the members are either one- or two-dimensional in nature along their natural direction (beam, column, truss, plate). That means that the formation of the stiffness matrix of the members in structural engineering is mostly one- or two-dimensional analysis, and the three-dimensional consideration comes from the assemblage of the global stiffness matrix.
3. Due to the three-dimensional nature of most geotechnical problems, the nature of the geomaterials and the various construction activities involved, it is very difficult or impossible to solve a geotechnical problem precisely, even for simple problems.
4. Analytical solutions are available for some simple cases in geotechnical engineering. Usually, these equations (particularly those from elasticity solutions) are very long and tedious even for simple problems, if available. To be honest, the authors never use calculators with these long equations. Although many design tables, charts and figures have been prepared to aid engineers in carrying out the calculations, their uses are very limited in practical purposes. For many real construction works, the geometry, loading conditions and ground conditions deviate greatly from the basic assumptions for these design tables, charts and figures, and the applicability of these design aids is questionable. However, the authors have noted that some engineers still adopt these design aids with various simplifications in their own problems, with the

DOI: 10.1201/9781032650586-1

adoption of a high factor of safety as the sacrifice. As the assessors for professional engineers in Hong Kong, the authors have asked engineers about the acceptability of such practice, and most engineers fail to answer this with sufficient justifications. The authors do not object to the use of simplifying assumptions in using various equations or design aids in the analysis and design, but stress that engineers should have a good concept of these aids before using them.

5. There are many uncertainties behind the geological conditions, soil/rock properties, ground water table and pore water pressure, loading, construction processes and workmanship in geotechnical projects. A very good understanding of these factors is vital to the good analysis and design of a geotechnical project.

In view of the nature of geotechnical engineering, there are four major principles for the analysis and design of geotechnical work:

1. Rule of thumb: reliance on past experience. For some temporary works or even permanent works, some engineers work out the solutions simply based on past experience without detailed assessment. This approach can be useful, particularly for some emergency works, but will not be applicable to new structures or situations.
2. Statistics: many engineers assess the field test results, settlement, bearing capacity, etc. based on the statistics from previous projects, which can be viewed as an extension of the rule of thumb with supporting statistics. This approach is very useful for many geotechnical works, and the authors suggest engineers consider the statistics from previous works in the feasibility study as well as the detailed analysis and design works. Currently, such statistical analysis also provides some design parameters for the detailed analysis in (3) and (4) below.
3. Use of analytical solutions: for some simple problems. Many classical methods of analysis rely on different assumptions which are the approximations of the actual situation. Design graphs and tables are also commonly used to replace the use of long analytical equations. This approach is limited to simple works and is not suitable for complicated conditions.
4. Computational method: rely on the use of computer programs with less assumptions. With the power of many modern commercial programs, many geotechnical construction works can be modeled easily. In many cases, engineers will still slightly simplify the numerical model so as to ease the mesh generation and the modeling of the problems. The use of computer programs needs knowledge and judgment, and engineers should remember that rubbish input will provide rubbish output. More importantly, there are various problems for which engineers should be aware of in using various computer programs.

In addition, the adoption of field monitoring for some complicated or sensitive geotechnical construction works is highly recommended, due to the highly uncertain nature of some projects. This will provide an additional safety factor as well as data for statistical analysis and research works. Currently, the authors are working on the use of the Internet of Things (IoT) to monitor slopes, retaining walls, dangerous buildings and debris flow barriers, which will be further elaborated later.

For illustration, the authors would like to describe a jack-in tunneling project in soft clay. For most geotechnical problems, particularly those related to real-life problems, analytical solutions are usually not available. The authors have carried out many research works and large-scale practical projects, and, in general, most of these works are complicated in geometry, applications of loadings, construction sequences, material behaviors, ground water conditions as well as other factors. As a good illustration, the construction of the Airport Link project in Brisbane, Australia, is a good example (Cheng et al. 2019). This project is located beneath the railway embankment of the North Coast Railway line adjacent to Kalinga Park, and the site comprises a thick layer of soft clay. The Airport Link, which is one of the most complex roads and tunnel engineering feats in Queensland's history, will be the first major motorway linking Brisbane city to the northern suburbs and airport precinct. The Link is a 6.7-km toll road, mainly underground, connecting the Clem 7 Tunnel, Inner City Bypass and local road network at Bowen Hills, to the northern arterials of Gympie Road and Stafford Road at Kedron, Sandgate Road and the East West Arterial leading to the airport. At one of the project sites, the tunnel section under the QR railway embankment at Toombul was constructed using the box jacking technique. The significant size of the launch box required 85,000 m^3 of spoil to be excavated under the railway embankment. Headwalls, canopy tubes and sidewall nails were constructed to retain the railway embankment for the excavation of the jacking shafts. The challenging ground conditions and requirements for the present project require the combination of innovative ground support, construction methods and detailed and realistic analysis for the proper execution of the works. In this project, the site is mostly comprised of soft clays which are susceptible to ground settlement problems during construction, and a typical section is shown in Figure 1.1. The standard penetration test (SPT) value for the soft clay is less than 10, whereas the CPT friction ratio for soft clay ranges between 2% and 4% with a mean pore pressure of approximately 0.12 MPa (see also Table 1.1). The SPT value for the firm clay is approximately 20, whereas the friction ratio for firm clay ranges between 4–8% with a mean pore pressure of approximately 0.38 MPa. The railway has to be maintained in operation during the whole construction to ensure transportation, and the settlement of the soft clay must be maintained at a low level with minimal disturbance to the railway track. This is technically a very difficult problem, and the original construction proposal was to inject a large amount of grout into the ground to stabilize it prior to excavation. However, the cost of the original scheme was extremely high so that a more economical alternative was considered. Ground improvement works underneath the QR railway embankment were hence required for the stability consideration during the box jacking stages. A trapezoidal jet grout block constructed immediately behind the headwall was used as a gravity-type retaining wall to reduce the earth pressures on the piled headwall. A smaller jet grout block was provided at the northwest of the final jacked box location and was used as an anchorage to the northern sidewall nails. A low-strength grout wall was installed west of the railway to provide a water cut-off for the TBM launch box. The grout wall was also used in the jacking scheme design to provide adequate anchorage to the geonails at the receiving pit side, eliminating an approximately 10 m length of nail with significant time and cost savings. The resulting "nail-anchored" western grout wall could then be used to maintain slope stability, enabling initial excavations in the cut and cover receiving pit to commence early.

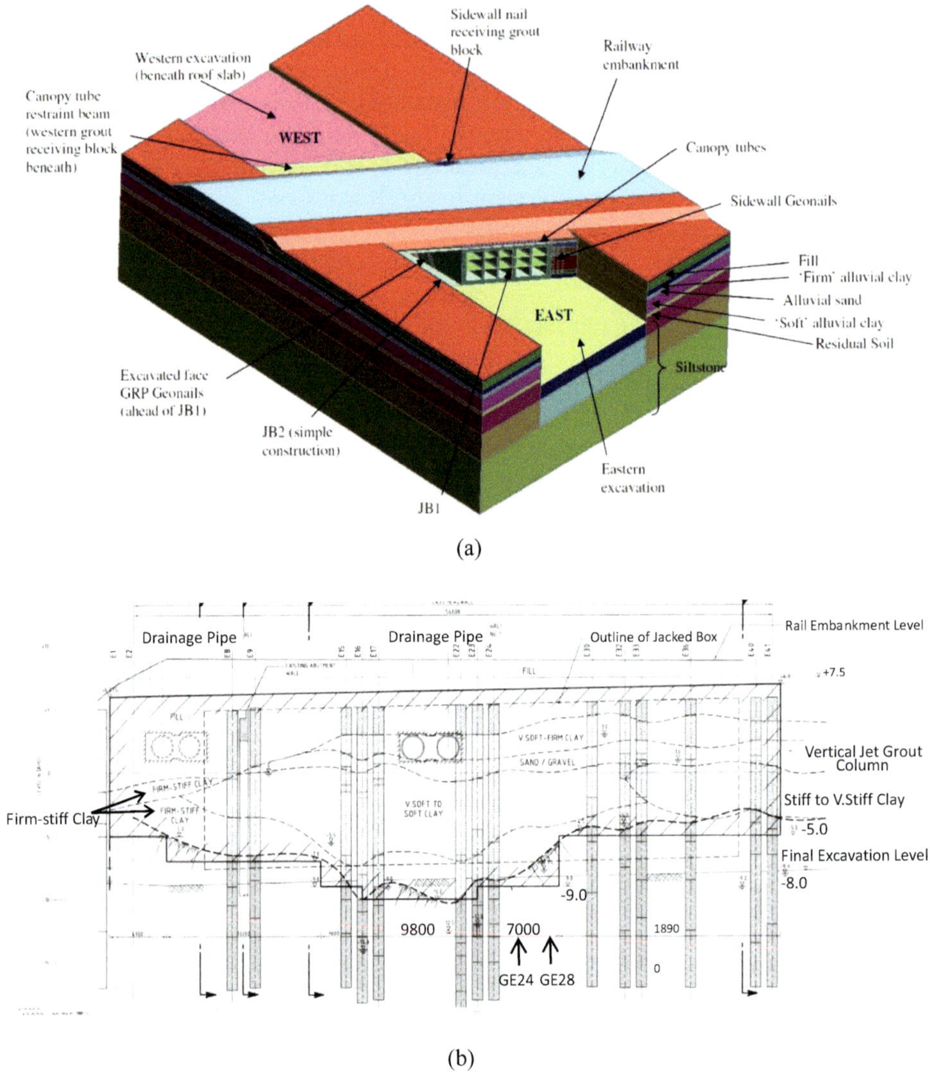

Figure 1.1 (a) Geological condition for the tunnel project in soft clay; (b) a typical section of the tunnel work.

Table 1.1 Average properties of ground soil (Young's modulus determined from dilatometer, vane shear and CPT tests)

Soil	*Undrained shear strength (kPa)*	*Young's modulus (MPa)*	*Water content (%)*	*Plasticity index*
Soft clay	20	6	57	25
Firm clay	37	20	46	45

In order to optimize the ground improvement design, a combined fracture grouting and GFRP soil nails ground improvement scheme was proposed by the authors as the alternative solution (Cheng et al. 2013), and the cost of the alternative scheme was critically reduced to 50% of the original scheme. In the past, fracture grouting was mainly adopted for compensation grouting, and the combined use of fracture grouting and a composite GFRP soil nails scheme for soil improvement was adopted in this project as pioneering work. The present project received the Fleming Award in 2011 and Ground Engineering Award in Technical Excellent in 2012 in Australia for satisfactory performance under such difficult conditions.

The proposed scheme was to utilize approximately 1.0 m grid size geonails and face shield to stabilize the soft clay faced slope in advance of the box jacking operation. With the ground improvement from the geonails and protection from the canopy tubes, a 60-degree cut slope was formed 0.2 m in advance of the box, with the box advanced gradually behind the excavation face. The basic principle of the combined geonail and face shield scheme was to obtain some face supports from a mining shield and, therefore, to reduce the number of geonails required for face stability. Sufficient nails should be installed to provide suitable size shield working compartments, which is considered to be approximately 3–4 m^2.

For the present project, the authors adopted three-dimensional analysis using the program Flac3D, and the number of elements used in the analysis exceeded one million (Cheng et al. 2019) which was generated by the use of Patran. The construction of the jacked box was modeled in a realistic manner, and the computer modeling sequences are shown in Table 1.2. For the analysis of the tunneling process, a three-dimensional analysis was performed as shown in Figure 1.2. The partial factors of safety adopted in the analysis were based on the Australian code AS5100:3. The soil model for different types of clay adopted was the elasto-plastic Cam-Clay model (with a partial factor of safety 0.65), while the Mohr-Coulomb model was adopted for sand, fill and rock. Canopy tubes and headwall piles were modeled as elasto-plastic material (with a partial factor of safety 0.6) with a maximum prescribed bending capacity. The geonail was modeled in a way similar to that by Wei and Cheng (2009) (partial factor of safety 0.65 to bond strength and 0.8 to tensile strength). The numerical model was set up based on the jacking drawing series. The computer model needed to consider the complicated ground conditions, the installation and removal of the geonails (due to cutting of nails during excavation), the excavation with a slope surface at the tunnel face, the formation of yield zones (with revision of the stabilization measures and re-analysis), and the installation of structural supports. The soil nails were modeled in a way similar to that by Wei and Cheng (2009).

The numerical models were established based on the arrangement of canopy tubes, side-wall nails and the front face geo-nails. Compartments were formed by temporary concrete blade walls at about 3.5 m c/c and two intermediate working platforms with hydraulic steel tables in front. Removable breasting plates were fixed to the steel tables, which could be pushed against the slope face to provide positive face pressure. The compartment sizes were determined mainly from the practical construction point of view, which was about 3.5 m × 3.5 m for the present project. During the box jacking excavation, arching of the clay within each compartment was required to maintain stability of the clay face within the compartment, therefore, bearing and frictional supporting force were induced to maintain

Table 1.2 Modeling of the construction process

S. phase no.	Description	Remarks
1	Initial condition (soil only)	Using "gravity load" approach
2	Initial condition (railway surcharge)	Activate railway surcharge
3	Construct piling platform	Filling to piling platform levels
4	Apply working load	Activate 10 kPa working load
5	Install trapezoidal grout mass and grout blocks at receiving side	Change soil to grout properties
6	Install headwall piles and excavate to existing ground level	Activate headwall piles and excavate to approximately +5 m AHD
7	Construct permanent roof slab of CC421 and restraint beams at the west side	Activate the spring support at west side
8	Install canopy tube from east side and connect with the restrain beam at west side	Activate canopy tube elements
9	Install geonails to stabilize west side grout block and commence with excavation in CC421	Activate the soil nail element (cable element) at west side
10	Excavated to –4.5 m ADH in CC422 adjacent to south-west grout block	Excavate to –4.5 m ADH
11	Install fan-shaped sidewall nails from western embankment slope	Activate fan-shaped pile elements at south side
12	Install anchors and walers to CC422 side walls as excavation proceeds	
13	Fill canopy tubes and connect them to the restrain beam	Change properties of canopy tube
14	West: Dig to canopy soffit. East: Excavate to +4.8 m AHD	Excavation initiated at east side
15	Connect canopy to east headwall piles and excavate at east to +3.3 m AHD	Canopy tube is activated but NOT connected to the headwall pile at this stage
16	Install first geonail and then dewater and excavate to +1.8 m AHD	The canopy tube is connected to the headwall pile at this stage
17	Install geonail to +2.8, dewater and excavate to –1.2 m	Activate geonails within +5.3 m AHD and +2.8 m AHD and change the properties of the soil layer to the improved soil mass
18	Install geonail down to –0.2 m, dewater and excavate to –4.2 m	Activate geonails within +2.8 m AHD and +0.2 m AHD and change the properties of the soil layer to the improved soil mass
19	Install geonail down to –3.2 m, dewater and excavate to –6.0 m	Activate geonails within +0.2 m AHD and –3.2 m AHD and change the properties of the soil layer to the improved soil mass
20	Install geonail to –5.2 m, dig and dewater to –8.2 m AHD	Activate geonails within –3.2 m AHD and –5.2 m AHD and change the properties of the soil layer to the improved soil mass
21	Install geonail to –7.17 m (1 m above FEL), dig and dewater to –8.7 m (500 mm over-excavation)	Activate geonails within –5.2 m AHD and –8.7 m AHD and change the properties of the soil layer to the improved soil mass. This stage is for ultimate checking, not for movement prediction
22	Cast jacking slab	

Table 1.2 (*Cont.*)

S. phase no.	*Description*	*Remarks*
23	JB2 jacked in 1/4	Box modeled with slope crest at about ¼ of the jacked box alignment, carry out the over excavation south end
24	JB2 jacked in 1/2	Box modeled with slope crest at about ½ of the jacked box alignment, carry out the over excavation middle of the jacking zone
25	JB1 jacked in 3/4	Box modelled with slope crest at about ¾ of the jacked box alignment, carry out the over excavation middle of the jacking zone
26	JB1 jacked to the end	Box modeled with slope crest at about the end of the jacked box alignment carry out the over-excavation north side of the jacking zone
23	JB1 jacked in 1/4	Box modeled with slope crest at about ¼ of the jacked box alignment. Excavate maximum 200 mm in advance of cutting shield and advance box in stages until in final position
24	JB1 jacked in 1/2	Box modeled with slope crest at about ½ of the jacked box alignment. Excavate maximum 200 mm in advance of cutting shield and advance box in stages until in final position
25	JB1 jacked in 3/4	Box modeled with slope crest at about ¾ of the jacked box alignment. Excavate maximum 200 mm in advance of cutting shield and advance box in stages until in final position
26	JB1 jacked to the end	Box modeled with slope crest at about at the end of the jacked box alignment. Excavate maximum 200 mm in advance of cutting shield and advance box in stages until in final position

the face stability. Based on this numerical model, the two-way arching effect of the soils between compartment walls could be modeled.

In order to obtain the individual effect of bulk excavation and jacking operation on the canopy tube and railway, the displacements were reset in the analyses before the jacking operation. During simulation of the jacking process, the displacements were reset in each excavation step to obtain the displacement change in each of the jacking advancements, and the procedures for the numerical modeling are given in Table 1.2.

The whole numerical simulation of each box jacking operation was divided into four stages in each box jacking operation. The shield embedment and 200 mm face excavation were simulated during the box jacking operation. When the box hits the culverts or timber piles, the obstruction should be removed before the advancement of the shield. A 1700 mm overcut should be allowed in the excavation process. The numerical analyses are divided into two jacking processes (JB1 and JB2). Each jacking process is subdivided into five stages. Stages 1–5 represent the JB2 jacking process and stages 6–10 represent the JB1 jacking process. The five stages' excavations are located at 0 m, 12.5 m, 25 m, 37 m away from the headwall and 1 m away from the west grout block. Each stage is subdivided into five steps. As the area required for over-excavation in JB1 is much less than that in JB2, the effect of over-excavation at JB1 will not be simulated. As the rapid loading

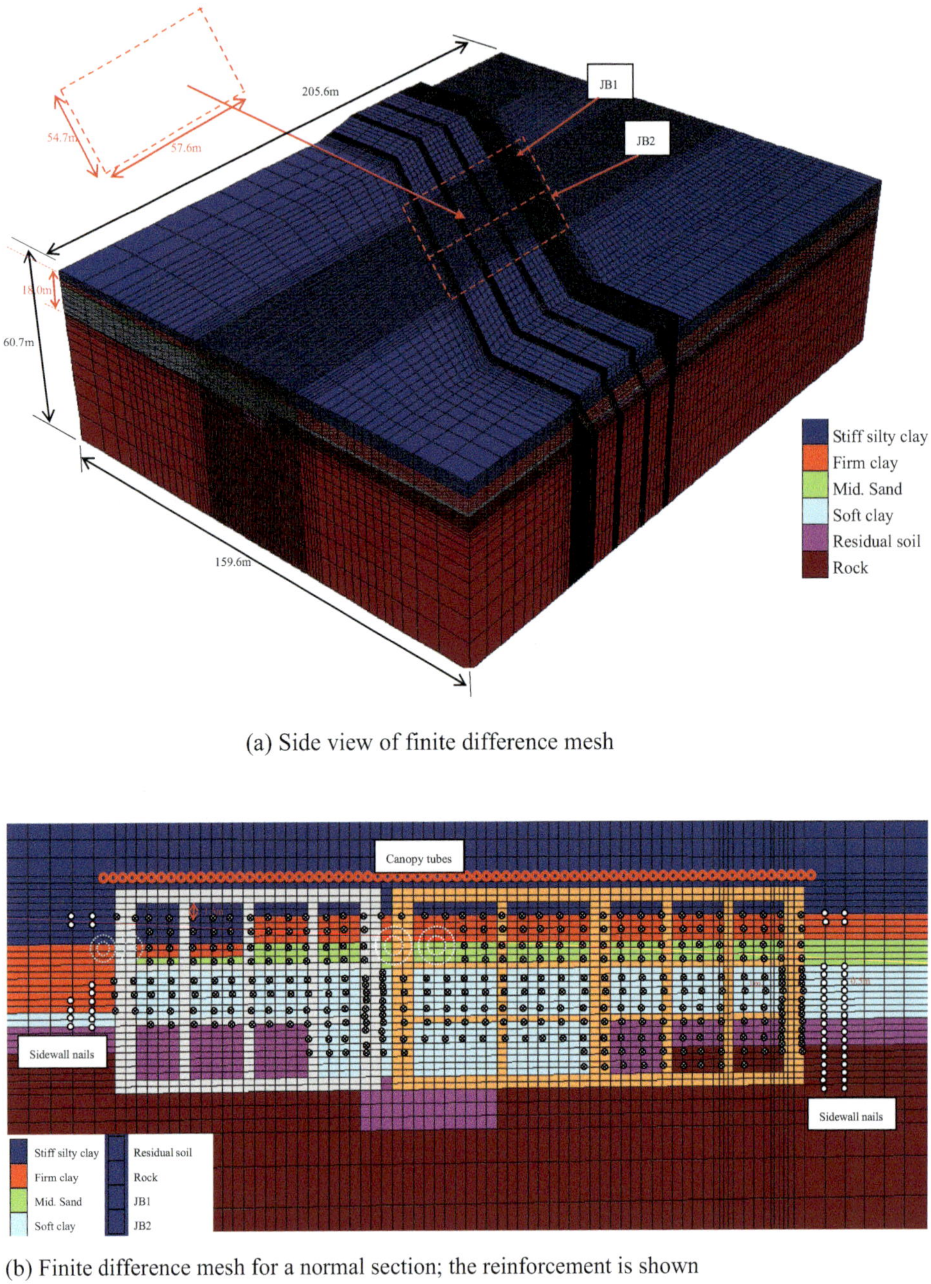

Figure 1.2 Three-dimensional finite difference mesh for jacked tunnel construction in soft clay.

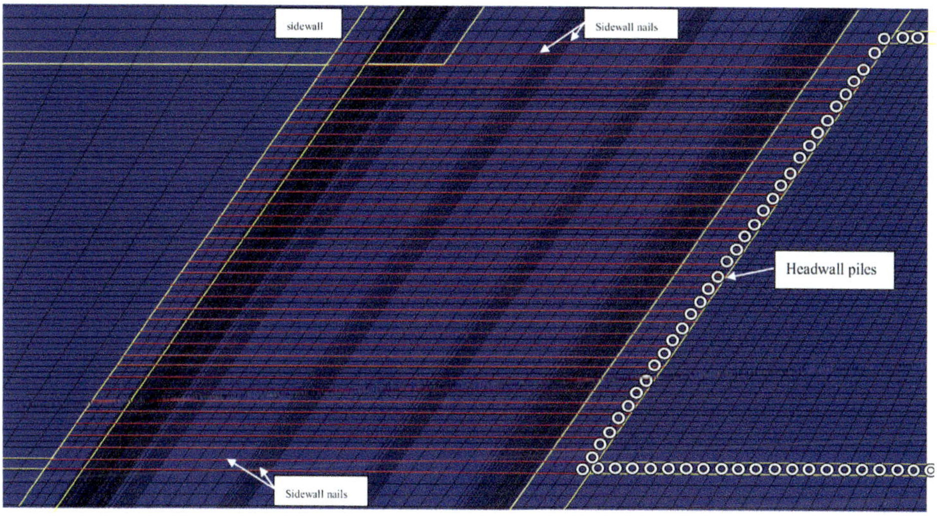

(c) Finite difference mesh; the headwall piles and side nails are shown

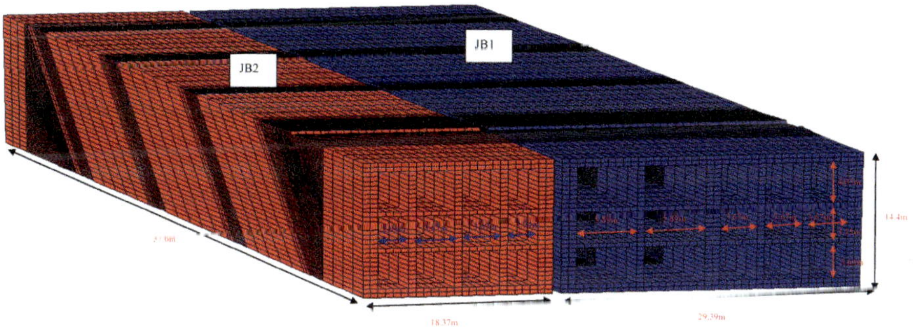

(d) Finite difference mesh during construction

Figure 1.2 (Continued)

and excavation sequence is demonstrated in the jacking processes, water seepage into the jacking is considered to be insignificant to the design. Therefore, undrained analyses are more suitable to reflect the behavior during the jacking operation.

Based on the three-dimensional finite difference analysis that involved approximately 1 million elements (Figure 1.2) to model the entire construction sequences, including the stresses, ground settlement, loadings in canopy tubes settlement derived from the canopy tube installation, the bulk excavation in the shaft and the ground improvement study (i.e., Geonail). Based on the results from numerical modeling, the stabilization measures are revised several times in order to maintain a balance between cost and stability of construction. Since there are vast amounts of computational results from the numerical analysis, only some selected results are shown in Figure 1.3 for illustration.

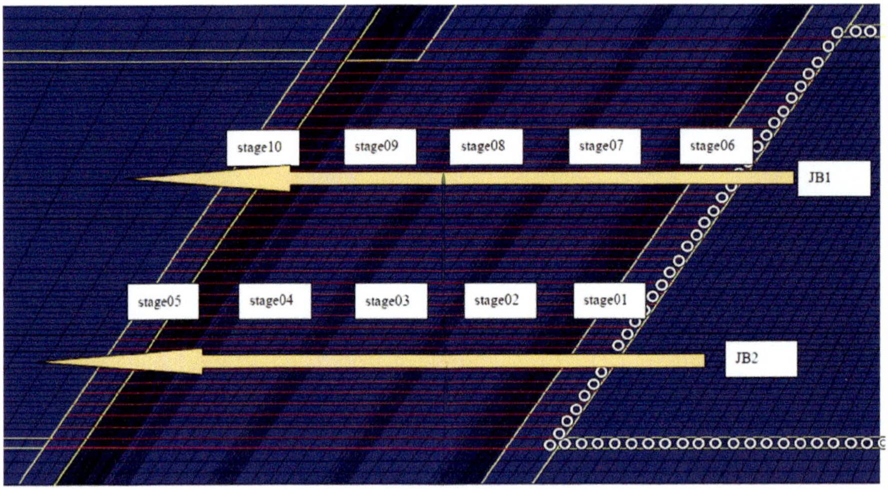

(e) Box jacking sequence in numerical modelling

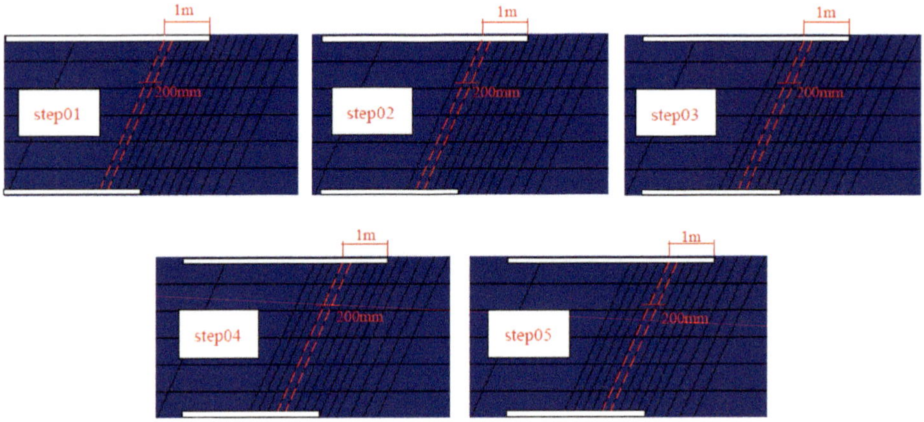

(f) Numerical simulations steps in each jacking stage

Figure 1.2 (Continued)

From Table 1.2, and Figures 1.2 and 1.3, it is clear that the analysis of the present problem must rely on the use of numerical methods. No analytical solution can provide the stress and displacement under all construction stages. In fact, most practical problems that the authors have encountered require the use of numerical methods and computer programs. Most engineers have adopted various engineering programs without questioning the daily works. In particular, as the assessor for the professional engineers

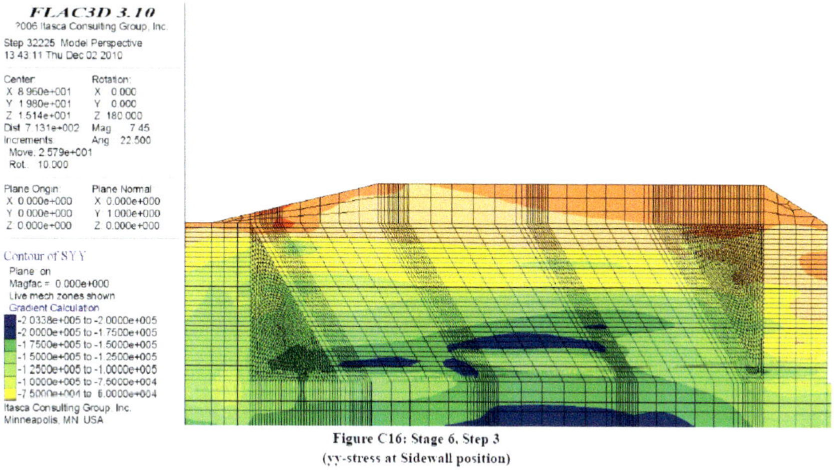

(a) σ_{yy} at sidewall position for stage 6, step 3

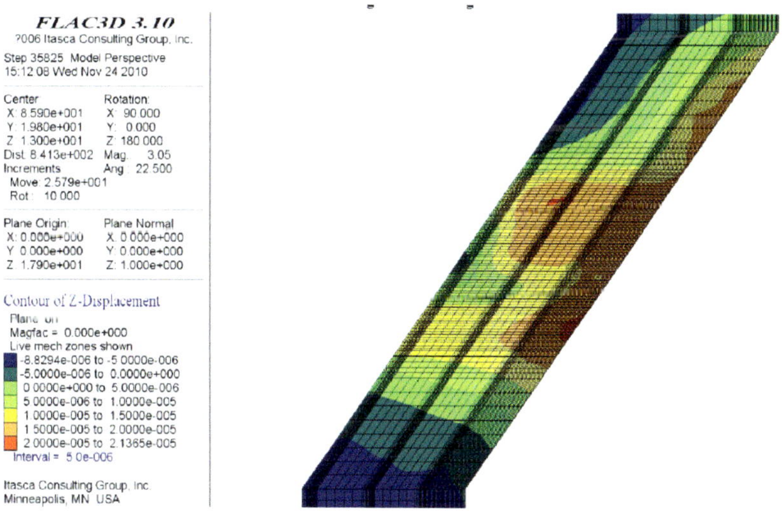

(b) Ground surface settlement at stage 10

Figure 1.3 Some results of analysis for the jacked tunnel.

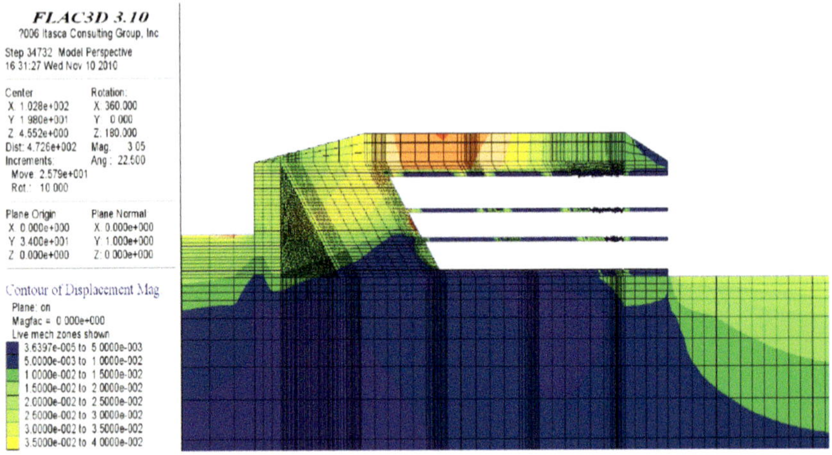

(c) Displacement contour at stage 8

Figure 1.3 (Continued)

examination for the Hong Kong Institution of Engineers, the authors have found that many engineers lack a basic understanding of the theory and limitations behind many engineering programs. It appears that some engineers are aware that unreasonable results may result from the numerical analyses, and some tricks have been developed to overcome the surprising results (without any theoretical basis). However, most of the problems from computer programs are simply neglected, as most engineers lack the ability to assess the acceptability of numerical results from different computer programs.

The present problem also reveals an important issue in geotechnical analysis and design. The design soil parameters are determined by considering laboratory tests and field tests (CPT, VST, dilatometer). For interpretation of the field test results before and after ground treatment, the soil parameters are obtained by the use of statistical interpretation. These design parameters are then fed into a sophisticated computer program for analysis. In this respect, no matter how advanced the theory in the soil constitutive model or the numerical modeling as adopted by the computer program is, the accuracy of the analysis is heavily limited by the accuracy of the various design parameters.

1.2 PROBLEMS IN COMPUTATIONAL ANALYSIS

Currently, many engineers adopt the use of different commercial programs for the analysis and design of geotechnical works without adequate background on the basic theory or numerical implementation. Although this situation appear to be unavoidable, there are also cases where adequate knowledge and judgment are required for the interpretation of computer output. Most engineering programs are based on finite element/difference methods, distinct element methods, meshless methods, limit equilibrium/analysis methods, and solutions of different types of governing differential equations (elasticity, plasticity, visco-plastic, dynamic, impact). Most engineers are unfamiliar with the basics of the numerical methods, the methods of implementations and the limitations of the numerical methods/programs. In fact, to a certain extent, the methods of implementation and the limitations of numerical methods are related. In many internal studies using different commercial numerical programs, the authors sometimes found noticeable or even completely different results with different programs or the same program with different default settings for a given problem, and this situation is not uncommon. For a problem with an unknown solution, how an engineer assesses the acceptability of the computer results is a difficult issue that needs serious attention. In several technical meetings in the Hong King Institution of Engineers, the authors have discussed with engineers the appreciation of the limitations of daily-used engineering programs. If two computer programs can produce significantly different results, how an engineer determines the acceptability of the results requires deeper knowledge about the basics of the numerical methods and implementations. Interestingly, the authors like to ask students a question: "Different answers can be obtained from different commercial programs. Which results should be accepted, and why should those results be accepted?" In general, the authors challenge their students (undergraduate and graduate students) every year with this question, and this question is almost never answered accurately.

The authors have carried out various internal studies on the acceptability of some major commercial geotechnical programs, and many surprising results have been found. In addition, the authors have also received many problem cases from engineers, and the assessment of these computer analysis results requires adequate fundamental knowledge in engineering as well as some knowledge of the procedures in the numerical implementation. In general, it is not easy to assess these problem cases. Some basic principles in engineering can however help to assess the acceptability of the numerical analysis. For example, the presence of a jump in the shear force for a raft foundation, as given in Figures 1.7 and 1.8, is considered to be unacceptable (even though the results come a world-famous program), as there is no point load or support at that jump location. The examples in this chapter illustrate how the authors assess the problems of different commercial programs, and the underlying principles include the use of standard problems with known solutions, use of basic engineering principles, and the adoption of different programs or even methods. It is also true that many numerical problems cannot be assessed easily, and there is also a strong need to investigate the reasons behind these numerical problems.

An interesting example is an engineering report submitted by an engineer for the professional examination in Hong Kong. Part of the design involve a beam simply supported at two ends with several point loads and distributed load. The engineer adopted a computer program instead of normal hand calculation in the analysis and design of such a simple problem. A very small but non-zero moment is obtained at the two ends of the beam, which contradicts the basic concept that the moments must be zero at simply supported ends. The authors regret that the engineer who is well experienced with the use of computer software for engineering analysis fails to explain the reason for such a phenomenon, even though this problem is very minor in nature. Another interesting case is the slope stability problem, as shown in Figure 1.4. An engineer adopted a commercial program for the analysis of the slope using a Janbu simplified method. A safety factor of 0.7 is given by the commercial program which is a surprise to the engineer. The engineer then changes the default initial factor of safety from 1.0 to 1.4, and a converged safety factor of 1.4 is obtained. The problem was later studied by the authors, and it was found that the first answer would give unreasonable internal forces which are not checked by the slope stability program (actually, most commercial programs do not check for this important issue). The authors had the opportunity to talk with that slope stability program developer about 10 years ago and mentioned such a phenomenon to them. To the author's disappointment, the software developer replied that the engineer should judge and accept the results instead of relying on the software developer to assess the acceptability of the results, while the acceptability of the results can actually be checked easily by the consistency and acceptability of the internal forces (a simple task). The authors were also very disappointed that many engineers and some researchers (mostly research students) simply believe the results from computer programs without adequate assessment. Although it is not easy to assess the acceptability of the computer results, the results from any computer program should be accepted with care. The authors have developed various engineering programs (include some commercial programs) for education and research purposes, and find that such experience will enhance the understanding of the various problems in numerical modeling, implementation and computations. Interestingly, the authors also like to ask students to do some small projects with different commercial programs on different types of problems. Sometimes, noticeable or even completely different results are obtained from different programs using the same input parameters. The authors also like to ask students a question: "Which set of results should be accepted, and why?" In general, the authors challenge the students (undergraduate and graduate students) every year with such

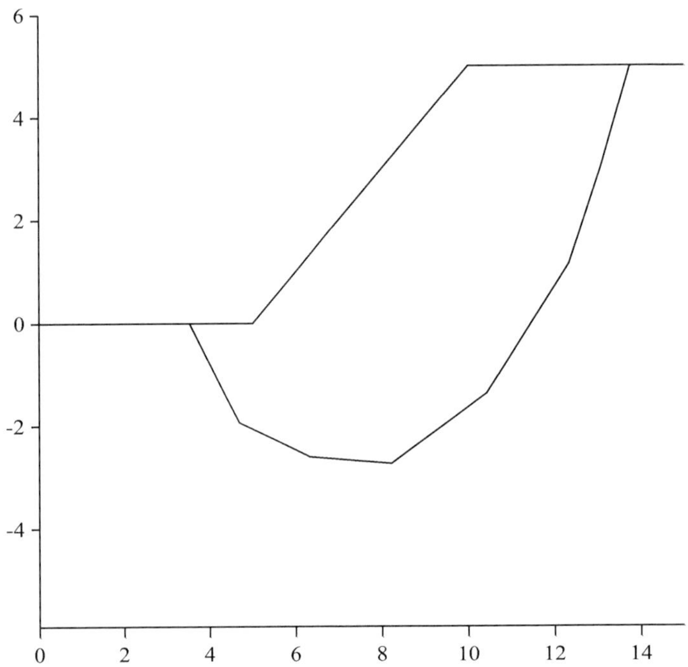

Figure 1.4 A trial slip surface with two different safety factors arising from the same commercial program using Janbu's simplified method.

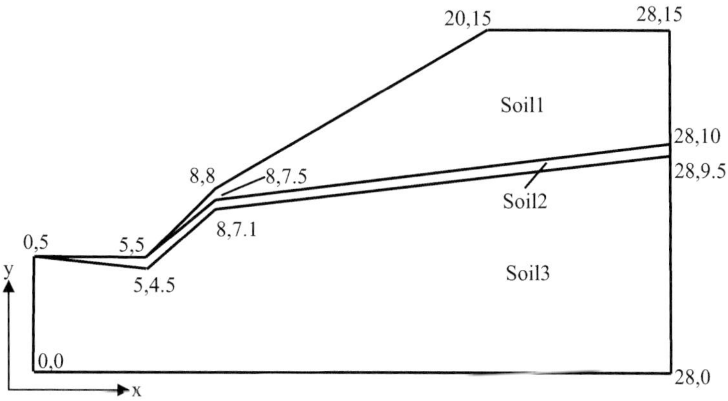

Figure 1.5 Strange results from SRM analysis of a slope with a soft band.

Table 1.3 Soil properties for Figure 1.5

Soil name	*Cohesion (kPa)*	*Friction angle (degree)*	*Density (kN/m³)*	*Elastic modulus (MPa)*	*Poisson ratio*
Soil 1	20	35	19	14	0.3
Soil 2	0	25	19	14	0.3
Soil 3	10	35	19	14	0.3

a question, and this question is almost never answered properly. In the following, some problems as revealed by small projects using some commercial programs are discussed.

Another interesting problem using the finite element strength reduction method (SRM) is shown in Figure 1.5, while the soil properties are shown in Table 1.3. This problem is considered by four different programs using three different domain sizes by SRM (Cheng et al. 2007). The critical solution using the Spencer method is shown in Figure 1.6 which looks reasonable, and the result is not affected by the domain size. On the other hand, the results by SRM as shown in Table 1.4 are disappointing, as it is not easy to conclude which result is reasonable. For some of the programs in Table 1.4, it is also interesting to find that different versions of the same program can give different analysis results. For engineers or researchers, what can they do with such results? During teaching, the authors always emphasize to students not to believe blindly in any computer program without prior knowledge and judgment. Besides these cases, the authors have also encountered different problems in programs for excavation and lateral analysis, pile driving signal analysis, and plate and shell analysis.

Another interesting problem case is given by engineers from the Housing Department of Hong Kong (Cheng and Law 2008). With reference to the raft foundation, as shown in Figure 1.7, due to the transfer of loading from the use of the super-structure structural analysis program, the mesh design for the raft foundation becomes highly irregular with a combination of three- and four-node thick plate elements. From the use of a world-famous

Tables 1.4 FOS by SRM from different programs when c' for soft band is 0. The values in each cell are based on SRM1 (zero dilation) and SRM2 (dilation angle equal to friction angle) respectively (min. FOS = 0.927 from Spencer analysis)

Program/FOS	*12 m domain*	*20 m domain*	*28 m domain*
Flac3D	1.03/1.03	1.30/1.28	1.64/1.61
Phase	0.77/0.85	0.84/1.06	0.87/1.37
Plaxis	0.82/0.94	0.85/0.97	0.86/0.97
Flac2D	No solution	No solution	No solution

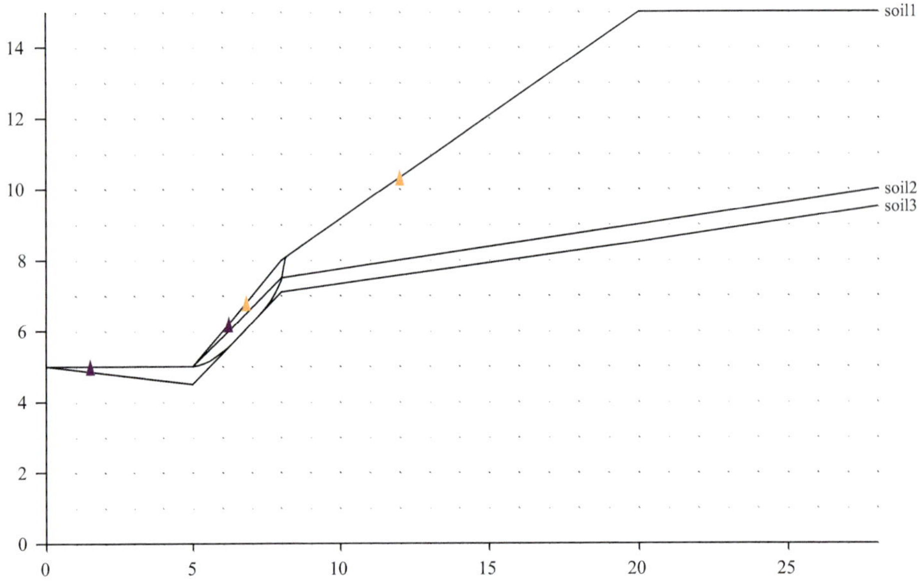

Figure 1.6 Critical slip surface by Spencer method.

finite element program, there are three places in total where surprising results are obtained, and an example is provided in Figure 1.8. For the given design strip, a very sharp jump in the shear force is obtained at a location where there is no point force or support. Conceptually, this is impossible but numerically possible. This problem is detected when the engineers find that the shear stress is so large that the shear capacity of the section is exceeded. Actually, the authors later found that besides the jump of shear forces, jump of moment can also occur at a location where there is nothing! As shown in Figure 1.9, a simple problem is one where the mesh is a combination of three- and four-nodes thick plate element is created. A simple uniform distributed load on the plate plus a line load along line 2 is applied to the problem, while no load is applied on lines 1 and 3. Some surprising jumps in the moment are obtained as shown in Figures 1.10–1.17. For Figures 1.10(a) and 1.10(b), they are actually the same problem, but two additional lines are added to the problem to make the mesh more distorted in Figure 1.10(a).

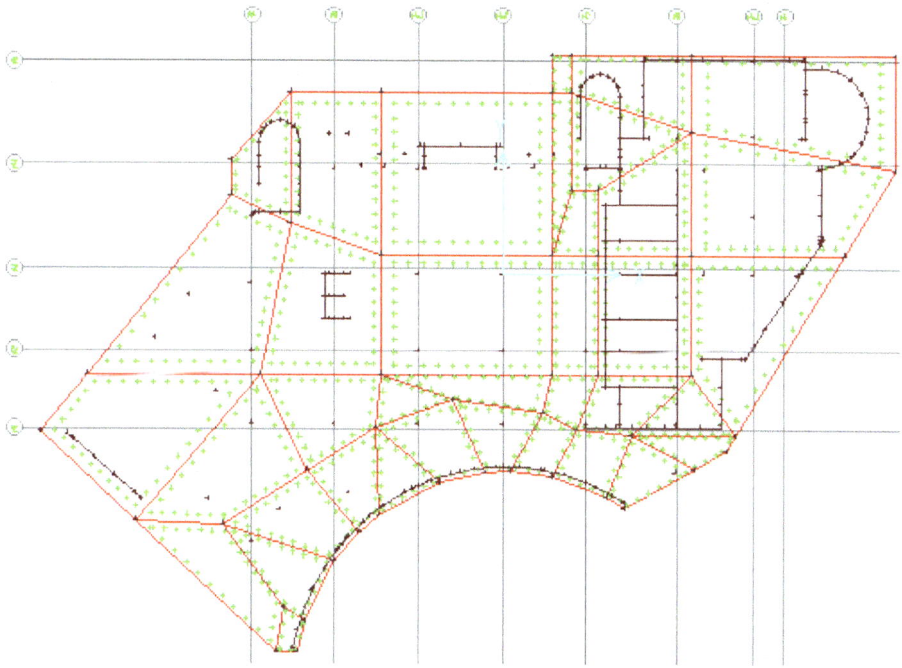

Figure 1.7 A raft foundation modeled by irregular mesh and combination of three- and four-node thick plate element.

In fact, the authors have created many similar surprising results using various commercial programs, and many of these programs are world-famous and have been used by many engineers and researchers. Sense and judgment in the interpretation of the results are extremely important. For the problems as shown in Figures 1.4–1.15, the authors managed to find the causes and overcome them in the self-developed plate programs. Cheng (2018) provided a series of technical problems for the DLO and limit equilibrium slope stability analysis commercial programs. Without experience in programming these problems, readers may not be aware of the technical issues and the reasons for these problems. Also, the authors like to explain to students that if we want a detailed understanding of an engineering problem, the best way is to program the problem into a software. During the programming, readers will find many questions which they may never have previously thought of, and many of these questions are not described in any textbooks or research papers.

If we focus on the local effects, some interesting results are observed. When the same problem using a regular mesh and irregular mesh is considered, there are great differences in the moment at the point (7.2, 6.3) as shown in Figure 1.16. The differences are noticeable, even when a fine irregular mesh is used. On the other hand, for the problem as shown in Figure 1.7, more than 10,000 elements are already used by engineers, which is usually more than adequate for normal problems, with surprising results still being obtained! For

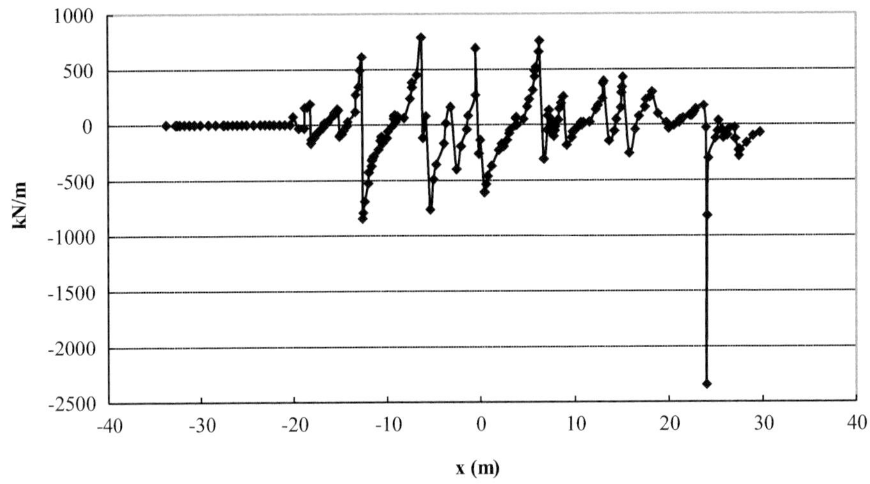

Figure 1.8 Shear jump from the use of thick plate finite element analysis.

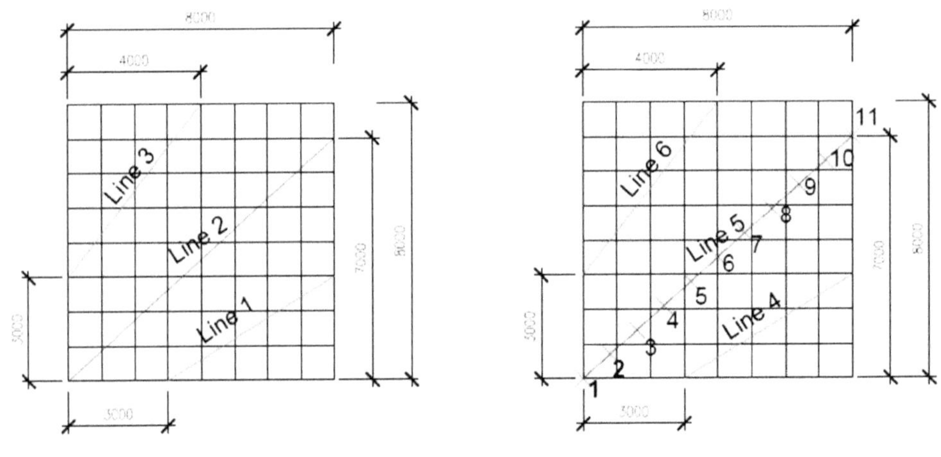

Figure 1.9 A simple problem where the mesh is a combination of three- and four-node thick plate element.

a problem without a known solution, how an engineer assesses the acceptability of the computer results can be a difficult issue. After a series of analyses into the local effects, the authors found that the use of triangular and adjoining quadrilateral elements is the cause of the incompatibility of the internal forces. This is due to the order of interpolation functions used in the two elements, and the derivatives of the displacement along the common side are not the same for the adjoining elements. On the other hand, the moment is a function of the derivative of the displacement, and the results on the moment can be very poor under such conditions. A mesh with a mixture of different elements can sometimes make mesh generation easier, but it will create noticeable numerical inaccuracy

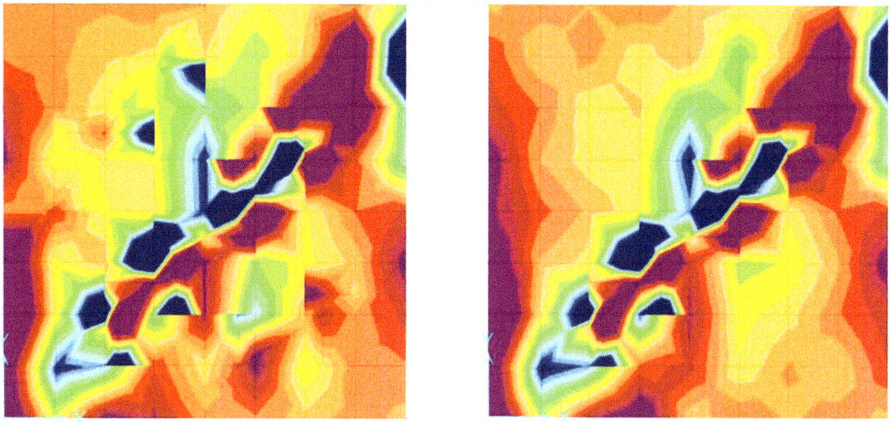

(a) (b)

Figure 1.10 Surprising results for Mx at lines 3 and 6, where only a line load is applied at the same problem.

for some problems, particularly those problems governed by a higher order differential equation.

The authors have found many numerical problems in different commercial programs over the years. Some commercial programs have actually updated their programs based on the problem cases discovered by the authors. Nevertheless, the authors can still find problem cases from many recent commercial programs, particularly for those involving nonlinear analysis. More problem cases will be provided in later chapters. For the plate analysis problems as shown above, Cheng et al. (2008, 2020) developed pseudo eight- and nine-node elements which behave as four nodes externally to the engineers. This approach can attain very high accuracy but possesses the advantage of simplicity in actual applications. The importance of a good thick plate element in foundation analysis and design will be further illustrated in Chapter 4.

1.3 SITE INVESTIGATION METHODS

Before the discussion about typical site investigation methods, the authors would like to discuss the purposes of site investigation. The types of subsurface information required for design include, but are not limited to, the following:

1. Areal extent, depth and thickness of each identifiable soil/rock stratum.
2. Description of the soil, rock and other materials. A consistent description should be adopted by all the engineers in a country so as to avoid ambiguity. In Hong Kong, this is governed by the GEO Guide 3 (Eurocode EN ISO 14688, 14689), Guide to Soil and Rock Description, and similar guidelines or codes are available in different countries. In general, the guidelines are similar among different countries.

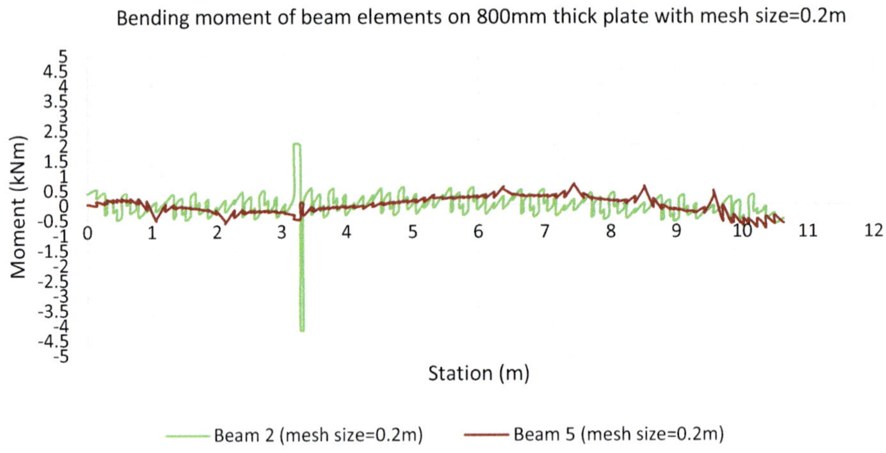

(a) Mesh size=0.2,

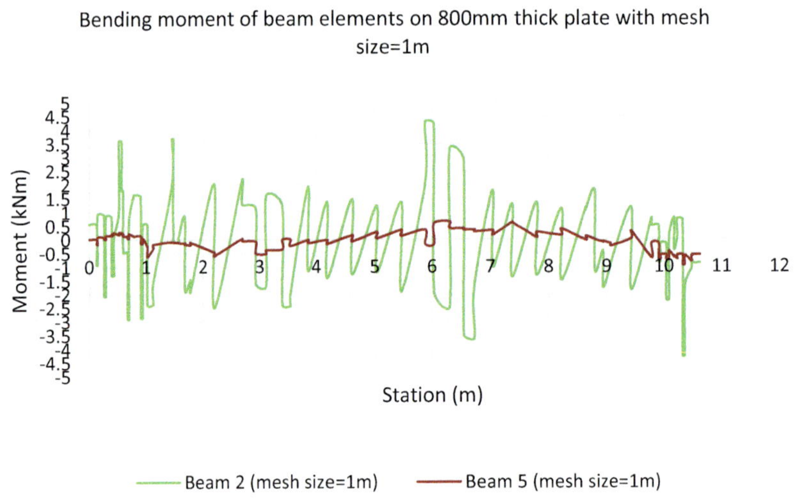

(b) Mesh size = 1.0m

Figure 1.11 Surprising result on Mx along lines 2 and 5.

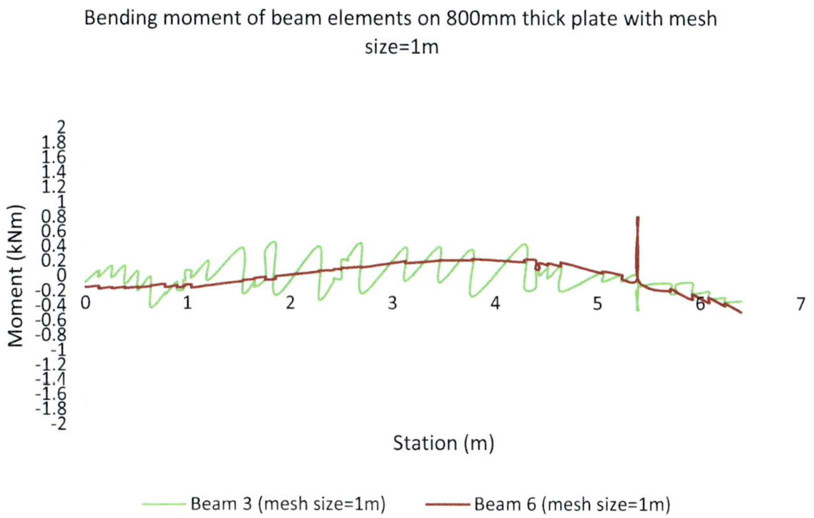

(a) Mesh size=1.0m

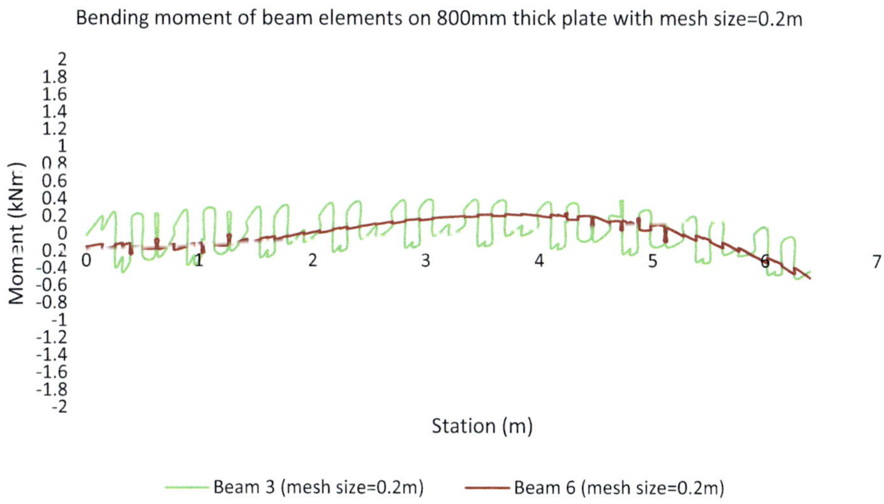

(b) Mesh size = 0.2m

Figure 1.12 Surprising result on Mx along lines 3 and 6.

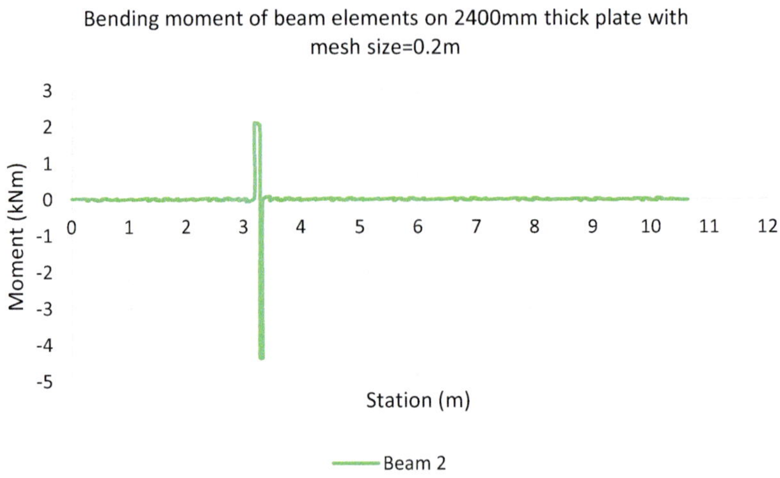

Figure 1.13 Surprising result on Mx along line 2.

3. Depth to top of rockhead and the characteristics of the rock. A clear definition about the definition of rock/boulder as well as the location of bedrock is required. In the Code of Practice for foundation in Hong Kong 2017, it requires a continuous 5 m drilling in rock for the definition of "bedrock", but it is also found that there are actually boulders with a size greater than 5 m, though these are not commonly found. The costs for drilling/excavation in soil and rock differ by a very great amount, and the presence of highly decomposed rock which is somewhere between rock and soil is frequently a source of argument in the measurement.
4. Location of groundwater or other fluid, and the fluctuation of the levels over time.
5. Engineering properties of soil and rock.
6. Presence of discontinuities or weak zone. A thin soft band in soil is sometimes found, which may not be detected due to its limited thickness.

Based on the site investigation results, the suitability of the proposed development can be assessed, and detailed analysis and design can be carried out subsequently. Usually, the cost for site investigation is about 1–2% of the total construction cost, but it can be very important for some projects. The authors would like to quote several cases for illustration. For the Tai Yau Building in Hong Kong, very few boulders were detected during the site investigation process (boreholes are not close to the site boundary). A sheet pile retaining wall was proposed by the engineers for the basement construction in this project. During the actual construction, many boulders were found around the perimeter of the site, and small-diameter boring rigs were used to break through the boulders for driving of the sheet pile, with a subsequent major increase in both the construction cost and time. The construction and collapse of the Mass Transit Railway at Admiralty, Hong Kong, is another case worth consideration. The rock tunnel was constructed by the drill and blast method

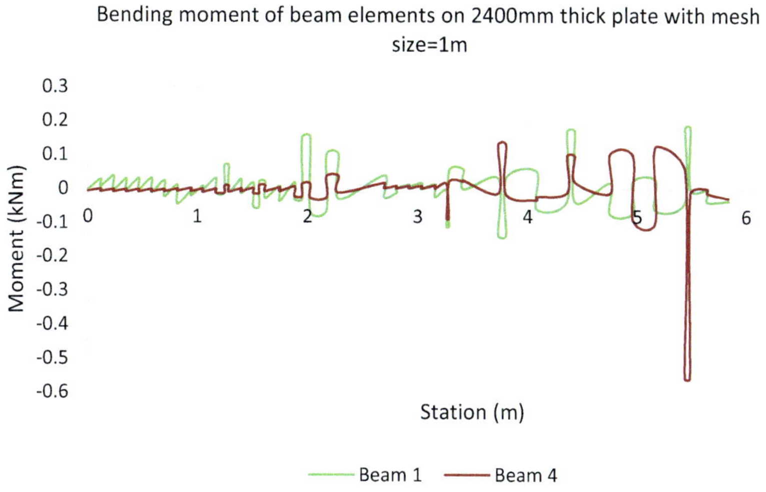

(a) The results for a thicker plate along lines 1,4

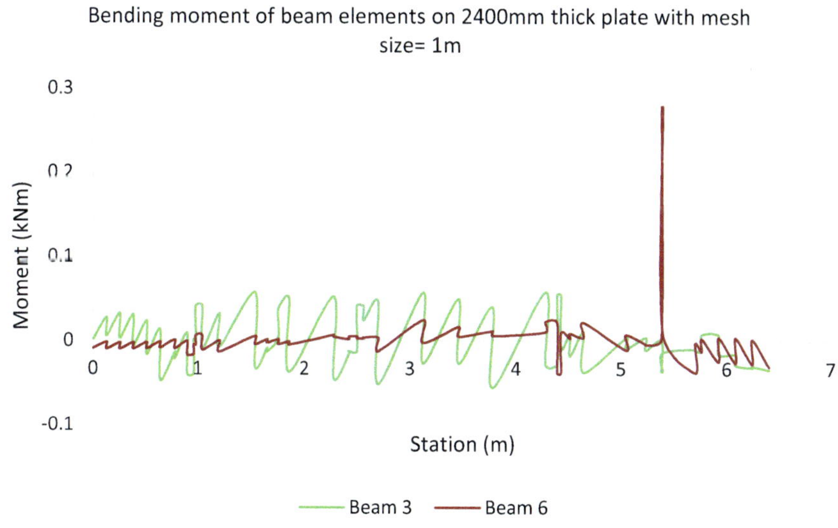

(b) The results for a thicker plate along lines 3, 6

Figure 1.14 Effect of thicker plate for the problem in Figure 1.9.

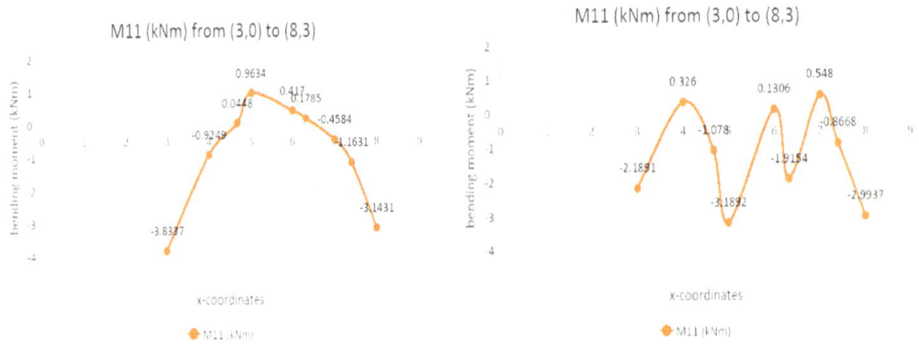

(a) Regular mesh (b) Irregular mesh

Figure 1.15 M11 from (3,0) to (8,3) for the regular and irregular mesh, where line moment is applied along line 2 or line 5, and no loading on other locations.

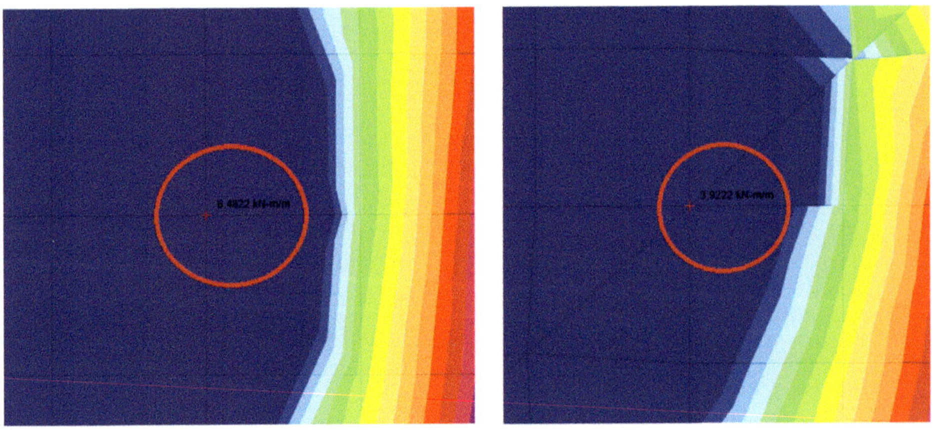

Regular mesh pattern (7.2,6.3): Irregular mesh pattern (7.2,6.3):

Moment = 6.4822kNm/m 3.922kNm/m

Figure 1.16 Result of moment at the same point to illustrate importance of mesh design.

in moderately decomposed granite in this project. An unknown but critical fault zone has caused a major collapse of the tunnel with a large and deep opening to the ground surface of the nearby road. Fei Tsui Road slope failure in Hong Kong is another good example in that the thickness of the soft band was so thin that it was not identified from the site investigation. Upon a heavy rain, the slope failed along the weak planes.

The extent of the investigation depends primarily upon the magnitude and nature of the proposed works and the nature of the site. A question frequently asked by many engineers is the number of boreholes that are required for a project. So far, there is no universal rule or guideline to this question. The second commonly asked question is the design of the borehole layout and the depth of exploration. If the layout of the structure has not been established when the site investigation is conducted, a uniformly spaced grid boreholes pattern is recommended. As per Eurocode 7 specifications, the spacing of normal exploration stations creating a grid ranges from 20 to 40 m. For linear structures (pipelines, tunnels, retaining walls), a spacing of 20 to 200 m is commonly adopted by many engineers (about 10 m spacing for the Mass Transit construction in Hong Kong). For the foundations of bridge piers, two to six boreholes is typical for each foundation, while a spacing of 25 to 75 m can be adopted for dams or similar constructions. Otherwise, boreholes should be positioned close to the suggested foundation, especially in the case where the depth of bearing stratum is varied. For pile foundations, Eurocode 7 specifies that borehole depth should extend 5 m or five times the shaft diameter below the pile toe. For the large-diameter bore pile in Hong Kong which requires pre-drill for each pile, the borehole depth is taken as 5 m below the pile tip to confirm that the bedrock level is reached. For shallow foundations, Eurocode 7 specifies that the depth of exploration below the foundation should be at least 6 m or three times smaller than the width of the foundation. For raft foundations, the depth of exploration should be at least 1.5 times smaller than the width of the foundation. For roads, the depth of exploration below the foundation should be at least 2 m while for trenches, it will be 2 m or 1.5 times the width of the excavation. For tunnels and caverns, the depth of exploration should lie between the width of excavation to twice the width of the excavation. The actual spacing or depth of excavation for different cases is given in Eurocode 7 and other similar guidelines and will not be repeated here. Even when guidelines can be found, engineers should exercise their knowledge and judgment to select a value suitable for their own projects. For example, if the presence of major discontinuity for a tunnel project is expected, the spacing should be adjusted. Codes or guidelines cannot replace engineering knowledge and judgment for geotechnical construction!

Due to dense urban development, construction activities can often affect adjacent properties, and this problem is frequently encountered in Hong Kong and many other highly developed cities. Site investigations should cover all the factors that may affect adjacent properties. Records of ground levels, groundwater levels and relevant particulars of adjacent properties should be made before, during and after construction. Where damage to existing structures is a possibility, adequate photographic records should be obtained. For the construction of the Mass Transit Railway in Hong Kong, Cheng has spent some time in preparing the assessment reports, records of all existing cracks and other special considerations for the building adjacent to the tunnel. The assessment reports also form the basis for the claims from the people adjacent to the construction sites. For the Sheung Wan concourse of the Mass Transit Railway, a right-angle corner of the concourse has to be removed to avoid conflict with some utilities which cannot be removed.

There are several stages in a site investigation project which include the following:

1. Desk study

   a. Collection and analysis of information relevant to the site, which includes the immediate environment and the proposed development. Much useful information can be obtained through this study. In Hong Kong and many developed cities, much useful information can be gathered through the government and university libraries, public works departments, consultants and contractor firms' libraries. Many geological maps, aerial photos and survey plans can also be gathered to gain further knowledge about the planned location. The interpretation and direct use of such old investigation data need care because the design/construction standards may have changed, and field practice dated 20 or more years previously may not be sufficiently reliable.

b. Define the types and amount of subsurface investigation required to establish the parameters for design.
c. Features to look for in a desk study include: (1) main rock and soil types present on site; (2) main geological structures, e.g. faults; (3) surface features, e.g. valleys, terraces; (4) groundwater conditions; (5) nearby developments, e.g. foundation of adjacent buildings; and (6) existence of any potential hazards, e.g. landslip, subsidence, toxic waste.

2. Site reconnaissance

It is also advised to visit the actual site after the desk study, though the authors are aware that some engineers will simply believe in the results from the desk study and prepare the site investigation tender document directly. The features to explore during site reconnaissance include: (1) topographical features; (2) subsurface strata that can be revealed in cut slopes, pits, quarries or natural exposures which provide data on the material and mass characteristics of soils and rocks, particularly the jointing characteristics, fault pattern, weathering profile, location of existing slip surfaces or weak zones; (3) surface depressions which may include sinkholes in limestone, pipes in chalk, depressions from quarries or underground mine workings; (4) surface drainage pattern; and (5) stable angle of existing natural or man-made slopes, availability and types of suitable construction materials, signs of distress.

It should also be noted that the actual site conditions may differ from the results in the desk study, particularly for areas with frequent development. The authors have good experience from the Kornhill Development project in Hong Kong, where the survey plan (prepared more than 10 years ago) differs significantly from the actual site condition.

3. Planning of site investigation works

With the information gathered from the above two stages, engineers can design the actual site investigation works. The method of exploration may include trial pits, boring or geophysical exploration methods.

4. Contract preparation and tendering

The engineer will prepare the design, tender document, specifications, bills of quantities and other documents/requirements in preparing the tender. Depending on the size of the project, a meeting between the contractors and engineer may be held in some cases. The contractor is strongly advised to visit the site before finalization of the tender prices and quantities. Some engineers like to provide provisional items in the tender document for works which may or may not be carried out, and the final decision will depend on the site investigation results.

5. Actual site investigation, field and laboratory tests

The actual site investigation works will be carried out by the contractor, and will be supervised by the engineer. The core boxes need to be prepared and transported carefully to prevent any disturbance to the samples. The field and laboratory tests should be carried out by accredited companies (for Hong Kong and some other countries). Currently, the laboratories must be Hoklas accredited in preparing the field and laboratory tests in Hong Kong. The advantage of Hoklas accreditation is recognition of the test results in many countries.

6. Preparation of the site investigation report and geotechnical report

The contractor will prepare the site investigation reports which will include a typical borehole log, and the classification of the soil/rock should be prepared by a recognized geotechnical engineer. The borehole logs will also include various in situ test results, location of water table and other useful information. In addition, photographs (movies in some cases) of various soil/rock types will be prepared. Upon receiving the site investigation reports, many engineers prepare an overall geotechnical design report. The design engineers will usually use the geotechnical design report directly in the design process.

1.3.1 Types of site investigation methods

1.3.1.1 Trial pits and surface stripping

Trial pits (typical 1.2 × 1.2m) are usually hand dug to a depth of about 3 m (see Figure 1.17), though a depth of 5 m has been experienced by the authors. The sides of the pit should be protected against collapse in order to protect personnel working in the pit. The spacing and stiffness of the shoring should be sufficient to prevent soil/water load as well as any other loadings induced from construction plants and stock materials. Backhoe falling into the trial pit has happened in Hong Kong in the past. Trial pits allows detailed investigation of the in situ condition of the ground in both lateral and vertical directions. Good-quality block samples can be retrieved while the conditions of the in situ soil can be examined visually. Occasionally, in situ tests (typical plate load tests) may be carried out within trial pits. The logs of the trial pits should be supplemented with color photographs and the results of field tests such as in situ density tests or Schmidt hammer, hand penetrometer and hand shear vane index tests.

The advantages of a trial pit are:

a) Speed, mobility, flexibility, economical (cost escalates rapidly with pit depth).
b) Allows examination of shallow slip surfaces, in situ structural details and determination of horizontal variability.
c) Allows more intensive in situ testing, useful for soil derived from in situ rock weathering and colluvium.

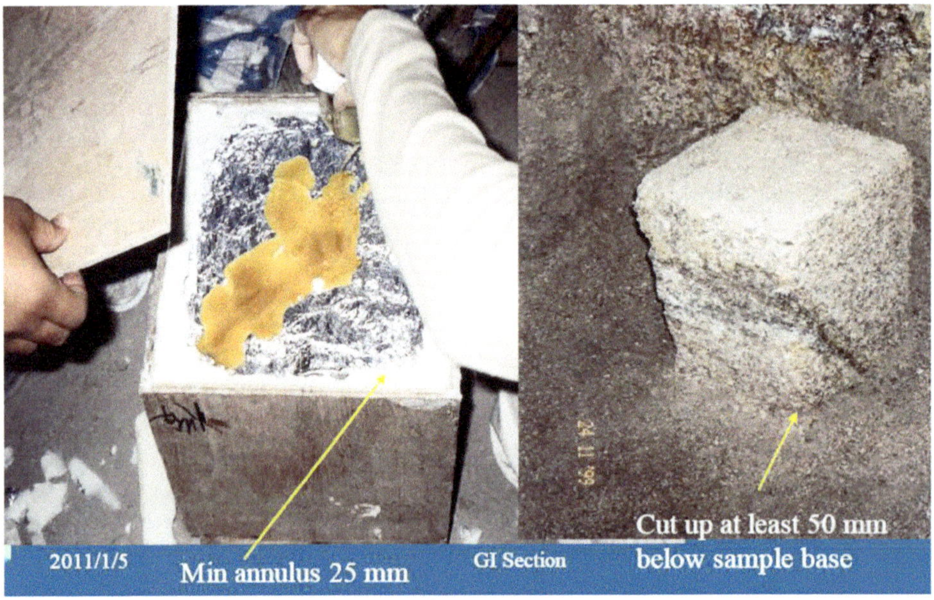

Figure 1.17 A typical block sample collected within a trial pit in Hong Kong.

The disadvantages of the trial pit are:

a) Limited working depth.
b) Existing services are at greater risk.
c) Dangers to workers if any utility is damaged, and such has happened in Hong Kong in the past.
d) There are no standard specifications or "normal practice" for trial pit exploration.

During sampling in trial pits, which is typically in 0.5 m intervals, disturbed samples recovered by Open-Drive or U100 sampler or undisturbed blocked samples (around 250–300 mm cube) are collected. Trial pit exploration is particularly suitable for sampling relict joints or shear planes, which is difficult for other exploration methods.

Trial pits can be extended into trenches or slope surface stripping for investigation. Surface stripping is commonly used in Hong Kong to investigate both natural and man-made slopes, and it is generally a 0.5 m wide strip, extending from the crest to the toe. Bamboo scaffolding is commonly provided for inspection, logging and reinstatement.

1.3.1.2 Boring explorations

The depth of a trial pit is limited to 6 m, to the best of the authors' knowledge. For deeper explorations, a boring method is commonly adopted. Typical boring methods include auger boring, wash boring, rotary boring and percussion boring. The suitability of the boring method will depend on the depth of exploration, type of soil, quality of exploration and cost consideration.

Many of the most frequently used and special-purpose sampling and in situ testing methods may be carried out in boring exploration, and the depth of exploration can extend to more than 200 m for rotary boring. On the other hand, site accessibility and labor costs may be a problem for rough terrain or inaccessible locations, and light rigs or a steel platform may be required, which will be expensive and time consuming when mobilization, supply of water or flushing medium are also considered.

1.3.1.2.1 Auger boring

Auger boring is mostly carried out for exploration in clay without a casing. For general exploration, helical or post-hole type, the hole is usually around 100 mm in diameter (200 mm is generally the maximum size). Auger can be hand-driven auger or power-driven continuous flight auger, and is popular in some countries for low-rise buildings exploration. It is not suitable to be used for the following conditions: (1) hard or unstable soil; (2) presence of rock fragments, boulders; (3) presence of high groundwater table; or (4) exploration beyond 30 m, as the power as well as maintenance of verticality can be difficult to manage. The soil sample is collected by withdrawal of the drill rod and collected from between the auger blade (Figure 1.18).

Figure 1.18 A typical auger boring rig for small diameter pile construction (similar to that for site investigation, but smaller).

1.3.1.2.2 Percussion boring

Percussion boring relies on the free fall of the cutter to cut into soil, and is an ideal and handy tool which can be operated by machine or human effort. The requirement on the headroom is low, and there is less of a verticality problem as compared with auger boring. For the construction of bore pile or diaphragm wall, the percussion method is widely adopted. In general, there are two types of percussion cutters: clay cutters and sand shell cutters. The clay cutter is usually a simple cylinder, while the sand cutter may have a valve flap and a hole for dissipation of pore water pressure. The use of percussion is also limited by the presence of boulders or rock fragments.

1.3.1.2.3 Rotary boring

Most boring exploration in Hong Kong is rotary boring, because most of the buildings in Hong Kong are tall while the depth of the bedrock can be quite low. In rotary boring, the drill bit or casing shoe rotates at the bottom of the borehole, and drilling fluid is pumped down to the bit through hollow drill rods, which lubricates the bit and flushes the drill debris up the borehole. The flushing medium may be water or drilling mud (water with clay or bentonite) for stabilizing the stability during opening. Use of air foam as a flushing medium enables increased core recovery in colluvium and in situ weathered soil. Rotary boring can go to great depths, but it is more expensive than other methods. The typical sizes of casings for rotary boring are given in Table 1.5.

There are two major types of rotary drilling in practice, rotary open-hole drilling and rotary core drilling, which depend on the quality of the samples that are required.

1.3.1.2.4 Open-hole drilling

For open-hole drilling, the drill bit removes all the material within the borehole. Water is usually used for drilling in relatively good soil, and the cuttings are brought to the surface by the flushing medium. This type of boring method is more suitable for wash boring or boring for grouting works, and high water pressure is used to flush the spoil out of the hole. The structure of the soil will be destroyed by this operation, hence this approach is not commonly adopted for ground investigation purposes.

1.3.1.2.5 Rotary core drilling

In core drilling, an annular bit is fixed to the outer rotating tube of a core-barrel. The assembly cuts a core within the inner stationary core-barrel tube, and the core is brought

Table 1.5 Typical casing size for rotary boring

Casing	*Casing size (mm) ID*	*Casing size (mm) OD*
N	76	89
H	102	114
P	128	140
S	156	168

(a) Vertical drill (b) inclined drilling

Figure 1.19 Typical rotary boring.

back to the ground surface. Usually, a casing is required to support the unstable ground during rotary drilling. There is a wide range of rotary drilling rigs with different weights and power ratings (typically 10–50 horsepower). A rig is normally skid mounted, as shown in Figure 1.19, which is capable of stable drilling with a string rotation of up to 1,500–2,000 rpm. Drilling rigs should be mounted on a stable platform (usually with wedges) to support a force of 12–15 kN. In choosing a suitable rig, the expected depth and diameter of the hole, any special possible casing requirements and site access have to be considered. The ram stroke length should also be greater than 600 mm for normal efficient operation.

For the flushing medium, water is most commonly used. In the case of very loose soil, drilling mud (usually water with clay or bentonite) may be used to stabilize the side and bottom of the borehole. It can also help to reduce soil disturbance during sampling, however, the permeability of the sample may be affected by the drilling mud. The authors have experienced that even with the use of drilling mud, samples could not be recovered in very loose alluvium in the Tai Koo Reservoir project in Hong Kong. No sample can be recovered for about 3 m length, despite many methods having been used to stabilize the borehole and to collect the sample. Besides drilling mud, air foam can also be used as a flushing medium to increase the core recovery and quality of samples in colluvium and soils derived from in situ rock weathering. Compared with the use of water as a flushing medium, air foam is better at maintaining the cuttings in suspension, and the low uphole velocity and low volume of water utilized help to reduce the disturbance of the core and the surrounding ground. The air foam also resists percolation into open fissures and stabilizes the borehole

Figure 1.20 A typical rock cutting drill.

walls. However, the polymer stabilizer in air foam will coat the walls of the hole so that an in situ permeability test will not be appropriate.

When boring has reached the rock head or when a boulder is encountered, the use of a rock drill similar to that in Figure 1.20 is required.

1.3.1.2.6 Wash boring

Wash boring is similar to rotary boring in some aspects. The spoils are, however, removed by high water pressure and the structure of the soil will be completely destroyed so that no soil sample can be recovered. Wash boring usually is not specified directly in a construction contract, but is a by-product during grouting work. For the Mass Transit construction in Hong Kong, boreholes are spaced at intervals of 10–20 m. From the colors of the spoil and the rock head level as determined from the drilling before grouting, additional information is obtained to supplement the borehole investigations. The properties of the soil and rock can also be estimated from the results at the surrounding boreholes.

1.3.1.2.7 Horizontal/directional boring

For the site investigation of the Mass Transit Railway Hong Kong Line, many inclined drilling works are carried out, as a major portion of the tunnel alignment is belowground. Horizontal and directional boring has been adopted for the Mass Transit Railway and other projects in Hong Kong (Route 8), though this is not common. Horizontal boring can provide continuous core samples along the tunnel alignment, which is important for the tunnel construction. Directional coring comprises three components: planning, steerable drilling and coring orientation survey. Steerable drilling is controlled by a steerable core barrel with wireline operating system. Coring survey is usually carried out by a miniature electronic multi-shot instrument with the time interval specified by the drilling specialist. The typical drilling rate is about 9–15 m/day in Hong Kong, while the rate can be increased to 25–35 m/day for straight drilling with wireline system.

1.4 SOIL AND ROCK SAMPLING

The sampling of soil and rock is crucial to the identification and testing of the soil and rock parameters which are required for the analysis and design of a project. In many cases, the engineers have not paid adequate attention and supervision to the sampling works, as these works are generally taken up by operators and technicians. Many poor practices in soil sampling have been reported in the past 30–40 years, but the situation has improved in recent years. There are several areas for which engineers should be aware of during the site investigation process. The number of samples to be collected for laboratory testing will depend on the specification in each country, and GeoSpec 3 is used in Hong Kong for the requirements of typical laboratory tests.

1.4.1 Sampling quality

There are generally four main techniques for obtaining samples, after Hvorslev (1949):

1. Taking disturbed samples from the drill tools or excavating equipment directly;
2. Driven sampling by a tube or split tube sampler using a static or dynamic force;
3. Rotary sampling in which a cutter is rotated into the ground to collect a core sample;
4. Hand block sampling from a trial pit, shaft or heading.

Soil and rock samples are required to be collected for visual examination as well as laboratory tests. For rock samples, there is usually no major difficulty, and this is discussed in greater depth later. Since the strength of soil depends on the condition of the soil sample, disturbance of the soil sample can lead to inaccurate determination of the soil parameters. In most construction contracts, undisturbed samples (100 mm diameter appears to be the most popular size) are usually specified. There is actually no way to prevent disturbance to the soil sample, as the original soil mass which is subjected to the in situ stress is taken out to the ground level without any confining stress. For good soil, this condition is basically similar to the unloading condition in a triaxial test. After the soil sample is placed in the triaxial cell and loaded by the initial cell pressure, the stress path will be generally similar to the reloading path. After the application of ram loading, the stress path will resemble the normal compression line and the soil parameters as determined will be basically similar to the in situ soil parameters. There are, however, also many loose soils which will not reproduce the in situ soil parameters in the triaxial test. Engineers should therefore look at the laboratory test results as well as the in situ test results for assessing the properties of soil. In Hong Kong, the UK and Eurocode 7, the soil sample quality is divided into five classes (not for rock samples), and a typical relation between the sample quality and the properties of soil is given in Table 1.6. In Eurocode, three

Table 1.6 Sample quality and the relevant soil properties

Sample quality	*Soil properties that can be determined*
1	Strength, deformation, soil classification, moisture content, compressibility, density
2	Soil classification, moisture and density
3	Soil classification, moisture
4	Soil classification
5	Only approximate identification of soil

sampling method categories are considered (EN ISO 22475-1), depending on the desired sample quality, as follows:

1. Category A sampling methods: samples of quality classes 1–5 can be obtained;
2. Category B sampling methods: samples of quality classes 3–5 can be obtained;
3. Category C sampling methods: only samples of quality class 5 can be obtained.

The quality of the samples that can be recovered depends on the care taken in the operation (a factor which is commonly neglected), soil types and the types of samplers. In general, soil derived from in situ weathering can give good-quality soil samples, while fill/alluvium/colluvium/marine deposit types result in high- to low-quality samples. The type of samplers and drilling fluid may also affect the quality of samples. Cheng has come across a case where the soil (high moisture content, at Tai Koo Shing, Hong Kong) is so loose that no sample can be recovered, despite various methods having been tried. For the five sample qualities described above, classes 1 and 2 can be viewed as undisturbed samples.

For sampling saprolitic soils, contamination and scour of the core caused by the drilling fluid are experienced in traditional core drilling. This problem is solved by the use of air-foam as the flushing medium, and excellent core quality and recoveries can be obtained, even for colluvium which is one of the most difficult soil types to sample (Phillipson and Chipp 1982). Polymer drill fluids of varying viscosities can also be used to obtain high-quality soil samples and can be used to any depths even below the water table, while foam flush has a maximum practical limit of around 50 m and is largely ineffective below the water table.

For soils derived primarily from the chemical decomposition of the parent bedrock, the characteristics of the soils will depend on the parent rock type. A rock decomposition grade classification is used by some engineers (GEO 2019), which is divided into six grades as follows:

(1) Fresh rock: no visible sign of decomposition. Fresh rock is not commonly found within areas of construction interest, as the depth to fresh rock may be so deep that foundation in fresh rock will be highly expensive. The strength of fresh rock is usually very high.
(2) Slightly decomposed: fresh rock with stains at joint surfaces. The strength of slightly decomposed rock is still high and it cannot be broken by a hammer easily. For practical construction works, foundations may terminate at slightly decomposed rock which is sometimes located not deep to the ground surface.
(3) Moderately decomposed: rock which is completely strained, but is still strong enough to withstand the strike of a hammer. Most pile foundations terminate in

this region, due to the cost of construction and the relatively high bearing capacity of this zone.

(4) Highly decomposed: completely discolored and can be broken by hammer strike, but the basic rock structure remains retained. This region can be considered as a transition between rock and soil, and the thickness of this region is usually limited in many projects. In Hong Kong, almost no pile foundations terminate in this region. In fact, engineers treat this zone as soil, although there is a clear distinction between highly decomposed material and soil.

(5) Completely decomposed: soil which still retains the rock texture, and can be broken by hand. The soil parameters and other strength properties for this grade are usually high.

(6) Residual soil: the original rock structure is completely destroyed, and it can be crumbled by hand easily. The strength properties for this grade are low, and slope failures occur commonly within this region.

The class of sample that can be obtained depends on the soil type, decomposition grade, sampling method, sampler and the use of special techniques.

1.4.2 Samplers

1.4.2.1 Block samples

Block samples are formed by hand in trial pits and excavations. They are usually taken in fill, soils derived from in situ rock weathering and colluvium in order to obtain samples with the least possible disturbance. Generally, block samples can be taken as undisturbed samples with which virtually all the standard tests for soil can be performed (Figure 1.21).

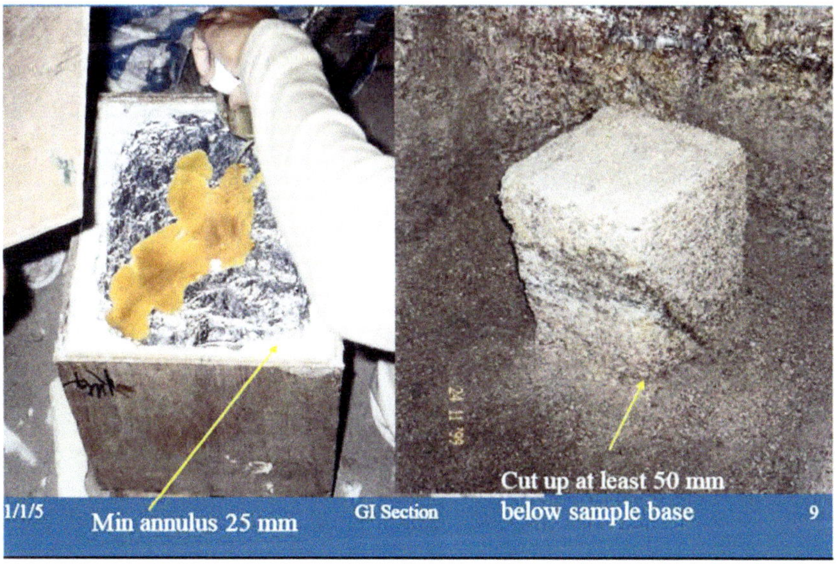

Figure 1.21 Block sampling.

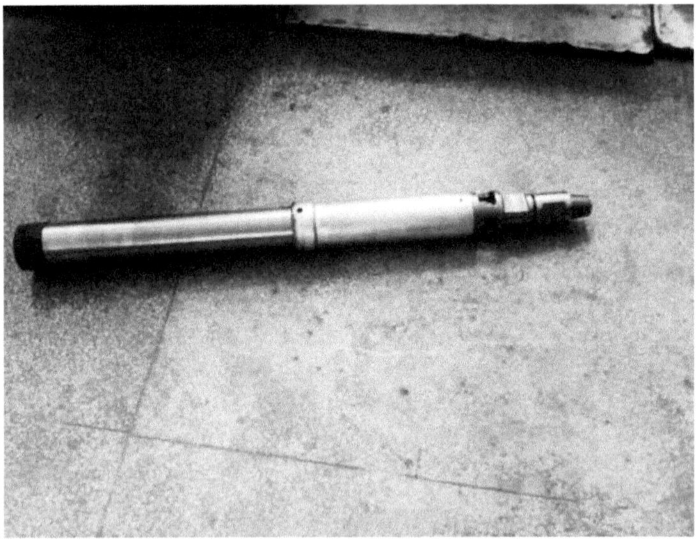

Figure 1.22 An open tube sample.

1.4.2.2 Open-tube sampler

An open-tube sampler is an open tube with a non-return valve that permits the escape of air or water as the sample enters the tube, as shown in Figure 1.22, and assists in retaining the sample when the tool is withdrawn from the ground. The sampler should cause as little remolding and disturbance as possible when being forced into the ground. The degree of disturbance is controlled by three major features: the cutting shoe, the inside wall friction and the non-return valve.

The cutting shoe

(a) Inside clearance: The internal diameter of the cutting shoe, Dc, should be slightly less the sample tube diameter, Ds, to give inside clearance, and the difference is typically about 1% of the diameter of the sampler. This clearance will reduce the friction from the inside wall during pushing of the sampler.
(b) Outside clearance: The outside diameter of the cutting shoe, Dw, should be slightly greater than that of the tube, Dt, to give outside clearance. This clearance will help the withdrawal of the sampler from the ground.
(c) Area ratio: The area ratio is the volume of the soil displaced by the sampler relative to the volume of the sample. This ratio should be as small as possible to reduce disturbance during sampling. This ratio is about 30% for the general purpose 100 mm diameter sampler, and about 10% for a thin-walled sampler.

Non-return valve

The provision of a non-return valve allows air and water to escape quickly and easily during driving the sampler, and will also assist in retaining the sample during withdrawal of the sampler.

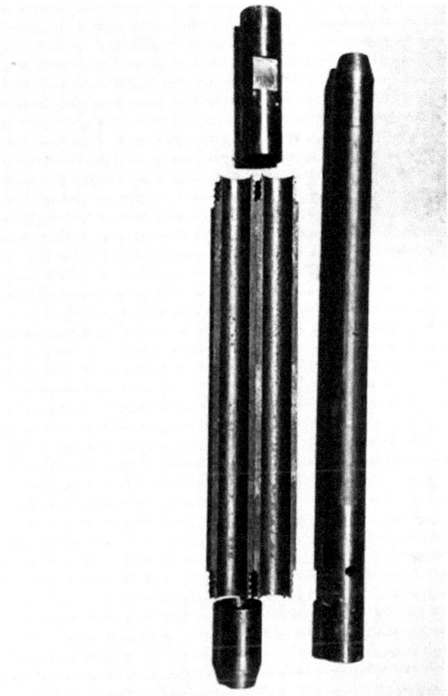

Figure 1.23 A typical split barrel standard penetration test sampler.

For the open-tube sampler, the best sample quality that can be obtained is approximately class 2.

1.4.2.3 Split barrel standard penetration test sampler

The split barrel sampler can be split into two pieces as shown in Figure 1.23, and is typically used in the standard penetration test as described in BS 5930, Eurocode 7. The split barrel sampler has an internal diameter of 35 mm, external diameter of 51 mm and an area ratio of about 100%. The cutting head is usually 25–50 mm in diameter, while the length of the sampler is usually around 610 mm. The quality of the sample that can be recovered is generally class 3 or 4 samples.

1.4.2.4 Thin-walled stationary piston sampler

The thin-walled stationary piston sampler consists of a thin-walled sample tube with a close-fitting sliding piston similar to that shown in Figure 1.24, which is slightly coned at its lower face. The piston is fixed to separate rods which pass through a sliding joint in the drive head and up inside the hollow rods. The diameter of the sample is typically 75 or 100 mm, but a sampler with diameter up to 250 mm is also used for special soil conditions. One hundred mm diameter, 1 m long piston sampling tubes with an area ratio less than 10% are used in the Hong Kong Airport Runway Project. The thin-walled

Figure 1.24 Some typical piston samplers.

sampler is usually used in low-strength fine soils and may give class 1 samples in silt and clay. It is able to recover samples below the disturbed zone, which is an advantage over the open-tube sampler. In addition, special piston samplers have also been designed for use in stiff clays (Rowe 1972). Large-diameter (254 mm) piston samples were taken during the ground investigation for the Container Terminal 9 Development in 2000 and Contract No. HK/13/02-Wan Chai Development Phase 2 in 2003. The samples were used for Rowe cell consolidation tests. This type of sample and the corresponding laboratory tests are generally expensive, and are used only for large-scale projects in Hong Kong.

1.4.2.5 Rotary core samples

For rotary cores, single-, double- or triple-tube core barrels can be used. The single-tube barrel is now seldom used.

For a double-tube core-barrel, shown in Figure 1.25, the inner tube is mounted on bearings and does not rotate against the core, and it can be used in fresh to moderately decomposed rocks. In highly fractured rocks, the use of a double-tube barrel can result in a jumble of rock fragments in the core box. Logging and measurement of fracture state indices will become difficult then, and the use of a core extruder is recommended in such conditions.

Figure 1.25 A typical double tube barrel used in site investigation.

The triple-tube core-barrel consists of detachable liners within the inner barrel that protect the core from drilling fluid and damage during extrusion, which is shown in Figures 1.26 and 1.27. It is suitable for use in fresh to moderately decomposed rock and some stronger highly decomposed materials, and is also particularly useful for highly fractured and jointed rock, as the split liners facilitate the retention of core with the joint system relatively undisturbed. For soils derived from in situ rock weathering, a triple-tube core-barrel is usually fitted with a retractable shoe. The cutting shoe and the connected inner barrel projects ahead of the drill bit and retracts when the drilling pressure increases in harder materials. This will greatly reduce the possibility of direct contact between the drilling fluid and the core. In Hong Kong, a Mazier core-barrel has an inner plastic liner which protects the sample during transportation to the laboratory. The 74-mm diameter core from the Mazier is compatible with the commonly used laboratory triaxial testing apparatus and is commonly specified. The Mazier has a tungsten carbide-tipped cutting shoe and is not suitable for fresh to moderately decomposed rock. In these cases, a core-barrel with a diamond-impregnated drill bit has to be used to advance the hole.

1.5 PRESENTATION OF SITE INVESTIGATION RESULTS AND GEOTECHNICAL INVESTIGATION REPORTS

After the borehole exploration and in situ/laboratory tests, rock samples are collected and arranged in a core-box for future examination, as shown in Figure 1.28c. The detailed information gathered from each borehole will be presented in a graphical form called the boring log, as shown below.

Generally, a borehole log similar to that in Figure 1.28 consists of the following information:

1. Name of the drilling contractor.
2. Drill hole number, location and coordinates/level of the borehole.

Figure 1.26 A typical triple tube barrel (Mazier) used in site investigation.

Figure 1.27 Non-retractable trip tube barrel.

3. Date of boring.
4. Subsurface stratification, with description of the soil/rock, boulders and other features. The colors, soil type, approximate soil classification, stiffness and the presence of gravel/boulders, shells or other materials will also be described.
5. Water table and the date of measurement, use of casing and mud losses and others.
6. In situ tests results with depths, e.g. SPT.
7. In the case of rock coring, type of core barrel, the actual length of coring, length of core recovery, RQD and other important features.
8. Driller's name, job description (optional).

Based on the various borehole records, geotechnical engineers prepare an overall report prior to the geotechnical design. This report typically gives an overall geotechnical view of the site, soil/rock properties, water table, discontinuities, assessment of the laboratory and in situ test results, and recommended design parameters for soil/rock. In addition, many engineers will also use a contouring program (SURFER is the most popular in Hong Kong) to prepare the contour for different soil layering and rock head. A typical rock head contour for a project in Hong Kong is given in Figure 1.29. The design engineers will then carry out the design works based on this geotechnical report, with reference to the actual borehole records if necessary. These are some of the actual design procedures used in some consultant firms.

1.6 LABORATORY TESTS VS. IN SITU TESTS

For collections and testing of samples, there are different requirements for different countries. In Hong Kong, this is governed by the Geospec 3 Model Specification for Soil Testing. Various laboratory tests may be specified by the engineers, which depend on the planned development. In general, these include moisture content, Atterberg limits, particle density, particle size distribution, chemical substances, dry-density moisture content relation, compressibility/consolidation, California bearing ratio (CBR), shear strengths based on direct shear and triaxial tests, unconfined compression test, ring shear test, Brazilian test and point load index test. The detailed procedures for testing can be found in BS 1377 and Geospec 3, Eurocode 7 and various ASTM standards. In Hong Kong, the laboratory and technicians carrying out the tests have to be accredited by authority (HOKLAS in Hong Kong) to ensure the quality of the testing.

There are various advantages and limitations in the laboratory tests, and some of the more important issues are discussed below.

Advantages of laboratory tests include:

1. Well-defined boundary conditions so that analysis of the test results is easier.
2. Strictly controlled drainage conditions.
3. Preselected and well-defined stress paths are followed during the tests, and a more complete constitutive model can be developed.
4. In principle, uniform strain fields (this assumption is acceptable for small strain levels only and soils which do not exhibit strain-softening behavior) are imposed on the specimens, which allow the application of continuum mechanics theories to the interpretation of test results.
5. Soil nature and physical features are positively identified.

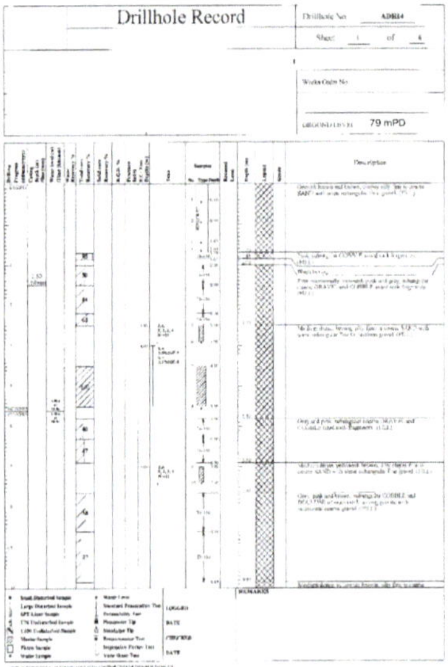

(a) first page of borehole log

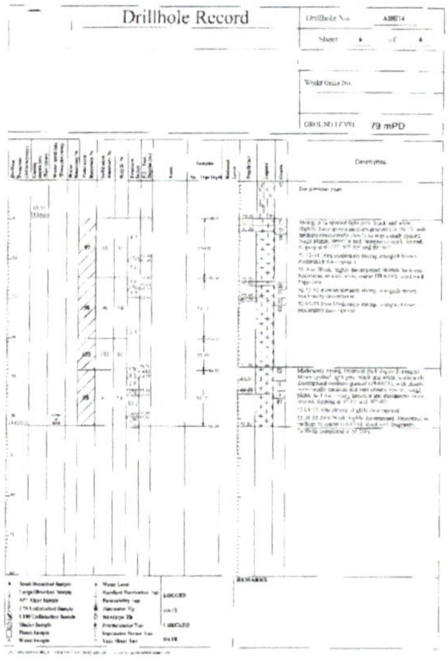

(b) last page of borehole log

(c) core box for rock sample

Figure 1.28 The first and last pages of a typical borehole log (Hong Kong).

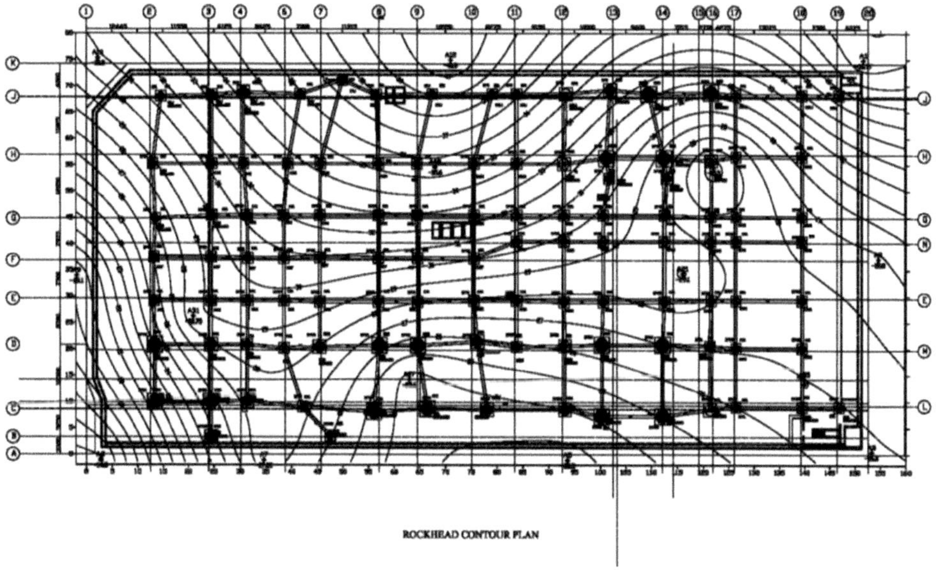

Figure 1.29 Contour for rock head or soil layer prepared for the geotechnical design.

Limitations of laboratory tests include:

1. In cohesive soils, the effects of unavoidable sample disturbance in even so-called "high-quality" undisturbed samples are sometimes difficult to nullify.
2. In cohesionless soils, undisturbed sampling remains an unsolved problem in everyday practice.
3. The small volume of laboratory specimens cannot reflect the macrofabrics and inhomogeneities of natural soil deposits. This leads to doubts as to what extent the field behavior of a large soil mass can be successfully modeled by small-scale laboratory tests. This is particularly important for the determination of the permeability of soil, as the results from laboratory and field tests usually differ several fold. Currently, most engineers rely on the results from the field permeability tests.
4. The factors causing the formation of shear planes during the testing of laboratory specimens are still very poorly understood. Shear planes are frequently associated with such phenomena as induced shear stresses, soil volume changes, non-uniformity of laboratory specimens and consequent non-uniformity of laboratory specimens and consequent non-uniform strain distributions, boundary and kinematic constraints, and stress concentrations imposed by the laboratory apparatus. It must be emphasized that once a shear plane has developed in a laboratory specimen, deformations are concentrated along this plane and displacements and stresses measured at the specimen boundaries are consequently no longer a function of the stress–strain behavior of the tested material.
5. In principle, the discontinuous nature of information obtained from laboratory tests may lead to erroneous modeling of the behavior of a large soil mass.

6. In general terms, soil explorations based on the laboratory testing of soil samples from borings are likely to be more expensive and time-consuming than explorations which make use of in situ testing techniques.

In view of the various limitations for laboratory tests, where there is not a need for expensive laboratory tests for some low-rise light structures, the use of in situ tests are also favored by some engineers. Actually, some engineers have more faith in in situ tests than laboratory tests, based on their engineering experience.

1.7 IN SITU TESTS

In view of the various limitations and the time/expense of laboratory tests, in situ tests are usually specified in site investigations. Actually, some experienced engineers do not specify any laboratory test in the investigation process, but rely totally on in situ tests for some simple, light structures.
Advantages of in situ tests include:

1. A larger volume of soil is tested than is usually done in most laboratory tests; hence in situ tests should in principle reflect more accurately the influence of the macrofabric on the measured soil characteristics.
2. Some in situ test devices can produce a continuous record of the soil profile, and this is particularly important when there is a thin layer of material which may be missed during the investigation.
3. In situ tests can be carried out in soil deposits in which undisturbed sampling is impossible or unreliable. Examples include cohesionless soils, soils with highly developed macrofabrics, intensively layered and/or heterogeneous soils, and highly fissured clays.
4. The soils are tested in their natural environment, which may not be preserved in laboratory tests. For example, the most successful attempts to measure the existing initial total in situ horizontal stress are the recent developments in situ techniques, e.g., self-boring pressure meter (SBP), flat dilatometer (DMT), Iowa stepped blade (ISB), spade-like total stress cells (TSC).
5. In general terms, soil exploration by means of in situ techniques is more economical and less time consuming than investigations based on laboratory tests.

Limitations of in situ tests include:

1. Boundary conditions in terms of stresses and/or strains are, with the possible exception of the self-boring pressuremeter (SBP), poorly defined, and a rational interpretation of in situ tests is very difficult.
2. Drainage conditions during the tests are generally unknown, which makes it uncertain if the derived soil characteristics reflect undrained, drained or partially drained behavior. In this respect, quasi-static cone penetration tests with pore pressure measurements (CPTU) and SBP tests (also with pore pressure measurements), when properly programmed, help to minimize this problem.

3. The degree of disturbance caused by advancing the device in the ground and its influence on the test results is generally (with the possible exception of the SBP) large but of unknown magnitude.
4. Modes of deformation and failure imposed on the surrounding soil are generally different from those of civil engineering structures; furthermore, they are frequently not well established, as, for example, in the field vane (FV) test.
5. The strain fields are nonuniform and strain rates are higher than those applied in laboratory tests or those which are anticipated in the foundation on structure.
6. With the exception of the SPT, the nature of the tested soil is not directly identified by in situ tests.

For most cases, the use of SPT, vane shear test (VST) and cone penetration test (CPT) will be sufficient, though there are many other in situ tests available. The use of a pressuremeter (PMT), dilatometer (DMT), resistivity probe and many other in situ tests are less common, and are used only for limited cases. There has been rapid development of in situ tests, and further information is provided by Lutenegger (2021).

1.7.1 Standard penetration test (SPT)

This test was developed around 1927 and is the most commonly used in situ test in Hong Kong. In this test, the split-spoon samplers are used to obtain soil samples that are generally disturbed but still representative. When a borehole is formed to a specific depth, the drill tools are removed and the sampler is lowered to the bottom of the borehole. The sampler is driven into the soil by hammer blows to an anvil at the top of the drill rod. The standard weight of the hammer is 63.5 kg and the height of free fall is 762 mm. The potential energy is hence constant and is standardized for this test.

The number of blows required for penetration of three 150-mm (6-in.) intervals is recorded. The number of blows required for the last two intervals (300 mm) is added to give the standard penetration number N at that depth. The first 150 mm is not counted, to avoid the effect of disturbance due to borehole formation. The number of blows to penetrate 150 mm is recorded over a distance of 450 mm. The number of blows for the last 300 mm penetration is added to give the SPT value N. The top 150 mm result is neglected because of disturbance from drilling. If, for example, the numbers of blows to penetrate three intervals of 150 mm are four, five and seven respectively, then N = 5 + 7. Note that only the integer is quoted for the penetration, which is sometimes not appropriate, but this appears to be the practice all over the world. If the resistance is too great, the test should preferably stop at N = 50, but local practice in Hong Kong will continue up to N = 200 (or even slightly higher) which may damage the sampler. The SPT values are commonly recorded in the borehole log, as shown in Figure 1.27.

The SPT value is limited by the efficiency of the machine in performing the tests. Machines that were used 40–50 years ago tended to have a lower efficiency, while the modern machines have greater efficiency. Greater efficiency means that more energy can be delivered to the SPT sampler for penetration, hence the apparent N value will be lower for modern machines. From experiments, it is found that the efficiency of the machine can vary from 55% to 75%, and a reference standard is now chosen as 70%. The N value should be corrected to this reference for comparison. For example, if N for 55% efficiency is denoted by N55, then N70 is calculated as N_{55} *55/70.

Overburden stress is an important factor in the bearing capacity of soil, which will be demonstrated in Chapter 2. For soil with the same soil properties, the N value will hence depend on the depth for test. For a fair comparison and interpretation, the N value should be adjusted for the overburden stress. A formula which is commonly used is that by Liao and Whitman (1986), where the overburden pressure is expressed in kPa.

$$C_N = \sqrt{\frac{95.76}{\sigma'_v}} \quad \text{and } N_{cor} = C_N N \tag{1.1}$$

N is commonly used for assessing the soil parameters, and some typical formulae are given below (Hatanaka and Uchida 1996, Peck, Hanson, and Thornburn 1974, Schmertmann 1975):

$$\varphi = \sqrt{20N_{cor}} + 20,\ 27 + 0.3N \text{ or } \varphi = \tan^{-1}\left[\frac{N_F}{12.2 + 20.3\left(\frac{\sigma'_v}{p_a}\right)}\right]^{0.34} \text{ for granular soil} \tag{1.2}$$

$$C_u = K_1 N \text{ in kPa} \tag{1.3}$$

$$E_s = K_2 N \text{ in MPa} \tag{1.4}$$

For the proportionality constants K_1 and K_2, K_1 usually varies between 3.5–6.5 kPa, while K_2 usually varies between 1.0–2.1. In Hong Kong, K_2 is usually taken as 1.0 to 1.5 for fill and alluvium and 1.5 to 2.0 for completely decomposed rock. In general, engineers should look up the suitable local statistics in the interpretation, as there can be wide differences for these proportionality constants between different places. Besides these soil properties, there are empirical formulae for relative density, consolidation coefficient or other soil parameters in the literature. Generally speaking, the levels of confidence are not as high for these as for other soil parameters, hence many engineers will not adopt such empirical correlations for engineering uses. An approximate relation between the N value and relative density of soil as used in Hong Kong (similar to those used in other countries) is given in Tables 1.7 and 1.8.

It should be noted that both the overburden stress and efficiency corrections are not performed in Hong Kong, and an efficiency of 60% has commonly been used for many design figures in the past. Correction for the length of rod is performed in some countries. Tests have, however, indicated that the efficiency of impact actually increases with the length of rod, and this correction may not be necessary.

The N value is used by many engineers to assess the soil parameters, stability of soil mass, bearing capacity and other properties. Even though theoretical relations between N and some soil properties have been established by some researchers, the authors tend to

Table 1.7 Approximate relation between N and relative density of soil, after GeoGuide 3

SPT N	*Approximate relative density (%)*
0–5	0–5
5–20	5–30
10–30	30–60
30–50	60–95

Table 1.8 Compactness and consistency of soil, after GeoGuide 3

Soil type	*Descriptive term*	*SPT N*
Sand and gravel	Very loose	0–4
	Loose	4–10
	Medium dense	10–30
	Dense	30–50
	Very dense	> 50
Silt and clay	Descriptive term	Undrained shear strength (kPa)
	Very soft	< 20
	Soft	20–40
	Firm	40–75
	stiff	75–150
	Very stiff	> 150

not rely on those theoretical relations, but stick to the use of statistical interpretation. In general, the Young's modulus, friction angle and cohesive strength can be relatively accurately reflected.

In using SPT in assessing the soil properties, the following limitations should be noted:

1. The design equations, tables or figures are approximate, and may not be applicable to general types of soil. In general, only statistical results obtained locally should be adopted for a particular project.
2. N values obtained from a borehole may vary widely for various reasons.
3. In soil that contains large boulders and gravel, SPT numbers may be erratic and unreliable. Although approximate, with correct interpretation, the standard penetration test provides a good evaluation of soil properties.
4. The primary sources of errors in standard penetration tests are inadequate cleaning of the borehole, careless measurement of the blow count, eccentric hammer strikes and inadequate maintenance of the water head in the borehole.

1.7.2 Vane shear test (VST)

The vane shear test is used during drilling operation to determine the in situ undrained shear strength (C_u) of clay soils – particularly soft clays. It has considerable application

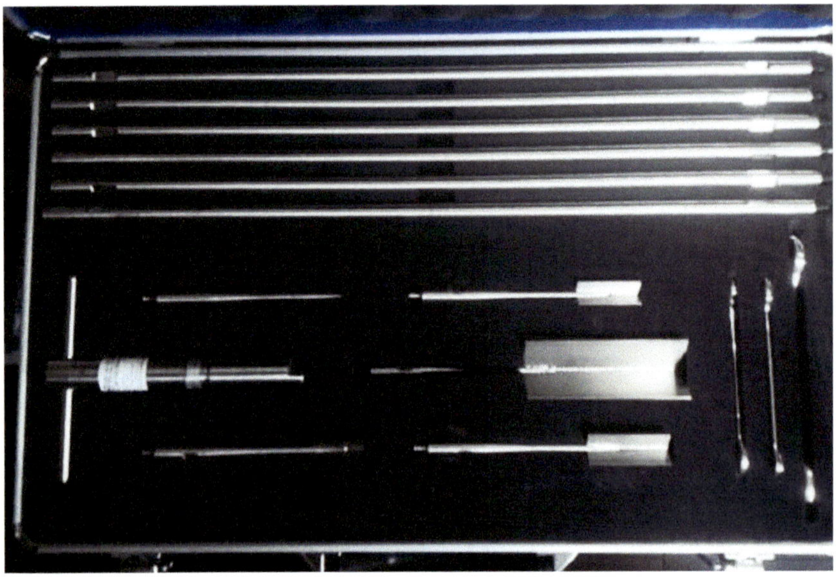

Figure 1.30 Vanes for VST.

in offshore oil exploration. In fact, the airport at Hong Kong is built on marine clay, and VST was used to assess the in situ undrained shear strength (C_u) of clay soils during the site investigations of the Hong Kong Airport and the Zhuhai-Hong Kong-Macau Bridge. In the vane shear test, the test apparatus consist of four blades on the end of a rod, as shown in Figure 1.30. The vane can be either rectangular or tapered. The vanes of the apparatus are pushed into the soil at the bottom of a borehole without disturbing the soil appreciably. Torque is applied at the top of the rod to rotate the vanes at a standard rate of 0.1°/s, and the test is usually completed in 5–10 minutes. In the Hong Kong Airport third runway project, the standard VST rotates at 0.2°/s, while the high-quality test rotates at 0.1°/s. There is no wait period for the standard test, while a wait period of 5 minutes is maintained for high-quality tests. For the remolded test, 12 revolutions are applied in the standard test, while 30 revolutions are applied for high-quality tests. This rotation will induce a cylindrical failure mass surrounding the vane. The most commonly used vane diameter ranges from 38.1 mm to 92.1 mm, while the height of the vane is double that of the diameter. Vanes with different height to diameter ratios are available, but they are less commonly used in practice. The thickness of the blade is usually 1.6–3.2 mm.

The measured torque can be related to the undrained shear strength of an anisotropic clay in the following way. Consider the failure mass in Figure 1.31, shear stress is mobilized at the side and the two ends of the cylindrical failure mass. For the side of the cylinder, the torque required to generate failure T_1 and the undrained shear strength S_v is given by

$$T_1 = \pi dh\, S_v{*}d/2 \tag{1.5}$$

For the bottom and top of the failure mass, refer to Figure 1.31b,

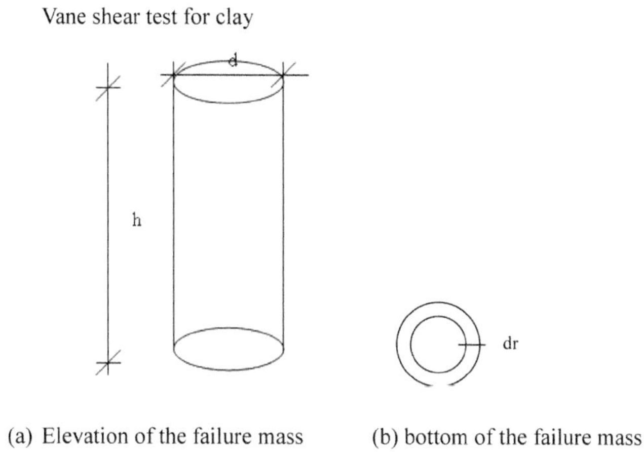

Figure 1.31 Failure mass in a vane shear test.

$$T_2 = 2\int_0^{\frac{d}{2}} 2\pi r \cdot dr \cdot S_h \cdot r = 4\pi S_h \int_0^{\frac{d}{2}} r^2 \cdot dr = \frac{\pi d^3}{6} S_h \tag{1.6}$$

Hence the measure torque T is given by

$$T = \frac{\pi d^2 h}{2} S_v + \frac{\pi d^3}{6} S_h \tag{1.7}$$

Obviously, T_1 and T_2 cannot be determined separately. This equation can be solved by assuming either isotropic soil condition or using two vanes to determine the undrained shear strengths in the horizontal and vertical directions. In most projects in Hong Kong and elsewhere, the anisotropy is usually neglected in practice. This may be not be good practice, as anisotropy can be important in some soils. In using the undrained shear strength from VST for design, one important point must be considered. The test is completed within several minutes, while the earth structures may remain in place for years. For clay, there is an important aspect, which is creep. For actual design purposes, the undrained shear strength values obtained from field vane shear [cu (VST)] are usually too high and should be reduced for the effect of creep. Cu should be reduced by multiplying with a creep factor, and a typical formula is proposed by Bjerrum (1972) as:

$$\lambda = 1.7 - 0.54 \log 10\ (PI) \tag{1.8}$$

It should be noted that there is no universal formula for the creep factor, and the curve fitting by Bjerrum (1972) is actually poor. This factor is also difficult to be assessed, as it takes a very long time to carry out the corresponding laboratory test. Interestingly, it appears that no one has ever talked about the possibility of anisotropy of the creep factor.

The advantages of VST include:

1. Fast and reliable for clay;
2. Wide database for interpretation;
3. No need for empirical correlation.

The limitations of VST include:

1. Presence of cobbles or boulders;
2. Uncertainty with the creep factor.

There is a common question raised by some engineers in the interpretation of VST results. For example, for the VST result at Macau, as shown in Figure 1.32, whether the maximum recorded reading or the residual value should be used for the design. With a closer look into the results, it is found that the maximum shear stress occurs at a high rotation, which corresponds to a rather high strain. For the actual structure, such a displacement is unacceptable. If the displacement of the actual structure is controlled, the maximum shear stress from VST should actually be used.

1.7.3 Cone penetration test (CPT)

The cone penetration test (CPT) is a versatile sounding method that can be used to determine the materials in a soil profile and estimate their engineering properties. This test is also called the static penetration test, and no borehole is necessary. It is applicable to a wide range of soil types, except gravel deposits or stiff cohesive materials. It has been used in many large-scale projects in Hong Kong but is less popular than SPT in Hong Kong. For the third runway project in Hong Kong, both conventional CPT as well as CPT with dissipation tests and seismic CPT were carried out to investigate the properties of the marine deposit. Vane shear tests were also carried out in this project to correlate the results with CPT tests.

For the CPT cone, the apex angle is 60° with a base area of 10 cm^2, which is pushed into the ground at a steady rate of about 10–20 mm/s, and the resistance to penetration is measured continuously. A typical CPT testing device is shown in Figure 1.33. There are generally two major types of cones: mechanical CPT and electrical CPTU (with pore water pressure measurement). A continuous result is usually possible with the CPT test.

The cone penetrometer measures:

1. The cone resistance (q_c), which is equal to the vertical force applied to the cone tip divided by its horizontally projected area.
2. The sleeve friction resistance (f_s), which is the resistance measured by the sleeve located above the cone with the local soil surrounding it. The frictional resistance is equal to the vertical force applied to the sleeve divided its surface area.
3. The pore water pressure (u).
4. the friction ratio f_r, which is defined as the ratio between friction resistance and cone resistance.

VANE SHEAR TEST RECORD SHEET

Contractor :

Contract No. : DE/2017/01

Project :

Depth of vane below ground level : 13.10 m

Tested / Supervised by : H.K. Poon

Drillhole No. : G2 — Sheet 2 of 2

Date of test : 2018/10/10

Ground Level : +3.970 mPD

Co-ordinates : E 840992.257 N 818604.408

Level of test : -9.130 mPD

Checked by : Henry Yip

EQUIPMENT DETAILS :

Torque head — Number: G001; Cali. date: 2018/7/24

Vane — Length (L): 110 mm; Diameter (D): 55 mm; Ratio (L/D): 2

DRILLHOLE DETAILS :

Diameter: 114 mm; Depth of hole: 12.60 m; Depth of casing: 12.60 m; Depth to water: 1.07 m

Shear Stress (S) = N x C

Where:

N = dial gauge reading corrected for shaft friction (if any)

C = calibration factor obtained from calibration graph

Peak undrained shear strength : 52.62 kPa

Remoulded undrained shear strength : 17.24 kPa

Sensitivity : 3.05

Angle degrees	Elapsed time s	Torque (N) divisions	Shear Strength kPa
UNDISTURBED TEST			
145	725	32.00	29.03
150	750	32.00	29.03
155	775	32.00	29.03
DISTURBED (REMOULDED) TEST			
75	375	19.00	17.24
80	400	19.00	17.24
85	425	19.00	17.24
90	450	19.00	17.24

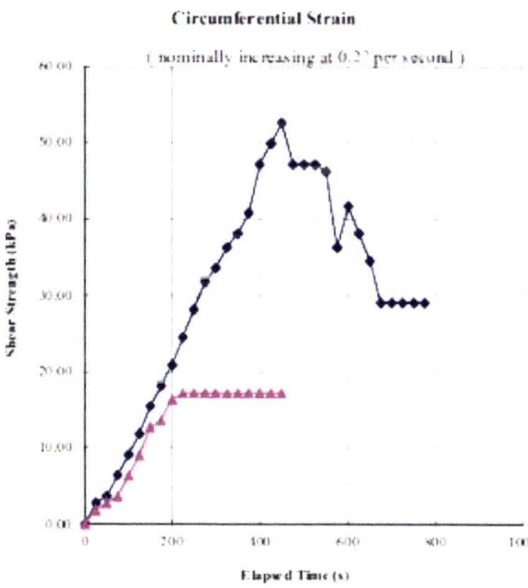

LEGEND :

Undisturbed test (■)

Disturbed (Remoulded) test (▲)

TEST PROCEDURE :

1. Insert vane
2. Wait 5 minutes
3. Conduct undisturbed test rotating at 0.2° per second
4. Rotate vane 12 times
5. Wait 5 minutes
6. Conduct disturbed test rotating vane at 0.2° per second

REMARKS :

Figure 1.32 VST results in marine clay, in situ test and test on remolded sample.

Figure 1.33 Typical CPT testing device.

For interpretation of the CPT results, various correlations are used by engineers. According to Mayne and Kemper (1988), for the undrained shear strength S_u, it is commonly taken to be

$$S_u = (q_c - \sigma'_v)/N_K, \quad \sigma'_v \text{ is the effective overburden stress at the point under consideration} \tag{1.9}$$

N_k is a factor which varies between 15–16 for normally consolidated clay and 18–20 for over-consolidated clay. For calibration of this factor, a commonly used approach is to carry out laboratory tests for some samples. With the N_k value established based on some samples, it is taken to be applicable to other locations and depths for a typical project. The friction angle of soil is usually taken as:

$$\phi = \tan^{-1}\left[0.1 + 0.38\log\left(\frac{q_c}{\sigma'_v}\right)\right] \tag{1.10}$$

The Young's modulus of soil is usually taken to be proportional to q_c, and the proportional constant is typically taken as 2.5–3.5. For the volume compressibility of clay, it is commonly taken $1/\alpha q_c$, while α typically ranges from 5 to 7.5. Again, relevant statistical results should be consulted instead of relying on typical results. One important factor which is called the friction ratio is also usually used, which is defined as the percentage ratio between f_c/q_c. In general, this ratio is low for sandy soil and greater for clayey soil. As shown in Figure 1.34, the top soil from ground to a depth of 2 m, and the soil at a depth between 10–12m are clayey in nature, as compared with the rest of the soil mass. There are several design figures which can be related to this friction ratio and the soil classification.

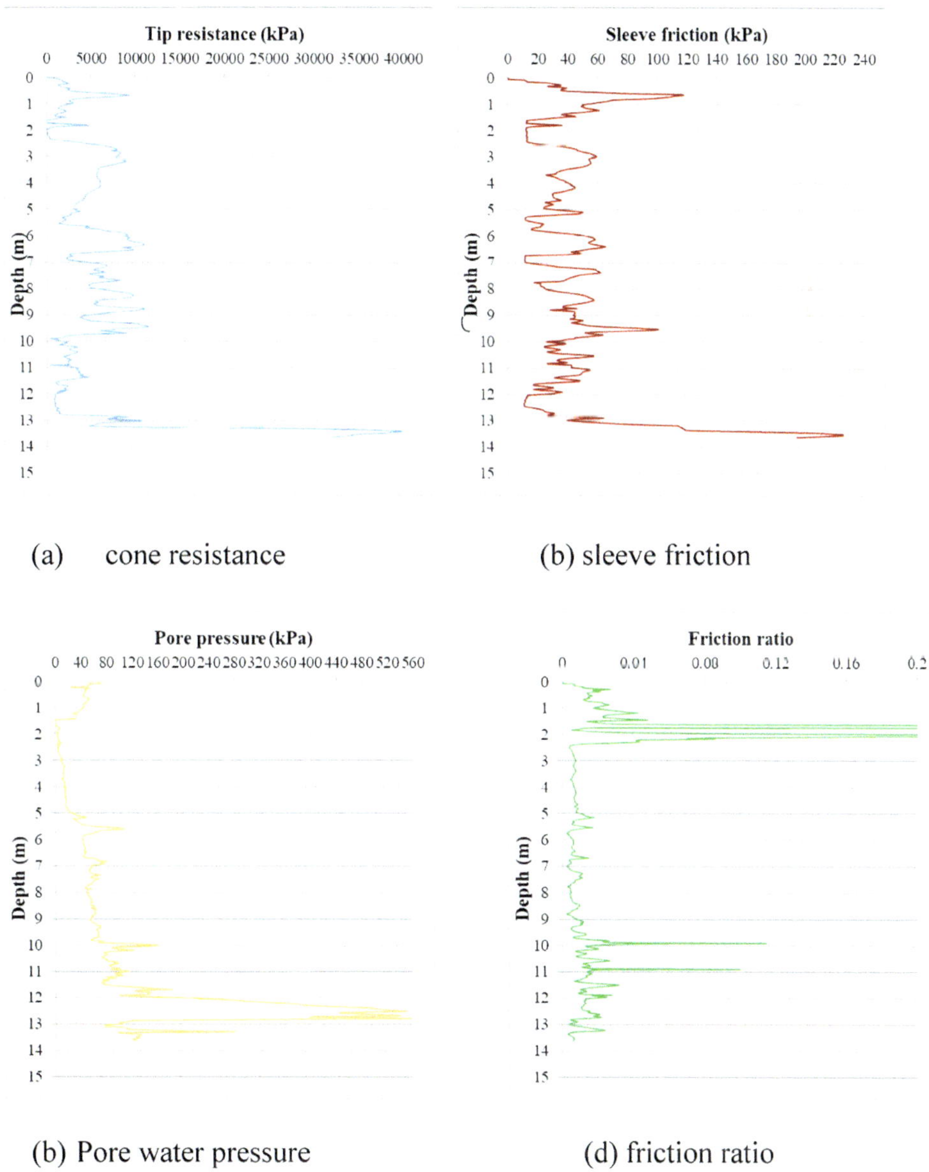

Figure 1.34 CPTU results.

Since CPT cannot recover any soil sample, these design figures are useful to engineers in assessing the properties of the soil. Again, there are several similar figures in the literature, and the differences between design figures are minor. The authors, however, have some doubts on some of these design figures which can give very fine soil classifications. Personally, the authors doubt the precise soil classification obtained from such design figures. In the literature, there are many correlation relations established between various soil parameters and CPT tests. Engineers should always refer to the published local results for consultation.

The advantages of CPT include:

1. More reliable and consistent results and faster than SPT.
2. Can be used for most type of soil conditions.
3. Continuous results, and large amount of data available for interpretation (see Figure 1.34).
4. Relatively cheap if used in great numbers.
5. Avoid human error.

The limitations of CPT include:

1. Not suitable in the presence of cobbles or boulder.
2. Samples cannot be recovered.
3. Needs calibration in some cases.

The piezocone was used on the Hong Kong Chek Lap Kok Airport project tool for checking the compaction of the fill material. After a series of trial tests, the compaction of the fill was considered to be acceptable if it fulfilled the following criteria:

1. The friction ratio was less than 45%;
2. The deviation between the pore pressure and hydrostatic pressure was no more than 0.2 bar;
3. The cone resistance was not less than 4 MPa; or the fines content in the gradings were less than 20% when the vibrocore samples were taken.

CPT is also commonly used to assess the effectiveness of backfill compaction or grouting. In Figure 1.35, the CPT values before and after grouting for a project in Australia are shown. It can be seen that the grouting has achieved a good improvement in the soil properties after grouting, and this can be tested easily and quickly using the CPT tests. Such an approach is also used for many reclamation works (West Kowloon reclamation and many others) and the third airport runway construction in Hong Kong.

1.7.4 Pressuremeter test (PMT)

The pressuremeter test, as shown in Figure 1.36, is an in situ test conducted in a borehole. It was originally developed by Menard (1956) to measure the strength and deformability of soil. The Menard-type PMT essentially consists of a probe with three cells. The top and bottom ones are *guard cells* and the middle one is the *measuring cell*. The use of guard cells

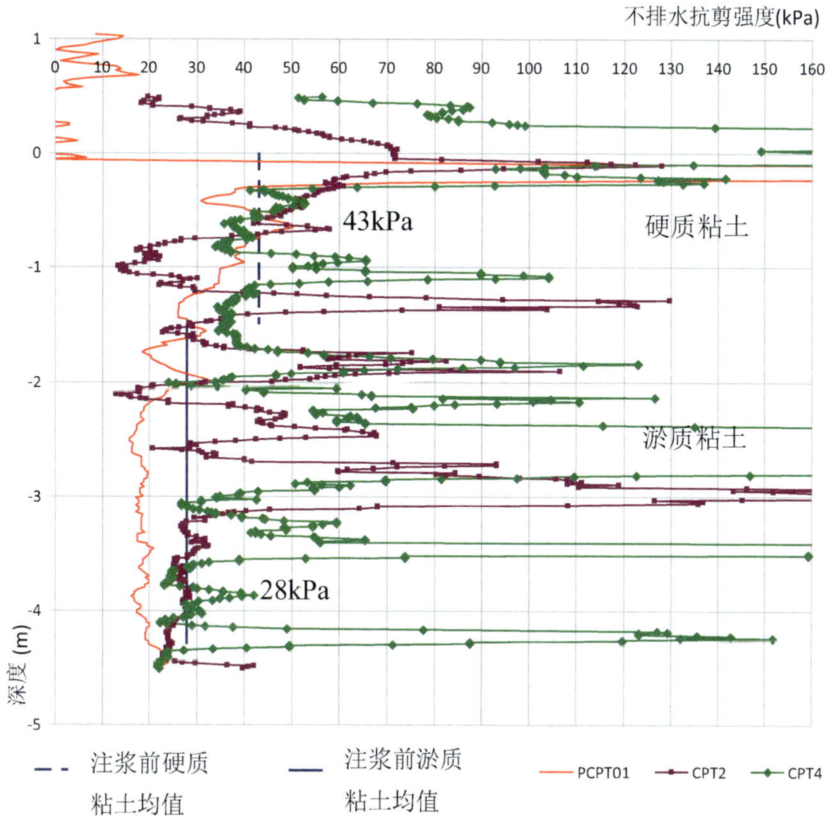

Figure 1.35 Application of CPTU results in ground treatment work.

at the top and bottom produces a near uniform displacement of the measuring cell, so that the relation between the constant displacement and the soil properties can be established. The Menard type probe is conducted in a pre-bored hole, while the self-boring type probe goes along with the boring. The pre-bore hole should have a diameter that is between 1.03–1.2 times the nominal diameter of the probe. The probe that is most commonly used has a diameter of 58 mm and a length of 420 mm. The probe cells can be expanded either by liquid or gas. The guard cells are expanded to reduce the end-condition effect on the measuring cell so that the deformation within the testing probe can be considered as uniform.

The original cell volume is measured and the probe is inserted into the borehole. Pressure is applied in increments and the volume expansion of the cell or the increase in the radial diameter is measured. This is continued until the soil fails or until the ultimate pressure limit of the device is reached. In practice, the test will usually stop when the volume of expansion is 100%. After completion of the test, the probe is deflated and advanced for testing at another depth.

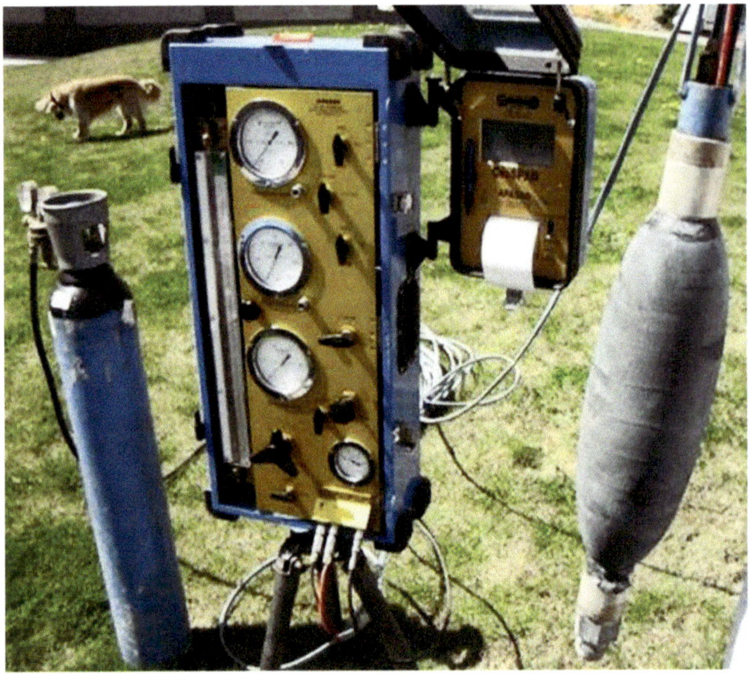

Figure 1.36 Pressuremeter test device.

The results of the pressuremeter test are expressed in a graphical form of pressure versus volume, as shown in Figure 1.37. In this figure, zone I represents the reloading portion during which the soil around the borehole is pushed back into the initial state (that is, the state it was in before drilling). The pressure, p_0, represents the in situ total horizontal stress. Zone II represents a pseudo-elastic zone in which the cell volume versus cell pressure is practically linear. The pressure, p_f, represents the creep, or yield, pressure. The zone marked III is the plastic zone. The pressure, p_l, represents the limit pressure. In order to overcome the difficulty of preparing the borehole to the proper size, self-boring pressuremeters (SBPMT) have also been developed. This is a more expensive type of testing probe but provides more accurate results. For the self-boring pressuremeter, zone I actually does not exist.

The pressuremeter modulus, E_p, of the soil is determined using the theory of expansion of an infinitely thick cylinder. Based on the linear part of the test results, the pressuremeter modulus E_m is given by

$$E_m = 2(1+\mu)(V_c + v_m)\left(\frac{\Delta p}{\Delta v}\right) \tag{1.11}$$

where $v_m = \dfrac{V_0 + V_f}{2}$, $\Delta p = p_f - p_0$, V_c = initial volume of the probe, V_0 is explained in Figure 1.37 and V_f is the volume of the probe at the end of the elastic range.

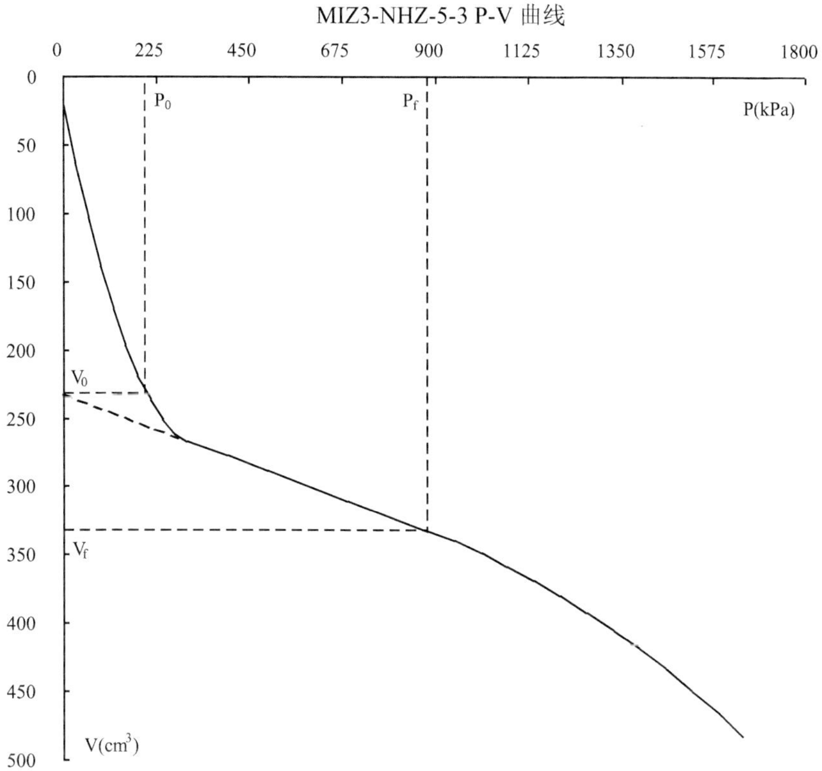

试验结果	P_0=203kPa	V_0=233cm^3	E=24.22MPa
	P_f=884kPa	V_f=332cm^3	E_m=25.39MPa
	P_L=2694kPa	P_f-P_0=681kPa	P_L-P_0=2491kPa

Figure 1.37 A pressuremeter test in China.

The limit pressure, p_f, is usually obtained by extrapolation and not by direct measurement to avoid damage to the probe which is expensive. It should be noted that no volume change is involved in the pressuremeter test, hence it is actually a shear test by nature.

The horizontal pressure, p_h, which is the start of the linear part of the test result, is usually taken as the in situ horizontal stress, hence the approximate at-rest earth pressure coefficient K_0 can be established accordingly. Correlations between various soil parameters and the results obtained from the pressuremeter tests have been developed by various investigators. Kulhawy and Mayne (1990) proposed that

$$p_c = 0.45p_l \tag{1.12}$$

where p_c = preconsolidation pressure. Based on the cavity expansion theory, Baguelin et al. (1978) proposed that

$$c_u = \frac{(p_l - p_0)}{N_p} \tag{1.13}$$

where c_u = undrained shear strength of a clay.

$$N_p = 1 + \ln\left(\frac{E_p}{3c_u}\right) \tag{1.14}$$

Typical values of N_p vary between 5–12, with an average of about 8.5. Ohya et al. (1982) correlated E_p with the field standard penetration number (N_F) for sand and clay as follows:

$$\text{Clay: } E_p\left(\text{kN/m}^2\right) = 1930 N_F^{0.63} \tag{1.15}$$

$$\text{Sand: } E_p\left(\text{kN/m}^2\right) = 908 N_F^{0.66} \tag{1.16}$$

Again, the correlation relations as given above are based on the statistics of some specific soils, and they cannot be applied to general cases.

Self-boring pressuremeter tests have been conducted in Hong Kong on several projects. For Contract No. HK/13/02 in 2003, the Wan Chai Development Phase 2 ground investigation, a total of 129 tests were undertaken at 46 marine drillhole locations. The tests were carried out in soft marine and variable alluvial deposits as well as some tests in granite derived saprolite. Two types of equipment manufactured by Cambridge were used: the conventional Self Boring Pressuremeter (SBP) and the Weak Rock Self Boring Pressuremeter (WRSBP). These pieces of equipment are essentially the same, but the WRSBP is fitted with a stronger membrane and the cutting system is arranged to slightly drill oversized.

1.7.5 Dilatometer test (DMT)

A dilatometer is a stainless steel blade with a flat, circular steel membrane mounted flush on one side, as shown in Figure 1.38, which was developed by Marchetti (1980). The blade is connected to a control unit on the ground surface by a pneumatic-electrical tube which transmits gas pressure and electricity during rod insertion. A gas tank which is connected to the control unit by a pneumatic cable supplies the gas pressure to expand the membrane. The control unit is equipped with pressure gauges, a pressure regulator and vent valves. The test procedure is given in Eurocode 7 (1997) "Geotechnical Design. Part 3: Design assisted by field tests, Flat Dilatometer Test (DMT)" and other similar specifications.

The DMT starts with insertion of the dilatometer. At the point under testing, the membrane will be inflated, and two readings are taken in about 1 minute:

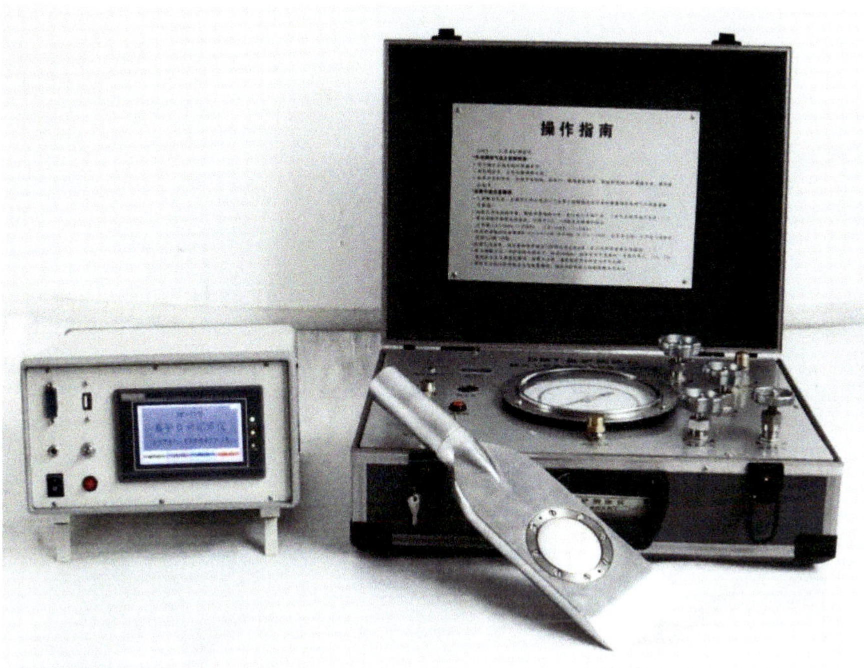

Figure 1.38 Dilatometer testing device.

1) The A-pressure p1 (kPa) at which the membrane just moves against the soil (termed "lift-off");
2) The B-pressure p2 at which the center of the membrane has moved by 1.1 mm against the soil;
3) An optional third reading, p3 (termed "closing pressure"), can be taken by slowly deflating the membrane soon after B is reached. According to Schmertmann (1986), this latter reading can be related to excess pore pressure.

The pressure readings A and B must then be corrected by the values ΔA and ΔB determined by calibration, to take into account the membrane stiffness, and converted into p1 and p2. The detailed procedure for setting up the test is provided by Marchetti (1980).

DMT can be applied to extremely soft soils to hard soils/soft rocks, for soils where the grains are small compared to the membrane diameter (60 mm), but is not suitable for use in gravels. The DMT readings are highly accurate, even in extremely soft/nearly liquid soils, and the blade is very robust and can withstand up to 250 kN of pushing force. The undrained shear strength for clays which can be tested by DMT ranged from 2–4 kPa up to 1000 kPa, while the range for moduli M is from 0.4 MPa up to 400 MPa.

According to Marchetti (1980),

$$d = \frac{2D(p_l - p_0)}{\pi}\left(\frac{1-\mu^2}{E_s}\right) = 34.7(p_l - p_0) \quad \text{for d=1.1mm} \tag{1.17}$$

The lateral stress and material indices (K_D, I_D) have been defined by Marchetti (1980) as

$$K_D = \frac{p_1 - u}{p_0'} \quad (1.18)$$

$$I_D = \frac{p_2 - p_1}{p_2 - u} \quad (1.19)$$

where the existing overburden stress p_0 has to be determined by the knowledge of the unit weight, and the water table has to be known. Based on these indices, some empirical correlations and design figures for the relations between various soil parameters and DMT results are established, which are however not generally universal. Some examples for the over-consolidation ratio (OCR) and the at-rest earth pressure coefficient K_0 are given:

$$OCR = 0.24K_D^{1.32} \quad (1.20)$$

$$K_0 = \left(\frac{K_D}{1.5}\right)^{0.47} - 0.6 \text{ in clay} \quad (1.21)$$

$$E_s = (1 - \mu^2)E_D \quad (1.22)$$

$$C_u = 0.35p_0'(0.47K_D)^{1.14} \quad (1.23)$$

$$\phi' = 28 + 14.6\log K_D - 2.1(\log K_D)^2 \quad (1.24)$$

Many soil classification design figures and other correlations between soil parameters and DMT results are available in the literature. These design tools can be useful to engineers, provided that they are suitable for the type of soil under consideration.

Trial tests were conducted by Schnaid et al. (2000) on a site at Kowloon Bay adjacent to the old Kai Tak Airport Site. A total of nine drillholes were undertaken in which standard penetration test (SPT), self-boring pressuremeter (SBP) and Marchetti dilatometer (DMT) tests were undertaken. Except for the trial tests, DMT appears to be not used for practical engineering projects in Hong Kong, though it is accepted for engineering investigations in many other countries.

1.7.6 Other in situ tests

For a typical borehole, there are some other in situ tests which may be carried out, which include the installation of piezometer/standpipe, and the test for the water table in soil, packer (lugeon) test and impression packer test. In a typical piezometer, as shown in Figure 1.39, a perforated piezometer tip is installed and the borehole is filled with sand filter. A dip meter is lowered into the standpipe within the piezometer, and there is an

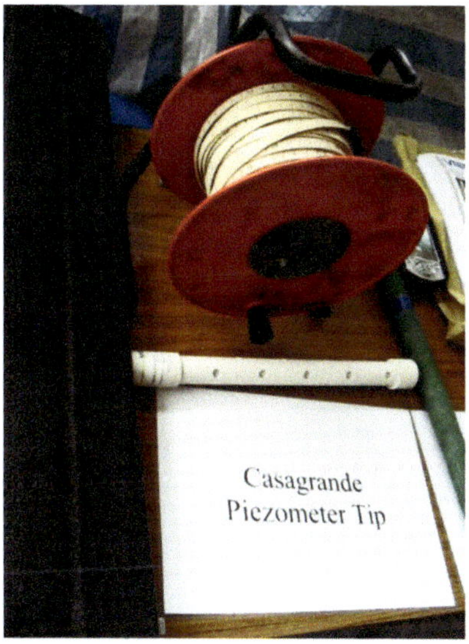

(a) Casagrande tip

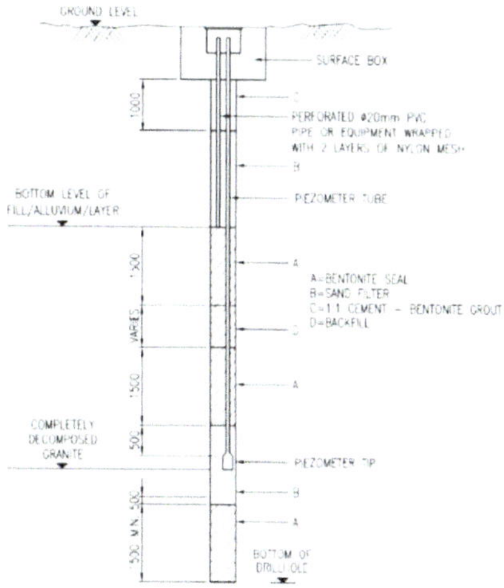

(b) typical piezometer and standpipe

Figure 1.39 Typical details of a piezometer.

incomplete circuit at the tip of the meter. When the dip meter reaches the water table, the circuit is completed, and sound will be emitted, hence the location of the water table is detected. It should however be noted that there may be false alarms (experienced by the authors) in the field, and technicians should lift the dip meter up and down several times to confirm the water table. Besides the ground water table, the falling head test is also commonly carried out to assess the permeability of a large volume of soil. It should be noted that the permeability obtained from a field falling head test usually differs with that resulting from laboratory tests by several times, as the field value represents the average permeability of a large volume of soil mass. Hvorslev (1951) and BS 5930 provide guidelines for the assessment of the field permeability test, and the permeability is given by

$$k = A \ln (h_t/h_0)/FT \text{ or } k = A/FT \tag{1.25}$$

where k is the permeability, A is the cross-sectional area of the borehole, T is taken at the time where $\ln (h_t/h_0)$, and h_t and h_0 are the height of the water table above the original water table at time t and the start of the test, respectively, and F is a factor which is given in BS5930. For a test with casing and where the bottom of the borehole is not at the rock head or an impermeable surface, F can be taken as

$$F = \frac{2\pi L}{ln\left[\frac{L}{D} + \sqrt{1 + \left(\frac{L}{D}\right)2}\right]} \tag{1.26}$$

where L and D are the length and diameter of the section of borehole without casing. The results for a constant head test in Hong Kong are shown in Figure 1.40.

For seepage in rock, since there is no continuous seepage path in rock, the concept of permeability is different from that for soil. The amount of water injection under a given pressure can give an indication of the degree of fracture. There are two types of packer tests for seepage in rock: single packer and double packer, as shown in Figure 1.41.

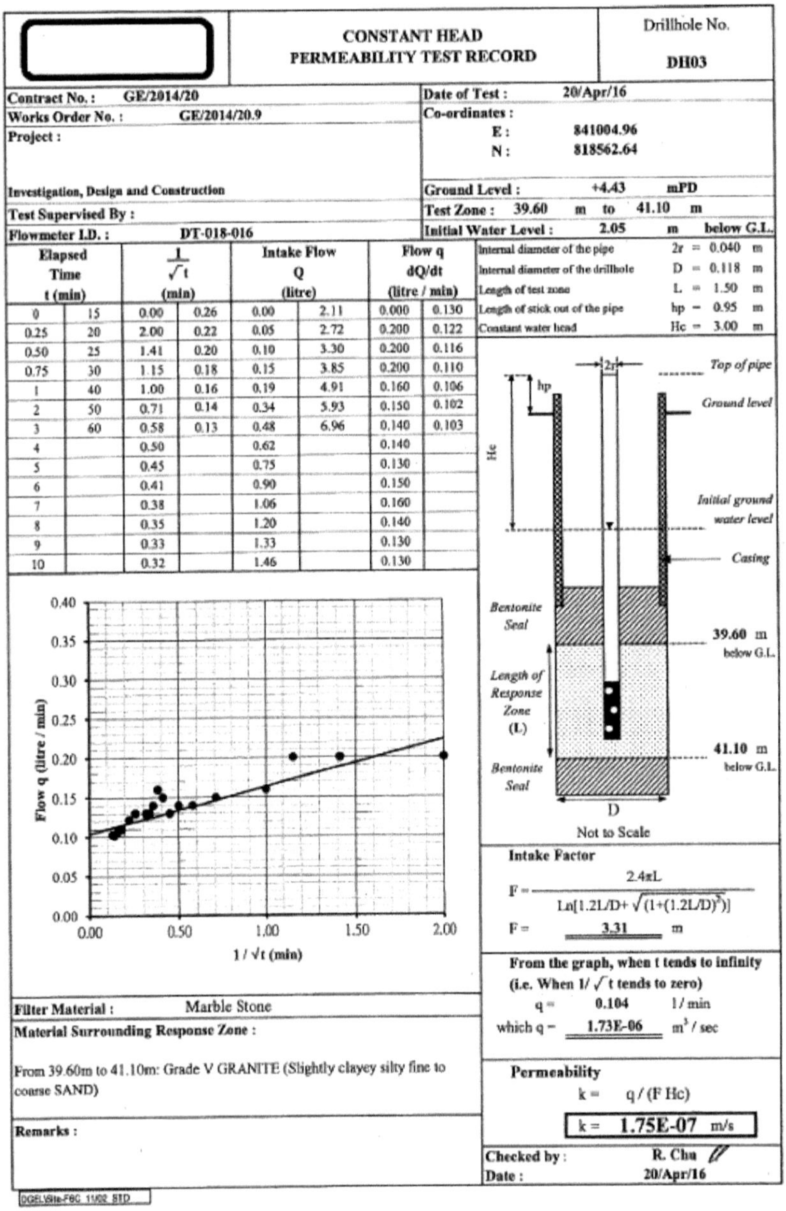

CONSTANT HEAD PERMEABILITY TEST RECORD

Drillhole No. **DH03**

Contract No. : GE/2014/20
Works Order No. : GE/2014/20.9
Project : Investigation, Design and Construction
Test Supervised By :
Flowmeter I.D. : DT-018-016

Date of Test : 20/Apr/16
Co-ordinates : E: 841004.96 N: 818562.64
Ground Level : +4.43 mPD
Test Zone : 39.60 m to 41.10 m
Initial Water Level : 2.05 m below G.L.

Elapsed Time t (min)		1/√t (min)		Intake Flow Q (litre)		Flow q dQ/dt (litre / min)	
0	15	0.00	0.26	0.00	2.11	0.000	0.130
0.25	20	2.00	0.22	0.05	2.72	0.200	0.122
0.50	25	1.41	0.20	0.10	3.30	0.200	0.116
0.75	30	1.15	0.18	0.15	3.85	0.200	0.110
1	40	1.00	0.16	0.19	4.91	0.160	0.106
2	50	0.71	0.14	0.34	5.93	0.150	0.102
3	60	0.58	0.13	0.48	6.96	0.140	0.103
4		0.50		0.62		0.140	
5		0.45		0.75		0.130	
6		0.41		0.90		0.150	
7		0.38		1.06		0.160	
8		0.35		1.20		0.140	
9		0.33		1.33		0.130	
10		0.32		1.46		0.130	

Internal diameter of the pipe	2r = 0.040 m
Internal diameter of the drillhole	D = 0.118 m
Length of test zone	L = 1.50 m
Length of stick out of the pipe	hp = 0.95 m
Constant water head	Hc = 3.00 m

Intake Factor

$F = \dfrac{2.4\pi L}{\ln[1.2L/D + \sqrt{(1+(1.2L/D)^2)}]}$

F = 3.31 m

From the graph, when t tends to infinity (i.e. When 1/√t tends to zero)

q = 0.104 l / min

which q = 1.73E-06 m³ / sec

Permeability

k = q / (F Hc)

k = **1.75E-07** m/s

Filter Material : Marble Stone

Material Surrounding Response Zone :

From 39.60m to 41.10m: Grade V GRANITE (Slightly clayey silty fine to coarse SAND)

Remarks :

Checked by : R. Chu
Date : 20/Apr/16

DGEL\Site-F6C 11/02 STD

Figure 1.40 Constant head test for the determination of permeability of soil.

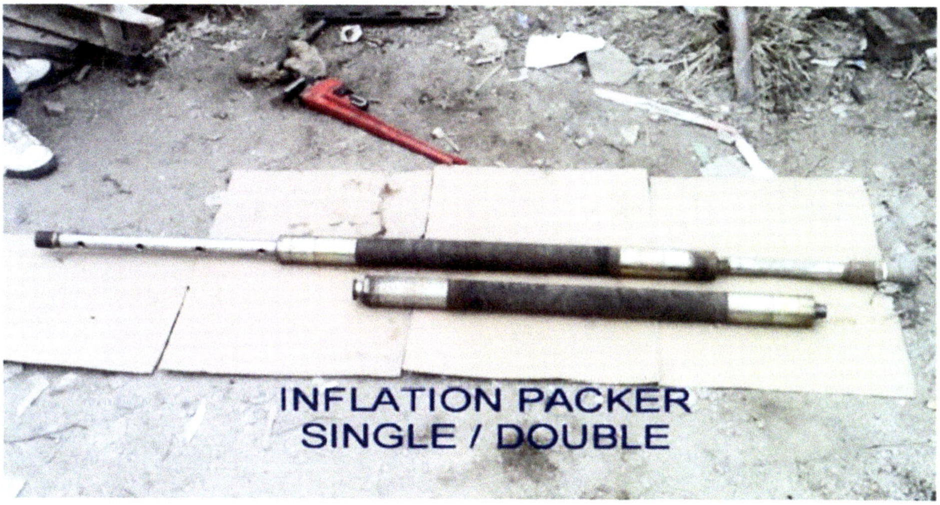

Figure 1.41 Packer test in Macau.

From the rock core on the ground surface, it is not possible to construct the 3D joint orientation. In the impression packer test as shown in Figure 1.42, a thin thermoplastic film is pressed to the surface of the rock by the internal packer. The marks of the joints/fractures are left on the film and are used to re-construct the 3D discontinuity profile. Alternatively, a small camera is used to capture the condition of the rock surface. This has been quite commonly adopted recently.

1.8 GEOPHYSICAL EXPLORATION

The classical site investigation with an in situ test can only provide the information for a limited area, which may be sufficient for many construction works. For exploration of a large area, the use of bore explorations will be very expensive, and bore exploration will be necessary only in some critical locations. To obtain reasonable geological information without expensive bore exploration, geophysical exploration techniques are an alternative. This permits a rapid and economical evaluation of the approximate underground conditions and some soil properties which will be sufficient for the preliminary assessment of the site condition. There are many geophysical exploration techniques available at present, and some of the more important methods include the seismic refraction test, cross-hole or single hole shear wave test, resistivity/magnetic survey and others. Interested readers can consult the various references suggested in this chapter.

The seismic cone penetration cone is a CPT or CPTU cone that is equipped with geophone sensors. The seismic cone penetration testing was undertaken for the Tung Chung New Town Development in 1995. This test is a fast and cost-effective technique to determine the in situ seismic wave velocities. In the seismic test method, the velocities of the body waves are measured (shear wave and primary pressure wave) by a wave source on the ground surface and an array of geophones in an in situ seismic cone penetrometer.

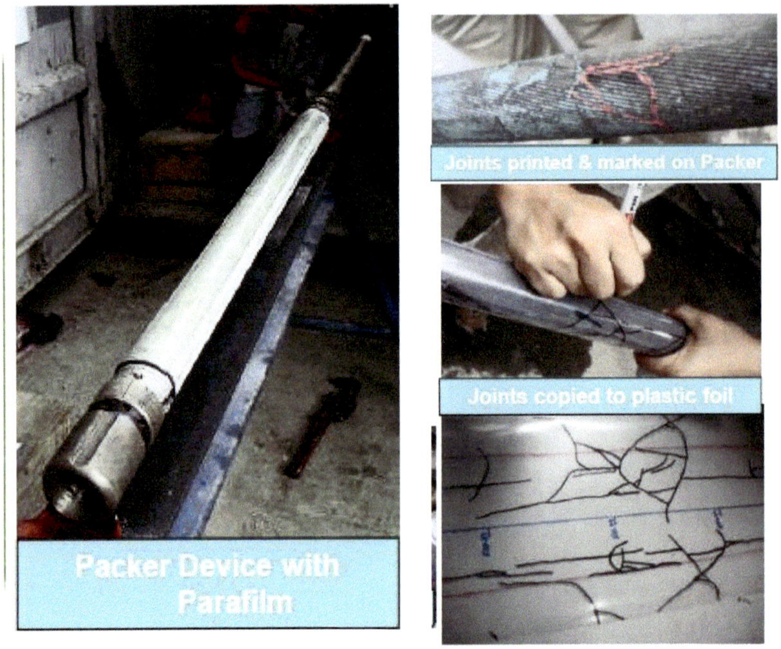

Figure 1.42 Impression packer test.

The seismic wave velocities give an indication of ground characteristics, including the low-strain shear modulus and Poisson's ratio. Furthermore, earthquake-resistant design of structures according to Eurocode 8 requires classification of the stratigraphic profiles based on the seismic wave velocities. In addition, Eurocode 8 recommends cone penetration tests (CPTs) for stratigraphic profiling and for the assessment of soil liquefaction potential. Seismic wave velocities also allow correlation of seismic profiling data and geotechnical borehole and the CPT results.

1.9 ROCK AS ENGINEERING MATERIAL

In many developed countries, tall buildings are generally supported on piles. If the loadings are high, the use of bore pile founded on rock is common. The stability of rock slopes and caverns is also an important topic in geotechnical engineering. In this section, some basic knowledge of rock as an engineering material will be discussed, and readers are strongly advised to consult other references for details of engineering geology. It is also generally accepted that unforeseen ground condition is one of the main cause for a delay/increase in cost and failures (Chan and Kumaraswamy 1995). The failure of the Hong Kong Admiralty to Wanchai running tunnel in rock caused by an unknown fault and the sharp increase in cost for the Hong Kong Section of the Guangzhou–Shenzhen–Hong Kong Express Rail Link due to unexpected high rock level are examples. A detailed knowledge of engineering geology is highly beneficial for some large-scale geotechnical projects.

In general, fractures in rock may govern the stability of near-surface structures, and also the in situ stresses may govern the stability of deep excavations. Examples include dam foundation, rock cavern, bore pile foundation, slopes, mining and other similar constructions. The effects will further be influenced by factors such as wet or dry condition, cold or hot, stable or squeezing and others. Marble clasts are generally not interconnected and dissolution of marble clasts is localized, leading to honeycomb weathering of the rock or even caverns in marble. The consideration of rock mass is different from that for soil. There are five major areas that influence the geological factors of rocks and rock masses:

1. Intact mass;
2. Discontinuities and rock structure;
3. In situ pre-existing rock stress;
4. Pore fluids and water flow;
5. Time.

1.9.1 Brief discussion of rock types

Before the discussion of these issues, we provide a brief discussion of the rock types that are most commonly found.

1.9.1.1 Igneous rock

Igneous rocks form from hot, molten rock. The magma originates deep within the Earth and rises toward the surface. Igneous rocks are further sub-divided into intrusive or extrusive rocks, depending on the location in the Earth where the magma solidifies.

A) Intrusive, or plutonic, igneous rocks are formed when the rising magma is trapped deep within the Earth and cools very slowly over thousands or even millions of years until it finally solidifies. This very slow cooling allows the individual mineral grains to grow and form relatively large crystals. Intrusive rocks have a coarse-grained texture with interlocking minerals.
B) Extrusive, or volcanic, igneous rocks are produced when magma erupts close to the Earth's surface. The erupted magma cools and solidifies relatively quickly when it is exposed to the atmosphere. Lava and tuff are two common types of volcanic rocks.

Two important igneous rocks are used to illustrate the characteristics of these groups. Granitic rock (intrusive) contains predominantly feldspar and quartz minerals, with some amphiboles, and micas. It can occur as plutons, dykes or sills. Individual minerals can generally be identified with the naked eye. Minerals are crystalline and show an interlocking texture. Granitic rock is strong, and is an ideal material for concrete production. Tuff (extrusive) contains fragments of minerals, glass, pumice and/or pre-existing rocks, and is classified on the basis of the relative components of the various fragments. The fragments are generally angular and broken. Tuff is commonly dark gray in color when the rock is unweathered. The rock may show a welding structure, while some tuff is columnar-jointed.

1.9.1.2 Sedimentary rock

Sedimentary rocks are formed from the eroded fragments of rocks, with the possibility of skeletal fragments of plants or organisms, and the transformation of sediments into a rock is termed as lithification. The amount of water expulsion during the transformation determines the porosity of the rock. The water content may be reduced by 80% for mudstone, while for sandstone, more voids will be retained. Sedimentary rocks usually demonstrate distinctive layering or bedding, and can be further sub-divided into three groups, including clastic, biological and chemical. Bedding is the primary feature in sedimentary rock. Adjacent sedimentary beds can however have substantially different weathering, composition and engineering properties. Bedding surfaces are usually sub-planar and sub-horizontal in their original depositional setting. However, post-depositional deformation and uplift have re-orientated most bedding planes, and many beddings are steeply dipping or sub-vertical. Graded bedding is found with settling out of the heavier grains with the fine- to medium-grained bedding at the top.

Clastic sedimentary rock is composed of rock and mineral grains eroded from rocks, and individual grains are held together by a cement that is commonly composed of quartz or calcite minerals. Clastic sedimentary rocks may contain fossils, bedding may be also present, and is defined by variations in the texture and composition of the constituent grains that are systematically arranged in layers. Biological sedimentary rock is formed by the accumulation of large amounts of dead plants or organisms which are compressed and cemented to form rock. Chemical sedimentary rock is formed by chemical precipitation from solutions, which begins when water passes through rock and dissolves and carries some of the minerals away. The minerals are finally deposited or precipitated, and rock salts are examples of chemical sedimentary rocks.

1.9.1.3 Metamorphic rocks

Metamorphic rock is formed when rock is subjected to high temperature, high pressure, hot and mineral-rich fluid, or a combination of these conditions. Metamorphic rock is usually formed deep within the Earth. There are two major types of metamorphic rocks.

Foliated metamorphic rock exhibits a platy or sheet-like structure, and the foliation develops when platy or prismatic minerals within the rock are compressed and aligned under extreme pressure. Examples of this group include slate, schist and gneiss.

Non-foliated metamorphic rock displays a massive structure, and it can be formed by contact metamorphism that occurs around intrusive igneous rocks. The existing rocks that come into contact with the intruding igneous rocks are baked by the heat, and the mineral structures of the pre-existing rocks are changed without intense pressure. Quartzite and marble are some typical examples of non-foliated metamorphic rocks.

1.9.2 Joints and discontinuity in rock

Joints are fractures or cracks in rocks. The patterns of joints generally have a characteristic geometry and a regular spacing. There are three major types of joints:

1. *Tectonic joints* are associated with regional tectonic deformation which may be formed under shear or tension. They facilitate the infiltration of groundwater and

the development of linear depressions in the rockhead. Tectonic joints associated with igneous intrusions generally are more localized, and form impersistent discontinuities that peter-out away from the contact zones. Tectonic joints formed under tension generally have rougher surfaces as compared with that formed under shear.

2. *Stress relief joints* usually develop in rocks close to the ground surface, as a result of the relaxation of confining pressure (overburden) followed by erosion of the overlying layers. For large-scale joints subparallel to the topography, they are called sheeting joints. An example is a major slope failure along a sheeting joint in Sai Kung, and three-dimensional slope stability has been carried out by Cheng. Smaller scale curved or concentric joints are called exfoliation joints. In general, the joint spacing in coarse-grained rocks is wider than that in fine-grained rocks. Stress relief joints in granitic rocks may be very persistent, extending for several hundreds of meters. On a local scale, they may facilitate the formation of exfoliation joints associated with corestone development.
3. *Cooling joints* develop from the cooling and contraction in granitic and volcanic rocks following their emplacement. They are typically perpendicular to the cooling surface and may form hexagonal columns, such as a famous type of rock in Sai Kung, Hong Kong.

Faults and the associated structures are critical in many engineering projects. Brittle faults are the most recent stage of movement for a variety of rock types and structures, and the fault zone is often annealed by secondary minerals. Angular rock and mineral fragments in a dominantly rich-clay matrix with sharp boundaries to the fault zones are commonly found. It is characterized by a high-level brittle type of deformation. In Hong Kong, the orientation of fault planes commonly dips steeply. The movement direction can usually be identified from the tectonic structures. A fault gouge in a fault zone can be composed of extremely weak material, however, strong quartz lenses may also be present in the fault gouge. Fault gouge zones usually have very sharp contacts with the host rock, and the contact may display grooves or mineral growths parallel to the fault movement. Fault breccia may also be by fragmentation of rock through brittle failure induced by tectonic movements. Slickensides are formed by the movement of two rock masses against each other to produce a highly polished surface. Ductile faults or shear zones are formed with no brittle failure. The fault displays many structures including foliation, S-C fabrics, folded mineral veins, mineral elongations and slickensides, which indicate the direction of movement. In extreme ductile faults, the rocks are significantly sheared to form very fine-grained, highly foliated rocks called mylonites.

Foliations are penetrative, very closely spaced, planar to wavy tectonic discontinuities formed during folding or faulting. In Hong Kong, foliations are found in most of the dynamically metamorphosed sedimentary rock as well as granite and volcanic rocks close to the faults. The foliation planes are commonly defined by the alignment of mica, and rock can split easily along foliation planes. This provides information for the orientation of the regional or local stress orientations and movement. Foliated rocks can be extremely fissile and friable, and this can increase the depth and intensity of weathering. It is the obvious plane of weakness in rock.

Folds are formed in a compression tectonic environment which can affect the whole rock sequence or just narrow zones. Some folds are, however, formed by slumping of sediments shortly after their deposition. This can provide information about the regional stress field.

Discontinuity is the separation in rock continuum with effectively zero tensile strength. Very often, the process by which a discontinuity is formed will determine its geometrical and mechanical properties. A particularly large and persistent discontinuity could critically affect the stability of the rock slope and underground excavation. Some well-known rock slope and tunnel failures in Hong Kong were caused by the presence of discontinuities. For example, the large-scale collapse of the Hong Kong Admiralty MTR in rock was caused by a discontinuity which was not found during the site investigation. A good understanding of the geometrical, mechanical and hydrological properties of discontinuities will be important for the design and construction of rock slopes, tunnels and foundations. Some of the critical considerations for discontinuity include the following:

1. *Spacing and frequency*: spacing of the discontinuity and the frequency which is the reciprocal of spacing. This is a very important factor for the global rock mass.
2. *Orientation, dip direction/dip angle*: for the design and construction of rock slopes and tunnels, this is an important consideration.
3. *Persistence, size and shape*: the extent of the discontinuity can be an important factor in the overall planning of the construction.
4. *Roughness*: the roughness is important in the stability of the rock mass.
5. *Aperture*: the aperture is another important consideration for construction. For the Fortress Hill MTR rock tunnel construction in Hong Kong, the amount of water coming out from the roof and shoulder was very significant (like a mild raining), and the main reason for this was the large aperture of the discontinuity.
6. *Discontinuity sets*: the number of discontinuity sets that characterize a particular rock mass geometry have to be determined, for the stability analysis of the rock mass.
7. *Block size*: this can be important for the design of rock bolts in rock slope and tunnel constructions.
8. *Infilling*: infill materials are usually soil, disintegrated rock or minerals which are weaker than the parent rock. This can be important for the stability of tunnel or rock slope.
9. *Seepage*: seepage can be an important engineering feature, which is a consideration for rock tunneling. For example, the amount of seepage during the rock tunneling construction of the Admiralty station tunnel of the HK MTR was minor, and no special precautions were required. For the Tin Hau-Fortress running tunnel of the HK MTR, the amount of seepage was very significant, and special arrangements to divert the water had to be implemented.
10. *Fracture state*: some typical indices are used for the quantitative description of the rock mass, which include the total core recovery, solid core recovery, rock quality designation and fracture index.

Karst is an important landform in Hong Kong and many other places, and some discussions about pile construction in karst zones are provided in Chapter 6. Dissolution of calcium carbonate in ground water with the formation of cavities is found in Yuen Long and surrounding areas, as well as some regions in Shenzhen. Most of the cavities are, however, partially or completely filled by wash-in sediments or collapse materials from the roof of the cavity in Hong Kong. Large cavities form within substantial underground drainage systems during high water flow, and such underground drainage has posed great engineering problems in Shenzhen.

The structural geology can greatly affect the design and construction of many tunnels, slopes and foundations, and more discussion on this can be found in Mukherjee (2015) and Groshong (2006).

1.9.3 Description of rock

The description of rock gives an indication of its likely engineering properties. An experienced engineer can give a quick estimation of the bearing capacity of rock foundation. The description of rock should include the name, descriptive terms for the strength, color, texture or structure, grain size, state of weathering and alteration, presence of discontinuities and other characteristics as appropriate. Following Hawkins (1984), the main characteristics that should be described include:

(a) Strength (material): which is usually based on the unconfined compression test or the point load index test.
(b) Color (material): whether the rock is dry or wet, and the descriptions should be supplemented by color photographs.
(c) Texture and fabric (material) and the structure (mass): texture refers to the general physical appearance of a rock such as the size and shape of the component grains or crystals, and the relationships between these aspects, while fabric refers specifically to the arrangement of the constituent grains or crystals in the rock
(d) State of and alteration (material and mass): weathering has a very significant effect on the engineering properties of rock. Most engineering projects in Hong Kong encounter substantial thicknesses of weathered rock, which may vary significantly on the degree of weathering over relatively short distances. Cheng has worked on several tunnel constructions in Hong Kong, and the importance of weathering on the stability of tunnels is obvious. In general, weathering is classified into six scales, from fresh rock (grade 1) to residual soil (grade 6). Rock is usually defined as materials from grades 1 to 4, but some engineers also define grade 4 materials (highly decomposed) as soil, and sometimes this is a source of dispute for the BQ rating.
(e) Rock name, including grain size (material).
(f) Discontinuities (mass).
(g) Any important additional geological information.

There are some important indices which are determined during the site investigation:

(a) Core recovery: the ratio of the recovery length of rock core against the length of the barrel.
(b) Rock quality designation: same as the core recovery, except that for rock fragments less than 100 mm in length, these rock fragments will be discarded.
(c) Fracture index: number of fractures per unit length.

In Hong Kong, the bearing capacity of rock is determined by the core recovery, which is given in Table 2.2 of the Hong Kong Code of Practice for Foundations. RQD is more used for assessing the stability of rock tunnels during construction and slope stability. RQD is also an important factor in the rock mass rating system, and the more common rating systems include the Terzaghi rock classification, rock structure rating (RSR), rock mass rating (RMR), Q system, geological strength index (GSI) and others. The indices are based on statistics and observations, and have been used by engineers for various projects around the world. Although these indices are empirical, they are useful for the preliminary design of rock tunnels and slopes.

In general, RQD can be related to the quality of rock mass as follows:

- <25% very poor;
- 25–50% poor;
- 50–75% fair;
- 75–90% good;
- 90–100% excellent.

1.9.4 Test for rock specimens

For laboratory tests, the unconfined compression test as shown in Figure 1.43 is the most important test for rock specimens. Some typical failures during the unconfined compression tests are shown in Figure 1.44. The procedures for the UCS are:

Figure 1.43 Laboratory uniaxial compression test.

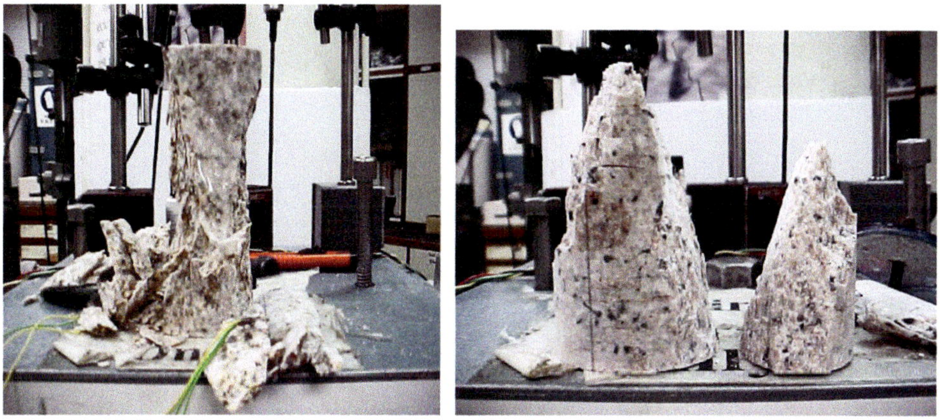

(a) Low end constraints (b) high end constraints

Figure 1.44 Unconfined compression test results.

1. The specimen shall be tested at a moisture content as close to field conditions as possible.
2. The L/D ratio should preferably be 2–3. If L/D < 2, then correction shall be applied.
3. The diameter of the specimen must be greater than 10 times the diameter of the largest grain size. Preferable diameter is 45 mm; in no case should the diameter of the core sample be less than 35 mm.
4. The specimen should be prepared in such a manner that the cylindrical surface shall be smooth and free from irregularities.
5. The diameter of the test specimen shall be measured to the nearest 0.1 mm by averaging two diameters measured at right angles to each other at about the upper height, the mid height and the lower height of the specimen.
6. Clean the surfaces of the two bearing discs and the test specimen. Place the specimen on the lower disc.
7. Carefully align the axis of the specimen with the center of thrust of the spherical seat.
8. Apply load continuously at a constant stress rate within the limits of 0.5–1.0 MPa/s.
9. Record the maximum load on the specimen at failure in N within 1% accuracy.
10. At least five specimens are required to give a representative value. The UCS of the specimen is calculated by dividing the maximum load carried by the specimen during the test, by the average cross-sectional area.

For a quick and cheaper assessment, the point load index test can be carried out. The point load index test, as shown in Figure 1.45, is used to provide an indirect estimate of uniaxial compressive strength, which is favored for simple quick site tests. The test is based on the ISRM suggested method for determining point load strength (ISRM Commission 1985). The procedures for this test are as follows:

(a) Laboratory test (b) field test

Figure 1.45 Point load index test.

1. The diametral test is conducted on rock core sample. A minimum of 10 test specimens are required to determine the average value of the point load strength index.
2. This test can be conducted on the core specimens which are completely dry or after soaking for 7 days.
3. Measure the total length (l) and diameter (d) of the core specimen. Specimens of $l/d = 1.5$ are considered suitable for this test.
4. Place the specimen horizontally between two platens in such a way that the distance between the contact point and the nearest free end (L) is at least 0.75 times the diameter of the core (d).
5. Measure the distance between two platen contact points (D) with the help of the scale attached with the loading frame. (Note: in the case of a diametral test, the diameter of the core (d) and the distance between two platens (D) will be the same.)
6. Apply load to the core specimen such that failure occurs within 10–60 s. Record the failure load (P).

From the measured value of the force, P, at which the test specimen breaks, an uncorrected point load index, I_s, is calculated as

$$I_s = \frac{P}{D_e^2} \tag{1.27}$$

where D_e is the equivalent core diameter. The index, I_s, varies with D_e and so size correction must be applied in order to obtain a unique point load strength index for a particular rock

(a) Typical Brazilian test for rock sample (b) failure start at centre and propagates towards the two ends

Figure 1.46 Brazilian test for rock sample.

sample for use in strength classification. Wherever possible, it is preferable to carry out diametral tests on 50–55 mm diameter specimens. The size-corrected point load strength index, $I_{s(50)}$, is defined as the value of I_s that would have been measured in a diametral test with D = 50 mm. The results of several series of tests carried out by a number of investigators show that the value of I_s determined in a test of equivalent diameter, D_e, may be converted to an $I_{s(50)}$ value by the relation

$$I_{s(50)} = Is\left(\frac{50}{D_e}\right)^{0.45} \tag{1.28}$$

Based on series of tests by Broch and Franklin (1972) and others, a correlation between the point load index and the uniaxial compressive strength is established, which is given by:

$$\text{UCS} \approx (22-24)I_{s(50)} \tag{1.29}$$

There are some limitations for the point load index test. For very soft rocks, and highly anisotropic rocks or rocks containing marked planes of weakness, spurious results may be obtained. A high degree of scatter is a general feature of point load test results, and large numbers of tests (often in excess of 100) are required in order to obtain reliable indices.

For the tensile strength of rock, which is also important in some cases, it is obtained in a similar way to that for concrete, and is called the Brazilian test, as shown in Figure 1.46. The Brazilian test is a geotechnical laboratory test for indirect measurement of the tensile strength of rocks. Due to its simplicity and efficiency, it is among the most commonly used laboratory testing methods in geotechnical investigation in rocks. In the Brazilian test, a disc-shaped specimen of the rock is loaded by two opposing normal strip loads at the disc periphery. The specimen diameter should preferably be not less than the NX core size (54 mm), or at least 10 times the average grain size. The thickness/diameter ratio should be 0.5–0.6. The load is continuously increased at a constant rate until failure of the sample occurs within a few minutes. The loading rate depends on the material and may be from 10 to 50 kN/min. At failure, the tensile strength of the rock is calculated as $2P/(\pi DL)$.

1.10 COMMON TYPES OF ROCK AND MINERALS

The engineering properties of different types of rock can vary significantly. Since the type of rock has important engineering implications, some of the rock and mineral types that may be useful to engineers are shown in Figures 1.47 and 1.48. These figures can be used for quick reference if required. The details of each type of rock and mineral should be referred to through Bonewitz (2012), Meyer (2017) and other texts for reference. More detailed discussions about the engineering properties of rock are given by Zhang (2017).

1.10.1 Common types of rocks

Gabbro

Shale

Schist

Tuff

Figure 1.47 Some common types of rocks.

Rhyolite

Gravel

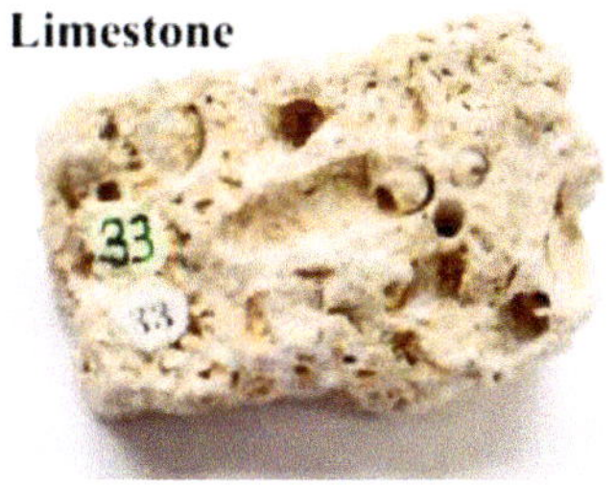

Limestone

Quartzite

Granite

Conglomerate

Dolomite limestone

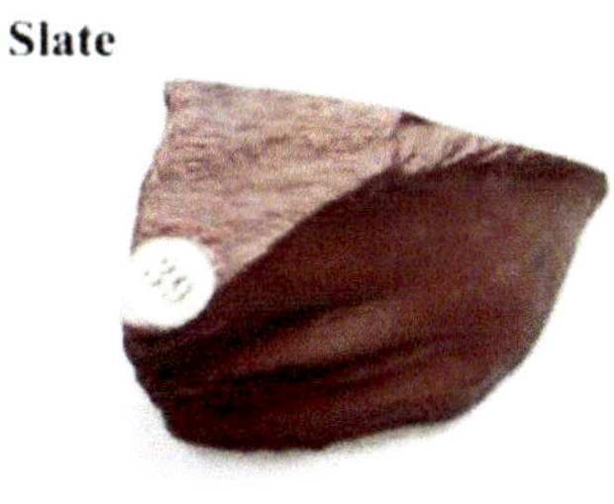

Slate

Figure 1.47 (Continued)

Syenite

Red Sandstone

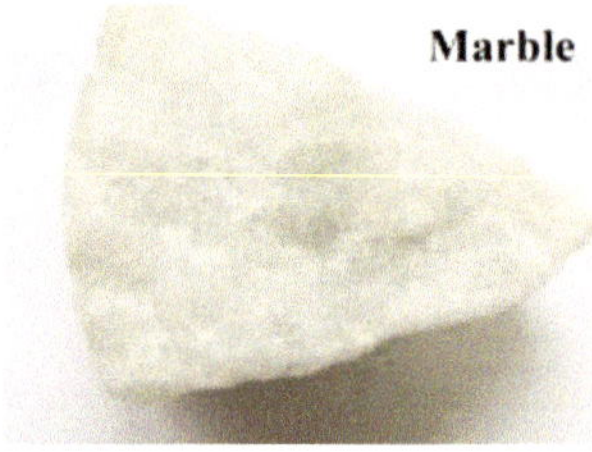

Marble

Basalt

Sandstone

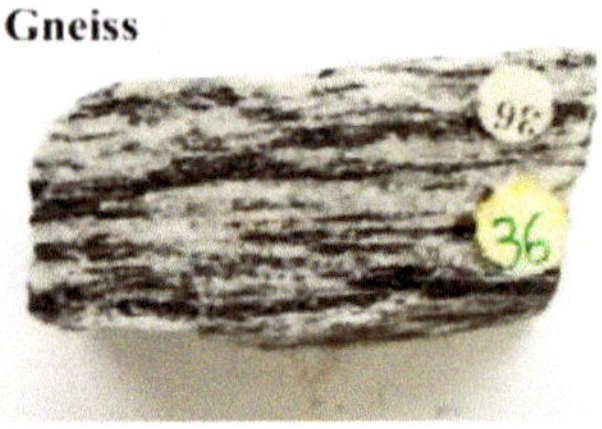

Gneiss

Phyllite

Feldsparphyric rhyolite

Mafic dyke

Figure 1.47 (Continued)

Table 1.9 Elastic properties of some common rock

	E (GPa)	*Poisson ratio*
Igneous		
Granite	30–70	0.1–0.3
Diorite	30–100	0.2–0.3
Gabbro	40–100	0.2–0.35
Rhyolite	10–50	0.2–0.4
Andesite	10–70	0.2–0.35
Basalt	40–80	0.1–0.3
Sedimentary		
Conglomerate	10–90	0.1–0.4
Sandstone	15–50	0.05–0.4
Shale	5–30	0.05–0.3
Mudstone	5–70	0.15
Dolomite	30–70	0.1–0.35
Limestone	20–70	0.1–0.35
Metamorphic		
Gneiss	30–80	0.1–0.3
Schist	5–60	0.15–0.25
Phyllite	10–85	0.25
Slate	20–90	0.2–0.3
Marble	30–70	0.1–0.35
Quartzite	50–90	0.1–0.35

The dynamic modulus is usually 1.1 to 1.4 times the static modulus, though there are some reported cases with a ratio as high as 2.8. The values in Table 1.9 are sufficient for typical engineering design, but should be verified by field or laboratory tests if possible.

1.10.2 Common types of minerals

Figure 1.48 Some common types of minerals.

Figure 1.48 (Continued)

Appendix A: Cavity expansion analysis for pressuremeter test

The elastic properties of rock can vary a lot, and some suggested values are given in Table 1.9. The dynamic modulus is usually 1.1 to 1.4 times the static modulus, though there are some reported cases with a ratio as high as 2.8. The values in Table 1.9 are sufficient for typical engineering design, but should be verified by field or laboratory tests if possible.

Equation of equilibrium in axi-symmetric case:

$$\frac{d\sigma_r}{dr} + \frac{\sigma_r - \sigma_\theta}{r} = 0 \tag{A1}$$

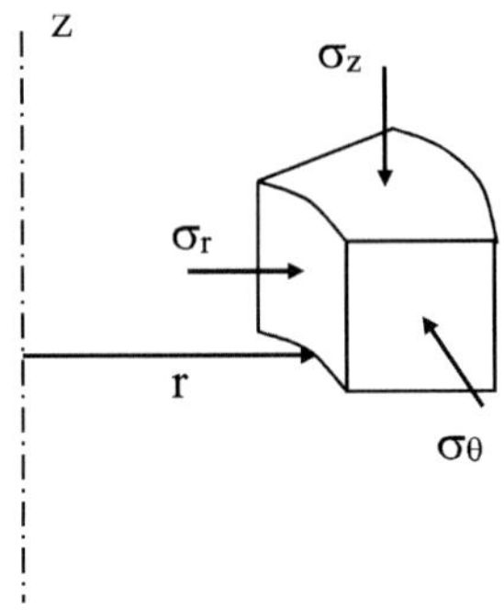

Figure A1 Stresses at an element using cylindrical coordinates.

Geometric or the strain equation is given by:

$$\xi_r = \frac{du}{dr} \tag{A2}$$

$$\xi_\theta = \frac{u}{r} \tag{A3}$$

<u>Elastic range</u>

According to the theory of elasticity, we can take the stress function φ as

$$\varphi = K \text{ In } r \tag{A4}$$

where K is a constant. The radial and circumferential stresses for axi-symmetric problems are given by:

$$\sigma_r = \frac{1}{r}\frac{d\varphi}{dr} = \frac{K}{r^2} \tag{A5}$$

$$\sigma_\theta = \frac{d^2\varphi}{dr^2} = -\frac{K}{r^2} \tag{A6}$$

Since $\sigma_r + \sigma_\theta = 0$, therefore, the volumetric strain is zero in the elastic stage, and this test is actually a shear test instead of a compression test. Let p_o denote the original horizontal stress before the test and Δp the increase in stress due to the test.

$$\text{At } r = r_o, \quad \Delta\sigma_r = \Delta p = \frac{K}{r^2} \tag{A7}$$

$$\therefore \qquad K = \Delta p\, r_o^{\,2} \tag{A8}$$

$$\text{Hence, } \sigma_r = p_o + \frac{\Delta p r_o^{\,2}}{r^2} \tag{A9}$$

$$\sigma_\theta = p_o - \frac{\Delta p r_o^{\,2}}{r^2} \tag{A10}$$

Within the elastic range, the radial and circumferential strain are given by

$$\xi_r = \frac{1-v^2}{E}\left(\Delta\sigma_r - \frac{v}{1-v}\Delta\sigma_\theta\right) \tag{A11}$$

$$\xi_\theta = \frac{1-v^2}{E}\left(\Delta\sigma_\theta - \frac{v}{1-v}\Delta\sigma_r\right) \tag{A12}$$

Putting equations (A9) and (A10) into (A11) and (A12), we have

$$\xi_r = -\frac{\Delta p}{2G}\frac{r_o^{\,2}}{r^2} \tag{A13}$$

$$\xi_\theta = \frac{\Delta p}{2G}\frac{r_o^{\,2}}{r^2} \tag{A14}$$

From equations (A2), (A13) and (A14), the radial displacement is given by

$$u = \frac{\Delta p}{2G}\cdot\frac{r_o^{\,2}}{r} = \frac{\Delta\sigma_r}{2G}r \tag{A15}$$

The radial displacement u_o at $r = r_o$ is given by

$$u_o = \frac{\Delta p}{2G} r_o \tag{A16}$$

Eq. (A15) is frequently expressed in term of volume change ΔV in the form

$$\Delta V = \frac{\Delta p}{G} V \tag{A17}$$

Based on eq. (A17), the shear modulus of soil can be found easily from the linear portion (elastic range) of the test results. It should be noted that even when the soil mass is anisotropic, the G and E values obtained are independent of the vertical direction and are controlled by the horizontal direction only. Furthermore, the modulus obtained from the pressuremeter test is usually lower than that obtained from the load test because:

1. Tensile stress can develop due to the radial load but this situation seldom occurs for load tests;
2. Disturbance of soil mass;
3. The mean stress is kept constant in this test as compared with other tests which usually increase the mean stress during the tests. Since the modulus of soil is highly dependent on the mean stress, the modulus obtained from the pressuremeter test is hence usually lower than that from other tests.

A.1 Ultimate condition for clayey material

For clay: $\sigma_r - \sigma_\theta = 2Cu$ (A18)

$$\therefore \frac{d\sigma_r}{dr} + \frac{2Cu}{r} = 0 \quad \text{or} \quad \sigma_r = K - 2Cu \ln(r) \text{ and is the total stress} \tag{A19}$$

K is a constant to be determined. At r = Ru, which is the new radius of the hole at the ultimate condition, the stress is P_L.

$$\therefore K = P_L + 2Cu \ln R_u \tag{A20}$$

$$\text{Hence} \quad \sigma_r = P_L - 2Cu \ln \frac{r}{R_u} \tag{A21}$$

$$\text{Put (A21) in (A18)}, \quad \sigma_\theta = P_L - 2Cu\left(\ln\frac{r}{R_u} + 1\right) \tag{A22}$$

For clay, the volume change can be taken as 0 (Poisson ratio $v \approx 0.5$). If R_p is the radius to the elastic/plastic zone,

$$\pi R_u^2 - \pi R_o^2 = \pi R_p^2 - \pi\left(R_p - u_p\right)^2 \quad \text{(A23)}$$

where u_p is the displacement at the boundary of R_p

$$\pi R_u^{\,2} - \pi R_o^{\,2} \cong 2\pi R_p u_p \quad \text{(A24)}$$

At ultimate condition $R_u >> R_o$, hence neglect R_o as well

$$2u_p \frac{R_p}{R_u^{\,2}} = 1 \quad \text{(A25)}$$

From (A15), $$u_p = \frac{R_p\left(\sigma_p - p_o\right)}{2G} \quad (\sigma_p \text{ refer to total}) \quad \text{(A26)}$$

where σ_p = stress at $r = R_p$

From (A21) and (A26), $$\sigma_p = P_L - 2Cu \ln\frac{R_p}{R_u} \quad \text{(A27)}$$

$$\therefore u_p = \frac{R_p}{2G}\left(P_L - 2Cu \ln\frac{R_p}{R_u} - p_o\right) \quad \text{(A28)}$$

At $r = R_p$, $\sigma_r - \sigma_\theta = 2Cu$ and $\Delta\sigma_\theta = -\Delta\sigma_r = -\Delta\sigma_p$

Hence $\Delta\sigma_r = Cu$ and $\sigma_r = p_o + Cu$, $\sigma_\theta = p_o - Cu$ (A29)

From (A25) and (A26), $$\frac{2R_p^{\,2}}{R_u^{\,2}} \cdot \frac{\left(\sigma_p - p_o\right)}{2G} \cdot \sigma_p = 1 \quad \text{(A30)}$$

Put (A29) into (A30), $$\frac{R_p^{\,2}}{R_u^{\,2}} = \frac{G}{Cu} = \frac{G}{\left(\sigma_p - p_o\right)} \quad \text{(A31)}$$

From (A27) & (A31),

$$Cu = (P_L - p_o) - Cu \ln\frac{G}{Cu} \quad \text{or} \quad P_L - p_o = Cu\left(1 + \ln\frac{G}{Cu}\right) \tag{A32a}$$

If $\Delta V = V_o$, P_l is given by $P_l - p_o = Cu\left(1 + \ln\frac{G}{2Cu}\right)$ (A32b)

For Mohr-Coulomb material:

$$(\sigma_r - \sigma_\theta) = (\sigma_r + \sigma_\theta)\ \sin\varphi + 2C\cos\varphi \quad \text{or} \quad \sigma_\theta = \sigma_r K_a - 2C\sqrt{K_a} \tag{A33}$$

$$\sigma_r - \sigma_\theta = \sigma_r(1 - K_a) + 2C\sqrt{K_a} = \sigma_r\frac{2\sin\phi}{1+\sin\phi} + \frac{2C\cos\phi}{1+\sin\phi} \quad \text{put } \sigma_r = \sigma_r' + p_o \tag{A34}$$

$$\text{Hence} \quad \frac{d\sigma_r}{dr} + \frac{2\sin\phi}{1+\sin\phi}\frac{\sigma_r}{r} + \frac{2C\cos\phi}{1+\sin\phi}\cdot\frac{1}{r} = 0 \quad \sigma_r' \text{ is increment} \tag{A35}$$

$$\text{Or} \quad \frac{d\sigma_r}{dr} + A\frac{\sigma_r}{r} + \frac{B}{r} = 0 \quad A = \frac{2\sin\varphi}{1+\sin\varphi};\ B = \frac{2C\cos\phi}{1+\sin\phi} \tag{A36}$$

To solve the homogeneous part $\frac{d\sigma_r}{dr} + A\frac{\sigma_r}{r} = 0$

$$\sigma_r = Dr^{-A} \tag{A37}$$

To solve the non-homogeneous part, we get

$$\sigma_r = \frac{-B}{A} + \frac{D_1}{r^A} \tag{A38}$$

When $\sigma_r = P_L$, $r = R_u$ (radius of hole at ultimate)

$$\text{Hence} \quad D_1 = \left(P_L + \frac{B}{A}\right)R_u^{\ A} \tag{A39}$$

$$\text{Since} \frac{B}{A} = C\cot\phi$$

$$\therefore \text{eq. (A38) reduce to } \sigma_r = (P_L + c \cot \phi)\left(\frac{R_u}{r}\right)^{\frac{2\sin\phi}{1+\sin\phi}} - C \cot\phi \tag{A40}$$

$$\text{now} \quad \pi Ru^2 - \pi R_o^{\,2} = \pi R_p^{\,2} - \pi\left(R_p - u_p\right)^2 + \pi(R_p^{\,2} - R_u^{\,2})\Delta \tag{A41}$$

where Δ is the average volumetric strain. Neglect R_o^2 and u_p^2

$$Ru^2(1+\Delta) = 2u_p R_p + R_p^{\,2}\Delta \tag{A42}$$

$$\text{or} \quad 1+\Delta = 2u_p \frac{R_p}{R_u^{\,2}} + \frac{R_p^{\,2}}{R_u^{\,2}}\Delta \tag{A43}$$

Put $r = R_p$ into eq. (A40),

$$\sigma_r = (P_L + c \cot \phi)\left(\frac{R_u}{R_p}\right)^{\frac{2\sin\phi}{1+\sin\phi}} - C \cot\phi \tag{A44}$$

Put eq. (A28) into eq. (A43),

$$u_p = \frac{1}{2G} R_p \left[(P_L + C \cot \phi)\left(\frac{R_u}{R_p}\right)^{\frac{2\sin\phi}{1+\sin\phi}} - c \cot\phi - p_0 \right] \tag{A45}$$

Put eq. (A45) into eq. (A42),

$$1+\Delta = \frac{R_p^{\,2}}{GR_u^{\,2}}[(P_L + c \cot \phi)\left(\frac{R_u}{R_p}\right)^{\frac{2\sin\phi}{1+\sin\phi}} - c \cot\phi - p_0] + \frac{R_p^{\,2}}{R_u^{\,2}}\Delta \tag{A46}$$

At $r = R_p$, σ_r & σ_θ satisfy Mohr-Coulomb relation as

$$\sigma_r = \sigma_\theta \frac{1+\sin\phi}{1-\sin\phi} + \frac{2c \cos\phi}{1-\sin\phi} \tag{A47}$$

Furthermore it also satisfies $\sigma_r' = -\sigma_\theta'$

Put $\sigma_r = \sigma_r' + p_o, \quad \sigma_\theta = \sigma_\theta' + p_o$ (A48)

$$\rightarrow \sigma_r' = p_o \sin\phi + c \cos\phi \tag{A49}$$

$$\rightarrow \sigma_p = \sigma_r' + p_o = p_o(1+\sin\phi) + c \cos \phi \tag{A50}$$

Put eq. (A49) into eq. (A43),

$$(P_L + c \cot \phi)\left(\frac{R_u}{R_p}\right)^{\frac{2\sin\phi}{1+\sin\phi}} - c \cot\phi = C \cos\phi + p_o(1+\sin\phi) \tag{A51}$$

Put eq. (A51) into eq. (A45),

$$1+\Delta = \frac{R_p^{\ 2}}{GR_u^{\ 2}}[p_o \sin\phi + c \cos \phi] + \frac{R_p^{\ 2}}{R_u^{\ 2}}\Delta \tag{A52}$$

or $$1+\Delta = \frac{R_p^{\ 2}}{R_u^{\ 2}}\left[\frac{(c+p_o \tan\phi)\cos\phi}{G} + \Delta\right] \tag{A53}$$

where Δ is small and $\sqrt{1+\Delta} \approx 1$. Introduce rigidity index I_r as

$$I_r = \frac{G}{c + p_o \tan\phi} \tag{A54}$$

$$\frac{R_p^{\ 2}}{R_u^{\ 2}}\left(\frac{1}{I_r \sec\phi} + \Delta\right) = 1+\Delta \tag{A55}$$

$$\frac{R_p^{\ 2}}{R_u^{\ 2}} = \left(\frac{I_r \sec\phi}{I_r \sec\phi + 1}\right)(1+\Delta) \tag{A56}$$

or $$\frac{R_p}{R_u} = \sqrt{\frac{I_r \sec\phi\ (1+\Delta)}{1 + I_r \Delta \sec\phi}} \tag{A57}$$

The corrected rigidity index I_{rr} is defined as

$$I_{rr} = \frac{I_r\ (1+\Delta)}{1 + I_r\ \Delta\sec\phi} = \xi_r I_r \tag{A58}$$

Hence $$\frac{R_p}{R_u} = \sqrt{I_{rr}\ \sec\phi} \tag{A59}$$

From eq. (A51), $$P_L = (p_o + c\cot\phi)(1+\sin\phi)\left(\frac{R_p}{R_u}\right)^{\frac{2\sin\phi}{1+\sin\phi}} - c\cot\phi \tag{A60}$$

From eq. (A59), $$P_L = (p_o + c\cot\phi)(1+\sin\phi)(I_{rr}\ \sec\phi)^{\frac{\sin\phi}{1+\sin\phi}} - c\cot\phi \tag{A61}$$

Denote $$F_q = 1 + \sin\phi\ (I_{rr}\ \sec\phi)^{\frac{\sin\phi}{1+\sin\phi}} \tag{A62}$$

Eq. (A61) can be written as $$P_L = cF_c + p_o F_q \tag{A63}$$

where $$F_c = \left(F_q - 1\right)\cot\phi \tag{A64}$$

Eqs. (A63) and (A64) are very similar to the classical bearing capacity equation which is discussed in Chapter 2.

References

Baguelin F, Jezequel JF and Shields DH (1978), *The pressuremeter and foundation engineering*, Trans Tech Publications, Clausthal, Germany.

Bjerrum L (1972), Embankments on soft ground. *Proceedings of the Specialty Conference, American Society of Civil Engineers*, 2, 1–54.

Bonewitz RL (2012), *Nature guide rocks and minerals*, DK Publishing.

Broch E and Franklin JA (1972), The point load strength test. *International Journal of Rock Mechanics and Mining Sciences*, 9, 669–697.

Chan, WM and Kumaraswamy, MM (1995), Reasons for delay in civil engineering projects–the case of Hong Kong. *Transactions of the Hong Kong Institution of Engineers*, 2(3), 1–8.

Cheng YM (2018), *Slope stability and Reliability Analysis*, Nova Publishing, New York.

Cheng YM and Law CW (2008), Development of a new and efficient thick plate element. *Structural Engineering and Mechanics*, 29(3), 327–354.

Cheng YM and Wei WB (2007), Application of innovative GFRP pipe soil nail system in Hong Kong, *Key Engineering materials*, 353–358, 3006–3009.

Cheng YM, Au SK and Li N (2013), An innovative Geonail System for soft ground stabilization, *Soils and Foundation*, 53(2), 282–298.

Cheng YM, Au SK and Wong H (2019), Fracture grouting and geonails for soft soil tunnelling, *Geomechanics and Geoengineering*, https://doi.org/10.1080/17486025.2019.1573321

Eurocode, *Eurocode 7 Part 2–Geotechnical design*.

Hatanaka M and Uchida A (1996), Empirical correlation between penetration resistance and internal friction angle of sandy soils. *Soils and Foundations*, 36(4), 1–9. doi:10.3208/sandf.36.4_1

Hawkins AB (1984), Rock descriptions. *Proceedings of the 20th regional meeting of the engineering group of the geological society*, Guildford, UK, pp. 59–66. (Discussion, pp 66–72). (Published as *Site Investigation Practice: Assessing BS 5930*, edited by A.B. Hawkins. *Geological Society, Engineering Geology Special Publication* no. 2, 1986).

Hvorslev MJ (1949), *Subsurface exploration and sampling of soils for civil engineering purposes*, U.S. Army Corps of Engineers, Waterways Experiment Station, Vicksburg, MS.
Hvorslev MJ (1951), *Time lag and soil permeability in ground-water investigations*, Waterways Experiment Station, Corps of Engineers, Vicksburg, MS.
ISRM (1985), Suggested methods for determining point load strength. *International Journal of Rock Mechanics and Mining Sciences & Geomechanics Abstracts*, 22, 51–60.
Kulhawy FH and Mayne PW (1990), *Manual on estimating soil properties for foundation design*, Cornell University.
Liao and Whitman RV (1972), Overburden correction factors for SPT in sand. *Journal of Geotechnical Engineering, A.S.C.E.*, 112(3), 373–380.
Lutenegger AJ (2021), *In situ testing methods in geotechnical engineering*, CRC Press.
Marchetti S (1980), In situ tests by flat dilatometer, *ASCE Jnl GED*, 106(GT3), March, 299–321.
Mayne PW and Mitchell JK (1988), Profiling of overconsolidation ratio in clays by field vane. *Canadian Geotechnical Journal*, 25(1), 150–158.
Menard L (1956), *An apparatus for measuring the strength of soils in place*, master's thesis, University of Illinois, Urbana, Illinois.
Mukherjee S. (2015), *Atlas of structural geology*, Elsevier.
Ohya S, Imai T, and Matsubara M (1982), Relationships between *N* value by SPT and LLT pressuremeter results. *Proceedings, 2nd european symposium on penetration testing*, Vol. 1, Amsterdam, pp. 125–130.
Peck RB, Hanson WE and Thornburn TH (1974), *Foundation engineering*, 2nd ed., John Wiley, New York.
Rowe PW (1972), The relevance of soil fabric to site investigation practice. *Geotechnique*, 27, 195–300.
SchMertmann JH (1975), In-situ measurement of shear strength, State of the art paper, Session III. *Proceedings of the conference on Insitu-measurement of soil properties, special conference of the geotechnical Div., ASCE*, vol. 1, North Carolina State University, 57–138.
Schmertmann JH (1986), Suggested method for performing the flat dilatometer test. *Geotechnical Testing Journal*, ASTM, 9(2), 93–101.
Schnaid F, Ortigao JAR, Mantarous FM, Cuhha RP and MacGregor J (2000), Analysis of Self Boring Pressuremeter (SBPM) and Marcheltic dilatometer (DMT) tests in granite saprolite. *Canadian Geotechnical Journal*, 37, 796–810.
Wei WB and Cheng YM (2009), Soil nailed slope by strength reduction and limit equilibrium methods. *Computers and Geotechnics*, 37, 602–618.
Zhang LY (2017), *Engineering properties of rocks*, Elsevier.

Further reading

Ameratunga J, Sivakugan N and Das BM (2016), *Correlations of soil and rock properties in geotechnical engineering*, Springer.
Bond A and Harris A (2008), *Decoding eurocode 7*, Taylor & Francis.
BSI (1999), *BS5930 code of practice for site investigation*, BSI.
Bull JW (2003), *Numerical analysis and modelling in geomechanics*, Spon Press *Code of practice for foundation*, Buildings Department, 2017.
Carter M and Bently SP (2016), *Soil properties and their correlations*, 2nd edition, John Wiley.
Clayton CRI, Matthews MC and Simons NE (1995), *Site investigation*, John Wiley.
Elsami A, Moshfeghi S, MolaAbasi H and Eslami M (2020), *Piezocone and cone penetration test (CPTu and CPT) applications in foundation engineering*, Butterworth-Heinemann.
Everett ME (2013), *Near surface applied geophysics*, Cambridge University Press.
Fletcher CJN (2019), *Geology for ground engineering projects*, CRC Press.

Fugro Engineering Services Ltd (2004), *CPT–Simplified description of the use and design methods for CPTs in ground engineering*, Fugro Engineering Services Ltd.
GEO (2006), *Foundation design and construction*, Hong Kong SAR Government.
GEO (2007), *Engineering geology practice in HK, GEO*, Hong Kong SAR Government.
GEO (2007), *Hong Kong geology guidebook*, Hong Kong SAR Government.
GEO (2017), *Guide to site investigation*, Hong Kong SAR Government.
GEO (2017), *Guide to soil and rock description, GEO Guide 3*, Hong Kong SAR Government.
Groshong RH (2006), *3-D structural geology*, Springer Verlag.
Hazelton P and Murphy B (2016), *Interpreting soil test results*, 3rd ed., CSIRO Publishing.
HKIE (2017), *Hong Kong foundation handbook*, Hong Kong Institution of Engineers.
Hossain S, Kibria G and Khan S (2019), *Site investigation using resistivity imaging*, CRC Press.
Hunt RE (2005), *Geotechnical investigation handbook*, 2nd ed., CRC Press.
Hunt RE (2007), *Geotechnical investigation methods, a field guide to geotechnical engineers*, CRC Press.
International Society for Soil Mechanics and Geotechnical Engineering (ISSMGE) (2001), The Flat Dilatometer Test (DMT), in *Soil investigations, report of the ISSMGE, Technica1 Committee 16 on ground property characterization from in-situ testing*.
Kearey P, Brooks M and Hill I (2002), *An introduction to geophysical exploration*, 3rd ed., Blackwell Science.
Korbel P, and Novak M (2001), *The complete encyclopedia of minerals*, Grange Books.
Lees A. (2016), *Geotechnical finite element analysis, a practical guide*, ICE Publishing.
Look BG (2014), *Handbook of geotechnical investigation and design tables*, 2nd ed., CRC Press.
Lunne T., Robertson PK and Powell JJM (1997), *Cone penetration testing in geotechnical practice*, CRC Press.
Mair JR and Wood DM (1987), *Pressuremeter testing, methods and interpretation*, CIRIA.
Mayne PW (2007), *Cone penetration testing, A synthesis of highway practice*, Transportation Research Board, US.
Meigh AC (1987), *Cone penetration testing, methods and interpretation*, CIRIA, UK.
Meyer J (2017), *Rocks and rock formations*, Princeton University Press.
Monnet J (2015), *In situ tests in geotechnical engineering*, John Wiley.
Potts DM and Zdravkovic L (2001), *Finite element analysis in geotechnical engineering*, Thomas Telford.
Rahman MS and Ulker MBC (2018), *Modeling and computing for geotechnical engineers*, CRC Press.
Rajapakse R (2015), *Geotechnical engineering calculations and rules of thumb*, Elsevier.
Robertson PK and Cabal KL (2010), *Guide to cone penetration testing for geotechnical engineering*, Gregg Drilling & testing Inc.
Samui P, Dixon B and Bui DT (2021), *Basics of computational geophysics*, Elsevier.
US DOT–Briaud JL and Miran J (1992), *The flat dilatometer test*. Department of Transportation–Fed. Highway Administr., Washington, DC, Publ. No. FHWA-SA-91-044, pp. 102.
Verbrugge JC and Schroeder C (2018), *Geotechnical correlations for soils and rocks*, John Wiley.
Wood DM (2004), *Geotechnical modelling*, CRC Press.
中华人民共和国国家标准 (2001), GB 50021, 岩土工程勘察规范

Chapter 2

Ultimate limit state analysis of shallow foundations

2.1 GENERAL DESCRIPTIONS AND TYPES OF SHALLOW FOUNDATIONS

For a foundation to be termed a shallow foundation, a common understanding is that its founding depth is shallow, usually within 2–3 metres below ground. Some textbooks and design codes however, give some more explicit definitions. BS8004 and the Hong Kong Code of Practice for Foundations 2017 define shallow foundations as those with founding depths less than 3 m, while Bowles (1997) defines those with depth over breadth ratios less than one (but may be with somewhat more). Nevertheless, limiting founding depths to some 2–3 metres is more common for the definition of a shallow foundation.

A shallow foundation is an enlarged base (in plan area) for the support of columns or walls of the superstructure that spreads the stresses of the columns or walls to a wider founding stratum without failure so as to achieve the ultimate (bearing capacity) and serviceability (settlement, angular distortion, etc.) limit states. The determination of the geometry of the shallow foundation therefore depends on the loads of the columns and walls being supported and the strength and stiffness of the supporting soil or stratum. To be effective in spreading the stress, adequate stiffness of the foundation is also required, in addition to the strength. Therefore, a shallow foundation usually has a comparatively large structural thickness. In this respect, some engineers define a shallow foundation as a foundation which tries to spread the applied loadings to wider regions, while the pile foundation will resist the loading mainly by skin friction and end bearing without a major effect on the surrounding soil. The depth of the shallow foundation will not be a concern in the classification. Actually, settlement is the most critical issue for the design of a shallow foundation. If the ultimate capacity is approached (yield), settlement must be large. If settlement is small, the system will be far from yield. The control of bearing pressure can hence also control settlement, and vice versa. This approach has been actually used for many shallow foundation designs in the past, when computers were not readily available to most engineers. Furthermore, as far as the authors are aware, there are very few ultimate state failures for shallow foundations, but there are a vast amount of settlement problems for them.

Though the columns and walls of the superstructure can be constructed of different construction materials including steel, concrete or brickwork, shallow foundations are predominantly constructed of reinforced concrete as durability is an important design criterion, because the foundations are buried in soil where aggressive actions by corrosive

 DOI: 10.1201/9781032650586-2

materials can take place. In Hong Kong, the maximum concrete grade used for shallow foundations is typically limited to Grade 45 (f_{cu} =50 N/mm^2). Since many buildings in Hong Kong are tall, bar sizes of 32 mm, 40 mm and even 50 mm are commonly used in construction. The foundation is hence an expensive item in the overall construction cost, and saving in the foundation design can be an important issue.

In accordance with the geometries and functions of a shallow foundation, it can generally be classified as follows:

(1) A pad footing has a small plan length to breadth ratio. It usually supports a "point load" from a single column. Where necessary, a pedestal may be used to locally thicken the footing for punching shear strength enhancement. A pad footing is often used where the load from the column being supported is relatively small and/or the subgrade is relatively strong.

(2) A strip footing has a large plan length to breadth ratio. It usually supports a wall and/or a series of columns arranged in a line. When a line of columns in close proximity is to be supported, a strip footing is used to provide adequate bearing area and to achieve economy in excavation and construction.

(3) When two or more footings are combined together, which can be in the form of a large elongated footing or individual footings with tie-beams, this type of footing is called combined footing. The shape of this type of footing is usually rectangular on plan, but can be trapezoidal if there are major differences in the column loads. When there are more than two columns in a combined footing, the columns should be aligned along a common line, otherwise the raft footing will be used, as a combined footing is commonly considered as a one-dimensional foundation. The advantages of combined footing include: (1) load sharing among the columns is possible so as to reduce the risk of adverse loading on an individual column and (2) reduction of differential settlement.

(4) A raft (or mat) footing has a large plan area of any convenient shape (can include openings or even embedded beams). The thickness of the raft footing is usually constant throughout, but a locally thickened raft footing is also very common if there are local heavy point or line loads. It usually serves to support a number of scattered columns and walls. The use of a large plan area serves not only to mobilize a large bearing area to minimize stresses on the subgrade, but also to reduce the differential settlements among the walls and columns being supported. A raft footing is a two-dimensional structure, and the method of analysis will be more complicated than in the earlier cases. Currently, most engineers will adopt a computer method in the analysis, which will be discussed in a later chapter.

(5) A buoyancy foundation is a buried hollow box structure in the ground. It works on the principle of minimizing additional stresses on the founding stratum by displacing certain depths of soil. Ideally, the weight of the soil being displaced can be approximately equal to the weight of the superstructure so that the net stresses on the founding stratum remain unchanged. Therefore, the stratum will remain as it was without or with little undue settlements.

Diagrammatically, the five types of shallow foundations are illustrated in Figure 2.1.

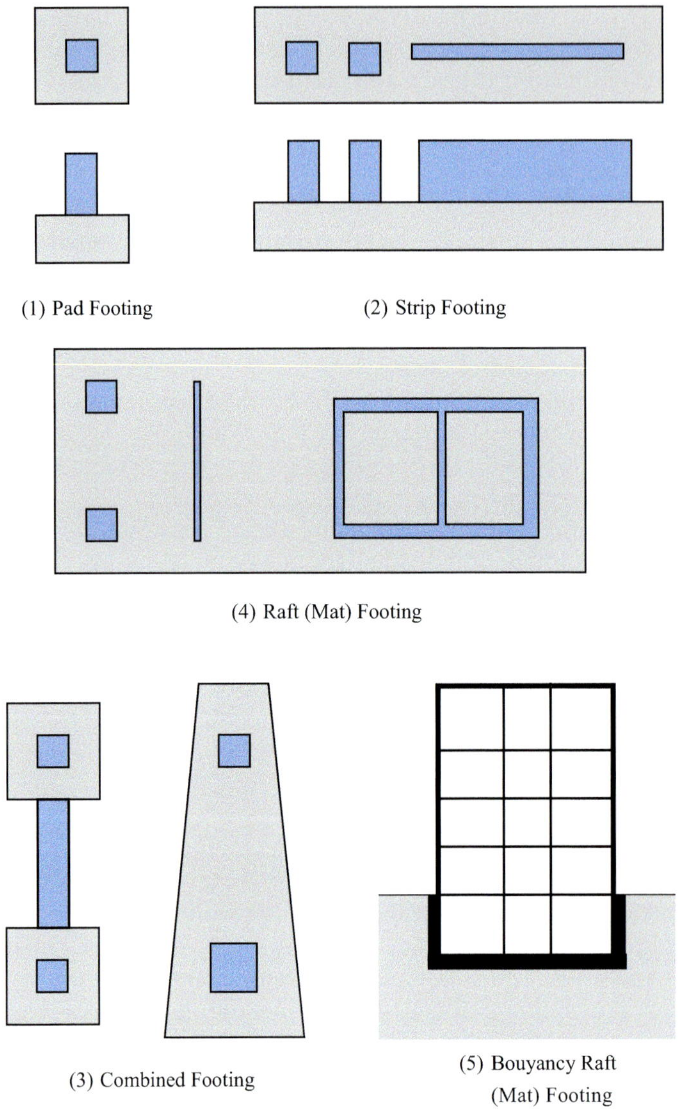

Figure 2.1 Diagrammatic illustration of different types of shallow footings.

2.2 FAILURE MODES OF SHALLOW FOUNDATIONS ON SOIL

Various laboratory tests have been performed by Vesic and Meyerhof and others on the bearing capacity and failure mechanisms under different conditions. These model test results form the background for the various empirical correction factors that have been developed for the bearing capacity determination, though many problems with these modification factors have been discussed by various researchers. Prior to the discussions on the bearing capacity of a shallow foundation, a review of the various failure modes of a shallow foundation on soil is first carried out. The modes are shear failure modes which can be classified as: (1) general shear failure; (2) local shear failure; and (3) punching shear failure, which are now discussed in turn as follows with reference to a long strip footing.

(1) General shear failure
This type of failure is found in dense and stiff soil with low compressibility characterized by well-defined and distinct failure surfaces developed between the edge of the footing and the ground, as shown in Figure 2.2. The failure starts by a state of plastic equilibrium being reached initially at the edges of the footing which spreads gradually downwards and outwards until the state of plastic equilibrium is developed throughout the soil above the failure surfaces. By the failure mode, continuous slipping and bulging of shear mass adjacent to the footing are visible as upheaval with a relatively long distance of disturbance beyond the footing edge. As the final failure mode usually takes place at one side, tilting of the footing often occurs. The failure is sudden and catastrophic with a pronounced peak in the pressure–settlement curve. The failure plane in a soil sample is a clear shear failure mode and the deformation will be due to relative movement along the failure plane.

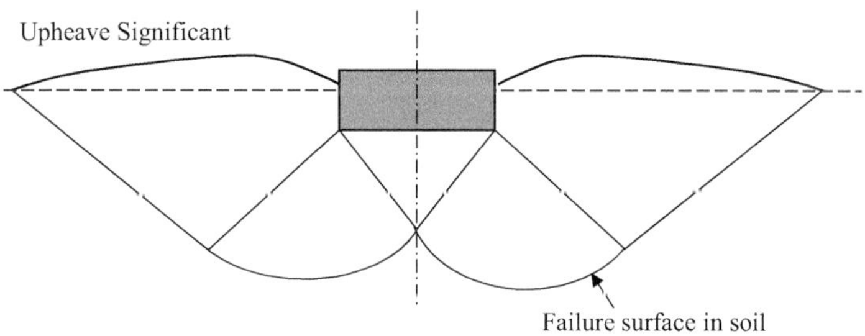

Figure 2.2(a) Failure surface of general shear failure mode.

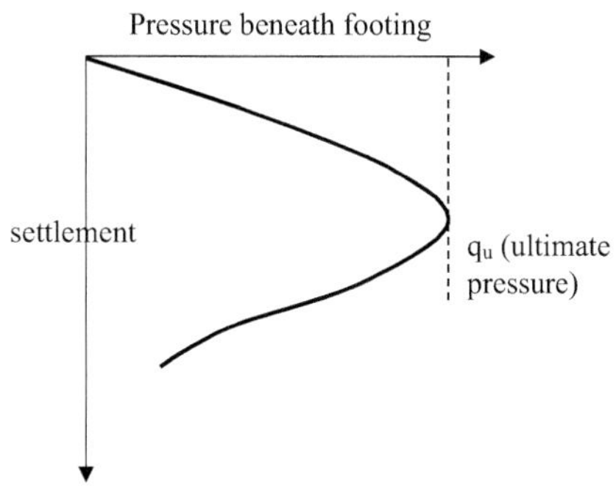

Figure 2.2(b) Pressure/settlement curve for general shear failure.

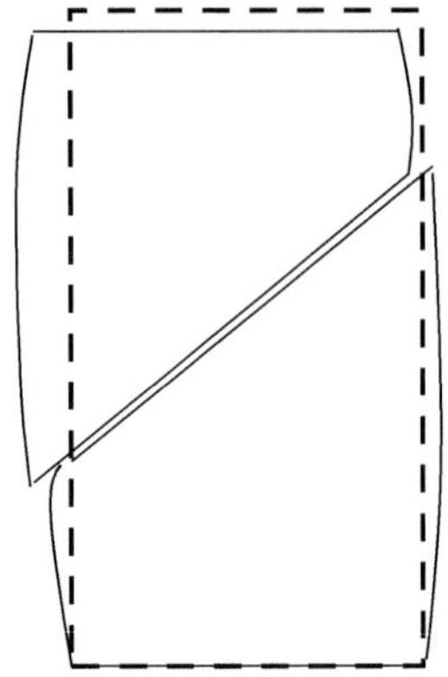

Figure 2.2(c) Failure mode of soil sample for general shear failure.

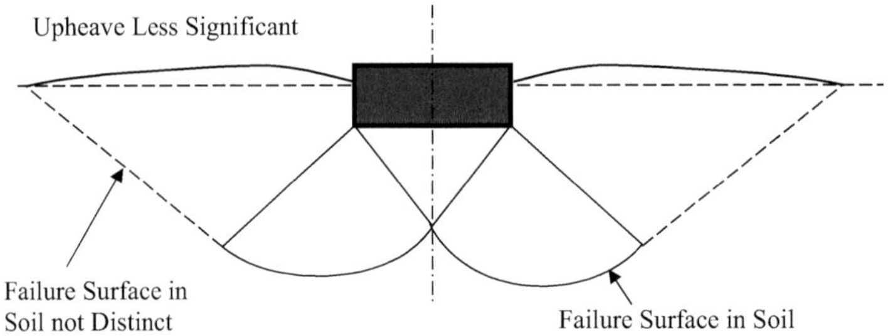

Figure 2.3(a) Failure surface of local shear failure mode.

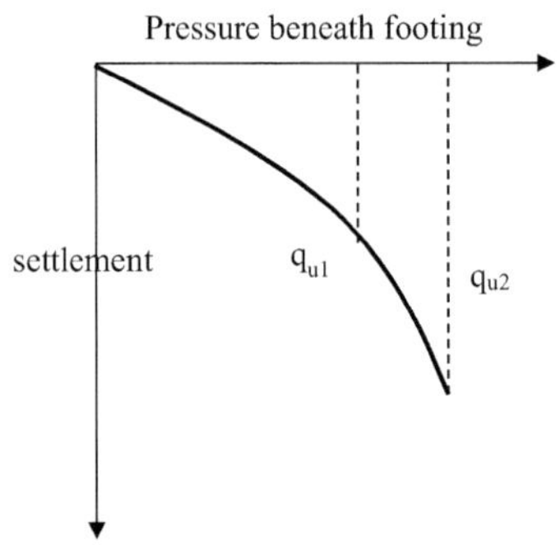

Figure 2.3(b) Pressure/settlement curve for local shear failure.

(2) Local shear failure
This type of failure is found in relatively loose and soft soil characterized by significant compression of soil below the footing and partial development of plastic equilibrium, as shown in Figure 2.3. The failure surface is not well defined and it

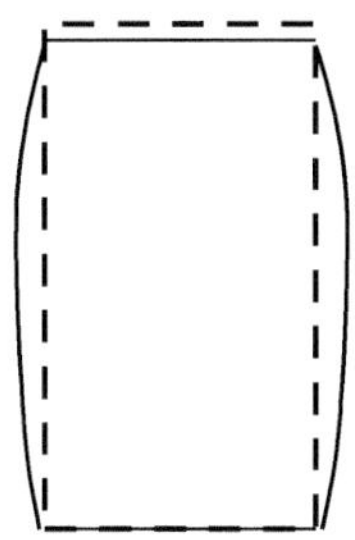

Figure 2.3(c) Failure mode of soil sample for local shear failure.

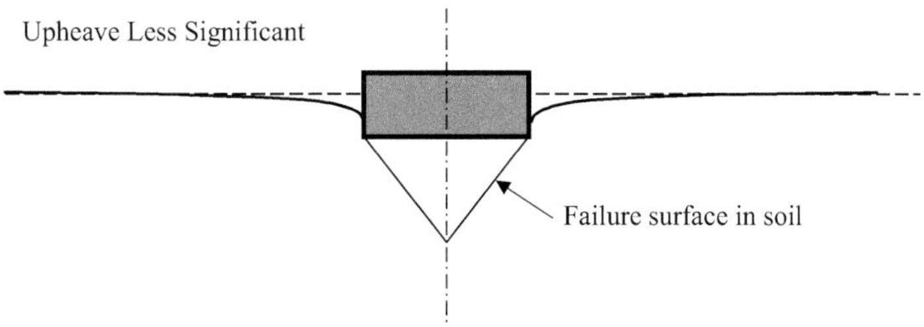

Figure 2.4(a) Failure surface of local shear failure mode.

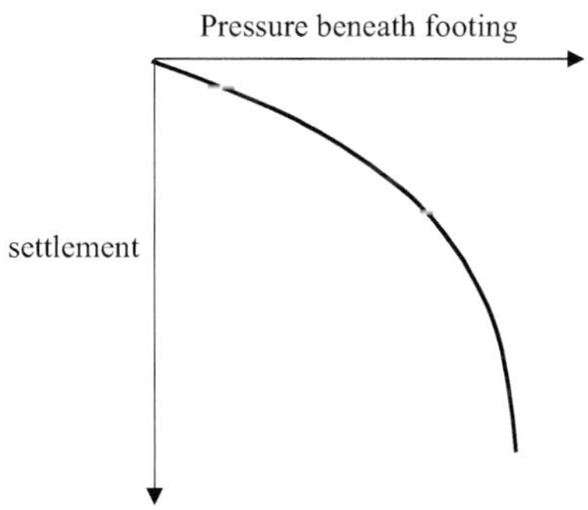

Figure 2.4(b) Pressure/settlement curve for local shear failure.

does not reach the ground surface. Only slight bulging of soil around the footing is observed. The failure is also accompanied by considerable settlement. Well-defined peak of stress and settlement is not found in the pressure–settlement curve in contrast with the general shear failure mode.

(3) Punching shear failure
This type of failure is found in loose and soft soil and at deeper elevations is characterized by relatively large and continuous displacement under shearing of soil in the vertical direction around the edges of the footing, as shown in Figure 2.4.

It can be also seen that the ultimate bearing capacity is not well defined. No heaving of the ground surface away from the edges of the footing and no tilting of the footing are observed. In general, this mode of failure depends on the compressibility of the soil and the depth of the footing relative to its breadth, and is more applicable for pile foundation problems.

Vesic has also demonstrated that the mode of failure is a function of the depth of the footing as well as the relative density of the founding soil. For these three failure modes, punching failure mode, which is more suitable for pile foundation, will be covered later. For the general shear failure mode, an analytical solution is available for limited cases which is used by many engineers in the analysis and design of shallow foundations. For the local shear failure mode, for which the compressibility of soil has to be considered, an approximate method has been suggested by Terzaghi (reduction of c' and ϕ'), while Vesic has proposed the concept of rigidity index which is based on the cavity expansion theory. In Hong Kong, the foundation code basically excludes the possibility of placing a shallow foundation on such compressible soil, due to the various uncertainties for the long-term foundation settlement.

2.3 BEARING CAPACITY OF SHALLOW FOUNDATIONS ON SOIL BY THE LIMIT EQUILIBRIUM METHOD

A shallow foundation is designed to withstand the imposed load without shear failure of the subgrade and excessive settlement. The ultimate bearing capacity of a footing is defined as the maximum pressure it can withstand without shear failure. As a safe design providing an adequate margin of safety, a safety factor usually with values ranging from 2–3 has to be applied, to which the applied pressure will be limited and the limited pressure is then termed the allowable bearing pressure or the allowable bearing capacity of the footing.

As the footing is often buried in soil for a certain depth, the weight of the soil above the bottom level of the footing is then exerting a pressure on the founding level of the footing, termed "overburden pressure", which will exhaust part of the bearing capacity of the subgrade. The magnitude of this "overburden pressure" is often known with high certainty. If we assign q_u and q_a to be the ultimate and allowable bearing capacities, respectively, q_o as the overburden pressure and F as the factor of safety, and the net ultimate and allowable bearing pressures become $q_u - q_o$ and $q_a - q_o$, we often relate these parameters by the following equation:

$$\frac{q_u - q_o}{q_a - q_o} = F \tag{2.1}$$

It can be easily understood that $q_o = \gamma D$ where γ is the unit weight of soil and D is the buried depth of the footing.

Making q_a subject of the formula in (2.1), we may list

$$q_a = \frac{q_u - q_0}{F} + q_0 \tag{2.2}$$

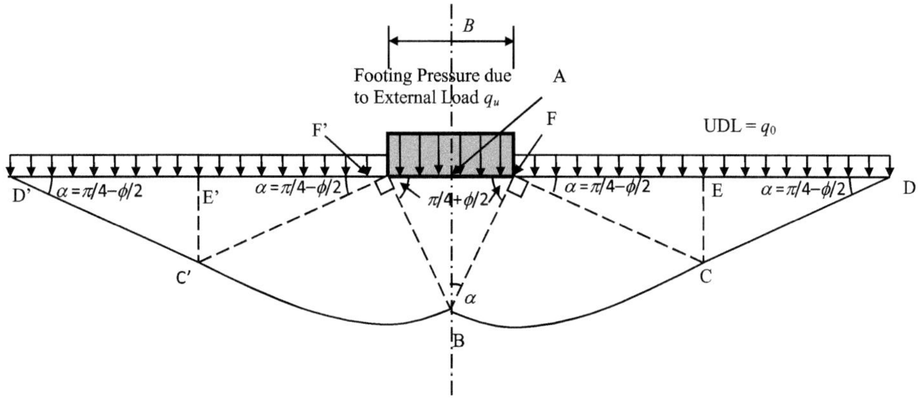

Figure 2.5 Assumed failure mechanism for ultimate bearing capacity of footing of infinite length (two-dimensional case) with zero unit weight of soil.

The bearing capacity of a footing can be studied in terms of the plasticity theory by which the soil beneath the footing reaches the plastic stage. Under the consideration of a perfectly plastic stress–strain relation, the derivation of the mathematical expression for ultimate bearing capacity is only realistic for soils of low compressibility, i.e. soils with "general shear failure mode", as illustrated in Figure 2.5.

This bearing capacity approach is formulated by the derivation of the bearing capacity factors which have been reported by Meyerhof (1963), Hansen (1970) and Vesic (1975a, 1975b). The failure mechanisms in these three methods are the same, as illustrated in Figure 2.5. For the simple case of a shallow foundation on the ground surface under a uniform pressure with weightless soil, two bearing capacity factors can be derived easily through the use of the classical soil mechanics limit equilibrium method. Consider Figure 2.5 for a strip footing (two-dimensional) resting on a mass of homogeneous and isotropic soil of angle of shearing resistance ϕ and cohesion c. At ultimate failure, the pressure beneath the footing reaches the ultimate bearing capacity value q_u and the soil mass F'FB is being pushed by the footing to reach a state of plastic equilibrium as an active Rankine zone. Therefore, the angles AFB and AF'B are both $\pi/4 + \phi/2$. On the other hand, zones FCD and F'C'D' are the passive Rankine zones being pushed outwards and therefore angles CFD and C'F'D' are $\pi/4 - \phi/2$. The zones FBC and F'B'C' are "transition zones", each enclosed by a logarithm spiral curve. Adopting the failure mechanism and using Figure 2.5, the following can be derived.

FB is the baseline of the radial zone taken as $r_0 = \dfrac{B}{2\sin\alpha}$ where $\pm = \pi/4 + \phi/2$ and r_0 is the length BE.

Within the radial zone FBC, the radial length is given a logarithm spiral curve by

$$r = r_0 e^{\theta\tan\phi} = \frac{B}{2\sin\alpha} e^{\theta\tan\phi}$$

where θ is measured from FB. The derivation of the equation is based on the principle of soil mechanics.

Length of AB is $\frac{B}{2}\cot\alpha$; Length of FC is $\frac{B}{2\sin\alpha}e^{\frac{\pi}{2}\tan\phi}$

Length of EC is $\frac{B}{2\sin\alpha}e^{\frac{\pi}{2}\tan\phi}\sin\alpha = \frac{B}{2}e^{\frac{\pi}{2}\tan\phi}$

Length of FE is $\frac{B}{2\sin\alpha}e^{\frac{\pi}{2}\tan\phi}\cos\alpha = \frac{B}{2}\cot\alpha \times e^{\frac{\pi}{2}\tan\phi}$

Ignoring the weight of the soil and the horizontal stress under the footing is minor stress, the uniform pressure on AB (zero unit weight of soil) is

$$P_{AB} = q_u K_a - 2c\sqrt{K_a} = q_u \tan^2\alpha - 2c \times \tan\alpha$$

where q_u is the ultimate stress immediately beneath the footing due to external load.

Uniform pressure on EC (zero unit weight of soil) is

$$P_{DC} = q_u K_p + 2c\sqrt{K_p} = q_u \cot^2\alpha + 2c \times \cot\alpha$$

Isolating the free body as shown in Figure 2.6 (as extracted from Figure 2.5), the applied forces are also shown.

Considering rotational equilibrium of the soil mass ABCEFA in Figure 2.6 and taking the moment about F, the overturning moments are given by

$$M_{AF} = q_u \frac{B}{2} \times \frac{B}{4} = q_u \frac{B^2}{8} \tag{2.3}$$

$$M_{AB} = P_{AB} \frac{B}{2}\cot\alpha \times \left(\frac{B}{4}\cot\alpha\right) = q_u \frac{B^2}{8} - \frac{B^2}{4} c \times \cot\alpha \tag{2.4}$$

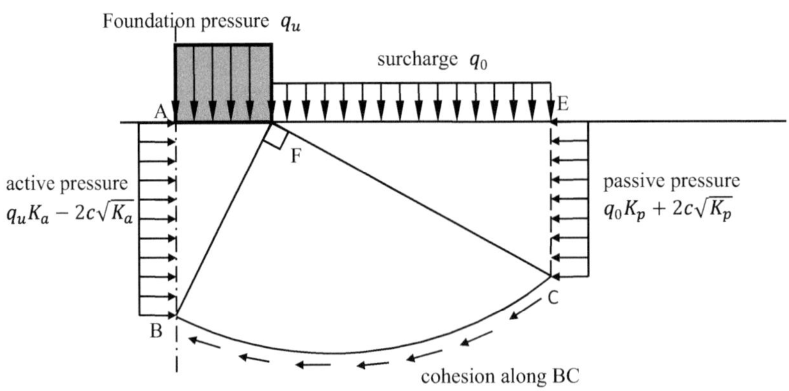

Figure 2.6 Assumed failure mechanism for ultimate bearing capacity of footing of infinite length.

The restoring moments are

$$M_{FE} = q_0 \frac{1}{2}\left(\frac{B}{2}\cot\alpha \times e^{\frac{\pi}{2}\tan\phi}\right)^2 = q_0 \frac{B^2}{8}\cot^2\alpha \times e^{\pi\tan\phi} \tag{2.5}$$

$$M_{EC} = \left(q_0\cot^2\alpha + 2c \times \cot\alpha\right)\frac{1}{2}\left(\frac{B}{2}e^{\frac{\pi}{2}\tan\phi}\right)^2$$

$$= q_0 \frac{B^2}{8}\cot^2\alpha \times e^{\pi\tan\phi} + c\frac{B^2}{4}\cot^2\alpha \times e^{\pi\tan\phi} \tag{2.6}$$

For the contribution of the spiral curve BC, from Figure 2.7, an elementary arc is under a normal stress σ_n and shear stress $\tau = c + \sigma_n\tan\phi$. As proven in Figure 2.6 where $\beta = \phi$, the component perpendicular to the radius to F effective in restoring rotational equilibrium is $ds\left(\tau\cos\phi - \sigma_n\sin\phi\right) = cds \times \cos\phi$, where ds is the length of the arc and $ds = \dfrac{rd\theta}{cos\phi}$. As the moment arm is r, the moment contributed is $dM_{BC} = crds\cos\phi = cr^2d\theta$. It should be noted that the actual stress distribution along the logspiral curve is not required or determined under the present analysis. The stress distribution along the logspiral curve has to be determined from the slip line analysis, which is briefly discussed later in this chapter.

$$\begin{aligned} \text{So for BC,} M_{BC} &= c\int_0^{\frac{\pi}{2}} r^2 d\theta = \int_0^{\frac{\pi}{2}}\left(\frac{B}{2\csc\alpha}e^{\theta\tan\phi}\right)^2 \\ d\theta &= \frac{cB^2\csc^2\alpha}{8\tan\phi}\left(e^{\pi\tan\phi} - 1\right) \end{aligned} \tag{2.7}$$

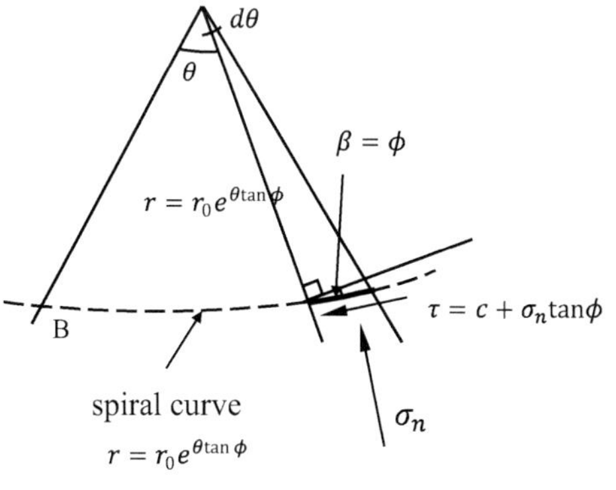

To find β,

$$\tan\beta = \frac{dr}{rd\theta} = \frac{1}{r}\frac{dr}{d\theta}$$

$$r = r_0 e^{\theta\tan\phi}$$

$$\Rightarrow \frac{dr}{d\theta} = r_0 e^{\theta\tan\phi}\tan\phi$$

$$\therefore \frac{1}{r}\frac{dr}{d\theta} = \frac{r_0 e^{\theta\tan\phi}\tan\phi}{r_0 e^{\theta\tan\phi}} = \tan\phi$$

$\Rightarrow \beta = \phi$ *(special feature of log-spiral curve)*

Figure 2.7 Proof of the contribution of the spiral BC.

Equating $M_{AF} + M_{AB} = M_{FE} + M_{EC} + M_{BC}$

$$q_u \frac{B^2}{8} + q_u \frac{B^2}{8} - \frac{B^2}{4} c \times \cot\alpha = q_0 \frac{B^2}{8} \cot^2\alpha \times e^{\pi\tan\phi} + q_0 \frac{B^2}{8} \cot^2\alpha \times e^{\pi\tan\phi} + c\frac{B^2}{4}\cot^2\alpha \times e^{\pi\tan\phi} + \frac{cB^2\csc^2\alpha}{8\tan\phi}\left(e^{\pi\tan\phi} - 1\right) \tag{2.8}$$

Simplifying Eq. (2.8)

$$q_u = q_0 \cot^2\alpha e^{\pi\tan\phi} + c\cot\alpha e^{\pi\tan\phi} + c\cot\alpha + \frac{c\csc^2\alpha}{2\tan\phi}\left(e^{\pi\tan\phi} - 1\right)$$

$$= q_0 e^{\pi\tan\phi}\tan^2\left(\frac{\pi}{4} + \frac{\phi}{2}\right) + c\left[\cot\alpha\left(e^{\pi\tan\phi} + 1\right) + \frac{\csc^2\alpha}{2\tan\phi}\left(e^{\pi\tan\phi} - 1\right)\right] \tag{2.9}$$

By trigonometry, it can be proven that

$$\cot\alpha\left(e^{\pi\tan\phi} + 1\right) + \frac{\csc^2\alpha}{2\tan\phi}\left(e^{\pi\tan\phi} - 1\right) = \left[e^{\pi\tan\phi}\tan^2\left(\frac{\pi}{4} + \frac{\phi}{2}\right) - 1\right]\cot\phi \tag{2.10}$$

$$\text{Putting } N_q = e^{\text{À}\tan\phi}\tan^2\left(\frac{\pi}{4} + \frac{\phi}{2}\right) \tag{2.11}$$

$$\text{and } N_c = \left[e^{\pi\tan\phi}\tan^2\left(\frac{\pi}{4} + \frac{\phi}{2}\right) - 1\right]\cot\phi \tag{2.12}$$

$$\text{then } N_c = \left[N_q - 1\right]\cot\phi \tag{2.13}$$

$$\text{and } q_u = q_0 N_q + cN_c \tag{2.14}$$

To include the effects of the weight of the soil mass ABCEFA, we need to cater for its weight and eccentricity. The practice is to express its effect as a factor over the parameter $0.5\gamma_s B$, where γ_s is the unit weight of the soil mass and allow a coefficient N_γ to express its effects in restoring rotational equilibrium. The exact derivation of N_γ is complicated, and cannot be expressed by a simple formula. Various researchers have put forward simplified expressions for this term. The expression put forward by Vesic as follows is now commonly used in the US and Hong Kong. In the UK, the Meyerhof expression is commonly adopted, while the Hansen expression is more commonly used in Europe. In fact, there are many similar expressions proposed by different researchers for N_γ, but the versions by

Vesic, Hansen and Meyerhof appear to be most popular among most engineers around the world. These expressions are all developed without detailed mathematical consideration.

$$\textit{Vesic} \quad N_\gamma = 2\left(N_q + 1\right)\tan\phi \tag{2.15a}$$

$$\textit{Hansen} \quad N_\gamma = 1.5N_c\tan^2\phi \tag{2.15b}$$

$$\textit{Meyerhof} \quad N_\gamma = \left(N_q - 1\right)\tan 1.4\phi \tag{2.15c}$$

Summing up, for an infinite strip footing of width *B*

$$q_u = 0.5\gamma_s BN_\gamma + q_0 N_q + cN_c \tag{2.16}$$

Where $N_q = e^{\pi\tan\phi}\tan^2\left(\frac{\pi}{4} + \frac{\phi}{2}\right)$

$N_c = \left(N_q - 1\right)\cot\phi$

$N_\gamma = 2\left(N_q + 1\right)\tan\phi$ for the Vesic method

N_q, N_c and N_γ are the bearing capacity factors depending only on the value of ϕ, and this approach of determining ultimate bearing capacity of footing is called the bearing capacity equation method. It should be noted that the failure mechanism proposed by Terzaghi is slightly different from that shown in Figure 2.5. The angle between the wedge and the foundation is taken as ϕ instead of $45+\phi/2$ as in Figure 2.5. When ϕ tends to 0, there is a problem with this failure mechanism. It is interesting to note that while the factors N_c, N_q are the same for the Vesic, Hansen and Meyerhof expressions, the expression for N_γ is not the same among these methods. Nevertheless, the differences between Eqs. (2.15a)–(2.15c) are not major. The more critical differences between the three methods are the modifications factors, and the differences can be major under some special cases (Cheng and Au 2005). Another important point to note is that the effects due to cohesion, surcharge and self-weight of soil are assumed that they can be superimposed. Cheng and Li (2017) demonstrated that such super-position is on the safe side and can be used for engineering design in general. The third important point to note is that the failure mechanism as given in Figure 2.6 is given by the plasticity solution under the condition of zero self-weight of soil. If the self-weight of soil is considered, the failure mechanism as given in Figure 2.6 will be an approximate solution only, which is discussed in more detail later.

The foregoing expressions are for a strip footing. A real footing foundation differs from Prandtl's mechanism in several aspects:

1. Usually it is not a 2D problem as the length of foundation is finite and the forces required for inducing failure at the two ends are not considered.
2. The foundation may be buried below ground level. This is commonly considered by assuming the ground level to be at the base of the foundation, while the soil above the base of the foundation is taken as a surcharge term $q_0 = \gamma D$ acting outside the foundation. The use of a surcharge term is not sufficient as the failure surface can go beyond the imaginary ground level.
3. The loadings acting on the foundation may not be a uniform distributed load (UDL), and there may be horizontal forces acting on the foundation.
4. Any other deviations from the ideal case as shown in Figure 2.5.

To cater for rectangular footing of finite dimensions ($L \times B$) or three dimensions, and other conditions, modification factors are applied to modify the bearing capacity factors. Summing up, factors to account for include: (i) width to length ratio of the footing; (ii) applied vertical load coupled with horizontal shear; (iii) tilting of the footing making an angle α with the horizontal; and (iv) adjacent slope with angle ω to the horizontal are proposed by various researchers. The most popular ones are listed in Table 2.1 with an illustration of α and ω given in Figure 2.8.

It should also be noted that if the resultant of the vertical load is eccentric to the footing with offsets from the centre of the footing by e_B and e_L along the dimension B and L, respectively, the effective dimensions of the footing are reduced to $B' = B - 2e_B$ and $L' = L - 2e_L$. This empirical approach has been used by most engineers.

D is the depth of the foundation, $k = D/B$ if it is less than or equal to 1.0, otherwise $k = \tan^{-1}(D/B)$.

Therefore, as a general formula, we may list the following for the bearing capacity of a rectangular footing with respect to the effective dimensions $L' \times B'$

$$q_u = 0.5\gamma_s B N_\gamma \zeta_{\gamma s}\zeta_{\gamma i}\zeta_{\gamma t}\zeta_{\gamma g}\zeta_{\gamma d} + q_0 N_q \zeta_{qs}\zeta_{qi}\zeta_{qt}\zeta_{qg}\zeta_{qd} + cN_c \zeta_{cs}\zeta_{ci}\zeta_{ct}\zeta_{cg}\zeta_{cd} \quad (2.17)$$

Alternatively, the total ultimate load that can be carried by the footing is

$$Q_u = q_u B' L' \quad (2.18)$$

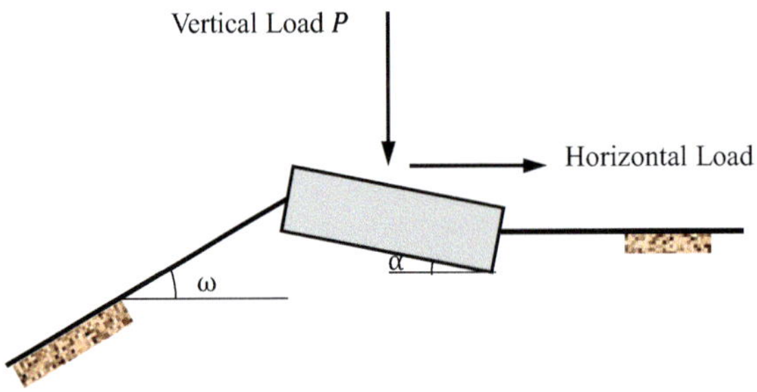

Figure 2.8 Illustration for use of modification factors to the bearing capacity factors.

Based on some model tests, Vesic and De Beer suggest using the reduction factor as $r = 1 - 0.25\ \log_{10}(0.5B)$, where B is in m. This reduction factor is particularly important for shallow soft foundations where the term $0.5rBN_r$ is great. This reduction factor is, however, not incorporated in many design codes. It should be noted that Meyerhof and Hansen have proposed different modification factors for the bearing capacity problems. These factors are based on the test results of a series of models with regression analysis dated about 50 years ago, and conservative results were finally adopted before coming up with these empirical formulae. Many researchers have doubts about the validity of these modification factors. In particular, the three-dimensional effect or shape factor for γ is less than 1.0, which is not consistent with all numerical analyses as more soil will be pushed up at the ends of the foundation. Nevertheless, many engineers are happy with these formulae instead of performing tedious numerical analysis, and viewing that the settlement is the more critical factor in the actual design, therefore such refinement appears to be of interest to researchers alone.

These equations are applicable to both granular and cohesive soil by substituting the respective ϕ and c values. For cohesive soil where $\phi = 0$, $N_q = 1$, $N_\gamma = 0$. $N_c = 2 + \pi$ by finding the limit of $(N_q - 1)\cot\phi$ when $\phi \to 0$ by methods such as the L'Hospital Rule.

The foregoing derivations for bearing capacity factors are for footings of rectangular plan shape. It can be seen that the basic factors depend only on ϕ and c. However, as the bearing capacity due to the triangular wedge of soil directly beneath the footing is $0.5\gamma_s BN_\gamma$, a greater B value will increase the overall bearing capacity. Therefore, it can roughly be stated that the bearing capacity of a rectangular footing increases with its breadth. Thus there is generally a conservative approach to determining the bearing capacity of a non-rectangular footing by drawing the largest possible inscribed rectangle within the perimeter of the footing and taking the breadth of the inscribed rectangle for calculation of the bearing capacity of the footing, as demonstrated in Figure 2.9.

However, if the modifying factors are involved which depend also on the B/L ratio, it may not be definite that the inscribed rectangle approach will lead to a conservative result. There are many cases where the modification factors are not sufficient or suitable. A good

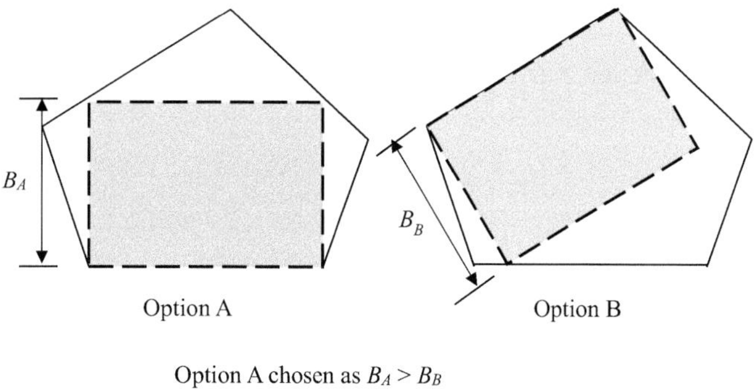

Figure 2.9 Approximating a non-rectangular footing by a rectangular one for determination of the bearing capacity.

example is a foundation close to the edge of a slope. Even though the ground slope factor is available, it is applied to the case where the ground slope is at the edge of the foundation, which is virtually impossible for a real foundation. Since this case is more critical, most engineers will carry out the calculation together with a plate load test to verify the actual bearing capacity of the foundation. Another critical issue in this case is that the term N_γ is derived based on a symmetric failure, while it will be an asymmetric failure in the present case. As discussed by the authors later, the N_γ term will be halved for this case, which means that the classical design approach may not be adequate for a slope adjacent to a slope.

There are several cases worthy of discussion. For a footing which is placed with a setback from the crest of a slope, the ground slope factor in Table 2.1 is not strictly applicable, as modification factors will be very conservative in the design if the failure mechanism is a symmetrical one. On the other hand, the footing may actually fail in an asymmetric mode. Towards this, Cheng and Au (2005) have developed a method based on the asymmetric failure mode and slip line method which has been demonstrated to give predictions of the test results by Shields as compared with that by Graham et al. (1988). The calculation is best carried out by the program SLIP developed by Cheng, and a demonstration copy can be obtained from Cheng at natureymc@yahoo.com.hk. Alternatively, an empirical approach has been adopted in Hong Kong. Take $x = 0$ for the crest of the slope, and assume that for a distance of 4B' from the crest of the slope, the slope has no effect on the bearing capacity of the footing which is denoted by q_1. If the footing is placed at the edge of the slope with $x = -D\cot\omega$ and bearing capacity q_2, the bearing capacity of any footing with a distance x away from the crest of the slope is given by a simple linear interpolation between q_1 and q_2.

Another problem is the use of the unit weight of soil in the bearing capacity equation. If the water table is well above or below the base of the footing, the unit weight of soil can be taken as γ' or γ without question. For the case where the water table is within a depth D_w (< B) below the base of the footing, an equivalent unit weight can be taken as $\gamma' + \gamma_w{}^* D_w/B$.

For footing on compressible soil, Terzaghi proposes to reduce c and $\tan\phi$ to two-thirds of their original values, while Vesic suggests using the compressibility factors (based on rectangular footing) for the analysis. Currently, it appears that the Terzaghi approach or the use of computer analysis are more popular among engineers.

For bearing capacity on layered soil, there is no rigorous equation for such a condition. Currently, most engineers will use the soil properties below the base of the footing for the design, which is usually conservative enough. Alternatively, some engineers adopt a geometric mean of the soil properties over a depth of B, but this is rare in practice. The conservative approach is generally sufficient, as the footing design will likely be controlled by the settlement instead of the bearing capacity.

Most of the bearing capacity factors are based on the condition of the smooth base of the foundation. For a rough base where the friction is considered, Cheng and Li (2017) determined the analytical formula for N_c and N_q, which will be discussed in a later chapter of this book. Kumar (2003) and Martin (2005) produced very consistent results based on limit analysis and slip line analysis, and the results can be given by the analytical solutions from Cheng and Li (2017).

Table 2.1 Bearing capacity factors and the modifying factors by Vesic

Basic bearing capacity factors	$N_q = e^{\pi\tan\phi}\tan^2\left(\frac{\pi}{4}+\frac{\phi}{2}\right)$
	$N_c = \left(N_q - 1\right)\cot\phi$
	$N_\gamma = 2\left(N_q + 1\right)\tan\phi$
Shape factors $\left(\text{depending on } \frac{B}{L}\right)$	$\zeta_{cs} = 1 + \frac{B}{L}\frac{N_q}{N_c}$
	$\zeta_{\gamma s} = 1 - 0.4\frac{B}{L}$
	$\zeta_{qs} = 1 + 0.4\frac{B}{L}\tan\phi$
Factors to modify vertical load P coupled with horizontal shear H	$\zeta_{ci} = \zeta_{qi} - \frac{1-\zeta_{qi}}{N_c\tan\phi}$
	$\zeta_{\gamma i} = \left(1 - \frac{H}{P + B'L'c\cot\phi}\right)^{m_i+1}$
	$\zeta_{qi} = \left(1 - \frac{II}{P + B'L'c \times \cot\phi}\right)^{m_i}$
	$m_i = \frac{2+\frac{B'}{L'}}{1+\frac{B'}{L'}}$ for load inclination along B'
	$m_i = \frac{2+\frac{L'}{B'}}{1+\frac{L'}{B'}}$ for load inclination along L'
Factors to modify tilting of the footing by an angle α to the horizontal	$\zeta_{ct} = \zeta_{qt} - \frac{1-\zeta_{qt}}{N_c\tan\phi}$
	$\zeta_{\gamma t} = (1 - \pm \times \tan\phi)^2$
	$\zeta_{qt} = \zeta_{\gamma t}$
Factors to modify adjacent sloping round of angle ω to the horizontal	$\zeta_{cg} = e^{-2\omega\tan\phi}$
	$\zeta_{\gamma g} = \zeta_{qg}$
	$\zeta_{qg} = (1 - \tan\omega)^2$ for $\omega \le 45°$
	$\zeta_{qg} = 0$ for $\omega > 45°$
Factors to account for the footing not on ground level (used in most countries, but not Hong Kong)	$\zeta_{cd} = 1 + 0.4k$
	$\zeta_{qd} = 1 + 2k\tan(1 - \sin\phi)^2$
	$\zeta_d = 1.0$

For the case of apparent earthquake load, Cheng and Au (2005) and Cheng (2007) determined the results for the bearing capacity and lateral earth pressure (reduced to a bearing capacity problem when the wall becomes horizontal) by a slip line approach. Yamamoto (2010) gave the bearing capacity factors by upper bound limit analysis, and the results are generally close to those of Cheng and Au (2005).

There is also an interesting application of the slope stability method in the bearing capacity problem (Cheng and Li 2017). For a foundation adjacent to an inclined slope, it can be viewed as a slope stability problem. The ultimate bearing capacity can be determined by varying the pressure until the minimum factor of safety of the slope is 1.0. For the classical bearing capacity problem without a slope, the same procedure can be applied by treating the problem as a horizontal slope. This procedure is limited by the use of the interslice force function f(x) in the slope stability analysis. Cheng et al. (2010) and Cheng and Li (2017) demonstrated that if f(x) is taken as a variable instead of the standard function (all standard functions give relatively poor results for bearing capacity determination) in all slope stability programs, by varying f(x) until the extremum of the factor of safety is determined, the exact bearing capacity factors can be determined by the slope stability method, which is the unification of the three major geotechnical problems by Cheng and Li (2017). Cheng and Li (2017) used the bearing capacity problem as an illustration of such unification.

Some engineers tend to rely on the use of SPT or CPT in assessing the bearing capacity of footing on soil, but the relations between the SPT or CPT values and the bearing capacity depend highly on the soil type and location, and such relations also vary between countries or even cities. No universal statistical relation can be applied for such a correlation, and engineers must refer to local building codes or practices for such interpretation. For bearing capacity on rock foundations, most engineers adopt presumed values given in building codes instead of relying on the apparent friction angle or cohesion.

There are many cases where the bearing capacity equation is difficult to apply. Nevertheless, there are now methods for determining the bearing capacity of an irregular plan size footing on soil by finite element analysis and pre-determined yield criteria. The method works by increasing the pressure from small values until the soil mass below yields to such an extent that no further bearing pressure can be added. The Coulomb failure criterion can be used as one of such criteria. The criterion can be defined by the following equation (2.19) which is originated from Figure 2.10.

$$\tau = c + \sigma \tan \phi \tag{2.19}$$

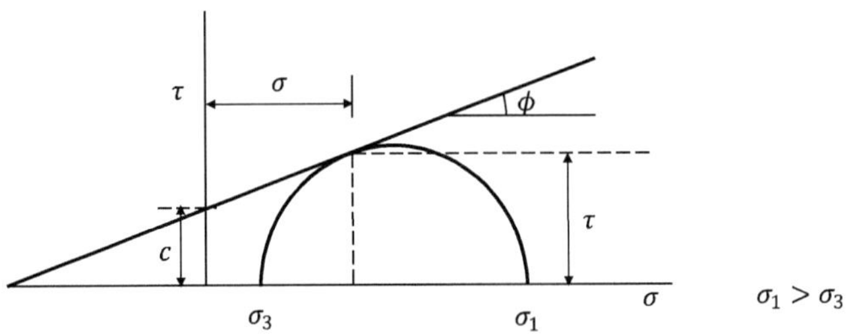

Figure 2.10 Coulomb failure criterion of soil.

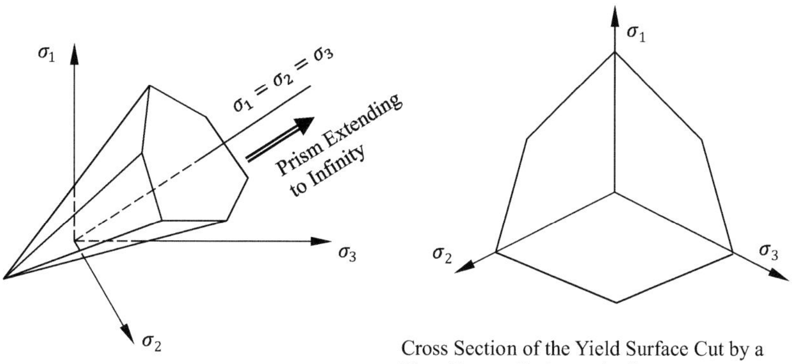

Cross Section of the Yield Surface Cut by a Plane Perpendicular to the line $\sigma_1 - \sigma_2 - \sigma_3$

Yield Surface in form of Hexagonal Prism Extending to Infinity

Figure 2.11 Yield surface of soil by Coulomb failure criterion.

Any stresses (σ, c) falling below the line $\tau = c + \sigma \tan\phi$ will not yield. The line is therefore a "yield line" for the soil.

Expressing Eq. (2.19) in terms of the principal stresses σ_1 and σ_3, we may list

$$\sigma_1(1-\sin\phi) - \sigma_3(1-\sin\phi) = 2c\cos\phi \tag{2.20}$$

Theoretically, σ_1 and σ_3 can be very large (as are σ and τ), so long as they fall below the yield line. Extending to a three-dimensional problem involving also σ_2, we may produce a "yield surface" as indicated in Figure 2.11. The yield surface is therefore a "hexagonal prism" which can in fact extend to infinity. Any stress state $(\sigma_1, \sigma_2, \sigma_3)$ outside the prism indicates failure and any stress on the surface indicates the yielding condition.

To apply the finite element analysis on the bearing capacity problem, the foundation can be modelled as flexible and uniform pressure is applied gradually until the system has reached the ultimate condition, which is usually a very large settlement. Sometimes, it may be difficult to control the increase in the pressure when the ultimate state is approached. Alternatively, the foundation can be modelled by assuming it to be rigid and prescribed displacement is applied. This approach is usually easier to control during the analysis, but the applied pressure beneath the foundation will not be uniform. The ultimate bearing capacity can then be obtained by an average of the stresses beneath the rigid foundation. In general, the difference between the two approaches is not significant.

2.4 APPLICATIONS OF BEARING CAPACITY FACTORS FOR SHALLOW FOUNDATION DESIGN ON SOIL

A worked example as shown in Figure 2.12 can illustrate the application of the various bearing capacity correction factors in Table 2.1.

$$B_{fe} = B_f - 2e_B = 5.4\,\text{m}$$

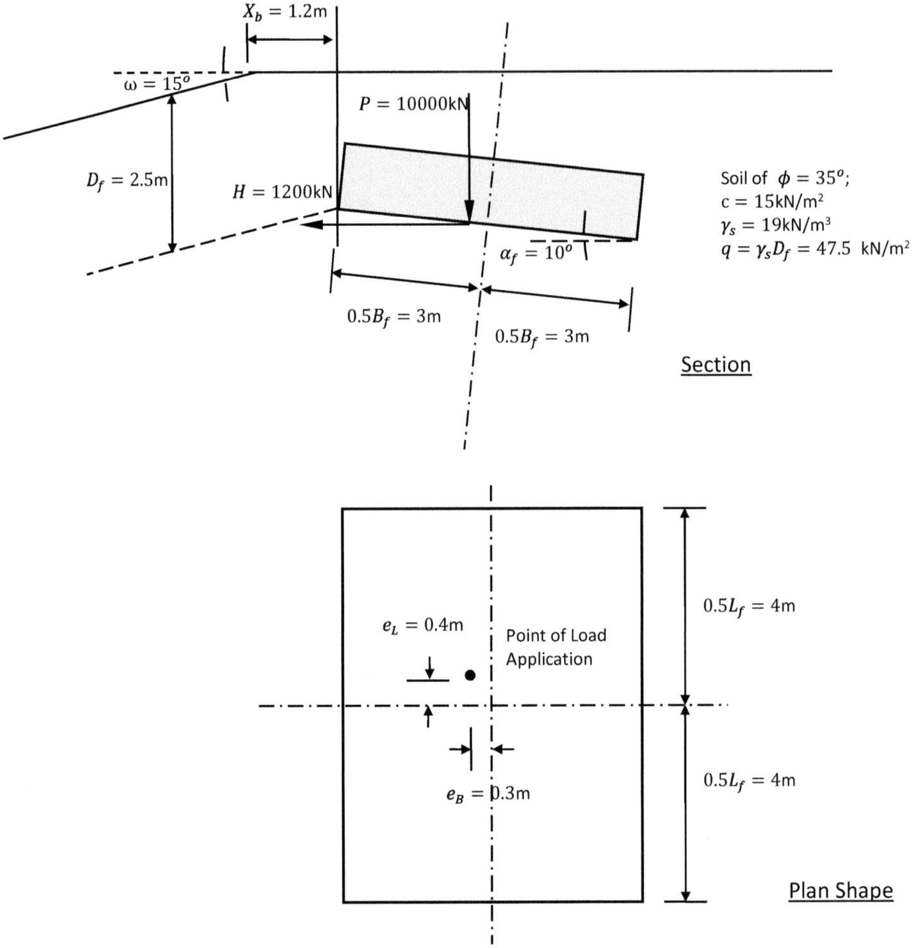

Figure 2.12 Worked example of determination of the ultimate bearing capacity of footing.

$$L_{fe} = L_f - 2e_L = 7.2\text{m}$$

Bearing capacity factors

$$N_q = e^{\pi \tan\phi} \tan^2\left(\frac{\phi}{2} + 45^o\right) = 33.296$$

$$N_c = \left(N_c - 1\right)\cot\phi = 46.124$$

$$N_\gamma = 2\left(N_q + 1\right)\tan\phi = 48.029$$

Shape factors

$$\zeta_{cs} = 1 + \frac{B_f}{L_f}\frac{N_q}{N_c} = 1.541$$

$$\zeta_{\gamma s} = 1 - 0.4\frac{B_f}{L_f} = 0.7$$

$$\zeta_{qs} = 1 + \frac{B_f}{L_f}\tan\phi = 1.525$$

Inclination factors

$$m_i = \frac{2 + B_{fe}/L_{fe}}{1 + B_{fe}/L_{fe}} = 1.571$$

$$\zeta_{qi} = \left(1 - \frac{H}{P + B_{fe}L_{fe}c\cot\phi}\right)^{m_i} = 0.832$$

$$\zeta_{ci} = \zeta_{qi} - \frac{1 - \zeta_{qi}}{N_c\tan\phi} = 0.826$$

$$\zeta_{\gamma i} = \left(1 - \frac{H}{P + B_{fe}L_{fe}c\cot\phi}\right)^{m_i+1} = 0.739$$

Tilt factors

$$\zeta_{\gamma t} = \left(1 - \alpha_f\tan\phi\right)^2 = 0.768$$

$$\zeta_{qt} = \zeta_{\gamma t} = 0.768$$

$$\zeta_{ct} = \zeta_{qt} - \frac{1-\zeta_{qt}}{N_c \tan\phi} = 0.761$$

Ground sloping factors

$$\zeta_{cg} = e^{-2\omega\tan\phi} = 0.687$$

$$\zeta_{qg} = (1-\tan\omega)^2 = 0.536$$

$$\zeta_{\gamma g} = \zeta_{qg} = 0.536$$

Therefore, the ultimate bearing capacity of the footing (Figure 2.12) is:

$$q_u = cN_c\zeta_{cs}\zeta_{ci}\zeta_{ct}\zeta_{cg} + 0.5B_{fe}\gamma_s\zeta_{\gamma s}\zeta_{\gamma i}\zeta_{\gamma t}\zeta_{\gamma g} + qN_q\zeta_{qs}\zeta_{qi}\zeta_{qt}\zeta_{qg}.$$

$$= 460.868 + 10.933 + 825.844 = 1297.645 \text{ kN/m}^2$$

2.5 USE OF DESIGN CODES

The bearing capacity equation method as discussed in Sections 2.3 and 2.4 has been incorporated in many national design codes and standards including Eurocode EC7, the Canadian Foundation Engineering Manual and the Hong Kong Foundation Code. Nevertheless, there is a simplified approach by prescribing "allowable bearing capacity values" in accordance with the quality of the soil and rock. The total allowable load that can be afforded by the footing will simply be the product of the allowable bearing capacity and the plan area of the footing. Allowable bearing capacities are listed in BS8004:1986, the Code of Practice for Foundations 2017 (Hong Kong) for soils and rock and BSEN 1997-1:2004 for rock. Instead of prescriptive values, the Eurocode has stated an approach with design net equivalent limit pressure from the pressuremeter test for soils. (The Eurocode has prescribed values for bearing capacities of rock as related to the strength of rock and the spacing of discontinuities.) Nevertheless, BS8004:1986 has commented on the use of presumed values for bearing capacity of footing as

> It is emphasized that the presumed bearing value should be used by the designer only for preliminary foundation design purposes and, in all cases, he should then review and, if necessary, amend his first design. This will frequently entail an estimate of settlements.

However, BS8004:2015 has changed the tone a bit to

> The design of many simple foundations has traditionally been checked against "allowable bearing pressures" which are normally very conservative estimates of the ultimate bearing resistance of the ground, selected on the basis of soil and rock descriptions.

Table 2.2 Presumed allowable vertical bearing pressure under foundations on horizontal ground

Category	*Description of rock or soil*	*Presumed allowable bearing pressure (kPa)*	
	Rock (granite and volcanic):		
1(a)	Fresh to slightly decomposed strong to very strong granite or volcanic rock of material weathering grade II or better, with 100% TCR of the designated grade which has a minimum uniaxial compressive strength of rock material (UCS) not less than 75 MPa (or an equivalent point load index strength PLI_{50} not less than 3 MPa)	10,000	
1(b)	Fresh to slightly decomposed strong granite or volcanic rock of material weathering grade II or better, and with not less than 95% TCR of the designated grade, which has a minimum uniaxial compressive strength of rock material (UCS) not less than 50 MPa (or an equivalent point load index strength PLI_{50} not less than 2 MPa)	7,500	
1(c)	Slightly to moderately decomposed moderately strong granite or volcanic rock of material weathering grade III or better, and with not less than 85% TCR of the designated grade, which has a minimum uniaxial compressive strength of rock material (UCS) not less than 25 MPa (or an equivalent point load index strength PLI_{50} not less than 1 MPa)	5,000	
1(d)	Moderately decomposed, moderately strong to moderately weak granite or volcanic rock of material weathering grade III or better, and with not less than 50% TCR of the designated grade	3,000	
	Meta-sedimentary rock:		
2(a)	Moderately decomposed, moderately strong to moderately weak meta-sedimentary rock of material weathering grade III or better, and with not less than 85% TCR of the designated grade	3,000	
	Non-cohesive soil (sands and gravels):	*Dry*	*Submerged*
3(a)	Very dense – SPT N-value >50	500	250
3(b)	Dense – SPT N-value 30–50; requires pick for excavation; 50 mm peg hard to drive	300	150
3(c)	Medium dense – SPT N-value 10–30	100	100
3(d)	Loose – SPT N-value 4–10, can be excavated with spade; 50 mm peg easily driven	<100	<50
	Cohesive soil (clays and silts):		
4(a)	Very stiff or hard–Undrained shear strength >150 kPa; can be indented by thumbnail	300	
4(b)	Stiff–Undrained shear strength 75–150 kPa; can be indented by thumb	150	
4(c)	Firm–Undrained shear strength 40–75 kPa; can be moulded by strong finger pressure	80	

Source: From *Code of Practice for Foundations 2017*, Hong Kong.

> The settlement of a spread foundation that has been designed using allowable bearing pressures is commonly assumed to be acceptable.

Therefore, the use of prescribed values for allowable bearing capacities of soil by various codes is quite established and they are often on the conservative side. As mentioned previously, most shallow foundation designs are controlled by settlement instead of the bearing capacity, hence a very precise value of the bearing capacity of shallow foundation is required only in limited cases.

A typical local design requirement is shown in Table 2.2, and there are different similar design guidelines in different places, for example, the use of RQD to determine the bearing capacity on rock foundation is used in some countries.

2.6 BEARING CAPACITY FROM PLASTICITY THEORY

At the ultimate condition, both equilibrium and yield conditions must be satisfied. Combining the Mohr-Coulomb yield criterion (which is generally adequate for soil) and the equilibrium equations, a set of hyperbolic partial differential equations of plastic equilibrium can be developed. In order to solve the governing partial differential equation, it is more convenient to transform the governing equations to curvilinear coordinates along the directions of the failure planes for mathematical convenience. Once the equations are solved, the failure modes with the corresponding systems of stresses will be automatically determined. The slip directions or slip lines constitute a network which is called the slip-line field. The governing equations can be solved with adequate boundary conditions to investigate the stresses at the ultimate condition, and the solution of the problem is commonly taken as the rigorous solution, as the solutions are either similar to those from other methods or better. Since the governing equations are written along the slip lines, the slip-line fields corresponding to the solutions are commonly considered as the failure mechanism of the governing problem. For example, the bearing capacity of footing and the lateral earth pressure behind a retaining wall are commonly analysed by the slip-line analysis, but not for the slope stability problem.

Kötter (1903) was the first to derive the slip-line equations for two-dimensional ultimate problems, while Prandtl (1920) was the first to obtain an analytical solution for footing by assuming the weight of soil to be negligible. His results were then applied by Reissner (1924) and Novotortsev (1938) to different problems on the bearing capacity of footing on weightless soil. The inclusion of soil weight in the solution to the governing partial differential equation is analytically impossible, and Sokolovskii (1965) proposed a finite difference approximation of the slip-line equations for which the accuracy can be further improved by an iteration scheme (Cheng 2002, 2003), and such iteration to update the coordinates of the grid points on the slip-line field has been demonstrated to be important for passive pressure evaluation. Sokolovskii (1965) solved many types of problems on the bearing capacity of footings, slopes, as well as the lateral earth pressure on retaining walls. De Jong (1957) developed a graphical procedure for the solutions which appear to be seldom used nowadays. There are other approximate solutions for the governing differential equations which include the applications of perturbation methods and series expansion methods, but these methods are not popular and versatile enough for more complicated problems and are seldom considered nowadays. More recent results and numerical techniques are given by Cheng and Au (2005) for the

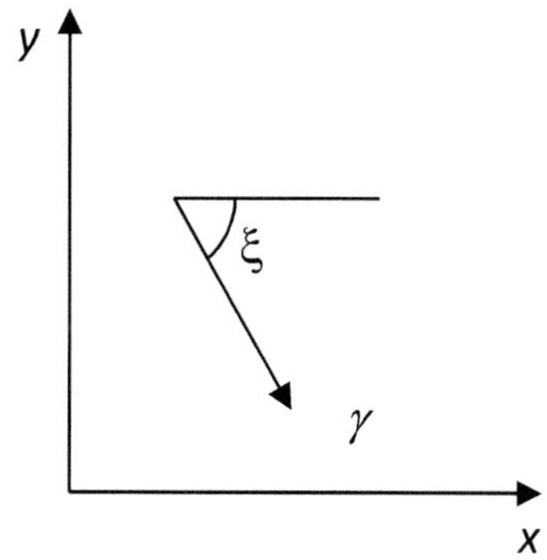

Figure 2.13 A general two-dimensional body force system with non-vertical gravity.

bearing capacity problem and Cheng (2003) and Cheng et al. (2007) for the lateral earth pressure problem.

Consider a general body force γ which may not align with XY axes as shown in Figure 2.13. Such consideration is commonly required when there is a rotation of axes applied to a problem or when earthquake load is considered, otherwise, ξ is usually 90°. Under the rotation of axes, the technique as proposed by Cheng (2003) is used, and ξ may not be 90° for a general case if there is a horizontal earthquake.

The stress equilibrium of an infinitesimal element gives:

$$\frac{\partial \sigma_x}{\partial x} + \frac{\partial \tau_{xy}}{\partial y} = \gamma \cos \xi$$
$$\frac{\partial \tau_{xy}}{\partial x} + \frac{\partial \sigma_y}{\partial y} = -\gamma \sin \xi \tag{2.21}$$

Let R and p represent the radius and the middle of the Mohr-circle of the stresses. From Figure 2.14,

$$p = (\sigma_1 + \sigma_3)/2, \ R = (\sigma_1 - \sigma_3)/\ 2 \tag{2.22}$$

At yield condition, from the Mohr-Coulomb relation:

$$\sigma_x = p + R \cos 2\theta \ \sigma_y = \mathrm{p} - \mathrm{R}\ \cos 2\theta \ \tau_{xy} = R\sin 2\theta \tag{2.23}$$

Putting Eq. (2.23) into Eq. (2.21) gives

$$\frac{\partial p}{\partial x}(1 + \sin\varphi\cos 2\theta) + \frac{\partial p}{\partial y}\sin\varphi\sin 2\theta + 2R\left(-\frac{\partial \theta}{\partial x}\sin 2\theta + \frac{\partial \theta}{\partial y}\cos 2\theta\right) = \gamma\cos\xi \tag{2.24}$$

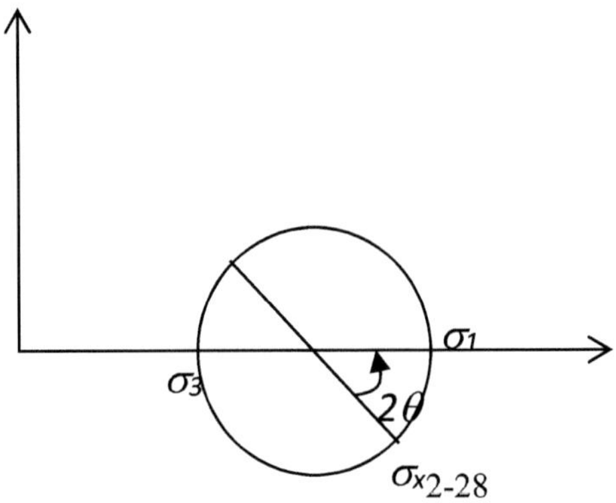

Figure 2.14 Mohr transformation of stresses.

$$\frac{\partial p}{\partial x}\sin\varphi\sin 2\theta+\frac{\partial p}{\partial y}(1-\sin\varphi\cos 2\theta)+2R\left(\frac{\partial\theta}{\partial x}\cos 2\theta+\frac{\partial\theta}{\partial y}\sin 2\theta\right)=-\gamma\sin\xi \tag{2.25}$$

where φ is the friction angle of soil. Multiply Eq. (2.2) by $\sin(\theta\pm\mu)$ and Eq. (2.25) by $-\cos(\theta\pm\mu)$, where $\mu=\frac{\pi}{4}-\frac{\varphi}{2}$, and then add up Eqs. (2.24) and (2.25) to get

$$\begin{aligned}&\frac{\partial p}{\partial x}\left[\sin(\theta\pm\mu)+\sin\varphi\sin(-\theta\pm\mu)\right]+\frac{\partial p}{\partial y}\left[-\cos(\theta\pm\mu)+\sin\varphi\cos(-\theta\pm\mu)\right]\\&\quad+2R\left[-\frac{\partial\theta}{\partial x}\cos(-\theta\pm\mu)+\frac{\partial\theta}{\partial y}\sin(-\theta\pm\mu)\right]=\gamma\sin(\theta+\xi\pm\mu)\end{aligned} \tag{2.26}$$

Using the formula

$$\begin{aligned}\sin(\theta\pm\mu)&=\pm\cos\left(\theta\pm\mu\mp\frac{\pi}{2}\right)=\pm\cos(\theta\mp\mu\mp\phi)\\&=\pm\cos(\theta\mp\mu)\cos\varphi+\sin\varphi\sin(\theta\mp\mu)\end{aligned} \tag{2.27}$$

$$\rightarrow\sin(\theta\pm\mu)+\sin\phi\,\sin(-\theta\pm\mu)=\pm\cos(\theta\mp\mu)\,\cos\phi \tag{2.28}$$

$$\begin{aligned}\rightarrow\cos(\theta\pm\mu)&=\sin\left(\frac{\pi}{2}-\mu\mp\theta\right)=\sin\left(\frac{\pi}{2}-2\mu\mp\theta+\mu\right)\\&=\sin(\varphi\mp\theta+\mu)=\sin[\varphi\mp(\theta\mp\mu)]\end{aligned}$$

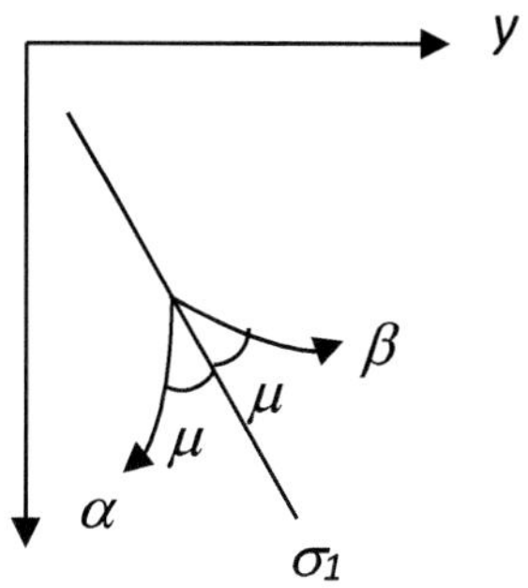

Figure 2.15 The α and β characteristic lines.

$$= \sin\varphi\cos(\theta\mp\mu)\mp\cos\varphi\sin(\theta\mp\mu) \tag{2.29}$$

$$\therefore\ \sin\varphi\cos(-\theta\pm\mu)-\cos(\theta\pm\mu)=\pm\sin(\theta\mp\mu)\cos\varphi \tag{2.30}$$

and $\sin(\theta+\xi\pm\mu)=\sin(\theta\mp\mu+\xi\pm 2\mu)$

$$=\sin(\theta\mp\mu)\cos(\xi\pm 2\mu)+\cos(\theta+\mu)\sin(\xi\pm 2\mu) \tag{2.31}$$

Therefore, Eq. (2.26) can be written as

$$\pm\cos(\theta\mp\mu)\cos\varphi\frac{\partial p}{\partial x}\pm\sin(\theta\mp\mu)\cos\varphi\frac{\partial p}{\partial y}-2R\left[\cos(\theta\mp\mu)\frac{\partial\theta}{\partial x}+\sin(\theta\mp\mu)\frac{\partial\theta}{\partial y}\right]$$

$$=\gamma\left[\sin(\theta\mp\mu)\cos(\xi\pm 2\mu)+\cos(\theta\mp\mu)\sin(\xi\pm 2\mu)\right] \tag{2.32}$$

Now a pair of lines defined by the α line and β line can be defined as the failure plane direction as shown in Figure 2.15.

The slopes of these two lines are:

$$\frac{dy}{dx}=\tan(\theta-\mu),\quad \frac{dy}{dx}=\tan(\theta+\mu) \tag{2.33}$$

The directional derivatives along these two lines are given by

$$\frac{\partial}{\partial S_\alpha}=\cos(\theta-\mu)\frac{\partial}{\partial x}+\sin(\theta-\mu)\frac{\partial}{\partial y}$$

$$\therefore \frac{\partial x}{\partial S_\alpha} = \cos(\theta - \mu), \quad \frac{\partial y}{\partial S_\alpha} = \sin(\theta - \mu) \tag{2.34}$$

$$\text{and} \quad \frac{\partial}{\partial S_\beta} = \cos(\theta + \mu)\frac{\partial}{\partial x} + \sin(\theta + \mu)\frac{\partial}{\partial y} \tag{2.35}$$

$$\therefore \frac{\partial x}{\partial S_\beta} = \cos(\theta + \mu), \quad \frac{\partial y}{\partial S_\beta} = \sin(\theta + \mu) \tag{2.36}$$

where S_α, S_β are the arc length. Since $\cos\varphi = \sin 2\mu$, from Eqs. (2.32) and (2.35), we can transform the two equations to axes along the failure surfaces as:

$$-\sin 2\mu \frac{\partial p}{\partial S_\alpha} + 2R\frac{\partial \theta}{\partial S_\alpha} + \gamma\left[\sin(\xi + 2\mu)\frac{\partial x}{\partial S_\alpha} + \cos(\xi + 2\mu)\frac{\partial y}{\partial S_\alpha}\right] = 0 \tag{2.37}$$

$$\sin 2\mu \frac{\partial p}{\partial S_\beta} + 2R\frac{\partial \theta}{\partial S_\beta} + \gamma\left[\sin(\xi - 2\mu)\frac{\partial x}{\partial S_\beta} + \cos(\xi - 2\mu)\frac{\partial y}{\partial S_\beta}\right] = 0 \tag{2.38}$$

Eqs. (2.24) and (2.25) are now transformed to Eqs. (2.37) and (2.38), which are the characteristic equations along the failure surfaces. Analytical solutions to Eqs. (2.37) and (2.38) can be evaluated if the unit weight of soil is 0, otherwise, the numerical method has to be adopted which requires complicated geometry with non-homogeneous conditions. There are three possible boundary conditions for the present problem: Cauchy type, Riemann type and mixed type. To solve the slip line equations, the iterative finite difference method is commonly used.

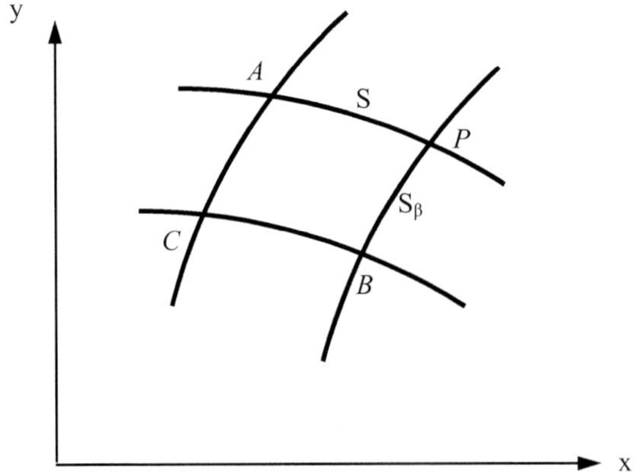

Figure 2.16 α and β lines and finite difference solution of the slip-line problem.

Cauchy type boundary condition: Starting from two points A and B with known solutions as shown in Figure 2.16, the results at point P along the intersection of α and β lines are given by the following relations.

Along α characteristic line :

$$R = \frac{R_P + R_A}{2}; \quad \theta = \frac{\theta_P + \theta_A}{2}; \quad \frac{\partial p}{\partial S_\alpha} = p_P - p_A; \tag{2.39}$$

$$\frac{\partial \theta}{\partial S_\alpha} = \theta_P - \theta_A; \quad \frac{\partial y}{\partial S_\alpha} = y_P - y_A; \quad \frac{\partial x}{\partial S_\alpha} = x_P - x_A$$

Along β characteristic line :

$$R = \frac{R_P + R_B}{2}; \quad \theta = \frac{\theta_P + \theta_B}{2}; \quad \frac{\partial p}{\partial S_\beta} = p_P - p_B; \tag{2.40}$$

$$\frac{\partial \theta}{\partial S_\beta} = \theta_P - \theta_B; \quad \frac{\partial y}{\partial S_\beta} = y_P - y_B; \quad \frac{\partial x}{\partial S_\beta} = x_P - x_B$$

If point P in Figure 2.16 is close to A and B, Eqs.(2.37) and (2.38) can be rewritten as:

$$-(p_P - p_A)\sin 2\mu + (R_P + R_A)(\theta_P - \theta_A) \approx -\gamma(y_P - y_A)\sin 2\mu - \gamma(x_P - x_A)\cos 2\mu \tag{2.41}$$

$$(p_P - p_B)\sin 2\mu + (R_P + R_B)(\theta_P - \theta_B) \approx -\gamma(y_P - y_B)\sin 2\mu - \gamma(x_P - x_B)\cos 2\mu \tag{2.42}$$

where $(x_P - x_A)$ and $(x_P - x_B)$ are given by Sokolovskii (1965) as

$$x_P - x_A \approx \tan\left(\frac{\theta_P + \theta_A}{2} - \mu\right)(y_P - y_A) \approx \left(\frac{dx}{dy}\right)_1 (y_P - y_A) \tag{2.43}$$

$$x_P - x_B \approx \tan\left(\frac{\theta_P + \theta_B}{2} + \mu\right)(y_P - y_B) \approx \left(\frac{dx}{dy}\right)_2 (y_P - y_B) \tag{2.44}$$

To solve for Eqs. (2.41) and (2.42), boundary conditions of known values of x, y, p and θ must be available on each of the slip lines. Such boundary conditions can usually be specified for a specific problem.

Riemann type problem: For the two segments CA and CB in Figure 2.16 on two corresponding α and β lines, the point P at the intersection of the slip lines passing through A and B can be initially estimated as:

$$x_P \approx x_A + x_B - x_C; \; y_P \approx y_A + y_B - y_C; \; p_P \approx p_A + p_B - p_C; \theta_P \approx \theta_A + \theta_B - \theta_C \tag{2.45}$$

In Eq. (2.45), the results for point C can be estimated as the average values of the corresponding results from points A and B. After obtaining the initial estimates for x, y and p, Eqs. (2.66) and (2.47) are used to determine the first iteration values of p_p and θ_p by substituting the initial estimates of x_p, y_p and R_p (= $p_p\sin\phi + c\cos\phi$) into Eqs. (2.41) and (2.42) which gives:

$$\theta_P = \frac{1}{R_A + R_B + 2R_p}\{[\gamma(y_A - y_B)\sin 2\mu - \gamma(2x_P - x_A - x_B)\cos 2\mu + (p_B - p_A)\sin 2\mu \quad (2.46)$$

$$+ (R_P + R_A)\theta_A + (R_P + R_B)\theta_B\}$$

$$p_P = \frac{1}{\sin 2\mu + (\theta_P - \theta_B)\sin\phi}\{[\sin 2\mu - (\theta_P - \theta_B)\sin\phi]p_B - 2c(\theta_P - \theta_B)\cos\phi \quad (2.47)$$

$$- \gamma(y_P - y_B)\sin 2\mu - \gamma(x_P + x_B)\cos 2\mu\}$$

$\left(\frac{dx}{dy}\right)_1$ and $\left(\frac{dx}{dy}\right)_2$ can be determined from Eqs. (2.43) and (4.44) while x_p and y_p can be determined from Eqs. (2.48) and (2.49).

$$y_P = \frac{x_B - x_A + \left(\frac{dx}{dy}\right)_1 y_A - \left(\frac{dx}{dy}\right)_2 y_B}{\left(\frac{dx}{dy}\right)_1 - \left(\frac{dx}{dy}\right)_2} \quad (2.48)$$

$$x_P = x_B + \left(\frac{dx}{dy}\right)_2 (y_P - y_B) \quad (2.49)$$

where x_P, y_P, θ_p and p_P obtained in the second step will be used in a refined analysis by using Eqs. (2.43), (2.44), (2.48) and (2.49) to produce a more accurate solution. The differences between x_P, y_P, θ_p and p_P before and after each iteration are checked. If the changes are less than 0.1% or any small value, convergence is assumed to be achieved. If the accuracy for a specific point is achieved, then the values for other points within the soil can be calculated in a similar manner. The importance of iteration in the solution of the finite difference equations will be illustrated in a later section.

A finite difference grid for numerical analysis is shown in Figure 2.18. It should be noted that due to the effect of soil weight, the slip lines are not straight lines. In Figure 2.18, j = 1 starts from the edge of the footing and increases towards the right. i starts from the edge of the footing also and increases in a radial direction. The calculations are divided into three parts: (1) the points at the second radial slip line where $i = 2$; (2) the points on the slip lines just below the boundary where $j = 2$; and (3) other points on the slip lines. The centre of the radial slip line has the serial numbers $i = 1, j = 1$. Starting from this point, the next outer radial slip line is $i = 2$ and so on. The maximum number value of i or the number of β lines is denoted by N_i. The number j increases as the slip line approaches the base of the footing. The maximum number of j in each radial slip line is denoted by N_k. The maximum number of points at the transition zone in each radial slip line is denoted by N_j. The authors have found that if N_j and N_k are set to be greater than 50, the results of analysis are practically unaffected by the values of N_j and N_k.

For the points where $i > 2$ and $j > 2$, the trial values of x, y, p at point P can be estimated from the corresponding values from points A, B and C on the corresponding α and β lines by Eq. (2.45). For the points at the second radial slip line where $i = 2$, the average value of previous two points A and B are used to estimate the y value as:

$$x_P = \frac{1}{2}(x_A + x_B)\ ;\ y_P = \frac{1}{2}(y_A + y_B);\ \theta_P = \frac{1}{2}(\theta_A + \theta_B)\ ;\ p_P = \frac{1}{2}(p_A + p_B) \tag{2.50}$$

For the points on the slip lines just below the boundary where $j = 2$, the initial estimates of the values at point P are given by the average values of the corresponding values at point A and B as:

$$\theta_P \approx \theta_B - \frac{\theta_{i=2,j=1}}{N_j + 1} \tag{2.51}$$

$$\left(\frac{dx}{dy}\right)_1 = \tan\left[\frac{1}{2}(\theta_P + \theta_A) - \mu\right] \tag{2.52}$$

$$\left(\frac{dx}{dy}\right)_2 = \tan\left[\frac{1}{2}(\theta_P + \theta_B) + \mu\right] \tag{2.53}$$

$$y_P \approx step * \tan(\pi/4 + \phi/2) + \frac{1}{2}(y_A + y_B) \tag{2.54}$$

$$x_P \approx x_B + \left(\frac{dx}{dy}\right)_2 (y_A - y_B) \tag{2.55}$$

$$p_P \approx \frac{1}{2}(p_A + p_B) \tag{2.56}$$

After obtaining the initial estimates for x, y and p, Eqs. (2.50)–(2.56) are used to determine the first iteration values of p_p and θ_p by substituting the initial guess of x_p, y_p, R_p ($=p_p\sin\varphi + c\cos\varphi$) into Eqs.(2.55) and (2.56) which gives:

$$\theta_p = \frac{1}{R_A + R_B + 2R_P}\{[-\gamma\sin(\varepsilon - 2\mu)(y_A - y_B) - \gamma\cos(\varepsilon - 2\mu)(2x_P - x_A - x_B) \\ + \sin 2\mu(p_B - p_A) + (R_P + R_A)\theta_A + (R_P + R_B)\theta_B\} \tag{2.57}$$

$$p_P = \frac{1}{\sin 2\mu + (\theta_P - \theta_B)\sin\phi}\{[\sin 2\mu - (\theta_P - \theta_B)\sin\phi]p_B - 2c(\theta_P - \theta_B)\cos\phi \\ - \gamma\sin(\varepsilon - 2\mu)(y_P - y_B) - \gamma\cos(\varepsilon - 2\mu)(x_P + x_B)\} \tag{2.58}$$

The refined values of x_p and y_p in the first iteration are then given by:

$$y_P = \frac{x_B - x_A + \left(\frac{dx}{dy}\right)_1 y_A - \left(\frac{dx}{dy}\right)_2 y_B}{\left(\frac{dx}{dy}\right)_1 - \left(\frac{dx}{dy}\right)_2} \tag{2.59}$$

$$x_P = x_B + \left(\frac{dx}{dy}\right)_2 (y_P - y_B) \tag{2.60}$$

The differences of x_P, y_P, θ_P and p_P, before and after each iteration analysis are checked. If the change is less than 0.1% (as used in the present study), convergence is assumed to be achieved. Based on trial-and-error analysis, it is found that a tolerance of 1% is actually sufficiently accurate for most cases but a smaller tolerance is adopted in the present study. If the accuracy for a specific point is achieved, then the values for other points within the soil can be calculated in a similar way. Otherwise, the newly obtained iteration values will be substituted into Eqs. (2.41) and (2.42) to determine the next iteration values. This process continues until accurate values satisfying the convergence criterion are achieved.

In general, for a bearing capacity problem, the pressure outside the foundation is assumed to be known, and the slip line equations are then solved towards the foundation. To determine the bearing capacity factors Nc and N_γ, a very small surcharge is applied outside the foundation, while for the bearing capacity factor Nq, a unit surcharge is usually applied. If the pressure beneath the foundation is obtained as q_{ult}, then it can be equated to cNc (with a practically zero q) or qNq to give the bearing capacity factors. For N_γ, the bearing pressure beneath the foundation will be obtained as triangular. For the interpretation, some researchers assume symmetric failure mode, while others assume asymmetric failure mode, and the bearing capacity for the asymmetric failure mode will then be only one-half of that for the symmetric failure mode.

2.6.1 Boundary conditions

In the case of a smooth horizontal footing base, the values y and θ are set to zero along this boundary. For the points below the footing, the values required for solution can be estimated from the corresponding values at points B, C and D at any two slip lines as:

At β characteristics line:

$$x_P - x_B = \left(\frac{dx}{dy}\right)_2 (y_P - y_B) \tag{2.61}$$

$$\frac{x_P - x_C}{x_D - x_C} = \frac{y_P - y_C}{y_D - y_C} \tag{2.62}$$

x_p and y_p can then be determined from the following

$$y_P = \frac{y_D - y_C}{x_D - x_C}\left[x_B - x_C + \left(\frac{dx}{dy}\right)_2 (y_P - y_B)\right] + y_C \tag{2.63}$$

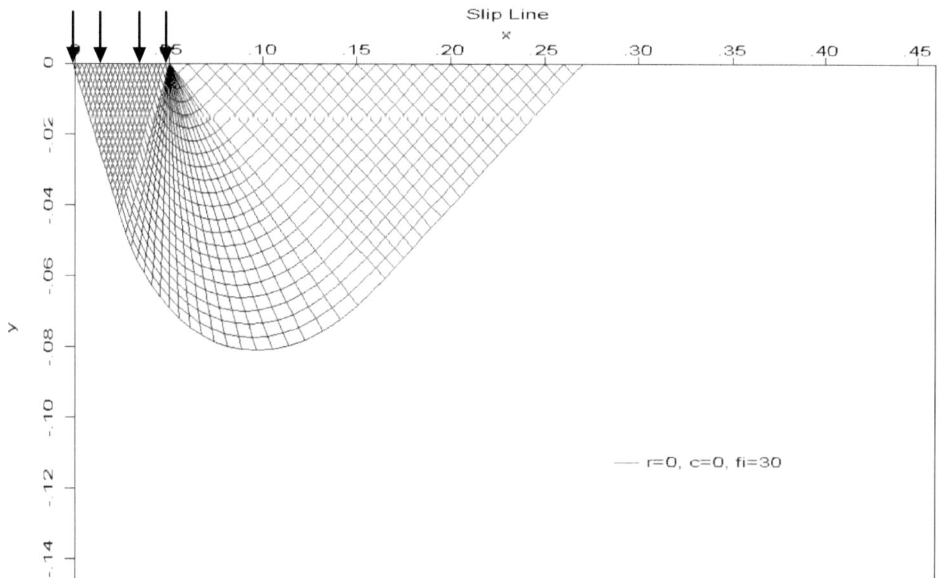

Figure 2.17 Typical solution for slip-line fields for N_c and N_q factors (uniform foundation load), unit weight of soil is 0.

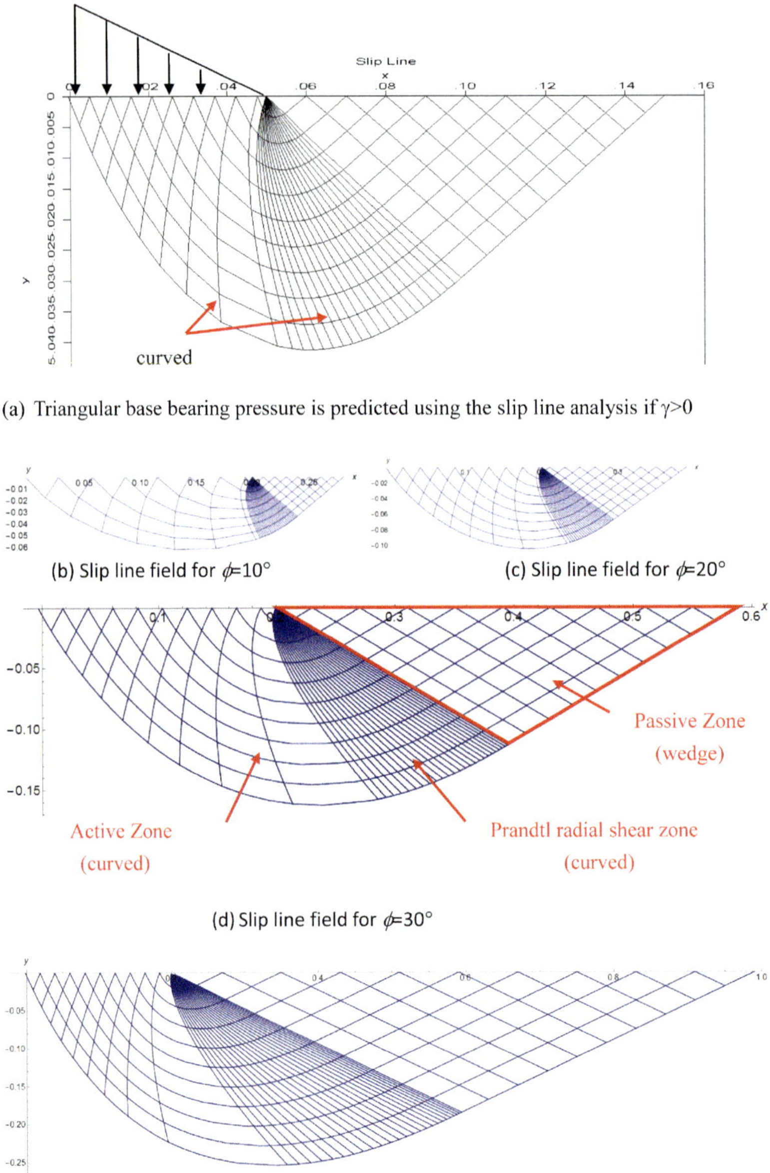

(a) Triangular base bearing pressure is predicted using the slip line analysis if $\gamma>0$

(b) Slip line field for ϕ=10°

(c) Slip line field for ϕ=20°

(d) Slip line field for ϕ=30°

(e) Slip line field for ϕ=40°

Figure 2.18 Typical slip lines for N_γ (triangular foundation load), ϕ=30°.

$$x_P = \left(\frac{dx}{dy}\right)_2 (y_P - y_B) + x_B \tag{2.64}$$

From the x,y obtained and the known value of q, p_p can be determined by Eq. (2.47).

2.6.2 Results of slip-line analysis

Based on the solution of the governing partial differential equations as given by Eqs. (2.42) and (2.42), Cheng has developed a program SLIP for the solution of bearing capacity problem and KA/KP for solution of a general lateral earthquake problem with consideration of the earthquake coefficient. The programs have been validated with other programs and published results, and extensions of these programs have been done so that these programs are more versatile for more complicated cases. Typical slip-line fields for bearing capacity factors N_c, N_q and N_γ are given in Figures 2.17 and 2.18. Based on such a known boundary condition (set q to practically 0 as the initial boundary condition for N_c and N_γ determination and 1 unit for N_q determination), the slip-line fields are evaluated from right to left. For N_c and N_q determination, it is assumed that the weight of soil is negligible, and a uniform distributed load (UDL) will be obtained at the base of the footing. Based on such conditions, the classical Prandtl's failure mechanism will be automatically obtained from the iterative finite difference analysis for N_c and N_q determination with the unit weight of soil being 0, and the failure mechanisms are the same for both bearing capacity factors. The failure mechanism as shown in Figure 2.17 is also exactly the same as the analytical solutions from Eqs. (2.37) and (2.38) if the unit weight of soil is 0 (Prandtl's mechanism). On the other hand, if the effect of the weight of soil is considered alone, a triangular pressure will be obtained at the base of the foundation, which is shown in Figure 2.18, and N_γ is determined by the average stress of this triangular pressure. The failure mechanism for this case deviates from the classical solution in two aspects: (1) the log-spiral zone will be distorted and is no longer a true log-spiral zone; and (2) the triangular zone underneath the foundation as shown in Figure 2.17 will also be distorted to that as shown in Figure 2.18. Interestingly, even

Table 2.3 Comparison of N_γ on level ground by various methods of analysis

ϕ\N_γ	*SLIP*	*Sokolovskii*	*Hansen*	*Meyerhof*	*Vesic*	*Chen*
10	0.56	0.560	0.4	0.4	1.2	0.72
20	3.16	3.16	2.9	2.9	5.4	3.45
30	15.31	15.3	15.1	15.7	22.4	15.2
40	86.6	86.5	79.4	93.6	109.3	81.79

Note: The values by SLIP are based on symmetric failure, while the results by Chen are based on associated flow rule

Table 2.4 Comparison of N_c on sloping ground with various methods of analysis (ϕ=30°)

ω\Nc	*SLIP*	*Eq. (2.32)*	*Vesic*	*Hansen*
10	24.33	24.32	24.66	28.09
20	19.57	19.57	20.15	26.04
30	15.68	15.68	16.48	24
40	12.5	12.5	13.46	21.94

Table 2.5 Comparison of N_q on sloping ground with various methods of analysis (ϕ=30°)

ω\Nq	*SLIP*	*Eq. (2.38)*	*Vesic*	*Hansen*
10	12.94	12.94	12.49	11.6
20	7.96	7.95	7.44	6.74
30	2.74	2.74	3.29	3.35
40	5.67	5.67	5.24	4.88

Table 2.6 N_γ under different earthquake coefficients by Cheng and Au (2005) for symmetric failure mode

$\phi \backslash k_h$	*0*	*0.05*	*0.1*	*0.15*	*0.2*
25	6.86	6.43	6.03	5.54	5.04
30	15.31	14.53	13.22	12.99	12.15
35	35.15	33.85	32.43	30.86	29.36
40	86.6	83.95	80.8	78.32	74.86

though the classical failure mechanism as shown in Figure 2.17 is different from that shown in Figure 2.18 for the effect of soil weight, the N_γ factor obtained by limit analysis using the mechanism in Figure 2.17 (Chen 1975) is still close to that obtained by slip-line analysis with the failure mechanism in Figure 2.18. The N_γ factors obtained by the slip-line method and limit analysis are relatively similar, but the differences between these factors and those by Meyerhof (1963), Hansen (1970) and Vesic (1973) are noticeable. While it is well known that N_c and N_q are the same among all methods for level ground, if the ground outside the foundation is sloping at an angle β, this understanding (using correction factors as compared with slip-line solutions) will no longer be true, as illustrated by Cheng and Au (2005). While the iterative finite difference solutions by program SLIP agree exactly with the analytical solution by Cheng and Au (2005), minor differences in these factors with those by Meyerhof (1963), Hansen (1970) and Vesic (1973) are noticed. It appears that the correction factors by Meyerhof, Hansen and Vesic are still acceptable for practical use, even though they are not the exact solutions. For the N_γ factor, the classical results by Vesic (1973) and Hansen (1970) have however over-estimated the bearing capacity and are not safe to be used. The results are shown in Tables 2.3–2.5. This is important for engineering design in Hong Kong, as there are many engineering problems in Hong Kong which rely on this factor on sloping ground. Cheng and Au (2005) further investigated the N_γ factor under earthquake conditions (by an equivalent pseudo-static earthquake coefficient) which is given in Table 2.6, while analytical solutions for the N_c and N_q factors are also provided by Cheng and Au (2005). Classically, the effects due to surcharge, cohesive strength and friction angle are directly superimposed in the bearing capacity and lateral earth pressure problems. Cheng (2002) has demonstrated that such a direct superposition approach is usually good enough for engineering use except for some very special cases. In the program SLIP or KA/KP by the authors, the global analysis or the analysis of individual effect can be considered, but there will be no corresponding coefficients in bearing capacity or lateral earth pressure then, as every factor is coupled with other factors for such cases. It should also be noted that the use of a logspiral curve as the transition is not rigorous if the self-weight of soil is not negligible (Li et al. 2018). Nevertheless, the use of a logspiral curve can be a good approximation for normal engineering design works.

The authors would like to discuss the effect of the weight of soil in mode details here. For the term N_γ, the rigorous plasticity solution by numerical solution of the slip-line equations requires the use of a computer program for which there is no available commercial program so far. Martin has offered his program ABC with less capabilities than SLIP which can be downloaded from his website. A good approximation of this term based on the use of logspiral mode and limit analysis is given by Chen as

$$N_\gamma = \frac{\tan\phi}{1+8\tan^2\phi}\left\{\tan\left(\tfrac{\pi}{4}+\tfrac{\phi}{2}\right)\left[3\tan^2\left(\tfrac{\pi}{4}+\tfrac{\phi}{2}\right)-1\right]e^{\left(\frac{3\pi}{2}\right)\tan\phi}+3-\tan^2\left(\tfrac{\pi}{4}+\tfrac{\phi}{2}\right)\right\} \tag{2.65}$$

Salgado and Lyamin have suggested the use of Eq. (2.66) for N_γ which is another good approximation

$$N_\gamma = [tan^2(45^\circ+\tfrac{\phi}{2})e^{\pi\tan\phi}-1]\tan(1.32\phi) \tag{2.66}$$

Finally, a very good approximation by simplified plasticity analysis is given by Eq. (2.67).

$$\begin{aligned} N_\gamma &= \tfrac{1}{4}\tan\left(\tfrac{1}{4}\pi+\tfrac{1}{2}\phi\right)\left[\tan\left(\tfrac{1}{4}\pi+\tfrac{1}{2}\phi\right)e^{\left(\frac{1}{2}3\pi\right)\tan\phi}-1\right] \\ &+\frac{3\sin\phi}{1+9\sin^2\phi}\left\{\left[\tan\left(\tfrac{1}{4}\pi+\tfrac{1}{2}\phi\right)-\frac{\cot\phi}{3}\right]e^{\left(\frac{1}{2}3\pi\right)\tan\phi}+\tan\left(\tfrac{1}{4}\pi+\tfrac{1}{2}\phi\right)\frac{\cot\phi}{3}+1\right\} \end{aligned} \tag{2.67}$$

The N_γ terms as given by Vesic, Hansen and Meyerhof, or the previous plasticity solutions, are all based on a symmetric failure mode. Since most of the foundation will carry a certain amount of moment and horizontal load, a more reasonable and conservative approach is to assume an asymmetric failure mode for the foundation. Under asymmetric failure mode, the N_γ term will be only 50% that of all the previous solutions, and this failure mode is adopted by Martin and given the program ABC. If we considered the unification of bearing capacity, lateral earth pressure and slope stability problem as given later, it will be seen that the N_γ term should be only 50% of all previous solutions, or the failure mode will be an asymmetric mode.

2.6.3 Refined bearing capacity factors from slip-line analysis

Based on the works by Cheng and Li (2017), the generalized bearing capacity factors N_c and N_q are given by

$$N_q = \frac{\cos\delta_1}{\cos\delta_2}\frac{\sin(\Delta_1+\delta_1)\sin\Delta_2}{\sin(\Delta_2-\delta_2)\sin\Delta_1}\exp\left[\left(2\alpha-\pi+\Delta_1+\delta_1+\Delta_2-\delta_2\right)\tan\phi\right] \quad (2.68)$$

$$N_c = \left\{\frac{\cos\delta_1}{\cos\delta_2}\frac{\sin(\Delta_1+\delta_1)\sin\Delta_2}{\sin(\Delta_2-\delta_2)\sin\Delta_1}\exp\left[\left(2\alpha-\pi+\Delta_1+\delta_1+\Delta_2-\delta_2\right)\tan\phi\right]-1\right\}\cot\phi \quad (2.69)$$

where $\alpha = 2\pi$ for level ground and $2\pi-\omega$ if there is a slope in front of the footing, $\Delta = \arcsin\frac{\sin\delta}{\sin\phi}$ and δ_1 and δ_2 refer to the base friction angle for the foundation side and outside the foundation. Cheng and Li (2017) further demonstrated that these two bearing capacity factors can be related to the active pressure terms *Kac* and *Kaq*. Eqs. (2.68) and (2.69) will reduce to Eqs. (2.11) and (2.12) if $\alpha = 2\pi$ and $\delta_1 = \delta_2 = 0$. Finally, bearing capacity, lateral earth pressure and slope stability problems are unified under one formulation by Cheng and Li (2017).

2.7 BEARING CAPACITY BY THE FINITE ELEMENT METHOD AND THREE-DIMENSIONAL STUDIES

The previous discussions are applicable for homogeneous conditions. For non-homogeneous ground conditions, the bearing capacities for limited cases have been derived (Das 2017). These approaches are tedious and limited in their use, as the soil layers must be horizontal for each layer. For practical applications, most engineers will simplify the support conditions to homogeneous conditions (usually conservative) for the analysis.

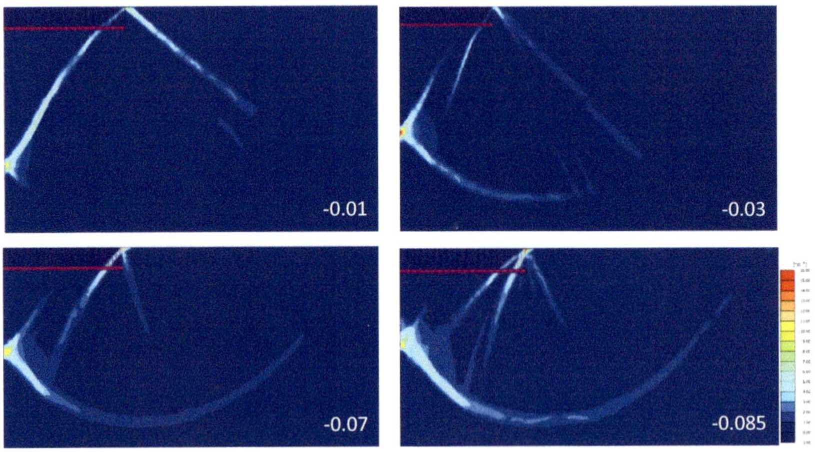

Figure 2.19 Bearing capacity failure mode from Plaxis.

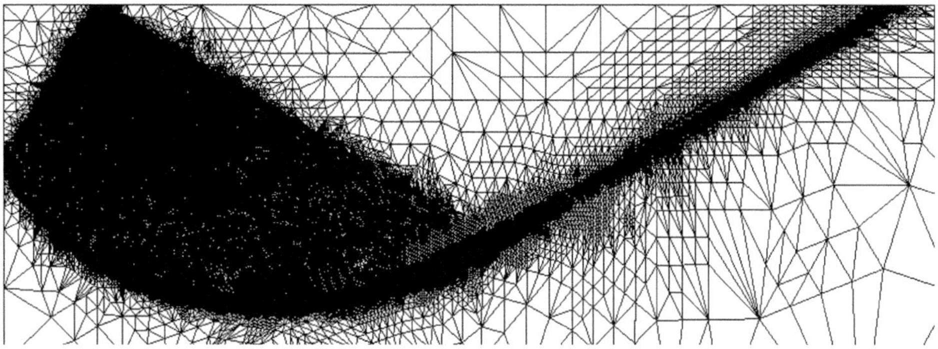

Figure 2.20 Finite element limit analysis of bearing capacity problem.

This approach is also favoured by the authors, as the bearing capacity is usually not the controlling factor in the design of the foundation.

For general complicated cases, many geotechnical finite element programs can be used for the analysis. The authors have used Adina, Phase, Plaxis and Flac for such purposes. In general, the collapse load from finite element analysis will be slightly higher than the analytical solution, but the differences are usually negligible for practical purposes. The critical issue is the failure mode as obtained from finite element analysis. Most of the finite element programs cannot provide a clear failure mode, that is, the wedges-logspiral mode. In particular, the logspiral part of the failure mode is seldom reproduced in most finite element programs. In many cases, the failure modes from finite element programs cannot form a continuous failure mechanism. In some cases, a failure mode cannot be seen clearly. With reference to Figure 2.19, the complete failure mode as given in Figure 2.5 is not clearly illustrated. The transition zone in this figure looks more like a circular arc than a logspiral curve. Similar results are obtained from all the programs that the authors have tested. In fact, the full failure mechanism as given in Figure 2.5 is obtained only in very limited cases by commercial finite element programs.

To overcome the limitations of the classical finite element analysis, finite element limit analysis can be adopted. As shown in Figure 2.20, the results were obtained by Cheng using a self-adaptive mesh refinement finite element limit analysis with error estimator control. While the results as shown in Figure 2.20 look acceptable, there are many practical difficulties in using this approach. There is no universal or commercial program available for finite element limit analysis, due to the various difficulties in mesh generation, error estimator control and self-adaptive mesh refinement analysis, for the design of a good mesh for general cases to capture the failure mode, time of analysis and other numerical difficulties.

Michalowski (2001) employed an upper bound limit analysis approach to study the bearing capacity of square and rectangular footings, and found that the computed shape factors differ greatly from the commonly used correction factors. Michalowski (2001) pointed out that the upper-bound estimates of limit loads for three-dimensional problems were likely to overestimate the true solutions by a larger margin (particularly for large ϕ) and suggested that an improvement in solutions for square and rectangular footings

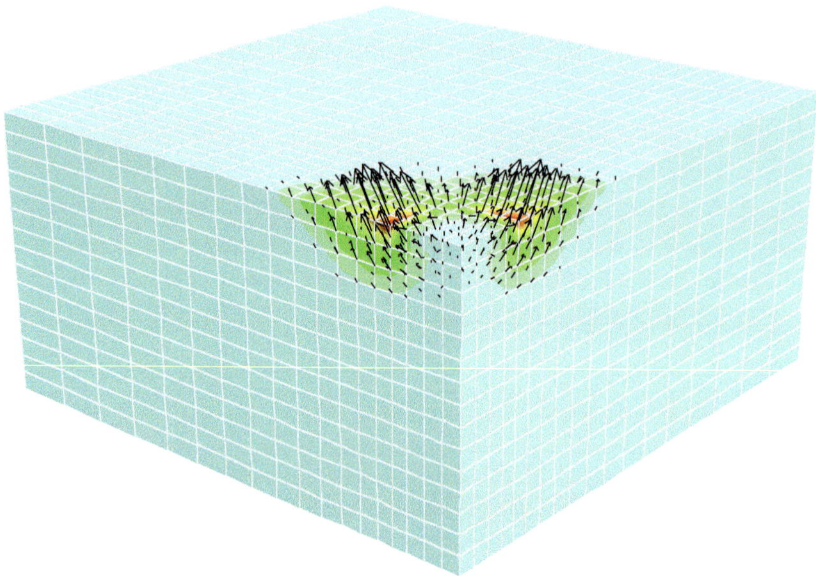

Figure 2.21 Three-dimensional analysis for bearing capacity problem: velocity field at collapse load for square footing ($\phi = 30^\circ$, $\psi = 0$).

could be obtained by employing numerical techniques, such as FEM. Michalowski's results (2001) are, however, surprising in that the three-dimensional results do not approach the plane strain solution, even if the L/B ratio is high and the multi-block mechanism with a limited number of blocks may be the reason for this phenomenon. In this respect, the correction factors as determined by Michalowski may not be accurate enough to reflect the behaviour of three-dimensional bearing capacity problem. Michalowski (2002) adopted FLAC^{3D} for the bearing capacity determination, and bearing capacity factors for $\phi = 30°$ are obtained from FLAC^{3D} which are much smaller than his results in 2001 (especially when ϕ is high) and are close to the results in the present study. In the present study, the authors found that three-dimensional results approach the plane strain results if L/B is high, which is in accordance with theory. Another limitation of the results by Michalowski (2001) is the adoption of associated flow rule in the analysis. Towards this, Cheng has applied both associated flow rule and non-associated flow rule for the present problem, as shown in Figure 2.21, and found that the dilation angle ψ is important only when ϕ is high. Also, it was found that the shape factor for N_γ, which is less than 1.0, is not given by three-dimensional analysis. This factor less than 1.0 is also conceptually impossible, as soil at the two ends of the footing is pushed up in the three-dimensional analysis. The numerical values in Tables 2.7 and 2.8 for associated and non-associated flow-rule analysis indicate that the results of Michalowski (2001) are over-estimated.

Table 2.7 Comparison of bearing capacity factor N_γ for square and rectangular footings ($c = 0, q = 0$ and $\psi = \phi$)

	L/B = 1		*L/B = 1.5*		*L/B = 2*		*L/B = 3*		*L/B = 5*		*L/B = 10*	
ϕ	*FLAC^{3D}*	*Michalowski*	*FLAC^{3D}*	*Michalowski*	*FLAC^{3D}*	*Michalowski*	*FLAC^{3D}*	*Michalowski*	*FLAC^{3D}*	*Michalowski*	*FLAC^{3D}*	*Michalowski*
10	0.76	0.778	0.84	0.923	0.87	0.978	0.92	1.046	1.00	1.076	1.04	1.098
15	1.64	2.560	1.80	2.623	1.86	2.654	1.92	2.679	2.08	2.692	2.15	2.699
20	3.72	7.160	4.01	6.878	4.10	6.641	4.22	6.431	4.54	6.246	4.59	6.111
25	8.34	20.399	8.69	17.863	8.86	17.073	9.06	15.873	9.58	14.788	9.64	14.094
30	19.16	59.895	20.13	49.789	20.28	46:203	20.33	41.646	21.15	36.945	21.08	33.756
35	49.94	189.640	50.85	151.367	51.01	134.184	49.65	116.138	50.52	100.130	49.28	85.140
40	142.61	671.694	144.51	502.152	140.41	417.755	135.22	353.455	134.44	279.149	130.23	234.531

Table 2.8 Comparison of bearing capacity factor N_γ for square and rectangular footings ($c = 0, q = 0$ and $\psi = \phi$)

	L/B = 1		L/B = 1.5		L/B = 2		L/B = 3		L/B = 5		L/B = 10	
ϕ	N_γ	s_γ	N_γ	s_γ	N_γ	s_γ	N_γ	s_γ	N_γ	s_γ	N_γ	s_γ
10	0.73	0.811	0.80	0.889	0.82	0.911	0.85	0.944	0.97	1.078	1.01	1.122
15	1.53	0.827	1.67	0.903	1.72	0.930	1.80	0.973	2.02	1.092	2.03	1.097
20	3.22	0.845	3.47	0.911	3.58	0.940	3.74	0.982	4.07	1.068	4.14	1.087
25	6.69	0.865	7.24	0.937	7.48	0.968	7.68	0.994	8.31	1.075	8.42	1.089
30	14.9	0.936	15.90	0.999	15.93	1.00	16.42	1.031	17.56	1.103	17.51	1.100
35	35.42	1.018	37.22	1.070	37.31	1.072	37.25	1.071	39.21	1.127	38.55	1.108
40	88.89	1.039	91.08	1.030	92.52	1.046	89.28	1.010	90.98	1.029	87.37	0.988

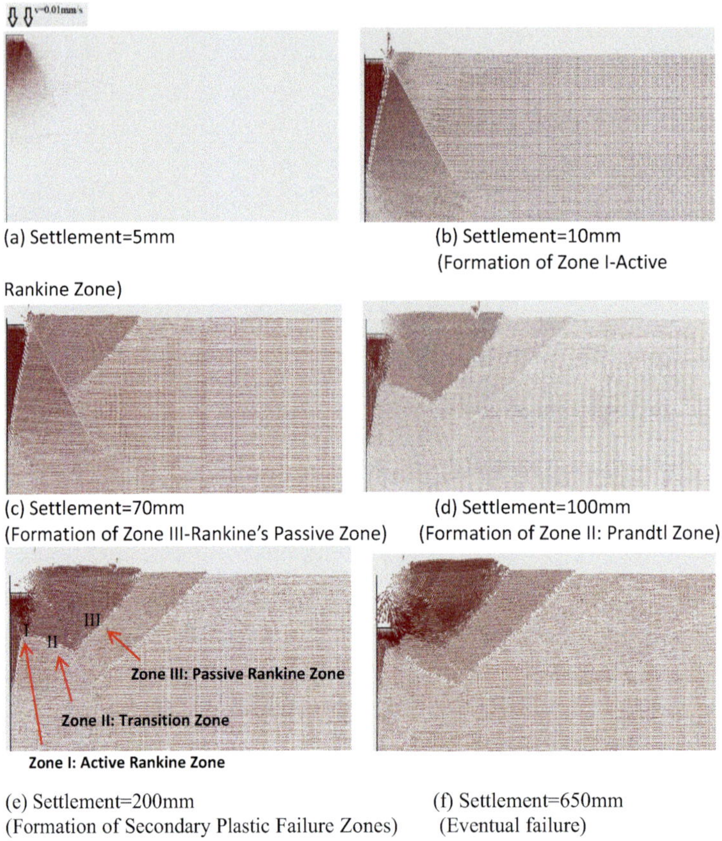

Figure 2.22 Failure progress of densely packed soil foundation for a 0.2 m width foundation (contact bond = 5 kN, $\varphi = 35°$).

(a) Instrumentation for load test on slope

(b) Bearing failure beneath bearing plate

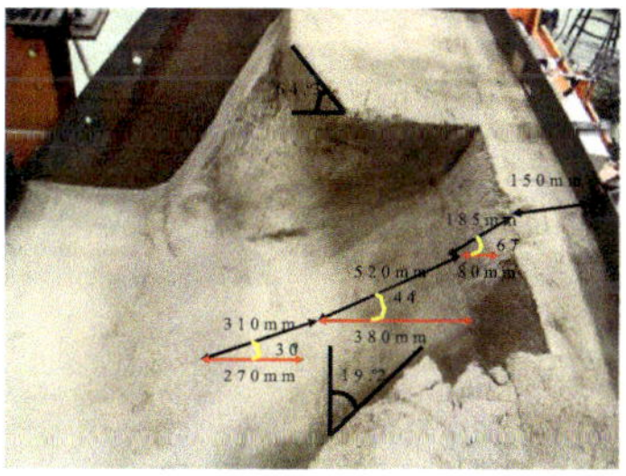

(c) Global three-dimensional failure

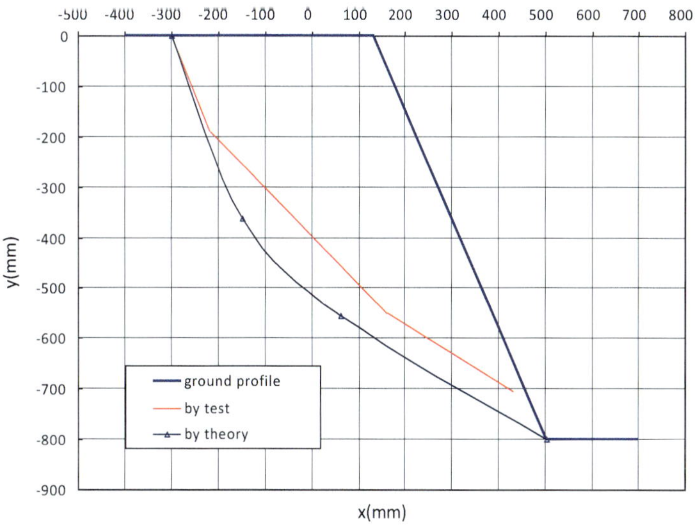

(d) Comparisons between the measure and the predicted failure surface profile at mid-section of failure

Figure 2.23 Laboratory bearing capacity test and investigation of failure surface.

2.8 BEARING CAPACITY BY DISTINCT ELEMENT METHOD

If distinct element analysis in the form of particle formulation is adopted, the failure mechanism is even more difficult to be captured. With respect to the particle flow program PFC analysis, as shown in Figure 2.22, the logspiral zone is actually obtained as a wedge zone from PFC.

To gain a better understanding of the failure mechanism, the authors have constructed a series of model tests with a slope in front of the footing, and the failed mass is removed for examination of the actual failure surfaces. In Figure 2.23, the hydraulic jack applies a local load on top of a 0.8 m high 65° inclined slope for which the soil is highly permeable, poorly graded river sand. The unit weight and the relative density of the river sand obtained in Hong Kong are $\gamma = 15.75$ kN/m^3 and 0.55. The soil parameters are $c' = 7$ kPa and $\varphi' = 35°$. The depth, breadth and height of the tank are 1.5 m, 1.85 m and 1.2 m, respectively. Five linear variable differential transducers (LVDTs) are set up to measure the displacement of soil at different locations, and the displacements at different vertical loads are monitored up to failure. The 10 mm thick steel bearing plate has a size B = 0.3 m and L = 0.644 m at 0.13 m away from the crest of the model, and an ultimate load of 35 kN is attained at a displacement of about 6 mm, as shown in Figure 2.24. As a result, the ultimate bearing capacity of the slope under the current soil properties, geometrical conditions and boundary conditions is 181.2 kPa, which gives a safety factor of 1.021 by the present method, and this value is very close to 1.0, which demonstrates that the result is reasonable. For the slope surface, the corresponding displacement at the maximum pressure is about 2 mm and 1 mm at the top and bottom of the slope, respectively. Beyond the peak load, the applied load decreases with increasing jack displacement. It is clear that the displacements of the slope are basically symmetrical. The failure surface of the present test is shown in Figures 2.23a and 2.23b, and the sectional view at the middle of the failure mass is shown in Figure 2.23c. In Figure 2.23a, after the maximum load is achieved, the load will decrease with increasing displacement. At this stage, the local triangular failure zone is fully developed, while the failure zones at the two ends of the plate are not clearly formed. When the applied load has decreased down to about 25 kN, the load maintained constant for a while and the failure zones at the two ends become visually clear. When the displacements are further increased, the applied load decreases further and the failure zone propagates towards the slope surface until the failure surface as shown in Figure 2.23b is obtained. For this test, there are several interesting phenomena worth discussing. The failure profile and cracks are firstly initiated beneath the footing, as shown in Figure 2.23b, which is a typical bearing capacity slope stability failure with a triangular failure zone noted clearly in Figure 2.5. As the load increased, the failure zone extended and propagated towards the toe of the slope, and the final failure surface is shown in Figure 2.24b. It is observed that the failure mechanism of the physical model test is hence a local triangular failure beneath the bearing plate, and the failure surface propagates towards the slope surface until a failure mechanism is formed. The classical failure mechanism for bearing capacity with an inclined slope based on that by Cheng and Au (2005) or Sokolovskii (1965) where a transition logspiral will appear after the triangular wedge underneath the footing is however not clearly developed. Instead, the transition zone appears to be another wedge which is followed by another wedge adjacent

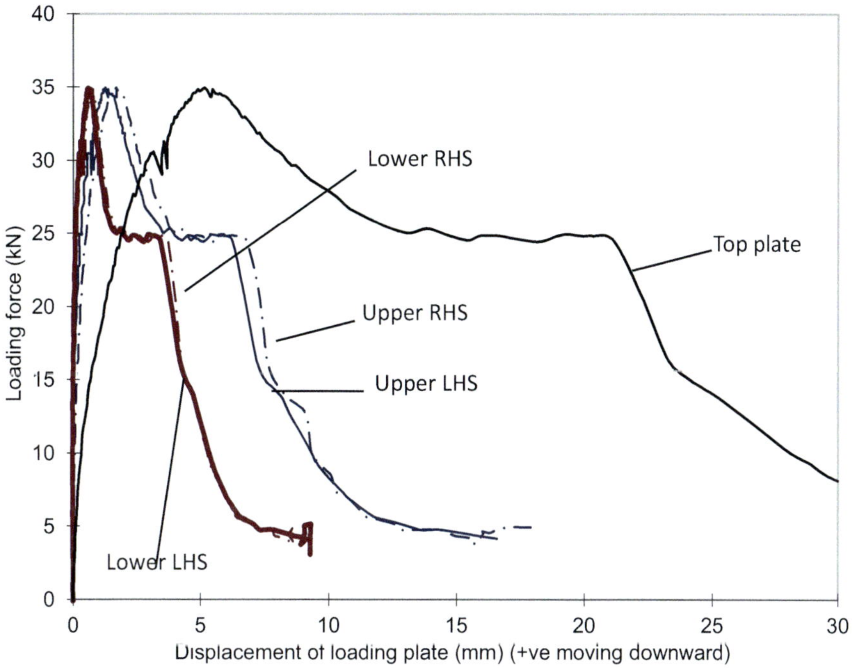

Figure 2.24 Loading force against the displacement of slope surface.

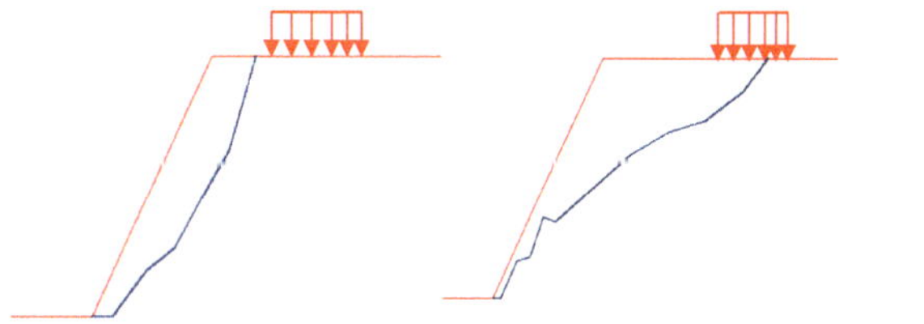

Figure 2.25 Failure mechanism with change of loading plate position.

to the transition zone. The actual failure mode in Figure 2.23c appears to match better with the results by DEM than by FEM or plasticity theory. If the position of the loading plate is varied as shown in Figure 2.25, then failure mechanisms from experiments still appear to be composed of three or more wedges instead of the classical failure mechanism. Yamamoto and Kusuda (2001) obtained a transition zone which is closer to a circular arc than a logspiral zone from the laboratory tests. In the experimental tests by Mwebesa et al. (2012), it was also found that the logspiral zone is not found in the bearing capacity test in Philippi Dune sand which has soil parameters of φ'=34° and c'= 6.7 kPa. The wedge underneath the footing is however found in all the literature that the authors are aware of. Other than that, it appears there are only limited test results where

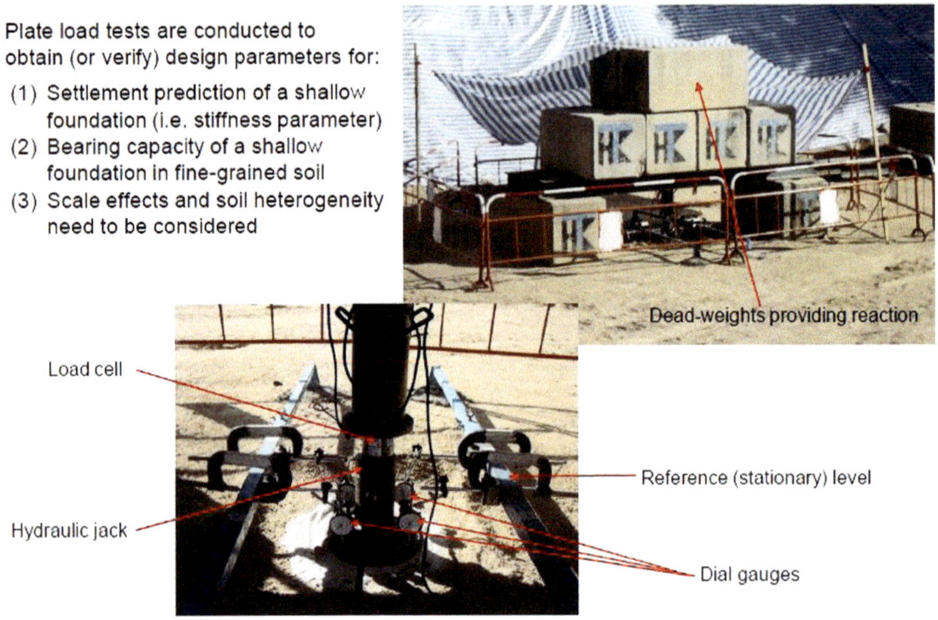

Figure 2.26 A typical plate load test in Hong Kong.

the logspiral transition is actually obtained in tests. Apparently, many researchers have jumped into the adoption of the logspiral mechanism and have overlooked that there are also many counter-examples in the literature.

2.9 PLATE LOAD TEST

When the ground condition is highly complicated, or when the footing is close to the edge of a slope, the use of the bearing capacity equation or prescribed bearing capacity in many foundation codes may not be appropriate. To overcome uncertainty in the analysis, the use of a plate load test (Figure 2.26) is commonly adopted by many engineers to ascertain the bearing capacity and the expected settlement for a proper design. The load applied on the plate actually comes from the action of the hydraulic jack instead of the kentledge, which is an issue not known by many young engineers. The hydraulic jack load is measured by a load cell, and the movement of the bearing plate is commonly monitored by dial gauges at the four corners of the plate. To ensure that the ground adjacent to the plate does not affect the result of the measurement, the dial gauges are usually attached to a reference beam with supports far from the testing plate. The relation between the expected settlement and bearing capacity for a plate (P) and a full foundation (F) is complicated. For plate load test in clay

$$q_u(F) = q_u(P) \tag{2.70}$$

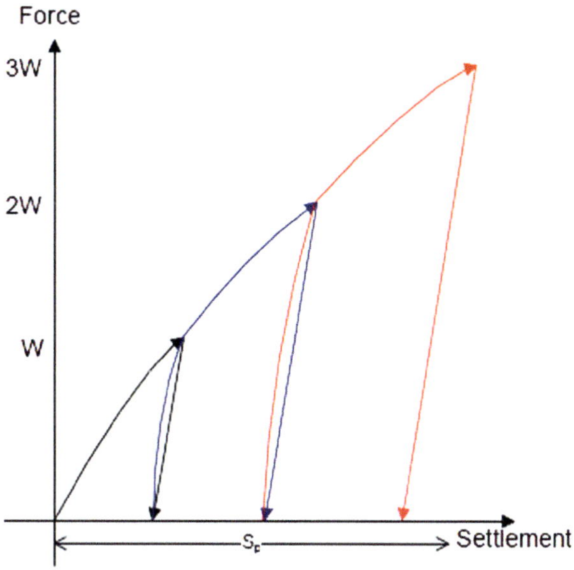

Figure 2.27 Typical load settlement relation in plate load test.

This equation implies that the ultimate bearing capacity in clay is virtually independent of the size of the plate. This is justified by the fact that $N_q = 1$ and $N_r = 0$ for clay. For test in sandy soils ($c = 0$) where the test is carried out on ground level with $q = 0$

$$q_{u(F)} = q_{u(P)} \cdot \frac{B_F}{B_P} \tag{2.71}$$

Eq. (2.71) is based on the basic bearing capacity equation $q_f = cN_c + qN_q + 0.5\gamma BN_\gamma$, and c and $q = 0$. Such extrapolation is actually not applicable when B is large or c is not zero.

For the plate load test, the loading plate should not be less than 300 mm for either a square or circular plate. Currently, a square or rectangular plate is more common than a circular plate in practice. The maximum test load is usually limited to 3W, where W is the working load, and the loading is applied in stages as shown in Figure 2.27. The test load should be applied through a hydraulic jack with a load cell for measuring the applied load in increments of 0.5W. Each stage should be maintained for at least 10 minutes or until the rate of settlement is less than 0.05 mm per 10 minutes, whichever is longer. The typical load settlement relation in a typical plate load test is shown in Figure 2.27. Some guidelines and procedures for conducting plate loading tests are given in BS EN 1997-1:2004 (BSI, 2004) and DD ENV 1997-3:2000 (BSI 2000). It is interesting to note that in many actual plate load test results, only a small part of the nonlinear relation is found (or even not found), which means that the bearing capacity equation together with the modification factors are usually quite conservative in nature.

Li (2007) discussed the problems of using a plate load test for verification of the bearing capacity of soil. Since the size of the testing plate ranges between 300 mm and 600 mm, which will be much smaller than the actual size of the footing, the initiation of failure in the plate load test does not necessarily imply that the footing will experience a similar failure at the same bearing pressure or settlement. In many cases, settlement is usually the controlling factor in determining the allowable bearing capacity. Usually, the plate load test is carried out at the founding level and above groundwater level (it can be below ground level). If the load test is to be carried out below the groundwater level soil, the water level should be lowered in order to prevent disturbance to the soil at the test level.

In the authors' view, it is not a meaningful exercise to verify the bearing capacity of soils using plate loading tests. There are already well-established bearing capacity theories for estimating the bearing capacity of footings. With a sufficiently large footing and a commonly adopted safety factor of 3, the bearing capacity is usually not an issue. The performance of footings on soils tends to be controlled by settlement. It will be more meaningful to use the plate loading test as a tool for estimating the stiffness of soils for settlement calculations.

If the plate loading test has to be used for verification of bearing capacity as a means of control of building safety, there are ways for a more rational interpretation of the plate loading test. The bearing capacity of footing depends on the footing size, but the bearing capacity factors do not. One logical approach for conducting the plate bearing test is to reduce the test pressure according to the size of the test plate on a pro-rata basis as discussed by Li (2007). This will not be repeated here.

A second approach is to shift the focus of a plate loading test to checking the validity of design values of N_γ and N_q used in bearing capacity calculations. The ultimate capacity of a test plate resting on the soils at the founding level is expressed as follows based on

$$q_{ult} = \tfrac{1}{2}\,\gamma B\, N_\gamma\, s_\gamma \qquad (2.72)$$

where B is now the size of the test plate. For a test plate resting on a level ground surface, which is usually the case, the value of the shape factor is 0.6. We will illustrate the approach by way of a worked example based on the following design parameters/assumptions:

a. Design allowable bearing capacity q_{all} = 400 kPa;
b. The footing is a 10 m square footing resting on a level surface;
c. Surcharge q_o = 20 kPa;
d. The soil is dry with a unit weight of γ = 20 kN/m^3;
e. There is no eccentricity of applied loading;
f. According to laboratory test/empirical correlation, a conservative estimate of the angle of shearing resistance ϕ = 33°;
g. Safety factor for net pressure = 3;
h. Size of test plate = 1 m.

For foundation design, the design process usually starts with a sufficiently conservative value of ϕ. The bearing capacity factors N_γ and N_q are then calculated to obtain an estimate of the ultimate bearing capacity q_{ult}. By adopting the safety factor of 3, the allowable bearing capacity q_{all} is then established based on soil mechanics theory. The actual design bearing capacity may be less than the computed allowable bearing capacity value based on soil mechanics calculations due to other design considerations.

Based on a value of ϕ = 33°, the allowable bearing capacity computed using the full bearing capacity equation is 1004 kPa. If a lesser design bearing capacity of 400 kPa is adopted, this will imply an actual design value of ϕ = 26.4° and the corresponding bearing

Figure 2.28 Photograph of a retaining wall resting on fill materials.

Source: Li and Lau (2012).

capacity factors are $N_\gamma = 13.2$ and $N_q = 12.3$. Based on a design value of $N_\gamma = 13.2$, the design ultimate bearing capacity corresponding with the conditions of a 1 m square test plate resting on a level ground is 79.2 kPa based on Eq. (2.70).

By observing whether the test plate will fail under a test load q_{test} equal to or larger than 79.2 kPa, one would verify the design values of ϕ, N_γ and N_q for bearing capacity calculations. The failure load of a test plate may be defined as the loading corresponding to a settlement equal to 10% of the size of the test plate, similar to a recommendation in BS8004 (BSI, 1986) for pile foundations. Using the elastic solutions for soils, one can also use the measured settlement of the test plate at working load to back-calculate the elastic modulus of soil E as described in GeoGuide 2 (GEO, 1987).

This approach of interpretation of the plate bearing test has been used by Li for verifying the bearing capacity for a few hundreds of metres of retaining walls for road-widening projects for the Tuen Mun Highway and Tolo Highways.

Figure 2.28 shows a 17.5 m high-gravity retaining wall founded on existing fill materials with a design bearing capacity of 450 kPa for a road-widening project in Hong Kong. The plate bearing test was used to verify the design values of the bearing capacity factor N_γ and deformation modulus adopted in the design of the retaining wall.

The original design of the retaining wall in Figure 2.28 was an L-shaped retaining wall supported by pile foundation. With the new approach for interpreting the plate loading test, it can be demonstrated rationally that the bearing stress and settlement of the gravity retaining wall are within acceptable limits. The piles can be safely deleted, leading to a significant cost saving for the project.

References

Bowles JE (1997), *Foundation analysis and design*, 5th ed., McGraw-Hill.

British Standard Institute (1986), *BS8004: 1986 code of practice for foundations.*

BSI (2000), *Eurocode 7. Geotechnical design.* Design assisted by field testing, Euro Code.

BSI (2004), *Eurocode 7. Geotechnical design.* General rules, Euro Code.

Chen WF (1975), *Limit analysis and soil plasticity*, Elsevier.

Cheng YM (2002), *Slip line solution and limit analysis for lateral earth pressure problem, the Ninth conference on computing in civil and building engineering*, April 3–5, Taipei, Taiwan, pp. 311–314.

Cheng YM (2003), Seismic lateral earth pressure coefficients by slip line method, *Computers and Geotechnics*, 30(8), 661–670.

Cheng YM and Au SK (2005), Slip line solution of bearing capacity problems with inclined ground, *Canadian Geotechnical Journal*, 42, 1232–1241.

Cheng YM and Li N (2017), Equivalence between bearing capacity, lateral earth pressure and slope stability problems by slip-line and extremum limit equilibrium methods, *International Journal of Geomechanics*, 17(12), ASCE, 04017113.

Cheng YM, Zhao ZH and Sun YJ (2010), Evaluation of interslice force function and discussion on convergence in slope stability analysis by the lower bound method, *Journal of Geotechnical and Geoenvironmental Engineering*, 136(8), ASCE, 1103–1113.

Cheng YM., Hu YY and Wei WB (2007), General Axi-symmetric active earth pressure by Methods of Characteristics – Theory and Numerical Formulation, *International Journal on Geomechanics*, 7(1), 1–15.

Das BM (2017), *Shallow foundations, bearing capacity and settlement*, 3rd ed., CRC Press.

De Jong DJG (1957), Graphical method for the determination of slip-line fields in soil mechanics, *Ingenior*, 69, 61–65.

GEO (1987), *Guide to site investigation.* Geotechnical Engineering Office, Hong Kong Government.

GEO *Foundation design and construction*, GEO Publication No.1/2006, HK SAR Government.

Graham J, Andrew M and Shields DH (1988), Stress characteristics for shallow footing in cohesionless slopes, *Canadian Geotechnical Journal*, 25(2), 238–249.

Hansen JB (1970), *A revised and extended formula for bearing capacity*, Danish Geotechnical Institute, Bulletin 28, Copenhagen.

Kotter F (1903), *Die Bestimmung des Druckes an gekemmten Gleitflchen, eine Aufgabe aus der Lehre vom Erddruck, Berlin Akad*, Wiss, Berlin, pp. 229–233.

Kumar J (2003), N_γ for rough strip footing using the method of characteristics. *Canadian Geotechnical Journal*, 40(3), 669.

Li KS (2007), *Bridging research and practice–the VLS experience*, Victor Li & Associates.

Li N, Cheng YM and Fung WH (2018), Transition zone in bearing capacity problem from plasticity method, discrete element method and laboratory tests, *Advances in Geoscience*, 2(1), 1–21.

Martin CM (2005), Exact bearing capacity calculations using the method of characteristics. *Proceedings of the 11th international conference of IACMAG*, Turin, 4, 441.

Meyerhof GG (1963), Some recent research on the bearing capacity of foundations, *CGJ*, 1(1), 16–26.

Michalowski RL (2001), Upper-bound load estimates on square and rectangular footings, *Geotechnique*, 51(9), 787–798

Mwebesa JM, Kalumba D and Kulabako R (2012), Simulating bearing capacity failure of surface loading on sand using COMSOL. *Second international conference on advances in engineering and technology*, 409–415.

Novotortsev VI (1938), Experience with the application of the theory of plasticity to problems of determination of the bearing capacity of foundation beds of structures, Izv, *Nauchno-Issled. Inst. Anzh. Geol*, 22, 115–128.

Prandtl L (1920), *Uber die Harte plasticher Korper, Nachrichten von der Koniglichen Gsellschaften*, Math-phys, Klasse, Gottingen, 74–85.

Rao NSVK (2011), *Foundation design theory and practice*, John Wiley.
Reissner H (1924), Zum Erddruckproblem In: Biezeno CB, Burgers JM (eds), *Proceedings of the first international congress for applied mechanics*, Delft, pp. 295–311.
Sokolovskii VV (1965), *Statics of granular media*, Pergamon Press.
Vesic AS (1973), Analysis of ultimate loads for shallow foundations, *Journal of the Soil Mechanics and Foundations Division, ACSE*, 99(SM1), 45–73.
Vesic AS (1975a), Bearing capacity of shallow foundations. In: HF Winterkorn and H Fang (eds), *Foundation engineering handbook*, Chapter 3, Van Nostrand Reinhold, New York, pp. 121–147.
Vesic AS (1975b), *Chapter 3: Foundation engineering handbook*, 1st ed., Winterkorn and Fang, Van Nostrand Reinhold, pp. 751.
Yamamoto K (2010), Seismic bearing capacity of shallow foundations using upper bound method. *International Journal of Geotechnical Engineering*, 4(2), 255.
Yamamoto K and Kusuda K (2001), Failure mechanisms and bearing capacities of reinforced foundations. *Geotextiles and Geomembranes*, 19(3), 127–162.
Yu H.S. (2006), *Plasticity and Geotechnics*, Springer.

Further reading

Baban TM (2016), *Shallow foundations, discussion and problem solving*, Wiley Blackwell.
Bowles JE (1996), *Foundation analysis and design*, 5th ed., The McGraw-Hill Companies, Inc.
Buildings Department (2017), *Code of practice for foundations 2017*, The Government of SAR, Hong Kong.
Canadian Geotechnical Society (2006), *Canadian foundation engineering manual*, 4th ed.
Das BM (2019), *Principles of foundation engineering*, Cengage Learning.
Day RW (2010), *Foundation engineering handbook*, 2nd ed., McGraw Hill.
Erol Tutumluer & Imad L Al-Qadi (ed.) (2009), *Bearing capacity of roads, railways and airfields, 1 and 2*, Taylor & Francis.
Huang AN and Yu HS (2018), *Foundation engineering analysis and design*, CRC Press.
Katzenbach R, Leppla S and Choudhury D (2017), *Foundation systems for highrise structures*, CRC Press.
Kimura T, Kasakabe O and Saitoh K (1985), Geotechnical model tests of bearing capacity problems in a centrifuge, *Geotechnique*, 35, 33–45.
Li N, and Cheng YM (2015), Laboratory and 3D-distinct element analysis of failure mechanism of slope under external surcharge, *Natural Hazards and Earth System Sciences*, 15, 35–43.
Li V, and Lau CK (2012), New developments in foundation design and construction in Hong Kong, *Proceedings of joint structural division annual seminar 2012–Contemporary design approaches and new technologies in structural engineering*, 63–77.
Meyerhof GG (1965), Some recent research on bearing capacity of foundations, *Canadian Geotechnical Journal*, 1(1), 16–26.
Saran S (2018), *Shallow foundations and soil constitutive laws*, CRC Press.
Shields DH, Scott JD, Bauer GE, Deschenes JH and Barsvary AK (1977), Bearing capacity of foundations near slopes, *Proceedings, 9th international conference on soil mechanics and foundations engineering*, 2, Tokyo, 715–720.
Terzaghi K (1943), *Theoretical soil mechanics*, Wiley, New York.
Tomlinson MJ (2001), *Foundation design and construction*, Pearson Education.
Vesic AS, Banks DC and Woodard JM (1965), Analysis of ultimate loads of shallow foundations, *Journal of the Soil Mechanics and Foundations Division, ASCE*, 99(SM), 45–73.
Yamamoto K, Lyamin AV, Abbo AJ, Sloan SW and Hira M. (2009), Bearing capacity and failure mechanism of different types of foundations on sand. *Soils and Foundations*, 49(2), 305–314.

Chapter 3

Serviceability limit state of shallow foundations

3.1 INTRODUCTION

The serviceability limit state of a shallow foundation is mainly related to its settlement under imposed load. The importance of settlement can easily be understood, as excessive settlement will not only create a serviceability problem, but will also adversely affect the integrity and stability of the structure. Additional bending moment and shear force can be generated from the displacement and relative displacement, and usually such displacements are not allowed in the design of buildings and bridges (continuous type). A good example is the housing problem at Tin Shui Wai, Hong Kong, where the significant relative displacement of the foundation prohibits the installation of lifts and caused a major engineering and political crisis in Hong Kong. Actually, there is virtually no ultimate limit state problem (Chapter 2) for foundations, as the ultimate bearing capacity will come much later than intolerable settlement and angular distortion of the superstructures.

Settlement of a shallow footing on soil is due to compressibility of the supporting soil or rock. Under the applied load from the structure which changes the "stress state" of the soil, the soil particles will re-arrange themselves by changing positions or even crushing, so that the whole soil mass has its void ratio and subsequently its volume reduced, leading to the settlement. As the soil particles re-arrange themselves by sliding, rolling and even crushing, leading to deformation of the whole mass which is largely unrecoverable, only very little of the settlement can be recovered upon removal of the load. Therefore, it can be claimed that the soil is hardly elastic. Nevertheless, it remains a practice (or as a convenient and yet conservative approach) to treat the soil as a "pseudo-elastic" material and estimate its "short-term" (in terms of hours or days) settlement by the elastic theory under the use of the elasticity parameter including Young's modulus and Poisson's ratio. Such an approach is applicable to coarse-grained soil. However, there is another type of settlement which takes months or years to complete, which is termed consolidation. Consolidation settlement is due to the expulsion of water from fine-grained soil which is very impermeable to water. The expulsion takes years to complete, resulting in consolidation settlement. Furthermore, there is another settlement caused by creep, which is critical mainly for clay. Creep is an important issue in airport construction. The construction of the first runway of the Hong Kong Chek Lap Kok airport cost about US$13 billion (1997 value), making it possibly the most expensive airport in the world. The main reason for such a high cost was to avoid creep, so that all the marine clay was removed by dredgers before reclamation.

 DOI: 10.1201/9781032650586-3

At the time of construction, half of the dredgers in the world were working in this project, and the high cost of this project was simply due to creep! The third runway of this airport cost about US$26 billion, and again, that is due to the potential consolidation and creep arising from the marine clay. There are many more examples where consolidation and creep play important roles in different types of construction, including the famous leaning tower of Pisa, ZhuZhai-Hong Kong-Macau bridge project, Hong Kong Chek Lap Kok International Airport, and others.

3.2 STRESS AND DISPLACEMENT DUE TO POINT LOAD, LINE LOAD AND OTHERS

The short-term (or immediate) settlement is discussed first, which largely involves elasticity theory. The total settlement in a soil stratum of depth H can be calculated by the following equation

$$S = \int_0^H \varepsilon_z dH \tag{3.1}$$

where ε_z is the strain in the Z (vertical) direction.

H is taken to the depth of "hard stratum", where further settlement is negligible, or it can be taken as infinity if the hard stratum is very deep. As such, the settlement is sitting on a semi-infinite space. However, if a few strata are encountered beneath the footing, the total settlement is

$$S = \sum_i^N \int_{H_{i-1}}^{H_i} \varepsilon_z dH \tag{3.2}$$

By the elastic theory,

$$\varepsilon_z = \frac{\sigma_z}{E_z} - \mu_x \frac{\sigma_x}{E_x} - \mu_y \frac{\sigma_y}{E_y} \tag{3.3}$$

where σ_z, is the normal stress in the vertical (Z direction)

σ_x, σ_y are the normal stresses in the lateral X and Y directions

E_x, E_y and E_z are the Young's moduli in the X, Y and Z directions

μ_x, μ_y are the Poisson's ratios in the X and Y directions.

Estimation of ε_z can be significantly simplified if the soil is taken as isotropic, which means $E_x = E_y = E_z = E$ and $\mu_x = \mu_y = \mu$. Eq. (3.3) then becomes:

$$\varepsilon_z = \frac{1}{E}\left[\sigma_z - \mu\left(\sigma_x + \sigma_y\right)\right] \tag{3.4}$$

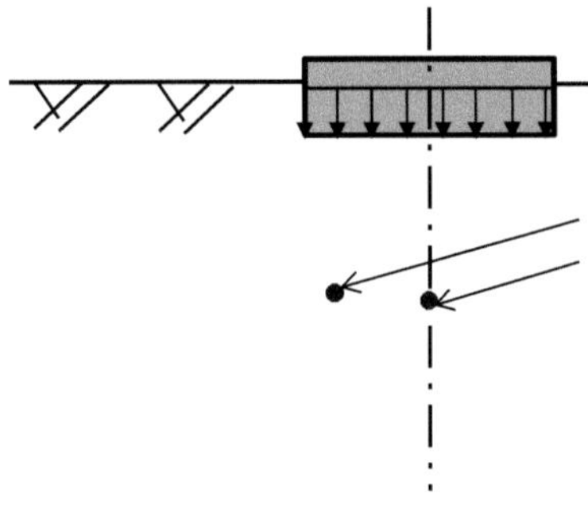

Figure 3.1 Illustration of estimation of settlement beneath footing.

The simplest case starts with determination of the stress components of a point within a semi-infinite homogeneous and isotropic medium with the point load P on the surface which is given by the famous Boussinesq equation with the symbols defined in Figure 3.2:

$$\sigma_z = \frac{3P}{2\pi z^2}\left(\frac{1}{1+(r/z)^2}\right)^{2.5} \tag{3.5}$$

$$\sigma_r = \frac{P}{2\pi}\left\{\frac{3r^2 z}{(r^2+z^2)^{2.5}} - \frac{1-2\mu}{r^2+z^2+z\sqrt{r^2+z^2}}\right\} \tag{3.6}$$

$$\sigma_\theta = \frac{-P}{2\pi}(1-2\mu)\left\{\frac{z}{(r^2+z^2)^{1.5}} - \frac{1}{r^2+z^2+z\sqrt{r^2+z^2}}\right\} \tag{3.7}$$

$$\tau_{rz} = \frac{3P}{2\pi}\left(\frac{rz^2}{(r^2+z^2)^{2.5}}\right) \tag{3.8}$$

τ_{rz} is the shear stress in the rz plane. By symmetry, $\tau_{r\theta} = \tau_{z\theta} = 0$.

There are several points about the Boussinesq equation which have to be noted. This fundamental equation applies for the case when the point load is located on the ground surface, otherwise, the more general Mindlin equation has to be used. At point (0,0,0), the vertical stress σ_z is infinity from Eq. (3.5). This result is obvious, as a point load has no size or zero sectional area. The pressure underneath the point load must hence be infinity, with infinite penetration. To overcome this discontinuity, a finite size distributed load q can be used to define a point load, but the fundamental solution Boussinesq equation will still apply by expressing P as qdA. The dA will cancel out the discontinuity at (0,0,0).

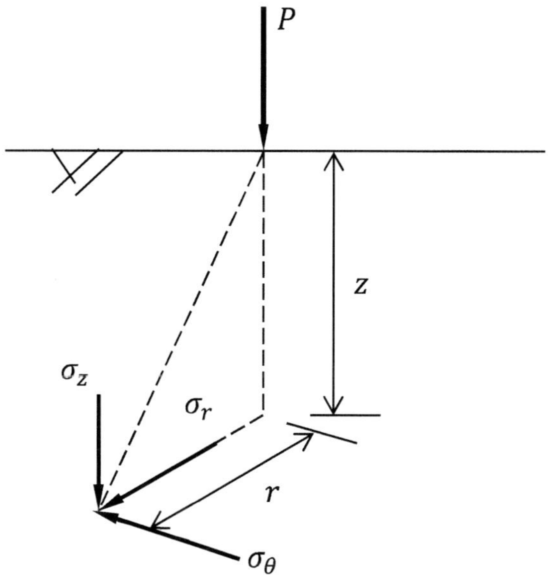

Figure 3.2 Boussinesq equation.

The plot of σ_z with z and r in a vertical plane containing the point load in form of contour is as shown in Figure 3.3.

As the stress distribution is radially symmetrical about the axis of the applied vertical load, a "bulb" form distribution for a constant stress will be created. The distribution is often described as a "stress bulb". This design figure can be used to obtain a quick assessment of the stresses below a foundation.

Eqs. (3.5)–(3.8) can be taken as a tool to calculate the stresses due to a patch of load of any shape or any load intensity distribution. For vertical stress, σ_v at any point of coordinate (x, y, z) as referred to a pre-determined coordinate system beneath the footing can be computed as

$$\sigma_v(x, y, z) = \int \frac{3}{2\pi z^2} \left\{ \frac{p(u,\ v)}{\left[1 + \left(\frac{\sqrt{(u-v)^2 + (v-y)^2}}{z} \right) \right]} \right\}^{5/2} dA \tag{3.9}$$

where $p(u, v)$ is the pressure exerted by the footing at a point of coordinate (u, v) as referred to the same coordinate system.

By Eq. (3.9), the following integral can be used to evaluate σ_v at a point at a coordinate (x, z) where x is the distance of the point from the centre of a circular footing of diameter B exerting uniform pressure of unity magnitude.

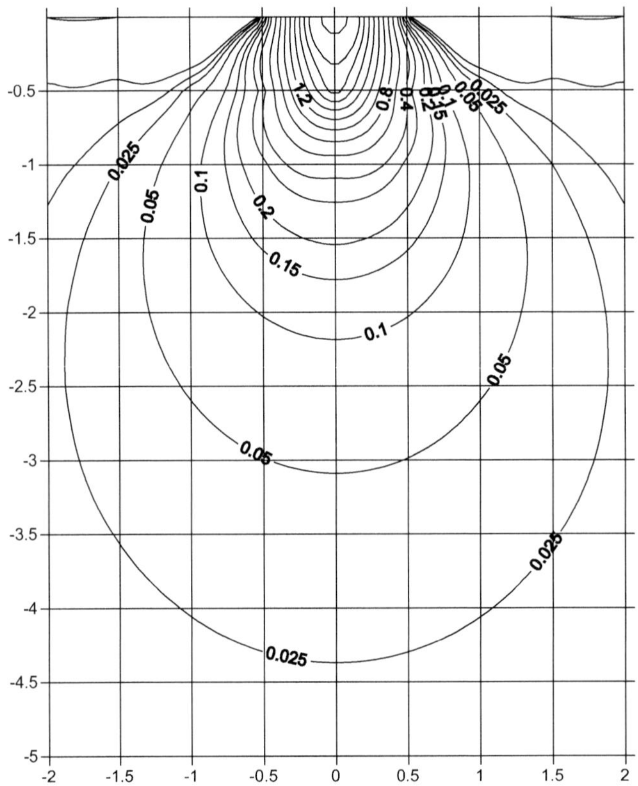

Figure 3.3 Vertical stress contour to Boussinesq equation (unit load).

$$\sigma_v(x,z) = \int_0^1 \int_0^{2\pi} \frac{3}{2\pi z^2} \left\{ \frac{1}{\left[1 + \left(\frac{\sqrt{r^2 + x^2 - 2xr\cos\theta}}{z} \right) \right]} \right\}^{5/2} d\theta dr$$

and the "stress bulb" (stress contour) at any vertical plane containing a diameter of the footing is as shown in Figure 3.4.

Similarly, for a square footing, $\sigma_v(x,z)$ for stress at any points within a vertical plane containing the centre line of the footing is

$$\sigma_v(x,z) = \int_{-1}^1 \int_{-1}^1 \frac{3}{2\pi z^2} \left\{ \frac{1}{\left[1 + \left(\frac{\sqrt{(u-x)^2 + v^2}}{z} \right) \right]} \right\}^{5/2} dudv \qquad (3.10)$$

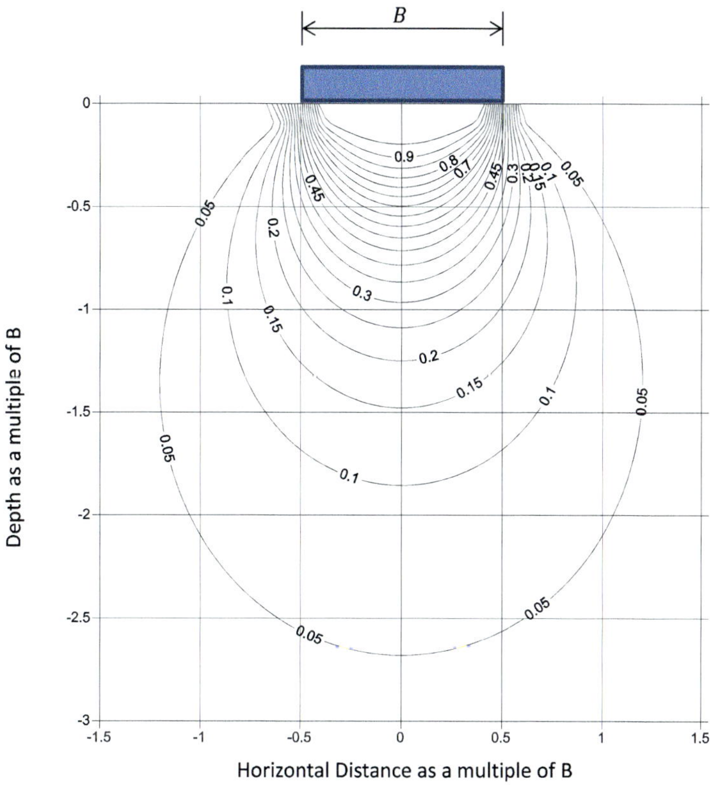

Figure 3.4 Contour of vertical stress beneath a circular footing (unit vertical pressure from footing).

The stress contour through the centre line of the footing is shown in Figure 3.5.

For a strip footing of length infinity, $\sigma_v(x,z)$ for stress at any points within a vertical plane perpendicular to the long length of the footing is

$$\sigma_v(x,z) = \int_{-\infty}^{\infty}\int_{-1}^{1} \frac{3}{2\pi z^2}\left\{\frac{1}{1+\left(\frac{\sqrt{(u-x)^2+v^2}}{z}\right)}\right\}^{5/2} dudv \tag{3.11}$$

The stress bulb is shown in Figure 3.6.

The stresses σ_r and σ_θ can be likewise computed according to Eqs. (3.6) and (3.7), and the strain at any point can therefore be calculated by Eq. (3.4) and the total settlement integrated by Eq. (3.1) or (3.2).

Using Eq. (3.9) and based on a coordinate system with the origin at a corner of a rectangular footing of plan dimension $B \times L$ which is applying uniform pressure of q on the

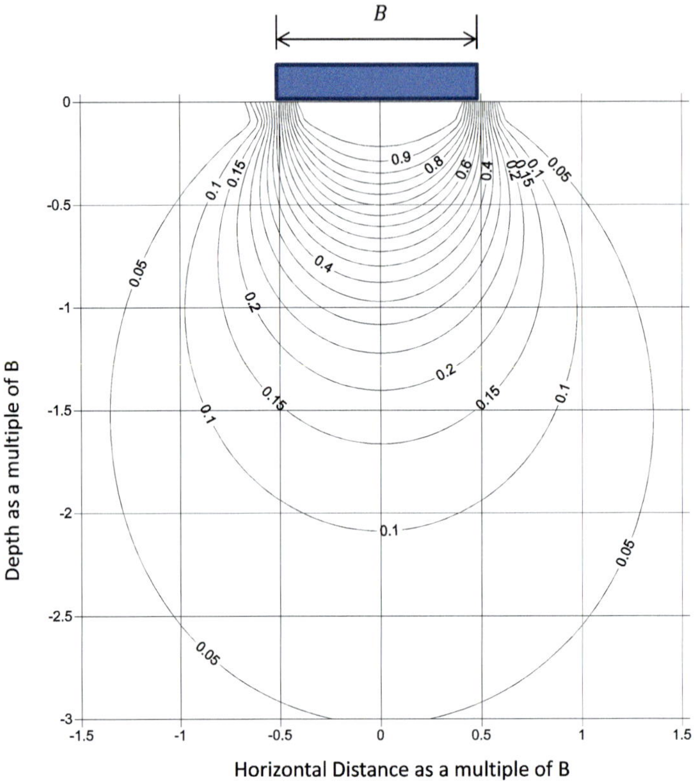

Figure 3.5 Contour of vertical stress beneath a square footing using normalized distance (unit vertical pressure from footing).

ground, the total vertical stress at a depth z beneath the corner due to the whole area is given by Eq. (3.11) as

$$\sigma_v = \int_{y=0}^{L}\int_{x=0}^{B} \frac{3q\,dx\,dy}{2\pi(x^2+y^2+z^2)^{2.5}} = pI_\sigma \tag{3.11}$$

where the coefficient I_σ can be worked out analytically as

$$I_\sigma = \frac{1}{4\pi}\left(\frac{2mn\sqrt{m^2+n^2+1}}{m^2+n^2+m^2n^2+1}\cdot\frac{m^2+n^2+2}{m^2+n^2+1}+\tan^{-1}\frac{2mn\sqrt{m^2+n^2+1}}{m^2+n^2+1-m^2n^2}\right) \tag{3.12}$$

with $m = \dfrac{B}{z}, n = \dfrac{L}{z}$

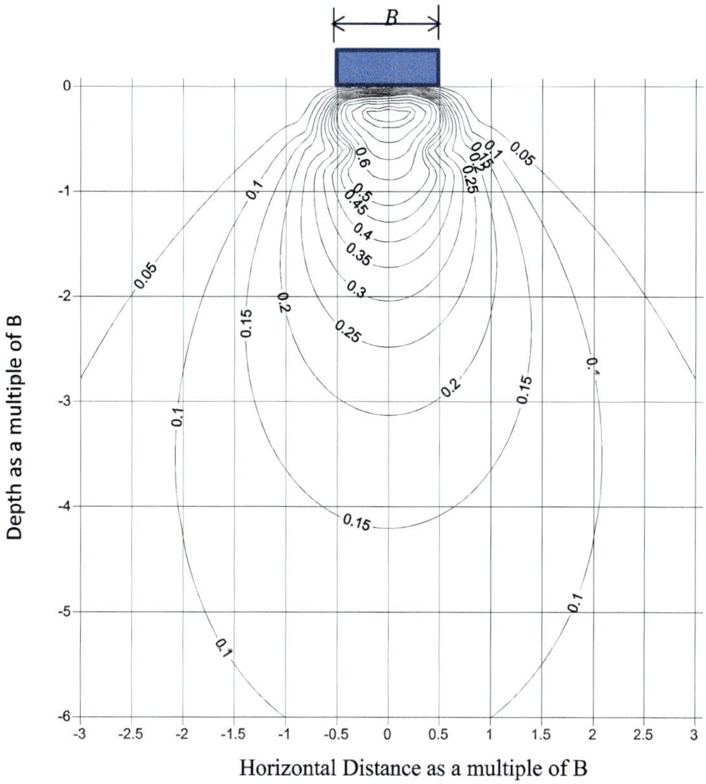

Figure 3.6 Contour of vertical stress beneath an infinitely long strip footing (unit vertical pressure from footing).

When m and n are large, $\tan^{-1}\dfrac{2mn\sqrt{m^2+n^2+1}}{m^2+n^2+1-m^2n^2}$ can be negative, then Eq. (3.12) takes the form

$$I_\sigma = \frac{1}{4\pi}\left(\frac{2mn\sqrt{m^2+n^2+1}}{m^2+n^2+m^2n^2+1}\cdot\frac{m^2+n^2+2}{m^2+n^2+1}+\pi+\tan^{-1}\frac{2mn\sqrt{m^2+n^2+1}}{m^2+n^2+1-m^2n^2}\right) \tag{3.13}$$

The values of I_σ are tabulated in Table 3.1 for ease of reference.

The reason why the corner is chosen is because of the ease in integration of Eq. (3.11) with the lower limits set at zero. Also, with the determination of the corner stresses, stress at any point can be worked out as per Figure 3.7. It should be noted that Table 3.1 is a symmetric one. This can be easily visualized by rotating a rectangular foundation by 90°. The stress at a point beneath the corner will not change by such a rotation of the foundation, hence this table must be symmetric by nature.

Table 3.1 Table of I_0 varying with m and n

n	m										
	0.1	*0.2*	*0.3*	*0.4*	*0.5*	*0.6*	*0.7*	*0.8*	*0.9*	*1.0*	*1.2*
0.1	0.0047	0.0092	0.0132	0.0168	0.0198	0.0222	0.0242	0.0258	0.0270	0.0279	0.0293
0.2	0.0092	0.0179	0.0259	0.0328	0.0387	0.0435	0.0473	0.0504	0.0528	0.0547	0.0573
0.3	0.0132	0.0259	0.0374	0.0474	0.0559	0.0629	0.0686	0.0731	0.0766	0.0794	0.0832
0.4	0.0168	0.0328	0.0474	0.0602	0.0711	0.0801	0.0873	0.0931	0.0977	0.1013	0.1063
0.5	0.0198	0.0387	0.0559	0.0711	0.0840	0.0947	0.1034	0.1103	0.1158	0.1202	0.1263
0.6	0.0222	0.0435	0.0629	0.0801	0.0947	0.1069	0.1168	0.1247	0.1311	0.1360	0.1431
0.7	0.0242	0.0473	0.0686	0.0873	0.1034	0.1168	0.1277	0.1365	0.1436	0.1491	0.1570
0.8	0.0258	0.0504	0.0731	0.0931	0.1103	0.1247	0.1365	0.1461	0.1537	0.1598	0.1684
0.9	0.0270	0.0528	0.0766	0.0977	0.1158	0.1311	0.1436	0.1537	0.1618	0.1684	0.1777
1.0	0.0279	0.0547	0.0794	0.1013	0.1202	0.1360	0.1491	0.1598	0.1684	0.1752	0.1851
1.2	0.0293	0.0573	0.0832	0.1063	0.1263	0.1431	0.1570	0.1684	0.1777	0.1851	0.1958
1.4	0.0301	0.0589	0.0856	0.1094	0.1300	0.1475	0.1620	0.1739	0.1836	0.1914	0.2028
1.6	0.0306	0.0599	0.0871	0.1114	0.1324	0.1503	0.1652	0.1774	0.1874	0.1955	0.2073
1.8	0.0309	0.0606	0.0880	0.1126	0.1339	0.1521	0.1672	0.1797	0.1899	0.1981	0.2103
2.0	0.0311	0.0610	0.0887	0.1134	0.1350	0.1533	0.1686	0.1812	0.1915	0.1999	0.2124
2.5	0.0314	0.0615	0.0895	0.1145	0.1363	0.1548	0.1704	0.1832	0.1937	0.2024	0.2151
3.0	0.0315	0.0618	0.0898	0.1150	0.1368	0.1555	0.1711	0.1841	0.1947	0.2034	0.2163
3.5	0.0316	0.0619	0.0900	0.1152	0.1371	0.1558	0.1715	0.1845	0.1952	0.2039	0.2169
4.0	0.0316	0.0619	0.0901	0.1153	0.1372	0.1560	0.1717	0.1847	0.1954	0.2042	0.2172
5.0	0.0316	0.0620	0.0901	0.1154	0.1374	0.1561	0.1718	0.1849	0.1956	0.2044	0.2174
6.0	0.0316	0.0620	0.0902	0.1154	0.1374	0.1562	0.1719	0.1850	0.1957	0.2045	0.2175

n	M									
	1.4	*1.6*	*1.8*	*2.0*	*2.5*	*3.0*	*3.5*	*4.0*	*5.0*	*6.0*
0.1	0.0301	0.0306	0.0309	0.0311	0.0314	0.0315	0.0316	0.0316	0.0316	0.0316
0.2	0.0589	0.0599	0.0606	0.0610	0.0615	0.0618	0.0619	0.0619	0.0620	0.0620
0.3	0.0856	0.0871	0.0880	0.0887	0.0895	0.0898	0.0900	0.0901	0.0901	0.0902
0.4	0.1094	0.1114	0.1126	0.1134	0.1145	0.1150	0.1152	0.1153	0.1154	0.1154
0.5	0.1300	0.1324	0.1339	0.1350	0.1363	0.1368	0.1371	0.1372	0.1374	0.1374
0.6	0.1475	0.1503	0.1521	0.1533	0.1548	0.1555	0.1558	0.1560	0.1561	0.1562
0.7	0.1620	0.1652	0.1672	0.1686	0.1704	0.1711	0.1715	0.1717	0.1718	0.1719
0.8	0.1739	0.1774	0.1797	0.1812	0.1832	0.1841	0.1845	0.1847	0.1849	0.1850
0.9	0.1836	0.1874	0.1899	0.1915	0.1937	0.1947	0.1952	0.1954	0.1956	0.1957
1.0	0.1914	0.1955	0.1981	0.1999	0.2024	0.2034	0.2039	0.2042	0.2044	0.2045
1.2	0.2028	0.2073	0.2103	0.2124	0.2151	0.2163	0.2169	0.2172	0.2175	0.2176
1.4	0.2102	0.2151	0.2184	0.2206	0.2236	0.2250	0.2257	0.2260	0.2263	0.2264
1.6	0.2151	0.2202	0.2237	0.2261	0.2294	0.2309	0.2316	0.2320	0.2324	0.2325
1.8	0.2184	0.2237	0.2274	0.2299	0.2334	0.2350	0.2357	0.2362	0.2366	0.2367
2.0	0.2206	0.2261	0.2299	0.2325	0.2361	0.2378	0.2387	0.2391	0.2395	0.2397
2.5	0.2236	0.2294	0.2334	0.2361	0.2401	0.2420	0.2429	0.2434	0.2439	0.2441
3.0	0.2250	0.2309	0.2350	0.2378	0.2420	0.2439	0.2450	0.2455	0.2461	0.2463
3.5	0.2257	0.2316	0.2357	0.2387	0.2429	0.2450	0.2460	0.2467	0.2472	0.2475
4.0	0.2260	0.2320	0.2362	0.2391	0.2434	0.2455	0.2467	0.2473	0.2479	0.2482
5.0	0.2263	0.2324	0.2366	0.2395	0.2439	0.2461	0.2472	0.2479	0.2486	0.2489
6.0	0.2264	0.2325	0.2367	0.2397	0.2441	0.2463	0.2475	0.2482	0.2489	0.2492

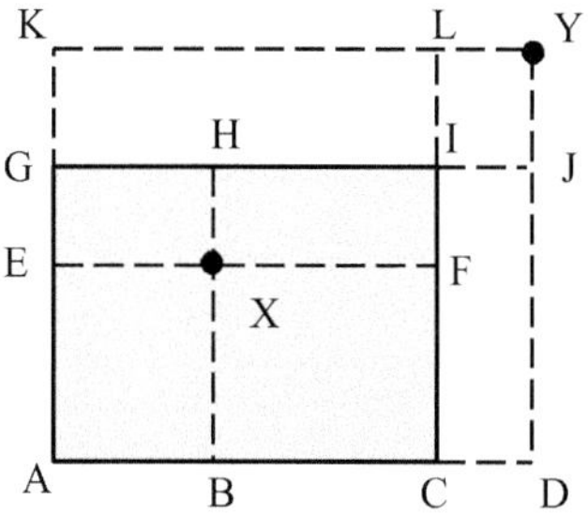

A rectangular footing ACIG is carrying uniform pressure. The stresses at depth z below points X and Y due to the footing can be calculated as addition and subtraction of the effects at a corner of each of the following rectangles :

(i) Point X : Sum of effects due to rectangles ABXE, BCFX, EXHG and XFIH;
(ii) Point Y : Effects of rectangles ADYK–CDYL – GJYK + IJYL

Figure 3.7 Illustration of determination of stress of a point due to a rectangular footing carrying uniform pressure.

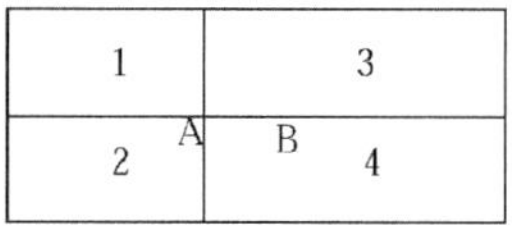

Figure 3.8 Example for calculation of stresses below a footing.

The lateral stresses can be obtained similarly. Apart from use in the determination of settlement, these formulae are useful in estimating stresses on soil beneath the footing.

Example: Consider a rectangular footing with a size of 4 × 6 m with a uniform distributed load of 100 kPa. Determine the stresses at a depth of 10 m below the ground surface, for the centre point B and a point A 2 × 2 m from the corner of the footing, as shown in Figure 3.8.

Solution:
Divide the rectangular footing into four equal 2 × 3 m rectangles. For point B in Figure 3.8, m = 3/10; n = 2/10, hence I_σ = 0.0259 from Table 3.1 (or a small program as used by the authors).

The vertical stress due to the ground surface load is hence 4 × 100 × 0.0259 =10.36 kPa (multiply by 4 because there are four equal rectangles).

For point A

Zone	*m (B/Z)*	*n (L/Z)*	I_σ
1	2/10=0.2	2/10=0.2	0.0179
2	2/10=0.2	2/10=0.2	0.0179
3	2/10=0.2	4/10=0.4	0.0328
4	2/10 = 0.2	4/10 = 0.4	0.0328
			Σ = 0.1014

The stress at point A is hence = 0.1014 × 100 = 10.14 kPa (slightly less than that at B).

In this example, there are several major limitations in the calculation:

1. The soil is assumed to be uniform and isotropic so that the Boussinesq equation can be used.
2. Rock exists at a depth of infinity for the Boussinesq equation.
3. The stiffness of the footing is neglected.
4. The loading is a uniform vertical pressure.

To overcome the various limitations as given above, the use of a computational method will be required, and the finite element method is the most widely adopted method for engineering programs. Alternatively, some engineers adopt a simplified approach for the stresses beneath a foundation. For a B × L rectangular footing with a uniform distributed load q on the ground surface, the loading is assumed to spread uniformly to a wider area of $(B + Z) \times (L + Z)$ at a depth Z, which is called the 1 in 2 load spread method by some engineers. The angle of spreading is also taken as 30° and 60° in some countries, but the 1 in 2 approach appears to be the most popular among engineers. The total load on the ground surface is qBL, which is equal to the stress $q(z) \times (B + Z) \times (L + Z)$ at depth Z. For the previous example, the stress should be 100 × 4 × 6/(14 × 16) = 10.71 kPa, which is basically similar to the results for points A and B using the Bousinessq equation. It should also be noted that if Z is small, the 1 in 2 method will give a stress lower than that from the Bousinessq equation.

Upon determination of the vertical and lateral stresses, the settlement at any point of the soil can be worked out using Eqs. (3.4) and (3.1). However, a more direct approach is based on the summation of effects on settlement due to point loads. A point load at the surface of an isotropic and homogeneous soil will produce a vertical strain in a point of coordinate (r,z) as

$$\varepsilon_z(r,z) = \frac{P}{2\pi E}\left[\frac{3(1+\mu)r^2 z}{(r^2+z^2)^{2.5}} - \frac{[3+\mu(1-2\mu)]z}{(r^2+z^2)^{1.5}}\right] \tag{3.14}$$

where P is the point load applied.

By integrating Eq. (3.14) for $\int_z^{\infty} \varepsilon_z(r,z)dz$ which is the settlement at (r,z) in a semi-finite mass medium, Eq. (3.15) is obtained as

$$\rho_z = \frac{P(1+\mu)}{2\pi E\sqrt{r^2+z^2}}\left[2(1-\mu) + \frac{z^2}{r^2+z^2}\right] \tag{3.15}$$

In general, the vertical strain of a point of coordinate (x,y,z) due to a patch of load on the surface of soil is

$$\varepsilon_z = \int \varepsilon_z q dA \tag{3.16}$$

where q is the pressure at an infinitesimal area dA so that qdA is the infinitesimal point load.

For a rectangular footing of plan dimensions $B \times L$ (with B as the lesser dimension) carrying a uniform pressure p with $dA = dxdy$, and with a hard stratum at height H beneath the soil surface, the total settlement of a corner is

$$\Delta H = \int_0^H \int_0^B \int_0^L \varepsilon_z dxdydz \tag{3.17}$$

Although ε_z is based on the Boussinesq equations on stresses for an elastic semi-infinite mass medium, it is assumed to be valid even if there is a hard stratum at some depth H, i.e. the stress distributions in accordance with Boussinesq equations is assumed to be the same even in the presence of a hard stratum.

For Eq. (3.17), the mathematics may sometimes be easier to work as the difference of two integrals, both involving infinity.

$$\Delta H = \int_0^\infty \int_0^B \int_0^L \varepsilon_z dxdydz - \int_H^\infty \int_0^B \int_0^L \varepsilon_z dxdydz \tag{3.18}$$

Eq. (3.18) can be solved analytically for the settlement of a corner of the rectangular footing as

$$\Delta H = qB\frac{1-\mu^2}{E_s}\left(I_1 + \frac{1-2\mu}{1-\mu}I_2\right)I_f \tag{3.19}$$

where E_s is the Young's modulus of soil
μ is the Poisson's ratio

$$I_1 = \frac{1}{\pi}\left[M\ln\frac{\left(1+\sqrt{M^2+1}\right)\sqrt{M^2+N^2}}{M\left(1+\sqrt{M^2+N^2+1}\right)} + \ln\frac{\left(M+\sqrt{M^2+1}\right)\sqrt{1+N^2}}{1+\sqrt{M^2+N^2+1}}\right]$$

$$I_2 = \frac{N}{2\pi}\tan^{-1}\left(\frac{M}{N\sqrt{M^2+N^2+1}}\right)$$

where $M = L/B, N = H/B$

In Eq. (3.19), I_f is an additional factor to account for the effect when the footing is buried at a depth D below the surface of the soil, bearing in mind that the original equations are based on Boussinesq's equations applying for point loads on top of the soil. I_f is based on the Fox equation with details as follows:

$$I_f = \frac{\sum_{i=1}^{5}\beta_i Y_i}{\left(\beta_1+\beta_2\right)Y_1} \tag{3.20}$$

where $\beta_1 = 3 - 4\mu$; $Y_1 = L\ln\left(\frac{R_4 + B}{L}\right) + B\ln\left(\frac{R_4 + L}{B}\right) - \frac{R_4^{\,3} - L^3 - B^3}{3LB}$

$$\beta_2 = 5 - 12\mu + 8\mu^2; Y_2 = L\ln\left(\frac{R_3 + B}{R_1}\right) + B\ln\left(\frac{R_3 + L}{R_2}\right) - \frac{R_3^{\,3} - R_2^{\,3} - R_1^{\,3} - R^3}{3LB}$$

$$\beta_3 = -4\mu(1 - 2\mu); Y_3 = \frac{R^2}{L}\ln\left[\frac{(B + R_2)R_1}{(B + R_3)R}\right] + \frac{R^2}{B}\ln\left[\frac{(L + R_1)R_2}{(L + R_3)R}\right]$$

$$\beta_4 = -1 + 4\mu - 8\mu^2; Y_4 = \frac{R^2(R_1 + R_2 - R_3 - R)}{LB}$$

$$\beta_5 = -4(1 - 2\mu)^2; Y_5 = R\tan^{-1}\left(\frac{LB}{RR_3}\right)$$

$$R = 2D;$$

$$R_1 = \sqrt{L^2 + R^2}; R_2 = \sqrt{B^2 + R^2};$$

$$R_3 = \sqrt{L^2 + B^2 + R^2}; R_4 = \sqrt{L^2 + B^2}$$

The I_f factors can be readily calculated with the help of spreadsheets. By dividing the numerator and denominator of eq. (3.19) by B, the Y and R (i.e., D) parameters can all turn to factors of B so that we can determine and plot I_f as the ratios of L/B; D/B. Typical plots for Poisson's ratios $\mu = 0.35$ and 0.5 are given in Figure 3.9.

In using the elastic settlement equation, it should be noted that the **effective loading** instead of total load, that is $q - D$, should be used, provided that q refers to the total pressure beneath the foundation. This is based on the rationale that the original overburden stress γD will not generate any elastic settlement. This approach is adopted in all textbooks and design codes that the authors are aware of. However, a problem in Hong Kong has caused the authors to rethink this problem. For a building with a raft foundation supporting an average load of 120 kPa, after demolition and re-construction works, the average loading on the raft footing will be about 160 kPa. What should be the design load for settlement calculation? Since the concept of effective loading has a major impact in this case, the authors view that 160 kPa should be used in the design calculation! The rationale behind this is that after demolition of the superstructure and existing raft footing, the soil will rebound to some extent, and the upheaval will be removed to the designed level for the new construction works. The new 160 kPa is imposed on this new configuration, and the settlement will be induced by this 160 kPa. However, using the unloading/reloading Young's modulus of soil until the previous maximum overburden pressure is exceeded,

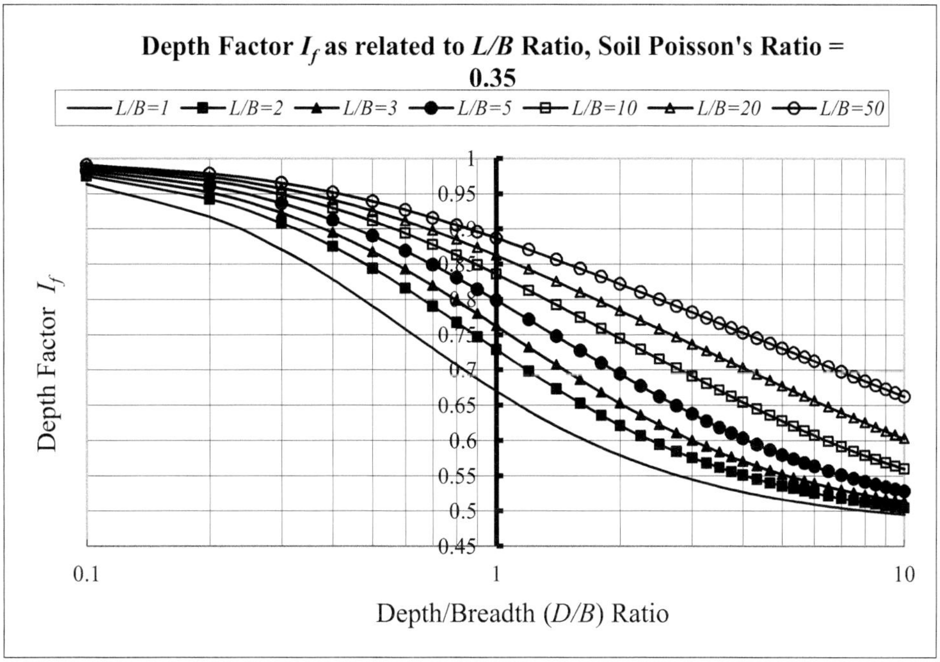

Figure 3.9(a) Plot of depth factor I_f with *L*/*B* and *D*/*B* ratios for μ = 0.35.

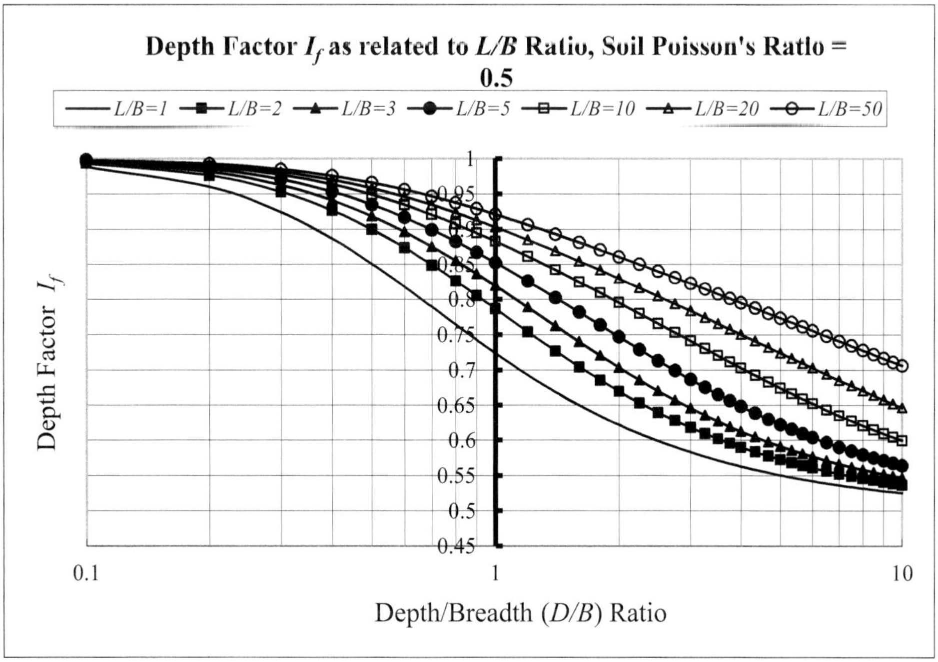

Figure 3.9(b) Plot of depth factor I_f with *L*/*B* and *D*/*B* ratios for μ = 0.5.

and the applied loading is located at a certain depth below the ground surface is a more realistic phenomenon. This should be the actual situation that is happening on site, and the full applied pressure should be adopted in design calculation, but using an unloading/reloading modulus.

For foundation on multi-layered soil, some engineers adopt the use of equivalent Young's modulus of soil, which is given by $E_{eq} = \Sigma E_i H_i / \Sigma H_i$. If the footing stiffness is considered, there are only minor changes to the settlement, as the settlement is mainly controlled by the settlement of the foundation soil. Hence, the use of a rigid footing cannot really reduce the amount of settlement. For rigid foundations, settlement is about 0.93 × settlement at centre of the flexible foundation. For the Young's modulus of soil, many engineers rely on the statistical relation with the SPT or CPT results, and an additional factor of safety is sometimes incorporated into the design codes in some countries. Some typical values are 10,000 kN/m² for fill, 10,000–15,000 kN/m² for alluvium, and 15,000–20,000 kN/m² for completely decomposed granite in Hong Kong, where a factor of safety of about 2 has been applied. For the Poisson's ratio, some typical values adopted by engineers are 0.3–0.4 for N equals 4–10, 0.2–0.35 for N equals 11–30 and 0.15–0.3 for N above 30.

With the application of Eqs. (3.18) and (3.19), the settlement of a corner of a rectangular footing can be determined and, with the principle as described by Figure 3.7, settlement at any point beneath the footing can be determined.

However, it can readily be seen from Eq. (3.18) that a footing applying uniformly distributed load on top of soil will result in different settlements at different parts of the footing. For a footing of plan dimensions 4 × 6 m applying a uniformly distributed load of 100 kPa on the top surface of soil of Young's modulus 50 MPa, Poisson's ratio = 0.35, and depth 32 m to hard stratum ($M = \frac{6}{4} = 1.5$, $N = \frac{32}{4} = 8$, $I_f = 0$), the settlement contour is shown in Figure 3.10, showing that the maximum settlement is at the centre of the footing.

However, it should be noted that the settlement shown in Figure 3.10 can only be true if the footing is a truly "flexible one", i.e. without stiffness. If a footing possesses some stiffness, the deflection will be affected, which has to be adjusted so that the deflection profile of the footing is identical to that of the soil. However, as the stiffnesses of the soil and the footing are different, the distribution of the load has to be adjusted until the two profiles under the adjusted loads match with each other. The determination is a tedious process and often results in very high stresses at the footing corners and edges. In theory, the stresses will become infinity at corners and edges if the footing is rigid, i.e. of infinite stiffness. As such, the soil will yield and there is a further redistribution of stresses and the soil behaves elasto-plastically.

Therefore, to be practical, the average settlement under the assumption that the footing is flexible is often taken as the settlement of the footing.

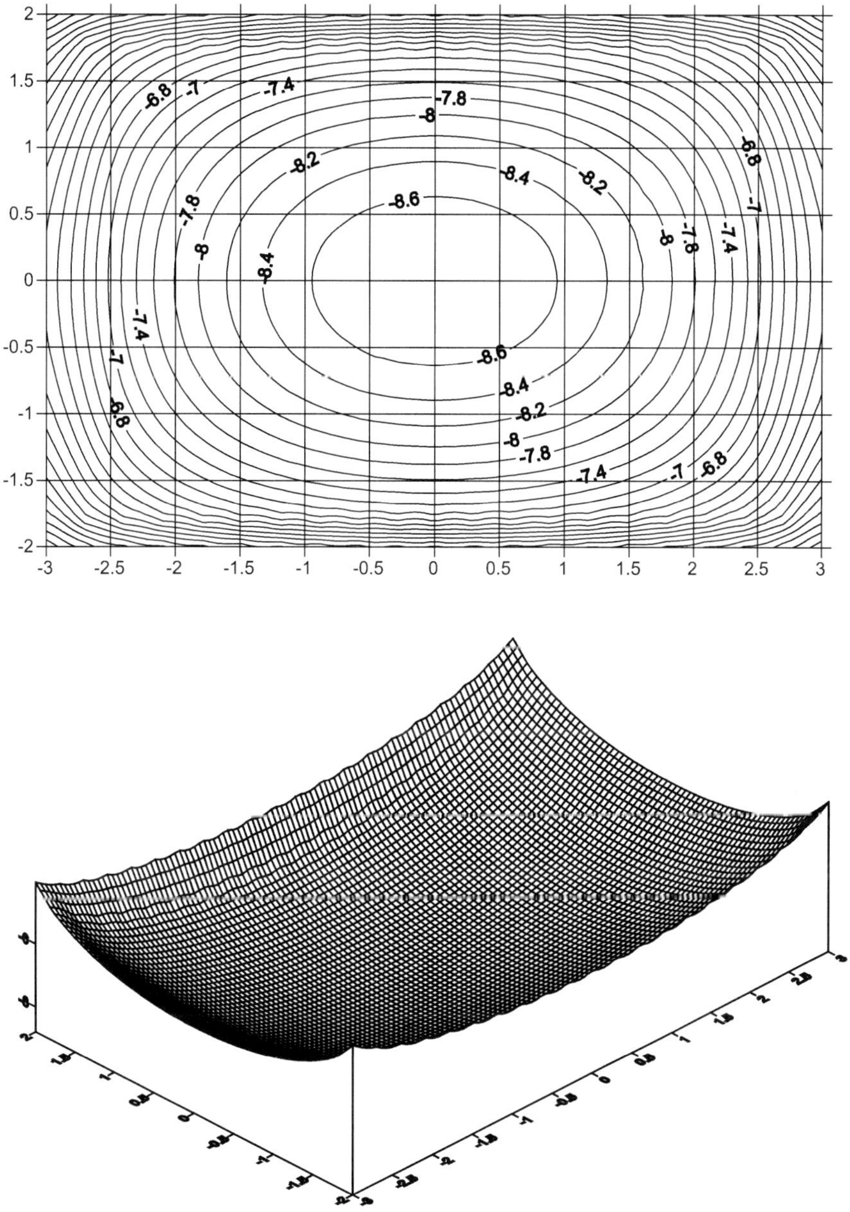

Figure 3.10 Settlement profile of a flexible footing carrying uniformly distributed load.

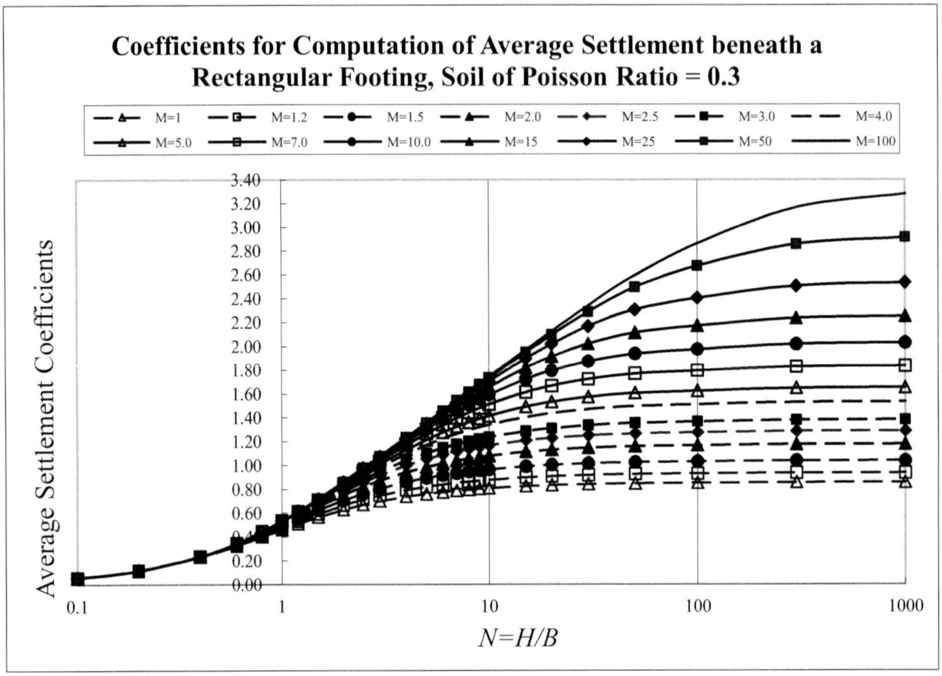

Figure 3.11(a) Variation of average settlement coefficient with *M* and *N* for soil of Poisson's ratio = 0.3.

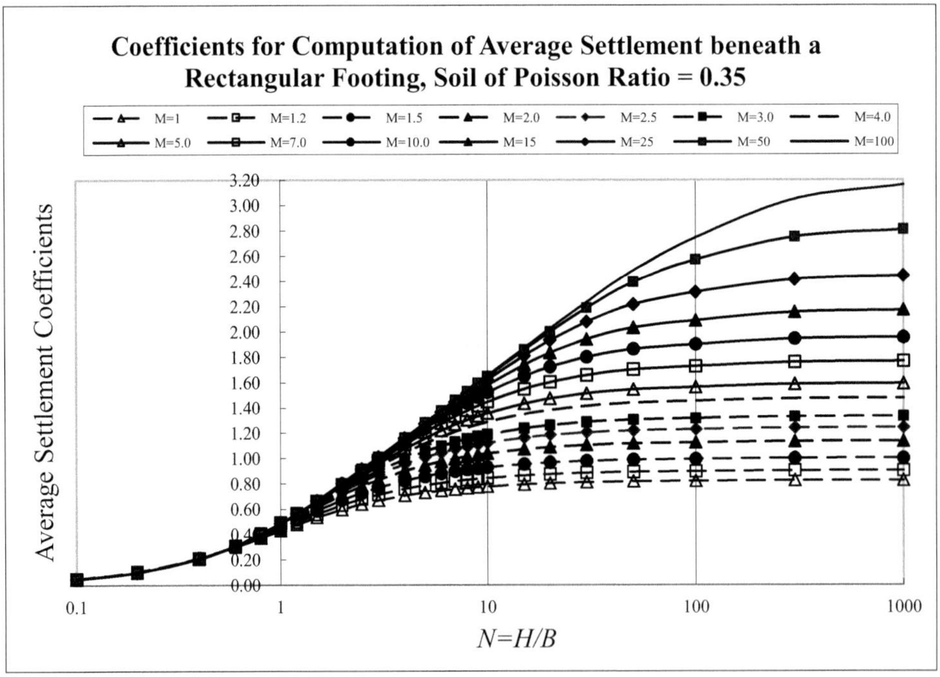

Figure 3.11(b) Variation of average settlement coefficient with *M* and *N* for soil of Poisson's ratio = 0.35.

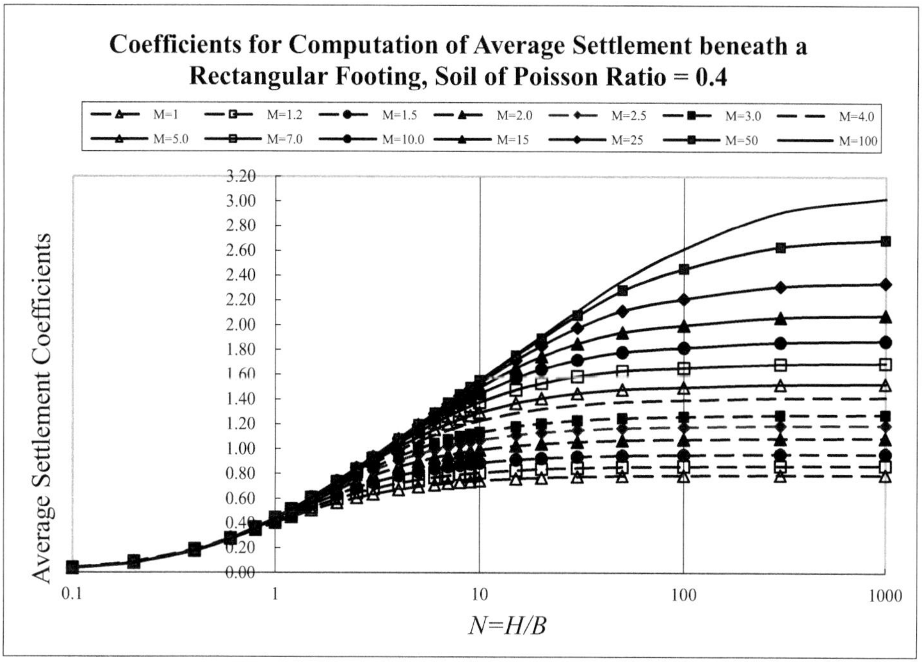

Figure 3.11(c) Variation of average settlement coefficient with *M* and *N* for soil of Poisson's ratio = 0.4.

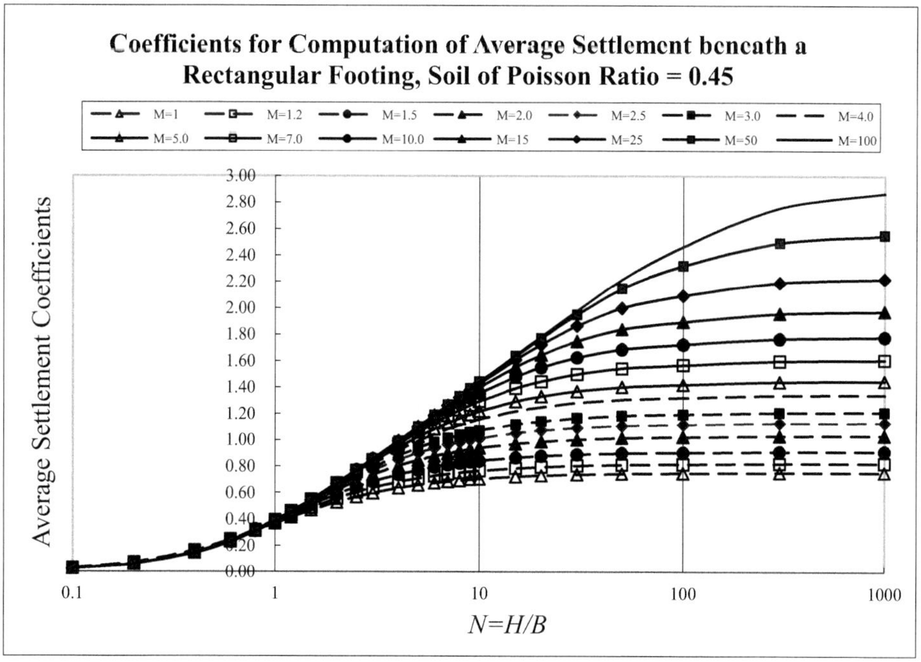

Figure 3.11(d) Variation of average settlement coefficient with *M* and *N* for soil of Poisson's ratio = 0.45.

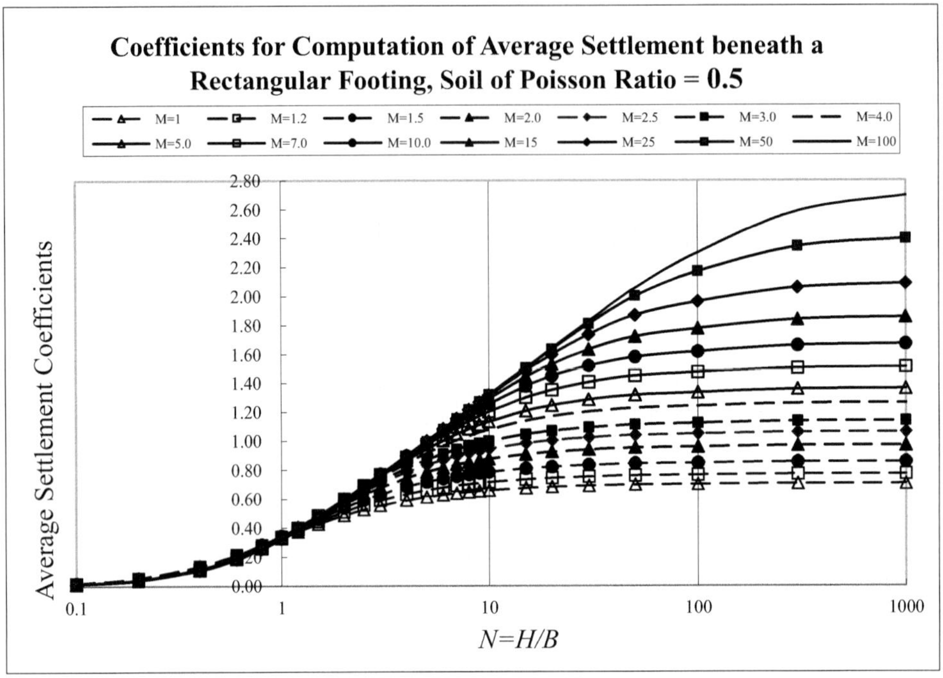

Figure 3.11(e) Variation of average settlement coefficient with *M* and *N* for soil of Poisson's ratio = 0.5.

Figure 3.11 gives the variation of the average settlement coefficient $I_\mu = (1-\mu^2)\left(I_1 + \frac{1-2\mu}{1-\mu}I_2\right)$. The results in Figure 3.11 multiplied by $\frac{qB}{E_s} \times I_f$ will give the average settlement available for the settlement estimation.

The afore-mentioned approach is based on the Boussinesq equation with point loads applied on the surface of the ground. Integration of point load effects with modification by the Fox equation have to be carried out to account for footings sunken in soil. However, Mindlin's equations can calculate effects of point loads within a semi-infinite soil medium so that their integration can be used to determine the effects of sunken footings. The equation for vertical settlement is listed as follows with the symbols explained in Figure 3.12.

$$\rho_z = \frac{P(1+\mu)}{8\pi E(1-\mu)}\left[\frac{3-4\mu}{R_1} + \frac{8(1-\mu)^2-(3-4\mu)}{R_2} + \frac{(z-c)^2}{R_1^{\ 3}} + \frac{(3-4\mu)(z+c)^2-2cz}{R_2^{\ 3}} + \frac{6cz(z+c)^2}{R_2^{\ 5}}\right]$$

$$= P \times I_{\rho-Mindlin} \qquad (3.21)$$

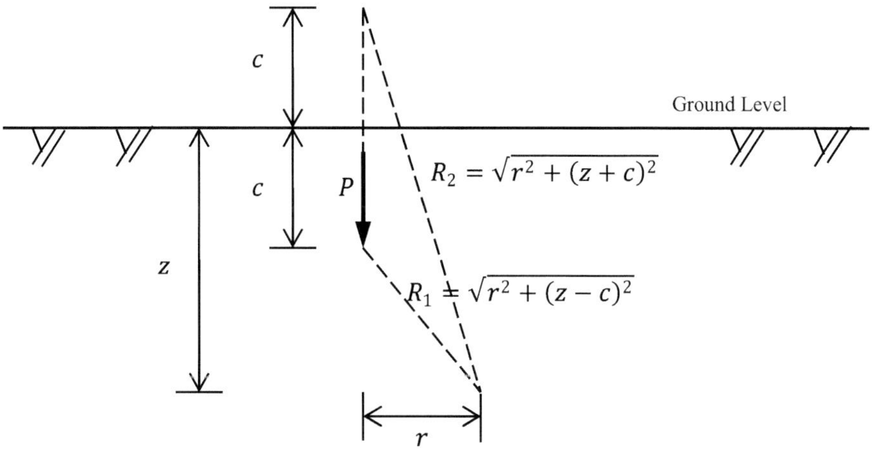

Figure 3.12 Defining symbols of Mindlin's equation.

It can be proved algebraically that Eq. (3.21) will be reduced to Eq. (3.15) when $c = 0$ in Eq. (3.21), showing that Eq. (3.15) is a particular case of Eq. (3.21).

The use of Eq. (3.21) can therefore be used by integration to find settlement at any point inside the soil medium due to a patch of load of pressure p by the following, as similar to that of Boussinesq equations.

$$\Delta H = \int p \times I_{\rho-Mindlin} dA \tag{3.22}$$

However, due to complexity of Eqs. (3.21) and (3.22), analytical solutions are extremely long and are not widely used, hence numerical methods (programs) are used for the solution.

There are Mindlin's equations for vertical and lateral stresses by which substitution into Eq. (3.4) can be used to calculate vertical strain and subsequently settlement as similar to Eqs. (3.16)–(3.19) in case there is a hard stratum at some depth below ground. However, the process is tedious.

The approach described above is for the determination of settlement of footing resting on homogeneous soil. However, the approach can be modified to calculate footing settlement on layered soil. Consider a footing resting on n successive layers of soil of Young's moduli $E_1, E_2, \ldots E_n$ and Poisson's ratio $\mu_1, \mu_2, \ldots \mu_n$. The total settlement of the footing is the sum of settlements due to all the n layers. Defining the function $S_i\left(E_j, \mu_j, D_j\right)$ for the calculation of the soil from top to the i-th level while the soil is the j-th layer of Young's modulus E_j and Poisson's ratio μ_j and depth D_j. The settlement of the i-th layer is

$$S_i = S_i\left(E_i, \mu_i, \sum_{k=1}^{j} D_k\right) - S_{i-1}\left(E_i, \mu_i, \sum_{k=1}^{j-1} D_k\right) \tag{3.23}$$

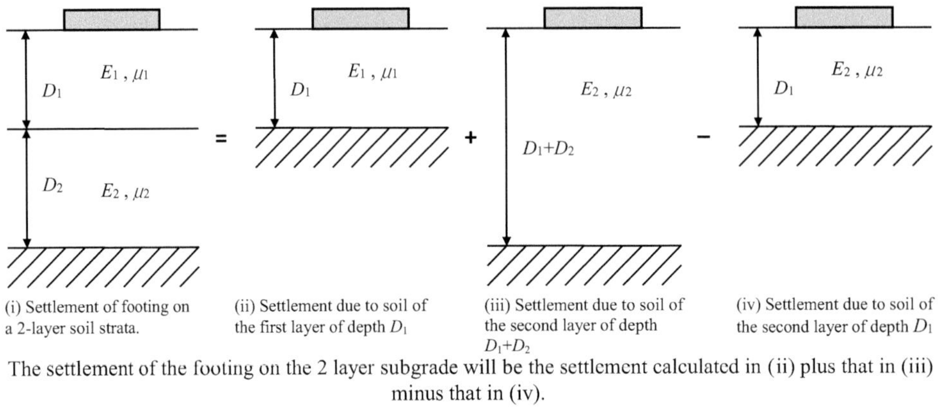

Figure 3.13 Illustration of calculation of footing settlement in two-layered subgrade.

Obviously $S_1 = S_1(E_1, \mu_1, D_1)$ and the total settlement is

$$S = \Sigma S_i \tag{3.24}$$

Figure 3.13 illustrates the approach for the simple case of two layers.

3.3 SETTLEMENT OF FOUNDATION FOR SIMPLE CASES

The following worked Example 3.2 illustrates the use of the afore-mentioned approach to find the immediate settlement of a footing on layered soil.

To find the average settlement of a sunken footing on a three-layered strata as shown in Figure 3.14, the following procedures can be adopted:

(i) Average settlement of the first layer of soil

$$M = \frac{L}{B} = \frac{12}{8} = 1.5; \quad N = \frac{h}{B} = \frac{8}{8} = 1.0, \mu = 0.4$$

Average settlement coefficient is 0.43 from Figure 3.11(c), so average settlement is

$$I \times \frac{q_0 B}{E} = 0.43 \times \frac{100 \times 8}{10000} = 0.0344 \text{ m} = 34.4 \text{ mm}$$

(ii) Average settlement of the second layer of soil
From the bottom level of the footing to the bottom level of the second layer of soil

$$M = 1.5; \quad N = \frac{h}{B} = \frac{8+6}{8} = 1.75, \mu = 0.35$$

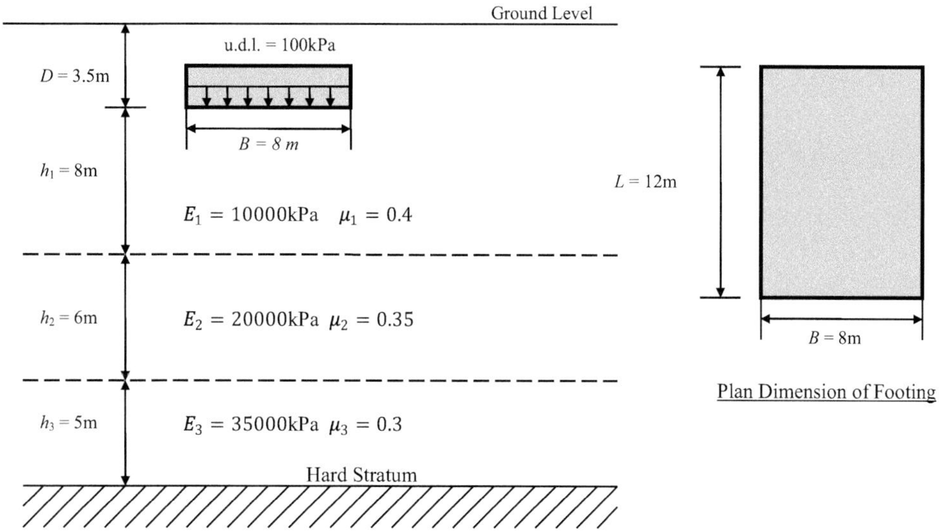

Figure 3.14 Worked Example 3.2.

Average settlement coefficient is 0.63 from Figure 3.11(b), so average settlement is

$$I \times \frac{q_0 B}{E} = 0.63 \times \frac{100 \times 8}{20000} = 0.0252 \text{ m} = 25.2 \text{ mm}$$

From the bottom level of the footing to the top level of the second layer of soil

$$M = 1.5; \quad N = \frac{h}{B} = \frac{8}{8} = 1.0, \mu = 0.35$$

Average settlement coefficient is 0.46 from Figure 3.11(b), so average settlement is

$$I \times \frac{q_0 B}{E} = 0.46 \times \frac{100 \times 8}{20000} = 0.0184 \text{ m} = 18.4 \text{ mm}$$

So the net average settlement of the second layer of soil is

25.2 – 18.4 = 6.8 mm

(iii) Average settlement of the third layer of soil
From the bottom level of the footing to the bottom level of the third layer of soil

$$M = 1.5; \quad N = \frac{h}{B} = \frac{8+6+5}{8} = 2.375, \mu = 0.3$$

Average settlement coefficient is 0.75 from Figure 3.11(a), so average settlement is

$$I \times \frac{q_0 B}{E} = 0.75 \times \frac{100 \times 8}{35000} = 0.0171 \text{ m} = 17.1 \text{ mm}$$

From the bottom level of the footing to the top level of the third layer of soil

$$M = 1.5; \quad N = \frac{h}{B} = \frac{8+6}{8} = 1.75, \mu = 0.3$$

Average settlement coefficient is 0.67 from Figure 3.11(a), so average settlement is

$$I \times \frac{q_0 B}{E} = 0.67 \times \frac{100 \times 8}{35000} = 0.0153 \text{ m} = 15.3 \text{ mm}$$

So the net average settlement of the third layer of soil is

17.1 – 15.3 = 1.8 mm

I_F as determined from Eq. (3.20) is 0.86

Therefore, the total average settlement is $0.86 \times (34.4 + 6.8 + 1.8) = 36.98$ mm.

Example 3.3: A 2 × 4m footing at 1.5 m below ground is loaded with a total UDL of 200 kPa. Determine the maximum differential elastic settlement along the long direction if the soil properties are E = 15,000 kPa, μ = 0.2 and bedrock is present at a depth of 7 m below ground, and unit weight of soil = 19 kN/m^3 (Figure 3.15).

Solution:

The effective loading q = 200 – 1.5 × 19 = 171.7 kPa

For point A, divide the rectangle into four smaller 1 × 2 m rectangles, M = 2/1, N = 5.5/1 (note that 6.5 m is used as H, which is the depth from the bottom of the footing to the bedrock), D/B = 0.75

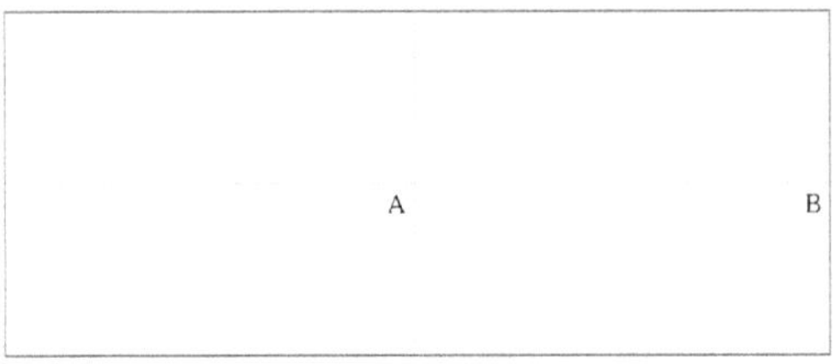

Figure 3.15 Calculation of the differential settlement for a simple footing.

$I_1 = 0.546$, $I_2 = 0.0535$, $I_f = 0.73$, hence

$$\Delta H_A = 4\times 171.5\times 1\times \frac{1-0.2^2}{15000}\left(0.546+\frac{1-2\times 0.2}{1-0.2}\times 0.0535\right)$$
$$\times 0.75\times 1000 = 19.28\,\text{mm}$$

$\Delta H_A = 19.28$ mm from Eq. (3.19) (note that the result has been multiplied by 4 when using the smaller rectangle for calculation).

For point B, divide the rectangle into two smaller 1 × 4 m rectangles, $M = 4/1$, $N = 5.5/1$

$I_1 = 0.584$, $I_2 = 0.0923$, $I_f = 0.73$,

$$\Delta H_{AB} = 2\times 171.5\times 1\times \frac{1-0.2^2}{15000}\left(0.584+\frac{1-2\times 0.2}{1-0.2}\times 0.0923\right)$$
$$\times 0.75\times 1000 = 10.73\,\text{mm}$$

hence $\Delta H_{AB} = 10.73$ mm from Eq. (3.19) (note that the result has been multiplied by 2).

The differential settlement is the distance between two points against the difference in the settlement, which is 1 in 2000/(19.28–10.73), or 1 in 235. It should be noted that the Hong Kong Foundation code and Euro Code 7 limit the differential settlement to 1 in 500, and similar values are used in other countries. The maximum allowable settlement of the foundation is less than 30 mm in the Hong Kong Foundation Code (or 50 mm in the Euro Code 7). In general, the maximum settlement is a less important factor in the foundation design, as the differential settlement will control the design in many cases. More importantly, the maximum settlement is important only for those cases where the utilities entering into a building cannot tolerate the settlement. The author has observed the effects of differential settlement on many old buildings (on raft foundation) around the Wan Chai area in Hong Kong which were built more than 50 years ago. Many of the walls, columns and beams are seriously cracked, with crack width wide enough for a finger. Such cracking is generated from the additional bending moment and shear force arising from the differential settlement, which were not considered in the design 50 years ago. The author has also observed some new building settlements at Wan Chai and Sheung Wan around the MTR construction. The maximum settlements amount to 250–300 mm, yet there is no cracking of the buildings which sit on piles. This is due to the fact that these buildings undergo rigid body movement during the settlement. As long as the utilities can tolerate the settlement, there are no major practical problems with these buildings.

3.4 CONSOLIDATION AND CREEP SETTLEMENT

As discussed in Section 3.1, in addition to "short-term" settlement, there is a relatively "long-term" settlement in fine-grained soil of low permeability to water. The process is termed "consolidation", by which the volume of a mass of fully saturated soil under pressure (exerted from a foundation) is reduced slowly due to the expulsion of water from its voids to the exterior through soil grains of low permeability. The process often takes years to complete until the excess pore water pressure set up by the increase in total stress has completely dissipated.

Consolidation settlement is the vertical displacement of the ground due to the volume change by the consolidation process. The significance of the soil consolidation process to a structure is that it can induce substantial settlements over the long term under the loads of the structure which is required to be estimated for both ultimate and serviceability design considerations.

The simplest consolidation theory is that based on the one-dimensional consolidation settlement equations using the compression index C_c and the coefficient of compression m_v.

Consider a stratum of saturated clayey soil of initial void ratio e_0 under external pressure from a structure leading to a change of void ratio of Δe, the vertical strain can be expressed as:

$$\varepsilon_z = \frac{\Delta e}{1+e_0} \tag{3.25}$$

where Δe is a function of stress on soil before the application of the foundation load σ_0, and σ_1 is the stress after application of the foundation load.

The settlement ΔH divided by the thickness of the clayey soil H is the strain ε_z, so

$$\varepsilon_z = \frac{\Delta H}{H} = \frac{\Delta e}{1+e_0} \tag{3.26}$$

Eq. (3.26) relates the change of void ratio to settlement of the clayey soil. Putting $\Delta e = e_0 - e$ where e is the void ratio after consolidation, Figure 3.16 shows a typical plot of e with effective stress σ for a saturated clay. The application of compression stress is first a "virgin" one by which the clayey soil is stressed to compressive stresses that it has never experienced before. Therefore, it is behaving as a "normally consolidated" soil. A plot of $e - \log\sigma'$ has shown a linear (or almost linear) relationship. However, when the stress is reduced at some point, the soil expands a little and its void ratio can only restore to a small amount of its original value. However, if stress is further increased, the void

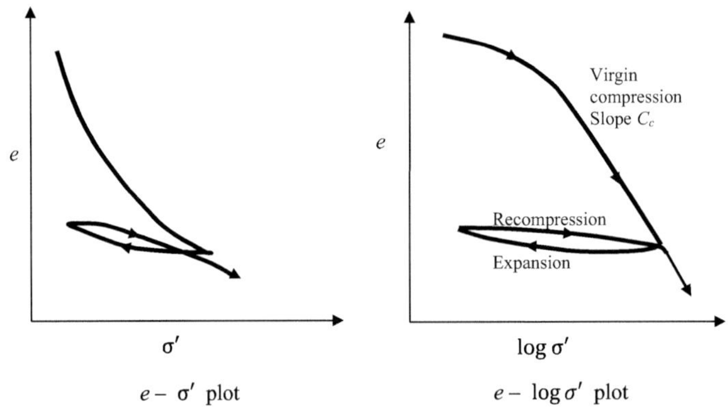

Figure 3.16 Typical plot of void ratio of clayey soil with effective stress.

ratio will restore to the value at the previous state of stress reduction at approximately the same previous stress level. A further increase in stress will result in a decrease in the void ratio by a compression slope approximately equal to the previous one.

In Figure 3.16, two parameters are defined:

(i) The coefficient of compressibility m_v as

$$m_v = \frac{1}{1+e_0}\left(\frac{e_0 - e_1}{\sigma_1' - \sigma_0'}\right) \tag{3.27}$$

where e_0 is the initial void ratio, σ_0' is the initial effective stress, and e_1 the void ratio at the increased stress of σ_1'. So m_v measures the volumetric percentage change of the soil per increase in effective stress. However, it should be noted that m_v is not a constant as per the $e - \sigma'$ plot shown in Figure 3.16.

i) The "compression index" C_c as

$$C_c = \frac{e_0 - e_1}{log\left(\sigma_1'/\sigma_0'\right)}. \tag{3.28}$$

which is taken as a constant at the straight line portion of the $e - \log\sigma'$ plot in Figure 3.16. Similarly, for the part on expansion, another index termed "expansion index" C_e can be defined.

In Figure 3.16, it can be seen that the change of void ratio (settlement similarly) is related to the stress history of the soil. During the first stage of the stress history (termed "virgin compression"), the soil is experiencing an increasing stress that it has never experienced before and the decrease in void ratio and settlement is comparatively significant. It is termed being "normally consolidated" and the soil is a normally consolidated soil. If the stress is then released, the expansion of the soil does not follow the original path as an elastic material but another path as shown by which the expansion is significantly smaller than the "virgin compression". Furthermore, if the increasing stress is re-applied gradually, the soil experiences stresses which are smaller than had been experienced previously. The decrease in void ratio and increase in settlement are comparatively small, and are of magnitudes similar to those of the previous expansion as shown in Figure 3.16. The soil is then an "over-consolidated" soil. After reaching the maximum stress it has experienced before, the rate of consolidation follows approximately the gradient of the $e - \log\sigma'$ path of the previous "normal consolidation". A method developed by Casagrande in estimating the "preconsolidation pressure", σ_c, is illustrated in Figure 3.17.

The procedure for estimating preconsolidation pressure in accordance with Figure 3.17 is as follows:

(i) On the consolidating curve ABC where B is the point with the greatest slope, the tangents DC on the straight portion and EBF are constructed;
(ii) The horizontal line BG is drawn;

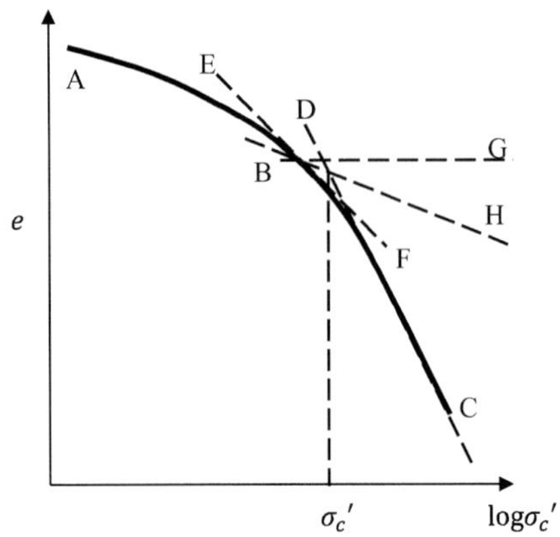

Figure 3.17 Estimation of preconsolidation pressure.

(iii) An angle bisector BH bisecting angle FBG is drawn;
(iv) The stress at the intersection of DC and BH is taken as the preconsolidation pressure σ_c'.

Therefore, from Figure 3.16, C_c can be taken as a material constant while m_v changes with stress.

Settlement of soil due to consolidation is

$$S_c = \int \varepsilon_z dz \tag{3.29}$$

where $\varepsilon_z = \dfrac{\Delta e}{1+e_0} = \dfrac{e_0 - e_1}{1+e_0}$ as shown in Eq. (3.26).

If stress is constant throughout a layer of clayey soil of depth H_c so that Eq. (3.29) becomes $S_c = \varepsilon_z H_c$, combining Eqs. (3.26), (3.28) and (3.29),

$$S_c = \frac{C_c H_c}{1+e_0} \log \frac{\sigma_1'}{\sigma_0'} \tag{3.30}$$

which is valid for the straight (or approximately straight) portions of the $e - \log \sigma'$ plot of Figure 3.16, i.e. the normally consolidated soil. If the equation is used to estimate consolidation settlement due to a newly added footing, σ_0' can be taken as the stress due to the original overburden soil and $\sigma_1' = \sigma_0' + \Delta\sigma$ where $\Delta\sigma$ is the additional stress created by the new footing.

If based on m_v which can be defined as $m_v = \dfrac{\Delta\varepsilon_v}{\Delta\sigma'}$ the settlement is

$$S_c = \int_0^H \varepsilon_v dz = \int_0^H m_v \Delta\sigma' dz \tag{3.31}$$

If $\Delta\sigma'$ is constant, or approximately constant with depth

$$S_c = m_v \Delta\sigma' H \tag{3.32}$$

However, even with a good estimation of C_c, settlement of a layer of clayey soil is often not straightforward as the stress changes with depth. An approximation is to find the average values of σ'_0 and σ'_1 of the whole clayey stratum if it is not too thick. A more accurate method is to divide the clayey stratum into a number of layers. σ'_0 and σ'_1 are determined in each layer at its mid-level for calculation of S_c of each layer. The total settlement is therefore the sum of the settlement of all layers.

While it is straightforward to calculate σ'_0, it is more complicated for $\Delta\sigma$. A simple method is to assume a uniform load spread from the footing, say in a gradient of 1 (horizontal):2 (vertical). Another approach is to adopt the principle of Boussinesq equations with the stress bulb load spread as shown in Figures 3.4–3.6. Both are not exact of course. However, they are acceptable under the current trade practice.

The following worked Example 3.4 illustrates the approaches to find the consolidation settlement of the normally consolidated clay at the vertical axis through the centre of the footing (Figure 3.18).

(i) The clay is normally consolidated clay. The average σ'_0 (as initial overburden pressure) and average σ'_1 are first determined.

$$\sigma'_0 = 3\times16.5 + 1.2\times(17.5-9.8) + 1.25\times(16-9.8) = 66.49\,\text{kN/m}^2.$$

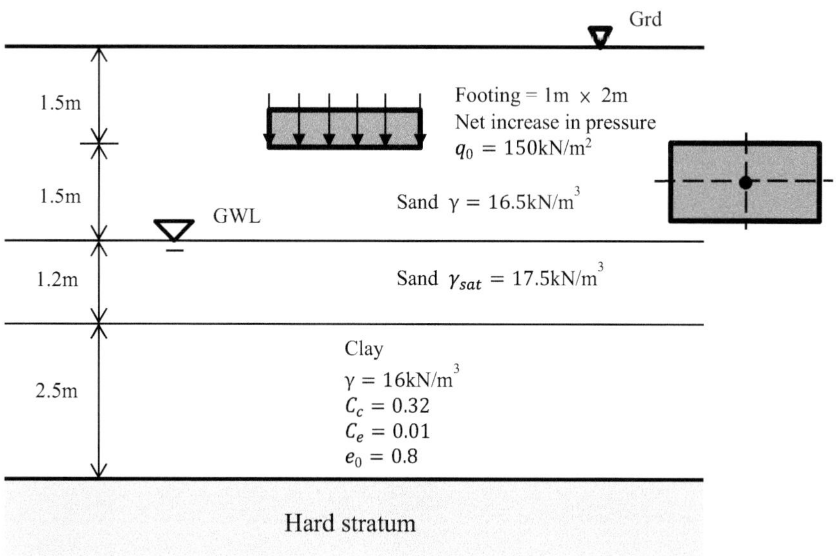

Figure 3.18 Worked Example 3.4 on settlement of normally consolidated clay and over-consolidated clay.

To find the average increase in stress due to the footing, the increases in stress at the top, mid-depth and bottom levels of the clay layer are first determined by Eq. (3.12), Eq. (3.13) or estimated by Table 3.1 by dividing the footing into four equal quadrants ($B = 0.5$ m, $L = 1$ m) and finding the stress at the mid-point of the footing. I_σ for each quadrant is:

So at top, $z = 2.7$ m, $m = \dfrac{B}{z} = 0.1852$; $n = \dfrac{L}{z} = 0.3704$, $I_\sigma = 0.0287$

At mid-depth, $z = 3.95$m, $m = \dfrac{B}{z} = 0.1266$; $n = \dfrac{L}{z} = 0.2532$, $I_\sigma = 0.0143$

At bottom, $z = 5.2$m, $m = \dfrac{B}{z} = 0.1266$; $n = \dfrac{L}{z} = 0.2532$, $I_\sigma = 0.0085$

The average value I_σ can be obtained by the trapezoidal rule as

$$I_{\sigma,ave1} = \frac{1}{4}(0.0287 + 2\times 0.0143 + 0.0085) = 0.0165$$

Or by the Simpson rule $I_{\sigma,ave2} = \dfrac{1}{6}(0.0287 + 4\times 0.0143 + 0.0085) = 0.0158$

Choosing $I_{\sigma,ave2} = 0.0158$, $\Delta\sigma' = 0.0158 \times 150 \times 4 = 9.48$ kN/m^2;

Therefore, the consolidation settlement by Eq. (3.30) is

$$S_c = \frac{C_c H_c}{1+e_0}\log\frac{\sigma_1'}{\sigma_0'} = \frac{0.32\times 2.5}{1+0.8}\log\frac{66.94+9.48}{66.94} = 0.0256 \text{ m} = 25.6 \text{ mm}$$

(ii) If the clay in worked Example 3.4 is over-consolidated with $\sigma_c' = 70$ kN/m^2, the consolidation settlement from the overburden pressure = 66.49 kN/m^2 to $\sigma_c' = 70$ kN/m^2 is

$$S_{c1} = \frac{C_e H_c}{1+e_0}\log\frac{\sigma_c'}{\sigma_0'} = \frac{0.1\times 2.5}{1+0.8}\log\frac{70}{66.94} = 0.0027 \text{m}$$

The settlement after passing the preconsolidation pressure is

$$S_{c2} = \frac{C_c H_c}{1+e_0}\log\frac{\sigma_1'}{\sigma_c'} = \frac{0.32\times 2.5}{1+0.8}\log\frac{76.42}{70} = 0.0169 \text{m}$$

Therefore, the total settlement is

$$S_{c1} + S_{c2} = 0.0196 \text{ m} = 19.6 \text{ mm}$$

Another example is a 2 × 2 m foundation supporting a total pressure of 230 kPa placed at a depth of 1.5 m below ground. If the water table is 3 m below ground, and there is a 2 m thick clay layer 5 m below the ground surface, determine the consolidation settlement if the unit weight of top soil is 18 kN/m³ above the water table and the saturated unit weight is 18.5 kN/m³. The soil parameters for the clay layer are initial void ratio = 0.8, saturated unit weight is 19.5 kN/m³, $p_c' = 90$ kPa.

$$q_{net} = 230 - 1.5\times 18 = 203\,\text{kPa}$$

$$p_0' = 3x18 + 2(18.5 - 9.81) + 1.0(19.5 - 9.81) = 81.07 \ \text{kPa}$$

$$\text{Top of clay: } M = N = 1/3.5, I_\sigma = 0.0344\times 4 = 0.138$$

$$\text{Middle of clay: } M = N = 1/4.5, I_\sigma - 0.0217\times 4 = 0.087$$

$$\text{Bottom of clay: } M = N = 1/5.5, I_\sigma = 0.015\times 4 = 0.06$$

$$\text{Mean increase in stress} = (0.138 + 4\times 0.087 + 0.06)/6\times 203 = 18.5 \ \text{kPa}$$

Since the final stress is $81.07 + 18.5 = 99.57 > p_c'(90)$ and $p_0' < 90$, therefore

$$\Delta\text{H} = \frac{0.1\times 2}{1+0.8}\log\frac{90}{81.07} + \frac{0.4\times 2}{1+0.8}\log\frac{99.57}{90} = 5.0 + 19.5 = 24.5 \ \text{mm}$$

It should be noted that analysis based on average σ_0' and σ_1' is applicable to a thin layer of clay. If the clay layer is thick, the clay layer should be divided into a number of sub-layers and settlement of each layer is calculated in accordance with its own σ_0' and σ_1' and the total settlement is the sum of the settlements of all layers. Varying values of C_c and e_0 of the sub-layers can be applied.

In addition to the primary settlement of soil due to the application of load from footing foundation as discussed in the foregoing, secondary settlement due to "creep" also occurs which is the continuing readjustment of the soil grains into a closer state under the compressive load even when the load is at a constant magnitude. This phenomenon is associated with both immediate and consolidation-type settlement, although it is usually not significant with immediate settlements. This readjustment occurs after the excessive pore pressure has dissipated and may continue for years.

A convenient correlation of this secondary compression represented by a secondary compression index C_a to the compression index C_c is

$$C_a = 0.032C_c \tag{3.33}$$

The index can be used similarly to C_c for estimation of secondary settlement. Some typical values for C_c can range from 0.1 to 10 (but are mostly less than 1.0), and many statistical relations have been established between the compression index and other soil indices.

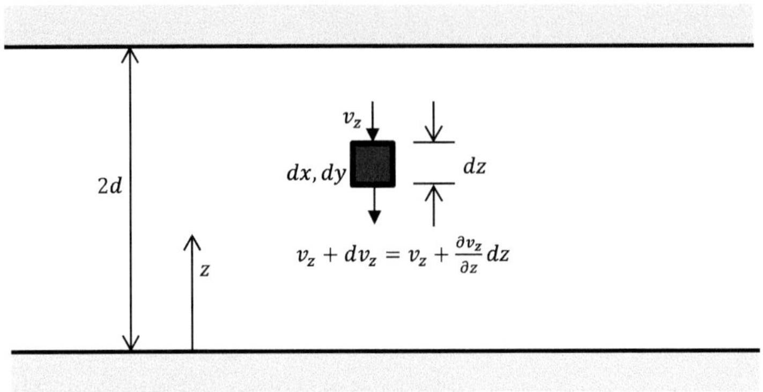

Figure 3.19 Deriving differential equation for consolidation.

A further discussion of the consolidation theory is given, following the theory developed by Terzaghi (1943). The following assumptions are made:

1. The soil is homogeneous;
2. The soil is fully saturated;
3. The soil particles and water are incompressible;
4. Compression and flow are one-dimensional (vertical). Two- or three-dimensional consolidation can be derived based on similar concept;
5. Strains are small (sometimes not too valid);
6. Darcy's law is valid at all hydraulic gradients;
7. The coefficient of permeability and the coefficient of volume compressibility remain constant throughout the process (which is strictly true for limited stress range);
8. There is a unique relationship, independent of time, between void ratio and effective stress. (Experimental results show that the relationship between void ratio and effective stress is not independent of time, creep effect.)

The theory relates the following three parameters:

1. The excess pore water pressure (u_e);
2. The depth (z) below the top of the soil layer;
3. The time (t) from the instantaneous application of a total stress increment.

Consider the element $dxdydz$ in Figure 3.19, by Darcy's law, the flow velocity

$$v_z = -ki_z = -k\frac{\partial h}{\partial z} = -k\frac{\partial\left(z + \frac{u_s + u_e}{\gamma_w}\right)}{\partial z} = -\frac{k}{\gamma_w}\frac{\partial u_e}{\partial z}$$

$$\because z + \frac{u_s}{\gamma_w} = h_s = \text{constant}(\text{static head})$$

$$v_z + dv_z = v_z + \frac{\partial v_z}{\partial z}dz = -\frac{k}{\gamma_w}\frac{\partial u_e}{\partial z} - \frac{k}{\gamma_w}\frac{\partial^2 u_e}{\partial z^2}dz$$

To satisfy the condition of water flow continuity, net volume of water out of the element = volume compression,

$$dxdy\left(v_z + dv_z\right)_{out} - dxdy\left(v_z\right)_{in} = -\frac{k}{\gamma_w}\frac{\partial^2 u_e}{\partial z^2}dxdydz = \frac{\partial V}{\partial t}$$

$$\frac{\partial V}{\partial t} = \frac{\partial\left(\varepsilon_v dxdydz\right)}{\partial t} = \frac{\partial \varepsilon_v}{\partial t}dxdydz = \frac{\partial\left(m_v \sigma'\right)}{\partial t}dxdydz = m_v\frac{\partial \sigma'}{\partial t}dxdydz$$

$$= m_v\frac{\partial\left[\sigma - \left(u_s + u_e\right)\right]}{\partial t}dxdydz = -m_v\frac{\partial u_e}{\partial t}dxdydz$$

$$\therefore \frac{\partial V}{\partial t} = -\frac{k}{\gamma_w}\frac{\partial^2 u_e}{\partial z^2}dxdydz = -m_v\frac{\partial u_e}{\partial t}dxdydz$$

$$\frac{k}{\gamma_w}\frac{\partial^2 u_e}{\partial z^2} = m_v\frac{\partial u_e}{\partial t} \Rightarrow \frac{\partial u_e}{\partial t} = c_v\frac{\partial^2 u_e}{\partial z^2} \tag{3.34}$$

where $c_v = \frac{k}{m_v \gamma_w}$

Strictly speaking m_v is stress dependent and is a constant under constant stress.

The total stress is assumed to be applied instantaneously. At time $t = 0$, the stress increment $\Delta\sigma$ will be taken up entirely by the pore water u_i which is an initial condition as

$$u_e = u_i \text{ for } 0 \leq z \leq 2d \text{ at } t = 0.$$

If the upper and lower boundaries of the soil are assumed to be free-draining, the permeability of the soil adjacent to each boundary is then very high compared with the soil. Water will therefore drain from the centre of the soil element to the two boundaries simultaneously so that the drainage path is half of the total soil thickness which becomes d. Thus the boundary conditions at any time after the application of $\Delta\sigma$ is

$$u_e = 0 \text{ for } z = 0 z = 2d \text{ at } t > 0$$

The solution to Eq. (3.34) by Taylor expansion and after incorporation of the above boundary conditions is

$$u_e = \sum_{n=1}^{n=\infty} \frac{2u_i}{n\pi}(1-\cos n\pi)\left(\sin\frac{n\pi z}{2d}\right)\exp\left(-\frac{n^2\pi^2 c_v t}{4d^2}\right) \tag{3.35}$$

When n is even, $(1-\cos n\pi)=0$, and when n is odd, $(1-\cos n\pi)=2$. Therefore, only odd values of n are used. So, using $n=2m+1$ and putting $M=\frac{\pi}{2}(2m+1)$ and $T_v=\frac{c_v t}{d^2}$ we may list

$$u_e = \sum_{m=0}^{m=\infty} \frac{2u_i}{M}\left(\sin\frac{Mz}{2d}\right)\exp\left(-M^2T_v\right) \tag{3.36}$$

It should be noted that though Eq. (3.34) is derived from the case that upper and lower layers of the clay of thickness $2d$ are free-draining, it is applicable to the case of a "half-closed layer" where one layer is impermeable with thickness d as illustrated in Figure 3.20.

Consider the degree of consolidation U_v that can be related to u_e and u_i as $U_v = 1-\frac{u_e}{u_i}$ and substituting Eq. (3.34) into it, we may list

$$U_v = 1-\sum_{m=0}^{m=\infty} \frac{2}{M}\left(\sin\frac{Mz}{2d}\right)\exp\left(-M^2T_v\right) \tag{3.37}$$

In practice, it is the average degree of consolidation over the depth of the layer as a whole that is of interest. Therefore, the average degree of consolidation at time t for constant u_i is given by

$$U_v = 1-\frac{(1/2d)\int_0^{2d} u_e dz}{u_i} = 1-\sum_{m=0}^{m=\infty}\frac{2}{M^2}\exp\left(-M^2T_v\right) \tag{3.38}$$

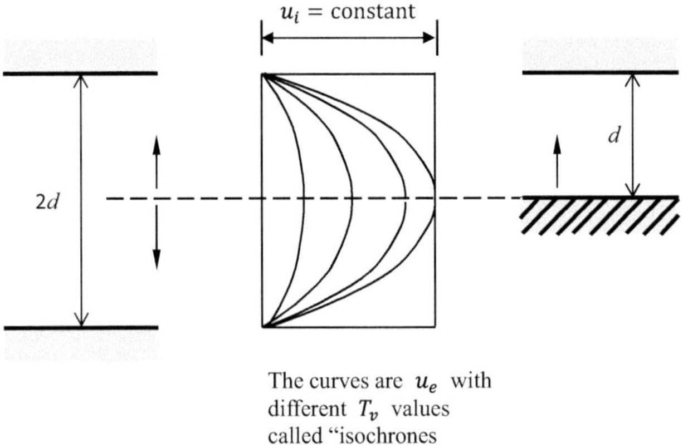

Figure 3.20 Plot of u_e curves under u_i constant with depth.

The relationship between U_v and T_v in Eq. (3.36) can be represented almost by the following empirical equations:

$$T_v = \begin{cases} \frac{\pi}{4} U_v^2 & U_v < 0.60 \\ -0.933\log(1 - U_v) & U_v > 0.60 \end{cases} \tag{3.39}$$

At the settlement at time t, S_t is the final settlement obtained by Eq. (3.30) multiplied by U_v in Eq. (3.36).

$$S_t = U_v S_c \tag{3.40}$$

The following worked Example 3.3 demonstrates estimation of settlement due to the time effect as per Terzaghi's approach.

A 3 m thick saturated clay layer with upper surface as a free drainage surface. The applied load increases to 50 kN/m² within half a year and remains constant thereafter. If $C_v = 0.68$ m²/yr, $m_v = 1.25$ m²/MN, determine the settlement at 3 months and 1 year.

As Terzaghi's approach consider loads to be applied instantaneously, the 3 months' time of application of load is approximated as load applied instantaneously at 1.5 months = 0.125 years.

Using Eq. (3.32), at time =∞,

$$S_c(\infty) = m_v \Delta\sigma' H = 1.25 \times 10^{-3} \times 50 \times 3 = 0.1875\text{m} = 187.5\text{mm};$$

$$\text{At 3 months' time}, T_v = \frac{c_v t}{d^2} = \frac{0.68 \times 0.125}{3^2} = 0.0094$$

$$\text{Using Eq. (3.37)}, \ U_v = \sqrt{0.0094 \times \frac{4}{\pi}} = 0.11$$

$$\text{Therefore settlement is } 0.11 \times 187.5 \times \frac{25}{50} = 10.3 \text{ mm (as only half of the load is applied).}$$

$$\text{At 1 year's time}, \ t = 1 - \frac{0.5}{2} = 0.75 \text{ years}, T_v = \frac{c_v t}{d^2} = \frac{0.68 \times 0.75}{3^2} = 0.057$$

$$\text{Using Eq. (3.37)}, U_v = \sqrt{0.0057 \times \frac{4}{\pi}} = 0.265$$

Therefore settlement is $0.265 \times 187.5 = 49.7$ mm.

For a soil system containing n layers with properties C_{vi} and thickness H_i, convert the system to one equivalent layer with equivalent properties as:

1. Select any layer i, say $i = 1$ as the reference.
2. Transform the thickness of remaining layers to

$$H_i = H_i \sqrt{\frac{C_{v1}}{C_{vi}}}$$

3. Calculate new thickness H as $\sum H_i'$.
4. Determine consolidation using C_{v1} and H.

The consolidation equations by Terzaghi can be found in virtually all textbooks on soil mechanics and foundation engineering, but have been found to not reflect the true consolidation phenomenon from many field measurement results in Hong Kong. The authors have been asked by engineers and students frequently on the use of the most appropriate consolidation equation among Eqs. (3.32) and (3.30). Actually, in Eq. (3.32), both m_v and K can change during the consolidation process, particularly for the volume compressibility m_v. K will also change due to the decrease in void ratio, but this term is in general less sensitive than the volume compressibility. Eq. (3.32) is less commonly used by engineers as compared with Eq. (3.30), as no equations or design charts are available for engineers to compute the time of consolidation. Eq. (3.30) is, however, a better equation to be adopted, as the compression index appears to be constant over a wide stress range, as compared with the volume compressibility. Actually, from Eq. (3.32), if the stress change is small, it can be approximated as

$$\Delta H = \frac{C_c H}{1+e_0} \log \frac{p_0' + \Delta p}{p_0'} \approx \frac{C_c H}{2.303(1+e_o)} \frac{\Delta p}{p_0'} \tag{3.41}$$

The volume compressibility can hence be approximated as

$$m_v = \frac{C_c}{2.303(1+e_o)p_0'} \tag{3.42}$$

Since the effective overburden stress can change by several times during the reclamation consolidation process, the volume compressibility will also change by several times. Even K has also decreased during the consolidation, the overall result is that C_v will also change significantly during the consolidation process. The poor prediction from Eq. (3.22) is hence not surprising. Actually, the field test results on reclamation settlement and pore pressure dissipation during reclamation in Hong Kong usually agree poorly with the classical Terzaghi consolidation equation. Cheng has also been asked by engineers on the poor performance of the Terzaghi equation on the reclamation work in Tseung Kwan O, Hong Kong. Considering that the effective overburden stress will change by several times while the consolidation settlement can exceed 1 m, it is not surprising that the classical consolidation equation may not work well in the prediction.

To overcome this problem, Eq. (3.34) can be rewritten as

$$\frac{\partial u}{\partial t} = C_v \frac{\partial^2 u}{\partial z^2} = \frac{K}{\gamma_w} \frac{2.303 p_o'(1+e_o)}{C_c} \frac{\partial^2 u}{\partial z^2} \tag{3.43}$$

To solve for Eq. (3.43), and hence the consolidation as well as the degree of consolidation, Cheng has adopted the Galerkin finite element formulation, where Eq. (3.43) is rewritten as

$$[C]\{\dot{u}\}+[K]\{u\}-\{F\}=0 \tag{3.44}$$

Eq. (3.44) can be solved by implicit finite difference scheme as

$$\{[C]+\theta\times\Delta t[K]\}\{u\}_b=\{[C]-(1-\theta)\Delta t[K]\}\{u\}_a+\Delta t\times\left[(1-\theta)\{F\}_a+\theta\{F\}_b\right] \tag{3.45}$$

where [C] = global capacitance matrix
[K] = global stiffness matrix
{F} is the equivalent force vector
$0 < \theta < 1$ specified by user

i. $\theta = 0$, forward difference method. The accuracy is low and a small time step is required but the steps in computation are simple. Unstable if Δt is large.
ii. $\theta = 0.5$, central difference method which is commonly used. Numerical oscillation can occur if Δt is large.
iii. $\theta = 2/3$, Galerkin difference method. Numerical oscillation can occur if Δt is large.
iv. $\theta = 1$, backward difference method. Unconditionally stable.

The time-step must be controlled carefully, otherwise, numerical oscillation may occur in the computation as shown. Based on Eq. (3.36), a computer program has been developed by Cheng where the compression index instead of the volume compressibility or coefficient of consolidation is used. In that program, the permeability is assumed to be constant, but the reader can modify this easily if the relation between the permeability and the void ratio is known (this is not implemented into the present code, as such a relation is generally not known for Hong Kong soil). The authors find that using such a simple one-dimensional consolidation program will be good enough without using the more complicated commercial geotechnical program. More importantly, interested readers can modify the program readily to incorporate any special constitutive model that is appropriate.

For sandy soil, the effect of creep is usually not important, though it is also considered in some countries. For some clays, besides the consolidation, the creep (secondary consolidation) can be a critical problem. Under unchanged loading and practically the end of the consolidation process, the settlement will keep increasing with time. It must be noted that creep starts at the beginning of consolidation process, and various means are used to separate these processes. However, many engineers prefer to separate creep from the consolidation process for ease of calculation, and the estimation of creep starts at the end of the primary consolidation process. The accurate determination of creep is very difficult, as the test for the creep parameters can be extremely time-consuming. Furthermore, the creep process is actually not a linear process. A classical creep equation is given by Eq. (3.46)

$$S_c=\frac{C_sH}{1+e_p}log\left(\frac{t_2}{t_1}\right) \tag{3.46}$$

where e_p is the void ratio at the end of primary consolidation, C_s is the creep (secondary consolidation) factor, and t_1 and t_2 are time for primary consolidation and time under consideration, respectively. Many engineers have criticized this equation, as, if t_2 tends to infinity, the creep settlement will tend to infinity. Creep must stabilize at some time, and this is the basis for many advanced creep models. Even so, many engineers still adopt this equation because t_2 which will be the expected life span of the construction work will actually be finite. Furthermore, due to the limitations of laboratory tests, parameters for more advanced constitutive models are not easily calibrated. From a practical point of view, there is no major harm in using this equation. The most important point in the determination of creep is not the type of equation, but the parameters for creep. Advanced creep models such as those by Cheng et al. (1996) and many others are hence not commonly adopted in practice.

There are many technical problems in reclamation, both for the construction and design. In general, consolidation and creep are the two most critical factors in marine clay. For construction, the typical construction method for reclamations over the in situ marine clay may include some of the following approaches:

1. Construction of the seawall over a trench dredged to the base of the soft marine clay.
2. Place a geotextile filter cloth on the sea bed (occasionally).
3. Place a sand drainage blanket in one or more layers.
4. Installation of vertical drains through the full thickness of the sand blanket and the soft deposit. In shallow water, the vertical band drains can only be installed after the first stage filling has raised the reclamation level above the high tide level.
5. Placement of the first stage of the reclamation fill. The thickness of the first-stage fill and slope of the leading edge being determined from stability analyses.
6. A delay period whilst the clay beneath the first-stage reclamation fill consolidates and gains in strength.
7. Placement of the second-stage reclamation fill, with thickness and slope being determined from stability analyses which adopt the increased strength of the marine clay.
8. A delay period whilst the clay beneath the second-stage reclamation fill consolidates and gains in strength.
9. Placement of the final reclamation fill and construction of surcharge mounds.
10. A delay period whilst the clay beneath the reclamation and surcharge fill consolidates.
11. Removal of the surcharge fill after a pre-determined degree of consolidation has been achieved, typically 90%. This is confirmed on site using monitoring data. By achieving an over-consolidated state, the magnitude of ongoing secondary compression settlement can be greatly reduced.

The specific construction sequence depends on a number of factors including the initial thickness and undrained strength of the soft marine clay layer, the water depth, fill type available, time for construction, etc. Additional steps such as vibro-compaction of sand fill or variations such as the use of stone columns to improve the characteristics of the marine clay have been adopted. The use of deep cement mixing (DCM) as shown in Figure 3.21 has been adopted for the third Hong Kong Airport runway project where the in situ marine clay will remain in place stabilized by high pressure cement grouting, following the major

Figure 3.21 DCM in third airport runway project in Hong Kong.

shift in the cellular cofferdam (maximum 6 m) associated with the Hong Kong ZhuHai Macau Bridge project which is induced by marine clay consolidation. For the artificial island in the Hong Kong ZhuHai Macau Bridge project, the use of vacuum preloading, lowering of the water table and preloading in the form of surcharge are used for accelerating the consolidation process.

There are also many problems or even failures for reclamation projects in Hong Kong, and mudwaves are the main cause of these problems. Many of the problems associated with mudwaves have been caused by failures which occurred during construction, primarily because the leading edge of the reclamation was too steep for the weak ground which underlies these reclamations. The long-term settlement problems which have resulted in damage to structures have been caused by incomplete consolidation of the soft clay deposit left in place beneath these reclamations. The settlement-related problems have occurred despite the installation of vertical drains and the use of surcharges to accelerate consolidation during construction.

3.5 AXI-SYMMETRIC CONSOLIDATION

In order to accelerate the process of consolidation, the use of a sand drain can be adopted. The horizontal and vertical seepage will have a compound effect which can greatly accelerate the consolidation which is beneficial to the early development of the land. One effect which should be considered is the "smear". A smear zone is created by the remoulding of

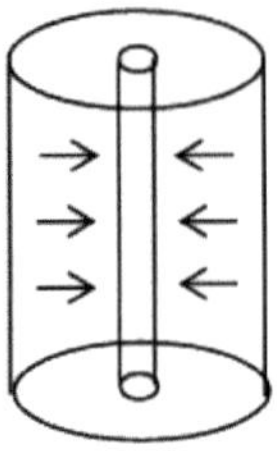

Figure 3.22 Axi-symmetric consolidation.

clay during the drilling or driving operation, which will result in a decrease in the permeability in the horizontal direction, as shown in Figure 3.22. The governing axi-symmetric consolidation equation will reduce to

$$\frac{\partial u}{\partial t} = C_v \frac{\partial^2 u}{\partial z^2} + C_r \left(\frac{\partial^2 u}{\partial \gamma^2} + \frac{1}{r} \frac{\partial u}{\partial \gamma} \right) \tag{3.47}$$

where $C_v = \dfrac{K_v}{m_v \gamma_w}$, $C_r = \dfrac{K_h}{m_v \gamma_w}$

The general solution to Eq. (3.47) can be expressed as (by Carrillo)

$$u(r,z,t) = u_r(r,t) \times u_z(z,t) \tag{3.48}$$

The average degree of consolidation u can be expressed as

$$1 - u = (1 - u_r)(1 - u_z) \tag{3.49}$$

where u_r is the radial degree of consolidation and u_z is the vertical degree of consolidation. Eq. (3.47) can be solved using three different approaches: free strain, equal strain and approximate solutions.

(1) Free strain solution
There is no restraint to the strain due to the consolidation process, hence the consolidation settlement will not be uniform. The solution to Eq. (3.47), which is given by Barron is difficult in general, which requires infinite series using first-order Bessel function and first-order Norman function and eigenvalues determination.

(2) Equal strain solution
Barron assumed that the strain is constant along a given horizontal plane. Obviously, this assumption is not correct, but the error is usually small, particularly when T_v is large. Without smear, the solution is given by

$$u_r = \frac{u_o}{r_e^2 F(n)} \left[r_e^2 \ln\left(\frac{r}{r_w} \right) - \frac{r^2 - r_w^2}{2} \right] e^{\frac{-2T_r}{F(n)}} \tag{3.50}$$

where $F(n) = \dfrac{n^2}{n^2 - 1}\ln(n) - \dfrac{3n^2 - 1}{4n^2}$

Solution with smear is given by

$$u_r(r,z) = u_o \sum_{m=1}^{\infty} \frac{1}{F+D}\left[\frac{K_h}{K_s}\left(\ln\frac{r}{r_w} - \frac{r^2 - r_w^{\,2}}{2r_e^{\,2}}\right) + D\right]\frac{2}{M}\sin\frac{Mz}{H}e^{-\beta mt} \tag{3.51}$$

= pore water pressure in smear zone

$$u_r(r,z) = u_o \sum_{m=1}^{\infty} \frac{1}{F+D}\left[\left(\ln\frac{r}{r_s} - \frac{r^2 - r_s^{\,2}}{2r_e^{\,2}}\right) + \frac{K_h}{K_s}\left(\ln s - \frac{s^2 - 1}{2n^2}\right) + D\right]\frac{2}{M}\sin\frac{Mz}{H}e^{-\beta mt} \tag{3.52}$$

= pore water pressure outside smear zone

Mean water pressure over the effective zone of a well is given by integration of the above equation with respect to r_w and takes the mean value.

$$\bar{u}(z,t) = u_o \sum_{m=1}^{\infty} \frac{2}{M}\sin\frac{Mz}{H}e^{-\beta mt} \tag{3.53}$$

Degree of consolidation is given by

$$u_r(z,t) = \left[1 - \sum_{m=1}^{\infty} \frac{2}{M}\sin\left(\frac{Mz}{H}\right)e^{-\beta mt}\right] \tag{3.54}$$

where $M = \dfrac{2m - 1}{2}\pi$, $m = 1,2,3......$

$$\beta_m = \left[\frac{M^2 C_{vr}}{H^2} + \frac{8C_h}{(F+D)d_e^{\,2}}\right]\alpha_E$$

$$\alpha_E = \frac{n^2 - S^2}{n^2 - 1} + \frac{S^2 - 1}{n^2 - 1}\frac{E_s'}{E_s} \qquad C_h = \frac{E_s K_h}{\gamma_w} \tag{3.55}$$

$$F = \left(\ln\frac{n}{S} + \frac{K_h}{K_s}\ln S - \frac{3}{4}\right)\frac{n^2}{n^2 - 1} + \frac{S^2}{n^2 - 1}\left(1 - \frac{K_h}{K_s}\right)\left(1 - \frac{S^2}{4n^2}\right) + \frac{1}{n^2 - 1}\left(1 - \frac{1}{4n^2}\right)\frac{K_h}{K_s}$$

$$G = \frac{K_h}{K_w}\left(\frac{H}{d_w}\right)^2; \quad D = \frac{8G(n^2 - 1)}{M^2 n^2} \tag{3.56}$$

K_w = permeability of drain
E_s, E_s' = Young's modulus of clay ad smear zone, respectively
d_e = effective zone diameter

$$n = \frac{de}{2r_w} = \frac{r_e}{r_w},\ r_w = \text{radius of well}$$

$$S = \frac{r_s}{r_w},\quad r_s = \text{radius of smear zone}$$

H = thickness of clay
K_h, K_s = horizontal permeability of clay and smear zone, respectively

To account for smear but using some design figures without smear, it is possible to adopt two approaches:

(a) Equivalent reduced drain diameter approach
$\ln(n) = \ln(n/s) + K_h/K_s \ln(s)$
(b) Equivalent reduced C_h approach
Equivalent $C_r = C_r{}^{*}F(n)/F_s(n)$

The degree of consolidation and the strain will be different for different layers using this solution, which is still tedious to be used.

The equivalent reduced Cr concept was applied to the reclamation test fill at Chek Lap Kok. The original undisturbed C_r was estimated to be 3.7 m²/year. With smear diameter ratio $s = 2$ and $K_h/K_s = 4.63$ (from a test), the equivalent C_r for the three vertical drain quadrants is determined and given in Table 3.2.

(3) Simplified solution by Hansbo

Hansbo proposes a simplified solution which assumes ξv and pore pressure u to be independent of depth (solution 2 depends on depth). This approximation is good enough for design use. For loading applied immediately, the radial degree of consolidation is given by

$$u_r = 1 - \exp\left(\frac{-2T_r}{m}\right) \tag{3.57}$$

$$\text{where } m = \frac{n^2}{n^2 - S^2}\ln\left(\frac{n}{S}\right) - \frac{3}{4} + \frac{S^2}{4n^2} + \frac{K_h}{K_s}\left(\frac{n^2 - S^2}{n^2}\right)\ln S$$

Table 3.2 Design parameters for Chek Lap Kok International Airport in Hong Kong

Band drain spacing (m)	*n*	*F(n)*	*F*	*Equivalent (m²/year)*
1.5	23.6	2.41	4.93	1.81
2	47	3.1	5.62	2.04
3 m sand drain	6.3	1.09	3.61	1.12

For the no-smear case, $r_s = r_w$, $K_h = K_s$, hence

$$m = \frac{n^2}{n^2 - 1}\ln n - \frac{3n^2 - 1}{4n^2} \tag{3.58}$$

For a ramp loading with no smear, the simplified equation can be written as

$$u_r \approx \frac{T_r - \frac{1}{A}\left[1 - \exp(-AT_r)\right]}{T_{ro}} \quad T_r < T_{r0} \tag{3.59}$$

$$\approx 1 - \frac{1}{AT_{ro}}\left[\exp(AT_{ro}) - 1\right]\exp(-AT_r) \quad T_r > T_{r0} \tag{3.60}$$

where $T_{ro} = \frac{C_r t_o}{r_e^2}$, $A = \frac{2}{m}$

Eq. (3.60) will reduce to Eq. (3.57) if T_{r0} tends to zero. Since the loadings are usually not applied instantaneously, the degree of consolidation as calculated will be greater than the actual one. Corrections can be done in the following ways as suggested by Terzaghi:

1. Determine the consolidation curve assuming instantaneous load. For a given time t, determine the degree of consolidation at $t/2$ as U. The corrected degree of consolidation is approximately given by $U' = U^* p/p_0$, where p_0 is the maximum load and p is the load corresponding to time t.
2. If loading is applied in several stages, the consolidation curve corresponding to each stage will be determined individually. The corrected degree of consolidation can be determined as $U' = U_1^* p_1/p_0 + U_2^* p_2/p_0$ ......

Henderson has suggested corrections which are similar to (2) above except that the radial and vertical consolidations are calculated individually.

3.5.1 Use of sand drain/wick drain

Sand drains, or more recently wick drains, are used to accelerate consolidation in clay. In general, the configuration is either a triangular or square pattern. The equivalent diameter d_e used in the consolidation calculation is given by

$$d_e = a\sqrt{\frac{2\sqrt{3}}{\pi}} = 1.05a \quad \text{traingaular pattern}$$

$$= a\sqrt{\frac{4}{\pi}} \quad \text{square pattern} \tag{3.61}$$

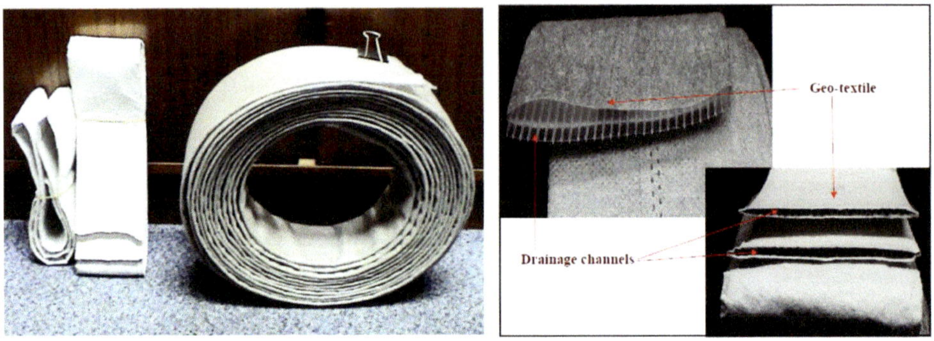

Figure 3.23 Examples of some wick drain.

where a is the spacing of the drains. The sectional area of those that are rectangular will be transformed to an equivalent circle for analysis. For loading which is applied instantaneously, the average degree of consolidation can be obtained from the radial and vertical consolidation as

$$u = 1 - (1 - u_v)(1 - u_r) \tag{3.62}$$

For ram loading, the above equation does not apply. Many engineers, however, still use it due to its simplicity.

Wick drains (Figure 3.23) and geodrains are usually faster in construction, and also better and cheaper than classical sand drains. For a rectangular wick drain, the equivalent radius of a circular drain is given by $(b+t)/\pi$. It is common practice to use a surcharge in conjunction with a wick drain to accelerate the consolidation settlement. In China, the use of vacuum preloading is a common practice. Since 1 atmosphere pressure is equal to nearly 10 m height of water, assuming an efficiency of 60%, the equivalent height of soil is about 3 m, which can save lots of transportation cost.

Construction of sand drain/wick drains

1. Ordinary sand drain with placement of sand: Ordinary rotary drill, vibroflotation, wash boring or auger drilling are adopted for forming the sand drain.
2. Sand drain formed by sand bag: It has been found that formation of good-quality but cheap and large-diameter sand drain is not easy due to necking. The formation process can be replaced by the use of a sand bag with a diameter of 7–12 cm. The bag is usually made of a geosynthetic with good strength.
3. Wick drain: This is the most common method as the drain is manufactured in a factory with good workmanship. The drain is installed by driving using a mandrel which will be retracted.

3.5.2 Vacuum preloading

Vacuum preloading is popular in China, and is a good solution as 1 atmosphere pressure is about 10 m height of water. Accounting for the efficiency, a 6 m height of water pressure

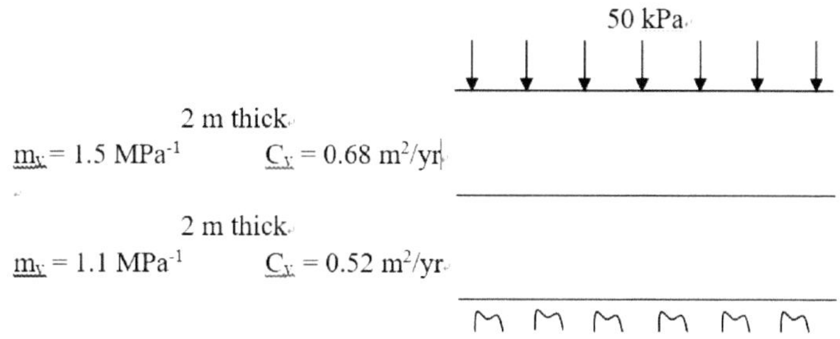

Figure 3.24 Example for axi-symmetric calculation.

can be achieved. It has the additional advantage that the loading can be applied immediately, which is highly efficient.

Example: Consider the problem as shown below. Determine the consolidation settlement after 1 year of construction. Assume d_w = 0.1 m, no smear, C_r = 0.3 m²/yr (av.), determine the corresponding consolidation settlement if sand strain is used (Figure 3.24).

Solution:
Layer 2 transformed to layer 1

$$H_2' = H_2 \times \sqrt{\frac{0.68}{0.52}} = 2.29\text{m}, \therefore \quad H = 2 + 2.29 = 4.29\text{m}$$

If p_0 = 50 kPa $\Delta H_{\infty} = 50 \times 1.5 \times 2 + 50 \times 1.1 \times 2$

$= 260$ mm

At time = 1 yr. $T_v = \dfrac{0.68 \times 1}{4.29^2} = 0.0369$

$$\therefore \quad u_v = \sqrt{0.0369 \times \frac{4}{\pi}} = 21.7\%$$

$\therefore \Delta H_{1\,yr} = 56.4$ mm

If square pattern sand drain with spacing 1.2 m is used

$$d_e = 1.2\sqrt{\frac{4}{\pi}} = 1.354\text{m}$$

Assume d_w = 0.1 m, no smear, C_r = 0.3 m²/yr (av.)

$$n = \frac{d_e}{d_w} = 13.54$$

$$m = \frac{13.54^2}{13.54^2 - 1}\ln 13.54 - \frac{3\times 13.54^2 - 1}{4\times 13.54^2} = 1.877$$

$$T_r = \frac{0.3\times 1}{0.677^2} = 0.655, \quad (0.677\,\text{is}\,r_e) \quad u_r = 1 - e^{\frac{-2T_r}{m}} = 0.50$$

$$\therefore \quad u = 1-(1-0.217)(1-0.50) = 0.61 > \frac{u_r}{u_v}$$

$\therefore \Delta H_{1\,yr}$ = 159 mm (consolidation settlement is accelerated)

Example: Determine the consolidation settlement for the case where the sand drain is not throughout the clay zone. C_r = 75.8 ×10⁻⁴ cm²/s, C_v = 7.18 ×10⁻⁴ cm²/s, d = 2.5 m. Pattern: triangular, d_w = 0.25 m, t = 1 month

$$d_e = 2.5\sqrt{\frac{2\sqrt{3}}{\pi}} = 2.625\text{m}$$

$$n = \frac{2.625}{0.25} = 10.5$$

$$m = \frac{n^2}{n^2 - 1}\ln(n) - \frac{3n^2 - 1}{4n^2} = 1.62$$

$$T_r = \frac{75.8\times 10^{-4}\times 30\times 24\times 60\times 60}{(131.25)^2}$$

$$= 1.14$$

$$\therefore \quad u_r = 1 - e^{-\frac{2T_r}{m}} = 0.755$$

(u_v is small and neglected)

For the 3 m without sand drain, consolidation is mainly in the vertical direction (Figure 3.25)

$$T_v = \frac{7.18\times 10^{-4}\times 30\times 20\times 60\times 60}{(300)^2} = 0.02$$

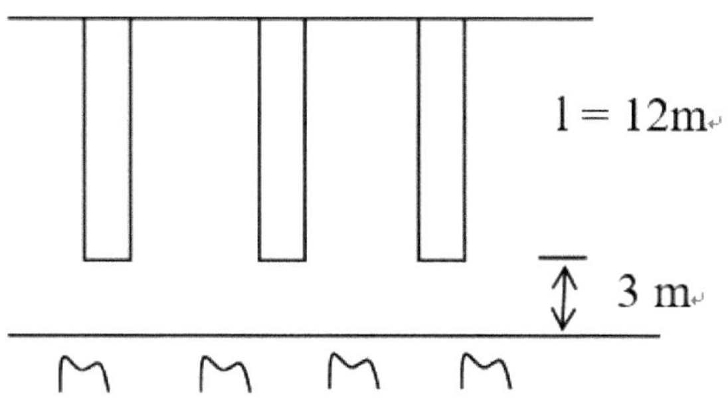

Figure 3.25 Another example for axi-symmetric calculation.

$$\therefore \quad u_v \approx \sqrt{0.02 \times \frac{4}{\pi}} = 0.16$$

$$\text{Overall } u \approx \frac{12}{15} \times 0.755 + \frac{3}{15} \times 0.16 = 0.636$$

3.6 EXTENSION TO TWO- AND THREE-DIMENSIONAL BIOT CONSOLIDATION

For a raft foundation on saturated clay, the use of 1-D consolidation is far from realistic as there is a spread of stress due to the three-dimensional effect. The use of 2-D or 3-D consolidation is required. Unlike 1-D consolidation, seepage can occur in different directions, and the stress changes include all the three normal and shear stresses. The treatment by Biot is given below.

From elasticity, the equilibrium equation of an infinitesimal element can be expressed as (equivalent to $\Sigma Fx=0$, $\Sigma Fy=0$, $\Sigma Fz=0$)

$$\begin{aligned}
&\frac{\partial \sigma'_{xx}}{\partial x} + \frac{\partial \tau_{xy}}{\partial y} + \frac{\partial \tau_{xx}}{\partial z} = fx \\
&\frac{\partial \tau_{xy}}{\partial x} + \frac{\partial \sigma'_{yy}}{\partial y} + \frac{\partial \tau_{yz}}{\partial z} = fy \\
&\frac{\partial \tau_{xz}}{\partial x} + \frac{\partial \tau_{yz}}{\partial y} + \frac{\partial \sigma'_{zz}}{\partial z} = fz
\end{aligned} \tag{3.63}$$

where compression is taken as positive and fx, fy and fz are body forces (Figure 3.26).

Strain compatibility equations are given by

$$\xi_x = -\frac{\partial u_x}{\partial x}, \xi_y = -\frac{\partial u_y}{\partial y}, \xi_z = -\frac{\partial u_z}{\partial z}$$

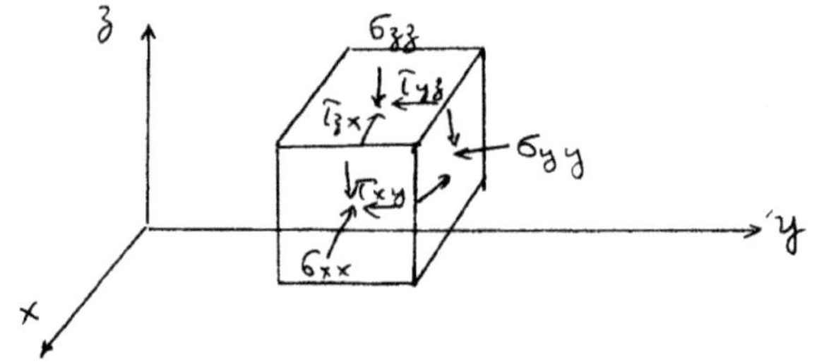

Figure 3.26 Stresses for a three-dimensional element.

$$\gamma_{xy} = -\left(\frac{\partial u_x}{\partial y} + \frac{\partial u_y}{\partial x}\right), \gamma_{yz} = -\left(\frac{\partial u_z}{\partial y} + \frac{\partial u_y}{\partial z}\right), \gamma_{zx} = -\left(\frac{\partial u_x}{\partial z} + \frac{\partial u_z}{\partial x}\right) \tag{3.64}$$

Darcy's law can be expressed as

$$v_x = -K_x \frac{\partial h}{\partial x}, v_y = -K_y \frac{\partial h}{\partial y}, v_z = -K_z \frac{\partial h}{\partial z} \tag{3.65}$$

where h = total head.

Similar to the 1-D condition, the difference between q_{in} and q_{out} will be equal to the change of volumetric strain, or

$$-\frac{\partial}{\partial x}\left(K_x \frac{\partial h}{\partial x}\right) - \frac{\partial}{\partial y}\left(K_y \frac{\partial h}{\partial y}\right) - \frac{\partial}{\partial z}\left(K_z \frac{\partial h}{\partial z}\right) = \frac{\partial \xi v}{\partial t} \tag{3.66}$$

For isotropic soil where $K_x = K_y = K_z = K$. Since

$$v_x = \frac{-K}{\gamma_w}\frac{\partial u}{\partial x}, v_y = \frac{-K}{\gamma_w}\frac{\partial u}{\partial y}, v_z = \frac{-K}{\gamma_w}\frac{\partial u}{\partial z}$$

$$\therefore -\frac{K}{\gamma_w}\left(\frac{\partial^2 u}{\partial x^2} + \frac{\partial^2 u}{\partial y^2} + \frac{\partial^2 u}{\partial z^2}\right) = \frac{\partial \xi v}{\partial t} \tag{3.67}$$

Based on elasticity,

$$\sigma'_x = 2G\left(\frac{v}{1-2v}\xi v + \xi x\right), \sigma'_y = 2G\left(\frac{v}{1-2v}\xi v + \xi y\right)$$

$$\sigma'_z = 2G\left(\frac{v}{1-2v}\xi v + \xi z\right), \tau_{xy} = G\gamma_{xy} \ldots\ldots \tag{3.68}$$

Putting Eq. (3.68) into Eqs. (3.66) and (3.63), we have

$$-G\nabla^2 u_x + \frac{G}{1-2v}\frac{\partial}{\partial x}\left(\frac{\partial u_x}{\partial x} + \frac{\partial u_y}{\partial y} + \frac{\partial u_z}{\partial z}\right) + \frac{\partial u}{\partial x} = 0$$

$$-G\nabla^2 u_y + \frac{G}{1-2v}\frac{\partial}{\partial y}\left(\frac{\partial u_x}{\partial x} + \frac{\partial u_y}{\partial y} + \frac{\partial u_z}{\partial z}\right) + \frac{\partial u}{\partial y} = 0$$

$$-G\nabla^2 u_z + \frac{G}{1-2v}\frac{\partial}{\partial z}\left(\frac{\partial u_x}{\partial x} + \frac{\partial u_y}{\partial y} + \frac{\partial u_z}{\partial z}\right) + \frac{\partial u}{\partial z} = -\gamma \tag{3.69}$$

$$\text{Since} \quad \xi_v = \xi_x + \xi_y + \xi_z = -\left(\frac{\partial u_x}{\partial x} + \frac{\partial u_y}{\partial y} + \frac{\partial u_z}{\partial z}\right) \tag{3.70}$$

∴ Eq. (3.67) can be rewritten as (continuity equation)

$$\frac{K}{\gamma_w}\nabla^2 u = \frac{\partial}{\partial t}\left(\frac{\partial u_x}{\partial x} + \frac{\partial u_y}{\partial y} + \frac{\partial u_z}{\partial z}\right) \tag{3.71}$$

Eqs. (3.69) and (3.71) form the governing equations and are termed as a couple analysis. For a 2-D problem where $\xi y = 0$, Eqs. (3.68) and (3.71) can be simplified as:

$$-G\nabla^2 u_x + \frac{G}{1-2v}\frac{\partial}{\partial x}\left(\frac{\partial u_x}{\partial x} + \frac{\partial u_z}{\partial z}\right) + \frac{\partial u}{\partial x} = 0$$

$$-G\nabla^2 u_z + \frac{G}{1-2v}\frac{\partial}{\partial z}\left(\frac{\partial u_x}{\partial x} + \frac{\partial u_z}{\partial z}\right) + \frac{\partial u}{\partial z} = -\gamma \tag{3.72}$$

$$\frac{K}{\gamma_w}\nabla^2 u = \frac{\partial}{\partial t}\left(\frac{\partial u_x}{\partial x} + \frac{\partial u_z}{\partial z}\right) \tag{3.73}$$

The above formulation is called Biot consolidation, which coupled the dissipation of pore water pressure with soil displacement. The Biot equation is very difficult to solve in general, and only complicated expressions can be obtained for simple cases. Even when analytical expressions exist for some simple cases, these expressions are difficult to evaluate by hand calculation, and computer analysis is necessary. Under general condition, the solution of Biot equation can only be solved by numerical method, and finite element method is the best method for this problem.

Terzaghi-Redulic assumes the mean pressure $(\sigma_x+\sigma_y+\sigma_z)/3$ remains unchanged during consolidation and simplified the Biot equation to

$$\xi v = \frac{p'}{K'} = \frac{1}{K'}(p-u) \tag{3.74}$$

where $p = \left(\sigma_x + \sigma_y + \sigma_z\right)/3$,

$K' = \dfrac{E}{3(1-2v)}$ = bulk modulus of volume compressibility

$$\frac{\partial \xi v}{\partial t} = \frac{1}{K'}\frac{\partial p'}{\partial t} = \frac{1}{K'}\left(\frac{\partial p}{\partial t} - \frac{\partial u}{\partial t}\right) \tag{3.75}$$

$$\text{Assume}\frac{\partial p}{\partial t} = 0, \frac{\partial \xi v}{\partial t} = -\frac{1}{K'}\frac{\partial u}{\partial t} \tag{3.76}$$

Putting Eq. (3.65) into Eq. (3.63)

$$\frac{K}{\gamma_w}\nabla^2 u = \frac{1}{K'}\frac{\partial u}{\partial t} \text{ or } C_{v3}\nabla^2 u = \frac{\partial u}{\partial t} \tag{3.77}$$

where $C_{v3} = \dfrac{K}{\gamma_w}K' = \dfrac{K}{\gamma_w}\dfrac{E}{3(1-2v)}$

For a 2-D problem, $\xi_y = 0$, Eq. (3.77) is reduced to

$$C_{v2}\nabla^2 u = \frac{\partial u}{\partial t} \text{where} C_{v2} = \frac{K}{\gamma_w}\frac{E}{2(1+v)(1-2v)} \tag{3.78}$$

There are some notable differences between the Biot consolidation and Terzaghi Redulic consolidation in the early stage of consolidation. In general, the pore water pressure can increase with time initially and then decrease afterward for Biot consolidation for some points (but not all locations). This phenomenon is called the Mandel-Cryer effect which does not occur for the Terzaghi's approach.

Biot consolidation is best analysed by finite element analysis as the analytical solution is very complicated to use. A Windows-based program, BIOT, written in Fortran 90 and Winteracter is shown below for teaching and research purposes. This program uses 8/9-node elements for greater accuracy, and is applicable for general non-homogeneous non-regular problems with elastic and Cam-clay model. The complete program code, including the Window interface, is too long to be included here. Interested readers can request the computer code by sending an email to Cheng at natureymc@yahoo.com.hk. A sample program output by BIOT is given below.

Pore pressure at time unit and node 14, 21, 41, 61 for a two-dimensional Biot consolidation problem:

Time	*14*	*21*	*41*	*61*
10	96.28	192.6	61.07	66.91
12	96.09	192.2	61.34	66.81
14	95.9	191.8	61.61	66.71
16	95.7	191.4	61.9	66.62
18	95.49	191	62.18	66.52
20	95.29	190.6	62.46	66.42
40	93.14	186.3	65.21	65.5
512	54.14	108.3	* 89.36	62.53
544	52.59	105.2	89.31	62.81
576	51.13	102.3	89.18	63.89
608	49.74	99.49	88.97	63.37
640	48.44	96.87	88.69	63.64

* It must be noted that the maximum applied load stops at time step 10 unit, while the maximum pore water pressure is achieved at time step 512. Such results actually match well with many reclamation field monitoring results in Hong Kong.

3.7 USE OF FOUNDATION CODES

Foundation settlement is important as it affects the performance of the superstructure, both in the ultimate and serviceability limit states. For the determination of shallow foundation settlements, the common national and international codes of practice for foundations either list the formulae as derived and discussed in the foregoing or ask the user to base their calculations on engineering principles. In some codes, empirical values for the parameters in the formulae such as Young's modulus of soil as related to SPTN values may be given. Or, more generally, the codes would simply ask the user to refer to laboratory or site tests. A brief discussion of these codes is given below.

The Eurocode EC7 (6.6.2) has defined three types of settlement, which are: (i) immediate; (ii) consolidation; and (iii) creep. The clause allows commonly recognized methods for the evaluation of the settlements. The Code also provides sample methods as specified in its Annex F. In the Annex, although detailed formulae are not given, basically the approaches for the three kinds of settlements are as indicated in the foregoing of this chapter. And yet it acknowledges that the calculations are not regarded as accurate, but only approximate indications.

The old British Code BS8004–1986 (2.1.2.3) only lists the different types of settlement as similar to EC7 with discussion of the nature but without an approach on the quantification of the settlement. The new version BS8004–2015 simply makes reference to EC7.

The Hong Kong Code of Practice for Foundation 2017 lists the formulae for settlement determination of coarse- and fine-grained soil and that of secondary settlements which are identical to the foregoing of this chapter.

The codes often stress that settlement and relative settlement should not incur detrimental effects on the structures, both in the ultimate and serviceability limit states. Some codes do not recommend limiting values, while some do. The Eurocode EC7 lists limiting values of structural deformation and foundation movement for normal routine structures in its Annex H which mainly comprises:

(i) Acceptable absolute settlement can be up to 50 mm;
(ii) Acceptable relative rotations (defined as relative settlements between two points divided by the distance between the two points) generally not to exceed 1/500 for "sagging mode" which should be halved for "hogging mode".

EC7 does not specify the loads (dead, live or wind) in giving rise to the settlement to which these limits apply. Presumably the loads should be dead load + live load.

The Code of Practice for Foundations 2017 (Hong Kong) 2.3.2(2) lists the following limits in general for checking of buildings or structures not particularly sensitive to movement:

(i) Absolute settlement of 30 mm due to dead + live loads;
(ii) Differential settlement/distance apart not to exceed 1:500 due to dead + live load;
(iii) Angular rotation again defined as differential settlement/distance apart not to exceed 1:500 due to wind or other transient loads.

For reclamation, the normally accepted residual settlement parameters adopted for general reclamations in Hong Kong are as follows:

1. Maximum residual settlement at 50 years after completion of the defects and maintenance liability period – 500 mm; and
2. Maximum differential settlement at 50 years after completion of the defects and maintenance liability period – 1:300.

By the Hong Kong Foundation Code, (ii) and (iii) can be considered separately. From a practical point of view, the code allows exemption from checking of (i) if the foundation is resting on rock. Exceedance of the above limits would, however, require the user to carry out assessment not only of the structures, but also on adjacent structures. The use of such criteria has a limitation of accumulation effect, which is also not clearly specified in nearly all the design codes. Consider a case when the construction has induced about 25 mm settlement after completion of the works. If a new construction starts immediately after the previous work and induced another 25 mm, should the new construction be considered as acceptable? Based on the code, the second construction has complied perfectly with the requirement, but the area has actually settled by a total 50 mm over a relatively short period. The situation will be further amplified if there is a third or fourth construction work nearby. This problem has happened for a settlement problem at Tin Shui Wai in Hong Kong, where the actual total settlement is very significant and has induced some problems to the surrounding area. Many design codes have not laid down clear guidelines about this problem, so the issue of accumulation effect can be critical in some cases.

3.8 COMPUTATIONAL METHODS

The analytical formulae as discussed previously bear many limitations, such as homogeneous half-space, isotropic elastic material, uniform shape foundation and simple uniformly distributed loadings. Other than these requirements, no analytical solutions are available. Under these simple conditions, analytical equations are available for the stress and settlement of circular or rectangular footings. The analytical equations (based on the basic Mindlin equation) are so long that the authors have never used them directly, although simple programs are prepared for applications. Currently, many computer programs are available for applications, and these programs can generally cater for arbitrary shape foundations with complicated loadings and constitutive models. Cheng has developed a program PLATE which is suitable for raft foundation, pile cap and transfer plate analysis and design. A simple and interesting illustration of the PLATE analysis is given in this section.

Consider a simple 4 × 2 m flexible foundation with zero stiffness on a soil with $E = 20{,}000$ kPa and $\mu = 0.2$ with 100 kPa is considered. This case is considered because an analytical solution is available for direct comparison. Based on the elastic solution as from Eq. (3.19) given in the previous section, the maximum settlement is given by

$M = 4/2 = 2$, $N = \infty$ (rock at infinity depth), $I_1 = 0.7659$, $I_2 = 0$

$$\Delta H - qB\frac{1-\mu^2}{E_s}(I_1 + \frac{1-2\mu}{1-\mu}I_2) = 4\times 100\times 1(1-0.22)(0.7659)/20{,}000 = 14.71 \text{ mm}$$

Using a mesh size of 0.2 m as shown below for the analysis, the maximum settlement is given by 14.34 mm from PLATE, with sample results shown in Figures 3.27 and 3.28. If the mesh size is further reduced to 0.1 m, the maximum settlement is given by 14.53, which is very close to the analytical solution. This result has illustrated that PLATE will converge to the classical elasticity solution, and a mesh with 200 elements is already adequate for engineering analysis. For practical engineering problems where thousands of elements are usually used, the accuracy of PLATE will be very good. The use of a computer program can consider the stiffness of the footing, which is not possible for Eq. (3.19).Furthermore, for non-uniform ground support, PLATE can accept borehole options to perform a true three-dimensional analysis using a two-dimensional configuration, as shown in Figure 3.28. It should also be noted that the stiffness of the foundation will mainly affect the differential settlement, while the maximum settlement is only slightly reduced by the use of a stiff foundation. This is due to the fact that the settlement mainly comes from the settlement of the ground and the applied loading, and the stiffness of the foundation will only affect the distribution of the settlement across the foundation. PLATE also has a borehole option where the three-dimensional underground condition can be defined as a series of boreholes, as shown in Figure 3.29. The equivalent interaction coefficient will be determined from the sub-base by a 3-D finite element analysis, and these coefficients are coupled to the 2-D plate analysis without the attention of engineers. More applications of settlement and internal force determination for a general footing problem are discussed in Chapter 4.

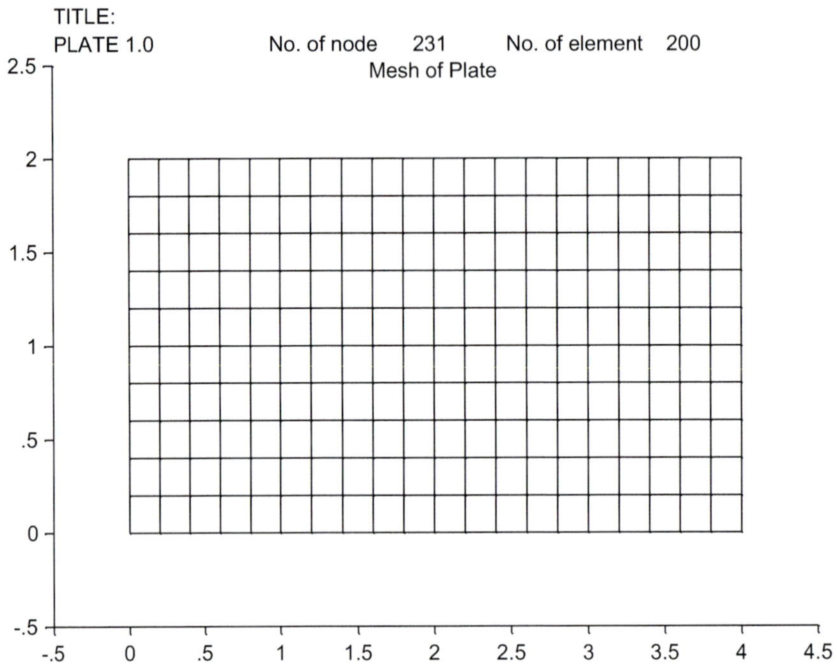

Figure 3.27 Mesh for the finite element analysis of the simple footing.

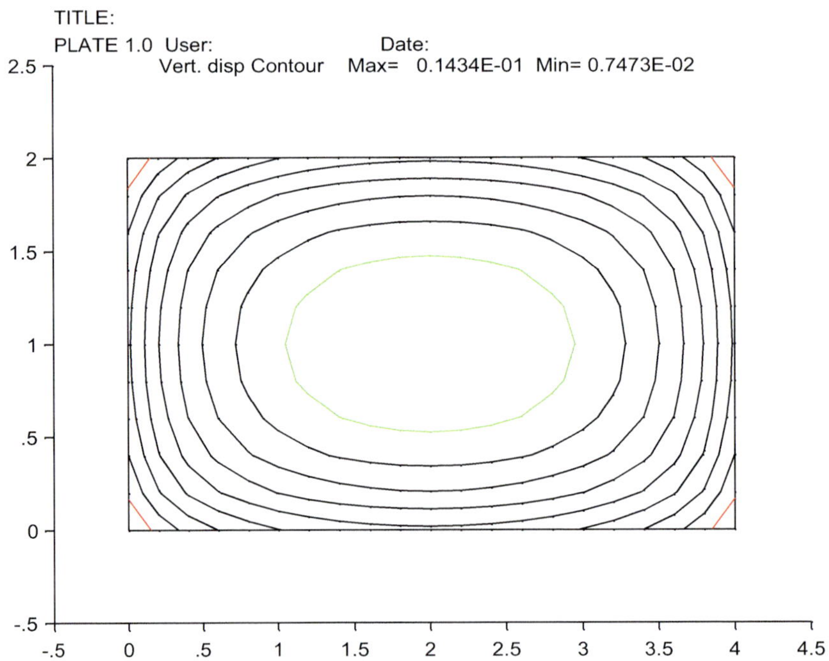

Figure 3.28 The surface for the simple flexible footing.

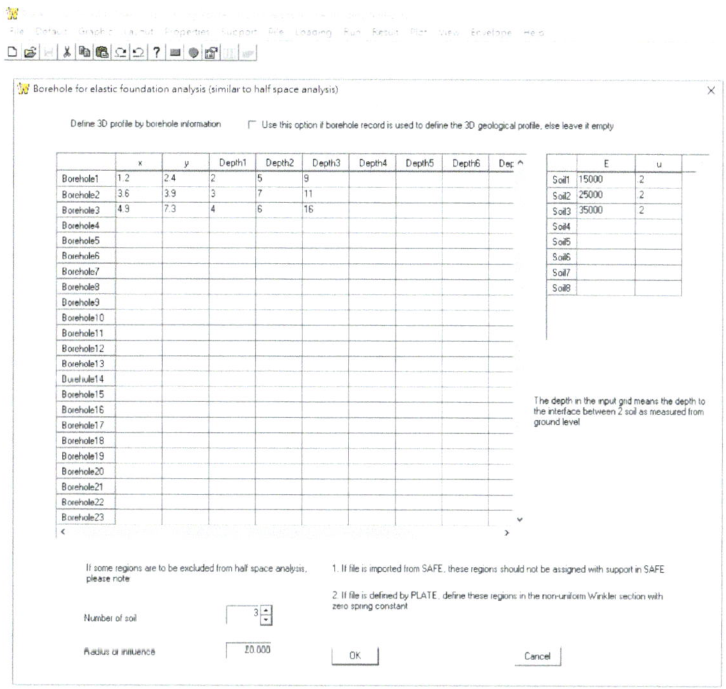

Figure 3.29 Defining the three-dimensional ground conditions using boreholes.

3.9 MONITORING OF SETTLEMENT

There are various uncertainties with consolidation/settlement of loose fill and soft marine clay, and many engineering problems have occurred in different projects in Hong Kong and many other countries. Even with great care in the analysis and design, it is still advisable to monitor the movement, stress/strain, pore water pressure, tilting, vibration, air quality, weather and various issues during the construction as well as long-term monitoring of the construction, for example, the settlement of the fill zone in the Hong Kong Link of the Hong Kong ZhuHai Macau bridge, third airport runway and other large-scale construction projects. Cheng is responsible for the monitoring works of some of these projects. For example, for the monitoring of the third airport runway project in Hong Kong, monitoring works in the form of inclinometers, magnetic probe extensometers, vibrating wire piezometers (VWPs), standpipes, settlement markers, deformation monitoring systems, vibration, tile plates and others are used with different monitoring frequencies. For the Hong Kong Link reclamation, GPS/GNSS is used for monitoring ground movement, as shown in Figure 3.30.

For IoT monitoring, which is now commonly used in construction, a platform is usually required by most engineers (there are only limited cases where API for data transfer are provided to the engineers but not a platform). A sample view of the platform is shown in Figure 3.31.

Figure 3.30 GPS/GNSS monitoring of ground movement for Hong Kong ZhuHai Macau bridge project.

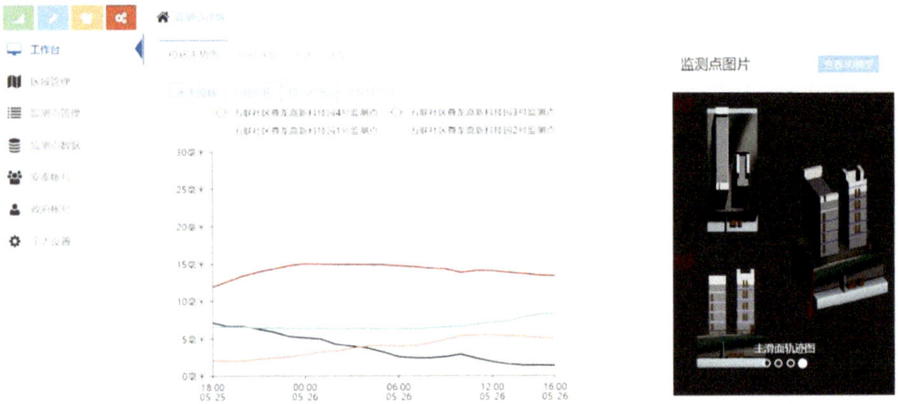

Figure 3.31 Platform for monitoring of settlement and ground water table for a building in Shenzhen, China.

In Hong Kong, the settlement and similar issues are usually responded to through an AAA monitoring plan, such as:

3.9.1 Alert level

1. RC to notify and submit written report to RSE/RGE.
2. RSE/RGE to assess the effect of movements and predict future movements.

3. RSE/RGE to agree with RC on a suitable action plan which may include the installation of additional check points and/or increasing monitoring frequency, and the remedial/mitigating measures to be taken upon reaching the "Alarm level".

3.9.2 Alarm level

1. RC to notify and submit updated report to RSE/RGE and notify the BD immediately.
2. RSE/RGE and RC to conduct joint site inspection to determine if any construction activities should be temporarily suspended.
3. RC to implement the necessary remedial/mitigating measures in accordance with the agreed action plan.
4. RSE/RGE and RC to discuss the instrument response and review the effectiveness of the response action.
5. RSE/RGE to agree with RC on the emergency plan detailing the measures to be taken upon reaching "Action level".
6. Any construction activities may be suspended if the response action has been implemented and on the advice of the RSE/RGE as necessary.
7. RSE/RGE to review the method statement of piling works to determine whether modification to the construction methods is required to prevent Action level from being reached.

3.9.3 Action level

1. All works that will cause ground movement are to be ceased.
2. RC to notify and carry out a joint site inspection with the RSE/RGE immediately. The BD and the relevant parties should be notified immediately.
3. RC to implement the necessary emergency measures in accordance with the agreed emergency plan.
4. RC to submit an incident report to RSE/RGE and the BD detailing the full history of movements and remedial/emergency measures implemented.
5. RSE/RGE to review the incident and agree with RC on further remedial and preventive measures to enable resumption of the suspended works.
6. Construction activities should not be resumed until the necessary remedial and preventive measures have been completed to the satisfaction of the BD.
7. If the trigger values and response action are revised, the amended plans should be submitted to the BD for approval. The suspended construction activities should not be resumed until the amended plans are approved by the BA and the consent is given.

In the case of a large-scale public construction, a public relation plan as shown may also be required:

1. Assign a Public Relationship (PR) Officer to response to any complaints.
2. Arrange meetings with the complainants from time to time.
3. Notify the occupants of the surrounding buildings of the construction.
4. Take record of any complaints (by the PR Officer).
5. PR Officer should respond quickly and react as soon as possible.

6. Identify the structural defects on the structural or non-structural elements.
7. Review the defect locations and carry out remedial works at the end of the works.

References

Baban TM (2016), *Shallow foundations, discussion and problem solving*, Wiley Blackwell.
British Standard Institute (1986), *BS8004: 1986 code of practice for foundations.*
Buildings Department (2017), *Code of practice for foundations 2017*. The Government of SAR, Hong Kong.
Canadian Geotechnical Society (2006), *Canadian foundation engineering manual*, 4th ed.
Cheng YM, Lu HL and Sun J (1996), Analysis of nonlinear creep in geotechnical engineering, *Chinese Journal of Geotechnical Engineering*, 18(5), 1–13.
Das BM (2017), *Shallow foundations, bearing capacity and settlement*, 3rd ed., CRC Press.
Das BM (2019), *Principles of foundation engineering*, Cengage Learning.
Day RW (2010), *Foundation engineering handbook*, 2nd ed., McGraw Hill.
Huang AN and Yu HS (2018), *Foundation engineering analysis and design*, CRC Press.
Powrie W (2014), *Soil mechanics, concepts and applications*, CRC Press.
Rahman MS and Ulker MBC (2018), *Modeling and computing for geotechnical engineering, An introduction*, CRC Press.
Rao NSVK (2011), *Foundation design theory and practice*, John Wiley.
Saran S (2018), *Shallow foundations and soil constitutive laws*, CRC Press.
Terzaghi K. (1943), *Theoretical soil mechanics*, John Wiley.
Tomlinson MJ (2001), *Foundation design and construction*, 7th ed., Pearson Education Ltd.

Further readings

Bowles JE (1996), *Foundation analysis and design*, 5th ed., The McGraw-Hill Companies, Inc.
Poulos H.G and Davis E.S. (1974), *Elastic solutions for soil and rock mechanics*, John Wiley and Sons.

Chapter 4

Analysis and design of footing, raft foundation and pile cap

The analysis of footing and raft foundations includes well-known soil–structure interaction problems, and many methods and computer programs have evolved in this respect. In this chapter, these important and expensive elements are discussed, together with some refined methods for the design of these structures.

4.1 USE OF CLASSICAL RIGID DESIGN METHOD FOR SIMPLE FOOTING

Before the extensive use of computer methods, the internal force in the out-of-plane direction of a raft foundation was often analysed by treating the raft foundation as a "beam" structure under the applied loads on top and the ground reactions from the ground below. Also, without computer methods, the ground reactions were often analysed by assuming linearly varying distribution of stresses in two directions to balance the applied loads based on the so-called "rigid cap (also apply to footing)" assumption. By the rigid foundation/cap assumption, the footing or the cap structure is assumed to be perfectly rigid, resulting in linearly varying strain (and therefore settlement) across its plan dimensions. As the ground reaction is assumed to be proportional to the settlement, a linearly distributed ground reaction results. By the assumption, at any point of coordinates (u,v), the stress in the vertical direction under the application of load P and biaxial moments M_u, and M_v acting in the U and V directions, which are principal axes of the plan section of the footing area, is determined by

$$\sigma_v = \frac{P}{A} + \frac{M_u}{I_u}u + \frac{M_v}{I_v}v \tag{4.1}$$

where A is the plan area; I_u and I_v are the second moment of area of the cross section along the U and V directions.

By the above assumption, the stiffness of the subgrade is considered constant at every point of the whole area of the raft footing and the stiffness of the subgrade is low compared with the stiffness of the footing. In addition, the stiffness at every point is "independent" of others. Its settlement is only affected by the load directly acting on it. The whole footing can be visualized as being supported by many evenly distributed identical and independent

DOI: 10.1201/9781032650586-4

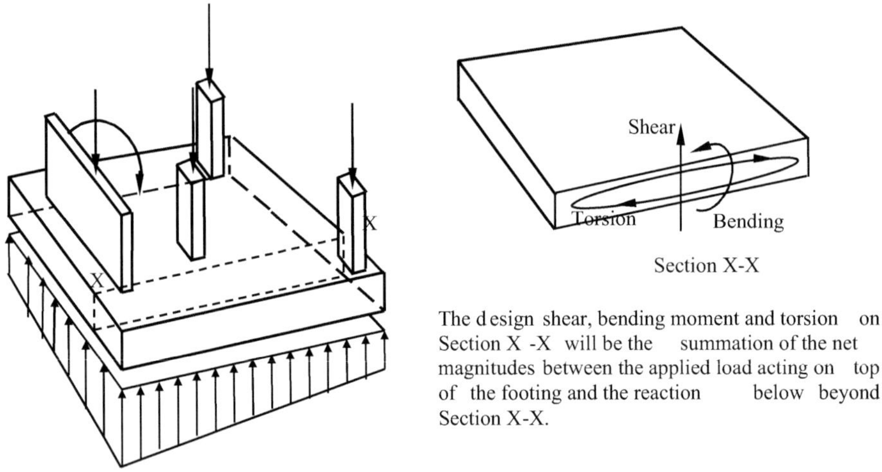

Figure 4.1 Demonstration of the determination of design forces on sections of footing by treating it as a beam structure.

springs which may be termed "Winkler's springs". The stiffnesses of the footing and the subgrade are not required in the analysis of the footing, which is clearly not a realistic assumption which is required at the time when there is no computer.

Being treated as a "beam" structure, the total shear force, bending moment and torsion of the footing at a section can then be calculated as illustrated in Figure 4.1.

Total shears, bending moments and torsions are calculated for various sections along the same direction and structural design is then carried out accordingly. Generally, the footing has to be analysed in two directions and structural design is carried out for the two directions independently. The reinforcement along the two orthogonal directions is then designed and placed in accordance with the two corresponding one-dimensional analysis. There are some obvious drawbacks or inaccuracies in adopting this method of analysis as discussed below:

a. By treating the footing structure as a single beam, only the average effects across the whole width of the footing can be captured, local effects which can deviate significantly from the average effect may occur leading to local under-design (shear lag effect is not considered);
b. Having unbalanced shear generally across the width of the footing in a section, some twisting of the section in the form of torsion may occur. The design for the torsion following that of the beam theory which is under the assumption of "spiral failure" is obviously not applicable to the footing as a plate or solid structure and can often result in over-design;
c. As a conservative design, it is a practice to design bending, shear and torsional reinforcement separately in two directions and superimpose the two sets of reinforcements. Very often, the reinforcements for the vertical shears in the two directions of the footing being simulated as a beam are superimposed, also, it is clear

that the vertical shear has thus been "doubly counted". The design method is obviously not realistic and often results in over-design in this aspect.

Despite the drawback and inaccuracies as discussed above, the method had been used for years before the computer method became extensively used. Nowadays, the classical method by idealizing the footing as a beam structure is not as acceptable except for simple cases. Consider Figure 4.2, the first approach is the rigid foundation method which is still used by many engineers. For the second approach which is practically a computer-based method, the subgrade is modelled as a series of individual springs, and the stiffness of the foundation is considered in the analysis. It should be noted that by putting high stiffness at the footing, method (b) will automatically reduce to method (a), as the basic assumption of method (b) is similar to that of method (a). Method (b) is currently the most popular method for the analysis of raft foundation, while method (a) is more popular for the simple single footing. Method (c) considers the continuity of the subgrade by elasticity theory (or elasto-plastic theory), which is obviously a more realistic modelling of the foundation, but is not popular among engineers. In fact, the assumption of the subgrade modulus in method (b) is a more difficult and unreliable parameter to be determined as compared with the simple elastic properties of the continuum in method (c). Method (c) is not popular among many engineers, and the main reason for this may be due to the fact that very few computer programs use method (c) in the analysis. The PLATE program developed by Cheng adopts both methods (b) and (c), and it is possible to mix methods (b) and (c) together in a single problem if the engineers really need to do so. Methods (b) and (c) are also called flexible methods, as they consider the stiffness of the foundation in the analysis.

In the classical rigid method, the plane section remaining plane is assumed, and the subgrade is treated as a classical beam under moment and axial force. The stresses acting on a beam will have a trapezoidal stress distribution in general. Classically, the moment on a footing is transformed to eccentricity e by assuming

$$Pe = M \tag{4.2}$$

Eq. (4.2) is valid only for a rigid footing, and should not be used if the footing is flexible enough. Consider Figure 4.2, where e is less than $L/6$, where L is the length of the footing along the bending direction. Assume a rigid footing using plane section remains plane, the base bearing pressure (assuming full contact) can be taken as linear and can be obtained from simple solid mechanics of eccentric axial load on a section as

$$q = \frac{P}{LB} \pm \frac{Pe \cdot X}{\frac{BL^3}{12}} \tag{4.3}$$

where B is the width of the beam normal to the bending direction, and q is pressure which takes a unit of kPa. At the two edges, the pressures are

$$q_{1,2} = \frac{P}{BL} \pm \frac{6Pe}{BL^2} \tag{4.4}$$

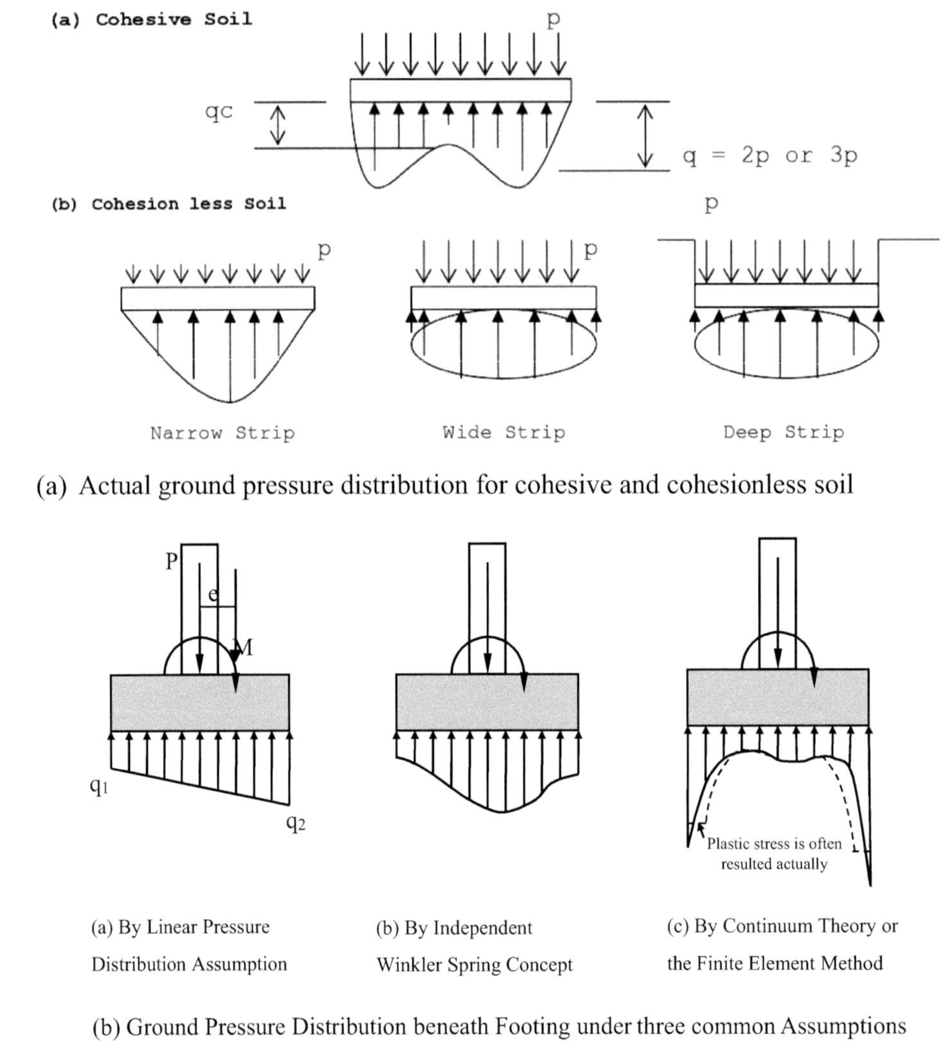

Figure 4.2 Ground pressure distribution by three major methods of analysis.

To make q similar to the uniform distributed load in classical 1-D structural mechanics problems, some engineers like to use qB (still denoted as q in this section) as q for the design, and this q takes a unit of kN/m. Hence

$$q_{1,2} = \frac{P}{L} \pm \frac{6Pe}{L^2} \tag{4.5}$$

If $e > L/6$, the smaller q value in Eq. (4.4) or (4.5) will be less than 0 or equivalently tension. Since soil cannot take tension in general, the assumption of a foundation in full

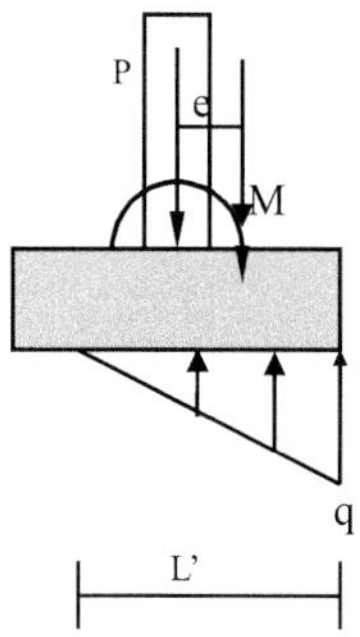

Figure 4.3 Base bearing pressure for rigid footing, when e > *L*/6.

contact with the subgrade will not be applicable. In this case, the contact length L' can be determined from Figure 4.3.

Consider vertical force equilibrium,

$$P = 0.5qL' \tag{4.6}$$

Taking the moment at the right end of the footing

$$\frac{1}{2}qL'\frac{L'}{3} = P(\frac{L}{2} - e) \tag{4.7}$$

Hence $$L' = 3\left(\frac{L}{2} - e\right) \text{ for } e > L/6 \tag{4.8}$$

It should be noted that for the design work in Hong Kong, tension at the base of footing is not allowed according to the building code, and engineers should enlarge the footing to ensure no-tension design. There is, however, no such requirement in other building codes. For the structural analysis and design of the footing, the foundation in Figure 4.4a is inverted, and the footing is now a classical beam which can be analysed easily. It should be noted that for the problem in Figure 4.4a, the two column loads are combined into one resultant column load, and the moments are then combined to an overall eccentricity when Eq. (4.4) or (4.5) is used. This assumption is obviously not valid when the distance between the two column loads is wide. For the case of Figure 4.4b, the beam is a structural indeterminate structure according to structural mechanics. However, it is statically determinate in the current problem, as the column loads are used to compute the trapezoidal base bearing pressure. If the column/support loads are known, the problem will then be statically determinate. For the case of biaxial bending, Eq. (4.1) should be used. If tension can be generated from high eccentricity, there can be several possibilities in the tension generation. Many engineers prefer to use a computer program which can carry out stress redistribution to eliminate tension at the base of the foundation (for example, PLATE by the author).

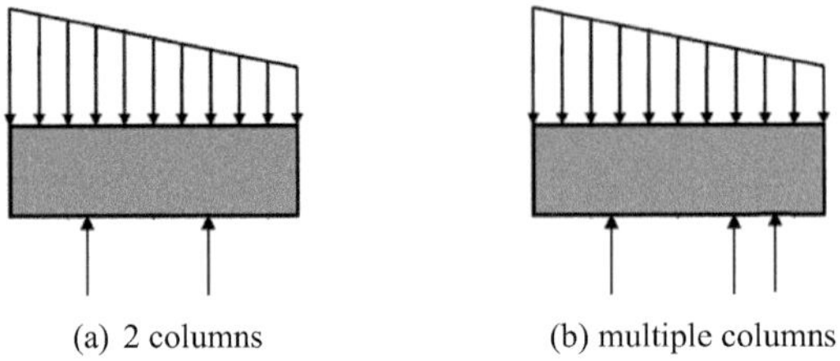

Figure 4.4 Structural analysis and design for the footing in Figure 4.2a.

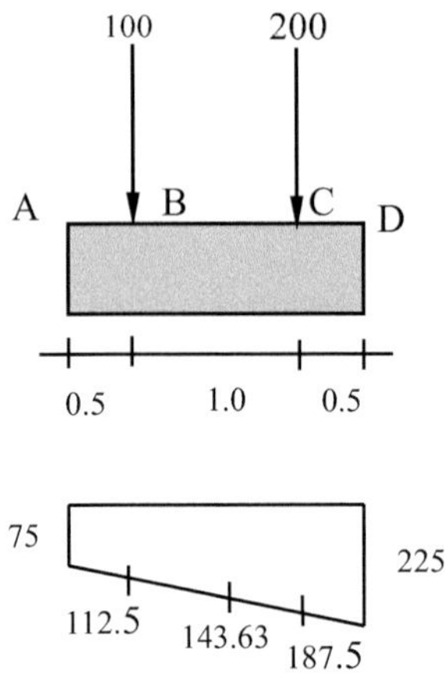

Figure 4.5 Example 1 for illustration of the rigid design method.

Example 1: Consider the problem as shown in Figure 4.5.

Determine the location of the resultant force of 300 kN and the location of the resultant force:

$$\text{Take moment about } A, 100\times 0.5 + 200\times 1.5 = 300\overline{X}$$

$\overline{X} = 1.167m$, which is the location of the resultant force from the left end. The eccentricity e is given by

$$e = \overline{X} - 2/2 = 0.167m \text{ towards the right of the centre line of the footing}$$

$$q = \frac{300}{2} \pm \frac{300\times 0.167\times 6}{2^2} = 150 \pm 75 \text{ kN/m, which is shown in Figure 4.5.}$$

SF at B = (75+112.5) × 0.5/2 = 46.88 kN (left); SF at B = 46.88 – 100 = 53.13 kN (right)

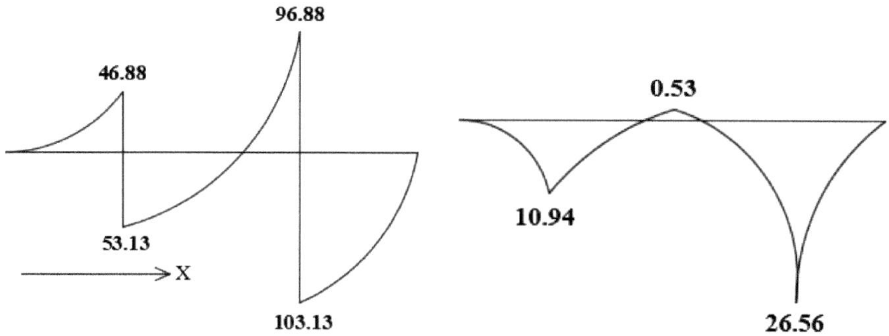

(a) Shear force diagram (b) Bending moment diagram

Figure 4.6 Bending moment and shear force for the footing in Figure 4.5.

SF at C =(187.5 + 225) × 0.5/2 = 103.13 kN (right); SF at C = 103.13 – 200 = 96.88 kN (left).

To determine the bending moment at B, divide the base pressure with a UDL + triangular load

BM at B = 75 × $0.5^2/2$ + 37.5 × 0.5/2 × 0.5/3 = 10.94 kN-m (note: 37.5 = 112.5 – 75 or x = 0.5)

BM at C = 187.5 × 0.5 × 0.5/2 + 37.5 × 0.5/2 × 0.5 × 2/3 = 26.56 kN-m

The point of zero shear is located at distance X from left, hence divide the trapezoidal pressure to a UDL + triangular load as:

$$\frac{1}{2}\cdot 75\cdot X^2 + 75X = 100, \text{ hence } X = 0.915 \text{ m}$$

The maximum hogging moment = 75*$0.915^2/2$ + (143.63 – 75)x 0.915/2 x 0.915/3-100 x (0.915 – 0.5) = –0.53 kN-m

The bending moment and shear force diagram for the footing in Figure 4.5 are given in Figure 4.6. Based on these results, the bending and shear reinforcement for the footing can be designed.

Example 2: For a 4 m long footing with two point loads 120 kN and 180 kN placed at 0.5 m and 3.5 m from the left of the footing, determine the maximum hogging moment using rigid analysis.

Solution:

For point of resultant vertical loads, 120x0.5 + 180x3.5 = 300X

$$\therefore X = 2.3 \text{ m} \Rightarrow e = 2.3 - 4/2 = 0.3 \text{ m to right of the centreline}$$

$$q = 300/4 \pm 6\text{x}300\text{x}0.3/4^2 = 75 \pm 33.75 = 108.75/41.25$$

For the point of zero shear:

$$41.25x + 0.5x16.875x^2 = 120 \text{ or } 8.4375x^2 + 41.25x–120 = 0 \ (16.875 = 33.75x2/4)$$

x = 2.05 m from the left for point of zero shear.

Maximum hogging moment = 41.25x2.05²/2 + 34.59x2.05/2x2.05/3 – 120x1.55 = –75.1 kN-m.

4.2 WINKLER'S SPRING MODEL FOR FOUNDATION ANALYSIS

The concept of Winkler's spring has been explained in Section 4.1. However, the term Winkler's spring was not popularly used in the classical method. Nevertheless, the term is more widely used in structural analysis of raft foundations using the computer method. Consider the plate load test result in Figure 4.7. If the settlement is not large, a linear relation between the applied pressure and plate settlement is generally obtained, and the linear part can extend to 100 mm for many plate load tests in Hong Kong, though the linear relation should not be considered to be valid in general. The slope of the pressure settlement relation is called the modulus of subgrade reaction, which is denoted by K_s.

When the strain is small, q vs. δ can be taken as linear and

$$K_s = q/\delta \text{ (unit = kN/m}^3\text{)} \tag{4.9}$$

Typical values of K_s can be found from GEO Guide 1.

$$(4.9) \times \text{A}: A \cdot K_s = q \cdot A/\delta = F/\delta$$

$$\text{Or } \delta = F/ A\, K_s \tag{4.10}$$

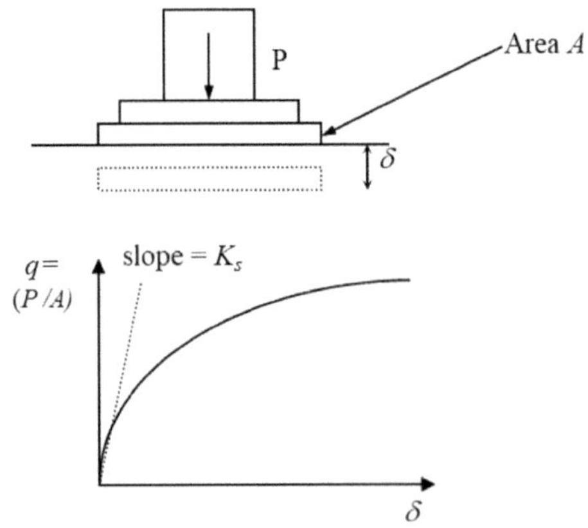

Figure 4.7 Results for a typical plate load test.

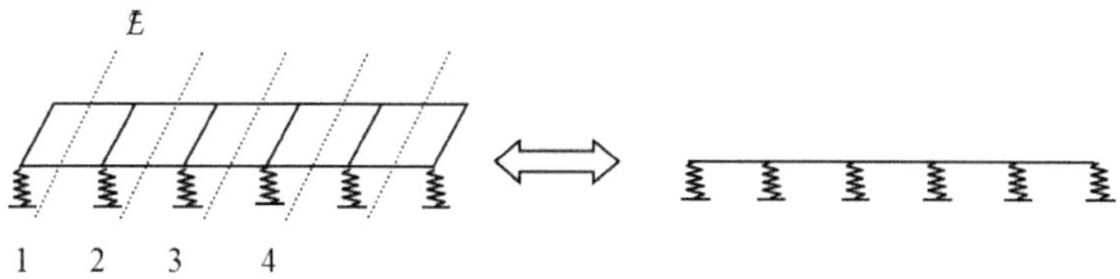

Figure 4.8 Idealization of the footing as a continuous beam with spring supports.

Hence $A{\cdot}K_s$ is conceptually similar to the classical 1-D spring, and $A{\cdot}K_s$ is taken as the spring constant. The response from the subgrade is hence a series of linear springs acting along the footing. If the width of the footing is much smaller than the length, it can be considered as one-dimensional. It can be analysed by the rigid design method as given before, or as a continuous beam on Winkler's spring as shown below. As shown in Figure 4.8, a footing is divided into a series of elements and nodes. In between the nodes, centre-lines are formed as shown in Figure 4.8. The area of footing between the centre-lines associated with each node is assigned as the area for computation of the equivalent spring. Obviously, there is no support between the nodal springs, which is physically incorrect, but if the number of nodal springs keeps on increasing, the analysis results will tend to converge and stabilize quickly. For a normal footing, the use of 10 elements may be sufficient to give a reasonable accuracy for the design. It can be seen that the effects of subgrade are lumped at the nodes only. Cut a centreline between nodes and the effective area between centrelines is used to compute $K = A_{eff} \cdot K_s$. K_s can be approximated as $0.65E/B(1-\mu^2)$ by Salvadurai, $E/B(1-\mu^2)$ by Vesic, or $\pi E/2B(1-\mu^2)\log_e(L/B)$. K_s is not really a constant parameter, as it is dependent on B according to the formulae by Vesic and Salvadurai or the field tests, however, it is usually considered to be constant for simplicity. It is interesting to note that the bending moment on the footing is not too sensitive to the precise value of K_s, but the settlement of the footing will be directly affected by the value of K_s.

The stiffness matrix of the beam is given by Eq. (4.11)

$$[K] = \begin{bmatrix} \frac{12EI}{L^3} & \frac{6EI}{L^2} & -\frac{12EI}{L^3} & \frac{6EI}{L^2} \\ \frac{6EI}{L^2} & \frac{4EI}{L} & -\frac{6EI}{L^2} & \frac{2EI}{L} \\ -\frac{12EI}{L^3} & -\frac{6EI}{L^2} & \frac{12EI}{L^3} & -\frac{6EI}{L^2} \\ \frac{6EI}{L^2} & \frac{2EI}{L} & -\frac{6EI}{L^2} & \frac{4EI}{L} \end{bmatrix} \tag{4.11}$$

The degree of freedom (DOF) vector is a four-element vector which includes the vertical displacement and rotations at the left and right ends. The stiffness matrix of all the members is formed, and has a size of $2n \times 2n$ where n is the number of nodes of the beam. We can put in the boundary condition and solve the matrix equation which will be the equivalent of the Winkler's spring at the nodes which affect only the vertical displacement. The nodal spring stiffness due to soil will be added in locations (1,1), (3,3), (5,5) ... (diagonal) of the global stiffness matrix, and odd numbers are superimposed because the even number entries are associated with rotation. Alternatively, a matrix due to the soil subgrade as given in Eq. (4.12) can be added to Eq. (4.11), which is based on the solution of the beam deflection equation. Using Eq. (4.12), there is no need to form the equivalent effective area and the subgrade nodal spring.

$$\left[K_{soil}\right] = \frac{k_s L}{420}\begin{bmatrix} 156 & 22L & 54 & -13L \\ 22L & 4L^2 & 13L & -3L^2 \\ 54 & 13L & 156 & -22L \\ -13L & -3L^2 & -22L & 4L^2 \end{bmatrix} \tag{4.12}$$

The basic beam deflection equation with Winkler's spring is (negative sign is put with $k_s y$ as the soil reaction is opposite to the direction of deflection)

$$EI\frac{d^4y}{dx^4} = q = -k_s' y \tag{4.13}$$

where $k_s' = k_s B$. If the conditions are uniform, solving the equations, a variable λ is introduced. The general solution to Eq. (4.13) is:

$$y = e^{\lambda x}(c_1 \cos\lambda x + c_2 \sin\lambda x) + e^{-\lambda x}(c_3 \cos\lambda x + c_4 \sin\lambda x) \tag{4.14}$$

$$\lambda = \sqrt[4]{\frac{k_s'}{4EI}} \quad or \quad \lambda L = \sqrt[4]{\frac{k_s' L^4}{4EI}} \tag{4.15}$$

λL is classically called the characteristic length, which is, however, not with current computers. Very long classical solutions have been derived for the case of point load and point moment in the past. The classical solution has several distinct disadvantages over the computational approach which are:

1. Assumes weightless beam (but weight will be a factor when footing tends to separate from the soil);
2. Difficult to remove soil effect when footing tends to separate from soil;
3. Difficult to account for boundary conditions of known rotation or deflection at selected points;
4. Difficult to apply multiple types of loads to a footing;
5. Difficult to change footing properties of I and B along member;
6. Difficult to allow for change in subgrade reaction along footing.

The advantage of using the matrix form in Eq. (4.11) instead Eq. (4.13) is that if the number of elements keeps on increasing, the matrix solution will tend to the differential equation solution. The limitations of the classical differential equation can be totally eliminated. It will be illustrated later that the continuous beam model with Winkler spring or elastic continuum can be replaced by the use of plate analysis, and numerical results will hence be provided only for plate analysis.

4.3 ANALYSIS OF RAFT FOUNDATION

The idealization of the footing by Figure 4.8 is not valid if the width of the footing is appreciable. In fact, if B is as large as L, there is no dominant bending direction for the problem. This problem can be further illustrated by the simply supported beam in Figure 4.9. For a 1-D beam, the maximum moment is $\omega L^2/8$, which is a well-known formula in classical structural mechanics. However, the width B is never considered in the derivation of this equation. During the reinforcement design, the factor $k = M/bd^2$ is commonly used in many reinforcement design codes, and the bending reinforcement is placed uniformly across the section. If the width is large, can we still carry out the design in this way? In fact, the factor can be rewritten as the factor $k = (M/b)d^2$ if the width is large. If M/b is not constant across the section, the use of the simple factor M/bd^2 can be misleading and unsafe. Cheng has carried out the design of a highway bridge, with the final bending steel increased by 50% for such non-uniform distribution of moment across the section of the bridge deck. Furthermore, the use of the simple beam model in Figure 4.8 does not consider the effect of torsion, which can be important if B is large enough. Consider case (b or c) in Figure 4.9, if all the loading is applied along half of the width of the beam, can engineers still use $\omega L^2/8$ and $k = M/bd^2$ in the reinforcement design?

For a two-dimensional raft foundation, the methods of analysis include:

1. Classical strip method based on the rigid analysis (simplified to 1-D hand calculation method), and analysis and design are carried out along the x and y directions

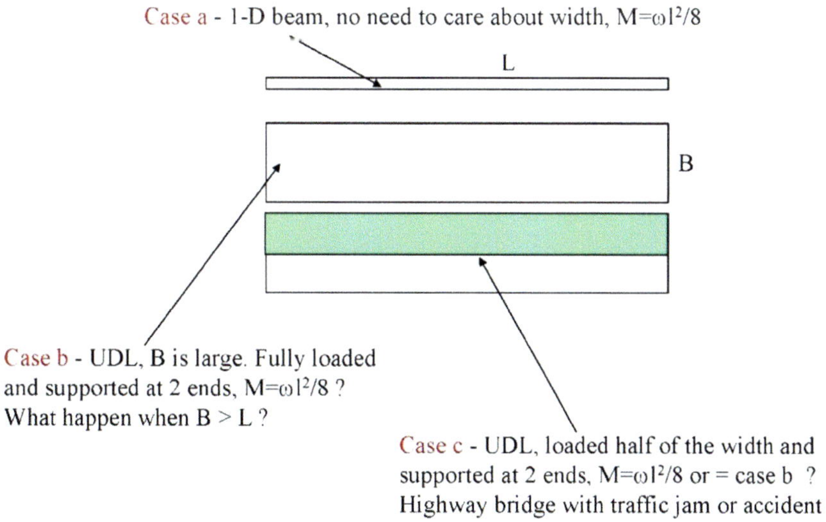

Figure 4.9 A simply supported beam with uniform distributed load ω, with the width of the beam not negligible.

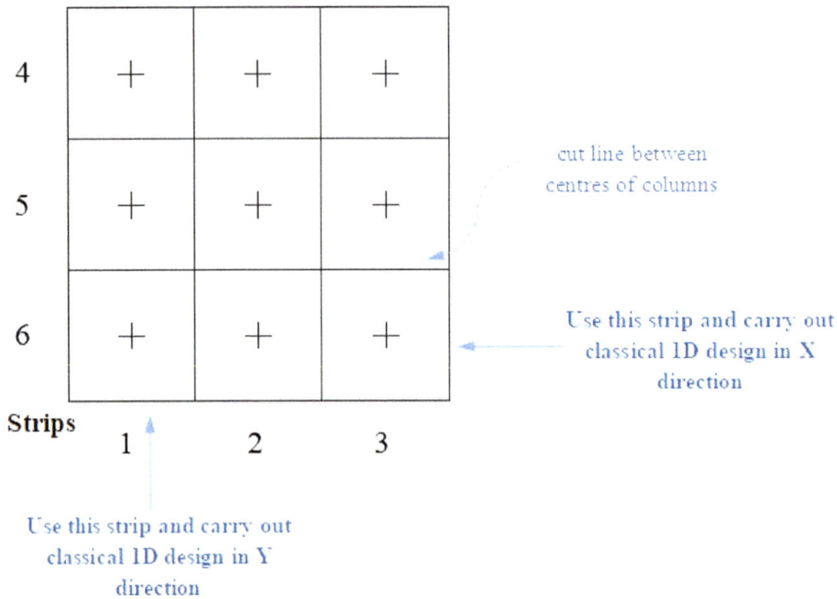

Figure 4.10 The use of the classical strip method for raft foundation analysis.

separately. As shown in Figure 4.10, three strips along the x direction and three strips along the y directions are considered. The bending and shear reinforcement from the two designs will be superimposed. Torsion is neglected in this design method. There are some major limitations to this method: (1) if the columns are spaced irregularly, it is very difficult to form a representative strip for analysis; (2) there will be displacement and rotation incompatibility between each adjoining strip. In view of that, some engineers neglect the width or length and carry out one 1-D analysis along the x and one along the y direction.

2. Differential equation method: The 2D plate bending on Winkler's spring problem has been solved for simple cases. A local analysis of the plate bending equation is also offered in some design codes. This method is now not commonly used, due to the availability of many computer programs.
3. Grillage method: The plate foundation is modelled as beams arranged in two major directions (usually orthogonal for ease of reinforcement design). Generally, low computer resources is required. This is an approximate method of analysis of 2-D slab-type structures. Irrespective of the number of grid beams used in the computer model, the area of each grid beam is actually zero, and the Poisson ratio is never reflected in such a model. The empty spaces between the grids are never filled in this model. If the Poisson ratio tends to 0, this model will be a good approximation of the raft foundation problem. This method was popular 30 years ago, and is still used by some engineers at present.
4. Finite element plate bending method: The actual raft foundation is modelled as an assembly of smaller plates which require higher computer resources, tedious

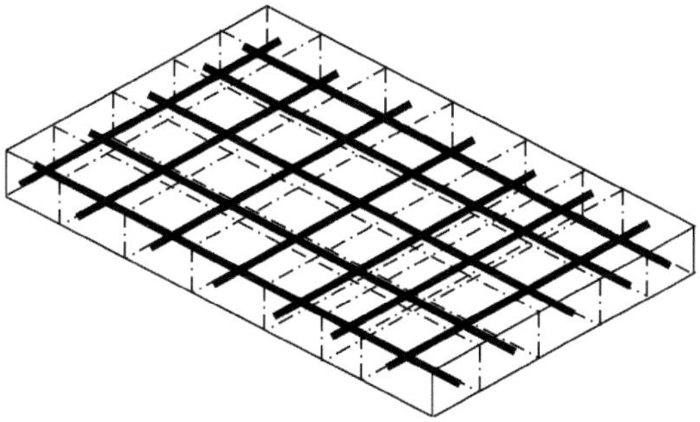

Figure 4.11 Simulation of a raft footing by a grillage beam system.

formulations, but more accurate modelling of the raft foundation. For most of the raft foundation, the thickness t of the plate is appreciable so that the thin plate model ($t/L > 0.1$) may not be a good representation of the actual foundation. At present, most engineers adopt this method for the analysis of raft foundation, using computer programs. More will be discussed on plate analysis in later sections.

The grillage beam simulation was popular when the finite element software was not yet popular. Basically, its formulation is under the simulation of the raft footing being cut into strips in two directions (often mutually perpendicular), with each strip simulated as a beam (line member) carrying the sectional properties in accordance with its cross-sectional dimensions. Figure 4.11 illustrates the simulation.

The stiffness matrix of a grid beam is given by Eq. (4.16). Eq. (4.16) is a special case of a space frame member, where only vertical displacement and rotation along x and y directions are considered. Each node therefore has three degrees of freedom, with a 6 × 6 stiffness matrix. Grid members need not be orthogonal to each other, and the average width of the member then will be used in determining the bending and torsional stiffnesses.

$$[K] = \begin{bmatrix} \frac{12EI}{L^3} & 0 & \frac{6EI}{L^2} & -\frac{12EI}{L^3} & 0 & \frac{6EI}{L^2} \\ 0 & \frac{GJ}{L} & 0 & 0 & -\frac{GJ}{L} & 0 \\ \frac{6EI}{L^2} & 0 & \frac{4EI}{L} & -\frac{6EI}{L^2} & 0 & \frac{2EI}{L} \\ -\frac{12EI}{L^3} & 0 & -\frac{6EI}{L^2} & \frac{12EI}{L^3} & 0 & -\frac{6EI}{L^2} \\ 0 & -\frac{GJ}{L} & 0 & 0 & \frac{GJ}{L} & 0 \\ \frac{6EI}{L^2} & 0 & \frac{2EI}{L} & -\frac{6EI}{L^2} & 0 & \frac{4EI}{L} \end{bmatrix} \tag{4.16}$$

The bending I and torsional J stiffnesses of each member are given by

$$I = \frac{bt^3}{12}, \quad J \approx bt^3\left[\frac{1}{3} - 0.21\frac{t}{b}\left(1 - \frac{t^4}{12b^4}\right)\right]\Omega \tag{4.17}$$

where $\Omega \approx 0.75\frac{L}{b} \leq 1.1$, and b is the width (or average width) of the grid member, and t is the thickness of the plate. In the assembly of the global stiffness matrix, it should be noted that the bending at one end of a member is coupled with the torsion of the member orthogonal to it. A typical grillage model in shown in Figure 4.12, where there are 25 nodes and 40 grid members. In general, grid members will be provided between the column lines and the edges of the foundation. For highly irregular spaced columns, it is difficult to define a good grillage system for the analysis. A complete Fortran program for the grillage analysis of raft foundation is provided by Cheng et al. (2020)

Centre-lines are formed between the grid members, and the effective area bounded by the centre-lines of each node is taken as the effective area for the calculation of the equivalent Winkler's spring. The global stiffness matrix of the grid model is assembled, and the problem can be solved with the application of the boundary condition, which is typically the Winkler springs. It can be easily understood that the method is more appropriate for regular plate structures where regular grillage of beams can be used to simulate the plate structure. For a footing of irregular plan shape and/or with openings, the grillage beam system may become irregular, resulting in an uneven arrangement.

Structural design can be based on member forces of the grillage beams as analysed. Though one may use the moments, shear forces and torsions obtained in analysis and carry out structural design accordingly by treating the raft footing as a series of beams and argue that we can arrive at structural adequacy, the following are, however, highlighted to indicate the shortcomings of the design method in terms of the design shear, moment and torsion:

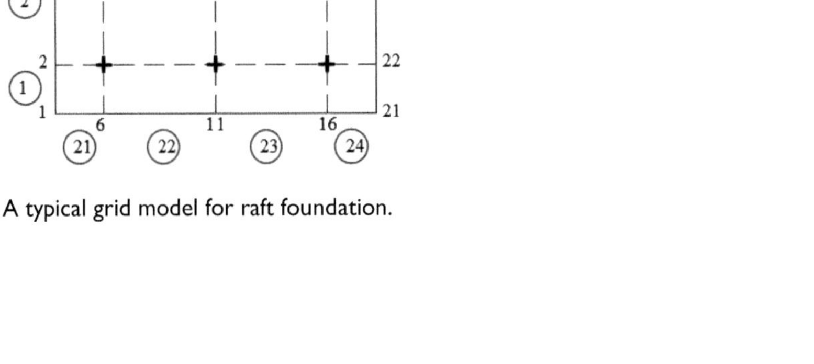

Figure 4.12 A typical grid model for raft foundation.

(i) Shear: As a plate structure, the maximum out-of-plane shear stress at any point is a single and unique value under a loading condition. But by the grillage analysis, two shear forces at the sections cutting the same point can be obtained by the two grillage beams running across the point and two shear stresses can therefore be obtained which are generally of different values. The value can only be taken as a value with which the true shear is of the same order. One may take the greater value for design purpose.

(ii) Moment and torsion: The moment and torsion obtained in a beam again depend very much on the divisions of the plate into beam elements. However, the bending moment in a beam running in one direction can be reduced or increased by the torsions of other beams running in the different direction along its length and vice versa for its torsion. Therefore, the moments and torsions of the beam are in "jumps" instead of in smooth profiles which should exist in a continuous plate structure to which the raft footing belongs. Also, the magnitudes of the moments and torsions depend very much on how the plate is divided into strips of beams. An approach to eliminate this phenomenon is to intentionally set the J values of the beams to zero. However, as the bending moments will then follow only the directions of the beams, this approach will force the bending moments in the slab to run only in the directions of the beams, which physically forces the principal moments to align with the directions of the beams. This method will distort the structural behaviour of the plate structure simulating the raft footing.

Summing up, there are quite a few limitations in the application of the method in simulating the structural behaviour of the structure realistically. With the advancement in structural analysis by simulating the raft footing as an assembly of plate elements or solid elements by the finite element methods which can simulate better the structural behaviour of the raft footing, the method is now gaining less popularity and in fact diminishing in use in the industry.

A typical output for a 4 m × 4 m raft foundation with several point loads around the centre as shown in Figure 4.12 from the grillage program by Cheng is shown below for illustration:

9 NODES 12 MEMBERS

Modulus of subgrade reaction = 10000 kN/m^3

Node	*X-ordinate*	*Y-ordinate*
1	0.000	0.000
2	0.000	2.000
3	0.000	4.000
4	2.000	0.000
5	2.000	2.000
6	2.000	4.000
7	4.000	0.000
8	4.000	2.000
9	4.000	4.000

Member	*Connectivity*			*EI*	*GJ*	*Length*
1	1	TO	2	0.20000E+07	0.20000E+07	0.20000E+01
2	2	TO	3	0.20000E+07	0.25000E+07	0.20000E+01
3	4	TO	5	0.20000E+07	0.20000E+07	0.20000E+01
4	5	TO	6	0.30000E+07	0.30000E+07	0.20000E+01
5	7	TO	8	0.25000E+07	0.32000E+07	0.20000E+01
6	8	TO	9	0.20000E+07	0.20000E+07	0.20000E+01
7	1	TO	4	0.20000E+07	0.20000E+07	0.20000E+01
8	4	TO	7	0.20000E+07	0.20000E+07	0.20000E+01
9	2	TO	5	0.20000E+07	0.20000E+07	0.20000E+01
10	5	TO	8	0.20000E+07	0.20000E+07	0.20000E+01
11	3	TO	6	0.20000E+07	0.20000E+07	0.20000E+01
12	6	TO	9	0.20000E+07	0.20000E+07	0.20000E+01

Result of grillage analysis

Joint	*Z-movement*	*X-rotation*	*Y-rotation*
1	0.13000E+03	−0.35070E+02	−0.27209E+02
2	0.65577E+02	−0.32900E+02	−0.14389E+02
3	0.00000E+00	−0.35663E+02	−0.29913E+01
4	0.19039E+03	−0.47882E+02	−0.29755E+02
5	0.94660E+02	−0.47563E+02	−0.14132E+02
6	−0.10318E+00	−0.47384E+02	−0.98634E+00
7	0.25000E+03	−0.60162E+02	−0.25959E+02
8	0.12507E+03	−0.62153E+02	−0.16277E+02
9	0.00000E+00	−0.59659E+02	−0.29228E+01

Member		*Moments for two ends*		*For nodes*	*Torque*
1	0.12812E+08	0.84708E+07	1 AND	2	−0.12820E+08
2	0.61929E+07	0.11720E+08	2 AND	3	−0.14247E+08
3	−0.53107E+06	−0.11678E+07	4 AND	5	−0.15623E+08
4	0.10943E+07	0.55539E+06	5 AND	6	−0.19719E+08
5	−0.12281E+08	−0.73024E+07	7 AND	8	−0.15490E+08
6	−0.72878E+07	−0.12276E+08	8 AND	9	−0.13355E+08
7	0.12820E+08	0.77266E+07	1 AND	4	0.12812E+08
8	0.78964E+07	0.15490E+08	4 AND	7	0.12281E+08
9	0.14277E+07	0.19414E+07	2 AND	5	0.14664E+08
10	0.21547E+07	−0.21353E+07	5 AND	8	0.14590E+08
11	−0.14247E+08	−0.10237E+08	3 AND	6	0.11720E+08
12	−0.94815E+07	−0.13355E+08	6 AND	9	0.12276E+08

Member	*Shear force*		*For nodes*	
1	−0.10641E+08	−0.10641E+08	1 and	2
2	−0.89566E+07	−0.89566E+07	2 and	3
3	0.84955E+06	0.84935E+06	4 and	5
4	−0.82470E+06	−0.82500E+06	5 and	6
5	0.97915E+07	0.97915E+07	7 and	8

Member	*Shear force*		*For nodes*	
6	0.97818E+07	0.97818E+07	8 and	9
7	−0.10273E+08	−0.10273E+08	1 and	4
8	−0.11693E+08	−0.11693E+08	4 and	7
9	−0.16845E+07	−0.16845E+07	2 and	5
10	−0.96898E+04	−0.96898E+04	5 and	8
11	0.12242E+08	0.12242E+08	3 and	6
12	0.11418E+08	0.11418E+08	6 and	9

From this grillage analysis output, it is noticed that the degrees of freedom are the vertical deflection, and rotation about the x and y directions. For each member, the lengths, EI and GJ, are defined. Based on the connectivity, the overall stiffness matrix is established. The boundary condition is given by the modulus of subgrade reaction, and the problem is then solved. The internal forces of each member are then evaluated by Eq. (4.16). It should be noted that the torque and shear are constant within each grid member, while the bending moment can vary along each member. With the development in a finite element program, grillage analysis is now less popular among engineers.

4.4 PLATE ANALYSIS OF RAFT FOUNDATION

With the development of the finite element method to formulate the stiffness matrix of a plate element, the structural analysis of a raft footing can be carried out by simulating the structure as an assembly of the plate element, as shown in Figure 4.13.

Each of the plate elements has basically three degrees of freedom at each node, which are the vertical settlement and out-of-plane rotations about two mutually perpendicular axes, say X and Y.

The basic mathematical derivation of the shears, bending moments, and twisting moments in plate bending are as follows, based on Figures 4.14 and 4.15. In Figure 4.15, the rectangular element of plan dimensions ∂x and ∂y is an infinitesimal element.

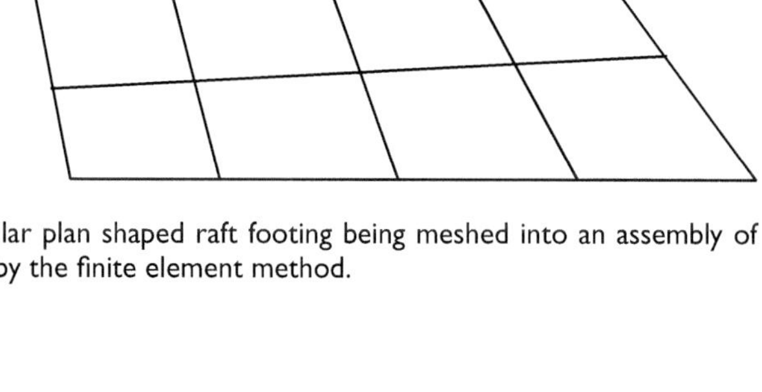

Figure 4.13 An irregular plan shaped raft footing being meshed into an assembly of plate elements to be analysed by the finite element method.

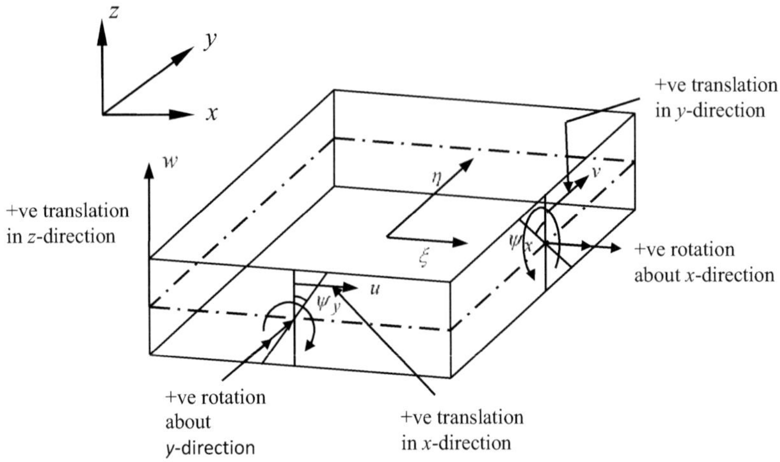

Figure 4.14 Plate bending theory for bending.

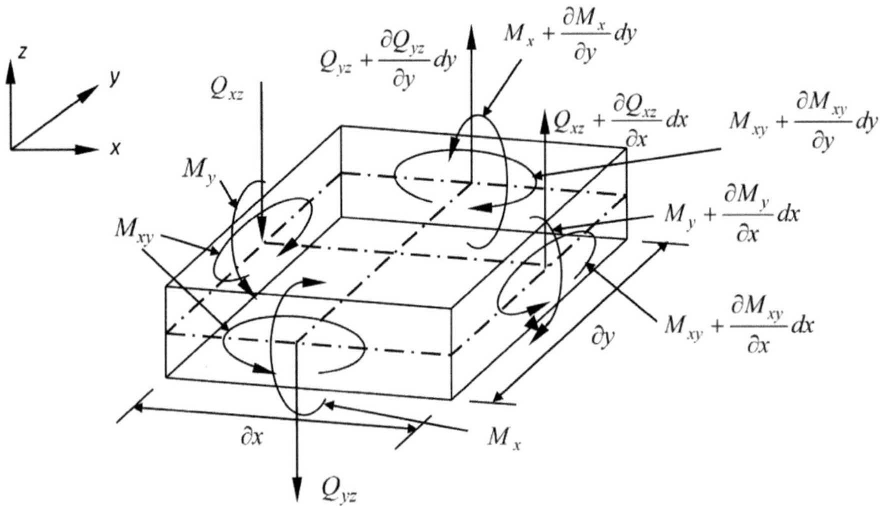

Figure 4.15 Plate bending theory for shear.

The fundamental stress–strain relationships as follows are listed:

$$\sigma_x = \frac{E}{1-\mu^2}\left(\varepsilon_x + \mu\varepsilon_y\right) \tag{4.18}$$

$$\sigma_y = \frac{E}{1-\mu^2}\left(\mu\varepsilon_x + \varepsilon_x\right) \tag{4.19}$$

$$\tau_{xy} = \frac{E}{2(1+\mu)}\gamma_{xy} \tag{4.20}$$

where E is the Young's modulus, and μ is the Poisson's ratio. The symbols σ, ε, τ, γ carry their usual meaning of direct stress, direct strain, shear stress, and shear strain, respectively, and the suffices indicate the directions.

The moments and shears are summed up as related to the vertical displacement of the plate w as follows:

$$M_x = \frac{Et^3}{12(1-\mu^2)}\left(\frac{\partial^2 w}{\partial y^2} + \mu\frac{\partial^2 w}{\partial x^2}\right) \tag{4.21}$$

$$M_y = \frac{Et^3}{12(1-\mu^2)}\left(\frac{-\partial^2 w}{\partial x^2} - \mu\frac{\partial^2 w}{\partial y^2}\right) \tag{4.22}$$

$$M_{xy} = \frac{Et^3}{12(1+\mu)}\left(\frac{-\partial^2 w}{\partial x \partial y}\right) \tag{4.23}$$

$$Q_{xz} = \frac{\partial M_x}{\partial x} + \frac{\partial M_{xy}}{\partial y} \tag{4.24}$$

$$Q_{yz} = \frac{-\partial M_x}{\partial y} + \frac{\partial M_{xy}}{\partial x} \tag{4.25}$$

$M_x, M_y, M_{xy}, Q_{xz}, Q_{yz}$ are defined in Figures 4.14 and 4.15. It should be noted that these quantities are moments and shears per unit width which are termed "stresses" in finite element terminology. t is the thickness of the plate element. Also, the derivations are based on flexural deformation of the plate only. Shear deformation has been ignored, though some advanced finite element formulations do include shear deformation which is considered more accurate for thick plate structures.

By considering equilibrium and strain compatibility, we can get the following equations relating shears, bending and twisting moments, and the strains by the plate bending theory.

$$\frac{\partial^4 w}{\partial x^4} + 2\frac{\partial^4 w}{\partial x^2 \partial y^2} + \frac{\partial^4 w}{\partial y^4} = \frac{q}{D} \tag{4.26}$$

$$\frac{\partial^2 M_x}{\partial x^2} + 2\frac{\partial^2 M_{xy}}{\partial x \partial y} + \frac{\partial^2 M_y}{\partial y^2} + q = 0 \tag{4.27}$$

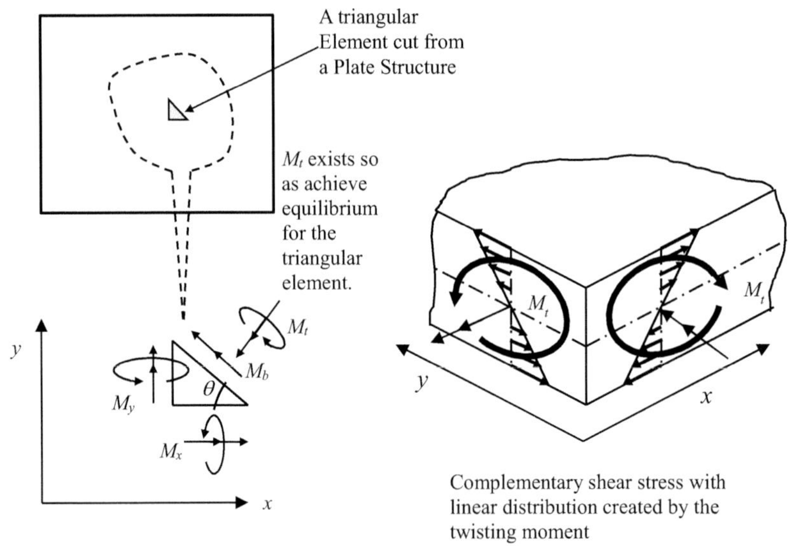

Figure 4.16 The occurrence and nature of the "twisting moment".

where q is the uniformly distributed load at the point (x, y) and $D = \dfrac{Et^3}{12(1-\mu^2)}$.

The finite element method can be based on the above plate bending theory and an assumed "shape function" for formulation of the stiffness matrix of an element. The stiffness matrix of the whole structure is formed as an assembly of the plate elements in the similar manner to that of the stiffness method.

While M_x, M_y producing flexural stresses can be easily understood, a parameter M_{xy} termed "twisting moment" (again per unit width) may be less familiar to some structural engineers. The existence of the twisting moment in a plate bending structure can be explained by Figure 4.16.

By Figure 4.16, the twisting moment M_t has to be in existence to achieve equilibrium of the triangular element. This M_t, unlike bending moment M_b producing flexural stress on the section (i.e. compressive stress on the top or bottom half and tensile stress on the other half), produces in-plane shear stress on the section instead, as shown in Figure 4.16.

Therefore, at any point of the plate structure, generally two bending moments in orthogonal directions bending moments of M_x, M_y and a twisting moment M_{xy} are acting in the x and y directions. However, the magnitudes of these moments will change at other directions, which is in complete analogy with the actions of direct stress and shear stress existing in an in-plane structure. For two moments M_x, M_y and about the x and y axes and the associated M_{xy}, the bending moment about a direction at an angle θ measuring anticlockwise from the x-axis is

$$M_b = M_x \cos^2\theta + M_y \sin^2\theta + 2M_{xy}\cos\theta\sin\theta \tag{4.28}$$

$$M_t = \left(M_y - M_x\right)\sin\theta\cos\theta + M_{xy}\left(\cos^2\theta - \sin^2\theta\right) \tag{4.29}$$

Again, in complete analogy with the in-plane stress problem, at a direction θ, M_t will vanish by Eq. (4.29), leaving two bending moments and these moments are analogous to the principal stresses in an in-plane problem where the associated shear stress becomes zero. Therefore, these moments are similarly termed "principal moments" and the directions for the principal moments are the "principal directions".

Theoretically, the designer may design for the principal moments in reinforced concrete structure including the raft and align the reinforcements in the principal directions. However, obviously, this is not practical as the principal directions change with locations and load cases. A practical design is based on the Wood Armer Equations (Wood 1968). The equations works on determining design bending moments at two pre-determined directions (usually orthogonal, say x and y) M_x^* and M_y^* that can cover bending moments generated by the bending moments in x and y directions coupled a twisting moment (i.e., M_x, M_y, M_{xy}) in all directions (by Eq. [4.28]). The equations work on the Johansen's yield criterion which gives the moment of resistance of the plate structure at an angle θ to the x axis by M_x^* and M_y^* as

$$M_\theta^* = M_x^*\cos^2\theta + M_y^*\sin^2\theta \tag{4.30}$$

which should then be not less than M_b in Eq. (4.14). So mathematically, we may list

$$M_x^*\cos^2\theta + M_y^*\sin^2\theta \geq M_x\cos^2\theta + M_y\sin^2\theta + 2M_{xy}\cos\theta\sin\theta \tag{4.31}$$

Based on Eq. (4.31), the optimal solutions of M_x^* and M_y^* are searched for various cases and the Wood Armer equations are derived. For these equations, sagging moments carry a positive sign and hogging moments carry a negative sign and M_x^* and M_y^* are the design moments. The designer will design the plate structure as having purely these bending moments.

For sagging moment, generally

$$M_x^* = M_x + \left|M_{xy}\right|; M_y^* = M_y + \left|M_{xy}\right| \tag{4.32}$$

If either M_x^* or M_y^* is found to be negative, then such a value is changecd to zero as:

$$M_x^* = M_x + \left|\frac{M_{xy}^{\ 2}}{M_y}\right| \text{with } M_y^* = 0 \text{ or } M_y^* = M_y + \left|\frac{M_{xy}^{\ 2}}{M_x}\right| \text{with } M_x^* = 0 \tag{4.33}$$

Similarly, for hogging moment, generally

$$M_x^* = M_x - |M_{xy}|; M_y^* = M_y - |M_{xy}| \tag{4.34}$$

If either M_x^* or M_y^* is found to be positive, then such a value is changed to zero as:

$$M_x^* = M_x - \left|\frac{M_{xy}^2}{M_y}\right| \text{with } M_y^* = 0 \text{ or } M_y^* = M_y - \left|\frac{M_{xy}^2}{M_x}\right| \text{with } M_x^* = 0 \tag{4.35}$$

The equations are simple to apply and have been widely used.

Worked Example 4.3

Consider two points in the raft:

Point A: $M_x = 7000\,\text{kNm/m}$; $M_y = 24000\,\text{kNm/m}$; $M_{xy} = 9000\,\text{kNm/m}$

For sagging:

$$M_x^* = M_x + |M_{xy}| = 7000 + 9000 = 16000\ \text{kNm/m};$$

$$M_y^* = M_y + |M_{xy}| = 24000 + 9000 = 33000\,\text{kNm/m}$$

For hogging:

$$M_x^* = M_x - |M_{xy}| = 7000 - 9000 = -2000\ \text{kNm/m} < 0;$$

$$M_y^* = M_y - |M_{xy}| = 24000 - 9000 = 15000\,\text{kNm/m} > 0;\ M_y^* = 0,$$

$$\text{So } M_x^* = M_x - \left|\frac{M_{xy}^2}{M_y}\right| = 7000 - \left|\frac{9000^2}{24000}\right| = 3625\ \text{kNm/m} > 0$$

So for sagging, $M_x^* = 16000\,\text{kNm/m}$; $M_y^* = 33000\,\text{kNm/m}$

for hogging, $M_x^* = 0\,\text{Nm/m}$; $M_y^* = 0\,\text{kNm/m}$

Point B: $M_x = 7000\,\text{kNm/m}$; $M_y = -24000\,\text{kNm/m}$; $M_{xy} = 9000\,\text{kNm/m}$

For sagging:

$$M_x^* = M_x + |M_{xy}| = 7000 + 9000 = 16000\,\text{kNm/m};$$

$$M_y^* = M_y + |M_{xy}| = -24000 + 9000 = -15000\,\text{kNm/m} < 0;$$

$$\text{So } M_y^* = 0 \text{ and } M_x^* = M_x + \left|\frac{M_{xy}^2}{M_y}\right| = 7000 + \left|\frac{9000^2}{24000}\right| = 10375$$

For hogging:

$$M_x^* = M_x - \left|M_{xy}\right| = 7000 - 9000 = -2000 \text{ kNm/m} < 0;$$

$$M_y^* = M_y - \left|M_{xy}\right| = -24000 - 9000 = -33000 \text{ kNm/m} > 0;\ M_y^* = 0,$$

$$\text{So } M_x^* = M_x - \left|\frac{M_{xy}^2}{M_y}\right| = 7000 - \left|\frac{9000^2}{24000}\right| = 3625 \text{ kNm/m} > 0$$

So for sagging, $M_x^* = 10375\,\text{kNm/m};\ M_y^* = 0\,\text{kNm/m}$

for hogging, $M_x^* = -2000\,\text{Nm/m};\ M_y^* = -33000\ \text{kNm/m}.$

The out-of-plane shear also needs be checked and add shear reinforcements if required. For analysis, shears Q_{xz} and Q_{yz} are formulated as indicated in Figure 4.14 which are the shear force (per unit width) on the faces parallel to the y and x axes, respectively. However, when the rectangular element in Figure 4.14 shrinks to a point, the shears Q_{xz} and Q_{yz} are the shear stresses at the point. Therefore, to find the maximum shear at the point, consider the triangular element having Q_{xz} and Q_{yz} as shown in Figure 4.17, to balance it can easily be shown that at $\theta = tan^{-1}\left(\frac{Q_{yz}}{Q_{xz}}\right)$, the vertical shear stress reaches its maximum value at

$$Q_{maxz} = \sqrt{Q_{xz}^2 + Q_{yz}^2} \tag{4.36}$$

Therefore, it can be concluded that in fact the point is having a shear stress of Q_{maxz} tending to cause shear failure at an orientation of $\theta = tan^{-1}\left(\frac{Q_{yz}}{Q_{xz}}\right)$ on plan.

The plate structure can accordingly be checked for the shear Q_{maxz} by the design code being used by the designer.

By the finite element analysis on the raft footing being simulated as an assembly of plate elements, the results of M_x, M_y , M_{xy}, Q_{xz}, Q_{yz} are determined which are usually outputted for all the nodes where the elements join together. The design of the raft footing can therefore be carried out in accordance with the approaches described above.

The finite element method based on plate is now a very popular method for the analysis of raft footings. To facilitate applications, the commercial software can perform "automeshing" of the footing and some have incorporated the Wood-Armer equations for direct

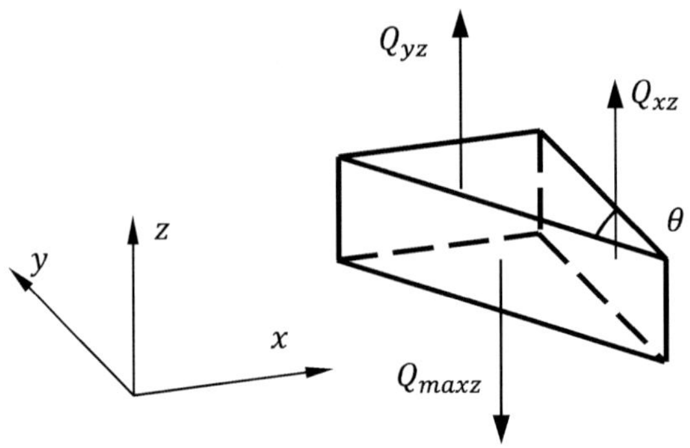

Figure 4.17 To find the maximum shear at a point in the plate bending model.

structural design. The determination of Q_{maxz} is also an easy job using this software. Stress check and subsequently reinforcement design can also be carried out readily.

It should be noted that for most of the raft foundation or pile cap, the thickness is not negligible, hence the use of the thin plate model as given above can only be viewed as an approximation of the thick plate theory. The thick plate theory is much more complicated in nature, and there are many problems for the present thick plate programs, which have been well covered by Cheng and Law (2008a).

4.5 REINFORCEMENT DESIGN FOR A THREE-DIMENSIONAL STRESS FIELD

Most of the reinforcement design method is based on the bending and shear force. For some specially shaped massive structures, it is difficult to define the meaning of the bending moment or shear force. Some construction works are also difficult to be modelled by a combination of solid and bending elements, hence some analyses are based on purely solid elements (a diaphragm wall deep excavation and a jacked tunnel by Cheng using more than 1 million solid elements). The bending and shear force can be recovered from the stresses for the reinforcement design. Alternatively, reinforcement design can be based on the stresses directly from the finite element analysis. Shape functions in the finite element method can be formed for three-dimensional "solids" (or "bricks") for analysis. Therefore, for a raft footing with comparable dimensions in its thickness and plan dimensions, its analysis as an assembly of brick elements should be able to capture the structural behaviour under applied loads more realistically. A modelling of a raft footing as an assembly of brick elements is shown in Figure 4.18.

However, there are six components of stresses obtained after analysis which generally comprise three direct stresses ($\sigma_x, \sigma_y, \sigma_z$) and three shear stresses ($\tau_{xy}, \tau_{yz}, \tau_{xz}$) along three mutually perpendicular axes as shown in Figure 4.19 for an element within a structure.

The structural design in accordance with the six components of stresses for a reinforced concrete footing analysed as an assembly of brick elements is, however, quite complicated.

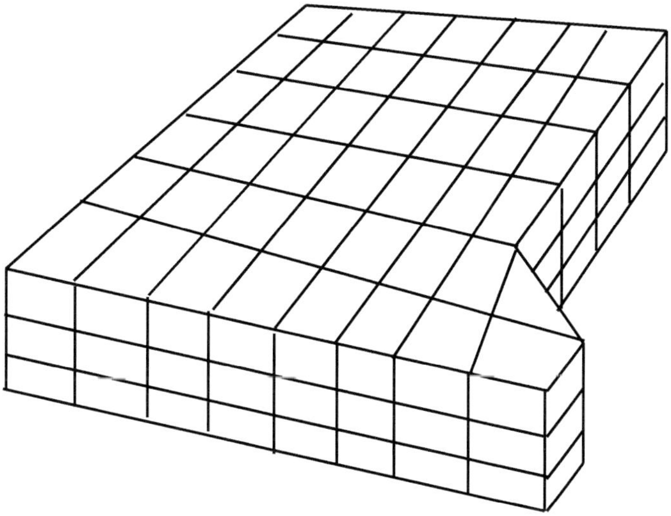

Figure 4.18 A raft footing being meshed to an assembly of bricks.

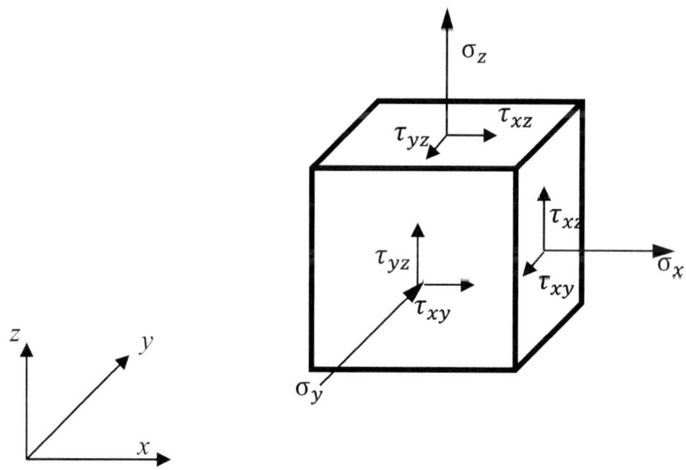

Figure 4.19 Three-dimensional stress.

It may be simpler to explain by starting with a two-dimensional problem having two direct stresses and a shear stress as $\sigma_x, \sigma_y, \tau_{xy}$.

Consider an element within a concrete structure having a direct stress up to the design concrete strength of f_c. The element is at the point of failure. Its stress state may be represented by points A and A' on a Mohr circle, as shown in Figure 4.20 (with negative direct stress for compression and positive for tension):

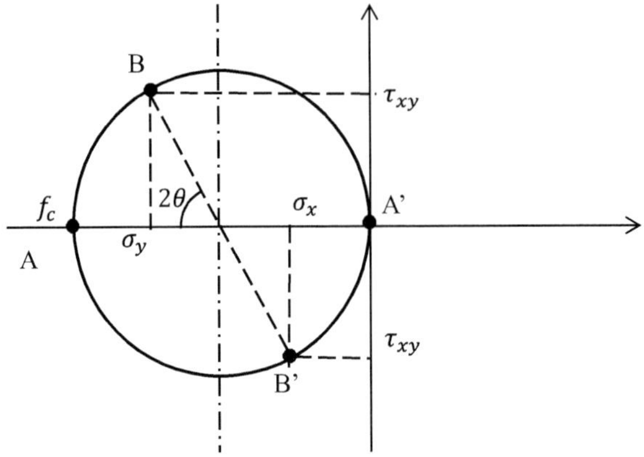

Figure 4.20 State of stresses in a two-dimensional stress field.

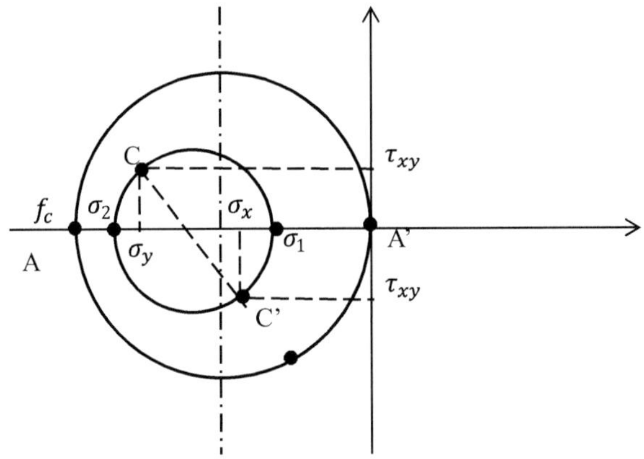

Figure 4.21 State of stresses in a two-dimensional stress field – inside strength circle.

The stress state represented by AA' is the "principal stresses" which are located at the opposite points of the diameter of the Mohr circle and shear stresses are zero. Nevertheless, it can be shown easily that for a plane at an angle θ to the original applied stress changes (represented by BB') represented by $\sigma_x, \sigma_y, \tau_{xy}$. AA' and BB' are created by the same stress state, only different orientations. Any stress states lying on a diameter of the same circle are the limiting failure state. Therefore, we may term the circle as the "strength circle".

Nevertheless, if the applied stress can be represented by a stress state CC' as in Figure 4.21, which lies completely within the concrete strength circle, its principal stresses are σ_1 and σ_2 less than f_c and greater than 0 (no tension), then no reinforcement is required.

However, for a stress state DD' outside the strength circle as shown in Figure 4.22, where the direct stresses are in tension, the concrete will obviously fail. If however, reinforcements are added that can take up the stresses represented by DE and D'E', the net concrete stress will then be represented by the Mohr circle EAE', and a safe design is achieved. It can easily be seen that the least tensile stresses required to be taken up by reinforcements are

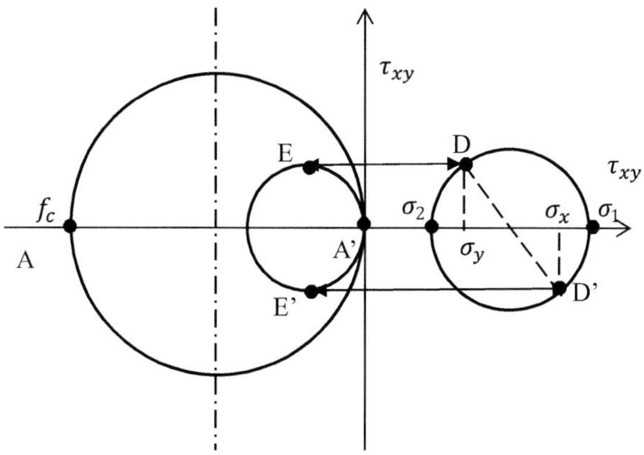

Figure 4.22 State of stresses in a two-dimensional stress field – outside strength circle.

$\sigma_x + |\tau_{xy}|$ and $\sigma_y + |\tau_{xy}|$, which are identical with that arrived at by Clark (1976), the derivation of which is based on equilibrium of forces.

Other scenarios of stresses combination, i.e. σ_x tensile and σ_y compressive, etc. can be similarly solved for required reinforcements and the results are identical to those of Clark (1976).

The design for three-dimensional stresses follows similar principles but more complicated. Law et al. (2007) derived an approach based on Clark's principle which involves seeking optimal solutions in solving 15 unknowns from 12 equations.

$$\sigma_x - \rho_x f_y = \sigma_1 l_x^2 + \sigma_2 m_x^2 + \sigma_3 n_x^2 \tag{4.37}$$

$$\sigma_y - \rho_y f_y = \sigma_1 l_y{}^2 + \sigma_2 m_y{}^2 + \sigma_3 n_y{}^2 \tag{4.38}$$

$$\sigma_y - \rho_y f_y = \sigma_1 l_y^2 + \sigma_2 m_y^2 + \sigma_3 n_y^2 \tag{4.39}$$

$$\tau_{xy} = \sigma_1 l_x l_y + \sigma_1 m_x m_y + \sigma_1 n_x n_y \tag{4.40}$$

$$\tau_{xz} = \sigma_1 l_x l_z + \sigma_1 m_x m_z + \sigma_1 n_x n_z \tag{4.41}$$

$$\tau_{yz} = \sigma_1 l_y l_z + \sigma_1 m_y m_z + \sigma_1 n_y n_z \tag{4.42}$$

$$l_x^2 + m_x^2 + n_x^2 = 1 \tag{4.43}$$

$$l_y{}^2 + m_y{}^2 + n_y{}^2 = 1 \tag{4.44}$$

$$l_z{}^2 + m_z{}^2 + n_z{}^2 = 1 \tag{4.45}$$

$$l_x l_y + m_x m_y + n_x n_y = 0 \quad (4.46)$$

$$l_x l_z + m_x m_z + n_x n_z = 0 \quad (4.47)$$

$$l_y l_z + m_y m_z + n_y n_z = 0 \quad (4.48)$$

From the above equations, ρ_x, ρ_y, and ρ_y are the required reinforcements and f_y is the design strength of reinforcement. σ_x, σ_y, σ_z, τ_{xy}, τ_{xz}, τ_{yz} are the applied direct stresses and shear stresses. σ_1, σ_2, and σ_3 are "principal stresses" under the action of the reinforcements but ignoring tensile strength of concrete. They will only be equal to the "principal stresses" calculated based on σ_x, σ_y, σ_z, τ_{xy}, τ_{xz}, τ_{yz} only in the absence of reinforcements and no tension. l_x, m_x, n_x etc. are the direction cosines of the plane where σ_x, σ_y, σ_z act with respect to the directions of σ_1, σ_2, σ_3. Figure 4.23 defines the symbols. In the figure, the UVW axes are the "principal planes" with respect to XYZ axes.

As there are more unknowns than equations, many solutions do exist. A systematic search process has been established by Law et al. to search for the optimal solution. Nevertheless, a simpler approach is that by Foster et al. (2003) which is briefly described in the following paragraphs.

The six components of stress σ_x, σ_y, σ_z, τ_{xy}, τ_{xz}, τ_{yz} can be completely stated by a tensor as

$$\sigma_{ij} = \begin{bmatrix} \sigma_x & \tau_{xy} & \tau_{xz} \\ \tau_{xy} & \sigma_y & \tau_{yz} \\ \tau_{xz} & \tau_{yz} & \sigma_z \end{bmatrix} \quad (4.49)$$

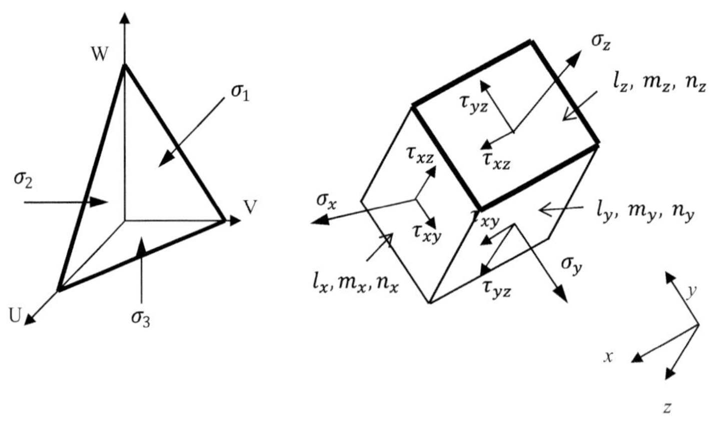

Figure 4.23 Definition of symbols and equilibrium conditions for the approach by Law et al.

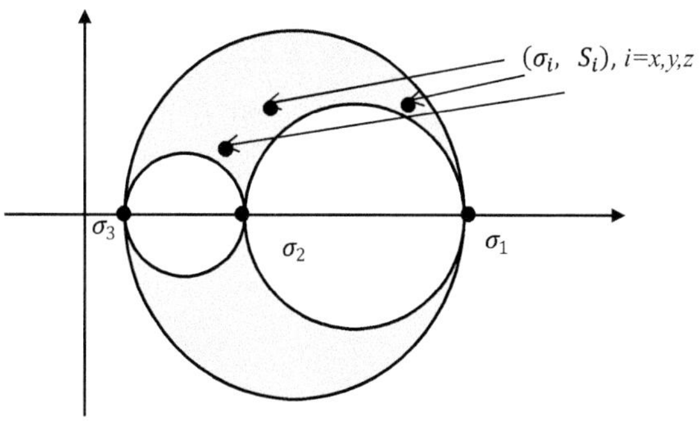

Figure 4.24 Three-dimensional stress state represented by three Mohr circles.

Similarly to the two-dimensional stress problem, there exists an orientation of xyz such that the shear stresses are zero and the direct stresses become the principal stresses. It is well known that the principal stresses (three numbers on three mutually perpendicular planes), $\sigma_1, \sigma_2,\ \sigma_3$, are equal to the eigenvalues of the tensor in Eq. (4.49). With $\sigma_1 > \sigma_2 > \sigma_3$, three Mohr circles are plotted as shown in Figure 4.24.

If the shear stresses on the xy plane comprising τ_{xz} and τ_{yz} are vector summed together to form S_z, i.e., $S_z = \sqrt{\tau_{xz}^{\ 2} + \tau_{yz}^{\ 2}}$; and similarly for the other two planes,

$$S_x = \sqrt{\tau_{xy}^{\ 2} + \tau_{xz}^{\ 2}}; S_y = \sqrt{\tau_{xy}^{\ 2} + \tau_{yz}^{\ 2}}; S_z = \sqrt{\tau_{xz}^{\ 2} + \tau_{yz}^{\ 2}} \tag{4.50}$$

The stress states of $(\sigma_x,\ S_x)$, $(\sigma_y,\ S_y)$, and $(\sigma_z,\ S_z)$, instead of lying on the Mohr circles, now exist on the shaded region outside the smaller Mohr circles.

Similarly, a Mohr circle can be constructed for the concrete. Depending on the applied stress, the concrete compressive stresses may not need to reach the maximum value of f_c. It will be at the designer's discretion to predetermine the concrete principal stresses and reinforcements added to bring the $(\sigma_i,\ S_i)$ states to within the maximum value of f_c.

In addition, the following equations relating to the six components of stresses to the principal stresses are used for solution:

$$\sigma_x + \sigma_y + \sigma_z = \sigma_1 + \sigma_2 + \sigma_3 \tag{4.51}$$

$$\sigma_x\sigma_y + \sigma_x\sigma_z + \sigma_y\sigma_z - \left(\tau_{xy}^2 + \tau_{xz}^2 + \tau_{yz}^2\right) = \sigma_1\sigma_2 + \sigma_1\sigma_3 + \sigma_2\sigma_3 \tag{4.52}$$

$$\sigma_x\sigma_y\sigma_z + 2\tau_{xy}\tau_{xz}\tau_{yz} - \left(\sigma_x\tau_{yz}^2 + \sigma_y\tau_{xz}^2 + \sigma_z\tau_{yz}^2\right) = \sigma_1\sigma_2\sigma_3 \tag{4.53}$$

These three equations can relate to the applied stresses and also to the concrete stresses.

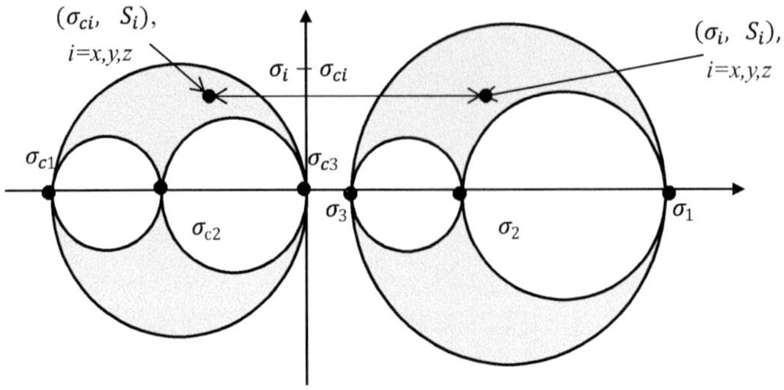

Figure 4.25 Three-dimensional stress state represented by three Mohr circles with tension.

Therefore, the design is based on choosing a scenario for the state of concrete, say $\sigma_{c3} = 0$ or $\sigma_{c3} = f_c$ to construct the concrete Mohr circles. Then Eqs. (4.36)–(4.38) are applied to calculate $\sigma_{cx}, \sigma_{cy}, \sigma_{cz}$ with the same S_x, S_y, S_z. Then the required reinforcement will be used to take up the stresses $\sigma_x - \sigma_{cx}, \sigma_y - \sigma_{cy}, \sigma_z - \sigma_{cz}$ as illustrated in Figure 4.25. The approach is similar to that illustrated for two-dimensional stress in Figure 4.22.

Although design to three-dimensional stress analysed by the finite element method is most realistic and is now within the capabilities of computer hardware and software, it is not yet very popular as many designers are not yet familiar with design to three-dimensional stress. In addition, while the design by plate model will result in reinforcements at the top and bottom faces of the raft footing only, design to three-dimensional stresses may result in reinforcements also at mid-depths of the raft footing if conventional elastic analysis is assumed in the finite element formulation, which may create reinforcement bars fixing difficulties on site. Nevertheless, plastic analysis where no tensile stress is to be taken can be assumed at locations where the designer does not wish to add reinforcements. Therefore, the analysis will allow the concrete to crack at these locations and the excess stresses will be taken up by concrete and reinforcement bars at other locations.

4.6 ANALYSIS BY THE STRUT-AND-TIE MODEL

The plate bending theory as described in the previous section is based on a basic assumption of linear strain and stress profiles across any cross section of the plate, which is more commonly known as "plane remains plane". Truly speaking, this assumption is only valid for "thin" sections, that is, when the thickness of the plate is small compared with the lateral dimensions. However, for a raft footing of considerable depth, the actual structural behaviour may deviate significantly from the "plane remains plane" phenomenon. Strut-and-tie models with strut-and-tie actions within the plate may be more appropriate. However, while the strut-and-tie model to be assumed in a pile cap can be quite clear, that in a raft footing is more ambiguous, as explained in Figure 4.26.

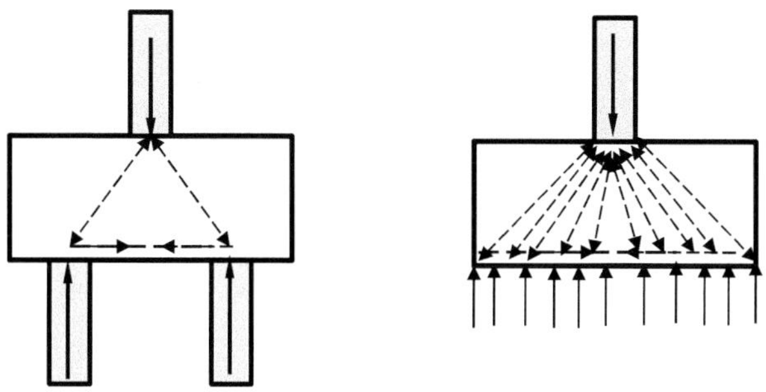

Figure 4.26 Strut-and-tie actions in pile cap and footing.

A strut-and-tie model (STM) is formed by predetermination of the stress paths by structural sense and analysed accordingly. Though determination may be simple or unique for simple cases, there can be many possible models for complicated cases. So the method is not popular for irregular footings with irregularly spaced columns and walls. STM is based on the lower-bound theorem of limit analysis which is a safe solution. The method is based on the equilibrium approach and failure criteria. In this model, the complex stress distribution is idealized as a truss carrying the imposed loading through the structure to its supports. Such idealization has also been verified by Cheng and Law (2006) using 3D finite element analysis of a pile cap.

The STM is an extension of the truss model, and the major difference between the two methods is that STM is a set of forces in equilibrium which do not form a stable truss system. STM is an idealization of the stress resultants derived from the flow of forces within concrete. In an STM model, a strut represents a concrete stress field with prevailing compression in the direction of the strut. On the other hand, a tie typically represents tension reinforcement. Figure 4.27A illustrates this phenomenon via a pile cap supported on two large diameter bored piles (LDBP) with 0.5m radius.

Both the bending theory and truss analogy are commonly used for the servicing as well as ultimate design of pile caps. The use of bending theory however prevails, as most engineers are familiar with bending theory. Furthermore, if the pile layout is irregular or the super-structure loadings are complicated, it is also not easy to define a proper strut and tie model. In this chapter, the authors investigate the applicability of the bending theory and the "truss" concept in pile cap analysis and design by performing three-dimensional stress analysis. It is not surprising to find that, under servicing conditions where the stress state is generally within the elastic limit and the local effect from the piles is significant, both the concepts of pure bending and "truss action" may not be good representations of the actual stress distribution within the cap structure. Consider the 2 m thick pile cap shown in Figure 4.28.

Three-dimensional contour plots of the vertical stresses at various levels are given in Figure 4.29. From the results at $z = 0$ and $z = 2.0$ m, it can be clearly seen that a strong strut model exists in this problem. A STM model can be established in Figure 4.30, and this model can avoid the problem of thin and thick assumption in the classical FEM approach.

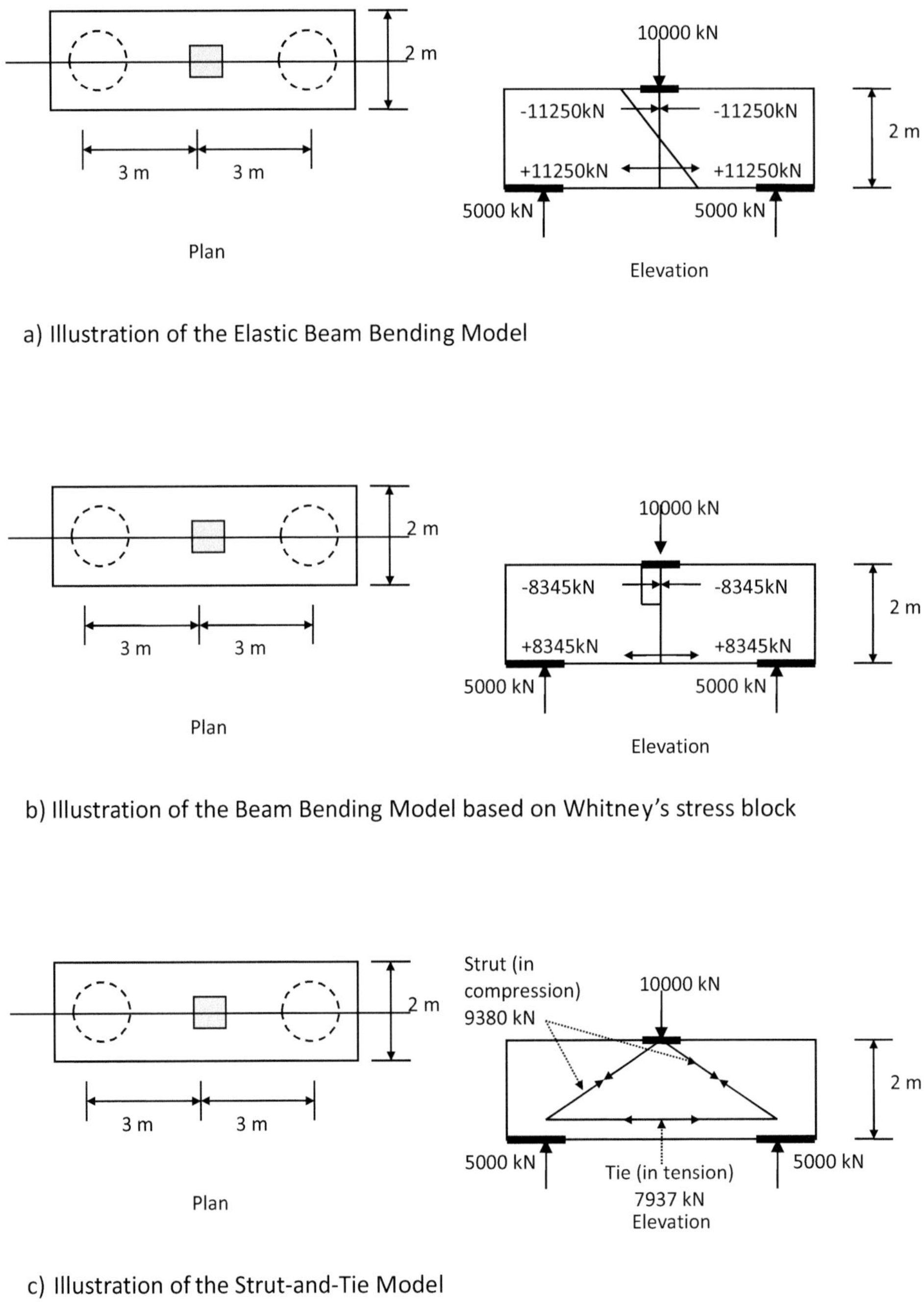

Figure 4.27 Different theories for the analysis of pile cap/pile raft.

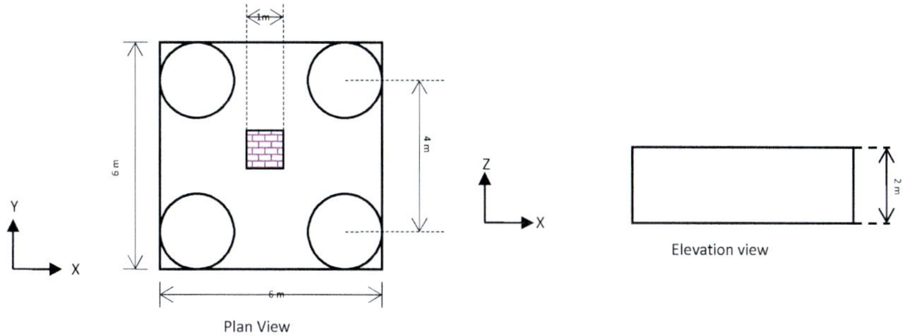

Figure 4.28 A pile cap supported by four large-diameter bore piles.

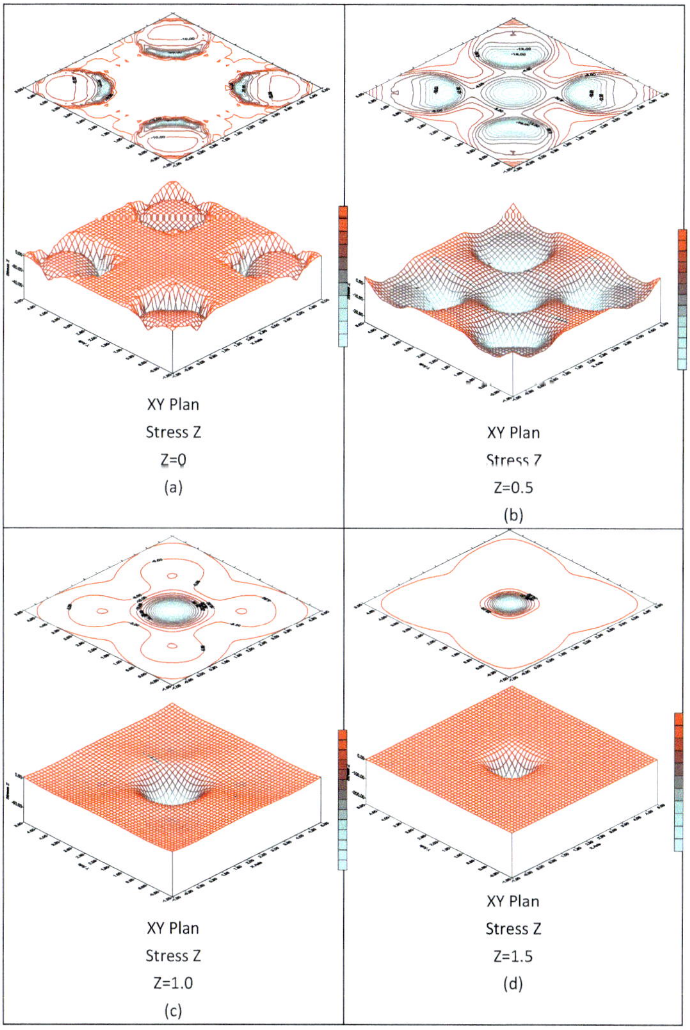

Figure 4.29 Compressive stress contours of the LDBP model.

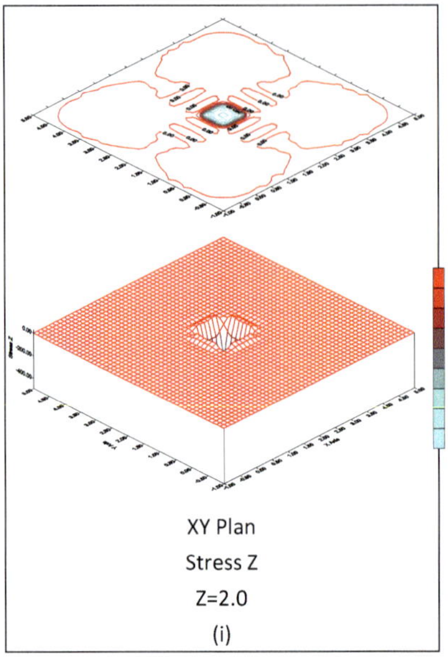

Figure 4.29 (Continued)

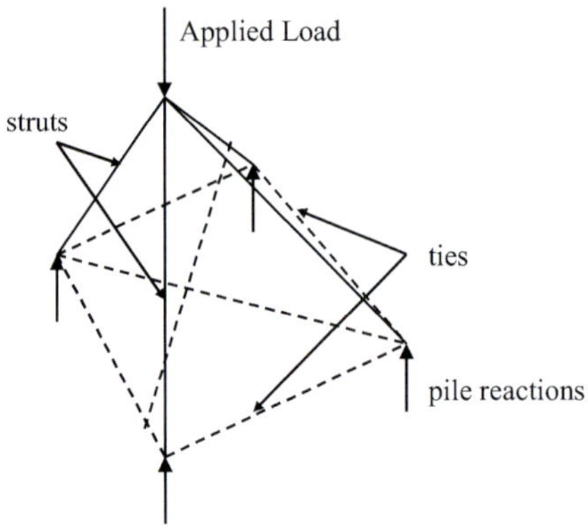

Figure 4.30 STM model for the problem in Figure 4.28.

4.7 CONTINUUM SUBGRADE MODEL

In this method, the structure of the raft footing is simulated by a grillage or plate model which can be quite accurate and realistic. The support is then simulated as a series of independent elastic springs, the stiffness of which can be based on the stiffness of the ground. The Winkler's spring model suffers from the limitation that for a beam with a uniformly distributed load using the Winkler's model, the moment and shear on the beam will tend to zero if the number of elements keeps increasing! Such a result is obviously not correct, but is a direct consequence of the individual spring assumption in the Winkler's model. Soil or rock is a continuum, and discontinuity may exist in the founding medium. The use of a simple Winkler's spring for a general case on a nonhomogeneous founding medium is not realistic, and there is great difficulty in determining the suitable Winkler spring constant for a general case.

The estimation of the ground stiffness can be in accordance with the continuum theory based on the integration form of the Boussinesq or Mindlin's equations involving parameters of the supporting soil or rock which are the Young's modulus and the Poisson's ratio. As such, a more realistic estimation can be arrived at as compared with the classical approach. A worked example can demonstrate the estimation, as shown in the worked Example 3.2.

The total average settlement is 36.98 mm under a pressure of 100 kPa. Therefore, the settlement stiffness is 100/0.03698 = 2704 kPa/m. Physically, this means that the subgrade will settle for 1 m under a pressure of 2704 kPa. This "Winkler's spring" can generally be inputted into the mathematical model simulating a raft foundation for structural analysis.

However, there is an obvious discrepancy in employing the continuum theory for determination of Winkler's spring as per the approach described above. This is because even under a uniformly distributed load on a homogeneous subgrade, the settlement is non-uniform, with the greatest value occurring in the middle of the load patch, as demonstrated in Figure 4.31. However, a uniformly distributed load acting on soil simulated by constant Winkler's springs will result in uniform settlement, as shown in Figure 4.31.

Despite the behaviour demonstrated in Figure 4.31(a), which may be described as being unrealistic as compared with the continuum theory, the analysis based on Winkler's spring by the computer method still marks an advancement in structural analysis as compared with the classical method. Apart from giving more realistic settlements, more realistic stress distribution beneath the footing can also result from higher stresses beneath heavy loads from walls or columns above, which can also be obtained with the use of "flexible footing".

Nevertheless, some more tedious simulations of support conditions can be carried out to account for the interaction of the support points. The structure is first idealized as being supported by a large number of support points simulating the continuum supports and the loads acting on the structure are also simulated as a series of point loads acting on the structure. Then, instead of assuming the settlement of any point support being only due to the load directly acting on it as a "Winkler's spring", the settlement of a point i is not only due to the point load P_i acting directly on it, but also due to loads acting on other points. The effect is first taken as elastic, by which the settlement is proportional to the

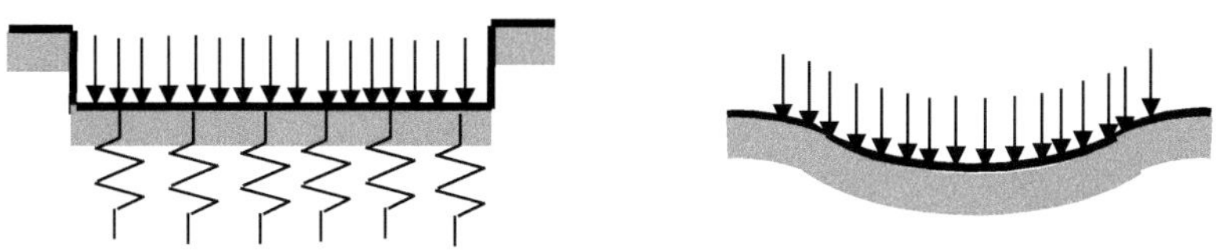

(a) Constant settlement under UDL on Winkler's spring

(b) Varying settlement even under UDL by the Continuum Theory

Figure 4.31 Different settlement profile as per Winkler's spring assumption and continuum theory.

load creating it. By means of the Boussinesq or Mindlin's equations, the settlement of a point i due to a point load P_j at some distance apart can be expressed as

$$\delta_{ij} = I_{ij} P_j \tag{4.54}$$

where I_{ij} is the interaction coefficient between the points i and j that can be calculated based on the Boussinesq or Mindlin's equations. From Boussinesq equation:

$$\mathrm{I}_{ij} = \frac{1-\mu^2}{\pi E_s r} \tag{4.55}$$

where r is the distance between points i and j.

The total settlement of point i is the summation of all the δ_{ij}. That is,

$$\delta_i = \Sigma\,\delta_{ij} = \Sigma\, I_{ij} P_j \tag{4.56}$$

where j. can be equal to or different from i.

The use of Eq. (4.56) will result in "interaction" of the supports, and is considered as the flexibility of the subgrade. The matrix of the flexibility coefficient [f] can then be established, which will give the equivalent stiffness matrix of the subgrade [k_s] by matrix inversion, and $[f]^{-1} = [k_s]$. [k_s] is a full matrix which is superimposed to the stiffness of the beam or plate to give the overall stiffness matrix of the whole system, which will yield the displacements a internal forces under the prescribed loading. In the superposition of [k_s] with the foundation stiffness, it must be noted that [k_s] affects only the vertical displacement component during the assemblage of the global stiffness matrix. It should also be noted that [k_s] is already the basic boundary condition to the problem. A structural designer will find that in formulating the stiffness matrix for a structure for analysis, the Winkler's spring approach only modifies the diagonal entries of the stiffness matrix of the raft footing structure. For a large problem with more than 100,000 nodes, the computer memory and time required for such coupled analysis can be demanding, and this is possibly the reason why very few computer programs implement this option for the foundation analysis. The PLATE program however includes such an option in the analysis. In addition, Cheng has added a borehole option so that engineers can define the ground conditions by borehole information and the soil properties of each layer of soil. Three-dimensional finite element analyses will then be automatically performed to determine the flexibility matrix [f] and hence [k_s]. It should also be noted that the influence coefficients for points far away from each other are small and negligible. To enhance the efficiency in computation without critical loss of accuracy, PLATE allows users to define a radius of influence so that the influence coefficient will be exactly zero if the distance between two points is greater than this influence zone. This will reduce the global stiffness matrix from a full matrix to a banded matrix.

4.8 COMPUTER MODELLING OF COMPLICATED RAFT FOUNDATIONS

Currently, most engineers rely on the use of a finite element plate program for the analysis and design of raft foundations. In the operation of the commercial program, engineers usually need to choose the option of thin or thick plate analysis. Over the years, Cheng has received many questions from engineers regarding strange results from the computer analysis. After serious review, Cheng and Law (2008a) found that the plate elements available in some famous commercial programs suffer from various limitations, if the mesh is highly irregular and distorted. In view of that, Cheng and Law have developed a pseudo eight-node thin/thick plate element, and the four internal nodes are statically condensed so that it behaves as a quadrilateral element. Such an element possess very high accuracy even for distorted mesh and also retains ease of use for practical problems. Cheng and Law have added many options which are not readily in many commercial software for engineering use. Some of the special features of the PLATE program by Cheng include:

1. Unlike some other commercial programs, the user need not choose between the thick and thin plate options. The plate element by Cheng and Law (2008a) has been verified to give very high accuracy for both thin and thick conditions.
2. No grid is required! Only coordinates are used to define everything.
3. Prescribed fixed or prescribed boundary conditions can be defined.
4. Winkler, piles and elastic half space support conditions can be modelled.
5. Contour lines for the results of an individual load case or load combination, result envelopes which are the maximum and minimum values out of all the load combinations can be viewed and generated. The user can also cut any section along any rotation to view the results.
6. Reinforcement design to Code of Practice for Structural Use of Concrete 2004, Hong Kong. Wood-Armer design for steel reinforcement is also available as an option.
7. Uniform elastic half space ground support or stratified layers of soil ground support defined by boreholes are accepted. This is a unique feature not available in any other existing commercial program.
8. For pile cap on pile group, the interaction matrix is generated automatically. This is a unique feature of PLATE compared with other commercial programs!
9. Line moment is automatically transformed to point load internally.
10. Reinforcement design at inclined axes is possible, a feature not available in other programs!
11. Punching shear is automatically checked in the design.
12. Correct shear reinforcement design, which considers shear enhancement automatically. The design practice by engineers using Fx and Fy is not correct, and Vmax is used in PLATE for shear reinforcement design, with reduced shear reinforcement in general.
13. Improved design strip bending moment, more economical, and avoids some strange mistakes in commercial programs.

For illustration, a 6 m long beam simply supported at the two ends with a uniform distributed load (UDL) is considered. The UDL is 10 kN/m, and the width of the beam is

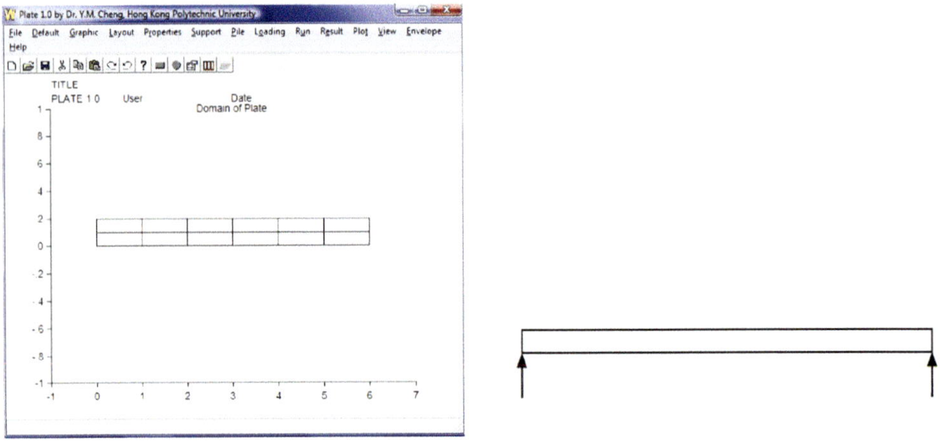

Figure 4.32 A simply supported beam is modelled by plate analysis.

0.2 m. The maximum moment on the beam is $M_{max} = \omega l^2/8 = 45$ kN-m. Using PLATE as shown in Figure 4.32, the UDL on the plate is 10/0.2 = 50 kPa.

With 24 elements, Mx at the centre = 224.23 kN-m/m, equivalent global 1D moment = 224.23 × 0.2 = 44.85 kN-m.

With 48 elements, Mx at the centre = 225, global moment = 225 × 0.2 = 45 kN-m = 1D resulting in $\omega l^2/8$.

The results indicated that if the width of the plate was small enough, the results from a 2-D plate analysis will reduce to the classical 1-D structural analysis.

Another interesting case is a raft foundation design in Hong Kong as shown in Figure 4.33. The point and line loads are given in Figures 4.33a,b, and the base bearing pressure is given in Figure 4.33c.

PLATE has some interesting options not available in other programs, as shown in Figures 4.34 and 4.35. In Figure 4.34, Cheng has implemented the elastic half space option using Eq. (4.55), with an additional option of radius of influence. Beyond this distance, f_{ij} will be taken as 0. This option is useful because it is noted that, for a fine mesh, over-interaction may sometimes occur. Also, from actual observations, the radius of influence of any point load is finite. Furthermore, Cheng has also implemented the borehole option as shown in Figure 4.36, which will be useful to the analysis of footing on complex ground conditions without the use of true 3-D analysis.

In addition to these features, the grillage program as discussed previously has been incorporated into PLATE, and grid members can be added to the plate structure if they exist. A special issue about the definition of a moment along a shear wall should be discussed in this section. Classically, a moment has no size, which is sufficiently good for columns. For the shear wall and core wall which take a large wind moment, the moment is carried by the long shear wall along its length. The length of a shear wall is long and cannot be modelled as a point moment. On the other hand, no program in the world allows a size to be assigned to a moment. To overcome this dilemma, the moment can be transformed to stress along the wall by the formula $\sigma = P/A + Mx/Ix$ for shear wall and $\sigma = P/A + Mx/Ix + My/Iy$

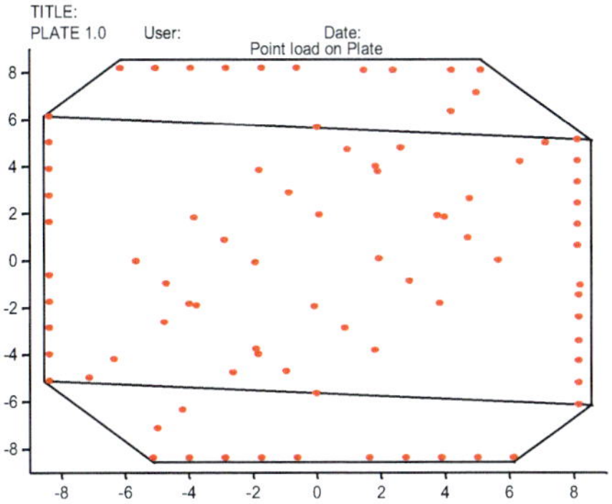

(a) Point loads

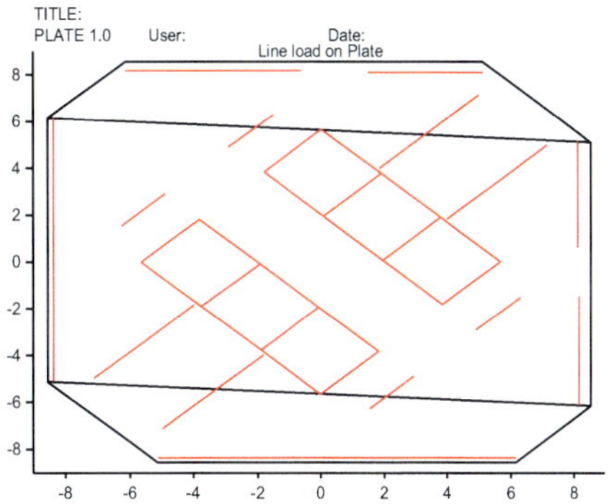

(b) line loads

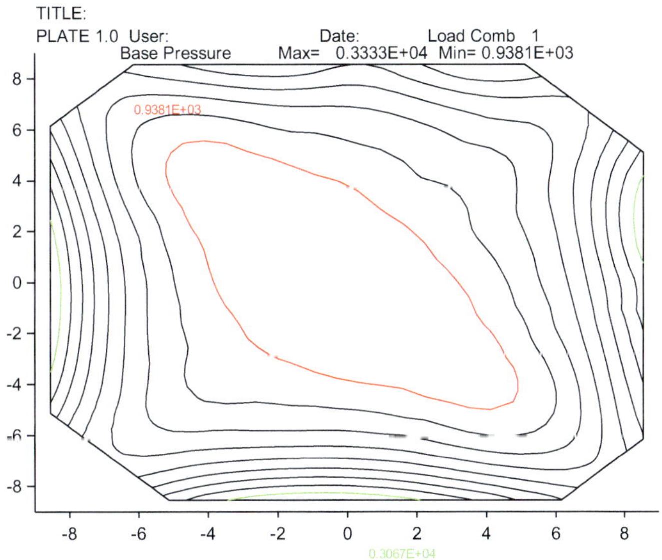

Base bearing pressure on the raft foundation (c)

Figure 4.33 A raft foundation design in Hong Kong using PLATE.

for core wall. A trapezoidal stress distribution will be obtained for the stress on the shear wall, which can then be distributed as point loads along the nodes. Along the side of a plate element where the stress is evaluated, the total area of the loading becomes point loads at the two ends of an edge distributed according to Figure 4.37. Through such an approach, the total vertical load and moment will be preserved. In the limit where there are many point loads in a computer model, the overall effect will then be equivalent to a line moment, and the moment can be given a size. In Figure 4.38, a size can be defined to a line moment, and the transformation of a line moment to point loads will be automatic.

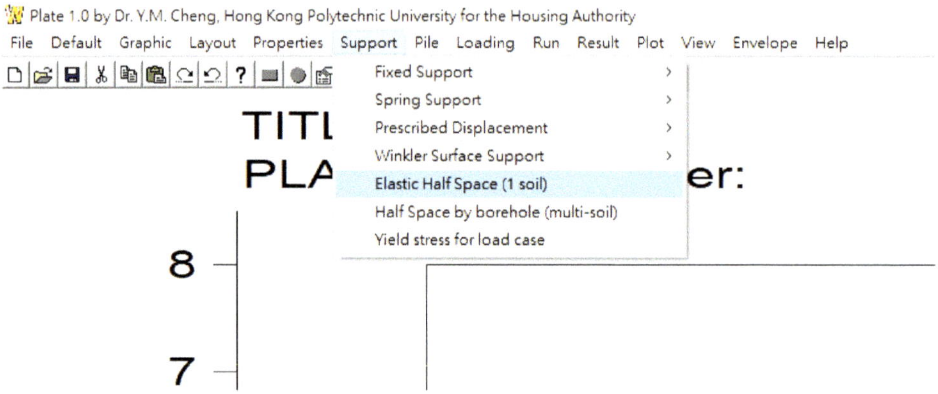

Figure 4.34 Various boundary conditions available in PLATE.

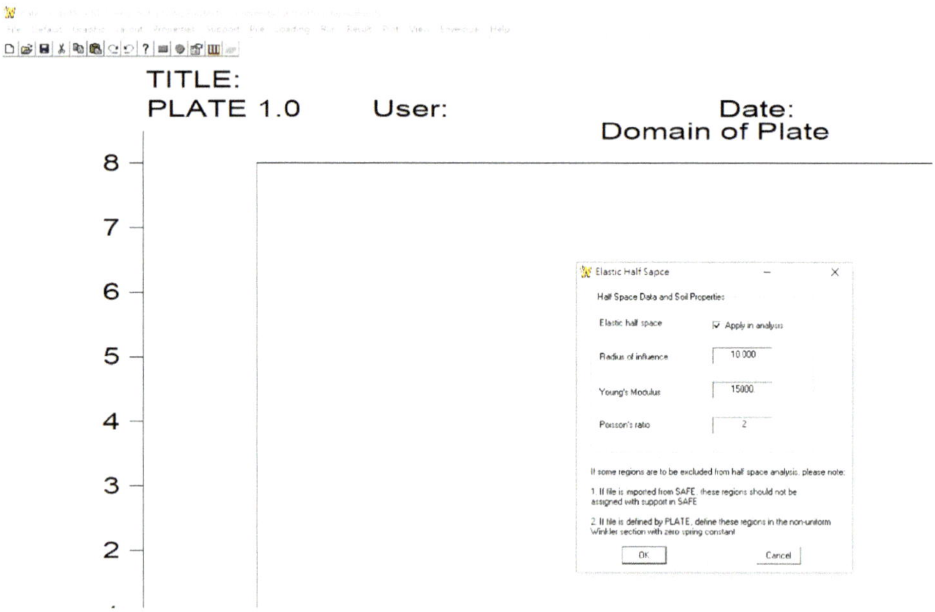

Figure 4.35 Elastic half space option in PLATE.

An example on distribution of line moment is given in Figure 4.39.

Two shear walls are shown in Figure 4.39. Each 4 m long shear wall carries an axial load of 5000 kN and moment of 1000 kN-m about the centre and along the wall. Determine the equivalent point loads for the shear wall, using five points (thickness of wall is 1 unit, exact value not required).

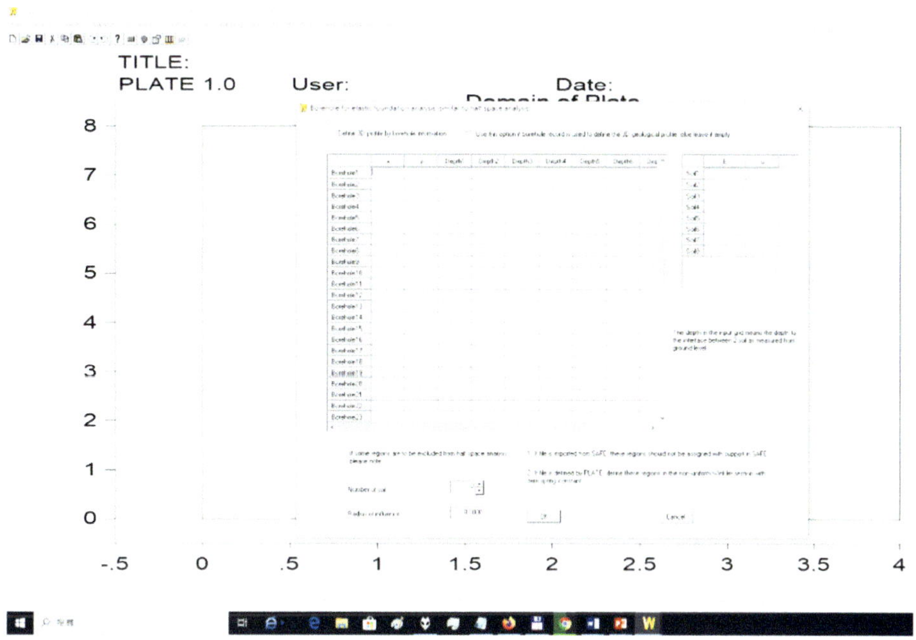

Figure 4.36 Borehole option in PLATE.

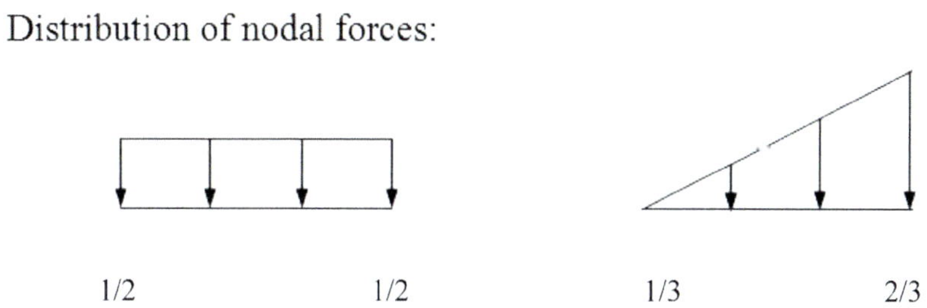

Figure 4.37 Distribution of stresses along a shear wall arising from the moment as point loads.

Solution (Figure 4.40):

Axial stress on shear wall = 5000/4 = 1250 kN/m
Maximum bending stress on shear wall = $1000 \times 2/(4^3/12) = 375$ kN/m
Therefore, maximum/minimum stress on shear wall = 1625/875 kN/m
For span 1–2, the distributed load is 1625 at the left and 1437.5 at the right
This distributed load will be transformed to two point loads at nodes 1 and 2
Equivalent load at node 1 = $1437.5 \times 0.5 + 0.5 \times 187.5 \times 2/3 = 781.25$ kN

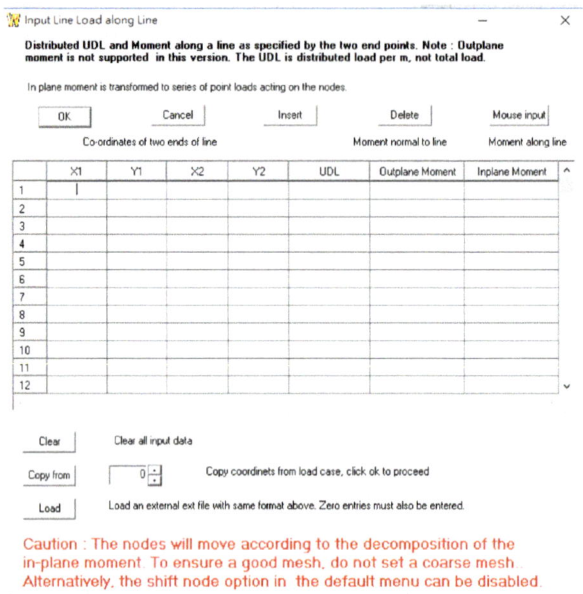

Figure 4.38 A size can be assigned to a moment in the updated PLATE.

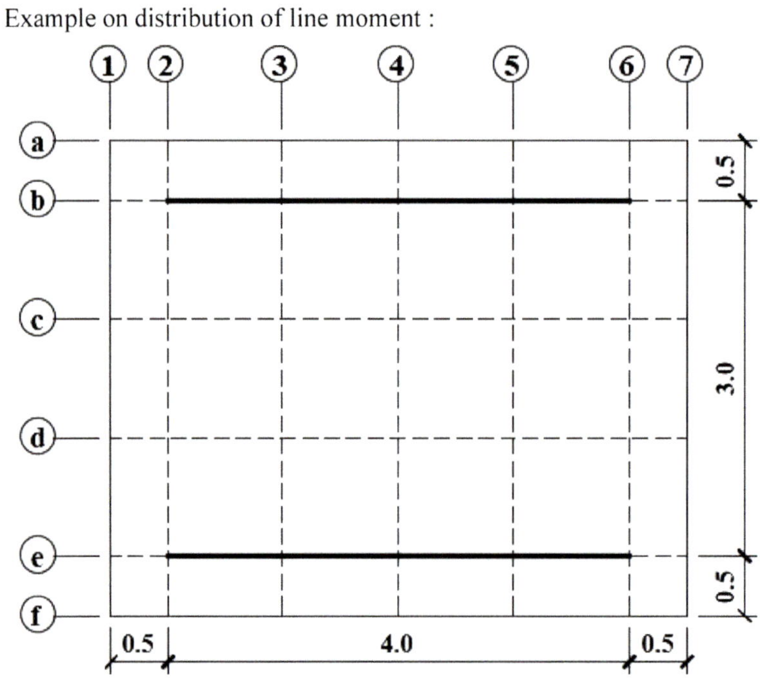

Figure 4.39 A raft foundation with two shear walls.

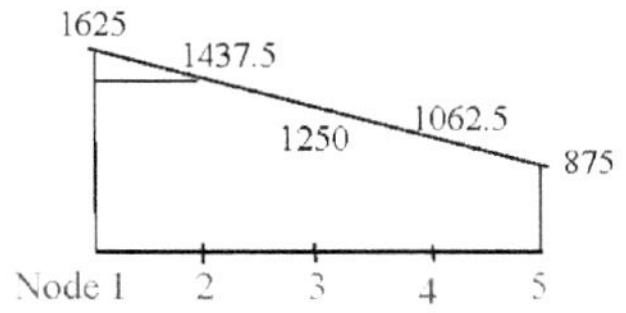

Figure 4.40 Stresses along the shear wall.

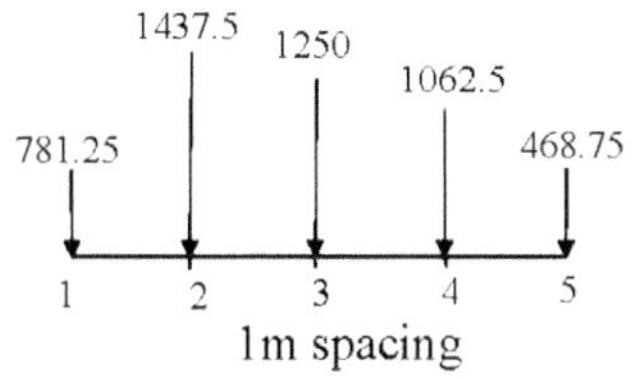

Figure 4.41 Equivalent point loads to represent the shear wall vertical loading and moment.

Equivalent load at node 2 = 1437.5 × 0.5 + 0.5 × 187.5 × 1/3 + 1250 × 0.5 + 0.5 × 187.5 × 2/3 = 1437.5 kN = 1437.5 × 1 (this applies when the spacings of nodes are uniform) (Figure 4.41)

Equivalent load at node 3 = 1250 kN = 1250 × 1
Equivalent load at node 4 = 1062.5 kN
Equivalent load at node 5 = 468.75 kN
Adding up these five point loads = 5000 kN
Take the moment at the centre = 1000 kN-m (781.25 × 2 + 1437.5 – 1062.5 – 468.75 × 2)

For pile cap or raft foundations, most engineers simply apply the loadings from the superstructure and neglect the effect of the superstructure in the foundation analysis. This rationale is based on the assumption of a hinge connection between the columns and pile cap which is incorrect, but this will greatly simplify the work load required to carry out the foundation analysis and design. In the updated PLATE, this limitation can be overcome by two approaches. The addition of grillage elements and nodal springs can partly represent the effect of walls and columns. For a more realistic analysis, the flexibility or stiffness matrix $[K_{ij}]$ of the superstructure can be determined from a structural analysis program. This stiffness matrix can then be coupled into the PLATE analysis program for a more refined analysis. Readers interested in program PLATE can obtain a demonstration copy from Cheng at natureymc@yahoo.com.hk.

Illustration

An 8 m × 0.2 m footing is loaded with the following point loads (Figure 4.42). Determine the bending moment on the footing, if k_s = 15000 kN/m³. E_{conc} = 2 × 107 kPa, Poisson ratio = 0.2, footing thickness = 0.2 m, unit weight of concrete = 24.5 kN/m³.

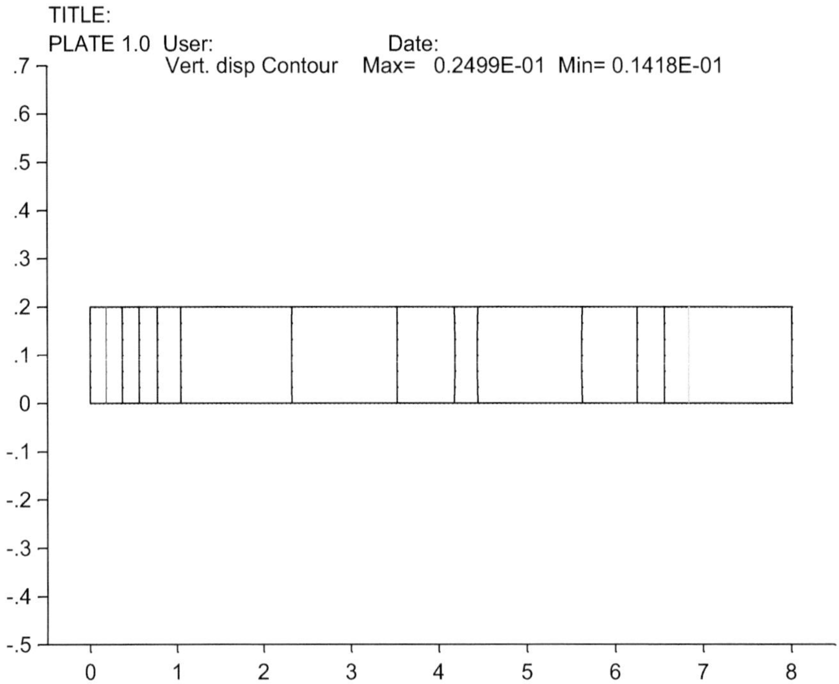

Figure 4.42 Settlement of the footing under four point loads.

X = 1 m, P = 120 kN, Mx = 10 kN-m
X = 3 m, P = 130 kN, Mx = 15 kN-m
X = 5 m, P = 110 kN, Mx = 17 kN-m
X = 7 m, P = 140 kN, Mx = 20 kN-m

The maximum and minimum settlements of the footing are 24.99 mm and 14.18 mm, respectively, using 1127 nodes and 960 elements in the analysis. The bending moment and shear force along the long axis are given in Figures 4.43 and 4.44. From this example, it is clear that there is not a strong need to use 1-D analysis or even the classical rigid design method. It is also interesting to note that if the subgrade modulus is increased by 10 times, only the hogging moment is slightly affected, which is shown in Figure 4.45. If the elastic half space option is used with E_s = 15000 kN/m^2, Poisson ratio = 0.2, the results of the analysis are shown in Figure 4.46. For this problem, the change in the bending moment is not significant. On the other hand, if the loading is a simple uniform distributed loading, the elastic half space option has a great difference with the Winkler spring option, as shown in Figures 4.47 and 4.48. As discussed earlier, the Winkler spring option will give zero bending moment and shear force in this case. The very small values slightly different from 0 in Figure 4.48 represent the small truncation errors during the calculation. If uniform distributed loading is the dominant load, the use of the Winkler spring model is not a good representation of the actual behaviour of the foundation.

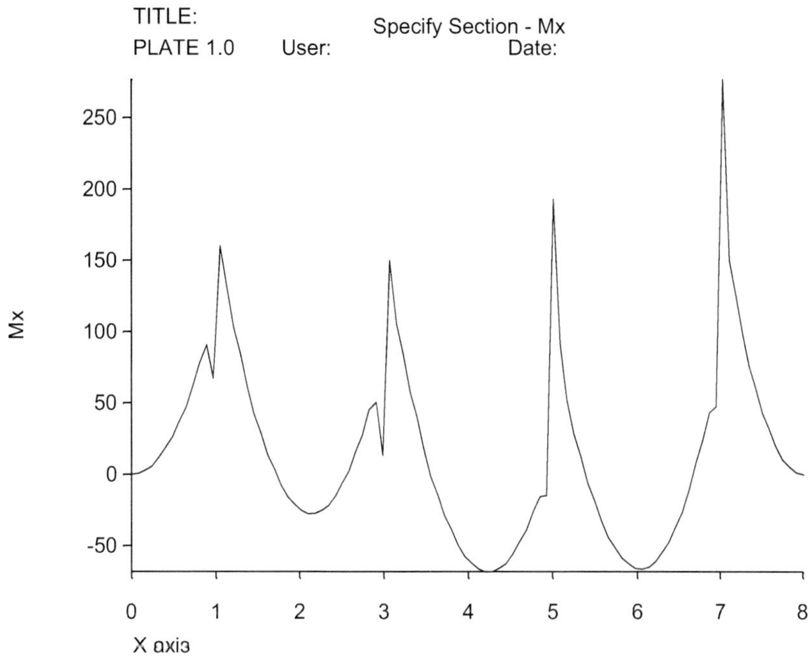

Figure 4.43 Moment along the long axis for the problem in Figure 4.42.

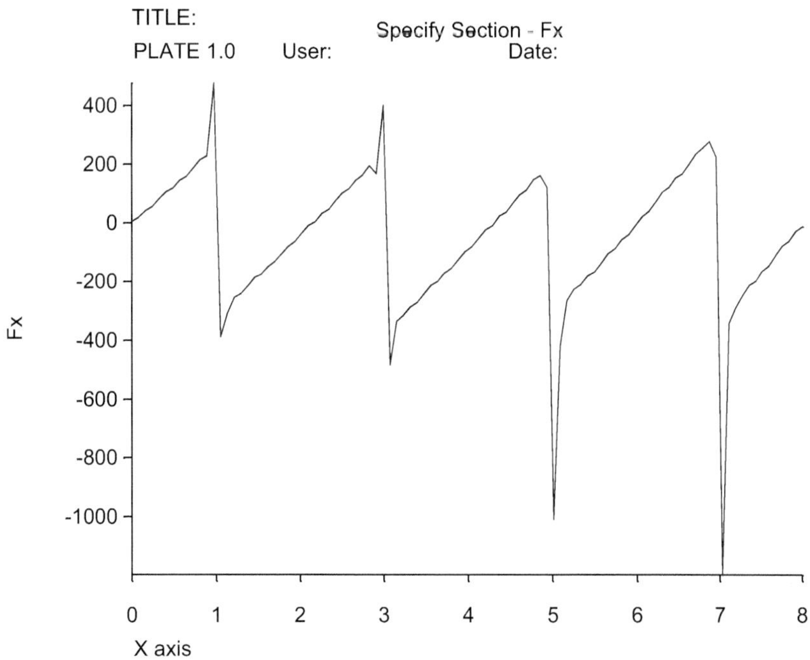

Figure 4.44 Shear force for the problem in Figure 4.42.

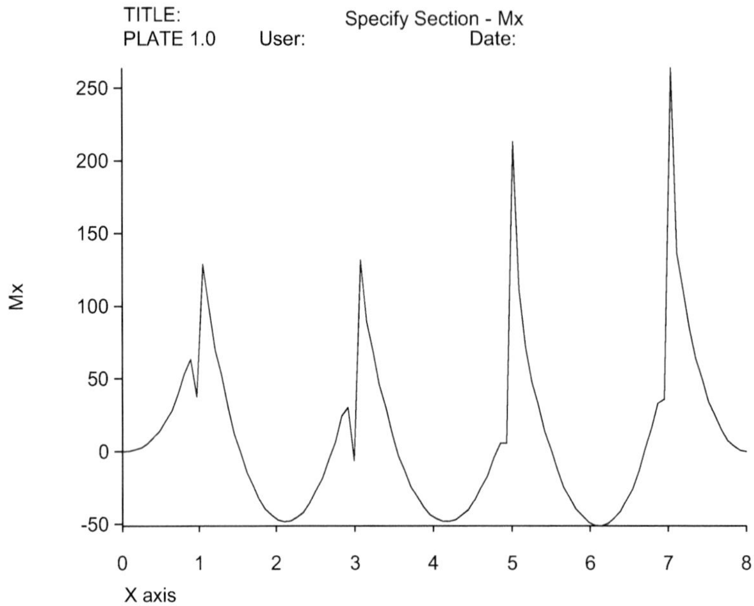

Figure 4.45 Moment along the long axis for the problem in Figure 4.42, k_s increased by 10 times.

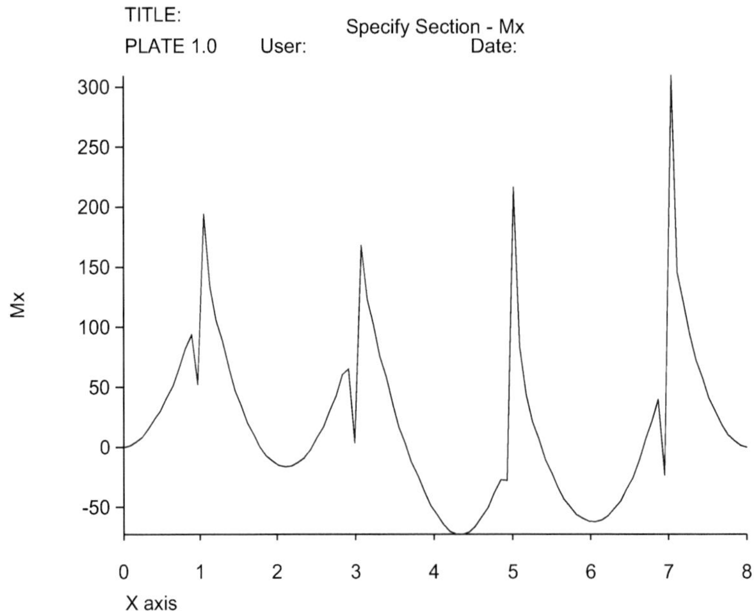

Figure 4.46 Moment along the long axis for the problem in Figure 4.42 using the elastic half space option.

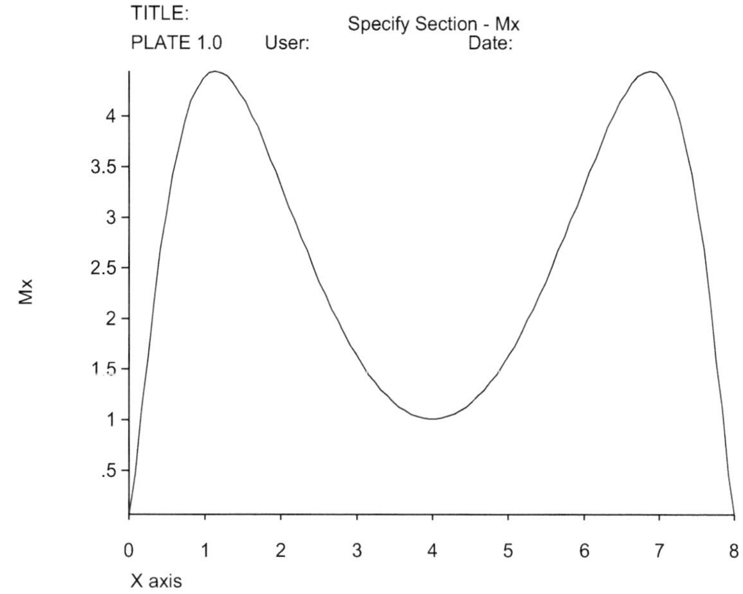

Figure 4.47 Moment along the long axis for the problem in Figure 4.42 using the elastic half space option, 100 kPa uniform distributed load, no concrete self-weight.

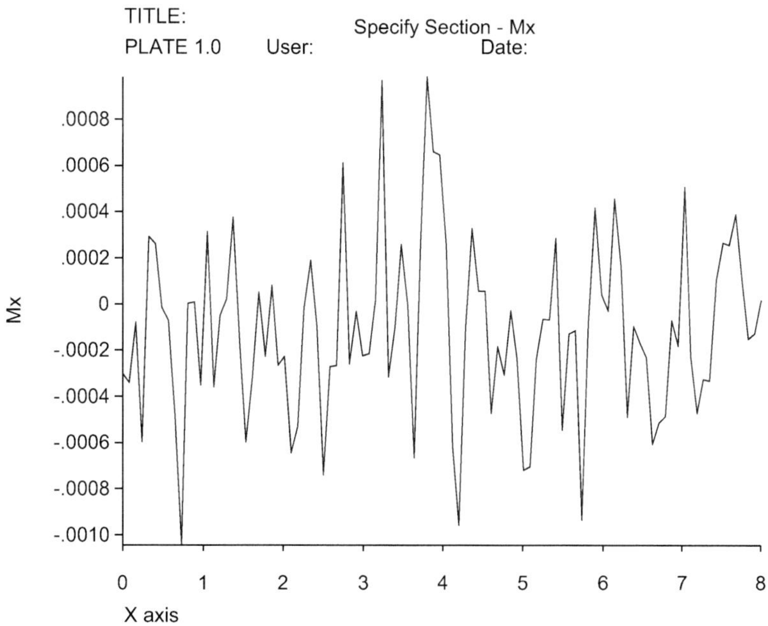

Figure 4.48 Moment along the long axis for the problem in Figure 4.42 using the Winkler spring model, 100 kPa uniform distributed load, no concrete self-weight.

The degrees of freedom at each node are the same for the grillage and plate methods, yet there are major differences in the internal forces. Consider a typical output from the plate analysis. The first entry is the element number, while the second entry is the Gaussian point number. Each element has four (or nine) Gaussian integration points in PLATE. Each Gaussian point is associated with five internal forces, *Mxx*, *Myy*, *Mxy*, *Fxz*, and *Fyz*. In general, these values will vary with locations, which are different from the grillage method. The bending and shear in each plate element are bi-directional, which is different from the 1-D formulation in the grillage method. In the PLATE program, the nodal internal forces are extrapolated from the Gaussian point results (which have the super-convergence property) through the use of the interpolation functions. Each element will give slightly different nodal internal forces at the same node, and the average of these contributions will then give a very high accuracy of the internal forces.

Element		*x/y of Gaussian*		*Mxx*	*Myy*	*Mxy*	*Fxz*	*Fyz*
1	1	0.0106	0.0070	0.1251E+00	0.2744E-01	0.7535E-02	0.1097E+02	0.4665E+01
1	2	0.0394	0.0070	0.4297E+00	0.2972E-01	0.8361E-02	0.1097E+02	0.4261E+01
1	3	0.0394	0.0263	0.4419E+00	0.1109E+00	−0.8965E-02	0.1059E+02	0.4261E+01
1	4	0.0106	0.0263	0.1501E+00	0.1186E+00	−0.9792E-02	0.1059E+02	0.4665E+01

4.9 ELASTO-PLASTIC ANALYSIS OF RAFT FOUNDATIONS

As shown in Figure 4.48, the use of the classical Winkler's spring model will create the embarrassing situation of zero internal forces if a uniform distributed load is applied on a raft foundation, no matter the magnitude of the applied load. The use of the elastic half space or bore-hole appears to be a more reasonable model for analysis, but another situation will then occur for this case. For a rigid foundation where the stiffness can be very high, it is found that the stresses at the corners of the foundation will tend to infinity, if the foundation stiffness tends to infinity, which is shown in Figure 4.49.

The results of the analysis are, in fact, consistent with the work of researchers including Borowicka (1939) for an infinitely long rigid strip and Muki (1961) for a circular rigid disc. Both are on an elastic semi-infinite subgrade and indicate infinite stress at the edges of the footing in their analytical approaches. In addition, Butterfield and Banerjee (1971) have shown similar high stresses at the edges and corners of a rigid disc embedded in a semi-infinite subgrade in his work based on a numerical approach. Muki's formula for determining contact surface vertical stress σ_z at a distance r from the centre of a rigid disc of radius a under a symmetrical load of P is quoted for reference as follows:

$$\sigma_z = \frac{P}{2\pi a\sqrt{a^2 - r^2}} \qquad (4.57)$$

which clearly indicates that σ_z will become infinity at the edge where $r = a$. In this respect, the more rigorous "rigid" analyses from various researchers are different from the engineer's rigid design method which is based on the classical beam bending equation. However, one should not regard the numerical method as useless even if infinite stresses are obtained.

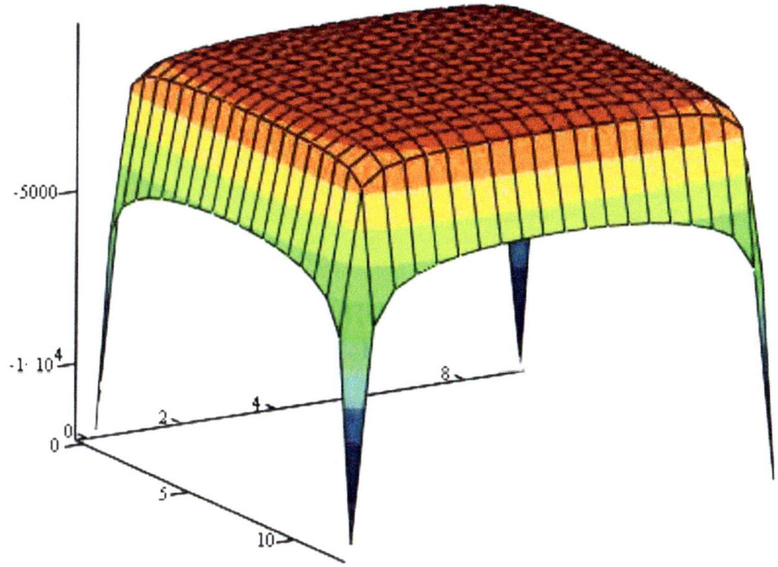

Figure 4.49 The bearing stress from a stiff foundation under a uniform distributed load.

This is because infinite bearing stress does not really exist. The subgrade will "yield" at some level of stress and the stresses over the whole footing will then be redistributed. For the current example, if the subgrade yields at 5000 kPa, i.e. the stress remains at 5000 kPa even with further settlement, then by elasto-plastic analysis with details as described later, the stress surface profile analysed by the 30×30 grid will be as shown in Figure 4.50, where the stresses near the corners stay at 5000 kPa. In this respect, the numerical approach is even better than the theory of elasticity as the plastic behaviour of the underlying material can be readily considered. The yielding of the support can be considered easily using the PLATE program by Cheng. The importance of foundation yielding will be illustrated by an example using PLATE.

A rectangular plan shape concrete (grade 40) footing of dimensions and loads as shown in Figure 4.51 is first analysed elastically. The footing is of overall plan dimensions 21.6 m × 18 m and the thickness is 2 m. The footing rests on a subgrade which is an elastic semi-infinite mass of Young's modulus 1 GPa and Poisson ratio 0.3.

The structure is idealized as an assembly of 20 × 20 = 400 plate bending finite elements. Two load cases are considered, as indicated in Table 4.1, each with an additional uniformly distributed load of 83 kN/m^2 (including own weight of footing, 1.5 m soil and 5 kN/m^2 live load) on the surface of the footing.

Total loads of the two load cases are identical, with loads per Load Case 1 symmetrical about both X and Y axes, whilst that of Load Case 2 is symmetrical about the Y axis only with an additional bending moment to the positive Y direction. By elastic analysis, the bearing stress surface profiles are as shown in Figures 4.52(a) and 4.52(b).

As anticipated, there are locations in both load cases exhibiting very high bearing stresses. If the limiting stress is set at 5000 kPa, the subgrade stresses redistribute as shown in Figures 4.53(a) and 4.53(b) for the two load cases. It is clear that stresses become plastic

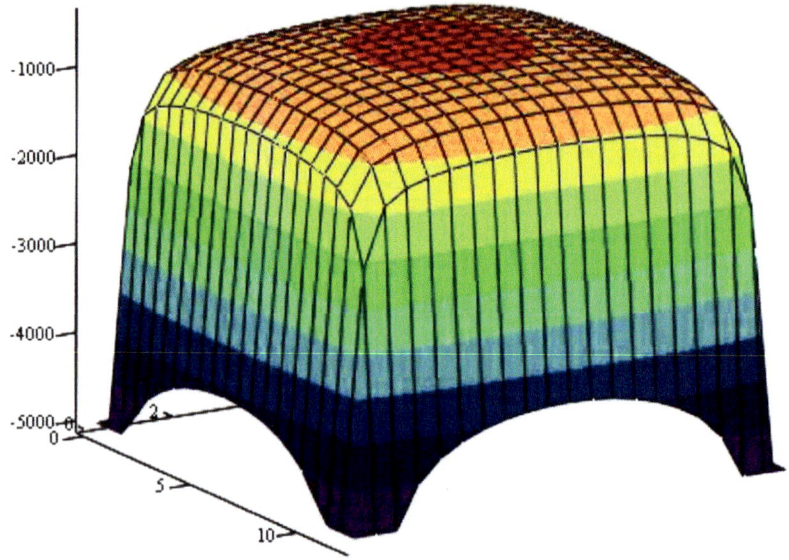

Figure 4.50 Bearing stress distribution, for a yield bearing stress of 5000 kPa.

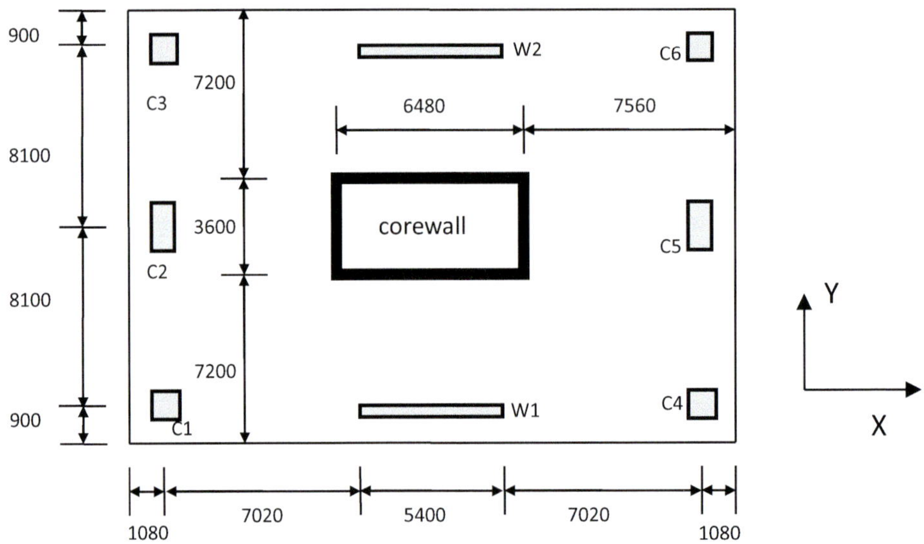

Figure 4.51 Plan of flexible footing analysed by subgrade structure interaction.

Table 4.1 Loading schedule of the footing in Figure 4.51

	Column load (kN)			*Wall load (kN)*		*Corewall load*	
	C1, C3	*C4, C6*	*C2, C5*	*W1*	*W2*	*Axial load (kN)*	*Moment about X axis (kNm)*
Load Case 1	60000	60000	90000	18000	18000	403200	0
Load Case 2	40000	80000	90000	12000	24000	403200	−185760

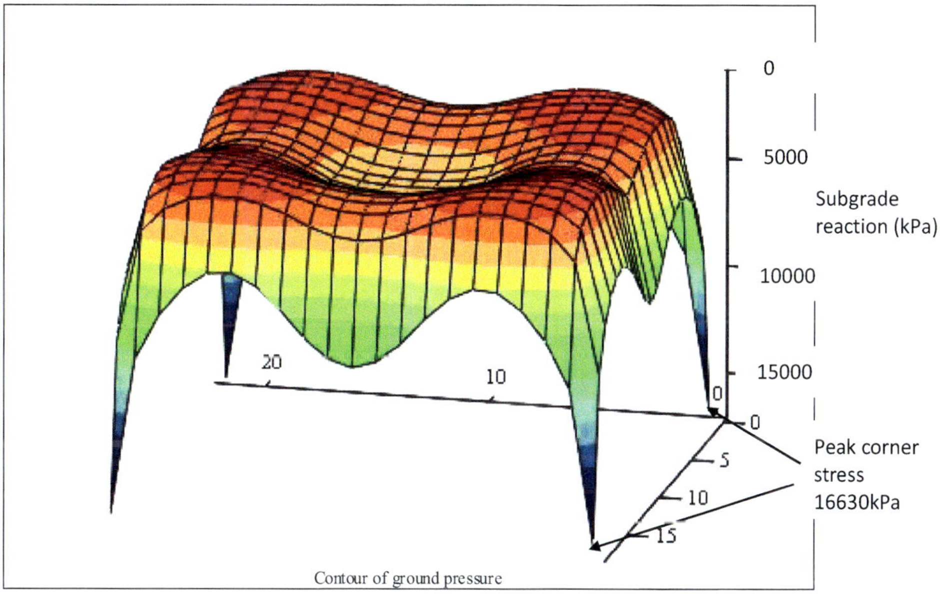

Figure 4.52(a) Ground bearing stress for Load Case 1 – elastic analysis.

along the edges of the footing for the two cases, with Load Case 2 having a greater extent of plastic stresses at the far end.

The structure is also analysed by the method of constant Winkler's spring against Load Case 1 and the bearing stresses distribution is plotted in Figure 4.54, which shows a much smaller variation of stress between 2517 kPa and 3333 kPa. No stress redistribution is necessary for this option as the plastic limit is not exceeded at any node. Comparison of the stress distribution between that with consideration of true subgrade structure behaviour under elasto-plastic behaviour (as worked out previously) along the diagonal of the footing is plotted in Figure 4.55. It is clear that the constant Winkler's spring approach has underestimated the stresses at the edges of the footing. To compensate for this underestimation, Bowles (1997) has suggested to double the end spring stiffnesses which appears to be popular among engineers in the US. The structure is then re-analysed with the subgrade spring along the doubled edges. Bearing stresses exceeding 5000 kPa are identified around

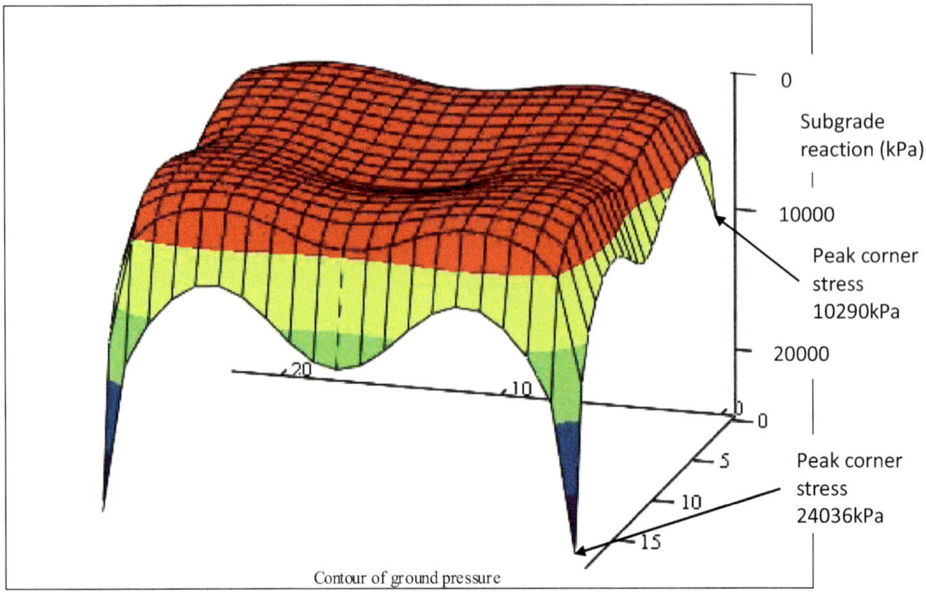

Figure 4.52(b) Ground bearing stress for Load Case 2 – elastic analysis.

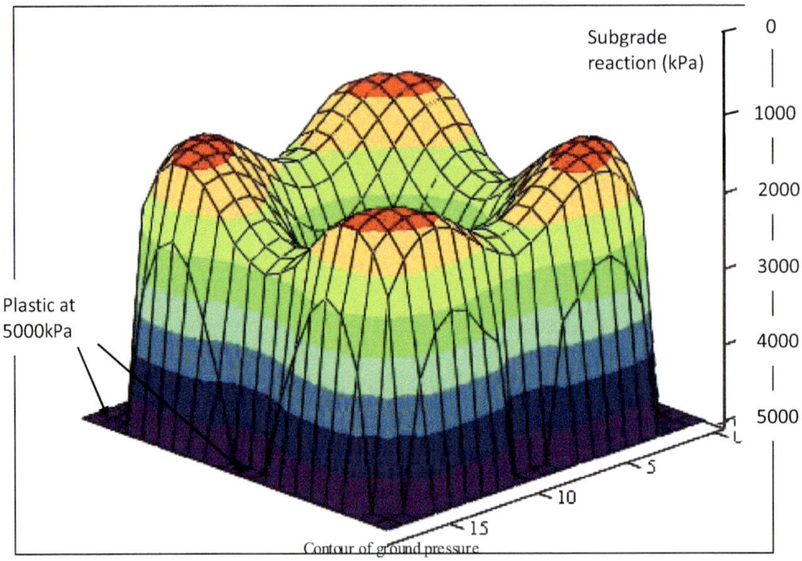

Figure 4.53(a) Ground bearing stress for Load Case 1 – elasto-plastic analysis only.

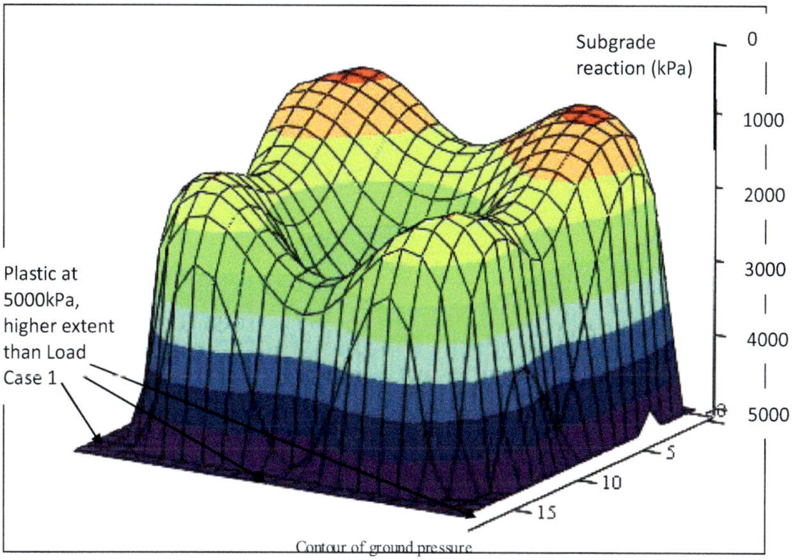

Figure 4.53(b) Ground bearing stress for Load Case 2 – elasto-plastic analysis only.

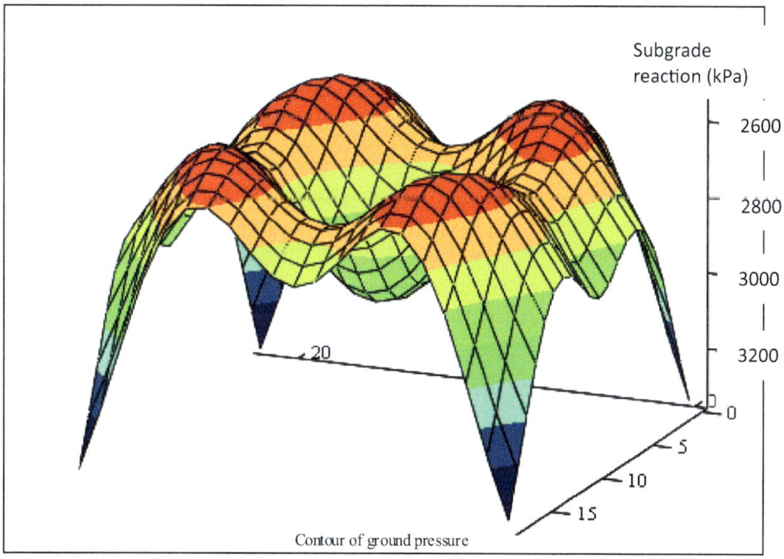

Figure 4.54 Ground bearing stress for Load Case 1 – constant Winkler's spring.

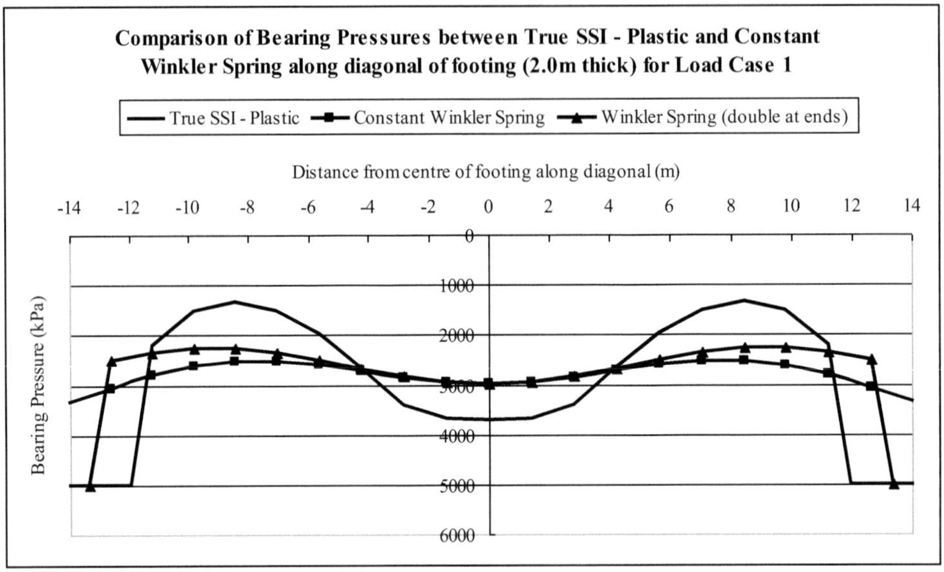

Figure 4.55 Comparison of bearing pressures – true subgrade structure interaction with plastic redistribution versus Winkler's spring approach.

the corners and elasto-plastic analysis is carried out. The stress profile along the diagonal are added to Figure 4.55.

By doubling the end subgrade stiffnesses, the stress profile is in closer agreement to that by the true subgrade structure interaction, though the simulation is still unsatisfactory. The suggestion by Bowles (1996) is, of course, only a rough guide as the increase in subgrade stiffness should depend on a number of parameters including the location of the loads, and dimensions and stiffness of the structure. Accurate analysis can only be achieved by comparing settlement profiles of the structure and that of the subgrade through various sets of subgrade stiffness which are normally obtained by iterations.

Through examination of the stress profiles of the three approaches: (i) true subgrade structure interaction (with elasto-plastic analysis); (ii) constant Winkler's spring; and (iii) Winkler's spring doubled at edges, we can roughly conclude that (i) will produce the least internal forces in the footing. This is because the upward stress on the footing by (i) is the lowest between the central portion and the edges where the corewall and the edge column are situated. By visualizing the corewall and edge columns as supports and the upward stress as loads, (i), having the lowest bearing stress between walls and columns, will produce the lowest internal forces (bending and shear) in the footing, as demonstrated in Figure 4.56.

Detailed designs by the Wood-Armer equations (Wood R.H. 1968) on Load Case 1 for the three models have been performed. The required total flexural reinforcements in the footing for (i), (ii) and (iii) are 2.52%, 3.23% and 2.96%, respectively, showing that the method with true subgrade structure interactions can achieve an economy of 22% in comparison with the constant Winkler's spring method. This result is not surprising in that redistribution of stresses by the interaction within the subgrade will help to reduce the

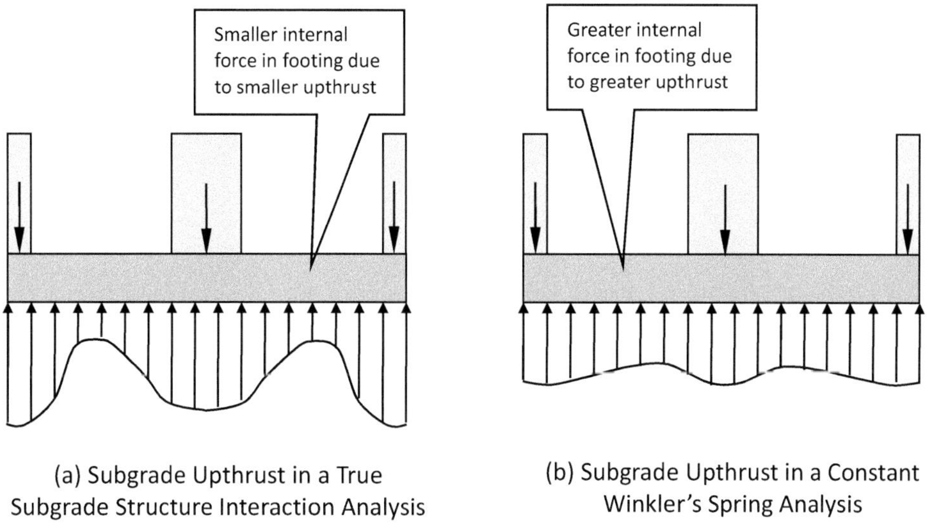

Figure 4.56 Demonstration of smaller internal forces in a true subgrade structure analysis.

internal forces. Layouts of the details of top flexural reinforcements in the X-directions are shown in Figures 4.57(a) and 4.57(b) for both approaches. It can be seen that (i) has significantly fewer top reinforcements than (ii) due to the lower upward reactions between walls and columns as explained above, which far outweighs the slightly greater quantities of bottom reinforcements in the same direction, thus achieving an overall saving in flexural reinforcements.

Since most engineers use commercial finite element programs for the analysis and design of raft foundations/pile caps, the authors would like to point out that many problems in the preprocessing, analysis and post-processing from finite element plate analysis programs have been found in different projects in Hong Kong. Cheng has delivered a short course to engineers about the various problems/limitations of commercial finite element programs with reference to many local projects. Consider Figure 4.58a, which is a raft foundation for a government project. The design strips are divided as given in Figure 4.58a also. Using the reinforcement design option in the program, the reinforcement design as given by the blue line in Figure 4.58b is obtained. There are two places where there is a sudden drop in the minimum reinforcement required which is absolutely unacceptable, as there is no support, point load or similar at those two locations. This is possibly due to the post-processing error in the program, where the equivalent nodal forces across the design section are used to give the reinforcement design. On the other hand, Cheng adopts a seven-point stress integration method (Simpson rule) to determine the strip design moment which is less affected by local problems. The pink line in Figure 4.58b is the result by Cheng which is reasonable and complies well with the moment contour in Figure 4.58a, where there is no sudden change in the moment. There are actually many different types of problems discovered by Cheng for plate analysis, and this is due to the nature of the governing equation for a plate bending problem being fourth order by nature.

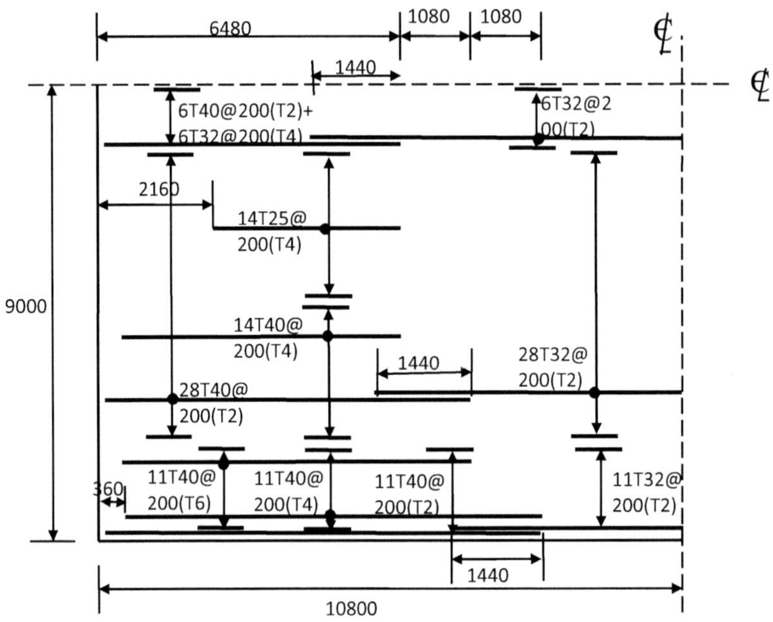

Figure 4.57(a) Top flexural steel in the X-direction analysed by the true subgrade structure interaction approach.

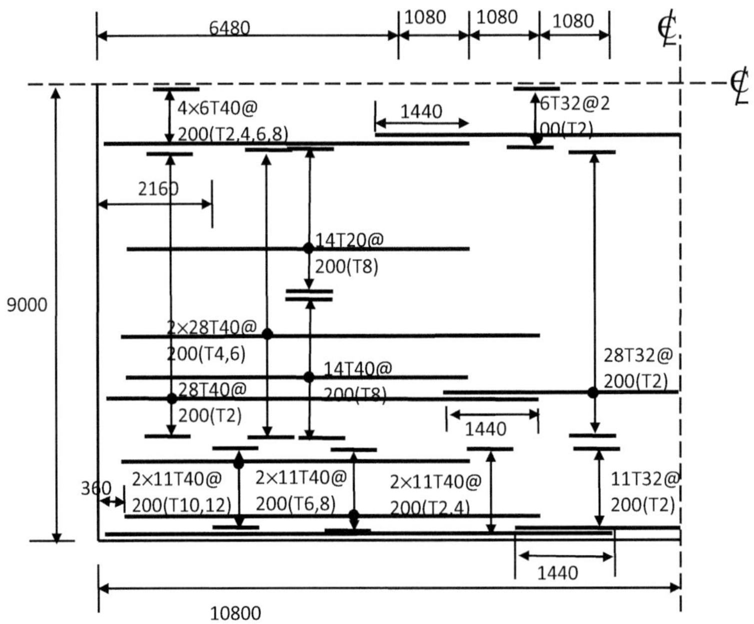

Figure 4.57(b) Top flexural steel in the X-direction analysed by the constant Winkler's spring approach.

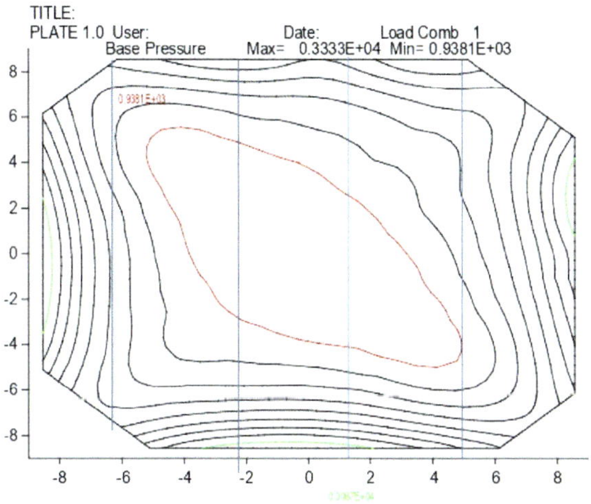

(a) Moment contour for a raft foundation in Hong Kong and the design strip

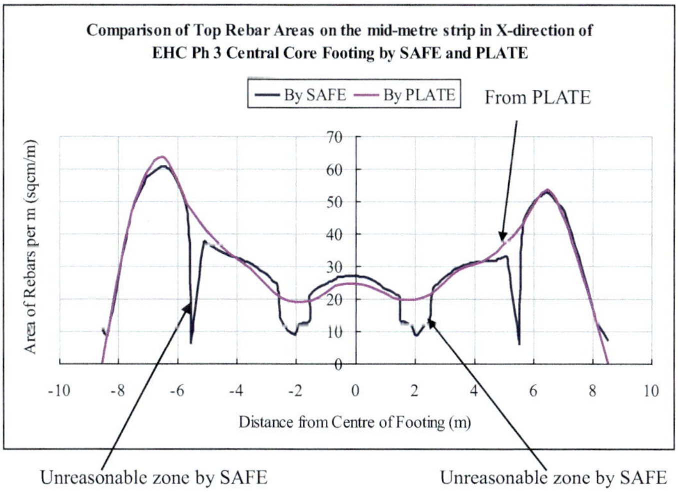

(b) Reinforcement design along a design strip by a famous program

Mx, My and Mxy per meter width are determined at 7 points

(c) Post-processing method by Cheng in PLATE

Figure 4.58 Strange result from the post-processing of a famous finite element program and the method adopted by Cheng in PLATE.

References

Borowicka H (1939), Druckverteilug unter elastischen Platten. *Ingenieur Archiv*, X(2), 113–125.

Butterfield R and Banerjee, PK (1971), A rigid disc embedded in an elastic half space. *Geotechnical Engineering*, 2(1), 35–52.

Cheng YM and Law CW (2006), Strut-and-tie actions in pile cap analysis–Elastic analysis, *HKIE Transaction*, 13, 1–10.

Cheng YM and Law CW (2008a), Development of a new and efficient thick plate element, *Structural Engineering and Mechanics*, 29(3), 327–354.

Cheng YM and Law CW (2008b), Shear jump in "plate bending" structures and other problems, *HKIE Transaction*, 15(3), 1–8.

Cheng YM, Wang JH, Li L and Fung WH (2020), *Frontier in civil engineering vol. 3: numerical methods in geotechnical engineering Part I*, Benjamin Press.

Clark LA (1976), The provision of tension and compression reinforcement to resist in-plane forces, *Magazine of Concrete Research*, 28(94), 3–12.

Foster SJ, Marti P and Mosjsilovic N (2003), Design of reinforced concrete solids using stress analysis. *ACI Structural Journal*, 100, 758–564.

Law CW, Cheng YM and Su RKL (2007), Approach for reinforcement design in reinforced concrete structures based on 3-dimensional stress field, *The HKIE Transactions*, 14(2).

Muki, R (1961), Asymmetric problems of the theory of elasticity for a semi-infinite solid and a thick plate. *Progress in Solid Mechanics*, Vol. 1, North Holland Publishing Co., Amsterdam.

Wood RH (1968), The reinforcement of slabs in accordance with a pre-determined field of moments, *Concrete*, 2, 69–76.

Further Readings

Bowles JE (1966), *Foundation analysis and design*, 5th ed., The McGraw-Hill Companies, Inc.

Cheng YM and Law CW (2005), Some problems in analysis and design of thin/thick plate, *HKIE Transaction*, 12, 1–8.

Cheng YM, Law CW and Yang Y (2012), Problems with some common plate bending elements and the development of a pseudo-higher order plate element, *HKIE Transaction*, 19(1), 12–22.

Cheng YM., Hu B, Wang JH and Li L. (2020), Development of a Pseudo Nine Node Thick Plate Element, *Structural Engineering International*, DOI: 10.1080/10168664.2020.1744502

Din E, Metwally E and Chen WF (2018), *Structural concrete, strut and tie models for unified design*, CRC Press.

Hemsley JA (2000), *Design applications of raft foundations*, Thomas Telford.

Law CW and Cheng YM (2006), Improved thick plate analysis and design, *HKIE Transaction*, 13, 49–58.

Tomlinson MJ (2001), *Foundation design and construction*, 7th ed., Pearson Education Ltd.

Tsudik E (2013), *Analysis of structures on elastic foundations*, J. Ross Publishing.

Chapter 5

Excavation and lateral support system (ELS)

For Hong Kong and many developed cities around the world, deep excavations are necessary for building foundations, basements, mass transit, subways and other construction works. In fact, ELS works are one of the major construction activities in Hong Kong, and many engineers are working on the design and construction of different ELS works. Many support systems have also been developed for different ground conditions. In Hong Kong, an ELS plan submission is required where the excavation is deeper than 4.5 m and longer than 5 m. In general, the submission includes:

1. Lateral support system, e.g. sheet piles, diaphragm wall, etc.
2. Strutting (or tie-back) layout.
3. Construction sequence.
4. Supporting geotechnical documents.
5. Realistic estimates of ground movements.
6. Assessment of effects of excavation and dewatering on adjoining structures.
7. Monitoring proposal.

5.1 TYPES OF RETAINING SYSTEMS

Due to the site constraints, time and cost of construction, ground and water table conditions, as well as the conditions of the adjoining buildings, many types of ELS have been developed, which include:

1. **Sheet pile wall** – This is versatile and cheap, water-tight, but cannot avoid utility. Sheet pile wall is relatively flexible, and is usually for bottom-up construction, e.g., Lok Fu MTR station in Hong Kong.
2. **Soldier pile wall** – This is restricted for use on stable ground and low water table conditions, but is cheap. Soldier pile wall is not water-tight, but can avoid utility (allowing utility to pass through the lagging), and is usually for bottom-up construction, e.g. Diamond Hill MTR station in Hong Kong.
3. **Caisson wall** – This is a stiff (temporary or permanent) wall which requires little machinery in the construction and can avoid utility. The construction of a hand-dug caisson is dangerous to workers and is seldom used currently. The caisson wall can be used for bottom-up as well as top-down construction, e.g., Choi Hung MTR station, Sun Plaza, in Hong Kong.

DOI: 10.1201/9781032650586-5

4. **Diaphragm wall** – This is a very common wall system in Hong Kong, and is usually permanent. It is a stiff wall which requires heavy machinery and care in the construction, and cannot avoid utility during construction. Examples include the Wan Chai Central Causeway, Sheung Wan MTR station, Times Square, in Hong Kong.
5. **Secant pile wall** – This is an expensive stiff permanent wall which requires heavy machinery in construction. Its construction cannot avoid utility, but it can provide a good water-tight wall, and is usually used for top-down construction. Examples include the Mongkok and Prince Edward MTR stations (rare in construction).
6. **Pipe pile wall** – This is versatile, flexible and cheap, but is not a water-tight wall and so it requires back grouting. This wall is currently very common in Hong Kong, but not other countries. It can avoid utility during construction, and is usually used for bottom-up construction, e.g. Nathan Centre, Hung Hom station, in Hong Kong
7. **PIP wall** – This is a wall formed by a grouted pile wall, usually with normal grouting pressure, and a special grout. Reinforcement cages as well as steel H have been used in the reinforcement of these piles in Hong Kong. Examples include the Jordan and Tsim Sha Tsui MTR stations, and Chartered Bank basement in Hong Kong. Due to the difficulty in controlling the quality and waterproofing, it is less common nowadays. Nevertheless, PIP walls can be cost-effective and a feasible solution if the water head and retaining height is not too high, with a limited working area.
8. **Jet pile wall** – This is a wall formed by a grout jetting process, where the grout pressure can achieve 100–200 bars (using double-tube or triple-tube technique). The soil structure is destroyed by such a high grouting pressure, and the cement grout can mix well with the soil to form a good-quality grouted wall. This type of wall is used mainly for soft clay, where the strength of soil is low. Examples include the Sheung Wan MTR station in Hong Kong, MRT, in Singapore.
9. **Mixed wall** – There are many cases where a combination of two or three types of the wall systems mentioned above can be adopted for economic construction. The most critical issue will then be the junctions between the different types of ELS walls. A good example is the Sun Plaza in Hong Kong, where the caisson wall system is mixed with the diaphragm wall system to deal with different ground conditions at different parts of the construction site. Another good example is the Nathan Centre in Hong Kong, where the sheet pile wall system is mixed with the pipe pile wall system to reduce the vibration induced to the tunnel of the Mass Transit Railway.

The choice of a suitable ELS is very important for a good ELS design and construction system. It should also be noted that many engineering problems have occurred during deep excavation, notably excessive wall and ground movement, stability of the bottom of excavation, difficulty due to ground water, and failure of the supporting system (mostly buckling failure of struts and welding).

In the analysis of the retaining wall, there are two major considerations. Most wall systems take the load directly and the internal forces span vertically, except for those with a distinct pile. In a soldier pile wall system, the horizontal lagging spans horizontally, and is simply supported on vertical piles. The support reaction from the lagging becomes the loading on the vertical pile. Wales are used to support the wall, which is usually considered as a continuous beam. The wales are further supported by struts or anchors, which are illustrated in Figure 5.1.

5.1.1 Sheet pile wall system

Cheng has worked on several major ELS works in Hong Kong in different types of ground conditions, and some practical considerations and precautions will be discussed in this chapter. Refer to Figures 5.1 and 5.2 for the sheet pile system, where it is noted that the sheet pile wall span vertically takes up the loading from the soil and water, which is supported by the wale which is practically a continuous beam. The wale is further supported by the strut to reduce the span length of the wale. There are several points to be considered in this design: (1) for the normal strut, it is noted that many engineers like to add inclined struts at the two sides of the normal strut so that a larger clear space is available between the strut for transportation of materials; (2) the normal strut can take up a very heavy load, in particular if the inclined struts are added, which transfer the loads to the normal strut; (3) buckling failure to the normal strut is a common failure problem for ELS works; (4) vertical king posts are used to reduce the effective length of the strut; (5) for some normal struts and wales, the loadings are so high that double steel sections or even concrete members may be used; (6) for the corner between two orthogonal sheet pile walls, corner inclined struts are commonly used (Figure 5.1b), and the strut will span between the two sheet pile walls; (7) for all the inclined struts, a horizontal force will be generated which is usually taken up by a fillet weld; (8) the bending moment on the wale can be very large, and some engineers may reduce the bending moment by considering the size of the strut in the analysis (instead of a point support), with a good example being the ELS design in Admiralty, Hong Kong; (9) a dewatering well may be required in or outside the excavation, depending on the depth of the excavation and the size of the site; (10) monitoring of the water table, settlement and adjoining buildings is required, otherwise disasters may result; (11) vibrators and drop/hydraulic hammers are usually used to install the sheet pile, while the use of a jacking method is generally less commonly used. The Giken system may be adopted when the headroom is limited, at the expense of high cost for the construction; (12) to create more working space, some engineers like to provide inclined struts supported by the pile caps; (13) preboring to place the sheet pile wall may be necessary to reduce vibration due to driving of the sheet pile; and (14) Cheng has come across several interesting cases, where heavy steel sections are used with wires to support several major water mains. The water mains are hence overhung during some MTR station constructions in Hong Kong.

Sheet pile wall comes in various types, with different shapes, stiffnesses and interlocks. The most common types of sheet pile in Hong Kong include the U-type, Z-type and H-type, with the U-type being the most popular among many construction projects. In extreme conditions, double layers of sheet pile wall can be used by providing horizontal ties between the two layers of sheet pile. It is possibly the most common ELS in Hong Kong and around the world, but there are also many practical problems in the sheet pile system. The most common problem with the use of a sheet pile wall is the presence of boulders, and small-diameter drilling rigs are required to break boulders for further driving of the sheet pile. Cheng has encountered a case where the sheet pile stopped on top of a large boulder, and beyond excavation to the bottom of that small section of sheet pile, a support similar to the soldier pile wall was used. This method is applicable only for stable ground conditions.

The design of a shoring system needs to consider the size, shape, utilities, construction process and other factors, and it is a challenging job to achieve an economical and flexible

(a) Bottom up sheet pile wall construction of the Hong Kong Polytechnic University Phase 8

(b) Typical strut/wale connection at different positions

Figure 5.1 Typical bottom-up construction using a sheet pile wall, wale and strut support.

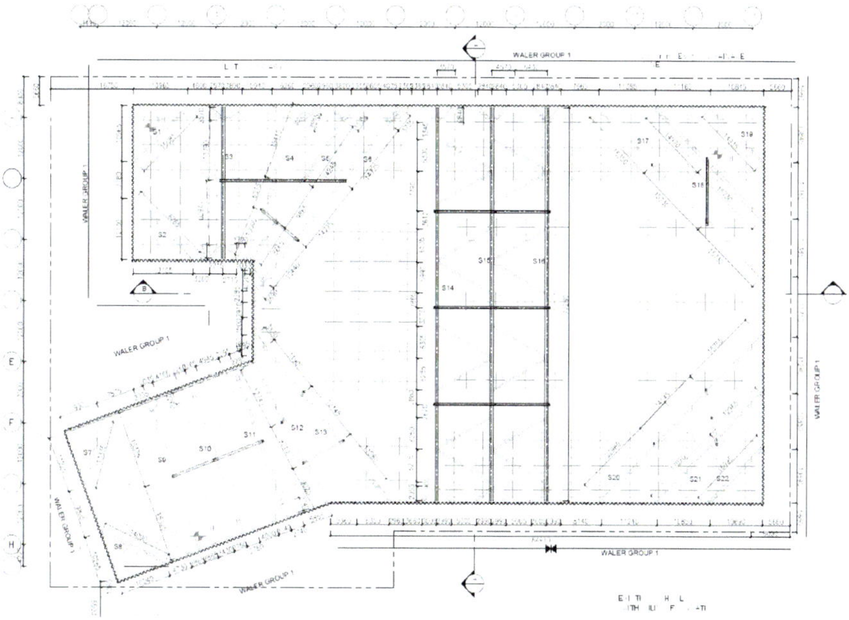

(c) a typical bottom up shoring system

(d) Concrete wales and struts used for ELS

Figure 5.1 (Continued)

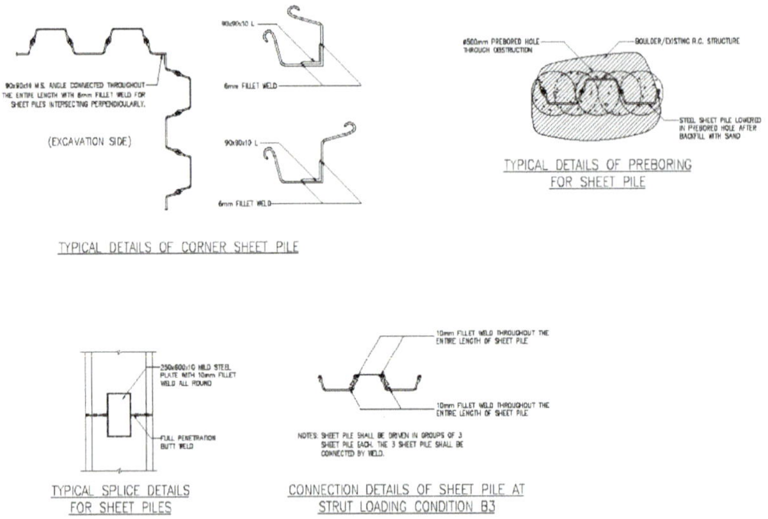

(a) Some details for sheet pile

(b) Press-in method for driving sheet pile to minimize sound and vibration

Figure 5.2 Interlocking sheet pile wall and some construction details.

system for the construction. A typical shoring system as shown in Figure 5.1d can illustrate the difficulty in designing a good shoring system, considering all the constraints. Sheet pile wall is economical and has good water-tightness, and its installation is easy. This system may, however, be limited by the presence of utilities, boulders or other hard materials, and the stiffness of the wall is generally low. The sheet pile wall is generally a temporary wall, with a permanent wall required to be constructed later.

Some typical guidelines for overcoming underground obstructions (mostly boulders) during sheet pile installation are:

1. Predrilled holes shall be sunk down to the proposed toe level of the sheet pile temporary casing to stabilize the drill hole.
2. The prebored drill holes should be backfilled with suitable granular materials during withdrawal of the casing.
3. The drive sheet pile in the backfill boreholes should be designed to the toe level.

According to Table NA.2 of NA to BSEN1993-5:2007, EA and EI for a sheet pile for one layer of strut under unfavourable conditions should be undertaken for sheet pile design. Similar requirements are also required in some countries, and engineers should note this. The loss of sheet pile thickness in fresh and sea water is also described in BSEN1993-5:2007.

5.1.2 Soldier pile wall system

For a soldier pile wall system, H section (Figure 5.3a) is the most commonly used pile. However, pipe piles or hand dug caissons have also been used in Hong Kong, and such systems are shown in Figure 5.3b. For the horizontal spanning lagging, it can be a wooden plank, steel sheeting or even in situ concrete (an access shaft is present at Causeway Bay in Hong Kong). In some cases, holes are left in the lagging for the utilities to pass through. The most critical problems for the soldier pile wall system are water leakage and the stability of the cutting face before lagging is installed. Back-grouting may be required in some cases.

A soldier pile wall system is economical and allows utilities to pass through the wall easily during construction without diversion. The suitability of this system is however limited by the stability of the open surface of soil before the installation of the lagging. Most soldier pile walls are temporary retaining walls, but there has been a tendency to make them permanent systems recently.

5.1.3 Caisson wall system

Hand dug caisson construction was invented in Hong Kong, and is used as a replacement for bore pile construction without heavy machinery. Due to the high accident rate in hand dug caisson construction, it is generally banned from construction in Hong Kong, though it is still used in China. As shown in Figure 5.4, the interlocking caisson wall forms a stiff and water-tight retaining wall, and it is relatively easy to deal with utilities

(a) Leakage through lagging in HK

(b) hand-dug caisson as the soldier pile

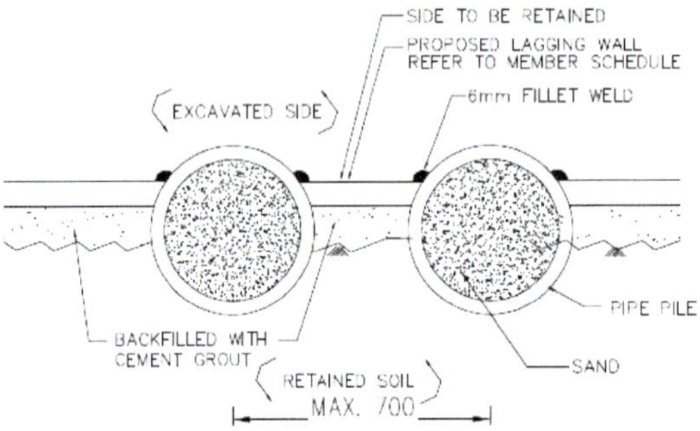

(c) Typical soldier pile wall details using pipe piles (common in Hong Kong)

Figure 5.3 Typical solder pile wall systems.

during construction. A caisson can be a temporary or permanent wall, but it appears that permanent walls are the more popular option. Except for the safety issue, it can be considered as a very flexible method of construction without the need for heavy plant. If a caisson is used in conjunction with lagging, as shown in Figure 5.3b, it can also be called a soldier pile wall system. Each caisson ring or excavation is about 0.8–1.0 m, and there is an overlap between adjoining rings. Normally, no reinforcement or only light reinforcement is required for the ring, which is typically 75 mm in thickness, as the ring is under pure compression due to its shape (arch action). After construction of the rings to the bottom, a reinforcement cage is put into the caisson and tremie concreting is carried out. For caisson construction in marine sand, alluvium, piping may be needed. Cheng has seen water and sand coming from the caisson linings in hand dug caissons in marine sand regions, and the depth of each ring construction is reduced to 0.5 m to maintain the stability of each ring excavation. Currently, the use of robots to construct hand-dug caissons is under investigation in Hong Kong, and it may be used in the future after a successful trial.

Figure 5.4 Caisson wall with tie-backs at Fortress, Hong Kong (another similar wall is at Kornhill, Hong Kong).

5.1.4 Diaphragm wall system

A diaphragm wall, as shown in Figure 5.5a, is a very popular construction method in Hong Kong, due to the ability for it to be constructed to a great depth without major vibration. The typical wall thickness ranges from 0.8 m to 1.5 m in Hong Kong, which is considered as stiff, and the panel size is typically 4–6 m. In the past, a grab (Figure 5.5c) was used to remove the soil within each panel, which is a relatively slow process. Currently, a hydrofraise is more commonly used to replace grab excavation, with a maximum excavation depth up to 100 m, and a typical hydrofraise is shown in Figure 5.5b. A typical hydrofraise has a rotation speed of 10–20 rpm, a cutting power of 3800 kg, with a circulation pump of about 300 m^3/h, a typical weight of 16–20 tons, and a power of about 365 HP. The typical construction sequence of a diaphragm wall is given below:

1. Guide wall construction – support the top soil, weight of reinforcement cage, and rotary grab.
2. Trenching by the use of bentonite and a grab or hydrofraise. In general, the excavation is carried out in three phases, left portion, right portion followed by the centre portions. The middle portion is usually 1–2 feet less than the grab or hydrofraise for excavation. As well as three passes, five-pass excavation is sometimes adopted. The stability of the trench is maintained by the use of bentonite with excess head.

3. Trench cleaning and stop ends fixing.
4. Reinforcement cage lowering, using the guide wall as a temporary support.
5. Tremie concreting.
6. Withdrawal of stop ends (some stop ends are permanent, for example, Sheung Wan MTR concourse). Some typical joint details are given in Figure 5.6, and a waterstop may also be left in place in some cases.
7. Obstructions due to boulders are commonly removed using chisels.
8. Toe-grouting as shown in Figure 5.8 is sometimes used to eliminate most of the seepage that may occur during excavation and dewatering inside the excavation. This can be useful to reduce ground settlement if the depth of excavation is high with a high water table and poor soil condition.

During panel excavation, stability is maintained by the bentonite slurry, which is a suspension. In general, the top level of bentonite should be 2 m above the ground water table so as to maintain a positive slurry head which can help to stabilize the panel. Bentonite is a suspension which forms a gel when it is not disturbed. In the slurry trench, some bentonite slurry infiltrates into the soil which will then become a jelly. This jelly wall separates the slurry from the soil and ground water table. If the net pressure/force from the slurry can balance the stress/force from soil and water, the panel excavation will be stable. It must be noted that this is a three-dimensional problem, as the width of the panel length is limited. An arching effect will be formed, so that the effective lateral pressure will be less than the classical active pressure. Bentonite slurry acts as a support fluid (affecting the stability of the slurry trench):

1. The bentonite suspension is montmorillonite group clay with exchangeable sodium cations (Na^+).
2. The action of bentonite in stabilizing the sides of bore holes is primarily due to the thixotropic property of bentonite suspension.
3. The bentonite suspension when undisturbed forms a jelly which when agitated becomes a fluid again. In the case of granular soils, the bentonite suspension penetrates into the sides under positive pressure and after a while forms a jelly. The bentonite suspension gets deposited on the sides of the hole, resulting in the formation of a filter cake in contact with soil against which the fluid pressure acts likes a jelly sheet pile wall as separation. In general, the density of the slurry is controlled between 1.02–1.2 g/cc. In theory, the denser the slurry, the better will be the stability. A very dense slurry will however create difficulties in the tremie concreting, hence the maximum density of the slurry must be controlled during the construction by continuous monitoring.
4. In the case of impervious clay, the bentonite does not penetrate into the soil, but deposits only a thin film on the surface of the hole.
5. The level of the supporting fluid must be above the ground water table to maintain a net force stabilizing the trench.

Bentonite recycling will involve the lowering of submersible pumps to the base of the trench and pumping to the surface and onwards to the desanding unit (Figure 5.5e), where coarse sand particles will be separated from the slurry where it will then be returned to the

Table 5.1 Acceptance criteria for bentonite testing

Bentonite properties	*Method of testing*	*Fresh slurry*	*Slurry during excavation*	*Slurry prior to placing concrete*	*Slurry prior to re-use*
Density (g/ml)	Mud balance	>1.015	1.08–1.15	≤1.15	≤1.30
Fluid loss and reside (ml)	Baroid filter press	≤25	<45	≤40	≤50
Viscosity (s)	March funnel	≤35	N/A	≤40	≤50
pH	Electric meter	7.5–10.5	8–11	8–11	7.5–11.5
Filter cake thickness (mm)	Baroid filter press	≤1	≤5	≤5	≤5
Sand content (%)	Decantation test	N/A	N/A	≤3	N/A
Frequency	–	Once per mixing day	Once for each shift during excavation	Once before placing concrete	At least three tests

silos or slurry pools for storage and re-use. Tests will be conducted before returning the bentonite to storage. The test results for acceptable slurry are given in Table 5.1.

The advantages of a diaphragm wall include:

1. Can be used for unfavourable ground and hydrological conditions.
2. Good control of water table during construction.
3. Less noise and vibration will be generated during construction.
4. Applicable for a great depth, with the world record for a diaphragm wall exceeding 100 m.
5. Stiff permanent wall, so that less strutting will be required, and maximum utilization of the site area.
6. Better control of ground settlement.
7. The wall need not be a linear element (Figure 5.5d).

The disadvantages include:

1. Due to the requirements of various construction plants, the site will be congested, and may cause inconvenience to the public.
2. Rough surface finish after exposure, due to tremie concreting.
3. Relatively expensive construction.
4. Treatment and disposal of slurry can create various problems.
5. Cannot avoid utilities, hence diversion of utilities is required before the construction.

For the reinforcement details, some typical details are shown in Figure 5.6. It is possible that boxing-out may be left in the panel for connection with structural elements or breaking out later. The diaphragm wall is not completely water-tight, and the connections between the panel can be made water proof with the use of a waterstop, as shown in Figures 5.6b and 5.6c. The use of grouting at the connection is also commonly used currently, as shown in Figure 5.6c. The reinforcement details between the wall and the permanent slab are

(a) Diaphragm wall with anchor support (b) typical hydrofraise

Figure 5.5 Diaphragm wall construction using a hydrofraise and cable grab.

shown in Figure 5.6d, which provides the commonly used details for top-down construction. To enhance the quality of the wall, a precast concrete panel (Figure 5.7) is sometimes used if the length of panel is not too large. To eliminate the movement of the bottom of the wall, a shear pin (Figure 5.9) is sometimes used.

5.1.4.1 Tests to the diaphragm wall

Interface coring may be carried out on the diaphragm wall panel to check the condition at the concrete and rock interface. A reservation tube as shown in Figure 5.10 can be fixed into the reinforcement cage at 1000 mm above the base level of the diaphragm wall. The core drilling shall be undertaken through the tube to 1000 mm below the concrete and rock interface.

A sonic test may also be carried out for diaphragm wall panels to check the integrity of the concrete panels. Reservation tubes shall be watertight steel tubes and securely fixed within the reinforcement cage at approximately 1.0–1.5 m spacing for the whole depth of each diaphragm wall panel. Tubes shall extend from between 0.2 m above the ground surface to within 0.1–0.2 m of the toe. The tubes will be filled with water and adequately capped and sealed at both ends before concreting.

The equipment for the sonic test consists of signal transmitter probes and signal receiver probes which are lowered into the tubes. The results of the sonic testing are displaced on a recording oscilloscope at the top of the diaphragm wall. The sonic test is carried out

c) Typical cable grab

(d) wall panel needs not be a straight in diaphragm wall construction

Figure 5.5 (Continued)

(e) Desilting unit

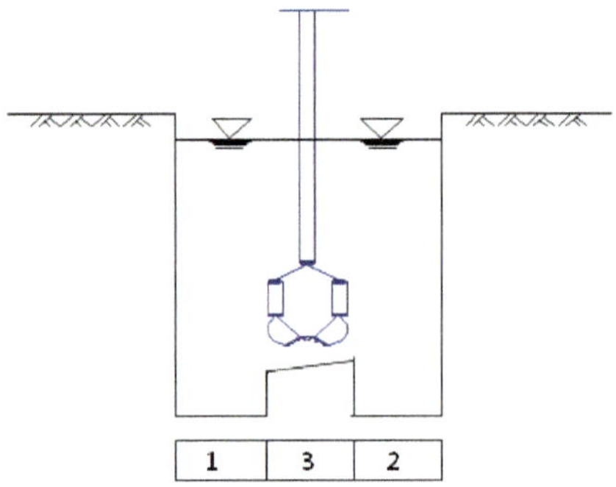

(f) Typical excavation sequence for diaphragm wall construction

Figure 5.5 (Continued)

at least 14 days after the concrete pour. The pipes are filled using a grout mix of similar strength compatible to that of the parent panel.

It is not easy to ensure complete water-tightness for the diaphragm wall, hence a maximum flow rate from the wall is usually specified in the specification. In the case of local poor-quality concrete with excessive inflow of water, epoxy grouting (commonly in the form of two-solution grout) under high pressure may be used to seal the voids within the concrete. This is commonly carried out in many mass transit diaphragm wall concourses in Hong Kong.

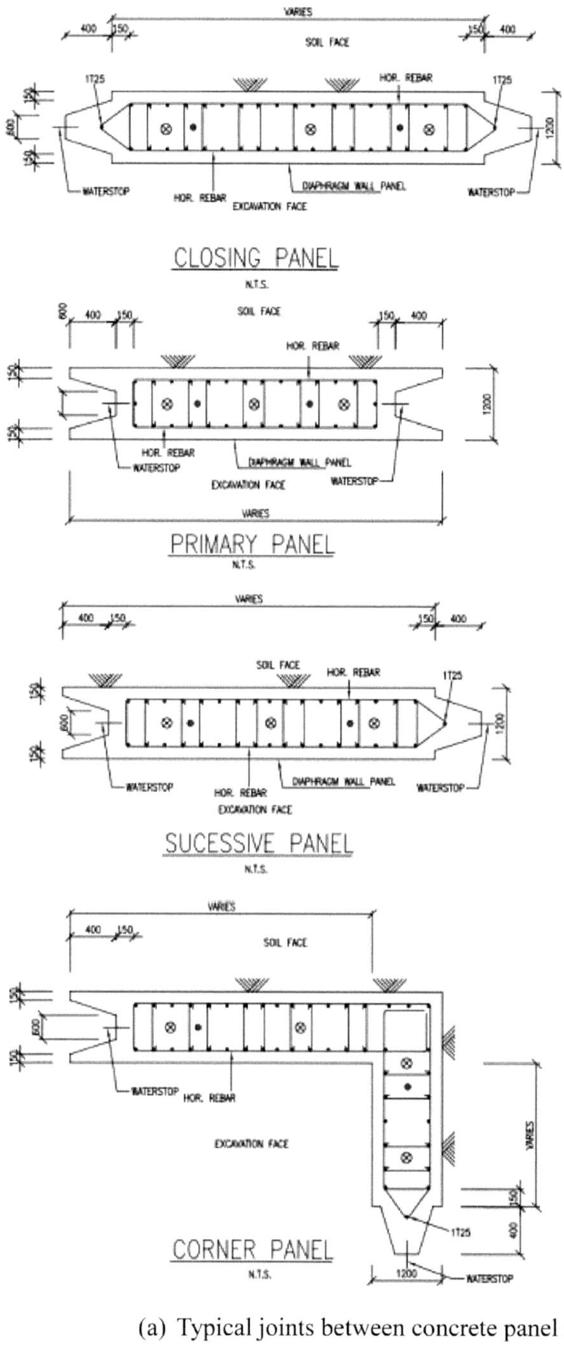

(a) Typical joints between concrete panel in diaphragm wall

Figure 5.6 Typical joint design in a diaphragm wall.

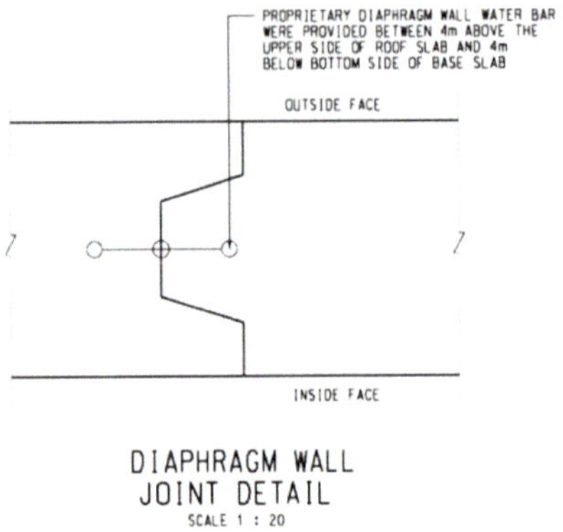

(b) Construction joint between panels, with proprietary water-stop

(c) Stop end with grout tubes

Figure 5.6 (Continued)

5.1.5 Secant pile wall system

For a secant pile wall, each unit is cut into the adjacent one to form a water-tight interlocking joint. This is an expensive construction and requires heavy machine. In the past, Benoto was used for the secant pile construction in Hong Kong (Mongkok and Prince Edward MTR station in Hong Kong), but it was replaced by modern heavy duty boring machines (RCD), which are shown in Figure 5.11. In Figure 5.11, some zig-zag signs are left on the surface of the secant pile, due to the alternate oscillation of the oscillator during extraction

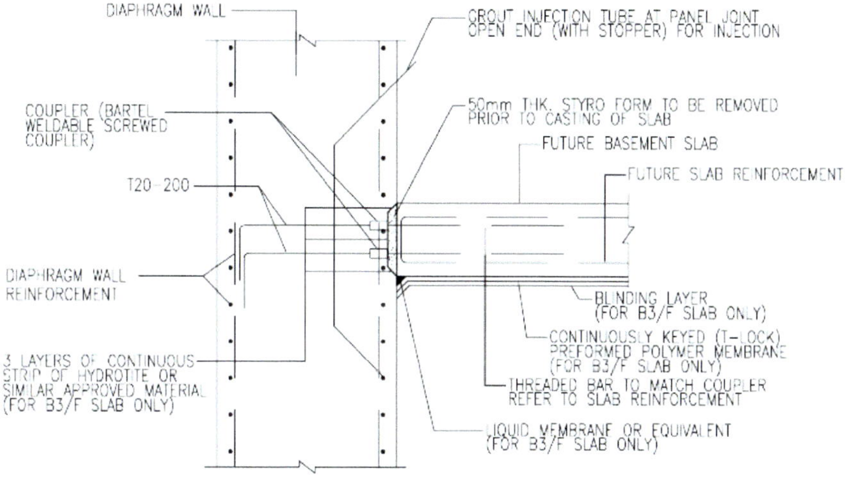

(d) connection of diaphragm wall with top slab and intermediate slab for top down construction (use of boxing out in diaphragm wall is to receive reinforcement from slab can also be used)

Figure 5.6 (Continued)

Figure 5.7 Use of a precast panel in a diaphragm wall construction.

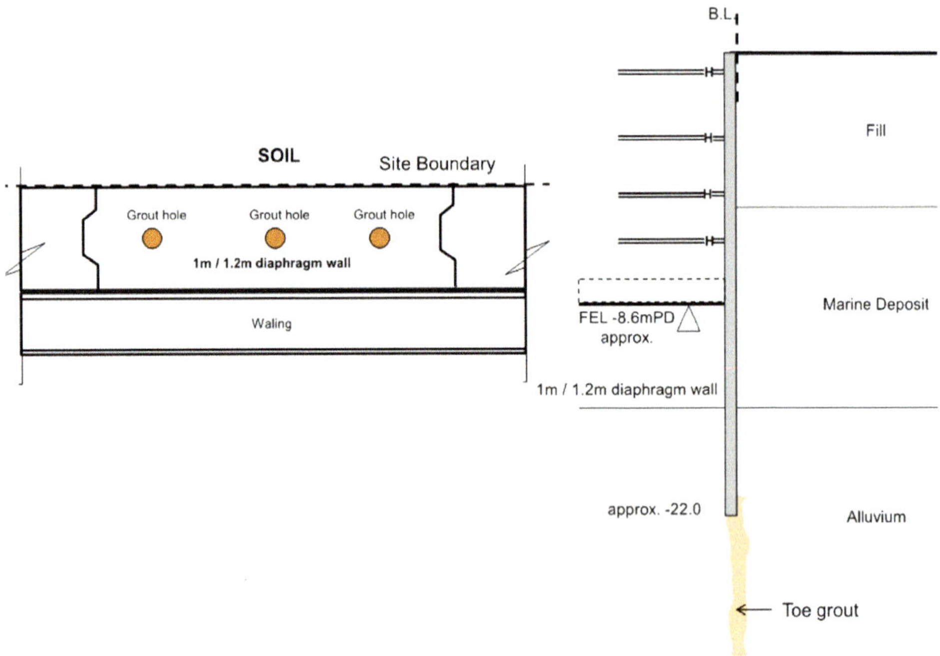

Figure 5.8 Typical details for toe-grouting in a diaphragm wall.

of the casing. A retarder may be used in the concrete for a secant pile, to facilitate cutting of the mother pile by the adjoining pile. There are two methods of secant pile construction: (1) all the odd-number piles are constructed before the even-number piles; (2) the piles are constructed sequentially (Figure 5.11c).

Secant piles are stiff with good watertightness, but the cost of construction is very high. Furthermore, the construction may be limited by the availability of RCD for the construction.

5.1.6 Pipe pile wall system

Pipe pile has been popular in Hong Kong in the last 20 years (but not in other countries). In the past, 150 mm pipes were commonly used, but 610 mm diameter pipe pile is very popular nowadays. The pipe piles are commonly installed by a small-diameter boring machine or ODEX with air flushing (see Chapter 6 for the updated system). In general, there are two major methods of installation: (1) the piles are constructed with very limited gap between piles; (2) a limited gap is left between the pipe piles, and the system may eventually be a soldier pile wall system with a steel plate lagging. Since there is no interlocking between the piles, back-grouting is required, which can be seen in Figure 5.12b. Pipe piles may be installed continuously or a small gap can be left between the piles, which are shown in Figure 5.13, with the choice depending on the cost of construction and stability of the open unsupported soil.

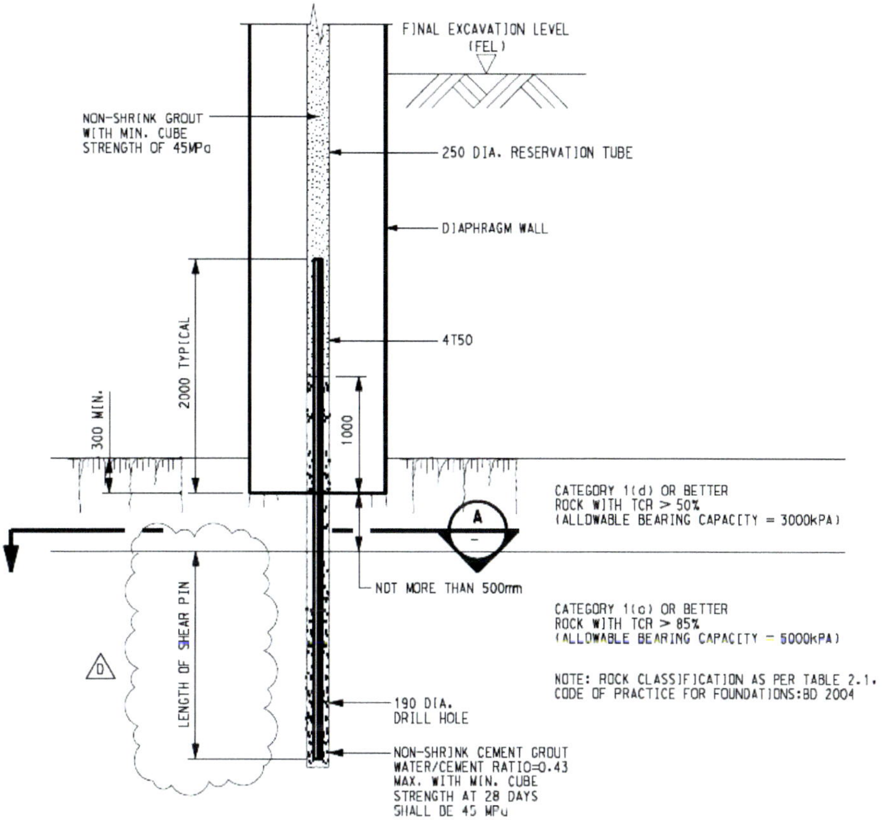

(a) Use of shear pin at bottom of diaphragm wall

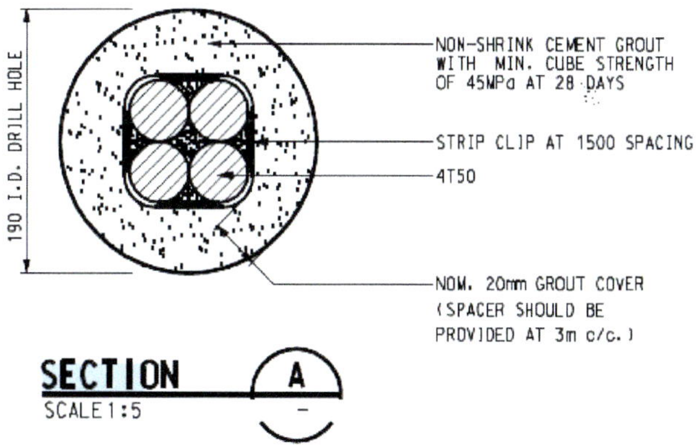

(b) Section A-A

Figure 5.9 Typical shear pin details.

Figure 5.10 Reservation tube fixed to a reinforcement cage for later testing.

A construction method statement for a pipe pile wall bottom-up project in Hong Kong is given below for reference:

1. Setting out for the pipe pile locations, and inclination and alignment of the pipe pile, will be monitored during construction.
2. A hole of 450 mm is sunk by a rotary percussion rig using the percussion ODEX drilling method (see Chapter 6) or ring bit drilling system (Figure 5.14) with air flushing from the existing ground to the bottom of the pipe pile. Hard materials and obstructions can be overcome using a pilot bit and concentric reamer.
3. Splicing of the OD 406 mm × 30 mm thick casing will be carried out by full penetration butt welding with a qualified welder during the drilling work.
4. For Tube a Manchette grouting, the pipe is 48 mm OD and 40 mm ID PVC with the grouting orifice covered with a rubber manchette at 350 mm spacing. The bottom end of the tube is capped. The tube is filled with water to reduce buoyancy and aid the installation. The spacing between the manchette pipe and the drill hole is grouted with a weak sleeve grout with a mix of bentonite:cement:water at 1:10:20 by weight. The sleeve grout is allowed to set for a minimum of one day prior to commencement of any further grouting.
5. Single- or double-stage back-grouting is required to prevent water flowing to the excavation. For double-stage grouting, see Section 5.4 for the procedures of the Tube a Manchettee grouting method. In general, the first grout mix will be cement

(a) Typical secant pile

(b) boring machine for secant pile

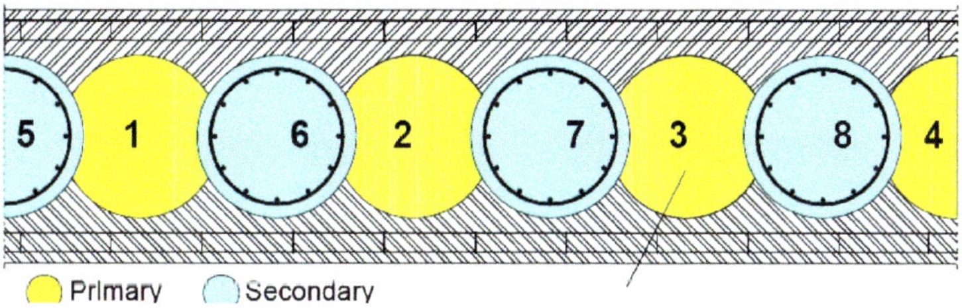

(c) another typical arrangement for secant pile wall, with primary and secondary piles (primary piles can be non-load bearing)

Figure 5.11 Secant pile construction.

bentonite grout with 350 kg cement, 35 kg bentonite and 600 litres of water per cubic metre of grout. The second sodium silicate grout is typically made up of 400 kg sodium silicate, 708 kg water with some admixture per each cubic metre of grout (based on several projects in Hong Kong).

5.1.7 PIP wall system

PIP has been used for the support of footbridges and lightweight structures in the past as a form of grouted piles. The PIP wall has been used for construction of some MTR stations in Hong Kong, but its application has now been replaced by the use of a pipe pile wall in Hong Kong. PIP piles are formed by rotating a continuous-flight, hollow-shaft auger into the ground. The flights of the auger are filled with soil as the auger is installed into the ground, and the soil on the auger is used to provide lateral support to maintain the stability of the hole. The diameter of the PIP is usually 450–550 mm diameter, and cement grout of 25 MPa with special admixture is injected (to improve the workability and compensate for the shrinkage effect) with sufficient pressure through the auger shaft, as the auger is being withdrawn, in such a way as to exert an upward pressure as well as a positive lateral pressure on the surrounding soil used for grouting of the pile. The grout is cast at least 600 mm above the specified cut-off level. A reinforcement cage (Figure 5.15)

(a) 610mm diameter pipe pile

(b) Grouting within pipe pile

Figure 5.12 610 mm diameter pipe pile used in Hong Kong.

or steel section (Tsim Sha Tsui and Jordan MTR stations in Hong Kong) can be used for the reinforcement of the pile. Two grout cubes of 100 mm size are made for each pile for unconfined compression testing. Additional sets of cubes should be made on any batch of grout used in piles. The maximum grout temperature shall not exceed 38°C. Adjacent piles closer than six pile diameters should not be placed within 24 hours of each other.

(a) Contiguous pipe pile with back grouting between piles

(b) pipe pile with a small gap

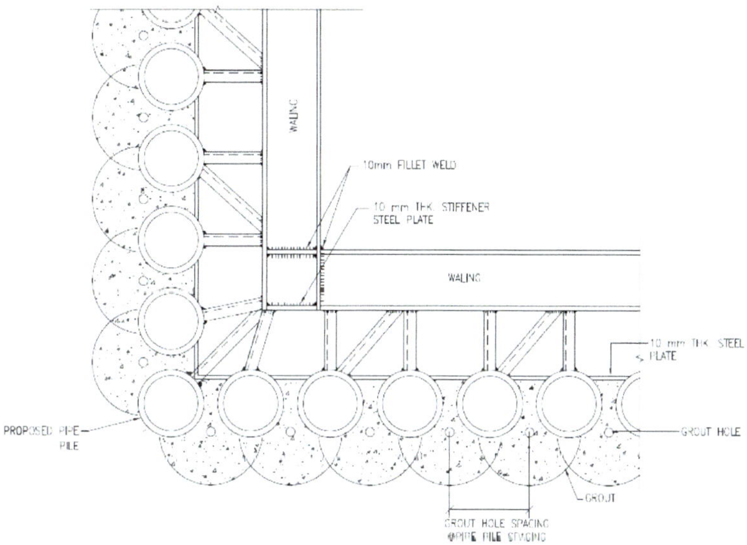

Details of pipe pile wall used in Hong Kong (c)

Figure 5.13 Pipe pile wall installation and back grouting.

For each PIP pile, install a single u-PVC access pipe with an internal diameter of 40 mm attaching to the centre of reinforcing steel cage. The length of access pipe should be equal to that of the reinforcing steel cage and filled with water. Sonic logging should be performed about 3 days after the pile has been installed. A pile integrity test using the pulse echo method/transient response method (vibration test) is carried out for 15% of the PIP piles installed.

The safe pile capacity P of PIP pile = $\mu N_{av} PeL + 5NA$ (5.1)

Where μ is taken as 0.7 normally, N_{av} is the average SPT over the pile shaft, Pe is the perimeter of the pile, L is the length of pile, N is the SPT at the pile toe and A is the sectional area of the pile.

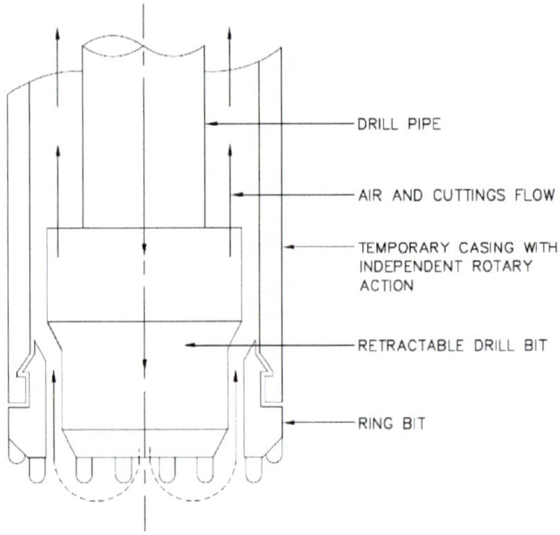

Figure 5.14 Ring bit drilling system (see also Chapter 6 for the concentric driving system).

Figure 5.15 PIP walls with a reinforcement cage.

5.2 METHODS OF EXCAVATION

For the excavation methods, there are two major types (Figures 5.16 and 5.17) with some hybrid forms.

5.2.1 Bottom-up construction

In bottom-up construction (Figure 5.16), the retaining wall (usually temporary) is constructed first. If king-posts are required, they are also installed. Excavation commences for a depth of 2–3m usually, which is followed by the installation of wales/struts or tie-backs. After that, excavation continues and the whole process is repeated until the final level is reached. The base slab/cap may then be constructed which is followed by the construction of the permanent wall. The support system will gradually be removed as the construction proceeds upward; these procedures are illustrated in Figures 5.1 and 5.16. King posts can help to reduce the slenderness ratio of the struts as well as providing support to the decking and are commonly constructed for storage of materials as well as vehicles. In general, the depth of construction using this approach is limited to avoid excessive wall and ground movement, though very deep bottom-up construction is also used in some cases.

5.2.2 Top-down construction

In top-down construction (Figure 5.17), the permanent retaining wall is constructed first. Barrettes, or bore piles, are also constructed, which eventually form the columns of the underground structures. The columns of many underground MTR stations are hence actually piles/barrettes during construction. Excavation commences for a depth of 2–3 m usually, followed by the installation of wales/struts or tie-backs. After that, the first roof slab is constructed, with at least two openings (for labourer and materials transportation). After that, excavation proceeds underneath the slab to form the second floor slab. There have been accidents in the MTR Admiralty station as the workers were underneath the roof slab while the blinding layer had not been completely removed. The falling concrete resulted in injury to some workers during the second layer of excavation. The construction of each floor proceeds from top to bottom, hence it is called a top-down method, and this method is used for most deep excavation works in Hong Kong. One distinct advantage of the top-down method is that the superstructure can be constructed in parallel with the underground construction. Top-down construction is more complicated and expensive than bottom-up construction, but it appears to be unavoidable for many deep excavation works in Hong Kong. The top-down construction process is illustrated in Figure 5.17. Many MTR concourses using a diaphragm wall, Sun Plaza, Times Square, Hong Kong Bank and other projects in Hong Kong, have been constructed using the top-down method. Using the Sun Plaza project as an illustration, the basements and superstructure were constructed simultaneously, and the major construction works for the top and bottom were completed at almost the same time. To the client, this may be an advantage in time saving, while costing more. For deep excavations, the top-down method appears to be preferred, as the more critical consideration is the wall/ground movement and the effects on surrounding buildings.

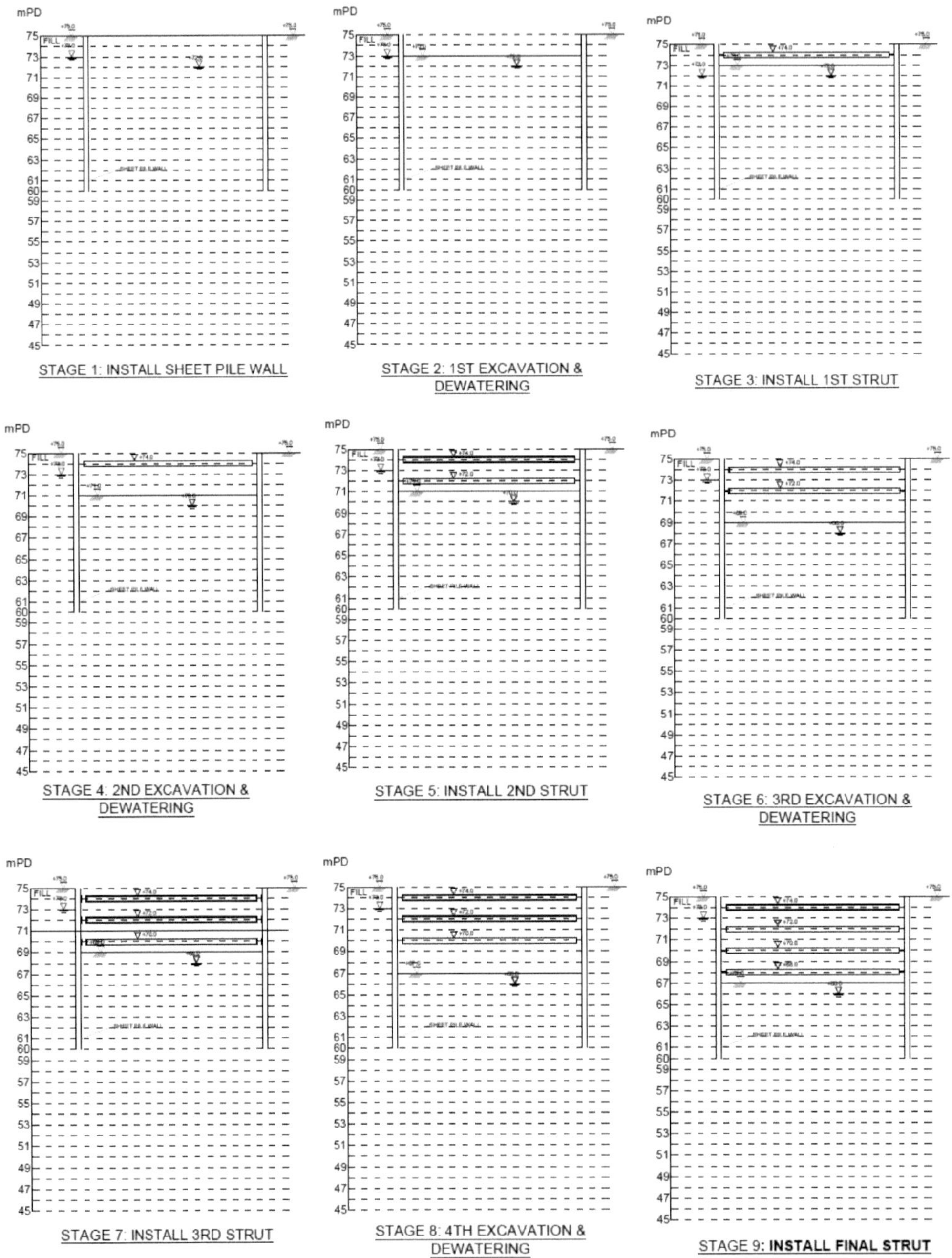

(a) Initial stages of bottom up construction

Figure 5.16 Illustration of bottom-up construction in a project in Hong Kong.

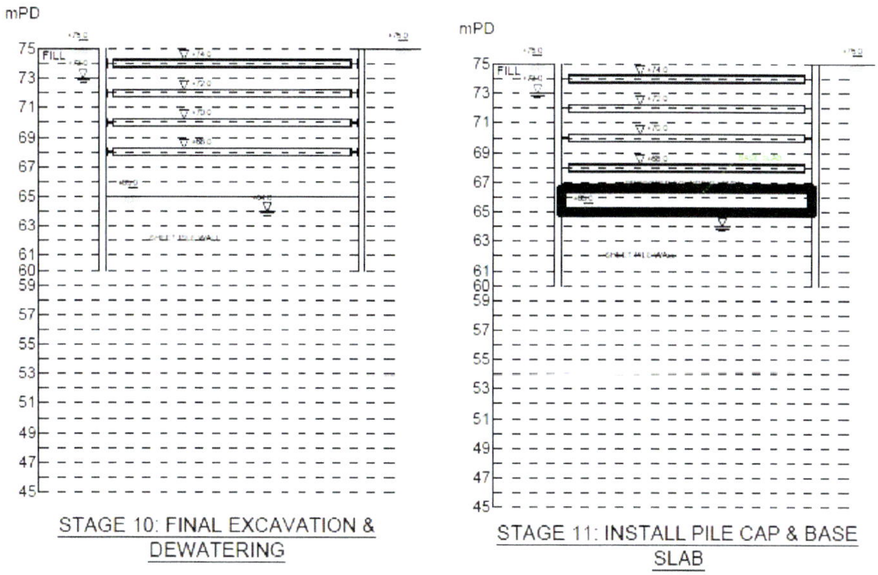

(b) Final stages of bottom up construction (internal piles can be constructed at this stage or at the start of the construction

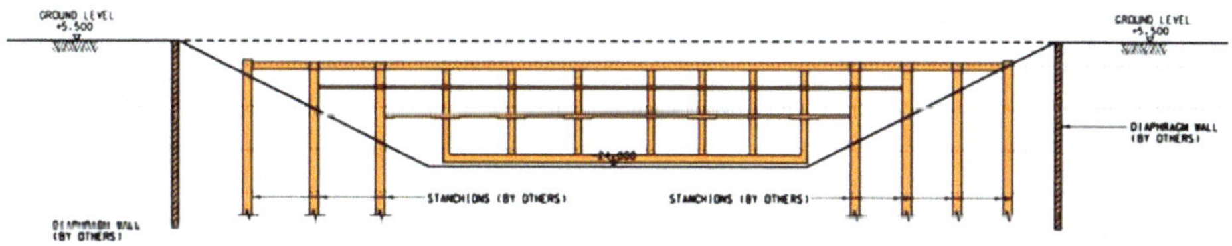

(c) Bottom up construction for West Kowloon express rail project

Figure 5.16 (Continued)

A pipe pile wall ELS bottom-up construction method statement for a project in Hong Kong is described below for reference. This excavation and lateral support plan primarily includes temporary soil/rock excavation and installation of lateral members. The general construction sequence is as follow:

1. Install the proposed pipe piles and king posts to the design levels, and apply grout curtain to the design depth;
2. Conduct a pumping test. It should be noted that drawdown analysis has been conducted for this project instead of assuming a stationary water table outside the pipe pile wall;

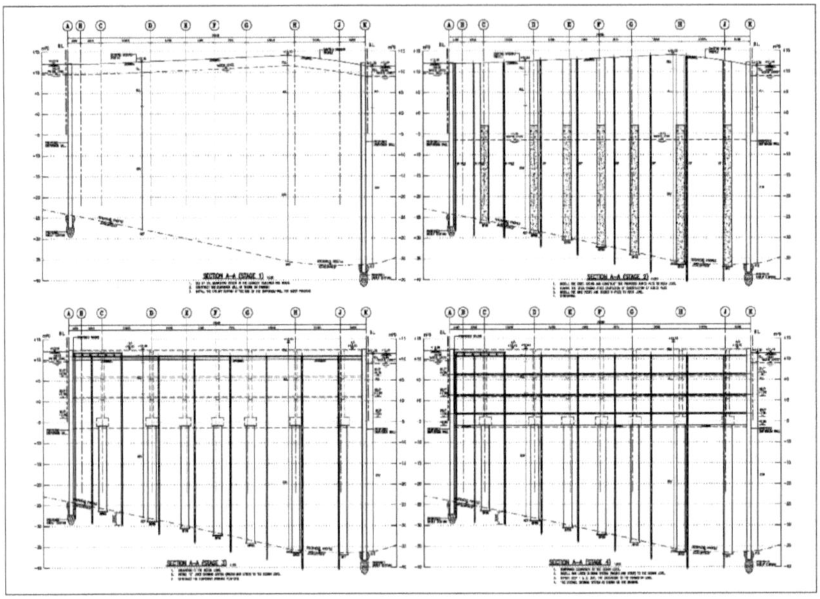

(a) Initial stage of top down construction

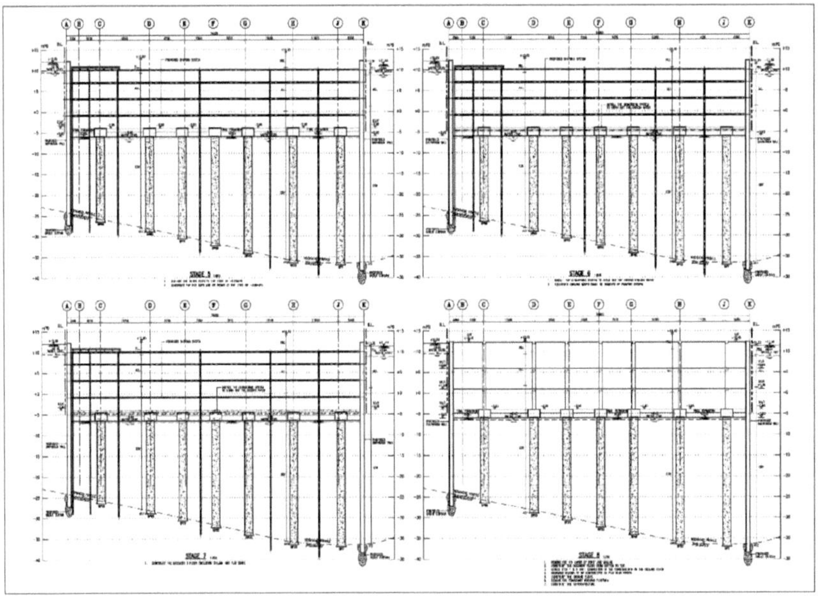

(b) Final stages of top down construction

Figure 5.17 Illustration of top-down construction (note that permanent slabs with openings are constructed).

3. Trim the existing retaining walls;
4. Construct an open cut slope down to the pipe pile wall at the road side;
5. Install the first and second layers of diagonal struts with preloading to prop against the existing retaining wall;
6. Trim the existing retaining walls to a width of 1.7 m;
7. Commence excavation from the ground surface;
8. Apply shotcrete to the pipe pile wall for every 1.0 m depth of excavation;
9. Further excavation and applying shotcrete where necessary until reaching 500 mm below the waling/strut level;
10. Install the proposed waling and strut;
11. Repeat steps (8) and (9) until all struts have been installed;
12. Form the final excavation profile;
13. Place 75 mm blinder layer;
14. Construct the proposed 2.5 m thick pile cap;
15. Dismantle the lower layer of struts except the diagonal one;
16. Trim 1.0 m off the pipe pile wall on the top at the existing retaining wall side;
17. Construct the proposed 1.0 m thick pile cap, E/M room and water tanks;
18. Dismantle the upper layer of struts except the diagonal one; and
19. Construct the upper floors of the superstructure.

The top-down construction sequences for a project in Hong Kong are given as:

1. Construct permanent/temporary retaining walls;
2. Construct prebored H piles;
3. Carry out dewatering test;
4. Carry out excavation phased with installation of strutting system down to soffit of the top slab, dewatering simultaneously as necessary;
5. Cast tunnel top slab;
6. Place water-proofing membranes on top slab;
7. Remove strutting system and back-fill to finished ground level, using the bottom slab as the strut;
8. Carry out cooling main laying;
9. Carry out excavation below the tunnel top slab phased with installation of the strutting system down to the final formation levels, dewatering simultaneously as necessary, using the extruded prebored H-piles as the support of the top slab and lateral buckling restraint to struts;
10. Place blinding and waterproofing membrane and cast tunnel base slab;
11. Remove bottom two strut layers (base slab now acting as a strut);
12. Construct rest of tunnel box walls without strut intrusion;
13. Decommission dewatering system and let the water return to normal.

For the wale and strut support, many variations exist in practice. For example, for deep excavation in poor soil, it is possible that double sections are required to resist the heavy loadings. The use of king posts to reduce the slenderness ratio of the strut is commonly used (Figures 5.1b and 5.17), unless the size of the excavation is small. In some cases, large-section concrete struts are required. In some cases, the struts are supported by the pile caps/footings/piles as shown in Figure 5.18, and this is also a common practice

(a) Support by multiple smaller pile caps

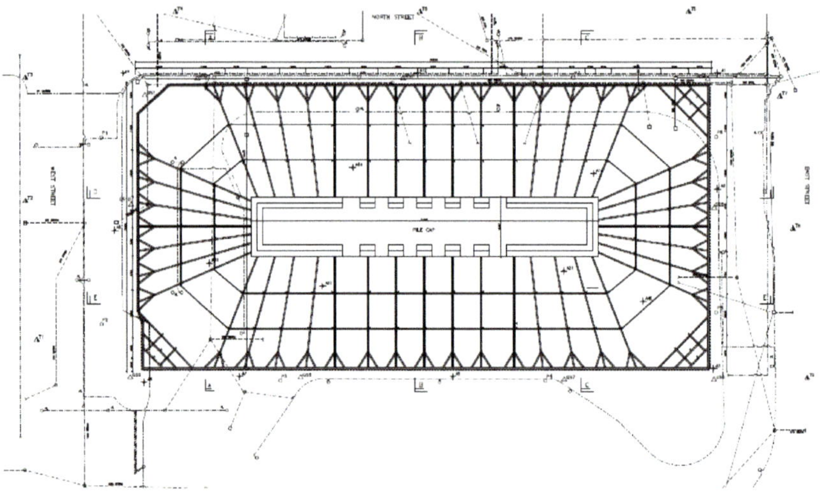

(b) Support by a large pile cap

Figure 5.18 Bottom-up construction with some struts supported by the caps, in Hong Kong.

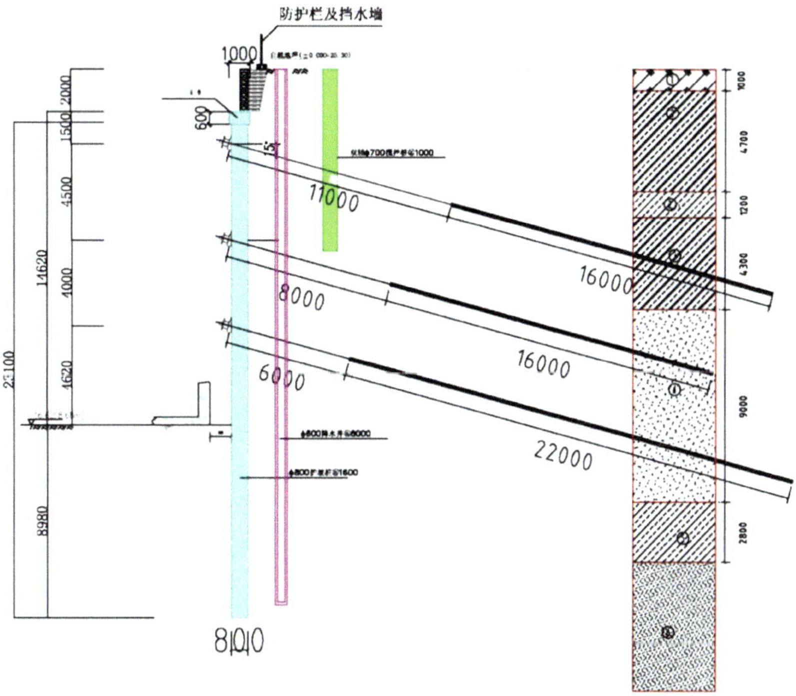

(a) An anchor wall for a deep excavation project in Beijing

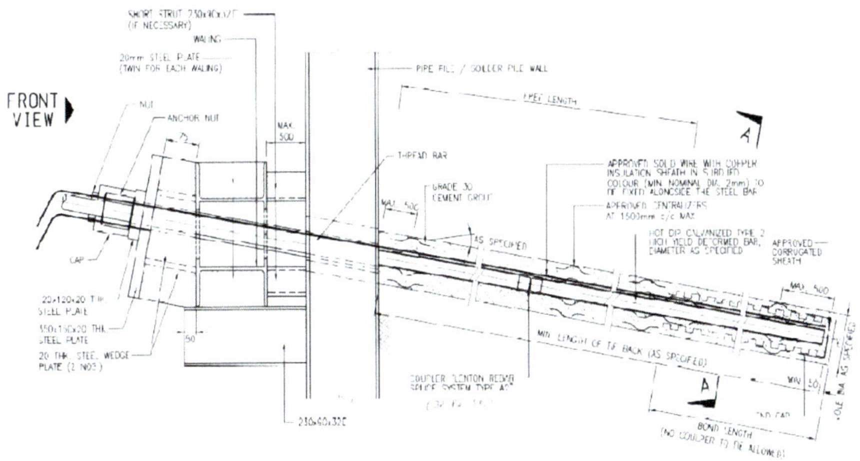

(b) An anchor used for deep excavation in Hong Kong

Figure 5.19 Use of an anchor system in deep excavation.

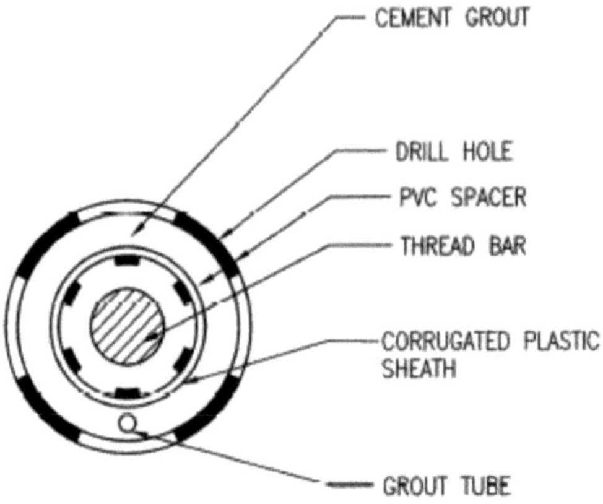

(c) Section AA for Fig.5.19b

Figure 5.19 (Continued)

in Hong Kong. Besides the use of struts, anchors are also used to support the walls. For example, the sheet pile wall of the Lok Fu MTR station in Hong Kong was supported by a series of temporary anchors during construction. There is a limitation to such an approach in Hong Kong, as the anchor will protrude to the surrounding areas which may not belong to the same client. Such lot boundary problems are very common in Hong Kong so that the use of anchor supports becomes less common, although it is a very common method in other countries. To control the lateral wall displacement, strut preloading is used for many very deep excavation projects in Hong Kong if a ground anchor is not allowed. Without the lot boundary restriction, the use of an anchor support as shown in Figure 5.19 can be adopted, which allows much greater flexibility in construction.

5.3 LATERAL EARTH PRESSURE FOR ELS SYSTEMS

For the proper design of an ELS system, several major factors have to be considered:

(1) Soil lateral earth pressure;
(2) Water pressure and seepage;
(3) Excavation and support method;
(4) Design of the retaining wall, shoring, dewatering;
(5) Assessment of the wall and ground movement, and any potential problems for the surrounding buildings and ground;
(6) Monitoring system for the water table, ground and building movements, development of cracks.

In general, the analysis and design of an ELS system is more difficult than other typical foundation works, as the analysis needs to consider the excavation process. For other foundation problems, usually only the final stage needs to be considered, while the temporary stages usually do not control the design. The problem is further complicated by: (1) the constitutive behaviour of soil under different stress paths; (2) insufficient time and tests, and the inability to calibrate a good constitutive model for most of practical projects; (3) many minor changes to the actual design may happen on site frequently; (4) the detailed soil–structure interaction in this problem is difficult to be modelled precisely by most commercial programs, and there is not a single commercial program with which the authors are totally satisfied; and (5) most of the analysis and design works are carried out using 2-D analysis, while the actual construction site is 3-D.

For the lateral earth pressure, there are many cases which should be noted by engineers. There are also various design methods and statistical relations which have been used with success. There is, however, an important concept which must be emphasized. For the full active and passive earth pressure coefficients which are used in routine design, many engin eers assume that the ultimate lateral earth pressure will be mobilized automatically. Many design methods are also based on the use of these coefficients without consideration of the actual lateral movement of the wall. For active pressure, it can be mobilized upon a small displacement, hence such an assumption may still be acceptable. For passive pressure, a relatively large strain is required to mobilize the full passive pressure, hence the use of the passive pressure coefficient in the design may not be appropriate. For example, in the free-earth support method, the maximum moment as calculated deviates from the experimental values, and an empirical moment reduction has been applied historically. Such an empirical reduction can be explained by the actual earth pressure lying between the full active and passive pressures, and the use of these two simple coefficients may leads to problems in calculation, as the free-earth method is statically determinate and should not give an incorrect answer.

In many building codes, a factor of safety of 2.0 is applied to the passive pressure if wall friction is used, and no factor of safety is applied if the wall friction is assumed to be zero. It should also be noted that *Kp* can be greatly increased by the use of wall friction, and the value of *Kp* after application of a factor of safety close to the traditional Rankine *Kp*. If the traditional Rankine *Kp* is used with a factor of safety of 2.0, the wall of *Kp* will be very small, and a very deep depth of penetration may be required. This case happens frequently in Hong Kong, where the depth of penetration is such that the sheet pile needs to penetrate into rock stratum. The use of wall friction is hence commonly adopted by many engineers.

There are many possible conditions for the lateral earth pressures, which will be discussed in the following sections.

5.3.1 At-rest earth pressure coefficient

Consider a deposit of soil formed by sedimentation in thin layers over a wide area. No lateral yield occurs as a result of the imposition of load upon it by the deposition of successive layers above. The in situ horizontal effective earth pressure σ_h in such a soil is known as the "earth pressure at rest". Theoretical and experimental determinations have shown that the value of the coefficient of earth pressure at rest (K_0) lies between the active and passive earth pressure coefficients, k_a and k_p. For the at-rest earth pressure coefficient of sand, k_0, Jaky (1944) developed the following theoretical equation for a granular material for normally consolidated state where the vertical effective stress has never been higher than at present:

$$k_0 = (1 - \sin\phi')\frac{\left(1 + \frac{2}{3}\sin\phi'\right)}{(1 + \sin\phi')} \approx (1 - \sin\phi') \tag{5.2}$$

Based on some limited tests in Hong Kong for sandy soil, k_0 in Eq. (5.2) should be multiplied by a factor of 0.95. Since a fine tuning is not a major factor, most engineers in Hong Kong (and also other countries) simply take $k_0 = (1 - \sin\phi)$ for sandy soil for normal design works. Mayne and Kulhawy (1982) and Meyerhof (1976) give an empirical equation for k_0 which takes into account the over-consolidation ratio (OCR) for a ground slope angle β as

$$k_0 = (1 - \sin\phi)OCR^{\sin\phi} \tag{5.3}$$

$$k_0 = (1 - \sin\phi)\sqrt{OCR}(1 + \sin\beta) \tag{5.4}$$

In Hong Kong, Eq. (5.4) is commonly used to estimate the at-rest earth pressure with an upslope angle. The use of a pressuremeter test to determine the at-rest earth pressure is seldom adopted in practice due to the high cost of the tests.

5.3.2 Rankine earth pressure

Classically, the Rankine equation is used for the active and passive pressures in the form:

$$\sigma'_a = \sigma'_v k_a - 2c\sqrt{k_a} + qk_a \tag{5.5}$$

$$\sigma'_p = \sigma'_v k_p + 2c\sqrt{k_p} + qk_p \tag{5.6}$$

$$\text{Where } k_a = \frac{(1 - \sin\phi)}{(1 + \sin\phi)} \qquad k_p = \frac{(1 + \sin\phi)}{(1 - \sin\phi)} \tag{5.7}$$

Eqs. (5.5)–(5.7) are based on the Mohr circle and the Mohr-Coulomb criterion, and are valid only for a vertical smooth wall with level back. It must be noted that: (1) the same lateral earth pressure coefficient applies for the three terms relating to the self-weight of soil, cohesive strength and surcharge are valid only for simple cases; and different coefficients will be obtained for the more general cases (see Appendix in Chapter 7 for more details); (2) the use of Eqs. (5.5) and (5.6) assume a direct superposition of the three effects, which is strictly not valid but is a good approximation, as demonstrated by Cheng (2003). For a smooth wall with an inclined ground (sloping angle β) at the back. Soubra mad Macuh (2002) has obtained the following:

$$P_{p,a} = \gamma z K_{p,a} \pm 2cK_{pc,ac} + qK_{p,a} \tag{5.8}$$

$$K_{pc,ac} = \sqrt{\frac{1 \pm \sin\phi}{1 \mp \sin\phi}}\cos\beta - c_q \cdot \frac{\cos\beta}{\cos\phi} \tag{5.9a}$$

$$c_q = 1 - \sqrt{\cos^2\beta + \sin^2\beta \cdot \frac{\left(\frac{c}{rz}\right)^2 - \left[\left(1+\frac{q}{rz}\right)\frac{\cos\beta}{\cos\phi}\right]^2}{\left[\frac{c}{rz} + \left(1+\frac{q}{rz}\right)\tan\phi\right]^2}} \tag{5.9b}$$

If wall cohesion c'_w is considered, Eq. (5.10) is commonly used in Hong Kong.

$$Kac = 2(Ka(1 + c'_w/c'))^{0.5} \qquad Kpc = 2(Kp(1 + c'_w/c'))^{0.5} \tag{5.10}$$

Rankine lateral earth pressure is totally based on the Mohr-Coulomb criteria (yield), and it is the lower bound solution instead of the exact solution. It is, however, on the safe side and is simple to be adopted for routine design. Rankine has also developed solutions for the case in Figure 5.20 for sandy soil which are given by

$$K_a = \cos\beta\frac{\cos\beta - \sqrt{\cos^2\beta - \cos^2\phi}}{\cos\beta + \sqrt{\cos^2\beta - \cos^2\phi}} \qquad K_p = \cos\beta\frac{\cos\beta - \sqrt{\cos^2\beta - \cos^2\phi}}{\cos\beta + \sqrt{\cos^2\beta - \cos^2\phi}} \tag{5.11}$$

Eq. (5.11) is given in some books and design codes, and has been used for design by some engineers. The authors, however, believe that these two equations are not correct in that the lateral stress is assumed to be parallel to the backfill with angle β, which is not possible for a smooth wall. If β tends to ϕ, k_a will be independent of β while k_p decreases with β! Rankine active pressure coefficient in Eq. (5.11) is a very good result which can be used safely. On the other hand, the Rankine passive equation in (5.11) and Table 5.2 is very poor, while the Coulomb passive coefficient can be highly unsafe in application. The limitations of Eq. (5.11) appear to not have been addressed previously. Actually, Rankine assumes the stress on every section is the same as the other to get Eq. (5.11), but this assumption is not valid for a section close to the wall, unless β is 0.

The Rankine solution is based on the lower bound method, and the most rigorous lower bound solution will be the solution by the slip line method (Cheng 2003). For

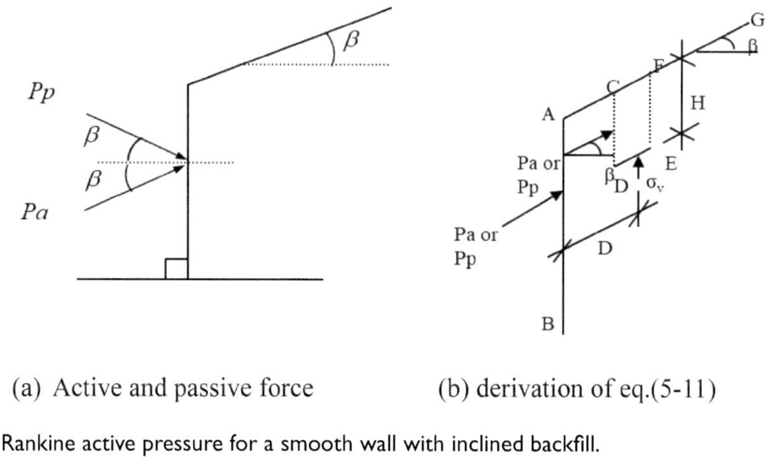

Figure 5.20 Rankine active pressure for a smooth wall with inclined backfill.

Table 5.2 Comparisons between passive earth pressure coefficients from Rankine's equation (5.11), Coulomb's equation and the slip-line method at fiction angle ϕ = 30°

		Coulomb/ Slip line					
β	*Rankine*	$\delta = 0°$	$\delta = 5°$	$\delta = 10°$	$\delta = 15°$	$\delta = 20°$	$\delta = 30°$
0	3.0	3.0/3.0	3.505/3.482	4.143/4.021	4.977/4.615	6.105/5.257	10.096/ 6.561
5	2.943	3.492/3.479	4.183/4.071	5.086/4.740	6.31/5.479	8.049/6.277	14.885/ 7.868
10	2.775	4.08/3.974	5.028/4.696	6.314/5.510	8.145/6.409	10.903/7.379	23.468/ 9.306
15	2.50	4.807/4.493	6.119/5.364	7.989/6.337	10.815/7.413	15.422/8.573	41.533/ 10.871
20	2.132	5.737/5.049	7.593/6.070	10.404/7.216	15.004/8.485	23.373/9.851	91.831/ 12.555
25	1.664	6.982/5.621	9.699/6.805	14.158/8.138	22.304/9.614	39.794/11.201	356.35/ 14.346
26	1.556	7.284/5.737	10.233/6.954	15.167/8.326	24.431/9.844	45.218/11.478	552.56/ 14.718
27	1.439	7.608/5.856	10.816/7.104	16.295/8.514	26.896/10.076	51.883/11.756	974.32/ 15.103
28	1.310	7.957/5.966	11.456/7.255	17.565/8.704	29.777/10.310	60.209/12.036	2173.25/ 15.43
29	1.159	8.334/6.081	12.161/7.406	19.001/8.894	33.178/10.543	70.813/12.316	8612.95/ 15.902
30	0.866	8.743/6.189	12.94/7.558	20.638/9.087	37.238/10.780	84.632/12.601	∞/16.191

the upper bound solution, Coulomb's solution is the most famous, but there are many limitations which need careful consideration. So far, Rankine's equation remains the most popular method in determining the lateral earth pressure, though the values are actually related to the ultimate values. It should also be noted that a factor of safety of 1.5 or 2.0 (Hong Kong) is adopted in many design codes for the passive earth pressure coefficient, no matter which method is used for determination of the coefficient. This factor of safety is considered as reasonable, as it takes a relatively large movement of the wall before the full passive pressure can be mobilized.

5.3.3 Coulomb earth pressure and Euro code

Coulomb's earth pressure is based on the assumption of a failure mechanism and is an upper bound solution. In general, a planar failure surface is assumed and this method is easy to be used and is applicable to more general problems, including earthquake loadings. The limitation of Coulomb's theory is that while the error in the active pressure is usually small, the error in the passive pressure can be significant if the wall friction is high. The active and passive Coulomb earth pressure coefficients are given by

$$K_a = \frac{\sin^2(\alpha + \phi)}{\sin^2\alpha \sin(\alpha - \delta)\left[1 + \sqrt{\dfrac{\sin(\phi + \delta)\sin(\phi - \beta)}{\sin(\alpha - \delta)\sin(\alpha + \beta)}}\right]^2} \tag{5.12}$$

$$K_p = \frac{\sin^2(\alpha - \phi)}{\sin^2\alpha \sin(\alpha + \delta)\left[1 - \sqrt{\frac{\sin(\phi + \delta)\sin(\phi + \beta)}{\sin(\alpha + \delta)\sin(\alpha + \beta)}}\right]^2} \quad (5.13)$$

Where δ is the wall friction and α is the inclination of the retaining wall back with respect to the horizontal direction for the back side of the retaining wall (=90° for a vertical wall). It must be noted that Coulomb's passive coefficient may be highly unconservative in some cases, and the use of a refined method or even Rankine's passive pressure coefficient may be required.

The best solutions for the active and passive earth pressures are given by the solution of the plasticity based slip line equations, and Cheng has developed two general programs, KA and KP, for such a purpose, with the consideration of pseudo earthquake coefficient. These demonstration programs can be obtained from Cheng for research or teaching at natureymc@yahoo.com.hk. These two partial differential equations can only be solved by iterative finite difference method. Rigorous solutions by KA and KP indicate that the log-spiral curve is a close approximation of the actual failure surface. Design figures for the lateral earth pressure in GEO Guide 1 and Design Manual 7 are based on the 1948 solutions, which have been found to be not accurate in some cases by the authors. Detailed solutions are provided by the updated earth pressure tables by Absi and Kerisel (1990), and the solutions by Absi and Kerisel (1990) are very close to the results by KA/KP, but without consideration of the earthquake coefficient. A good approximation for the static normal components of the active and passive earth coefficients is provided in Euro Code 7, which are

$$k_a = \frac{[1 - \sin\phi \sin(2m_w - \phi)]}{1 + \sin\phi\sin(2m_t - \phi)} e^{-2(m_t + \beta - m_w - \theta)\tan\phi} \quad (5.14)$$

$$k_p = \frac{[1 + \sin\phi \sin(2m_w + \phi)]}{1 - \sin\phi\sin(2m_t + \phi)} e^{2(m_t + \beta - m_w - \theta)\tan\phi} \quad (5.15)$$

$$\text{Where } 2m_t = \cos^{-1}\left(\frac{-\sin\beta}{\pm\sin\phi}\right) \mp \phi - \beta \text{ and } 2m_w = \cos^{-1}\left(\frac{\sin\delta}{\sin\phi}\right) \mp \phi \mp \delta \quad (5.16)$$

for passive and active cases, respectively, θ is the inclination of the wall from the vertical, and is 0 for a vertical wall. It should be noted that for m_t, if the term inside $\cos^{-1}$ is negative, take the second quadrant value. In Eqs. (5.14) and (5.15), radian instead of degree is used in the equations.

5.3.4 Discussion on 2-D earth pressure theory

Rankine and Coulomb earth pressures represent the upper and lower limits of the ultimate earth pressure, but they are the true solutions only for special cases. Yield alone is not necessarily the ULS, hence Rankine theory is the lower bound. Once an assumed failure mechanism is assumed in the upper bound analysis, a more critical solution may be available, hence Coulomb theory is the upper bound.

For k_p, the Coulomb equation is poor and design figures/tables should be used. For an active equation, the difference between the Coulomb active pressure and the rigorous solution is actually small. For Coulomb active pressure, as long as a failure surface is chosen, it is either the true solution or less critical solution. If it is a less critical solution, the force from failure should be less, hence we need to search for the maximum active pressure.

For Coulomb passive pressure, as long as a failure surface is chosen, it is either the true solution or less critical solution. If it is a less critical solution, then the soil mass is less easy to fail or equivalently greater passive pressure must be exerted to generate failure. Hence we need to search for the minimum passive pressure in Coulomb's method. Furthermore, the failure surface may not be a planar surface, and we need to try all possible failure shapes in the Coulomb analysis.

It must be noted that besides the soil lateral earth pressure, additional earth pressure can be induced from construction load, plants, foundations from adjacent structures, earthquakes, compaction processes, water pressure from breakage of water mains and other possible factors. Furthermore, the choices of soil parameters and factor of safety are also important in the design. In general, the wall friction should be limited, particularly for the passive pressure determination, as sufficient wall movement must be allowed to mobilize the full passive pressure. Many engineers adopt a maximum wall friction up to half of the soil friction angle in the design, while a factor of safety of 2.0 to the passive pressure coefficient is commonly applied. The effects of line load, point load, and patch load on retaining wall have been provided by Terzaghi, but the validity of such approaches has not been well proved.

5.3.5 3-D lateral earth pressure for slurry wall

In some cases, it is necessary to consider the 3-D effect, due to the limited amount and extent of the backfill. Some approximate formulae have been proposed in the past, but most of these formulae are not readily applicable for real problems, and most engineers still adopt the 2-D lateral earth pressure as discussed for the actual design work. For the slurry excavation in diaphragm wall, the 3-D effect in the form of arching is however important, and is usually considered in the actual design. Without the arching effect, the design of the slurry wall will be extremely conservative and expensive. Consider the yielding of a horizontal strip, as shown in Figure 5.21. The actual failure surfaces will be curves instead of straight lines, starting from the yield point to the ground surface. Assume the failure surfaces to be vertical for simplicity. Consider the equilibrium of the yielding slice of width B and unit length, then

$$B[\sigma_z + \Delta\sigma_z] = B\sigma_z + \Delta W - 2\Delta z\tau_f \tag{5.17}$$

Since $\Delta W = \gamma B\Delta z$, $\tau_f = c' + \sigma_x'\tan\phi$ and $\sigma_x' = K\sigma_z'$ (5.18)

$$B\Delta\sigma_z = \gamma B\Delta z - 2c'\Delta z - 2K\sigma_z\tan\phi\Delta z \tag{5.19}$$

$$\text{Or}\,\frac{\Delta\sigma_z}{\Delta z} = \gamma - \frac{2c'}{B} - \frac{2K\tan\phi'}{B}\sigma_z \tag{5.20}$$

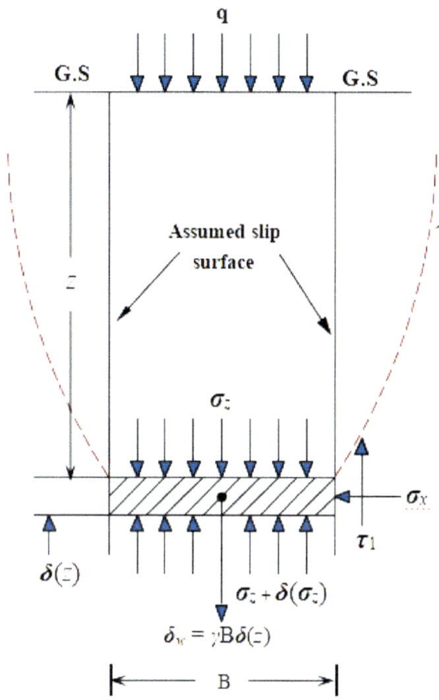

Figure 5.21 Arching in a slurry wall or silo problem.

At ground surface, the surcharge is q, hence integration of Eq. (5.20) will give

$$\sigma_z = \frac{B[\gamma - 2c'/B]}{2K\tan\phi'}\{1 - e^{-(2Kz/B)\tan\phi'}\} + qe^{-(2Kz/B)\tan\phi'} \tag{5.21}$$

If c' is zero, which is commonly adopted by most engineers, in the design, this will give

$$\sigma_h = \mathrm{A}k_a\sigma_v \text{ and } \mathrm{A} = \frac{1 - e^{-2nK_a\tan\phi'}}{2nk_a\tan\phi'} \tag{5.22}$$

Where $n = z/B$, and B is the panel length of the slurry trench

Besides Eq. (5.21), the method by Schneebeli (1964) is also used by some engineers. The arching effect serves to reduce the effective vertical pressure from pure overburden weight of the soil and surcharge by Eq. (5.23):

$$\sigma'_v = \frac{\gamma_s B}{\sin 2\phi}\left(1 - e^{-\sin 2\phi z/B}\right) + qe^{-\sin 2\phi z/B} \tag{5.23}$$

γ_s is the effective weight of soil;

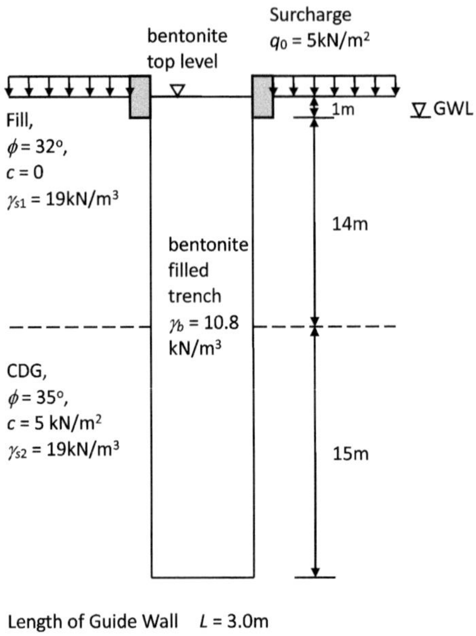

Figure 5.22 Soil profile for worked Example 1.

Eq. (5.23) applies only for homogeneous soil. For layered soil, σ_v' needs to be assessed by layers with q_0 being the vertical stress at the top level of the layer, and σ_h' is simply taken as $k_a \, x \, \sigma_v' - 2c\sqrt{k_a}$.

Consider Example 1 as shown in Figure 5.22, determine the factor of safety for the slurry wall during excavation under different stages.

Determinations of bentonite pressure and ground water pressure are straightforward. For example, at the level 6 m below ground:

Bentonite pressure is $p_b = 10.8 \times 6 = 64.8$ kN/m²;

Ground water pressure is $p_w = 10 \times 5 = 50$ kN/m².

Determination of soil active pressure is more difficult as the effective vertical pressure takes the "arching effect" into account:

(i) For the fill above GWL;
$q_0 = 10$ kN/m², the external surcharge, $\gamma_s = 19$ kN/m³;
At z=1, $\sigma_v' = 23.996$ kN/m²;
Using Eq. (5.23), $\sigma_h' = 7.999$ kN/m².

(ii) For the fill below GWL, $q_0 = 23.996$ kN/m², $\gamma_s = 19 - 10 = 9$ kN/m³; therefore, at the bottom level of the fill where $z = 10$; $\sigma_v' = 30.777$ kN/m²; and by Eq. (5.23), $\sigma_h' = 10.259$ kN/m².

(ii) For the CDG, at the top level, $\sigma_v' = 30.777$ kN/m² as obtained in (ii); $\sigma_h' = 3.134$ kN/m² for ϕ=35° and c = 5 kN/m².

(iv) For the CDG, at the bottom level, $q_0 = 30.777$kN/m²; z=12 m; $\sigma_h' = 28.78$ kN/m² and $\sigma_h' = 2.594$ kN/m².

Table 5.3 Calculation of the factor of safety during slurry trench excavation for Example 1 (σ_b is the fluid pressure due to bentonite), FOS = $(\sigma_b - \sigma_w)/\sigma_h$ (* supported by guide wall)

Depth(m)	*ϕ (degree)*	*$\gamma_s'(kN/m^3)$*	*σ_v' (kN/m^2)*	*$\sigma_h'(kN/m^2)$*	*$\sigma_w(kN/m^2)$*	*σ_b (kN/m^2)*	*$\sigma_b - \sigma_w$ (kN/m^2)*	*FOS*
0	32	19	5.000	1.536	0	0	0	0
1	32	19	20.124	6.183	0	10.8	10.8	1.75
1	32	9.2	20.124	6.183	0	10.8	10.8	1.75
2	32	9.2	22.864	7.025	9.8	21.6	11.8	1.68
3	32	9.2	24.894	7.649	19.6	32.4	12.8	1.67
4	32	9.2	26.399	8.111	29.4	43.2	13.8	1.70
5	32	9.2	27.515	8.454	39.2	54	14.8	1.75
6	32	9.2	28.341	8.708	49	64.8	15.8	1.81
7	32	9.2	28.954	8.896	58.8	75.6	16.8	1.89
8	32	9.2	29.408	9.036	68.6	86.4	17.8	1.97
9	32	9.2	29.745	9.139	78.4	97.2	18.8	2.06
10	32	9.2	29.994	9.216	88.2	108	19.8	2.15
11	32	9.2	30.179	9.273	98	118.8	20.8	2.24
12	32	9.2	30.316	9.315	107.8	129.6	21.8	2.34
13	32	9.2	30.417	9.346	117.6	140.4	22.8	2.44
14	32	9.2	30.492	9.369	127.4	151.2	23.8	2.54
15	32	9.2	30.548	9.386	137.2	162	24.8	2.64
15	35	9.2	30.548	3.073	137.2	162	24.8	8.07
16	35	9.2	30.232	2.987	147	172.8	25.8	8.64
17	35	9.2	30.000	2.924	156.8	183.6	26.8	9.17
18	35	9.2	29.831	2.878	166.6	194.4	27.8	9.66
19	35	9.2	29.708	2.845	176.4	205.2	28.8	10.12
20	35	9.2	29.617	2.820	186.2	216	29.8	10.57
21	35	9.2	29.551	2.802	196	226.8	30.8	10.99
22	35	9.2	29.503	2.789	205.8	237.6	31.8	11.40
23	35	9.2	29.467	2.780	215.6	248.4	32.8	11.80
24	35	9.2	29.442	2.773	225.4	259.2	33.8	12.19
25	35	9.2	29.423	2.768	235.2	270	34.8	12.57
26	35	9.2	29.409	2.764	245	280.8	35.8	12.95
27	35	9.2	29.399	2.761	254.8	291.6	36.8	13.33
28	35	9.2	29.391	2.759	264.6	302.4	37.8	13.70
29	35	9.2	29.386	2.758	274.4	313.2	38.8	14.07
30	35	9.2	29.382	2.757	284.2	324	39.8	14.44

The results are tabulated in Table 5.3. The following observations can be made:

(i) The vertical stresses and hence the horizontal pressure by the soil become more or less constant at greater depths instead of increasing linearly. The arching effect is thus very significant;

(ii) The factor of safety of 1.2 can be achieved at all levels, except at the top first metre which is safeguarded by the presence of the guide walls (Figure 5.23).

5.3.6 Axisymmetric lateral earth pressure

For bore pile, circular shaft or similar works, the shape of the excavation is circular instead of a plane strain problem. For such a condition, the intermediate stress plays an

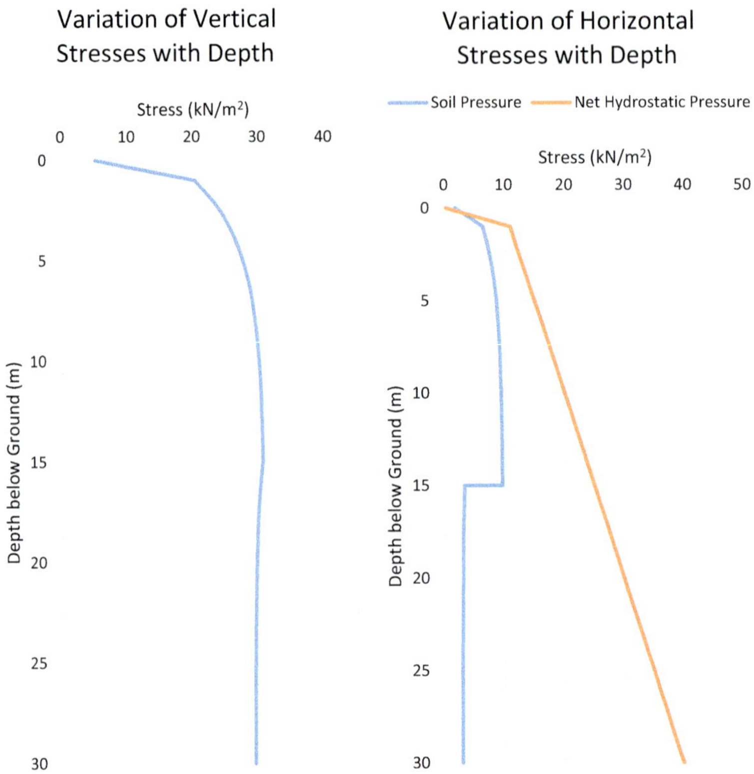

Figure 5.23 Graphical representation of variation of stresses with depth in worked Example 1.

important role in the lateral earth pressure determination. Berezantzev (1958) adopted the Haar-Von Karman's hypothesis where the tangential stress is assumed to be equal to the major principal stress and the failure surface is assumed to be a conical shape. Based on the simplified failure mechanism and Haar-Von Karman hypothesis, an analytical solution for active pressure with a circular cut was developed. At present, many engineers still prefer to adopt the plane strain active pressure coefficient in analysis and design instead of Berezantzev's solution because Berezantzev's theory gives a low value of axisymmetric active earth pressure which may be risky in practice. Steinfeld (1958) and Karafiath(1953) have assumed an axisymmetric Coulomb-type failure surface and the sliding mass is assumed to be a cone in determining the active earth pressure on the shaft lining. They obtain the total earth pressure in a way similar to the classical Coulomb's method. Lorenz (1966) adopted Steinfeld's theory to the active pressure where the tangential stress is neglected in the equilibrium condition. Based on these previous works on using limit equilibrium in axisymmetric problem, Prater (1977) adopted Coulomb's theory and introduced the tangential stress coefficient equal to K_0 and K_a in investigating the active pressure on shaft lining. Prater also viewed that Berezantzev's solution gave a very low value of active earth pressure because of Haar-Von Karman's hypothesis. The authors'

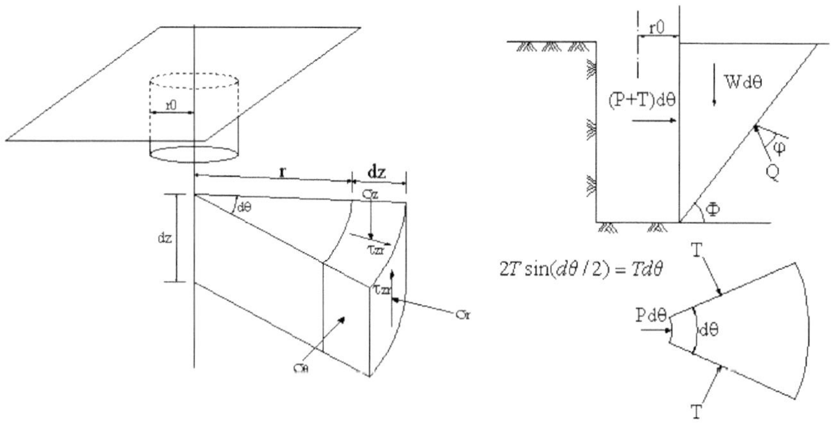

(a) Cylindrical coordinate system and stress (b) Assumed failure mechanism

Figure 5.24 Method of determining the axi-symmetric active earth pressure.

view is that the assumptions by Berezantzev and Prater may not be reasonable. Prater's results indicate that active pressure reaches 0 at a relatively shallow depth, which is shown in figures 1 and 9 of Prater's paper. Cheng and Liu (2007, 2008) introduced a general tangential stress coefficient and solved this problem by means of a rigorous slip-line approach and simplified approach.

For two-dimensional analysis, it is well known that active pressure by Coulomb's equation is close to the results by Caquot and Kerisel (1948) using a log-spiral curve or Sokolovskii (1965) using a method of characteristics. This means that a planar failure surface is a relatively good representation of the failure surface for active pressure determination. Berezantzev has used this fact and assumed a conical failure surface, as shown in Figure 5.24b, for the analysis.

Prater has adopted limit equilibrium in the analysis and the model is as shown in Figure 5.24b. The equilibrium equations for a toroidal element (Figure 5.24a) can be written in cylindrical coordinates r, θ, z as:

$$\frac{\partial \sigma_r}{\partial r} + \frac{\partial \tau_{rz}}{\partial z} + \frac{\sigma_r - \sigma_\theta}{r} = 0, \qquad \frac{\partial \tau_{rz}}{\partial r} + \frac{\partial \sigma_z}{\partial z} + \frac{\tau_{rz}}{r} = \gamma \tag{5.24}$$

The four stress components can be expressed in terms of the mean stress σ and the inclination angle ψ which is formed by extending the major principal stress to the r axis (Figure 5.25b) as:

$$\sigma_r = \sigma(1 + \sin\phi\cos 2\psi) - c \cdot \cot\phi \quad \sigma_z = \sigma(1 - \sin\phi\cos 2\psi) - c \cdot \cot\phi \tag{5.25}$$

$$\tau_{rz} = \sigma\sin\phi\sin 2\psi \quad \sigma_\theta = \lambda\sigma_1 = \lambda\sigma(1 + \sin\phi) - \lambda c \cdot \cot\phi \tag{5.26}$$

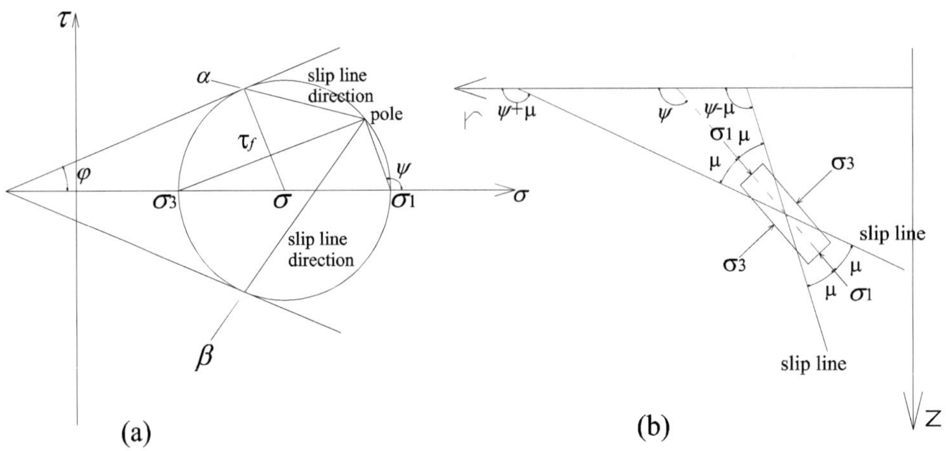

Figure 5.25 (a) Mohr circle under failure condition. (b) Sign convention and notation.

And $\sigma = (\sigma_1 + \sigma_3)/2 + c \cdot \cot\phi$, c is cohesive strength and the tangential coefficient λ is a ratio between σ_θ and σ_1. λ has been taken as 1.0 by most researchers in the past for simplicity, which is known as the Harr and Von Karman hypothesis but this ratio should lie somewhere between 1 and K_a (to be discussed later) and should be taken as a variable generally. Parter (1977) considered $\lambda = K_0 = 1 - \sin\phi$ and $\lambda = 1$ in his analysis with limit equilibrium method.

The characteristic lines for the solution are the α and β lines on which the shear strength is fully mobilized. The slopes of the slip lines are

$$\frac{dz}{dr} = \tan(\psi + m\mu) \text{ Where } \mu = \frac{\pi}{4} - \frac{\phi}{2} \tag{5.27}$$

and m takes the value –1 for an α line and +1 for a β line (shown in Figure 5.21b). These equilibrium and yield equations form a set of hyperbolic partial differential equations expressing the changes in stress along each characteristic line in terms of the changing inclination ψ and position (r, z)

$$\begin{aligned} &d\sigma + m2\sigma\tan\phi d\psi + \frac{(1-\lambda-\lambda\sin\phi)\sigma - c(1-\lambda)\cot\phi}{r}dr \\ &+ m\frac{\lambda(1+\sin\phi)\sigma + c(1-\lambda)\cot\phi}{r}\tan\phi dz = \gamma(m\tan\phi dr + dz) \end{aligned} \tag{5.28}$$

To simplify the study, the variables will be normalized with the circular shaft radius r_0 as

$$\Omega = \frac{\sigma}{r_0\gamma}, \quad R = \frac{r}{r_0}, \quad Z = \frac{z}{r_0}, \quad C = \frac{c}{r_0\gamma}, \quad Q_0 = \frac{q_0}{r_0\gamma} \tag{5.29}$$

where r_0 is the radius of the circular shaft, z is the vertical depth under consideration, γ is the unit weight of soil, and q_0 is the external surcharge. Substituting Eq. (5.29) into Eqs. (5.27) and (5.28), we have:

$$\frac{dZ}{dR} = \tan(\psi + m\mu) \quad \text{where } \mu = \frac{\pi}{4} - \frac{\phi}{2} \tag{5.30}$$

$$d\Omega + m2\Omega\tan\phi d\psi + \frac{(1-\lambda-\lambda\sin\phi)\Omega - C(1-\lambda)\cot\phi}{R}dR$$

$$+m\frac{\lambda(1+\sin\phi)\Omega + C(1-\lambda)\cot\phi}{R}\tan\phi dZ = m\tan\phi dR + dZ \tag{5.31}$$

If we assume the slip lines to be straight lines in the R–Z plane, which is an assumption made by Berezantaev, the inclination of β slip line is:

$$m = 1, \quad \psi = \frac{\pi}{2} = const \text{ and } dZ = \tan\left(\frac{3\pi}{4} - \frac{\phi}{2}\right)dR \tag{5.32}$$

Putting Eq. (5.32) into Eq. (5.28), we can obtain

$$\frac{d\Omega}{dR} - \left[\lambda\frac{(1+\sin\phi)^2}{\cos^2\phi} - 1\right]\frac{\Omega}{R} - \frac{C(1-\lambda)}{R}\frac{1+\sin\phi}{\sin\phi\cos\phi} = -\frac{1}{\cos\phi} \tag{5.33}$$

Let

$$\eta = \lambda\frac{(1+\sin\phi)^2}{\cos^2\phi} - 1 = \lambda\tan^2\left(\frac{\pi}{4}+\frac{\phi}{2}\right) - 1 \text{ and } \xi = \frac{1}{\eta}\frac{\lambda}{}\tan^2\left(\frac{\pi}{4}+\frac{\phi}{2}\right) + 1 \tag{5.34}$$

Putting Eq. (5.34) into Eq. (5.33), the solution of differential Eq. (5.33) is:

$$\Omega = aR^{\eta} + \frac{R}{(\eta-1)\cos\phi} - \frac{C(1-\lambda)(1+\sin\phi)}{\eta\sin\phi\cos\phi} \tag{5.35}$$

Eqs. (5.26) and (5.27) are normalized with the shaft radius r_0 and reduce to the following forms with the use of Eq. (5.32):

$$\Omega_R = \Omega(1-\sin\phi) - C\cdot\cot\phi\,\Omega_Z = \Omega(1+\sin\phi) - C\cdot\cot\phi \tag{5.36}$$

$$\Omega_\theta = \lambda\Omega(1+\sin\phi) - \lambda C\cdot\cot\phi \tag{5.37}$$

According to Eqs.(5.36), (5.37) and (5.35), the component of the stress tensor can be expressed as:

$$\Omega_R = a(1-\sin\phi)R^{\eta} + \frac{(1-\sin\phi)}{(\eta-1)\cos\phi}R - \cot\phi\frac{(1+\eta-\lambda)}{\eta}C \tag{5.38}$$

$$\Omega_Z = a(1+\sin\phi)R^{\eta} + \frac{(1+\sin\phi)}{(\eta-1)\cos\phi}R - \xi C\cdot\cot\phi \tag{5.39}$$

$$\Omega_\theta = ak(1+\sin\phi)R^{\eta} + \frac{(1+\sin\phi)}{(\eta-1)\cos\phi}kR - \xi kC\cdot\cot\phi \tag{5.40}$$

Here R_b denotes the intersect of the β slip line which passes through the point (1, Z) with the R axis, so

$$R_b = 1 + Z\cdot\tan\left(\frac{\pi}{4} - \frac{\phi}{2}\right) \tag{5.41}$$

Since $\Omega_z\big|_{R=R_b} = Q_0$, the integral constant can be determined as:

$$a = \frac{Q_0 + \xi C\cot\phi}{R_b^{\eta}(1+\sin\phi)} - \frac{1}{(\eta-1)\cos\phi R_b^{\eta-1}} \tag{5.42}$$

Putting Eq. (5.42) into Eqs. (5.38)–(5.40), these equations will be simplified as

$$\Omega_R = R\frac{\tan\left(\frac{\pi}{4}-\frac{\phi}{2}\right)}{\eta-1}\left[1-\left(\frac{R}{R_b}\right)^{\eta-1}\right] + Q_0\left(\frac{R}{R_b}\right)^{\eta}\tan^2\left(\frac{\pi}{4}-\frac{\phi}{2}\right)$$
$$+C\left[\xi\left(\frac{R}{R_b}\right)^{\eta}\tan^2\left(\frac{\pi}{4}-\frac{\phi}{2}\right) - \frac{1-\lambda+\eta}{\eta}\right]\cot\phi \tag{5.43}$$

$$\Omega_Z = R\frac{\tan\left(\frac{\pi}{4}+\frac{\phi}{2}\right)}{\eta-1}\left[1-\left(\frac{R}{R_b}\right)^{\eta-1}\right] + Q_0\left(\frac{R}{R_b}\right)^{\eta} + C\xi\left[\left(\frac{R}{R_b}\right)^{\eta} - 1\right]\cot\phi \tag{5.44}$$

$$\Omega_\theta = \lambda R\frac{\tan\left(\frac{\pi}{4}+\frac{\phi}{2}\right)}{\eta-1}\left[1-\left(\frac{R}{R_b}\right)^{\eta-1}\right] + \lambda Q_0\left(\frac{R}{R_b}\right)^{\eta} + \lambda\xi C\left[\left(\frac{R}{R_b}\right)^{\eta} - 1\right]\cot\phi \tag{5.45}$$

In Eq. (5.43), putting $R = 1$, the active earth pressure on the shaft lining is:

$$P_a = \gamma r_0\frac{\tan\left(\frac{\pi}{4}-\frac{\phi}{2}\right)}{\eta-1}\left(1-\frac{1}{R_b^{\eta-1}}\right) + Q_0\frac{\tan^2\left(\frac{\pi}{4}-\frac{\phi}{2}\right)}{R_b^{\eta}} - C\left[\frac{1-\lambda+\eta}{\eta} - \xi\frac{\tan^2\left(\frac{\pi}{4}-\frac{\varphi}{2}\right)}{R_b^{\eta}}\right]\cot\phi \tag{5.46}$$

Therefore, the actual active pressure can be obtained by Eq. (5.29) is:

$$p_a = \gamma r_0 \frac{\tan\left(\frac{\pi}{4}-\frac{\phi}{2}\right)}{\eta-1}\left(1-\frac{1}{R_b^{\eta-1}}\right)+q_0\frac{\tan^2\left(\frac{\pi}{4}-\frac{\phi}{2}\right)}{R_b^{\eta}}-c\left[\frac{1-\lambda+\eta}{\eta}-\xi\frac{\tan^2\left(\frac{\pi}{4}-\frac{\phi}{2}\right)}{R_b^{\eta}}\right]\cot\phi \quad (5.47)$$

where

$$R_b = 1+Z\cdot\tan\left(\frac{\pi}{4}-\frac{\phi}{2}\right)=1+\frac{z}{r_0}\cdot\tan\left(\frac{\pi}{4}-\frac{\phi}{2}\right) \quad (5.48)$$

When $\lambda = 1$, Eq. (5.34) becomes:

$$\eta=\tan^2\left(\frac{\pi}{4}+\frac{\phi}{2}\right)-1=2\tan\varphi\tan\left(\frac{\pi}{4}+\frac{\phi}{2}\right),\ \xi=1 \quad (5.49)$$

and Eq. (5.47) becomes:

$$p_a = \gamma r_0 \frac{\tan\left(\frac{\pi}{4}-\frac{\phi}{2}\right)}{\eta-1}\left(1-\frac{1}{R_b^{\eta-1}}\right)+q_0\frac{1}{R_b^{\eta}}\tan^2\left(\frac{\pi}{4}-\frac{\phi}{2}\right)-c\left[1-\frac{1}{R_b^{\eta}}\tan^2\left(\frac{\pi}{4}-\frac{\phi}{2}\right)\right]\cot\phi \quad (5.50)$$

Eq. (5–50) is the same as Berezentzav's original formula for axisymmetric active pressure. Eqs. (5.46) and (5.47) are more general, in which λ is now a variable. According to Eq. (5.50), the active earth pressure can be formulated as a linear combination of the effects from the self-weight of soil, external surcharge and cohesive strength of soil in the form:

$$p_a = K_{a\gamma}\gamma z + K_{aq}q_0 - K_{ac}c \quad (5.51)$$

where the active earth pressure coefficients $K_{a\gamma}$, K_{aq}, K_{ac} on shaft lining are defined as:

$$K_{a\gamma} = \frac{\tan\left(\frac{\pi}{4}-\frac{\phi}{2}\right)}{\eta-1}\left(\frac{r_0}{z}-\frac{r_0}{zR_b^{\eta-1}}\right)=\frac{\tan\left(\frac{\pi}{4}-\frac{\phi}{2}\right)}{\eta-1}\left\{\frac{1}{Z}-\frac{1}{Z\left[1+Z\tan\left(\frac{\pi}{4}-\frac{\phi}{2}\right)\right]^{\eta-1}}\right\} \quad (5.52)$$

$$K_{aq} = \frac{1}{R_b^{\eta}}\tan^2\left(\frac{\pi}{4}-\frac{\phi}{2}\right)=\tan^2\left(\frac{\pi}{4}-\frac{\phi}{2}\right)\frac{1}{\left[1+Z\tan\left(\frac{\pi}{4}-\frac{\phi}{2}\right)\right]^{\eta}} \quad (5.53)$$

$$K_{ac} = \left[\frac{1-\lambda+\eta}{\eta}-\frac{\xi}{R_b^{\eta}}\tan^2\left(\frac{\pi}{4}-\frac{\phi}{2}\right)\right]\cot\phi=\left\{\frac{1-\lambda+\eta}{\eta}-\frac{\xi\tan^2\left(\frac{\pi}{4}-\frac{\phi}{2}\right)}{\left[1+Z\tan\left(\frac{\pi}{4}-\frac{\phi}{2}\right)\right]^{\eta}}\right\}\cot\phi \quad (5.54)$$

Because of the effect of arching action, axisymmetric active earth pressure must be less than or equal to plane strain active pressure. For the first term in Eq. (5.50), η must be greater than 0 and hence:

$$\eta - 1 \geq -1 \tag{5.55}$$

Putting Eq. (5.55) into Eq. (5.34), we obtain

$$\lambda \geq \tan^2\left(45 - \frac{\phi}{2}\right) = K_a \tag{5.56}$$

Hence the tangential stress ratio λ will lie between 1.0 and K_a. Using Eqs. (5.46) and (5.47), the upper and lower estimates of the active pressure can be determined by putting λ = 1.0 and K_a.

Before excavation, the tangential stress can be taken as $K_0\sigma_v$. During excavation, σ_r will decrease while σ_θ will increase according to the theory of elasticity. λ will possibly be greater than K_0 but is less than or equal to 1.0. Since a greater λ value will give lower active pressure, for the design of circular cut, the authors suggest using $\lambda = K_0$ in Eqs. (5.52)–(5.54) for estimation of the active pressure. These equations can also be used for determining the lower and upper estimates of the active pressure. Engineers can consider both the lower/upper bounds of the active pressure and the active pressure by using $\lambda = K_0$ before the final adoption of a suitable active pressure for the design of a circular cut. The equations as given in this section are generally accurate enough for design purposes. A more realistic analysis by using the slip-line method has been provided by Cheng and Liu, and the computer program for the axisymmetric slip line solution is provided by Cheng (2020).

5.3.7 Method of trial wedge/slope analysis

In case of problems with an irregular ground surface, limited extent of the soil behind the retaining wall, non-uniform distributed load and/or consideration of wall adhesion, the trial wedge method is commonly used by engineers. Cheng has developed a spreadsheet for trial wedge analysis (or Culmann's method), which can be obtained at natureymc@yahoo.com.hk. Although the trial wedge method has been used for many years by engineers, the authors are not in favour of this method, as the point of action of the lateral load cannot be determined precisely. Some empirical formulae for the point of action for simple loading cases are available in different texts or design guides (GeoGuide 1, GEO 1993), however, the validity of such empirical formulae is hardly ever verified, and the effects of different soil layers are not considered in such empirical formulae.

The classical trial wedge method suffers from the limitation of a straight line failure surface. Slope stability analysis methods based on the limit equilibrium approach are upper bound methods, which are actually similar to the upper bound method in trial wedge analysis or the Coulomb's method without the limitation of a straight line failure surface. In the Janbu simplified approach, the basic assumptions are:

1. Upper bound limit equilibrium approach;
2. Mohr-Coulomb relation;
3. Force equilibrium.

In the classical earth pressure problems, the Coulomb earth pressure theory is also based on the upper bound approach with consideration of force equilibrium but not moment equilibrium. The assumption used in the Coulomb theory is actually the same as that in the Janbu simplified method, and so the two problems should actually be equivalent problems. Consider a 5 m height slope with a level back. The soil parameters are $c' = 0$, $\phi' = 30°$ and unit weight=20 kN/m^3. The total horizontal lateral pressure of 16.665 kN/m is found by trial and error when the minimum factor of safety is equal to 1.0. It should be noted that since force equilibrium is used in the Janbu simplified method, the use of point load, uniform distributed load or triangular load will be completely equivalent in the present analysis. An identical result will be obtained if triangular pressure is applied in this example. The correction factor f_o should not be used in this case because only force equilibrium is considered in the Coulomb mechanism, while f_o will correct for the inter-slice shear force and moment equilibrium for the Janbu simplified analysis.

The minimum factor of safety as found is 1.0001 and the critical failure surface is shown in Figure 5.26a and is completely equivalent to the classical Rankine solution. The total load on this slope is 16.665 × 5 = 83.325kN. The active pressure coefficient can be back calculated as

$$83.325 = 0.5\, K_a \times 20 \times 5^2 \quad \Rightarrow \quad K_a = 0.3333 = (1 - \sin\phi)/(1 + \sin\phi)$$

This result is exactly the same as for the Rankine solution or Coulomb solution. The critical failure surface from the non-circular search is also found for a plane surface incline at an angle of 60° with the horizontal direction, which is equivalent to 45° + $\phi/2$ from the Rankine solution. The next example is the same as the previous one with a 20° slope behind a 5m high slope. The total load on this slope for a minimum factor of safety of 1.0 is 110.25 kN. The active pressure coefficient can be calculated as

$$110.25 = 0.5\, K_a \times 20 \times 5^2 \quad \rightarrow \quad K_a = 0.441$$

This result is exactly the same as the Coulomb solution for a slope with 20° back, which is shown in Figure 5.26b. The failure surface is also found to incline at an angle of 52°. The angle of inclination for this case can be found from Design Manual 7 (see 7.2, page 65) which is given by:

$$\tan\theta = \tan\phi + (1 + \tan^2\phi - \tan\beta/\sin\phi/\cos\phi)^{0.5}$$

If we put in $\beta = 20°$, θ is obtained as 52°, and this is exactly the same as that obtained from the non-circular search as discussed earlier (see also the user guide of SLOPE 2000 for the detailed results).

The point of application of the active pressure can be determined from the moment balance of the failure mass and is found to be at one-third of the height of the retaining wall. The present technique can be extended for a retaining wall with a non-homogeneous backfill, and the point of application of the active pressure can be determined by the moment equilibrium between the base normal and shear forces, the weight of soil and the total active pressure from slope stability analysis.

For more complicated cases with an irregular ground surface, several soil layers, water, surface and/or below surface loadings, limited extent of soil behind the retaining wall, wall friction, earthquake loads as well as other conditions, the use of the Janbu simplified

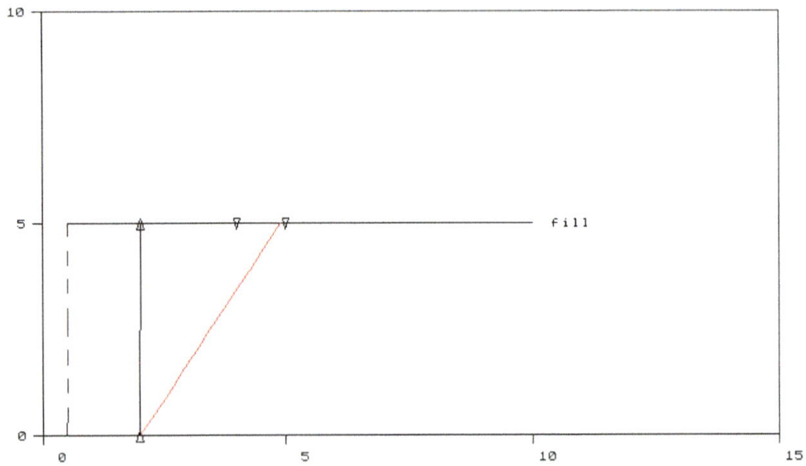

(a) Critical failure surface from Janbu simplified without f_0 based on non-circular search, completely equal to Rankine solution

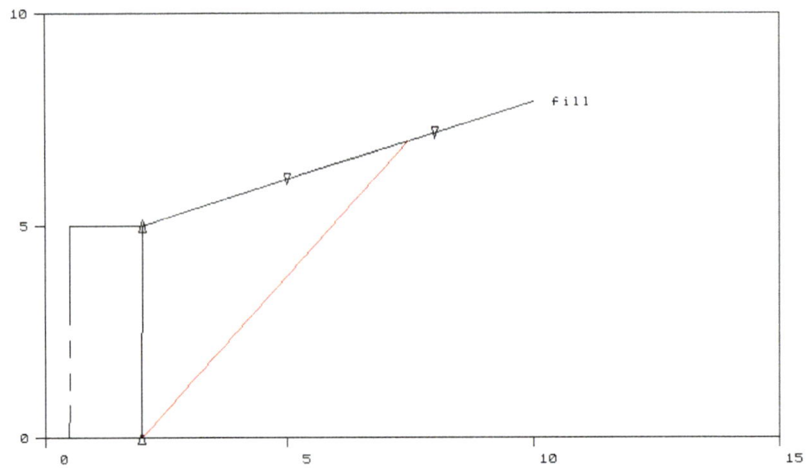

(b) Critical failure surface from Janbu simplified without f_0 based on non-circular search, completely equal to Coulomb solution.

Figure 5.26 Use of the slope stability method to replace the Rankine and Coulomb solution.

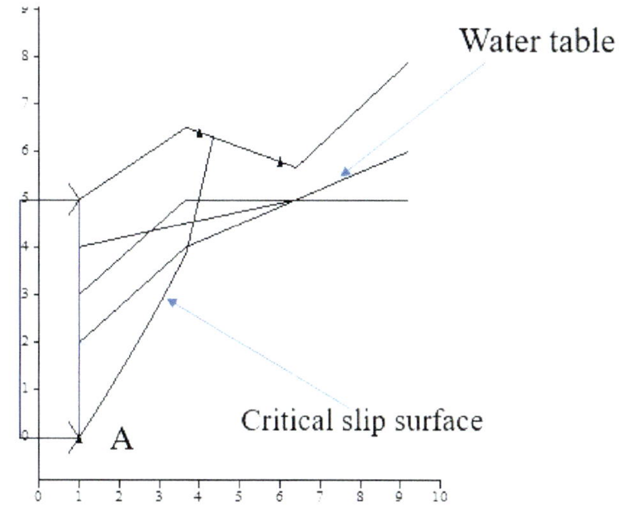

(a) Critical non-circular search using Janbu simplified method

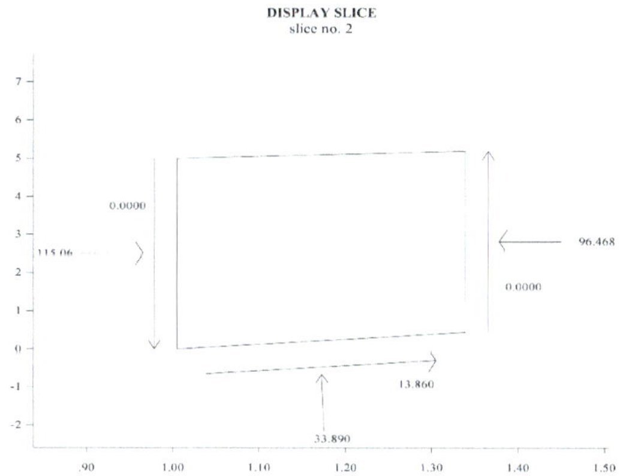

(b) Forces acting on a slice for the critical solution

Figure 5.27 Slope stability method to replace the trial wedge method.

method with the automatic non-circular search algorithm can be used. The active/passive pressure with wall friction can be prescribed and varied until the minimum factor of safety is 1.0. The effect of wall cohesion can also be prescribed (but not varied) in such analysis. Refer to Figures 5.27a and 5.27b, from the base normal and shear force and weight of soil mass, take the moment to point A. The total force F thrust line or centroid is the distance from H to A, and is found by equating with the moment from the total lateral force with respect to point A. If the effective force is required, deduct F_w from F to give F'. Line of thrust of F' is found by $F \times H = F' \times H' + F_w \times H_w$. This method is more rigorous than the trial wedge method, and actually easier to use with greater flexibility!

5.3.8 Lateral earth pressure for earthquakes and other cases

For an earthquake case, the use of pseudo-earthquake coefficients K_h/K_v is commonly adopted for design purposes. Classically, the Mononobe–Okabe (1929, 1926) active/passive pressure coefficients are commonly used:

$$K_a = \frac{\sin^2(\alpha + \phi - \theta)}{\sin^2\alpha\cos\theta\sin(\alpha - \delta - \theta)\left[1 + \sqrt{\dfrac{\sin(\phi + \delta)\sin(\phi - \beta - \theta)}{\sin(\alpha - \delta - \theta)\sin(\alpha + \beta)}}\right]^2} \quad (5.57)$$

$$K_p = \frac{\sin^2(\alpha - \phi + \theta)}{\sin^2\alpha\cos\theta\sin(\alpha + \delta + \theta)\left[1 - \sqrt{\dfrac{\sin(\phi + \delta)\sin(\phi + \beta - \theta)}{\sin(\alpha + \delta)\sin(\alpha + \beta + \theta)}}\right]^2} \quad (5.58)$$

where $\tan\theta = K_h/(1 - K_v)$, and $Pa/p = 0.5\gamma H^2(1 - K_v)Ka/p$. When the Mononobe-Okabe method, which is a simple extension of the Coulomb's method to earthquake condition, is used, the line of thrust is no longer at one-third of the height of the retaining wall. The simple total static active/passive earth pressure *Pa/p* and the additional component *Pae/Ppe* (total earthquake load-static component) are determined. The static component still takes a lever arm of one-third of the height of the retaining wall as measured from the base of the wall, while the additional component is assumed to be located at a distance of 0.6*H* from the base of the wall. Alternatively, the slip line approach by Cheng (2003), which is based on an axis transformation technique and coded in the program KA/KP, can be used.

Classically, the concept of an earth pressure envelope has been adopted in some cases and is based on the field measurement results by Terzaghi (1943) for some projects in the US and Germany. The authors however do not favour such approaches, as they are based on very old data with limited depth of excavation and conventional construction methods. More practically, most projects involve non-homogeneous soil layers and different soil types with water. The traditional equivalent earth pressure envelopes as proposed by Terzaghi cannot fit into most practical projects. With the use of computer software, such an earth pressure envelope is not required in the analysis and design. For interest, apparent earth pressure envelopes may appear from the computer results, which is not required to be defined by the engineers. The actual load imposed on a semi-vertical retaining wall is dependent on eight construction aspects:

1. The degree of relative compaction of the wall backfill within the active or at rest envelopes.
2. The structural stiffness of the retaining wall system and its ability to deflect outward in response to "active" earth loads.
3. The magnitude and inclination of thrust generated by soil loads due to either inclined bedding, filling, or finished grades behind the wall.
4. The inclination between the wall itself and the backfill materials.
5. The applicable skin friction at the wall/backfill interface.
6. The vertical and horizontal components of loads associated with structural tie-backs.

7. Live or dead surcharge loads above and/or behind the wall.
8. The degree of saturation of the wall backfill in the zone of active or at-rest earth pressure

5.4 GROUND WATER TABLE DURING EXCAVATION

The ground water table and water pressure are very important aspects in the design and construction of the ELS. Take a simple estimation for a 30 m depth excavation, and the water table is taken at ground level for simplicity. The effective vertical stress at a depth of 30 m is about 10 × 30 = 300 kPa, while the effective horizontal stress may be about 78 kPa (for ϕ = 36°). The water pressure may amount to about 300 kPa, which is several times the effective horizontal stress due to soil. Under such a high water pressure, water can seep through the diaphragm wall, and if the amount of seepage is higher than the value as specified in the particular specification, epoxy grouting is commonly applied to reduce the amount of seepage. For a permanent structure, 100% water-proofing is usually not specified, as it is very difficult to be achieved in practice. It is hence not surprising to find traces of water at the side walls of many MTR concourses in Hong Kong and other countries. For control of the ground water table during excavation, besides the use of retaining and cut-off walls, the exact ground water control method will depend on the soil type, depth of excavation and adjacent structures. In general, the more commonly adopted options are:

(1) Use of a well point for relatively shallow excavation – A well point system utilizes a vacuum pump which has a maximum uplift of about 7 m in practice. For greater depth, it may be required to carry out multi-stage well point dewatering, but this approach is largely limited by the availability of space, as a slope will be formed at the sides of the excavation. If the lot boundary is limited, that is equivalent to reduced usable space for the underground construction. For this reason, a well point system is seldom used in Hong Kong and many developed cities. In general, a single-stage well point system is limited to about 5 m depth excavation, while a two-stage well point system can be used for up to 10 m excavation. In general, a trench or sump will also be prepared to collect the water at the sides of the excavation. Besides the classical boring method to install the well point system, a self-jetting well point is also available.
(2) Use of deep well – A deep well with a diameter of about 300 mm and the use of a submersible pump with a capacity limited to 30 m (typical) is used when the depth of excavation is great. For the Sheung Wan Concourse in Hong Kong which Cheng has worked on, the maximum depth of the water table to be lowered was about 35m, an intermediate tank was hence placed at about 15 m depth below the ground surface, and a submersible pump was placed into the tank to carry out the second lift. Deep wells can be placed inside or outside the excavation, and there are cases where both internal and external deep wells are used to lower the ground water table. Due to the excessive drawdown that may occur during deep excavation, recharge wells (Figure 5.28b) are used for the Wan Chai and Sheung Wan MTR concourses to bring up the water table outside the retaining wall in order to reduce ground settlement. Some of the typical details for dewatering wells are shown in Figures 5.28a and 5.28c. A typical ground water control plan is shown in Figure 5.28d.
(3) Open excavation by ditch and sump – This is used for shallow excavation.
(4) Use of a grout curtain with complete cut-off – Dewatering can cause serious ground settlement, and settlement up to 350 mm and serious differential settlement causing

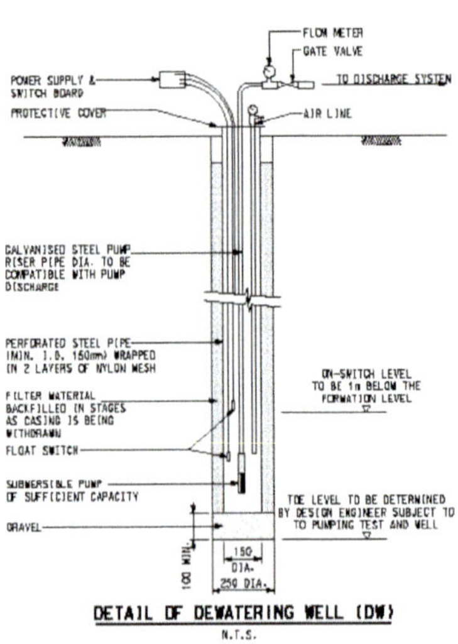

(a) Typical dewatering well

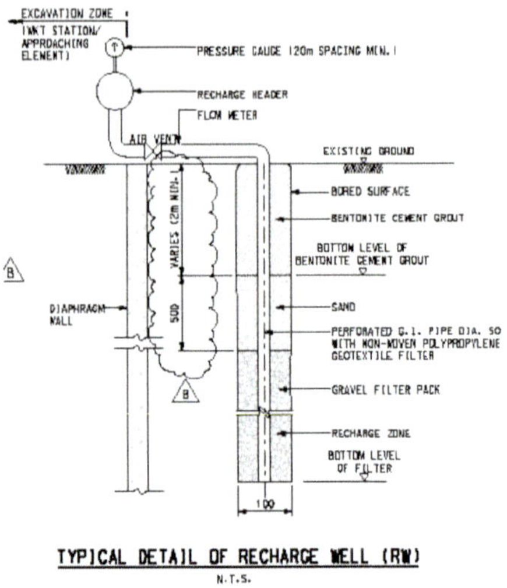

(b) typical recharge well

(c) dewatering well used for an ELS project in Beijing

Figure 5.28 Typical dewatering and recharge well in deep excavation.

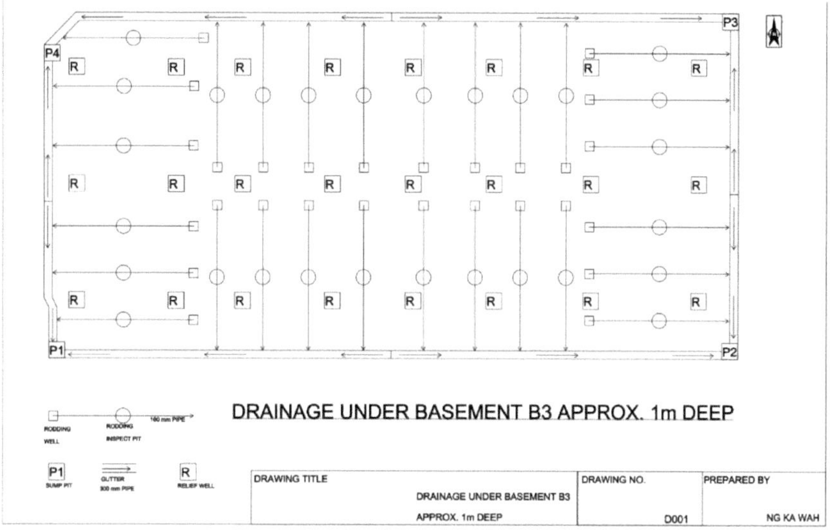

(d) Ground water control plan –dewatering and relief wells for a project in Hong Kong

Figure 5.28 (Continued)

serious cracking of old buildings have occurred in Hong Kong and many developed cities. To avoid the construction of a very deep diaphragm wall, some engineers prefer to construct a grout curtain wall beneath the diaphragm wall using the Tube a' Manchette method, as shown in Figure 5.29, and the lowering of the water table outside the retaining wall will then be greatly reduced. This approach is now commonly used in Hong Kong, as many buildings and MTR concourses in Hong Kong have a very deep basement, and many constructions are carried out in reclaimed land where the strength of the upper soils is not good.

(5) Electro-osmosis – The above methods are useful for well-drained soil. For low-permeability soil, it is difficult to use the pumping method directly. The use of electric voltage can help to increase the permeability in fine-gained soil, and the efficiency will increase with the increase in the voltage. Electro-osmosis is used for dewatering in fine soils, sediments and sludge, and can be combined with the well point in application. This method is expensive, and is used in some special cases only. Currently, some engineers prefer to use grouting plus a retaining wall to reduce seepage instead of using electro-osmosis.

A typical grout curtain installation procedure using the Tube a Manchette method for a large project in Hong Kong is described here for reference:

1. The grout curtain consists of one row of grout holes with 800 mm spacing. The grout curtain extends from the existing ground level to the required level.
2. The grout hole is 75 mm diameter for installation of a Tube a Manchette (TAM) with 48 mm OD. A grout pipe with grouting orifices is covered with rubber Manchette

Figure 5.29 Illustration of the grouting process through Tube a Manchette (TAM).

Table 5.4 Typical grout mix used for grout curtain

Solution A		*Solution B*		
Sodium silicate	*water*	*Cement*	*water*	*Approximate gel time (s)*
167 l	333 l	375 kg	375 l	30
167 l	333 l	273 kg	410 l	50
167 l	333 l	176 kg	440 l	90

at 300 mm spacing. A typical and similar arrangement is shown in Figure 5.26 for illustration. The bottom ends of the tubes are sealed.

3. The TAM should be lowered to the bottom of the hole and surrounded by a sleeve grout (similar processes are used for many MTR tunnel projects in Hong Kong). The mix proportion of the sleeve grout is 1:10:20 for bentonite, cement and water, respectively.
4. A curtain grout is carried out in increases from the bottom to the top of the TAM in successive phases using cement/bentonite grout followed by chemical grout. Each rise of the double packer should be 500 mm.
5. The first phase consists of cement bentonite grout, and the mix proportion is 350 kg cement, 35 kg bentonite and 870 litres of water to form 1m^3 of the grout.
6. The second and later phases (if required) consist of chemical grout with solutions A and B. The mix proportion and approximate gel times are shown in Table 5.4.
7. During grouting, the control criteria are: (1) if the grout pressure can be maintained for 5 minutes for a pressure of 2 × effective overburden pressure + 2 bar; (2) if the cement bentonite grout injection reaches 200 litres per linear metre rise or the chemical grout injection reaches 200 litres.

If a cut-off wall with a grout curtain is used in the construction, the water pressure inside and outside the excavation can be assumed to be hydrostatic. Otherwise, a seepage analysis should be carried out in general. A typical flow-net construction in a deep excavation problem is well covered in many classical soil mechanics textbooks and will not be described here. The authors seldom draw the flow-net manually, because we can never draw squares during flow-net drawing. The seepage problem is governed by the Laplace equation in Eq. (5.59), where K_x and K_y are the permeabilities of soil in the x- and y-directions. For a more general case, the principal directions may inclined at an angle θ with the horizontal direction, and this case can be considered easily. There are many finite element programs, both academic and commercial, for such purposes. Finite difference is efficient and fast in the solution, however the authors have noted that the grid spacing around the tip of the sheet pile can greatly affect the accuracy of the solution around the tip. Furthermore, for highly non-homogeneous and highly irregular problems, the finite difference is complicated. The use of the finite element method is more commonly used by engineers currently (Cheng 2020). Before the invention of the computer, the use of complex analysis and conformal mapping method were used for the solution of Eq. (5.59), and many analytical solutions have been developed for simple problems. The main problem of finite element/difference or analytical solutions is the effective zone, or the horizontal extent where the effect of seepage is negligible so that the simple hydrostatic boundary condition can be applied. In general, a horizontal effective distance twice the total head difference will be sufficiently accurate for practical purposes.

$$K_x \frac{\partial^2 \phi}{\partial x^2} + K_y \frac{\partial^2 \phi}{\partial y^2} = 0 \tag{5.59}$$

Alternatively, a simplified fragments approach has been proposed by Pavlovsky (1933). In this method, the Davidenkoff and Franke form factors ϕ_1 and ϕ_2 are determined from d_1/T_1 and d_2/T_2, with reference to Figure 5.30. Determine h_2 (total head loss in discharge area, or total head at tip of sheet pile) and q (flow rate) with respect to the total head difference H according to the following condition:

1. Trench having a width $2b$ and an infinite length

$$h_2 = H\frac{\phi_2}{\phi_1 + \phi_2} \qquad q = kH\,\frac{1}{\phi_1 + \phi_2} \tag{5.60}$$

2. Circular cofferdam of a diameter $2b$

$$h_2 = 1.3\,H\frac{\phi_2}{\phi_1 + \phi_2} \qquad q = 0.8kH\,\frac{1}{\phi_1 + \phi_2} \tag{5.61}$$

3. Square cofferdam of a side $2b$

$$h_2 = 1.3\,H\frac{\phi_2}{\phi_1 + \phi_2} \text{ at middle of side} \qquad h_2 = 1.7\,H\frac{\phi_2}{\phi_1 + \phi_2} \text{ at corner}$$

$$q = 0.75kH\,\frac{1}{\phi_1 + \phi_2} \tag{5.62}$$

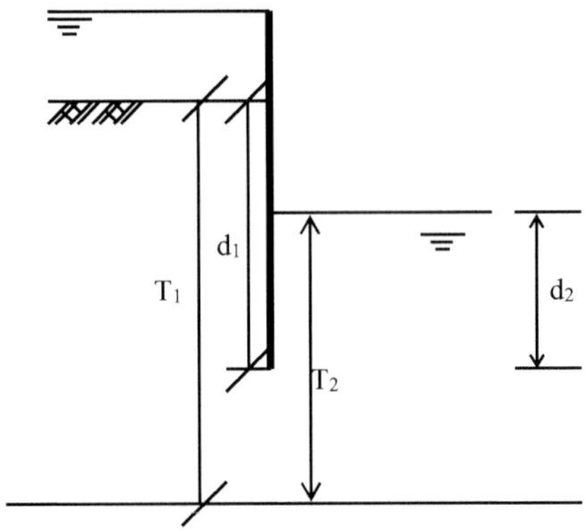

Figure 5.30 Methods of fragments by Pavlovsky.

The average hydraulic gradient on the downstream side is calculated from $i_{av}=h_2/l$, where is l is the length of the flow path. To the best of the authors' knowledge, the method of fragments is now seldom adopted in practice, and most engineers prefer to use a computer program for such purposes.

5.4.1 Free surface seepage flow

There are some problems where a free surface is present and the determination of the free surface is relatively difficult. The free surface is governed by $\phi = y$ $(u = 0)$ and the flow rate normal to the free surface is 0. The trial-and-error flow-net construction or iterative finite element analysis are the common techniques for this type of problem, but are commonly neglected by many engineers in routine analysis and design. Consider the boundary conditions for a typical seepage problem as shown in Figure 5.28. If the boundary condition along AB is a constant head, that will be a classical seepage problem, and the water table along AG will remain stationary during seepage analysis. If the total head along AB is not specified with the boundary condition $\phi = y$ along AG (also no flow across AG), then the precise position of AG will be an unknown which can be determined from iterative finite element analysis. Cheng and Tsui (1993) developed the method of the air element and the corresponding code which is very efficient for this problem.

For the problem in Figure 5.31, the finite element equation will be $[K]\{\phi\} = \{q\}$, where $\{q\}$ is the vector of the point source or sink. For the solution of the stream function, the stiffness matrix $[K]$ is formed by putting $K_x' = 1/K_x$ and $K_y' = 1/K_y$ together with the boundary condition in Figure 5.31b. For a steady-state problem where the head is prescribed along AG and FD, this matrix equation can be solved directly to give ϕ. If AG is assumed to be a free surface, then the boundary condition of AB is given as ϕ = constant and given by height AB, while the precise location of AG will be varied until $\phi = y$ and $u = 0$ along AG. The results of this analysis are typically illustrated by the seepage flow in Figure 5.32.

To illustrate the seepage process during deep excavation, two projects from Hong Kong are considered in Figures 5.33a and 5.33b. In Figure 5.33b, it is noted that the use of free surface seepage analysis will create greater changes in the effective stress on soil with

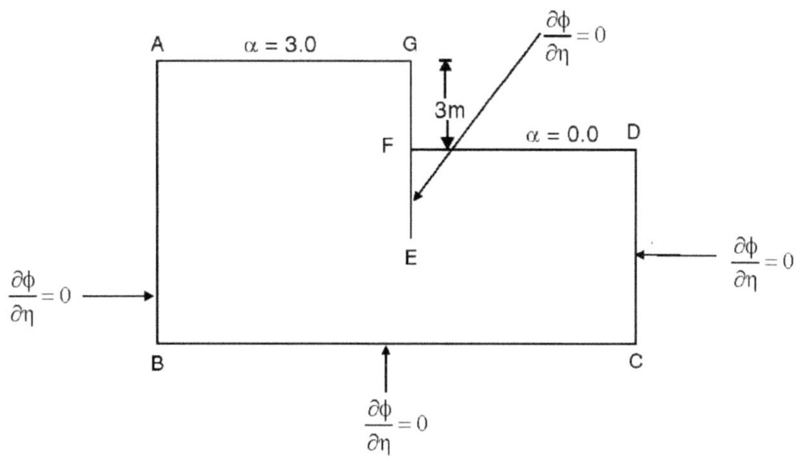

(a) Solution for the total head

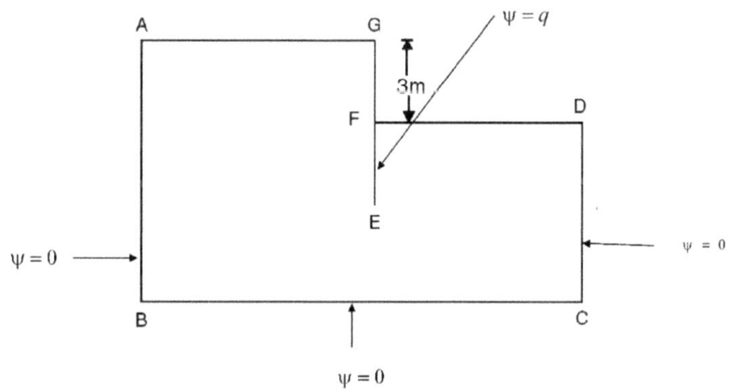

(b) Solution for the stream function (with the inverse of the permeability to replace the original permeability in the equivalent stiffness matrix)

Figure 5.31 Boundary condition for a seepage problem.

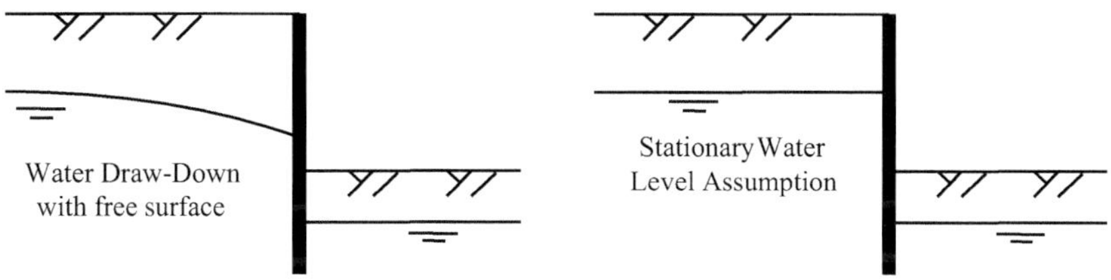

Figure 5.32 Water table for two different boundary conditions.

greater ground settlement. In the free surface seepage flow in Figure 5.33b, the distance of influence is commonly taken to be at least 2 × depth of excavation. From the field measurement results in Wan Chai and Sheung Wan MTR concourses construction, the drawdown of the water table outside the excavation is very significant, and amounts to around 10 m unless the recharge well is used. In view of the water table drawdown problem, some engineers specify the use of curtain grout below the diaphragm to the rock head level so as to reduce the seepage problem. Although some seepage will still come from the fractured rock at the rock head level, the amount of seepage will be greatly reduced. In many large-scale deep excavation projects, pumping tests are specified to confirm the water-tightness of the grout curtain and to estimate the permeabilities of soils. Some case studies in Hong Kong are given by Cheng et al. (2021).

While most commercial programs for seepage analysis are based on the finite element, classically, the finite difference method also has been used on some occasions. The authors have carried out an analysis similar to that in Figure 5.34, with the stationary water table outside the sheet pile, but with different soil layers near to the tip of the sheet pile. The grids are hence not uniform, particularly close to the toe of the sheet pile. The total head contour for the region inside the sheet pile (half section due to symmetry) is shown in Figure 5.34. Based on the total head, the velocity of flow is then computed. It is, however, noted that while in some regions the water flow upward inside the excavation is a reasonable result, there are also some regions where the water flows towards the bottom (rock boundary), which is definitely wrong. When the same problem is analysed by the finite element method, such a problem does not exist. In this respect, use of the finite element method is preferred over the finite difference method for seepage analysis.

Based on the finite element/difference method or equivalently flow-net construction or free surface construction, the net water pressure difference at the retaining wall for ELS is illustrated in Figure 5.35. In general, the actual water pressures at the two sides of the retaining wall are not required, and the net water pressure difference will be sufficient for the analysis and design of the ELS. As shown in Figure 5.35, the net water pressure difference from the finite element/flow-net is almost a triangular distribution, except at the tip of the sheet pile where there is a very rapid change in the water pressure. This result is similar between the classical confined flow and the free surface flow analysis. The blue line in Figure 5.35 is based on the free surface flow analysis, which has been observed in many deep excavation works by Cheng, and the water pressure is lower than the classical approach where the water table outside the retaining wall remains stationary during construction. A simplified approach is adopted in Hong Kong (see also Ou 2006), where the design water pressure difference is a triangular pressure, and the maximum water pressure is given by

$$u = \frac{2 \times HD}{2D + H} \times \gamma_W \tag{5.63}$$

It should be mentioned that the net water pressure difference at the tip of the retaining wall must be zero! Cheng has been asked by many engineers and students about this, as a trapezoidal water pressure difference assuming hydrostatic condition is used by engineers in

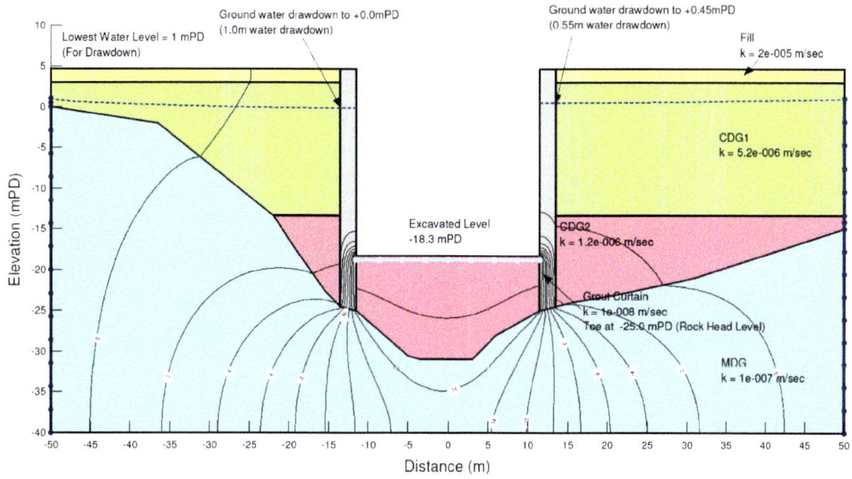

(a) Finite element seepage analysis of a deep excavation in Hong Kong, based on stationary water table outside retaining wall

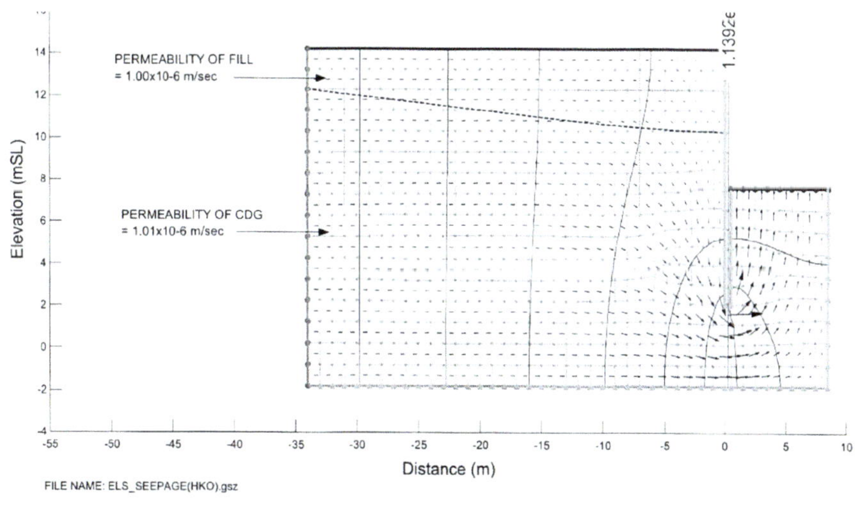

(b) Free surface seepage flow for another project in Hong Kong

Figure 5.33 Two different types of seepage analysis.

the actual design. At one point, there is only one water pressure, hence the water pressure difference at the tip of the sheet pile must be zero, but this simple fact has been neglected by some engineers. A detailed treatment of the seepage process in deep excavation problem is given by Cheng et al. (2021).

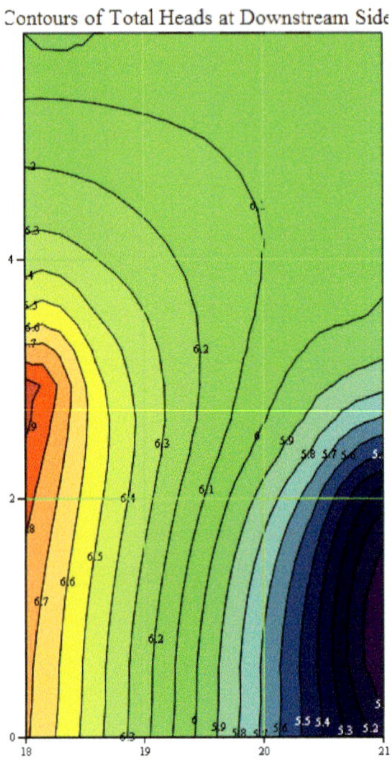

Figure 5.34 Some unacceptable results from finite difference seepage analysis.

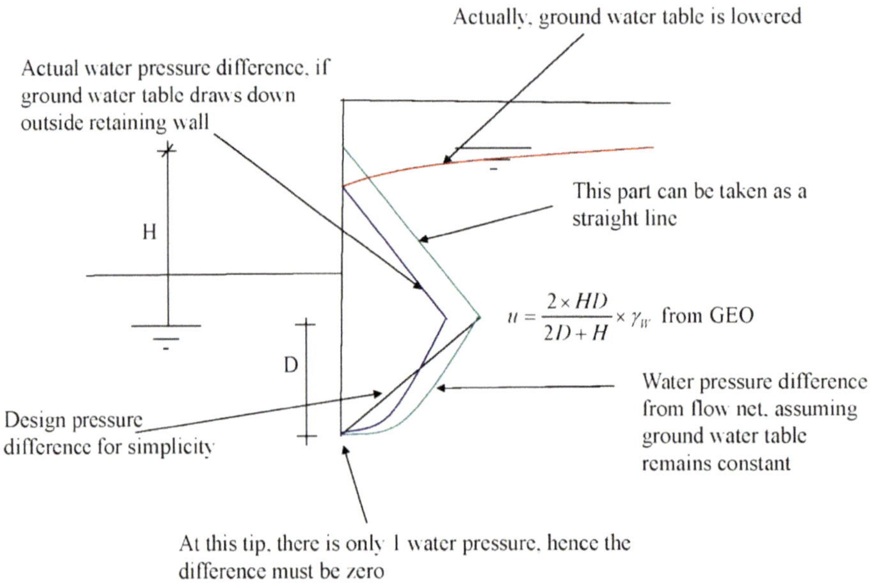

Figure 5.35 Design net water pressure difference on the retaining wall.

One of the difficult problems in seepage analysis is the radius of influence R_0, which is largely empirical. Currently, this value is usually removed from the statistical information. Alternatively, a simplified method is adopted by some engineers as:

$$R_0 = CH\sqrt{k} \tag{5.64}$$

Where C is a constant, H is the amount of drawdown and k is the permeability of soil as determined from the field test. C is commonly taken as 3000 for radial flow and 1500 for line flow to trenches or well points.

5.5 ANALYSIS AND DESIGN OF ELS SYSTEMS

For the analysis and design of ELS, there are several approaches:

1. The hand-calculation method is based on simple earth pressure coefficients and simplified analysis. The problem with hand-calculation is that the construction process is difficult to be considered, and no information about the wall and ground movement during the excavation can be obtained.
2. Subgrade reaction or pseudo finite element method – This is a simple one-dimensional analysis approach based on a computer program. This approach appears to be unpopular currently. Since the vertical component is not considered, no ground movement can be considered in the analysis.
3. Finite element approach – Many finite element/difference programs, both for 2D and 3D analysis, are adopted by most engineers in Hong Kong and other countries. In theory, the wall and ground movement can be obtained from the analysis, however, the estimation of the ground movement is usually not good, due to various limitations and difficulties in defining a realistic model for this soil–structure interaction problem.

5.5.1 Subgrade reaction model

As shown in Figure 5.36, a continuous beam is used to model the retaining wall, which is governed by the beam deflection equation as follows:

$$E_pI_p\frac{d^4v}{dz^4} + P\frac{d^2v}{dz^2} + K_hv = 0 \tag{5.65}$$

where v is the lateral displacement of the wall at depth z
E_pI_p is the flexural rigidity of the wall at depth z
p is the axial load in the wall at depth z which is usually 0
K_h is the elastic spring stiffness per unit length of the pile at depth z
P is usually 0, hence Eq. (5.66) can be formulated by the finite difference method for a node i as

$$E_pI_p\frac{v_{i-2} - 4v_{i-1} + 6v_i - 4v_{i+1} + v_{i+2}}{\delta^4} + P\frac{v_{i-1} - 2v_i + v_{i+1}}{\delta^2} + K_{h,i}v_i = 0 \tag{5.66}$$

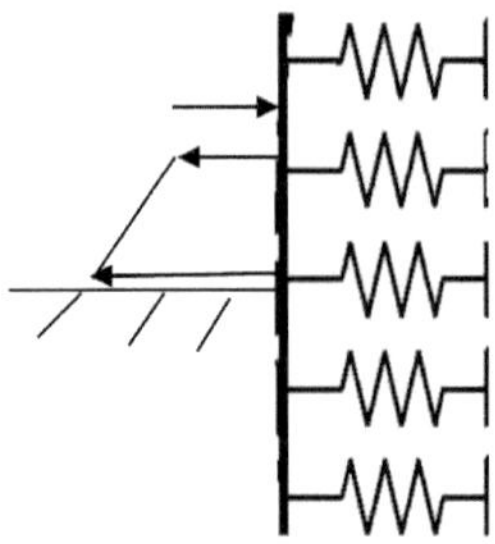

Figure 5.36 Subgrade reaction model for ELS.

where v_i is the lateral displacement of node i. The spring constant is commonly taken as $K_h = k_h B$ and $k_h = n_h \left(\frac{z}{B}\right)^n$ which is taken from the pile lateral load analysis. The spring forces are governed by the limiting active or passive pressure; hence the analysis is usually nonlinear if some of these soil springs yield during the analysis. The excavation process is modelled by the equivalent excavation pressure on the wall as well as the net water pressure difference between different stages. A strut or tie-back can be applied as the support in the model, and yielding or tension in the strut can be considered so that the strut will no longer function in the model, which can also be modelled. One of the critical problems is the use of the equivalent excavation forces. Either the original lateral earth pressure or the updated lateral earth pressure is used by commercial programs, and there will be some differences in the results of these analyses. The lateral stresses are also not affected by the release of the vertical stress during excavation, which is obviously not correct. Furthermore, some programs do not provide any lateral springs for the regions behind the excavation and above the excavation level, while other programs may provide lateral springs which can yield upon reaching the active or passive pressure limits as imposed by the users during the analysis. The programs PileWall by Cheng, Wallap and the code by Bowles can work in this way to model the excavation. The wall movement, shear and bending on the wall at different stages can then be obtained. The axial load on the wall can be considered also, and this will create a further increase in the lateral deflection and internal forces in the retaining wall, which can be considered by the program by Cheng. Currently, the subgrade reaction method is seldom adopted, due to the intrinsic problem of the subgrade reaction model/parameters and the inability to estimate the ground movement. A case where the 1-D subgrade reaction model is used for a project in Hong Kong is shown in Figure 5.37. A sample program PileWall with graphics support can be obtained from Cheng at natureymc@yahoo.com.hk.

Figure 5.38 show the typical deflection and moment on a long cantilever sheet pile, where the deflection at the sheet pile tip is minimal. If the sheet pile is not long enough, there will also be lateral movement at the sheet pile bottom. For staged excavation, typical plots of wall displacement and moments are given in Figure 5.39.

For illustration, some of the inputs and outputs for a large-scale project in Hong Kong are given below.

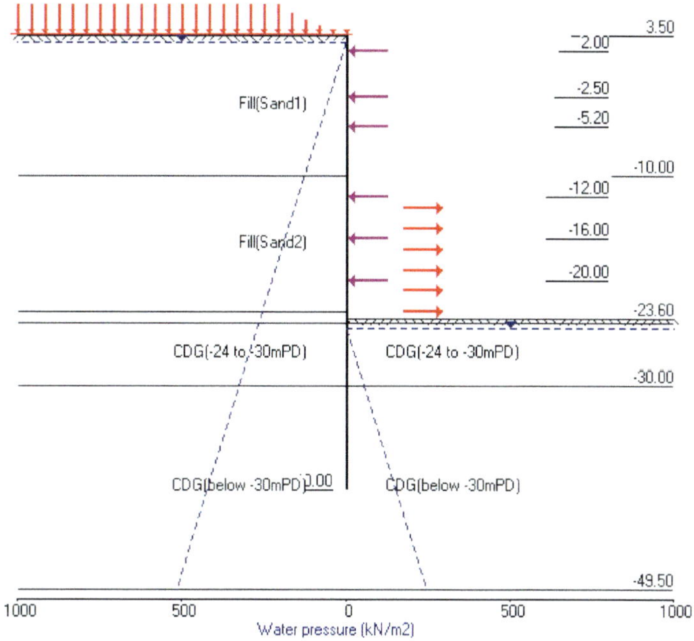

Figure 5.37 A typical design using a 1-D model for a project in Hong Kong.

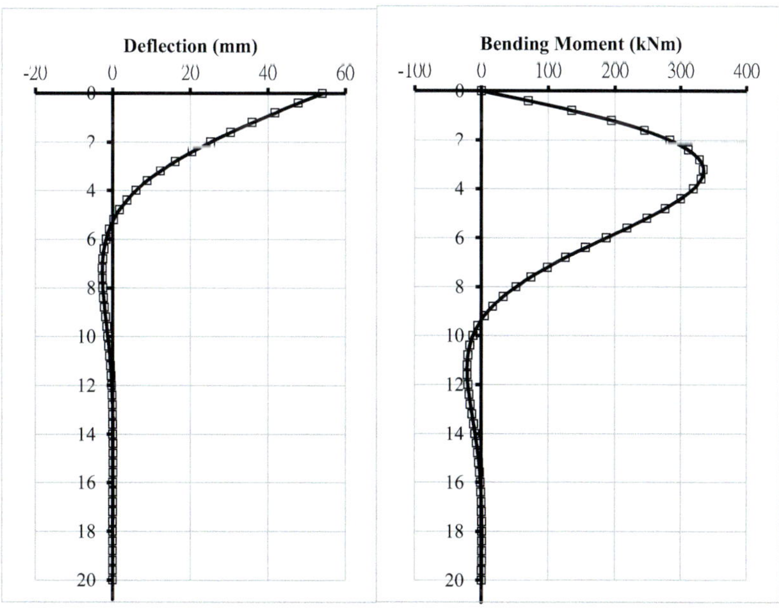

(a) Typical deflection and bending moment for cantilever retaining wall

Figure 5.38 Typical output for cantilever excavation for a project using a subgrade reaction model in Hong Kong.

SOIL PROFILE

Stratum Elevation of ----------------- Soil types -------------------

No.	Top of stratum	Active side	Passive side
1	3.50	2 Fill (Sand 1)	2 Fill (Sand 1)
2	–5.00	3 Fill (Sand 2)	3 Fill (Sand 2)
3	–9.00	4 MD (Clay)	4 MD (Clay)
4	–15.00	5 MD (Sand)	5 MD (Sand)
5	–17.00	7 All (Sand)	7 All (Sand)
6	–22.00	8 CDG (–22 to –30 mPD)	
		8 CDG (–22 to –30 mPD)	
7	–30.00	9 CDG (–30 to –40 mPD)	
		9 CDG (–30 to –40 mPD)	
8	–40.00	10 CDG (below –40 mPD)	
		10 CDG (below –40 mPD)	
9	–52.00	11 Rock	11 Rock m^3

SOIL PROPERTIES

Soil type	Bulk density (kN/ m^3)	E (kPa)	K_0	NC/OC (Nu)	K_a (Kac)	K_p (Kpc)	C (kPa)
1 not defined							
2 Fill (Sand 1)	19	20000	0.437	OC (0.3)	0.248 (0)	6.26 (0)	
3 Fill (Sand 2)	19	100000	0.437	OC (0.3)	0.248 (0)	6.26 (0)	
4 MD (Clay)	17.2	9200	1.0	OC (0.49)	1.0 (2.0)	1.0 (2.0)	23
5 MD (Sand)	19	20000	0.454	OC (0.3)	0.259 (0)	5.79 (0)	
6 not defined							
7 All (Sand)	19.5	25500	0.439	OC (0.3)	0.249 (0)	6.2 (0)	
8 CDG (–22 to –30 mPD)	19	56000	0.45	OC (0.3)	0.259 (1.017)	5.89 (4.854)	5
9 CDG (–30 to –40 mPD)	19	92000	0.438	OC (0.3)	0.249 (0.997)	6.23 (4.992)	5
10 CGD (below –40 mPD)	19	168000	0.438	OC (0.3)	0.249 (0.997)	6.23 (4.992)	5
11 Rock	23	1.75E+7	0.293	OC (0.3)	0.134 (0.866)	18.01 (17.011)	300
12 Compacted fill	19	20000	0.437	OC (0.3)	0.248 (0)	6.26 (0)	

Additional soil parameters associated with Ka and Kp

No.	Description	Ka: Soil friction angle	Ka: Wall adhesion coeff.	Ka: Back-fill angle	Kp: Soil friction angle	Kp: Wall adhesion coeff.	Kp: Back-fill angle
1	Not defined						
2	Fill(Sand1)	34.09	0.434	0.00	35.98	0.443	0.00
3	Fill(Sand2)	33.98	0.456	0.00	35.93	0.448	0.00
4	MD(Clay)	0.00	0.000	0.00	0.00	0.000	0.00
5	MD(Sand)	32.94	0.467	0.00	34.68	0.460	0.00
6	Not defined						
7	All(Sand)	33.07	0.612	0.00	34.36	0.612	0.00
8	CDG(-22 to -30mPD)	36.08	-0.003	0.00	45.21	0.000	0.00
9	CDG(-30 to -40mPD)	36.99	-0.003	0.00	46.33	0.000	0.00
10	CDG(below -40mPD)	36.99	-0.003	0.00	46.33	0.000	0.00
11	Rock	45.00	1.000	0.00	45.00	1.000	0.00
12	Compacted Fill	37.05	0.001	0.00	46.43	0.000	0.00

GROUND WATER CONDITIONS

Density of water = 9.810 kN/m3

	Active side	Passive side
Initial water table elevation	1.60	1.60

Automatic water pressure balancing at toe of wall : No

Water press. profile no.	Active: Point no.	Active: Elev. (m)	Active: Piezo elev. (m)	Active: Water press. (kN/m2)	Passive: Point no.	Passive: Elev. (m)	Passive: Piezo elev. (m)	Passive: Water press. (kN/m2)
1	1	1.60	1.60	0.0	1	-9.30	-9.30	0.0

WALL PROPERTIES

```
        Elevation of toe of wall =    -37.50
   Maximum finite element length =      2.40
       Youngs modulus of wall E = 2.4000E+07 kN/m2
     Moment of inertia of wall I = 0.10080 m4/m run
                             E.I = 2.4192E+06 kN.m2/m run
            Yield Moment of wall = Not defined
```

SURCHARGE LOADS

Surcharge no.	Elev.	Distance from wall	Length parallel to wall	Width perpend. to wall	Surcharge kN/m2 Near edge	Surcharge kN/m2 Far edge	Equiv. soil type
1	3.50	0.00(A)	50.00	2.00	10.00	=	N/A
2	3.50	2.00(A)	50.00	4.00	10.00	48.00	N/A
3	3.50	6.00(A)	50.00	500.00	58.00	=	N/A
4	3.50	-0.00(P)	50.00	50.00	20.00	=	N/A

Note: A = Active side, P = Passive side

A trapezoidal surcharge is defined by two values:
N = at edge near to wall, F = at edge far from wall

CONSTRUCTION STAGES

Construction stage no.	Stage description
1	Apply surcharge no.1 at elevation 3.50
2	Apply surcharge no.2 at elevation 3.50
3	Apply surcharge no.3 at elevation 3.50
4	Apply surcharge no.4 at elevation 3.50
5	Apply water pressure profile no.1

```
FACTORS OF SAFETY and ANALYSIS OPTIONS
   Type of structure    -   Retaining wall
   Stability analysis:
      Method of analysis  -  Burland-Potts
      Factor on passive for calculating wall depth =   1.50
      Factor on passive for calculating tie force  =   1.00

   Parameters for undrained strata:
      Minimum equivalent fluid density             =   5.00 kN/m3
      Maximum depth of water filled tension crack  =   0.00 m

   Bending moment and displacement calculation:
      Method  -  Subgrade reaction model using Influence Coefficients
      Open Tension Crack analysis? - No
      Non-linear Modulus Parameter (L) = 0 m

   Boundary conditions:
      Length of wall (normal to plane of analysis) = 100.00 m

      Width of excavation on active  side of wall  = 200.00 m
      Width of excavation on passive side of wall  = 22.00 m

      Distance to rigid boundary on active side  = 100.00 m
      Distance to rigid boundary on passive side = 11.00 m


OUTPUT OPTIONS

 Stage ------ Stage description ----------- ------- Output options -------
  no.                                       Displacement   Active,  Graph.
                                            Bending mom.   Passive  output
                                            Shear force   pressures
   1 Apply surcharge no.1 at elev. 3.50         Yes          Yes      Yes
   2 Apply surcharge no.2 at elev. 3.50         Yes          Yes      Yes
   3 Apply surcharge no.3 at elev. 3.50         Yes          Yes      Yes
   4 Apply surcharge no.4 at elev. 3.50         Yes          Yes      Yes
   5 Apply water pressure profile no.1          Yes          Yes      Yes
   * Summary output                             Yes           -       Yes
```

In the 1-D P-Y program by Cheng (Dos mode program), the stress–strain relation is shown in Figure 5.40, and an elastic limit "stroke" is defined which is sufficient to define the relation from active to passive. Loadings from staged excavation include the unbalanced soil and water pressure difference. The wall is defined by a beam element, while the support is the spring elements. The results from Figures 5.38 and 5.39 from the use of commercial programs can also be obtained by the P-Y program. Interested readers can obtain the Fortran code for P-Y from Cheng.

5.5.2 2-D/3-D finite element/difference method

Currently, 2-D and 3-D finite element programs are used by many engineers in the analysis and design of ELS. Some representative programs include Plaxis 2D/3D, DeepEx, Flac2D/3D, Frew and Soilstru by Cheng and others. Two different approaches to develop a program for 2D excavation are:

1. The mesh and node numbering remain unchanged, elements that are removed are assigned E = 0 or 1.0, excavated nodes are given prescribed displacement as 0 (Plaxis, etc.).
2. The mesh, number of elements and nodes keep changing with construction (Flac2D, 3D).

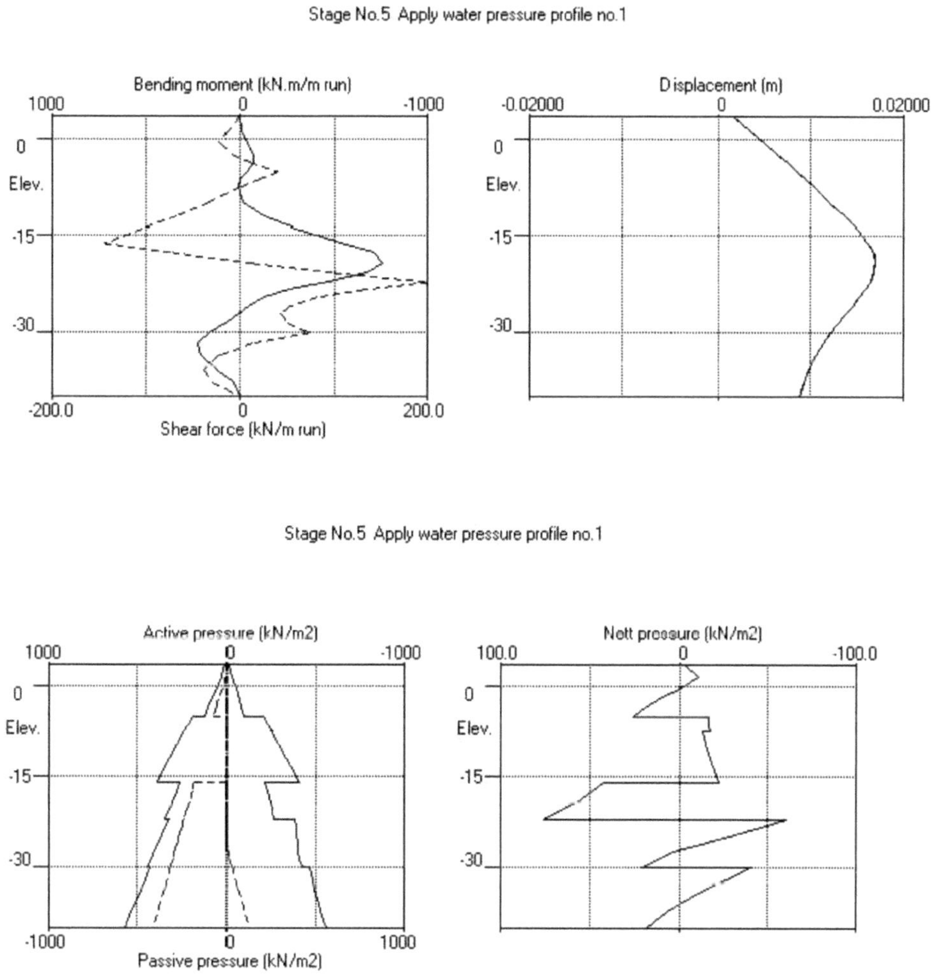

Figure 5.39 Typical output for strutted stages for a project using a subgrade reaction model in Hong Kong.

Modelling of ELS:

1. Form the mesh, apply the body forces by $\int[N]^T\{0,-\gamma\}dv$.
2. Form the horizontal stress by using $\sigma_h = K_0\sigma_v$.
3. Apply equal and opposite forces arising from the stresses on the excavated surfaces. Excavation forces:

$$\int_v [B]^T\{\sigma\}dv = \{F\} \quad \text{or} \quad \int[N]^T\{\sigma\}ds \tag{5.67}$$

4. See page forces: $\int[B]^T\{\Delta u\}dv$, Δu will be the nodal change in pore pressure.
5. Insert strut/anchor stiffness.

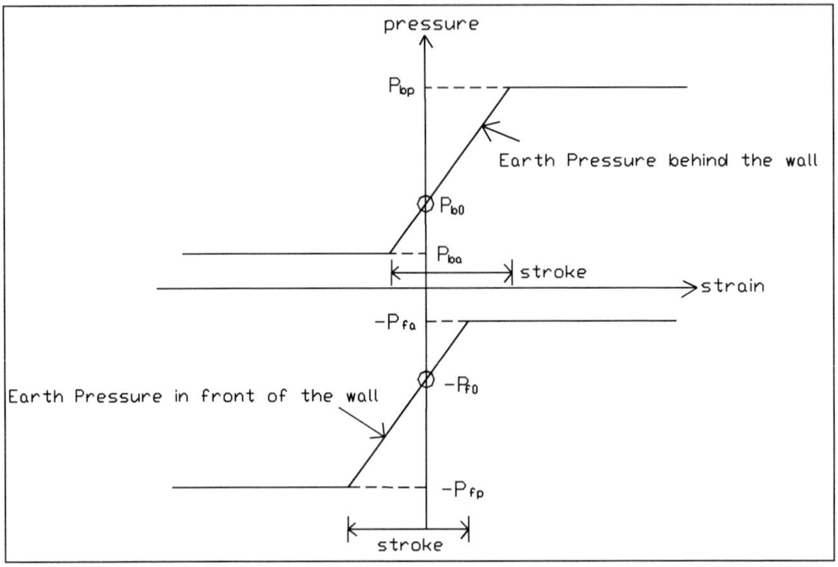

Figure 5.40 The stress–strain model used in the program P-Y by Cheng.

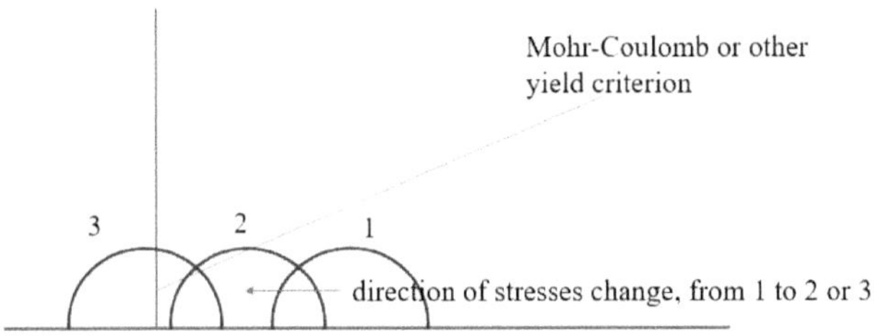

Figure 5.41 Correction of stresses during computation.

6. Solve the global equilibrium equation and check yielding of soil. Correct the stresses on every element by the Mohr–Coulomb relation. Perform iteration analysis until unbalanced force due to yielding of soil is 0. If the Mohr circle moves from 1 to 2 or 3 (2 or 3 represent the stress state after computation), we need to correct for the unbalanced stresses and bring the Mohr circle back to the yield criterion. Iteration continues until all soil elements are controlled within the yield criterion (Figure 5.41).

Limitations of ELS computer modelling:

1. Most commercial programs cannot model the stress transfer between the wall and soil well. Many strange results frequently result from Plaxis or other programs. Interface between wall and soil: zero thickness interface element or thin layer of material. Problems: (1) Stress transfer; (2) node compatibility not maintained after soil/wall move.

2. Analysis may be terminated by the program due to the formation of a local failure zone or negative element stiffness matrix (use of an elastic material in such a region to resume analysis by engineers). Commonly encountered by engineers.
3. Stress-free surface after analysis may not be stress free, and re-iteration should be done, but not in a commercial program.
4. Free surface seepage flow analysis is not commonly adopted.
5. the effect of water is only considered for the unbalanced water pressure on the retaining wall but not the whole soil mass. This is usually compensated for by using a drop in the water level and volume compressibility to compute additional ground settlement due to water drawdown.

Based on various problems for engineers, there are some further practical considerations for computer modelling of ELS:

1. Program Frew neglects the vertical component of movement, hence the element stiffness matrix in Frew is only one-quarter of that in other programs. This problem is, however, not so critical, as ELS is more controlled by the lateral force and movement. Furthermore, the effect of seepage and change in water pore pressure within the soil mass is not considered in Frew. Nevertheless, Frew can be viewed as an approximation which is good enough for design purposes.
2. Wall deflection is greatly controlled by the stiffness of the soil, but not the wall stiffness. A very stiff wall is not cost effective to control wall and ground movement.
3. Ground movement is difficult to be assessed with good accuracy. Ground movement estimated by the volume of wall deflection, Plaxis (or similar) or statistics can differ by a significant amount. Personal experience is that ground movement from FEM programs is usually not very good, due to the various complicated issues in ELS modelling. My personal preference is based on the statistics of 0.5–1.0 maximum horizontal wall movement.
4. What is the meaning of 25 mm ground movement limitation? Should the cumulative effect be considered? How important is this 25 mm? This 25 mm is a nightmare to many design engineers in Hong Kong. From the author's experience on MTR and other ELS in Hong Kong, the most critical problem is the tilting of buildings. If 25 mm is rigidly enforced, many MTR concourses would not be built. It is better to spend more effort on monitoring.
5. Field measurement indicates a ratio of 0.5–1% maximum wall movement for maximum ground settlement. No program can give a good estimation of ground settlement. There are similar statistics for Hong Kong, Taiwan and the US. See the GEO Guide and many internal reports within various consultants firms (MTR). For the Island line, 1% was measured by the author.
6. To reduce wall and ground movement, the excavation process can be divided into more processes. For example, excavation of 2 m depth of soil can be modelled by excavation of 0.5 m followed by 0.5 m, 0.5 m and 0.5 m in a total of four steps. A reduction in computed wall and ground movement is possible because ELS is nonlinear by nature in the analysis.

The elements to be excavated are usually assigned a very small stiffness, or the air element by Cheng and Tsui (1991) can be used. Except the program Frew, where the vertical

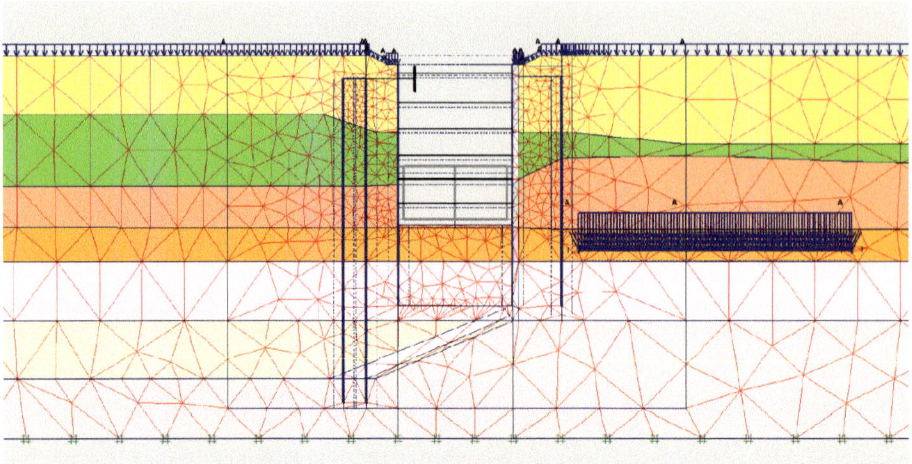

Figure 5.42 A typical 2-D finite element model for a project in Hong Kong.

component is not considered, all other programs will give predictions on the heave within the excavation, ground movement outside the retaining wall, wall movement, strut loads as well as other information. A typical finite element analysis for ELS is shown in Figure 5.42. There are two major points to consider in the finite element/difference analysis. The wall and soil movement may not be compatible in most cases, which is not well reflected in most finite element programs. A ground upheaval adjacent to the retaining wall is frequently obtained in the analysis (but not often in practice), which is due to release of the overburden stress as well as the ground heave within the excavation. Since the wall and soil movement are assumed to be compatible in most cases, the ground heave outside the excavation is usually over-estimated by most programs. Even with the use of an interface element between the walls and soil, such over-estimation still exists in general. The final result will be relatively great differences between the actual and the computed ground settlement, which has been frequently experienced by authors in many deep excavation projects (Cheng and Tsui 1989). Another important issue is that different programs tend to give different results, with the same input parameters. This is highly different from slope stability or other problems in which different programs will give similar results for the same problem. Cheng has carried out internal studies using several commercial programs, and found that the differences can range from minor to major. For example, the differences between Frew and Plaxis2D range from 2% to 26% for the wall bending moment, 1% to 42% for the wall shear force and 0% to 47% for the wall deflection for a hospital construction project. Such differences among different programs are not encountered among other types of problem. The predicted wall and ground movements also deviate from the actual measurement by several-fold, which can be attributed to the very conservative soil parameters used for computer analysis. Typical results for ground settlement for a hospital project in Hong Kong are given in Figure 5.43. The internal forces for this ELS system are given in Figure 5.44.

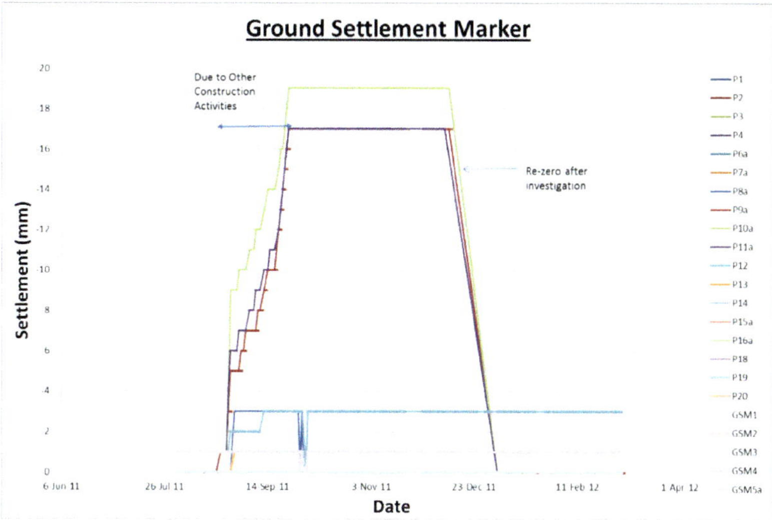

Figure 5.43 Ground settlement against time for a settlement marker in a hospital construction project in Hong Kong.

One important point which is sometimes neglected by engineers is the choice of the strut length in the strut stiffness calculation and computer modelling. Since the middle of the strut is the stationary point, only half of the strut length should be used in the strut stiffness determination.

Due to the variation in the soil boundaries, an interesting measurement has been recorded for an MTR station construction in Guangzhou, China. In this project, the highly decomposed granite (HDG) profile dips at an angle of about 30°, while the soils above the HDG are poor quality. There are great differences between the wall displacements for the north and south walls, which also implies that the axial shortening of the struts is significant. The authors strongly support monitoring those critical deep excavation projects to avoid excessive wall and ground movements. Another interesting, important and useful consideration is the use of 3-D ELS analysis, which is particularly important for struts close to corners. Due to the restraints from the corners, the wall deflection will be less than that for the middle section, but this effect is not commonly considered in actual design and construction. Referring to the 3-D ELS analysis for a project in Hong Kong, the maximum wall deflections below the first layer of struts by 2-D and 3-D analyses (from different companies) were 18.65 mm and 2.95 mm, respectively. The differences between 2-D and 3-D analyses and the outputs from different software are too significant to be overlooked.

The authors have also carried out a very simple problem for a rectangular excavation, as shown in Figure 5.45. Based on 2-D and 3-D excavation analyses, great differences were found between the two analyses, and the ground settlement estimation is given in Table 5.5. In general, it can be concluded that 3-D analysis will enable more economic design, reduced wall and ground movement, reduced internal forces in the retaining wall and reduced strut loads. A very large scale 3-D modelling of a very deep excavation project

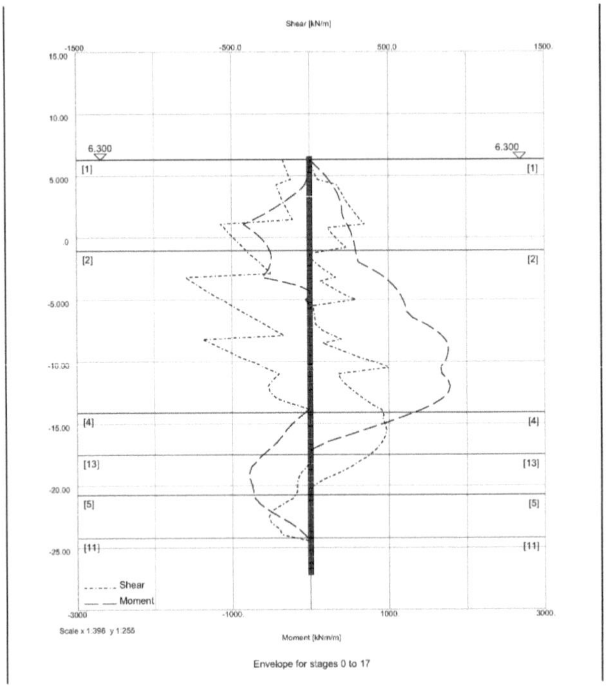

(a) Moment envelope on retaining wall, at end of construction

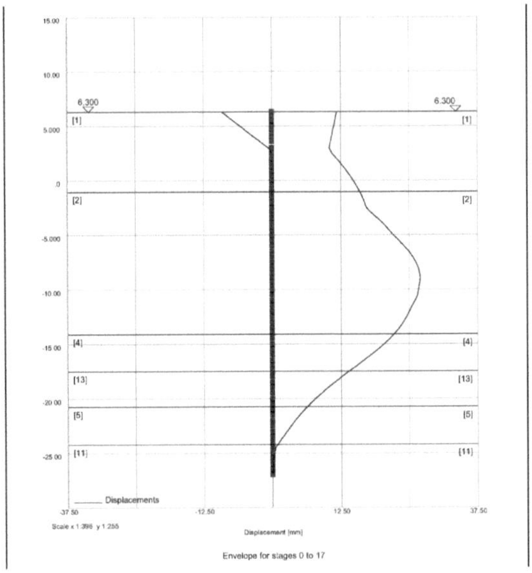

(b) Wall deflection, at end of construction

Figure 5.44 Deflection and bending moment on the diaphragm wall at the end of construction, with four layers of strut supports for a hospital project.

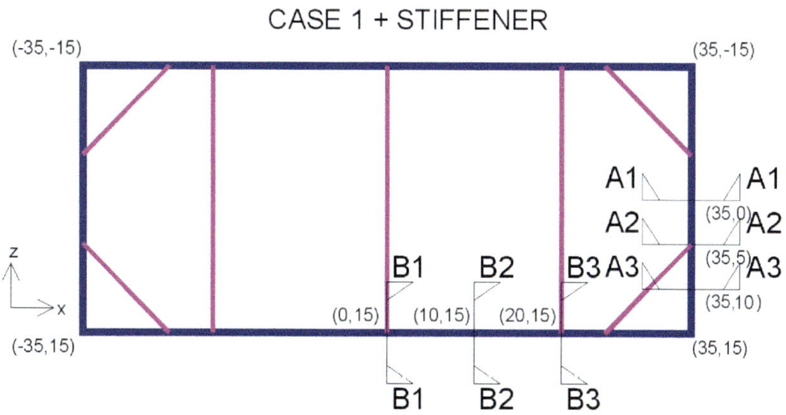

Figure 5.45 3-D excavation analysis against 2-D analysis.

Table 5.5 Comparisons of ground settlement prediction using 2-D and 3-D analysis

	2-D Analysis (mm)		*3-D Analysis (mm)*	
Case	*A-A*	*B-B*	*C-C*	*D-D*
1	13.64	17.46	2.87	.21
2	20.83	27.34	3.78	7.70
3	15.98	21.66	8.31	9.06
4	15.75	21.25	4.92	7.22
5	4.73	6.73	2.62	3.04
6	11.42	15.48	5.31	5.84

in Egypt by Cheng is shown in Figure 5.46, where more than 1 million elements were used in the computer model.

Some field observations have indicated that the maximum ground settlement does not occur immediately after the retaining wall, but is located at a distance approximately half the excavation depth away from the wall. This is due to the effect of heaving and stress relief during excavation. Such an observation can be estimated by the method after Milligan (1983), Nicholson (1987) and Rouke (1976). Another observation is that the ground settlement is usually very small after a distance of twice the excavation depth. Another useful observation is that the maximum ground settlement usually amounts to 50–100% of the maximum wall deflection (dependent on the quality of soil and workmanship of construction), and this relation is also commonly used to estimate the ground settlement in Hong Kong.

Besides the wall and ground movement, the 3-D effect is also important for the strut load determination. Based on the actual field measurements, it is generally found that the actual strut loads are less than the predicted values (Yau and Sum 2010). This phenomenon can

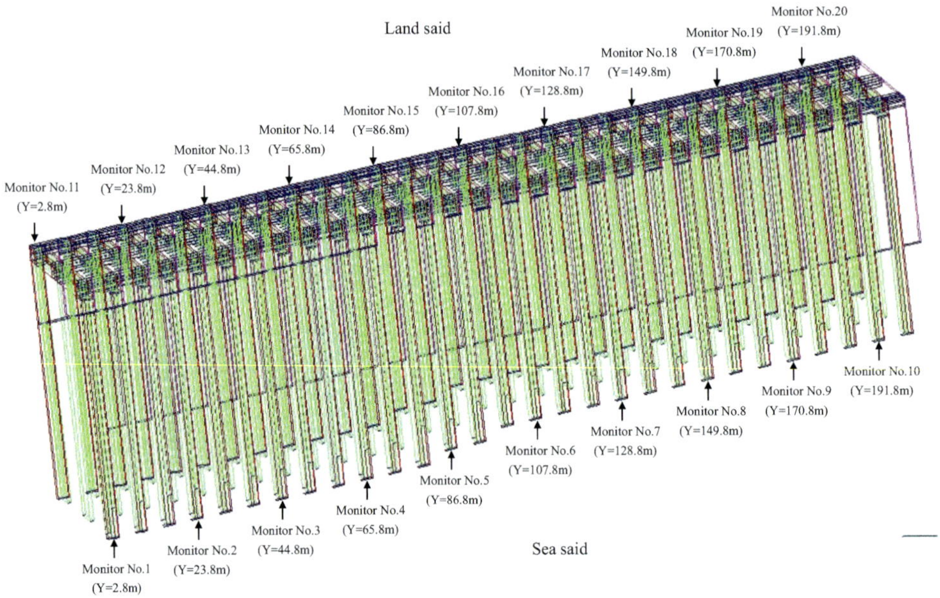

Figure 5.46 3-D analysis of a deep excavation in soft clay in Egypt by Cheng (more than 1 million elements and nodes).

be attributed to the 3-D and soil arching effect. Since tedious workings are required for 3-D deep excavation analysis, most routine designs are still based on 2-D analysis, while 3-D analysis is adopted for more critical projects.

The most important parameters towards a good analysis of the ELS problem are the soil parameters. Based on the use of $E = N$ (MPa) commonly used in Hong Kong, some comparisons between predicted and measured results have been studied. Pan et al. (2001) reported that the measured deflection was only about 50% of the predicted values for a diaphragm wall excavation for the Tseung Kwan O station and tunnel, where the ground condition consisted of 15–18 m of reclaimed fill with 8 m of soft marine clay. In a deep basement project in Central where the ground comprises 8–10 m of fill and 8 m of alluvium/marine deposit, the measured deflection was about 53% of the value predicted for the diaphragm wall excavation (Sze and Young (2003). Based on some back-analyses of deep excavation works, Chan (2003) found that an E/N value of 1.5 (MPa) for fill, alluvium and marine deposit and an E/N value of 2 (MPa) for CDG are considered as reasonable and conservative. Askew et al. (2006) reported that the maximum measured wall deflection was 74% of the predicted magnitude of a strutted pipe pile wall for the East Rail tunnel project. The measured strut forces were about 42–67% of the predicted values. Another pipe pile wall cofferdam for a youth centre in Chai Wan recorded strut loads measuring between 26% and 60% of those predicted. Similarly, the ground settlement from the analysis is usually found to be smaller than the predicted values. Occasionally, the predicted values can be smaller than the measured values, but such a condition is not common.

5.5.3 Classical method of analysis

Classically, ELS is analysed by simple and approximate methods, which are meant for design purposes instead of being rigorous. The authors only give some discussions and simple illustrations about the classical approach, as it has been largely replaced by the use of computational methods now. Furthermore, only the ultimate active and passive pressures are used in the calculation, while the actual earth pressure depends greatly on the wall displacement and construction process. The soil–structure interaction is totally neglected in the classical design method. Nevertheless, the design method as outlined below can give a satisfactory design in general.

5.5.3.1 Cantilever case

Consider an excavation in dry soil, where $\gamma = 18$ kN/m³ and $\phi = 33°$, $K_a = 0.295$, $K_p = 3.392$ (Rankine solution). It is assumed that the 3 m cantilever wall has deflected sufficiently that the full active and passive pressures are mobilized. From numerical analyses as well as experimental tests, it has been found that a large force exists close to the bottom of the sheet pile, and the force extends over only a small region close to the pile tip, as shown in Figure 5.47. For such a case, it is easy to determine the depth of penetration, shear force and bending moment by taking the moment about the sheet pile bottom.

Take the moment at the bottom of the retaining wall:

$$\frac{1}{6}K_a\gamma(D+3)^3 = \frac{1}{6}K_p\gamma D^3 \Rightarrow \left(\frac{D+3}{D}\right)^3 = \frac{K_p}{K_a} = 11.499$$

$D + 3 = 2.257D$, which gives $D = 2.39$ m

Since the extent of the point force is not considered in this analysis, it is a common practice to add 20% to D for the sake of safety. Hence the designed depth of penetration is increased to 2.86 m. Some engineers check this additional depth by considering a balance between active and passive pressures, but the authors' view is that this is not necessary for normal designs.

If water is present at a depth of 2 m below ground and at the dredge level for the two sides, take $\gamma_{sat} = 19.5$, $\gamma' = 19.5 - 9.81 = 9.69$ kN/m³.

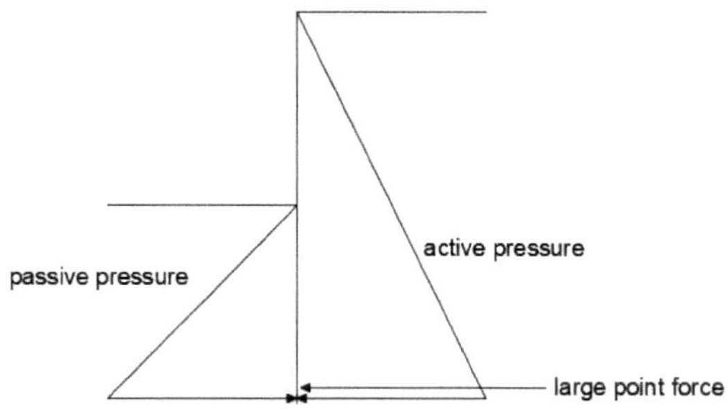

Figure 5.47 A simple cantilever sheet pile wall.

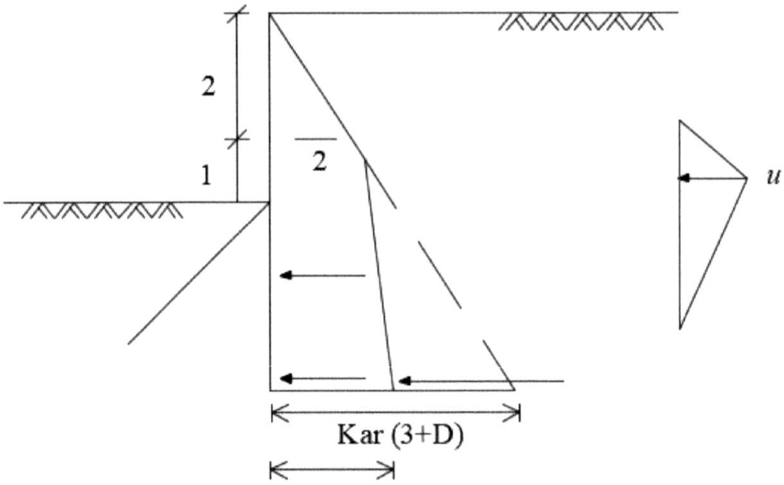

Figure 5.48 Differential water table for the problem in Figure 5.47.

For water pressure, it is only necessary to consider the difference between the pressures on the two sides of the sheet pile. This pressure difference can be determined with the use of flow-net construction, computer analysis or approximation. The net differences are usually approximated to a triangular pressure as shown for ease of hand calculation. The maximum value of the triangle is usually taken as $\gamma_w H$ or from Eq. (5.63). In the present case, the authors adopt Eq. (5.63) in the calculation, which considers the steady-state seepage condition. The earth and water pressures at the two sides of the retaining wall are given in Figure 5.48. The head difference H is 1 m in the present case.

$$u = \frac{2 \times HD}{2D + H} \times \gamma_W = \frac{2D}{2D+1} \times 9.81$$

It should be noted that some engineers simply adopt $u = \gamma_w H$ in Figure 5.48, for ease of computation. Taking the moment to the bottom of the sheet pile

$$\frac{1}{2}K_a\gamma 2^2\left(\frac{2}{3}+1+D\right)+\frac{1}{2}K_a 2\lambda(D+1)^2+\frac{1}{6}K_a\gamma'(D+1)^3+\frac{1}{2}u\left(\frac{1}{3}+D\right)+\frac{1}{2}uD\frac{2}{3}D=\frac{1}{6}K_p\gamma' D^3$$

or $0.885(D+3)^3 - 0.409(D+1)^3 + 4.91\frac{2D}{2D+1}\left(\frac{1}{3}+D\right) + 6.54\frac{D^3}{2D+1} = 5.478D^3$ By trial and error, $D = 3.5$ m, LHS = 257.3, RHS = 234.9; $D = 3.8$ m, LHS = 297.0, RHS = 300.

The solution with $D = 3.8$ m is considered to be adequate enough, and more digits for D is considered to be unnecessary, as it is impossible for the sheet pile to have such a precise depth in practice. The maximum bending moment occurs at the point where the shear force is zero. Assume it is located at a distance x below the dredge level.

$$\text{Without water } \frac{1}{2}K_a\gamma(x+3)^2 = \frac{1}{2}K_p\gamma x^2 \quad \Rightarrow \quad \left(\frac{x+3}{x}\right)^2 = \frac{Kp}{Ka} = 11.499$$

$x = 1.255$ m, hence

$$M_{\max} = \frac{1}{6}\times 0.295\times 18\times 4.255^3 - \frac{1}{6}\times 3.392\times 18\times 1.255^3 = 48.1 \text{ kNm/m}$$

With water (use 3.8 m depth of penetration in the calculation)

$$\text{u} = 8.67\,\text{kPa},\ \frac{1}{2}\times 0.295\times 18\times 2^2 + 10.62(1+x) + \frac{1}{2}\times 0.295\times 9.69(1+x)^2$$

$$+\frac{1}{2}\times 8.67\times 1 + \frac{1}{2}\left(8.67 + \frac{3.8-x}{3.8}8.67\right)x = \frac{1}{2}\times 3.392\times 9.69x^2$$

$$or \qquad 25.6 + 10.62x + 1.429(1+x)^2 + 8.67x - 1.14x^2 = 16.434x^2$$

$$16.145x^2 - 22.15x - 27.03 = 0 \quad \text{gives} \quad \text{x} = 2.15\text{m}$$

The maximum moment is given by the active pressure taking the moment to point of zero shear passive pressure:

$$M_{\max} = \frac{1}{2}\times 0.295\times 18\times 2^2\times\left(\frac{2}{3}+3.15\right) + 10.62\times 3.15^2/2 + \frac{1}{6}\times 0.295\times 9.69\times 3.15^3$$

$$+\frac{1}{2}\times 8.67\times\left(\frac{1}{3}+2.15\right) + 3.76\times 2.15^2/2 + \frac{1}{2}\times 4.91\times 2.15^2\frac{2}{3} - \frac{1}{6}\times 3.392\times 9.69\times 2.15^3$$

= 80.7 kNm/m. Choosing the FSPII type sheet pile from Japan which is commonly used in Hong Kong. Section modulus Z = 874 cm^3/m

$$\sigma = \frac{80.7\times 10^6}{874\times 10^3} = 92.3\text{N/mm}^2 \quad < \quad 165\text{N/mm}^2 \qquad \text{O.K}$$

5.5.3.2 Free/fixed earth method for one layer of struts

Assume the water pressure to be hydrostatic for simplicity and it is on the safe side ($u = 2 \times 9.81 = 19.62$ kPa). The pressures at the two sides of the sheet pile are given in Figure 5.49.

After installation of the first strut and excavation, the wall below the formation level can move, and the large point force at the bottom is released. The design pressure can then be taken as simple active and passive pressure for this case, which is called the free earth method. Take the moment about support for simplicity (can take the moment at the bottom or another point):

$$15.93(D+2)\left(\frac{D+2}{2}+1\right) + \frac{1}{2}\times 0.295\times 9.69(2+D)^2\left((D+2)\frac{2}{3}+1\right)$$

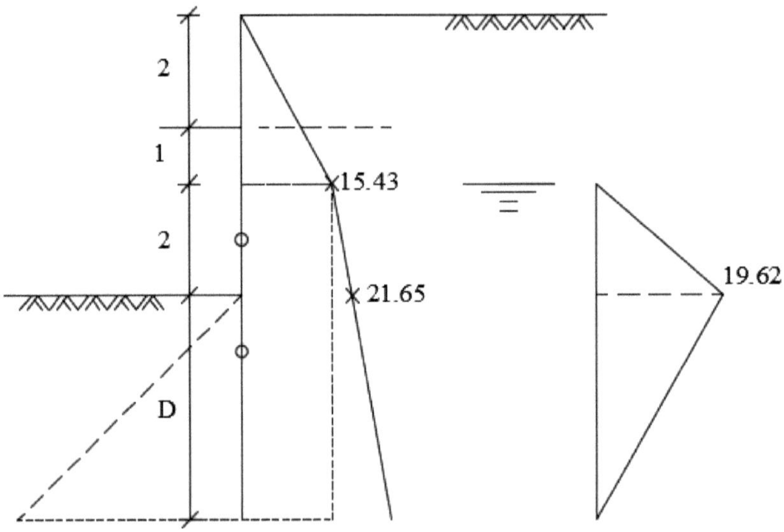

Figure 5.49 Free earth support analysis.

$$+\frac{1}{2}\times 19.62\times 2\times\left(\frac{4}{3}+1\right)+\frac{1}{2}\times 19.62\times D\times\left(\frac{D}{3}+3\right)=\frac{1}{2}\times 3.392\times 9.69\times D^2\left(\frac{2}{3}D+3\right)$$

$$15.93(D+2)\left(\frac{D}{2}+2\right)+1.43(2+D)^2\left(\frac{2}{3}D+2.333\right)$$

$$+45.78+9.810\left(\frac{D}{3}+3\right)=16.434D^2\left(\frac{2}{3}D+3\right)$$

Try $D = 2.5$: LHS = 432.2 RHS = 479.3

$D = 2.3$: LHS = 400.7 RHS = 394.1 O.K.

Add 20% to D, $D \approx 2.3 + 20\% = 2.76$ m. Anchor force is the difference between the total active and passive forces, hence

$$\text{Anchor/strut force} = \frac{1}{2}\times 3\times 15.93+15.93\times 4.3+\frac{1}{2}\times 0.295\times 9.69\times 4.3^2$$

$$+\frac{1}{2}\times 19.62\times 4.3-\frac{1}{2}\times 3.392\times 9.69\times 2.3^2 = 74.1\,\text{kN/m}$$

$$\text{Active pressure above the dredge level} = \frac{1}{2}\times 15.93\times 3+(15.93+21.65)+\frac{1}{2}\times 19.62\times 2$$

$$= 81.1\ \text{kN/m} \quad > \quad 74.1\ \text{kN/m}$$

∴ The point of zero shear is located above the dredge level, say distance x from ground level

$$\frac{1}{2}\times 15.93\times 3+15.93(x-3)+\frac{1}{2}\times 0.295\times 9.69\times (x-3)^2+\frac{1}{2}\times 9.81\times (x-3)^2=74.1$$

$$6.33(x-3)^2+15.93(x-3)-50.21=0, \text{so } x-3=1.83, \text{so } x=4.89\text{m}$$

$$M_{max}\ \frac{1}{2}\times 15.93\times 3\times 2.83+15.93\times 1.83^2/2+\frac{1}{6}\times 0.295\times 9.69\times 1.83^3$$

$$+\frac{1}{6}\times 9.81\times 1.83^3-74.1\times 2.83=-147.5kNm/m$$

Based on the series model test results, Rowe found that the maximum moment based on the above calculation deviates from the test results, and a moment reduction factor has been proposed by Rowe, which is not used in Hong Kong. According to the calculation shown above, the problem is statically determinate, and the result must be correct. Moment reduction in the free earth method is a superficial phenomenon, and is simply due to the use of the ultimate active and passive pressure coefficients which are not realistic. If computer analysis is used, there is no need to use the moment reduction coefficients. Furthermore, the moment reduction coefficients are based on mainly model tests, and the extension of these results to the various current retaining wall types is questionable.

Besides the free earth method, there is also a fixed earth method, which assumes that the sheet pile is long enough that a hinge is formed at a certain distance below the sheet pile. The location of the hinge is assumed to be known, and then the strut force can be determined. Several assumptions for the hinge location have been proposed previously, but there is not one which is universally used. In some of the ELS works in Hong Kong, some engineers adopt a distance of 0.5 m below the excavation level, which appears to be a rule of thumb practice. The authors do not prefer the fixed earth method, as the hinge location should depend on many factors which are not easily defined. Currently, an empirical approach is used in Hong Kong and China, for which the hinge point can be taken as the point of net zero pressure on the sheet pile. This assumption is obviously not correct, but other approaches do the same. In general, this approach is conservative, and can be used for some simple braced excavation design works. For the above example, suppose we adopt a designed depth of penetration *D* 10 m. Using Eq. (5.63) and Figure 5.50

$$u=\frac{2\times 2\times 10\times 9.81}{20+2}=17.84\ kPa$$

$$\text{stress at } RHS=\left[18\times 3+9.69(2+x)\right]\times 0.295+\frac{10-x}{10}\times 17.84$$

$$=9.69\times 3.392\times x=\text{stress at LHS}$$

$$32.868x=21.647+2.859x+17.84-1.784x\Rightarrow x=1.242m$$

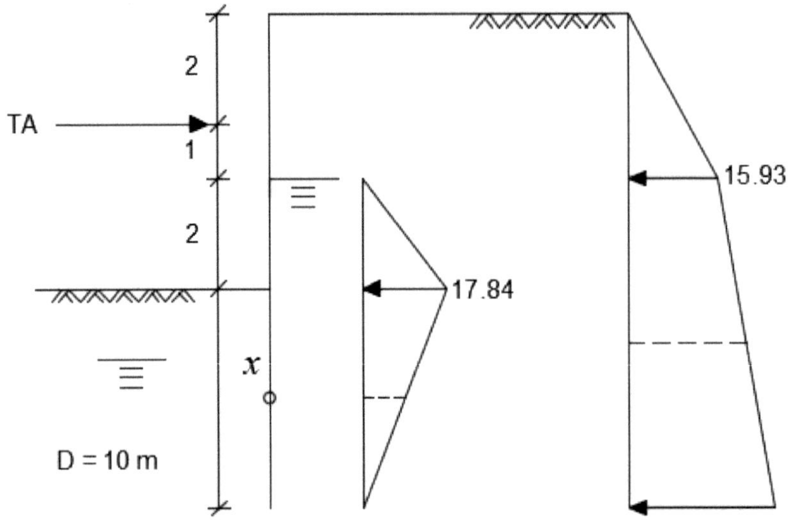

Figure 5.50 Fixed earth method for one layer of struts.

Take the moment at the hinge point

$$T_A \times 4.242 = \frac{1}{2} \times 15.93 \times 3 \times 4.242 + 15.93 \times 3.242^2/2 + \frac{1}{6} \times 0.295 \times 9.69 \times 3.242^3$$

$$+ \frac{1}{2} \times 17.84 \times 2 \times \left(\frac{2}{3} + 1.242\right) + 15.624 \times 1.242^2/2 + \frac{1}{2} \times 2.22 \times 1.242^2$$
$$\times \frac{2}{3} - \frac{1}{6} \times 9.69 \times 3.392x1.242^3 \Rightarrow T_A = 58.6kN/m$$

Excavate 3 m further as shown in Figure 5.51, assume T_A is remain unchanged in the second stage (unreasonable, but make this assumption)

$$u = \frac{2 \times 5 \times 7 \times 9.81}{14 + 5} = 36.14\ kPa$$

$$RHS = [18 \times 3 + 9.69(5 + x)] \times 0.295 + \frac{7 - x}{7} \times 36.14 = 9.69 \times 3.392x\ (LHS)$$

$$30.22 + 2.859x + 36.14 - 5.163x = 32.868x \Rightarrow x = 1.89m$$

$$\text{Hence } T_A \times 7.89 + T_B \times 4.89 = \frac{1}{2} \times 15.93 \times 3 \times 7.89 + 15.93 \times 6.89^2/2$$
$$+ \frac{1}{6} \times 0.295 \times 9.69 \times 6.89^3 + \frac{1}{2} \times 36.14 \times 5 \times \left(\frac{5}{3} + 1.89\right) + 26.38 \times 1.89^2/2$$
$$+ \frac{1}{3} \times 9.76 \times 1.89^2 - \frac{1}{6} \times 9.69 \times 3.392x1.89^3$$

$$\Rightarrow T_B = 123.4\ kN/m$$

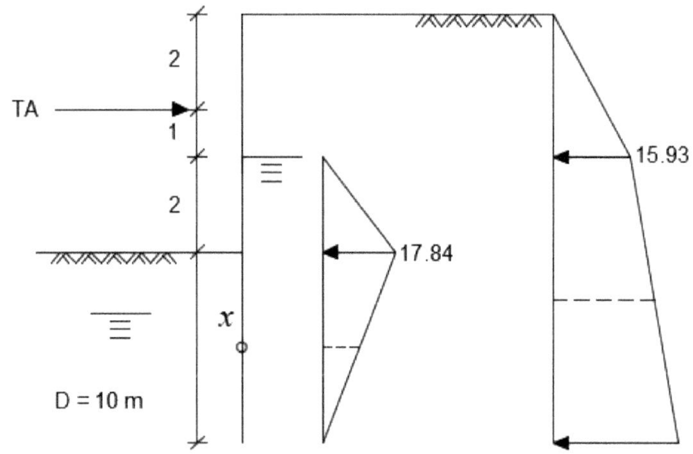

Figure 5.51 Fixed earth method for two layers of struts.

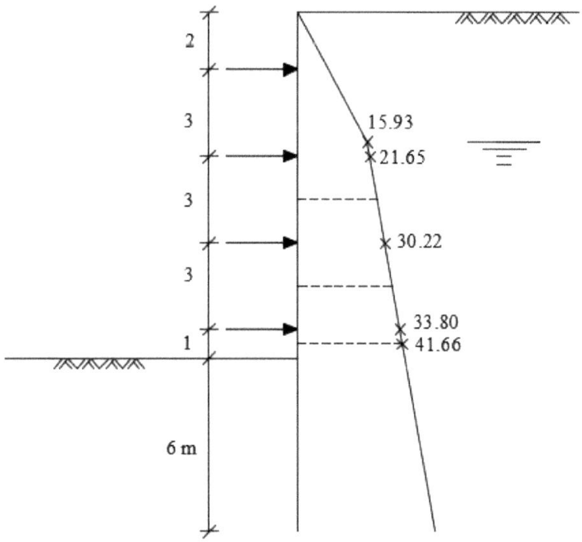

Figure 5.52 Design for the sheet pile.

After excavation to the full depth, the maximum moment on the sheet pile can be determined by assuming the hinge at the support locations. With reference to Figure 5.52, the trapezoidal pressure from soil and water can be averaged, and the maximum span moment will be:

Average load on span CD = (30.22 + 38.8)/2 + 9.81 × 6.5 = 98.28 kN/m

Design bending moment of sheet pile on span CD = 98.28 × 3^2/8 = 110.57 kN-m/m.

5.5.3.3 Depth of penetration required

The depth of penetration (push-in failure) required is determined by the piping and earth or moment requirement. According to GEO publication 1/90 and DM 7, with reference to Figure 5.53, the active side pressure Re which is measured from the mid-line between the lowest strut to the formation level is given by

Re = [(40.23+41.66)/2+(47.65+50.45)/2] x 0.5 = 45 kN/m

$$\text{Passive resistance } = \frac{1}{2}\times 3.392\times 9.69\times 6^2 = 591.6\ kN/m$$

Active force mid-line to bottom of sheet pile =

$$45 + 41.66\times 6 + \frac{1}{2}\times 0.295\times 9.69\times 6^2 + 50.45\times \frac{6}{2} = 497.8\text{kN/m} < 591.6\text{kN/m}$$

Alternatively, D can be estimated from the moment balance by taking the moment about the lowest strut D (using the pressure below the lowest strut) as:

$$\frac{1}{2}\times 38.8(D+1)^2 + \frac{1}{3}\times 0.295\times 9.69\times (D+1)^3 + 47.65\times \frac{1}{2} +$$

$$\frac{1}{2}\times 50.45D\times\left(\frac{D}{3}+1\right) = \frac{1}{2}\times 3.392\times 9.69\times D^2\left(\frac{2}{3}D+1\right)$$

Try D = 3.5: LHS = 694.8 RHS = 671

D = 4.0: LHS = 879 RHS = 964 $D \approx 3.6$ m

Piping is another critical condition to be checked in deep excavation. To avoid the use of computer software or the construction of a flow-net, a simplified method as shown in Figure 5.54 can be performed as

since $i_{crit} = (Gs - 1)/(1 + e) = \gamma'/\gamma w \approx 1.0$

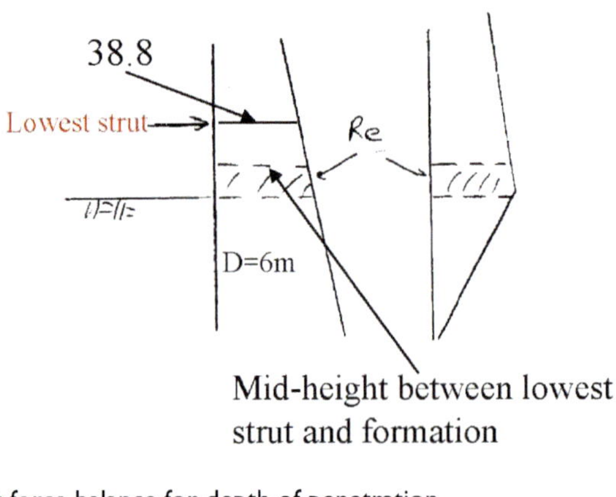

Figure 5.53 Check against force balance for depth of penetration.

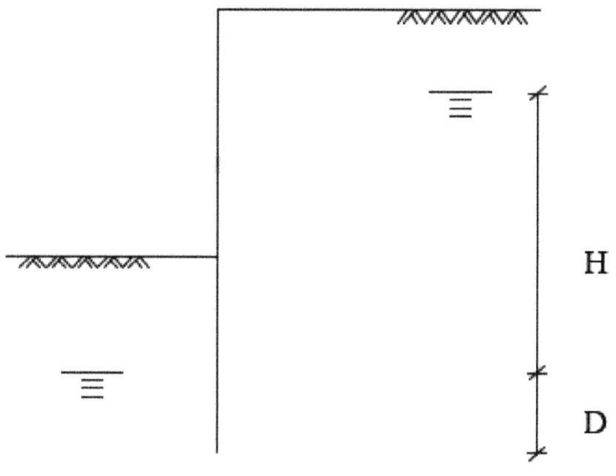

Figure 5.54 Simplified method for piping analysis.

For piping, average hydraulic gradient $= i = \dfrac{9}{9+2\times 6} = 0.43 < I_{crit}/1.5$ OK

In this approximate approach, the shortest flow path is used and the average hydraulic gradient is considered, which is the flow path adjacent to the wall. Alternatively, a com puter program like SEEP by Cheng, Modflow or Plaxis can be used for the seepage analysis. Consider a problem by Craig, where the head difference is 4 m, and the depth of penetration is 4.8 m. Using the classical soil mechanics approach, Craig (2004) obtained a factor of safety of 3.4 by considering flow-net construction. Using the simplified approach as outlined above, a factor of safety of 2.54 is obtained. In general, the piping problem is not critical towards many practical problems, and may be problematic for marine sand and alluvium only (in Hong Kong). Alternatively, some engineers carry out the design figures based on the design figures in DM7, which are established for homogeneous conditions only. In Figure 5.55, piping is observed for deep excavation projects in China. In fact, piping is also common for caisson construction in Hong Kong.

5.5.4 Equivalent earth pressure

Equivalent earth pressure envelopes have been proposed by Peck for excavation in sand, soft to medium clay and stiff clay. Such a pressure envelope is based on the field tests by Terzaghi and Peck in the US and Germany in the 1940s. The use of these simplified pressure envelopes is based on the use of an effective loading area concept, and is meant to design the strut load only, with an adequate factor of safety. Such pressure envelopes are not used for the design of the sheet pile, and should not be taken as the actual pressure at the retaining wall. For most of the complicated conditions in the presence of a water table and multi-layer of soil, there is no simple way to use these pressure envelopes. Furthermore, the database for these design figures may not reflect the current situation, hence they have not been used for design in Hong Kong for many years.

Figure 5.55 Piping causing basal failure in a deep excavation project in China.

5.6 GROUND SETTLEMENT

Ground settlement has been a great problem in many developed cities, particularly for Hong Kong where many deep excavations are carried out at reclaimed areas with a high ground water table and poor soil conditions. The maximum ground/building settlements that the authors have experienced in Hong Kong amount to 350 mm, hence recharge well and strut preloading are used to help reduce the wall and ground settlement. Estimating the ground settlement is not an easy task, as the settlement comes from the settlement induced by the wall installation and excavation process. For wall installation, the estimation of ground settlement relies on the use of statistical results, which depend on the soil condition, wall type, use of machines, quality of workmanship and other factors. For the excavation process, several methods are available. The authors have studied many ground settlement results due to excavations. In general, maximum ground settlement occurs at a certain distance behind the retaining wall. Several methods are available for the estimation of ground settlement during deep excavation.

5.6.1 A ground/wall settlement coefficient

Based on the field observations by Milligan (1983), Nicholson (1987) and Clough and O'Rouke, the maximum ground settlement is now commonly taken as 0.5–1.0 of the maximum wall movement for the design in Hong Kong. The exact factor will depend on the ground conditions. In general, such a correlation is found to be reasonable, based on the field data from Hong Kong, Taiwan, the US and elsewhere. For normal conditions, a factor of 0.5 is taken, while 1.0 is used mainly for excavation in poor ground conditions. Currently, this is one of the methods used in Hong Kong. As already mentioned, we could not find a program which can model well the interface between the soil and wall during the staged excavation. Based on the observations in several very deep excavations in Hong Kong, Cheng found that the wall does not move at the same magnitude as the soil, hence there is slipping along this interface which is not considered in most commercial programs. This is the most critical reason for the poor prediction in ground settlement by most commercial programs. On the other hand, prediction of the wall movement by commercial programs is relatively poor, hence many engineers prefer to use the settlement ratio in the design of deep excavations.

5.6.2 From finite element analysis

Many finite element programs can give a prediction of ground movement (except Frew which is semi-finite element). From many field observations, it is noted that the predictions from these programs are usually not good. The reasons for this are possibly due to the use of conservative soil parameters as well as inadequate modelling of the soil–structure interface. An upheaval adjacent to the wall is commonly obtained from finite element analysis, which is however not commonly observed on site. On the other hand, Cheng has observed an upheaval of up to 25 mm when the dewatering wells are turned-off after completion of the base slab construction of an MTR station (high water pressure exerted at the base slab).

5.6.3 Settlement classification by Peck

Peck (1969) produced a settlement profile estimation, where the influence zone extends to $2H$ for good conditions and $4H$ for poor conditions. The settlement ratio is divided into three zones, and the maximum settlement is taken as being adjacent to the wall, which is not supported by many of the current field measurement results. Since the database by Peck is more than 50 years old, these results are now usually used for reference instead of design.

5.6.4 Method by Bauer

Based on a series of field measurements, Bauer (1984) proposed a design method where the maximum ground settlement is taken as a ratio r_0 of the excavation height, which is a function of the relative density of the soil. The ground settlement is assumed to vary according to

$$S = S_0\left(\frac{x}{B}\right)^2 f_1 f_2 \quad B = 1.5H\tan\left(\frac{\pi}{4}-\frac{\phi}{2}\right) \tag{5.68}$$

where x is the distance from the wall, B is the distance of the influence zone, f_1 is a function of the workmanship and f_2 is a function of the construction difficulty. This method has been used in Hong Kong in the past, but has been largely replaced by the use of settlement coefficient. Furthermore, this method assumes the maximum ground settlement to be adjacent to the wall, which is not realistic, also, non-homogeneous conditions are not well covered in this design approach.

5.6.5 Method by Bowles

Based on the assumption of the volume of wall deflection, V_s equals the volume of ground settlement, which is a relatively reasonable assumption for good soil. Bowles (1996) proposed a method which has been used by some engineers in the past. Assume that V_s can be obtained by the subgrade reaction model (or finite element analysis), H_p can be obtained as B for $\phi = 0$ and $0.5H \tan(45° + \phi/2)$.

$$H_t = H + H_p \tag{5.69}$$

Distance of influence D is computed as

$$D = H_p \tan(45° - \phi/2) \tag{5.70}$$

The maximum ground settlement S_w adjacent to the wall and the ground settlement S at distance x from the wall are given as

$$S = \frac{2V_s}{D} \tag{5.71}$$

$$S = S_w \left(\frac{x}{D}\right)^2 \tag{5.72}$$

The methods by Bauer (1984) and Bowles (1996) are used when finite element programs are not available to engineers, but these are not commonly used nowadays. The use of the ground settlement coefficients is favoured by many engineers, as this method is simple and the accuracy is as good as the finite element analysis. Ou and Hsieh (2000) proposed a method which is practically equal to the settlement ratio approach together with a ground settlement envelope. The estimation of the ground settlement due to excavation, ground water lowering and diaphragm wall construction for a project in Hong Kong is shown in Figure 5.56. Precautions in the form of ground treatment or others can be assessed from this ground settlement estimation.

For settlement due to ground water drawdown, Cheng has observed that there is not always a clear relationship between the ground settlement and water table drawdown. Nevertheless, some engineers estimate the vertical ground settlement from drawdown of the water table as $m_v \sigma H$, where σ and H are the drawdown of water pressure, H is the thickness of soil under consideration, and m_v is the coefficient of volume compressibility which is sometimes taken as the inverse of E which is estimated as $1/N$ or $(1 - 2\mu)(1 + \mu)/E(1 - \mu)$.

Finally, a collection of all the settlements arising from retaining wall construction, wall deflection and ground water drawdown is determined, and the accumulated results are checked with the corresponding government regulation. If the maximum allowable

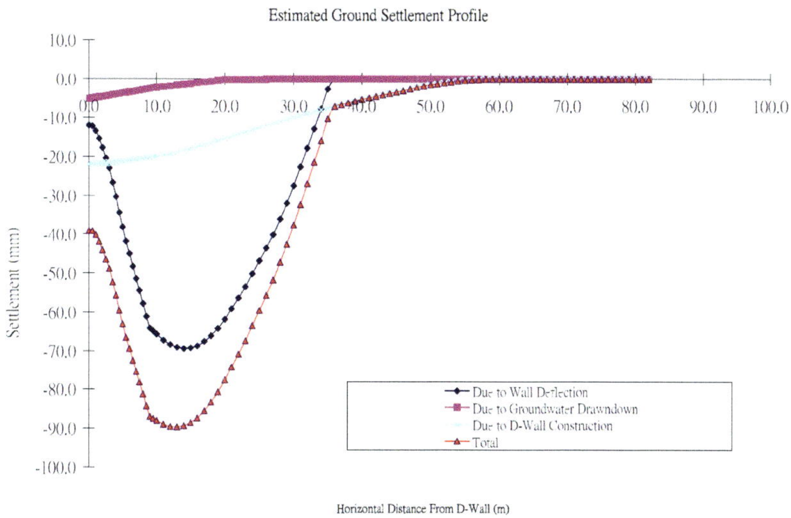

Figure 5.56 Determination of various settlements at the final stage for a project in Hong Kong.

settlement is exceeded, precaution measures must be applied, which may include ground treatment, strut preloading, use of recharge wells, underpinning to surrounding structures and/or others.

5.7 BASAL STABILITY PROBLEM IN CLAY

The basal stability problem is not common in sandy soil, but can happen easily in soft soil, as shown in Figure 5.57. For excavation in clay, if a rock or hard layer exists at a depth of more than $0.7B$, where B is the width of excavation, the factor of safety FS for the basal stability by Terzaghi is taken as

$$FS = \frac{5.7cB_1}{\gamma HB_1 + qB_1 - cH} \tag{5.73}$$

where q is the surcharge outside the retaining wall, and B_1 is $B/\sqrt{2}$. If a rock or hard layer exists at a depth of D which is less than $0.7B$, D will be used instead of $B1$ in Eq. (5.73). It should be noted that, under this method, the depth of penetration of the sheet pile has no contribution to the factor of safety, which is obviously incorrect. For sandy soil, checking of the piping and the force/moment balance as outlined previously will generally be sufficient. Some engineers adopt the limit equilibrium slope stability method in assessing the factor of safety against basal stability (Ou 2006). In general, a factor of safety of 1.5 should be applied for the basal stability. Basal instability has been recorded for a series of deep excavation projects in China. The use of jet grouting or the deep cement mixing method (see Chapter 3) is usually adopted to overcome the basal instability problems in soft soil.

Figure 5.57 Basal instability problem for a deep excavation project in FoShan, China.

5.8 MONITORING SCHEME

Excessive ground/building settlement, and damage to utilities and buildings have occurred in Hong Kong and many developed cities. To safeguard the public and surrounding buildings from damage, a well-considered monitoring scheme should be proposed, and a survey of the adjacent buildings and ground should be conducted prior to the construction. In the survey, the conditions of the existing buildings have to be examined and recorded in detail. Existing cracks should be monitored for assessing the ground movement as well as future court cases. In general, the water table and ground movement should be monitored twice every week. For deep excavations, an inclinometer is also commonly installed at the retaining wall as well as on the ground. A typical inclinometer is shown in Figure 5.58, which is commonly used in Hong Kong, and some typical settlement markers are shown in Figure 5.59.

An ELS monitoring scheme in Hong Kong involving ground settlement points, building settlement points, vibration and tilting monitoring, piezometer, with dewatering wells inside the excavation and recharge well outside the excavation is shown in Figure 5.60 (the AAA levels are 12, 18 and 25 mm for settlement, 7.5, 10 and 15 m/s for vibration, and 1:1000, 1:750 and 1:500 for tilting). The extent and requirements for monitoring depend on the depth of excavation, ground conditions and surrounding buildings.

Currently, Cheng is working on the development and implementation of IoT (Internet of Things) monitoring systems in slopes, retaining wall and buildings. In such a system,

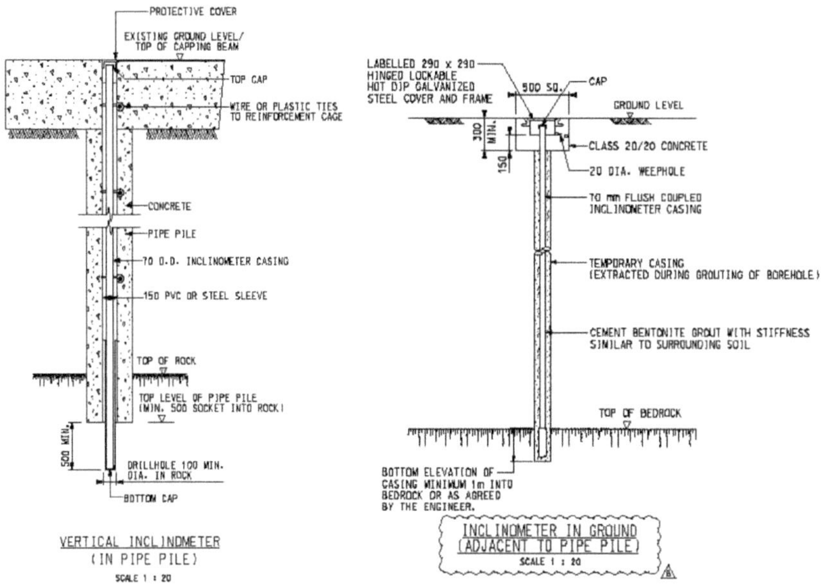

(a) Inclinometer in retaining wall (b) inclinometer in ground

Figure 5.58 Typical inclinometer for retaining wall and ground movement monitoring in deep excavation.

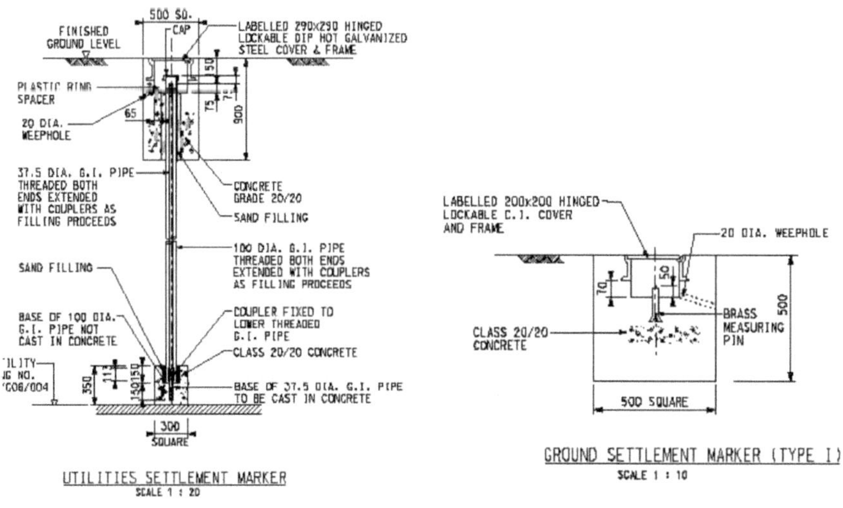

(a) Settlement marker at utilities (b) ground settlement marker

Figure 5.59 Typical monitoring devices for an ELS project (also piezometer/standpipes as shown in Chapter 1).

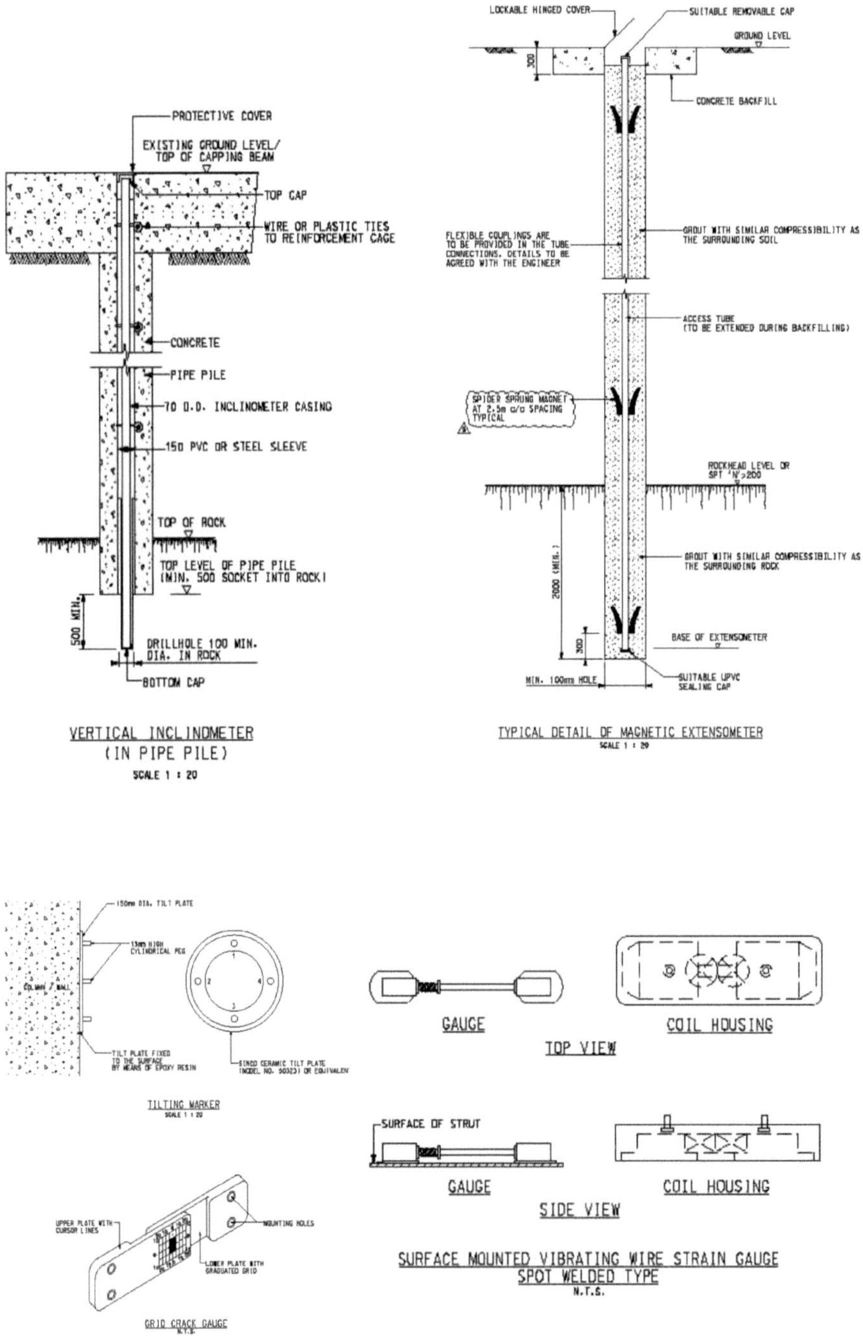

Figure 5.59 (Continued)

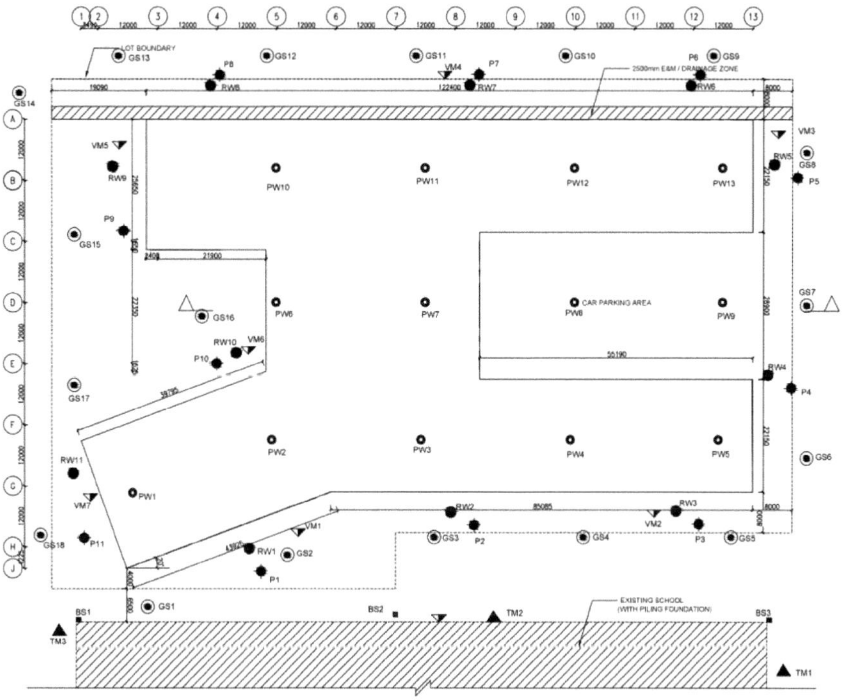

Figure 5.60 A typical monitoring plan for an ELS project in Hong Kong.

every sensor is powered by electricity instead of requiring manual checks as described above. Every sensor is equipped with a sim-card and functions as a mobile in sending out the measured results through the cloud. The server receives the results and assesses and sends out warnings according to guidelines built into the server system. A typical tiltmeter and GPS station installed by Cheng are shown in Figure 5.61, while a pore pressure transducer and IoT are shown in Figure 5.62. The concept of IoT monitoring is that these sensors take measurements of the inclination, vibration, tilting, pore pressure, crack width and other values regularly (hourly or any other value). The results are sent to the server through the sim-card in the IoT unit as shown in Figure 5.62c and received by the server. The engineers should build in the alarm level in the server system for assessment. Warnings will be issued to engineers automatically through email, SMS or other means. For the typical triggering levels used for ground settlement monitoring in Hong Kong, a maximum ground movement exceeding 12 mm is the alert level, which is followed by 18 mm and 25 mm for the alarm and warning levels. The complete information can also be viewed from a mobile app or web page, which greatly facilitates engineers in monitoring.

5.8.1 Importance of IoT monitoring and instantaneous analysis

With regard to Figure 5.63 for the Kwun Lung House retaining wall failure (Report GEO Report 103) and the large-scale slope failure in China (Figure 5.64), deaths can be reduced

(a) Tiltmeter with solar cell (b) GPS/GNSS station

Figure 5.61 Tiltmeter and GPS station.

(a) Electric pore pressure transducer (b) transducer in borehole (c) IOT unit

Figure 5.62 IoT pore pressure measurement and IoT unit.

or even eliminated if an IoT system is used for these two cases (and many other cases). Take the Kwun Lung House case for reference:

1. The consultant firm carried out many visual observations before the failure, and the wall was rated as "safe". The rapid change in the pore pressure within a short time and the development of a failure could not be assessed by routine inspection.

Figure H5 - Photograph of the Failure Scar Showing the Field Observation Points Described in Table H3

Figure 5.63 Kwun Lung House retaining wall failure that killed two people in 1994.

2. The retaining wall failed when the rainfall reduced.
3. The failure occur suddenly within a short time.
4. The failure occurred at midnight and resulted in the deaths of two people, no warning was be given, as no one had inspected the work.

Using the mobile app, it is possible to examine the pore pressure, tilting and displacement at any time and place, even while on holiday, at night time or during very heavy rainfall, without the need for on-site measurement. A warning signal system will also be issued automatically subject to the results from the cloud computing system. Using an IoT measurement and warning system, a sudden increase in pore pressure due to breakage of utilities and movement prior to failure can be automatically detected and assessed at any time, without the direct attention of an engineer, and a warning signal can be issued quickly. Currently, IoT monitoring is becoming increasingly popular in China, Hong Kong, Taiwan and many other countries. A typical mobile app by Cheng is illustrated in Figure 5.65

In the middle of Figure 5.65, the 3-D view of the site under monitoring is shown, with the locations of the sensors highlighted. Users can point to the sensor locations, and the detailed *instantaneous* field data will be shown. An overall assessment of the retaining wall is also shown on the mobile app. The IoT system is expensive in the short term, but is cost-saving in the long term, as no technicians are required to take the measurements, and no engineers are required to process the measured result. The advantage of this concept is clear: monitoring of the instantaneous field conditions at any time, in any location, by all authorized users.

Figure 5.64 Large-scale slope failure and subsequent debris flow in Shenzhen, China, in 2015.

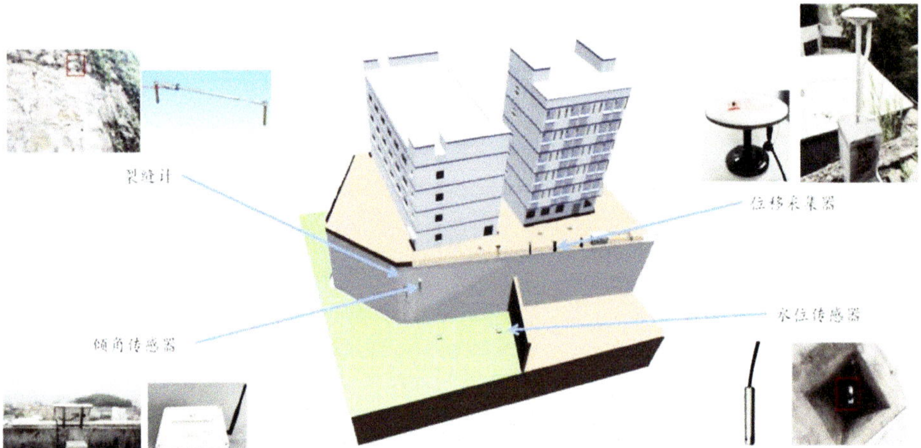

Figure 5.65 Mobile app for IoT monitoring of a retaining wall by Cheng.

The testing and monitoring works for the third runway of Hong Kong Airport provide a good illustration of the needs and importance for reclamation works. In this project, the sensors include a vertical inclinometer, magnetic probe extensometer, vibrating wire piezometer, standpipe, settlement marker, surface movement marker, temporary ground movement marker, automatic deformation monitoring system, utility monitoring point, tilt plate, master survey control points and vibro graph. These sensors are used for monitoring of the reclamation, seawall and existing structures.

References

Absi E and Kerisel J (1990), *Active and passive earth pressure tables*, Taylor & Francis, UK.

Askew I, Sein D, Frame A (2006), Improving design efficiency and safety in substructure building works in Hong Kong through observational method design. *Proceedings of the seminar on the observational method and its application* in Hong Kong, HKIE-HKGES-AGS(HK).

Bauer GE (1984), Movements associated with the construction of a deep excavation, In: *Proceedings of the 3rd international conference on ground movements and structures*, Cardiff, 694–706.

Berezantzev VG (1958), Earth pressure on the cylindrical retaining wall, *Proceedings of Brussels Conference on Earth Pressure Problems*, 2, 21–27.

Bowles JE (1996), *Foundation analysis and design*, 5th ed., The McGraw-Hill Companies, Inc.

Caquot A and Kerisel J (1948), *Tables for the calculation of passive pressure, active pressure and bearing capacity of foundations*, Gauthier-Villars, Paris, France.

Chan A (2003), Observations from excavations–A reflection. Proceedings of the seminar on case histories in geotechnical engineering in Hong Kong, HKIE Geotechnical Division.

Cheng YM (2003), Seismic lateral earth pressure coefficients by slip line method, *Computers and Geotechnics*, 30(8), 661–670.

Cheng YM and Tsui Y (1989), A fundamental study of braced excavation construction, *Computers and Geotechnics*, 8, 39–64.

Cheng YM and Tsui Y (1991), A simple method in the solution of free surface seepage problem, *Asian Pacific conference on computational mechanics*, Hong Kong, pp. 833–838.

Cheng YM and Tsui Y (1993), An effective method for the solution of free surface seepage flow problems, *Computers and Geotechnics*, 15, 47–62.

Cheng YM, Chen F, Yuan CF and Li L (2021), Seepage analysis for Construction, with applications to some projects in Hong Kong, *Geofluids*, 2021, https://doi.org/10.1155/2021/6623816

Cheng YM, Hu YY and Wei WB (2007), General axi-symmetric active earth pressure by methods of characteristics–Theory and numerical formulation, *International Journal on Geomechanics*, ASCE, 7(1), 1–15.

Cheng YM, Hu YY, Au SK and Wei WB (2008), Active pressure for circular cut with Berezantzev's and Prater's Theories, *Soils and Foundations*, 48(5), 621–632.

Cheng YM, Wang JH, Li L and Fung WH (2020), *Frontier in civil engineering vol.4 Numerical methods in geotechnical engineering Part II*, Benjamin Press.

Craig RF (2004), *Craig's soil mechanics*, 7th ed., Spon Press.

Jaky J (1944), The coefficient of earth pressure at rest, *Journal for the Society of Hungarian Architects and Engineers*, October, 355–358.

Karafiath L (1953), On some problems of earth pressure, *Acta Tech, Acad, Sci. Hung*, 328–357.

Lorenz H (1966), *Offene Senkkasten*. Grundbautaschenbuch. W. Ernst & Son, pp. 795–798.

Mayne PW and Kulhawy FH (1982), *Ko–OCR* relationships in soil, *Journal of the Geotechnical Engineering Division, ASCE*, 108(GT6), 851–872.

Meyerhof GG (1976), Bearing capacity and settlement of pile foundations, *JGED, ASCE*, 102(GT 3), 195–228.

Milligan GWE (1983), Soil deformations near anchor sheet pile walls, *Geotechnique*, 33(1), 41–55.

Nicholson DP (1987), The design and performance of retaining walls at Newton Station, *Proceeding of the Singapore Mass Rapid Transit Conference*, Singapore, 99, 147–154.

O'Rourke TD, Cording EJ and Boscardin M (1976), *The ground movements related to braced excavation and their influence on adjacent structures*. University of Illinois Report for U.S. Dept. of Transportation, Report No. DOT-TST-76T-22.

Ou CY (2006), *Deep excavation*, Taylor & Francis.

Ou CY and Hsieh PG (2000), *Prediction of ground surface settlement induced by deep excavation, Geotechnical Research Report No. GT200008*, Department of Construction Engineering, National Taiwan University of Science and Technology.

Pan, JKL, Pappin, JW, Cowan S, and Lam LWY (2001). An application of the observational method at Tsueng Kwan O Station and Tunnels. *Proceedings of the Seminar on Geotechnical Deformation and Movement*, HKIE Geotechnical Division.

Pavlovsky NN (1933), Motion of water under dams. Trans. 1st Congr. on Large Dams, Vol. 4., Stockholm, pp. 179–192.

Peck RB (1969), Deep excavation and tunneling in soft ground, *Proceedings of the 7th International Conference on soil Mechanics and Foundation Engineering*, Mexico City, State-of-the-Art Volume, pp. 225–290.

Prater EG (1977), An examination of some theories of earth pressure on shaft lining, *Canadian Geotechnical Journal*, 14(1), 91–106.

Schneebeli G (1964), Le stabilite des tranchees profondes forees en presence de boue. *Houille Blanche*, 19(7), 815–820.

Sokolovskii VV (1965), *Statics of granular media*, Pergamon Press.

Steinfeld DK (1958), Uber den Erddruck auf Schacht-und Brunnenwandungen. Contribution to the foundation engineering meeting, Hamburg. German Soc. of Soil Mech. Found. Eng., 111–126.

Sze, JWC and Young, STM 2003. Design and construction of a deep basement through an existing basement at Central. *Proceedings of the seminar on case histories in geotechnical engineering in Hong Kong*, HKIE Geotechnical Division.

Terzaghi K (1943), *Theoretical soil mechanics*, John Wiley.

Yau KF and Sum HL (2010), Some *observations associated with design and construction of excavation and lateral support in Mainland*, HKIE Geotechnical Seminar 2010, HKIE.

Further reading

Burland J (2004), *Design and construction of deep basements including cut-cover-structures*, Institute of Structural Engineers, UK.

Cashman PM and Preene M (2021), *Groundwater lowering in construction*, 3rd ed., CRC Press.

Cheng YM (2016), Rankines earth pressure coefficients for inclined ground reconsidered by slip line method, *Journal of Civil Engineering*, 1(1), 1–7.

Cheng YM and Hu YY (2005), Active earth pressure for circular shaft lining by simplified slip line solution with general tangential stress coefficient, *Chinese Journal of Geotechnical Engineering*, 27(1), 110–115.

Cheng YM, Hu YY and Wei WB (2007), General axi-symmetric active earth pressure by methods of characteristics–Theory and numerical formulation, *International Journal on Geomechanics, ASCE*, 7(1), 1–15.

Cheng YM, Hu YY, Au SK and Wei WB (2008), Active pressure for circular cut with Berezantzev's and Prater's theories, *Soils and Foundations*, 48(5), 621–632.

Cheng YM, Wang JH, Li L and Fung WH (2020), Frontier in civil engineering, vol.3, *Numerical methods and implementation in Geotechnical Engineering*, Benjamin Press, 2020.

Clayton CRI, Woods RI, Bond AJ and Milititsky J (2013), *Earth pressure and earth-retaining structures*, 3rd ed., CRC Press.

Clough GW and O'Rouke TD (1990), Construction induced movements of in situ walls, In: *Proceedings of the 4th national geotechnical conference*, Hawlin, Taiwan.

Desai CS and Zaman M (2014), *Advanced geotechnical engineering, soil-structure interaction using computer and material models*, CRC Press.

Endicott J (2020), *Deep excavations in soil*, CRC Press.

GCO (1990), *Review of design methods for excavations*, GCO Publication No.1/90, GCO, Hong Kong Government.

GEO (1993), *Guide to retaining wall design*, GeoGuide 1, 2nd ed., Geotechnical Engineering Office, Hong Kong.

Haque MI (2015), *Mechanics of groundwater in porous media*, CRC Press.
Henn RW (1996), *Practical guide to grouting of underground structures*, ASCE Press.
Hettler A and Kurrer KE (2020), *Earth pressure*, Ernst & Sohn.
Hsieh PG and Ou CY (1998), Shape of ground surface settlement profiles caused by excavation, *Canadian Geotechnical Journal*, 35, 1004–1017.
Kempfert HG and Gebreselassie B (2006), *Excavations and foundations in soft soils*, Springer Verlag.
Kovacs G (1981), *Seepage hydraulics*, Elsevier.
Lees A (2016), *Geotechnical finite element analysis*, ICE Publishing.
O'Rouke TD, Cording EJ and Boscardin MD (1976), The ground movements related to braced excavation and their influence on adjacent structures, *University of Illinois, Report for U.S. Dept. of Transportation*, Report No. DOTTST-76T-22.
Ou CY (2021), *Fundamentals of deep excavations*, CRC Press.
Powers JP, Corwin AB, Schmall PC, Kaeck WE and Herridge CJ (ed.) (2007), *Construction dewatering and groundwater control*, John Wiley.
Puller M (2003), *Deep excavations, a practical manual*, 2nd ed., Thomas Telford.
Smedt FD and Ziji W (2018), *Two and three-dimensional flow of groundwater*, CRC Press.
Soubra AH and Macuh B (2002), Active and passive earth pressure coefficients by a kinematical approach, *Geotechnical Engineering*, 155(2), 19–131.
Sterrett RJ (2007), *Ground water and wells*, 3rd edition, Johnson Screens.
Tonon F, Liu X and Wu W (2010), *Deep and underground excavations*, Geotechnical Special Publication No.l26, ASCE.
Toth J (2009), *Gravitational systems of groundwater flow*, Cambridge University Press.
Turner JM (2009), *Excavation systems, planning, design and safety*, McGraw Hill.
Zhang WG and Liu HL (2022), *Design of deep braced excavation and earth retaining systems under complex built environment*, Science Press, Beijing.

Chapter 6

Piling foundation design and construction

Piles are used to take up vertical loadings, lateral loads and tensions. Since piles are installed below ground, many types of piles have been developed for different ground and environmental conditions. There are many types of piles available in the construction industry, and it is impossible to include all types in this chapter.

6.1 CLASSIFICATION OF PILES

Piles are usually classified by the material type, method of installation or load transfer method. For load transfer, piles can be classified as end-bearing piles or friction piles according to the relative proportion between the skin friction and end-bearing. It should be noted that many end-bearing piles function as friction piles under working load conditions.

Classifications for different types of piles:

(a) End bearing and friction pile – This classification is based on the load transfer of the piles. If the main pile capacity is derived from friction along its shaft, it is termed a "friction pile". If the main pile capacity is derived from its end-bearing capacity on a hard stratum, it is termed an "end-bearing pile". It should be noted that the classification is mainly used for design purposes, as a pile's actual geotechnical capacity is a combination of both friction and end-bearing, and the ratio between the shaft friction and end-bearing capacity will vary with the applied loading. Many end-bearing piles function as friction piles under working load conditions, which is revealed by many static load tests in Hong Kong and other places.

(b) Percussive pile and non-percussive pile – This classification is based on the method of installation. A percussive pile is installed by percussive driving. A non-percussive pile is normally installed by forming a hole in the ground by methods such as auger or rotary cutting after which (i) a preformed pile such as a precast concrete pile or steel pile is placed; or (ii) wet concrete is poured into the hole to form the pile.

(c) Displacement pile, small displacement pile and non-displacement pile – This classification is based on the amount of displaced soil during installation. Displacement (or large displacement) pile involves relatively large soil displacement, and this type of pile is usually a solid pile or a hollow pile with closed ends. A small displacement pile creates relatively low displacement of soil during the installation, which

DOI: 10.1201/9781032650586-6

is commonly steel H or hollow section piles where the cross-sectional area is small. Non-displacement pile is formed by boring into the ground before the pile installation. The significance of soil displacement dictates the friction along the pile shaft and densification/loosening of the surrounding soil.

(d) By materials:
Timber pile – Seldom used nowadays, but many old structures are supported on timber piles.
Concrete pile – Usually in the form of precast piles (reinforced and more commonly prestressed) or cast in situ piles.
Steel pile – Mostly tubular or H pile, but piles in other shapes are sometimes used.
Composite pile – A combination of two or more structural materials, and most commonly tubular steel pile with concrete or cast in situ concrete pile with steel section.

(e) Mode of loading – Axially loaded, transverse or laterally loaded, moment resisting or combinations of these loadings under different load cases.

(f) Shape – Square (solid or hollow), octagonal (solid or hollow), circular (solid or hollow), fluted, H pipe and others

(g) Miscellaneous – Piles may have a constant cross-section or be tapered, and in some occasions, may have different sizes along their lengths. Many special piles are used in China to suit different ground conditions and to reduce the cost of construction. A special screw pile is shown in Figure 6.1, which appears to be not used in other countries.

Figure 6.1 A low-capacity screw pile used in China.

6.2 INSTALLATION OF PILES

Piles can either be prefabricated as a steel section pile or precast concrete pile prior to installation into the ground or formed by cast in situ concrete. There are many types of piles around the world, however this section only discusses the more commonly used piles in construction.

6.2.1 Prefabricated piles

A prefabricated pile (Figure 6.2) can be installed into the ground by percussive driving, and the pile capacity comprises the shaft friction and the end-bearing at its tip, or a combination of both capacities in most cases. Prefabricated pile is driven by a heavy hammer dropped from height. During the impact, energy is transferred from the kinetic energy of the hammer to cause energy dissipation along the shaft and base. To confirm the pile capacity, temporary compressions of pile and soil and permanent set of the pile are usually measured. H and tubular piles are the most popular steel piles in construction, but other shapes (usually high capacity) are also available. The use of steel tubular piles is popular in many earthquake zones, as the section modulus of the pile is the same for all directions of earthquake. There are also cases where a hole is bored with the insertion of prefabricated pile, but this is not a common practice.

Besides steel piles, precast concrete piles are also commonly used (Figure 6.2c). In the past, reinforced concrete piles were used for low-capacity purposes, and the shape of the reinforced concrete pile was usually square. Such low-capacity reinforced concrete pile is not popular now, but is still used in some countries where the load capacity requirement is not high. The main advantage of such pile is the low investment cost and the flexibility in production and transportation. Since the capacity of the reinforced concrete pile is low, it has now been largely replaced by precast prestressed concrete piles in many countries. The precast prestressed spun concrete pile (Figure 6.1c) with a diameter of 400–600 mm is manufactured using high-grade concrete (cube strength up to 78 MPa) with prestressed tendons embedded in the pile. The allowable capacity ranges from 1690 kN to 3500 kN per pile, and is commonly used for many building works in Hong Kong, Macau and China. Spinning of the pile shaft is involved during manufacturing of the pile. The prestressing force by the tendons serves to resist the transient tensions which may be induced during pile driving. It should be noted that the prestressing force does not reduce the **ultimate** pile capacity, as the applied load will shorten the pile and hence reduce the prestressing force in the pile. In Hong Kong, each pile is normally 12 m in length, and connection between piles is achieved by welding the steel joint plates at the ends of each pile section. Steel conical pile shoes are usually installed at the pile ends for pile driving. The common sizes of this pile type include 500 mm diameter with 100 mm thick walls and 500 mm diameter with 125 mm thick walls. The maximum compressive capacities of these two types of piles are 2300 kN and 2700 kN, respectively. Since the displacement for this pile is large, the driving sequence should be carefully arranged. Normally, driving is commenced for piles in the central portion of the site, and driving is progressed in an outward direction towards the periphery of the site. This type of pile is not suitable for sites where underground boulders and other obstructions exist, since the piles may crack under excessively hard driving which has been found from some piling projects.

(a) Steel H pile (b) pipe pile

(c) Precast pretensioned concrete pile

Figure 6.2 Some typical prefabricated driven piles.

There are other types of driving type piles, which are not commonly used nowadays, including the Franki pile, Vibro pile, pressure pile and others. While percussive driving may induce undesirable sound pollution and ground vibrations, a pile may be silently jacked into the ground, usually by high hydraulic pressure. Pile jacking is less efficient than percussive driving and is generally used when percussive piling is not allowed.

6.2.2 Socketed H pile

The prefabricated pile can be installed in a prebored hole formed by preboring or augering. Shaft friction is usually less significant for this type of pile, and the pile capacity is usually derived from the end-bearing with a hard stratum or the bond stress with the rock socket (Figures 6.3a,b). For the former case, the H-pile rests on rock, but this is less common in Hong Kong. In Hong Kong, piles socketed into rock are more common. A socketed H-pile is formed by preboring a hole through the soil stratum using an Odex (Overburden Drilling with EXcentric bit) drill or concentric driving system (Figure 6.4) and then into

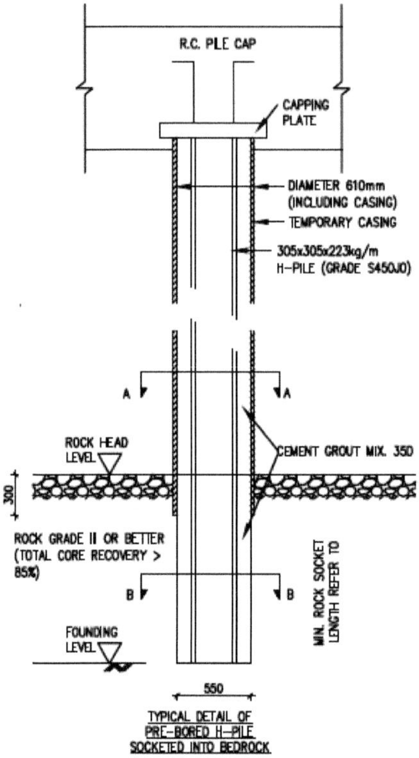

(a) typical socket H pile

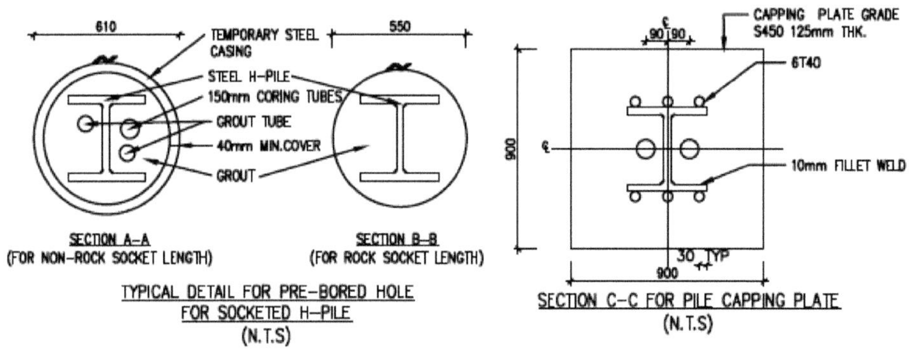

(b) Section A-A and B-B for Fig.6.3a

Figure 6.3 Socket H-pile construction.

(c) drilling to form socket H pile

Figure 6.3 (Continued)

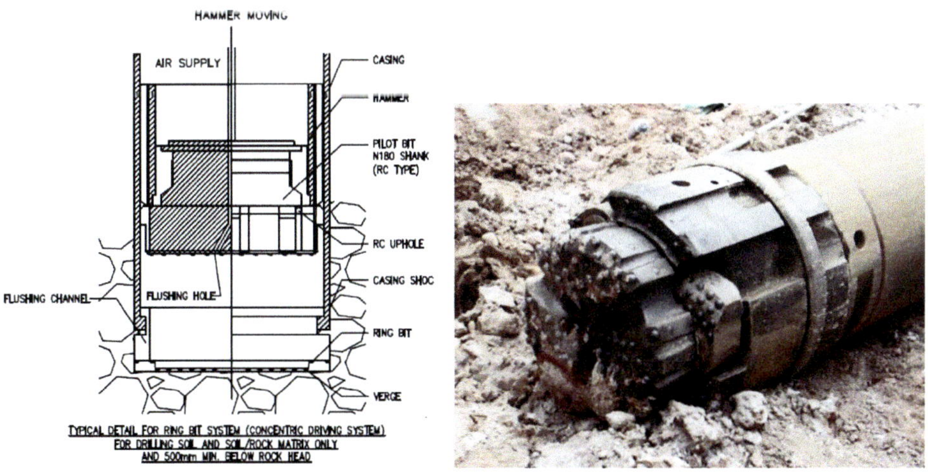

(a) Concentric driving system

(b) Odex drill

Figure 6.4 Driving system for prebored socket H-pile.

hard rock. The H-section steel pile is then installed into the hole and cement grout is injected into the hole which acts as a medium in bonding the pile to the rock by which the pile derives its geotechnical bearing capacity. It should be noted that while socketed H-pile is common in Hong Kong, it is seldom adopted in other countries. The reason behind the adoption in Hong Kong is to avoid the pile resting on underground boulders as well as providing pile capacity with a higher level of confidence (at the expense of high cost).

Steel piles (H or other shapes) are usually installed by driving, but the socketed H-pile is installed by a different method as described in the previous paragraph. Shear connectors are commonly installed along the socket H-pile shaft to increase the bonding between the cement grout and steel, and many field tests in Hong Kong have proved that the steel/cement bond stress is greatly increased by the use of shear connectors. The major difference between the socketed H-pile and another pile type called minipile is that the steel casing is normally extracted along with grouting, while the steel casing remains in the ground for minipiles after installation. Another difference is that socketed H-pile can be designed for bending which is not possible for the minipile. Different sizes of steel casings and steel sections are available for this pile type. One of the most common steel casing sizes is the 610 mm diameter hollow section, which is sufficient to cater for all different sizes of H-sections with adequate cover.

6.2.2.1 *Typical method statement for socketed H-pile construction in Hong Kong*

1. Drilling with the first temporary casing, alignment is checked during the installation.
2. Drilling works by concentric driving system or ODEX (Overburden Drilling Excentric) method, where the Odex drill bit cuts ahead of the steel casing with a diameter slightly larger than the casing. It should be noted that the ODEX system is not allowed to be used now, as there have been several problem cases caused by this system. For the Odex system, the casing can be hammered down easily and continuously. High-pressure air is applied through the flushing holes at the front of the drilling bit to push the soil up. Once the designed depth is reached, the drill bit is drilled in reverse order. For the concentric driving system, a ring bit system is used to install a 610 mm O/D casing with minimum disturbance to the surrounding. The casing is welded to the shoe to seal around the end of casing and to prevent air from passing outside the casing. The casing and casing shoe are carried down the hole as a whole at the same time. The pilot bit and ring bit are locked together by the bayonet coupling. The pilot bit is projected just a little ahead of the ring bit to avoid over-break. The pilot bit and ring bit rotate with the drill rod. The casing shoe and casing, however, do not rotate to minimize the disturbance to the surrounding. During drilling, only air is used as the flushing medium, and no dewatering is allowed. The maximum operation air pressure is usually limited to 17 bars. Upon reaching the rockhead, the pilot drill bit is replaced by the "down the hole" flat bit for drilling the rock socket.
3. Preparation of steel H-pile – shear studs and steel bars are commonly welded to the pile to enhance the bond stress.
4. Splicing of temporary casing/H-pile.
5. Obtaining rock samples after completion of drilling.
6. Preparation for grouting.

7. Air-lifting to clean the hole before grouting.
8. Cement grouting for the socketed pile. The minimum 28 days strength for the cement grout is typically 35 MPa, and the water cement ratio is usually limited to 0.45 maximum.
9. Extraction of temporary steel casing by vibrating hammer or hydraulic jack in some cases.
10. Grout testing work.

6.2.3 Minipile

A minipile usually comprises a few (four to seven – a high number is not recommended) steel reinforcing bars in a drill hole (not exceeding 400 mm) filled with grout. Its bearing capacity is usually derived from the bond stress with the rock socket. Due to its small size, the pile capacity is comparatively low. However, as the machinery required for construction is small and the speed of construction is fast, the pile is suitable for the construction of buildings in sites where access is difficult. In addition, due to the limited load-carrying capacity (1374 kN for 4T50 pile) of the pile and the requirement for minimal spacing (the greater of 750 mm and twice the outer diameter), the pile type can normally carry loads from buildings of up to 18 storeys.

For a minipile, four or more 50 mm diameter steel bars are installed in a prebored pile (Figure 6.5) through soil to rock, and the hole is then grouted by cement grout to achieve bonding with the rock for derivation of the bearing capacity.

6.2.3.1 Typical method statement for minipile construction in Hong Kong

1. Set up drilling rig for minipile, and working platform may be required for construction on slopes.
2. Drilling with alignment checking of minipile.
3. Splicing of steel casing.
4. Collecting rock sample at founding level.
5. Air lifting to clean the pile after completion of drilling.
6. Prefabrication of steel bars on ground.
7. Installation of steel bars.
8. Cement grouting.

6.2.4 Bored pile

A cast in situ concrete pile is often formed by preboring a hole in the ground down to the design level, and the hole is then concreted after the reinforcement cage installation. The bored pile size can range from 400 mm to 3500 mm diameter. By definition according to the British and Hong Kong standards, pile diameters above 750 mm are termed "large-diameter bored pile", otherwise it is termed "small-diameter bored pile". A bored pile can derive its bearing from friction along its pile shaft and/or end-bearing.

A bored pile is usually constructed by driving a steel casing into the ground and the soil within the casing is removed with the use of a grab or other means. The reverse circulation drilling (RCD) technique is now commonly used for many large-diameter piles; a typical RCD used for bored pile construction is shown in Figure 6.6c. RCD uses the

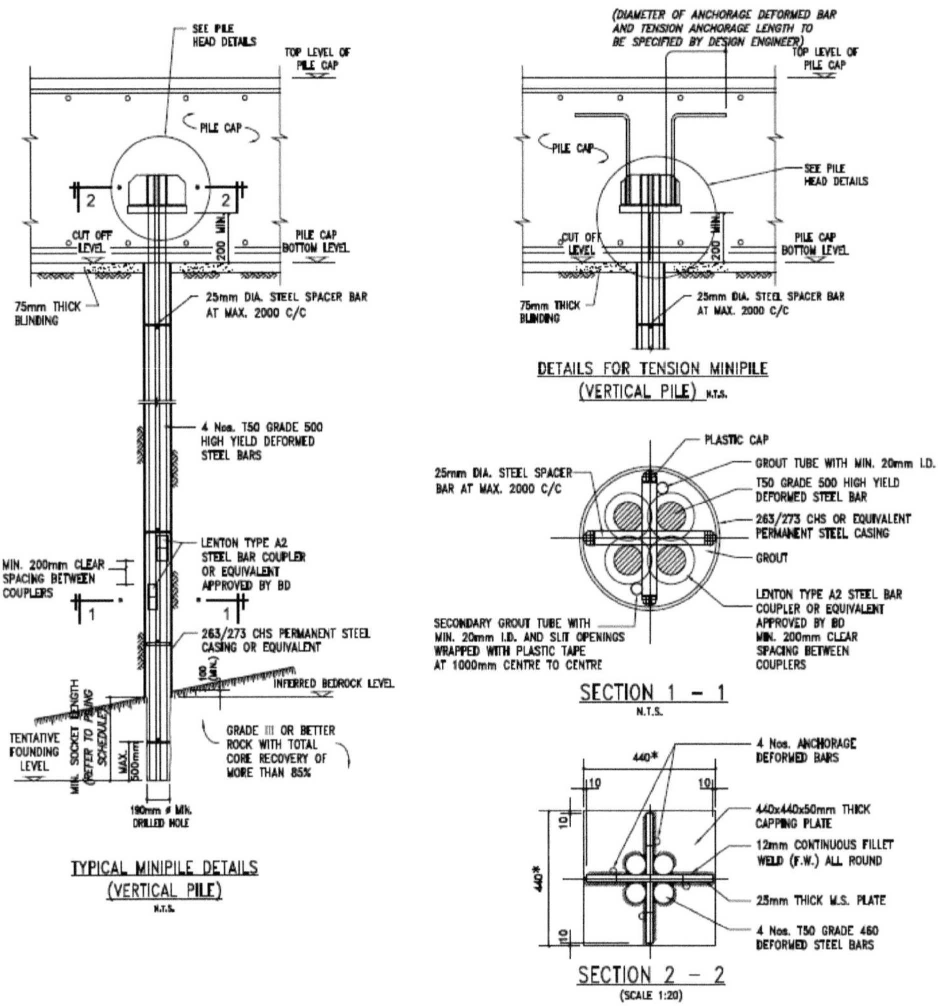

Figure 6.5 Typical details for a minipile in Hong Kong.

self-weight and rotary motion of the drill to crush hard materials. Crushed materials are then retrieved by circulation of fluids within the drillhole. This method effectively reduces the noise and vibration during the construction of rock sockets. It is also possible to bore the hole using drilling fluid without a casing, but there are also some reported unsuccessful cases. Concreting is then performed after installation of the reinforcement cage with the casing gradually withdrawn from the ground. Bell-out, as shown in Figure 6.7, is commonly used, which is based on the allowable stress on concrete being greater than that for rock with $A_s f_{cu} \geq A_b f_{rock}$.

The Koden test is an acoustic device originally used in checking overbreak of a diaphragm wall trench (details will be provided in a later section of this chapter). The application is now extended to checking verticality and bell-out of bored piles. An ultrasonic Koden test is usually used to check the integrity of the bell-out for the bored pile. Since

Figure 6.5 (Continued)

the diameter of the bored pile is usually large in Hong Kong, predrilling to confirm the founding level (instead of just relying on the site investigation) is practically mandatory nowadays.

Some of the common problems encountered during bored pile construction as reported by piling contractors include:

1. Inability to drive temporary casings to inferred rockhead level.
2. Cave-in of unprotected loose ground.
3. Tearing of temporary casings during chiselling.
4. Dropping the drill and bell-out bit.
5. Pile off centre and out of plumb.
6. Crooked pile shaft.
7. Necking in pile shaft.
8. Insufficient bell-out size.

During air-lifting and concreting, some common reported problems include:

1. Seepage of underground water, especially through rock joints.
2. Collapsed reinforcement cage.
3. Bursting of permanent casings.
4. Uplift of reinforcement cage and permanent casings.
5. Leakage and blockage of tremie pipes.

(a) use of grab

(b) use of auger

(c) Use of RCD

Figure 6.6 Different types of plants used for bored pile construction.

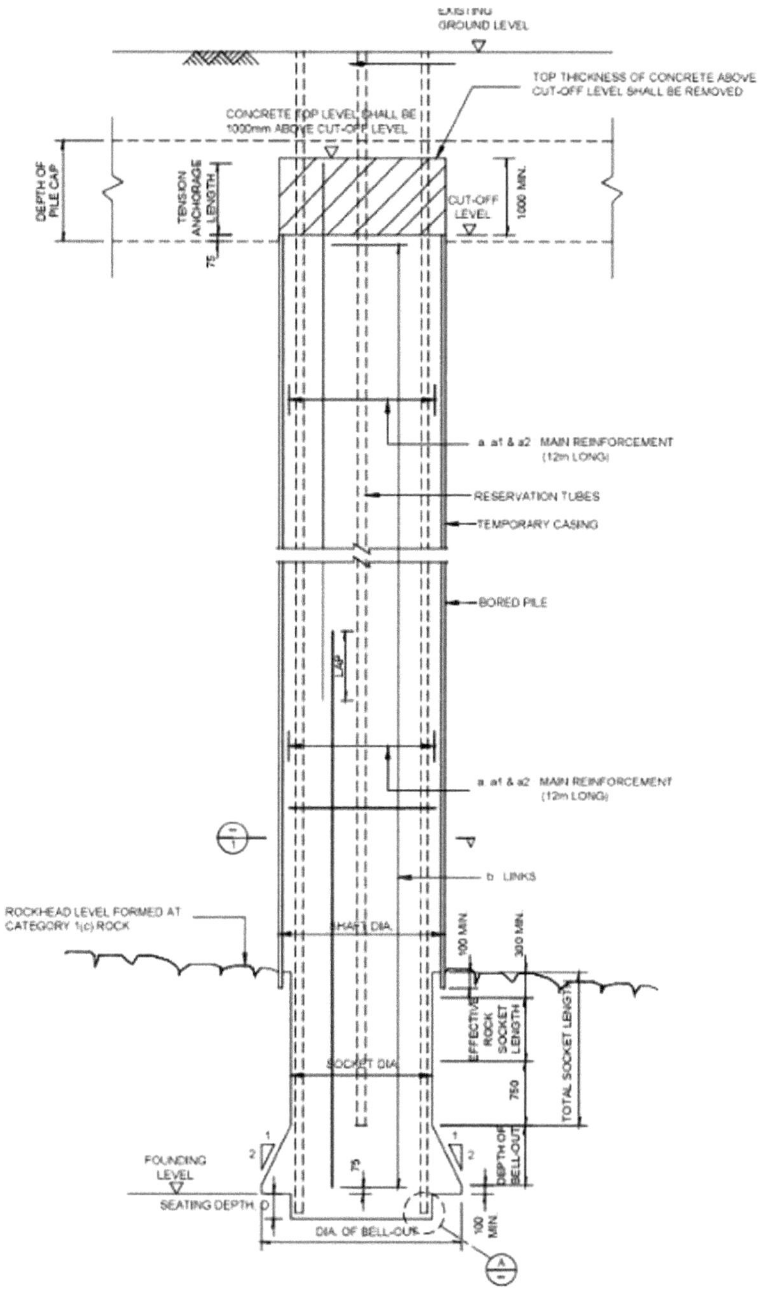

(a) Typical bored pile details with bell-out

Figure 6.7 Typical details for a large-diameter bored pile with bell-out.

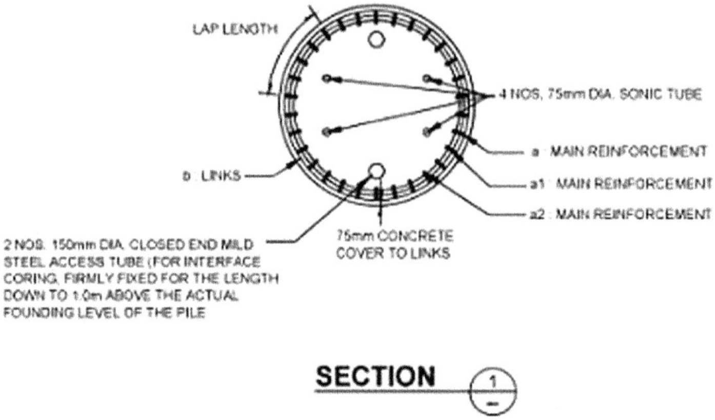

(b) Section 1-1 showing different tubes left-in place

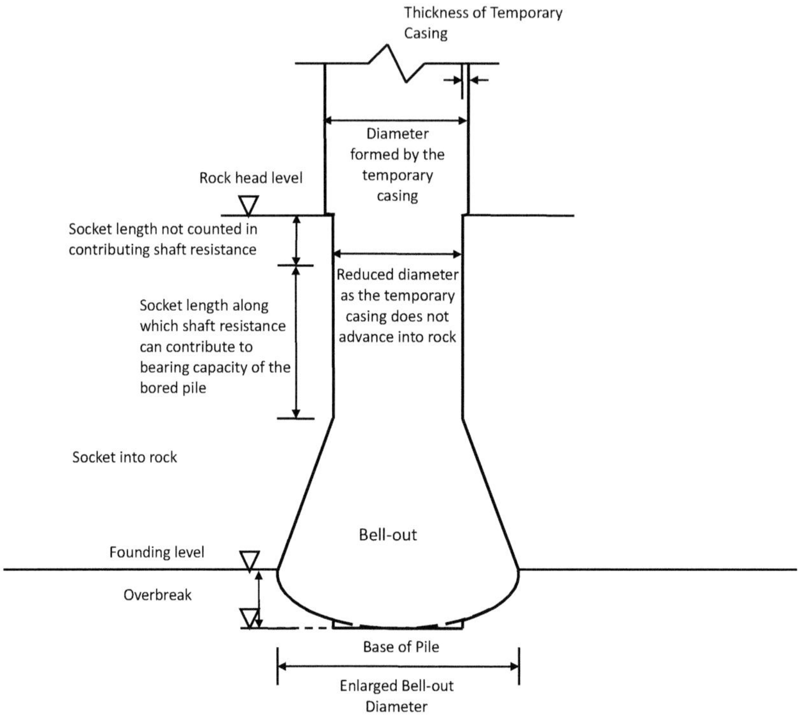

(c) bell-out with socket

Figure 6.7 (Continued)

6. Leakage and blockage of sonic tubes.
7. Inability to retrieve temporary casings during concreting.

Some other considerations during bored pile construction include:

1. It should be noted that the finished concrete top level should be at least 750 mm above the top cut-off level. The top 750 mm concrete for the bored pile is usually weak due to inadequate compaction by self-weight of the tremie concrete and is therefore removed to expose the pile with good-quality concrete to the pile cap.
2. If a permanent liner is required (usually for very poor soil which will collapse during casing extraction), the gap between the temporary and permanent liners should be backfilled with soil.
3. During construction, the alignment and verticality of each bored pile should be checked frequently. If the deviation exceeds the allowable limit, remedial actions must be carried out as soon as possible to rectify any problems.
4. In the case of obstruction, the use of a backhoe or hand-held machine may be used if the obstruction is located at a shallow depth. If the obstruction is found at depth, chiselling by rock chisels or under-reaming is commonly used.
5. Sonic access tubes are installed for most large-diameter bored piles in Hong Kong to assess the quality of the concrete. The sonic tubes which are capped with screw on steel watertight shoes are usually fixed to the reinforcement cage, and project 300 mm above the cut-off level of the pile.
6. In the case that there is a thin layer of sediment (segregation) at the pile–rock interface, remedial pressure grouting is usually carried out. The grout should be non-shrink cement grout with a minimum grouting pressure of 25 bars which is maintained for a minimum of 5 minutes. The sediment or defective concrete is cleaned by high-pressure water with a minimum jetting pressure of 200 bars prior to pressure grouting.

For pile foundations in a marble area with sinkholes, a modern technique is to use high water pressure to crack the vertical faces of the bored pile or barrette followed by injection of grout through the formed cracks. The grouting thus formed will enhance significantly the shaft friction of the pile. This type of pile is called "shaft grouted barrette" or "shaft grouted bored pile" which is discussed later, and this approach is popular in some marble areas in Hong Kong.

The typical procedures for shaft grouting are:

1. Crack the fresh concrete cover using a double packer and water within 24 hours of concrete casting.
2. Carry out shaft grouting for each manchette from bottom to top. The refusal grouting pressure is around 50 bars in Hong Kong.

If the intake cannot be achieved on some manchettes, the target intake immediately above, below or on its side should be increased.

It should be noted that tremie concreting is required for the construction of a bored pile or barrette. A reduction factor of 0.8 on the concrete cube strength is common for the design of these piles in Hong Kong.

(a) setting out of bored pile position

(b) mobilization of hydraulic oscillator

(c) setting up the temporary steel casing

(d) Removal of soil by grab

(e) connection of temporary steel casing

(f) use of chisel to break boulder

Figure 6.8 Typical procedures for the construction of a large-diameter bored pile.

For illustration, the complete procedures for a large-diameter bored pile construction in Hong Kong are given in Figure 6.8. The connection and lowering of heavy reinforcement cages are important, and the typical connection details for the reinforcement cages are also shown in Figure 6.9. In general, the temporary reinforcement, which is very heavy, will

(g) cutter to cut in rock

(h) A reamer to form the bell-out

(i) Reverse circulation to remove spoil

(j) cleaning of pile base

(k) Lowering of steel cage

(l) tremie concreting of bored pile

Figure 6.8 (Continued)

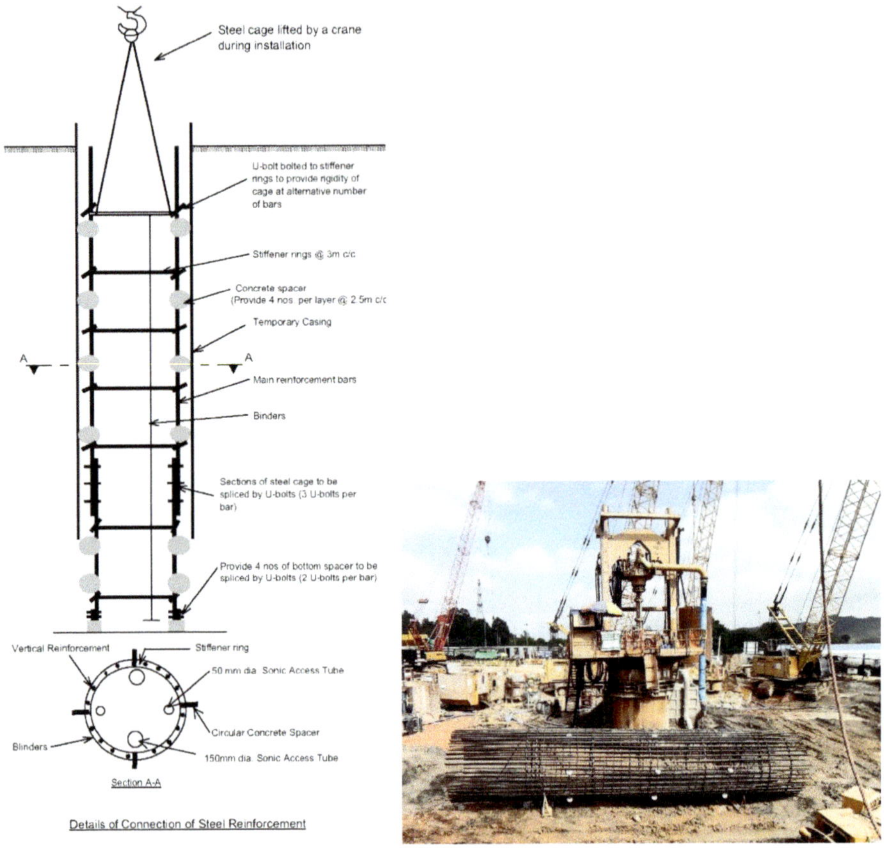

Figure 6.9 Connection between reinforcement cages by U-bolts.

be temporarily suspended on the casing before the insertion and connection of the next reinforcement cage by bolts. A similar method is used for lowering of the reinforcement cage in diaphragm wall construction.

A typical bored pile construction sequence using casing for a project in Hong Kong is as follows:

1. Set out the pile location in accordance with construction drawings and carry out the predrilling work to determine the rock head level and founding level of each bored pile.
2. Set out the pile location in accordance with construction drawings and reference points for commencement of bored pile construction. Allow time for RE checking prior to pitching of temporary casing.
3. All monitoring points shown on the piling layout plan are installed prior to the commencement of piling work. All monitoring points together with the existing one are maintained through-out the contract. The frequency of monitoring is in accordance with the approved plan.

4. Pitch temporary steel casing at the pile location and drive the casing down to the rockhead level of bedrock or hard stratum with a heavy-duty hydraulic oscillator/rotator. The steel casing forms a watertight seal to prevent collapse of soil into the unlined shaft at all times.
5. Excavate within the steel casing with a hammer grab working with a crane. Casings are coupled together by means of welding or bolting. The bottom casing has a thickened cutter shoe to facilitate penetration into hard ground.
6. The excavated materials are disposed to the approved government dumping area. The detailed disposal system and procedures are shown in the environmental management plan.
7. A surveyor will check the offset of the temporary casing after the first casing pitching and before airlifting. Moreover, the verticality of the casing is frequently checked using a spirit level.
8. In the case of excessive out-of-tolerance of the steel casing, the bored hole is backfilled and then the casing is re-installed in the correct location.
9. A positive water head is maintained within the temporary steel casing throughout the course of excavation to ensure stability of the pile and prevent excess pore water pressure. In addition, prior to the installation of RCD, the excavation level should always be kept at least 1 m above the casing toe level.
10. All underground obstructions or boulders encountered during the course of pile excavation should be removed by grabbing or RCD.
11. When the casing is required to penetrate through rock prior to reaching the bedrock level, a reverse circulation drilling method is adopted. The drilling rig is securely attached to the top of the casing, either by hydraulic clamps or special joints. A large-diameter drill bit, attached to a heavy-duty drill string, is rotated by means of a power swivel located on the top of the rig. The cuttings are airlifted into a large mud tank for settlement and subsequent removal from site. Water is returned to the pile shaft to balance any head of water in the ground. As drilling continues, additional drill strings are connected until the required level is reached.
12. A rock chip sample is collected for inspection and verification of the rock head and founding level. Rock socket is achieved by means of a reverse circulation drill. The rock head, founding and pile base level should be checked by the resident engineer.
13. The extent of additional length for formation of bell-out and the details of the bit are based on the on-site calibration record. The additional length is included during drilling of the rock socket.
14. The rock chip is collected at the inlet of the desilting tank from the discharged effluent of RCD.
15. The bell-out base is then formed using a calibrated expandable drill bit mounted onto a reverse circulation drill rig to form the desired bell-out base.
16. The size of the expandable drill bit must be checked and calibrated on the ground before construction.
17. The bell-out base is formed by applying the drilling motion coupled with the downward thrust to the drill rod from which the bell-out drill bit is expanded. This operation is continued until the downward displacement of the drill bit rod coincides with the value of the specific diameter of the bell-out base formed.

18. The bored piles are constructed in accordance with the tolerances stated in the specification.

The reinforcement cage as shown in Figure 6.9 is usually fabricated on an arc-shaped fabrication frame. Initially, several stiffener rings are erected on the arc-shaped frame. Rebars are fixed at the bottom of the stiffener rings with the use of U-bolts. Fixed tie-bracing to the stiffener rings is done using U-bolts at both ends of the cage, then sufficient bracings are added to the stiffener rings. Top rebars are fixed afterwards, followed by fixing of the side bars. Spacers are added to the reinforcement cage to ensure the provision of cover to the reinforcement.

The following is the method statement for construction of a large-diameter bored pile using drilling fluid:

1. Predrilling is conducted at a location in close proximity to the piles.
2. The survey and set out centre of bored piles should be submitted to the resident engineer before drilling.
3. To construct a bored pile of diameter D, a casing of diameter slightly less than D has to used.
4. Cores (minimum NX size) should be taken to at least 5 m below the proposed pile base level to prove the nature of the founding material.
5. The acceptable values of index parameters, such as total core recovery, unconfined compressive strength (or point load strength), RQD, joint spacing and nature of discontinuities and any infilling below the founding level must be determined in relation to the design method.
6. The chippings should be inspected carefully to confirm the nature of the materials when the proposed founding level is reached. For core sampling, a minimum of one to two cores should be taken for every 100 piles. Nevertheless, judgement should be exercised should more confirmatory cores be required.
7. In the case that the slurry-support technique is used, the technique involves excavation of a shaft under a drilling fluid using a reverse-circulation drill with rock roller bits for the sake of less weathered zones.
8. A reverse circulation drilling incorporating an under-reaming head can be used to form the bell-outs.
9. A suspension system is needed to control the reverse-circulation drills due to the need for a larger volume of drilling fluid needing to be treated prior to recharge into the bore.
10. Polymer fluid should be used.
11. Polymer fluid does not need special disposal measures.
12. Acceptable criteria for polymer fluid testing suggested by Beresford et al. (1987) should be fulfilled.
13. Use reverse-circulation drills to remove the soils around the obstruction by grabbing or airlifting and place lean mix concrete to encase the obstruction to facilitate subsequent drilling.
14. Use high-pressure water jetting to remove the loose sediments at the base. Then use pressurized grout to fill any voids.

15. A potential problem of trapping debris at the pile base can be reduced by lifting the tremie pie with hydraulically operated equipment.

In most cases, the bored pile is filled with water and underwater tremie concreting is required. The procedure is as follows:

1. Initially, the tremie pipe bottom level will be adjusted about 300 mm above the pile base level and securely handed on the casing top. A hopper will then be placed on top of the tremie pipe with support by the working platform.
2. In the initial concrete pouring stage, two hoppers are used. The bottom hopper is contained with cement grout and a hopper of concrete is placed on top. A plug of polystyrene chip is discharged inside the pipe ahead of discharge of cement grout to prevent the grout from mixing with the water in the pipe. The reinforcement cage is connected with a rope which acts as an indicator for uplifting of the cage. During the concreting and extraction of the casing, the foremen/supervisor closely monitors any movement of the cage.
3. Immediately following the cement grout, concrete is continually discharged into the tremie pipe. During the concreting operation, the bottom of the tremie pipe is kept immersed in the concrete to a depth of at least 3 m to maintain a seal. This is checked by dipping a measuring tape to the concreted surface and using the casing top as the reference. The pipe is lifted in stages as pouring proceeds.
4. The steel casing is extracted by an oscillator/rotator during concreting of the pile shaft. The toe of the casing is always maintained at least 5 m below the top of the concrete to prevent ingress of surrounding material.
5. Concrete is placed to a minimum of 1.0 m above the required cut-off level for the pile.
6. During concreting, logging is kept of deliveries, volumes and concrete levels. Slump test, temperature control and concrete cube making are also conducted.
7. After completion of the concrete pouring, a temporary casing is left in the bore hole until the initial setting of the cement grout/concrete.
8. Immediately after extraction of the temporary casing, the empty bore above the concreted level of the pile is backfilled with granular material using a backhoe.

6.2.5 PIP pile

PIP piles have been popular in Hong Kong over the past decade as low- to medium-capacity pile foundations. For example, the Jordan, Tsim Sha Tsui MTR stations and the Charter Bank basement have been constructed with the use of PIP piles. Apart from the use of a normal reinforcement cage, the use of a standard H section as the reinforcement of the PIP pile has been popular in the past (Jordan and Tsim Sha Tsui MTR station). Auger is commonly used for the construction of small-diameter bored piles or PIP piles. The size of the PIP piles is commonly in the order of 450–550 mm, as compared with the small-diameter bored piles which have a diameter of 800–1000 mm in Hong Kong.

6.2.5.1 Typical process for PIP construction (similar to many other small-diameter bored piles)

1. Setting out for pile.
2. Drilling with auger.
3. Extension of auger drilling rod.
4. Drilling completed to the required depth.
5. Grouting of pile, preparation for prefabricated steel cage.
6. Flow cone test for grout.
7. Initial setting time test for grout.
8. Grout cube compression test.
9. Completion of grouting work.
10. Installation of steel cage into the pile shaft.
11. Surveying for the top level of the steel bar.
12. Stabilizing the steel cage (in grout) using U-bolts.
13. Inspection of grout at the top level of the pile before backfilling.

Grouted piles in the form of PIP piles, chemical churning piles (CCPs), mixed-in-place piles (CFG) have been used for some low-capacity works, and actually such piles have been popular for footbridges and low-rise buildings in the past, and are popular in many countries including China, but not in Hong Kong.

6.2.6 Barrette

Barrette is actually a pile constructed by the diaphragm wall technique (see Chapter 5). Besides the normal rectangular section, T sections, L sections are X sections are also commonly adopted. For some MTR stations in Hong Kong, the barrette serves as the foundation as well as columns in some MTR concourses. Hence the connections between the barrette and slabs have to be carefully designed. There is an interesting case in Hong Kong. For the construction of the Central Plaza in Hong Kong, diaphragm wall and bored pile are specified in the drawings. To avoid mobilization of different plants, the contractor proposed to replace the bore pile with barrette with a major saving in the construction cost.

6.3 ANALYSIS AND DESIGN OF SINGLE PILE – VERTICAL LOAD

The design of a pile for the vertical load should consider both the structural capacity and the geotechnical requirement. In general, the structural capacity and hence the pile type is chosen first, which is followed by the geotechnical design for the length of pile, requirement of bell-out or socket length. The design of the pile is largely controlled by the local building code and practice, and there is no universal method of design. The examples below illustrate the design practice used in Hong Kong, which is based on a consideration of both the pile capacity and pile settlement; this design practice can be used for reference in other countries.

6.3.1 Steel pile by driving or jacking/bored and socketed

The maximum loading capacities (axial load) of steel pile are given by:

(installed by driving) $P = 0.3f_yA_p$

(installed by preboring or jacking) $P = 0.5f_yA_p$

Where f_y is the yield strength of the steel and A_p is the cross-sectional area of the pile. A lower factor of 0.3 is used for pile installed by driving to reduce the bearing capacity of the driven pile so as to minimize the use of heavy hammers and high drop heights, which can subsequently lead to high stresses during pile driving (dynamic load is different from static load, and the axial stress can be magnified during driving). Pile capacity is increased by 25% if the wind load is considered. For combined bending and axial load, the working stress can be increased by 50%. The minimum yield stresses of steel should be determined from the "design strength" of steelwork in accordance with Table 3.2 of the Code of Practice for the Structural Use of Steel 2011 (CoPSteel2011) or BSEN10025. The maximum allowable average bond stress between grout and steel H-pile should be 0.5 MPa for grout having a minimum cube strength of 30 MPa at 28 days and 0.45 MPa for grout having a minimum cube strength of 25 MPa at 28 days.

Thus, for the H-section S450 305×305×223 where the cross-sectional area is 284 cm^2, the yield stress is 430 MPa (with the flange and web thicknesses, respectively, at 30.4mm and 30.3mm), the axial load-carrying capacity for a DRIVEN H pile without wind is $0.3 \times 430 \times 28400 \times 10^{-3} = 3676$ kN (add 25% with wind). The design load-carrying capacity of a SOCKETTED H-pile S450 305×305×223 H-section of area 284 cm^2 without wind is $0.5 \times 430 \times 28400 \times 10^{-3} = 6106$ kN. It should be noted that due to the static load test of pile which is loaded to twice the design load, to avoid buckling of pile during the static load test which has happened but is rare, some engineers limit the ratio to 0.49 instead of 0.5 in the pile capacity determination.

To fully utilize the bearing capacity effected by the cross section, the minimum socket length into Category 1(c) rock or better in accordance with Table 2.1 of the Hong Kong Code of Practice for Foundation is determined as follows (suppose the diameter of the bored hole in the rock is 550 mm and the presumed bond or friction between the rock and grout according to Table 2.2 of the Code of Practice for Foundation is 700 kPa)

$$0.55\pi L \times 700 = 6106 \Rightarrow L = 5.048 \text{ m}$$

Design example for bearing capacity of rock-socketed H-pile

<u>Design condition:</u>

Diameter of pile = 0.55 m
Socket length ls = 4.5 m (assumed)
Grade of rock = II
Type of steel H-pile = 305 × 305 × 223
Sectional area of H-pile A = 28500 mm^2
Perimeter of pile L = 1727.88 mm

Table 6.1 Typical small-diameter bored pile capacity estimation

Depths below ground	*Soil description*	*Layer of depth (m)*	*SPT N-value from GI report*	*Design SPT N-value*	*Shaft resistance (kN/m²)*	*Shaft friction (kN)*
0–6 m	Fill or marine deposit	–	–	–	Neglected	–
6–21 m	Completely decomposed granite	1.50	18	18	86.4	68.9
		1.50	25	25	120	95.7
		1.50	36	36	172.8	137.8
		1.50	48	40	192	153.2
		1.50	57	40	192	153.2
		1.50	62	40	192	153.2
		1.50	78	40	192	153.2
		1.50	91	40	192	153.2
		1.50	104	40	192	153.2
		1.50	118	40	192	153.2
		Allowable friction resistance				1,374
		Allowable end-bearing resistance				40.5
		Total loading capacity of pile				1414

Material strength

Grade of grout = 30 MPa
Reduction factor for grout Ra = 0.85
Yield stress of steel H-pile = 430 MPa
Allowable stress/yield stress = 0.5
Allowable bond stress between rock and grout = 700 kPa (under compression)
Allowable bond stress between rock and grout = 350 kPa (under tension)
Allowable bond stress between steel and grout = 600 kPa (under compression)
Allowable bond stress between steel and grout = 600 kPa (under tension)

Calculation of bearing capacity

a) Bearing capacity of steel H-pile = $0.5 \times f_y A_p$ = 6106 kN
Assume provide 16 no. of H-pile, total load capacity of H-pile = 6106 × 16 = 97696 kN
Assume NSF up to 20% of the pile capacity = 18924 kN
Design load = Column load + NSF = 60000 +18924
= 78924 kN < 97696 kN OK!

6.3.2 Small-diameter bored pile

A small-diameter bored pile is a bored pile having a diameter not exceeding 750 mm. It is formed by boring a casing into the ground and subsequently filling the hole with concrete or grout. As the pile derives its bearing largely from skin friction, pressure grouting and pressurizing the concrete are favoured. Small-diameter bored pile is less popular in Hong

Kong due to its low capacity, which would result in a large number of piles for the Hong Kong buildings which are generally tall or very tall. Small-diameter bored pile is however still very popular in other countries.

The following design principles should be adopted in general:

1. The geotechnical capacity of a small-diameter bored pile is usually derived from shaft friction but the combined use of end-bearing and shaft friction in soil or rock can also be used;
2. The contribution of friction from any fill or marine deposit layer, if it exists, should usually not be included in the design unless it can be proved that there is no possibility of future consolidation;
3. Where pressure grouting or pressurizing the concrete is involved in construction, the empirical relationship of shaft friction of 4.8 N (kPa) is generally used for N values up to maximum of 40, usually with a factor of safety of 3, as described in Clause 6.4.5.3 of GEO Publication No. 1/2006. Otherwise, smaller values should be used with reference to other publications or tests;
4. The structural capacity of the small-diameter bored pile should also be considered, and in the case that it is less than the geotechnical capacity, the structural capacity will be the controlling factor;
5. The small-diameter bored piles can be used to resist lateral loads even when they are not raking piles.

A sample design calculation of a small-diameter bored pile with pressurizing concrete is presented as follows. The design data of the pile are:

Pile diameter = 508 mm
Shaft resistance taken as $\tau = 4.8$ N
Concrete strength = 20 Mpa
Factor of safety = 3 (as skin friction is not tested)
No. of piles under the same pile cap: less than four
Concrete mix of the pile is grade 20 where the permissible stress is 5 MPa
Capacity of the pile in accordance with its structural provision is

$$P = \frac{1}{4} \times 508^2 \pi \times 5 \times 10^{-3} = 1013kN$$

The geotechnical capacity of the pile is determined in accordance with the empirical relations that shaft friction is $\tau = 4.8$ N and end-bearing is $\sigma_b = 5$ N (pile terminated at soil). The SPTN values of the soil embedding the pile are given in Table 6.1, based on which the total shaft friction and end-bearing are calculated in Table 6.1, giving a sum of 1414 kN, which is greater than that by the structural provision of 1013 kN. Therefore, the load-carrying capacity of the pile is 1013 kN. For cast-in-place concrete foundations, the concrete strength should be reduced by 20% where groundwater is likely to be encountered during concreting or where concrete is placed underwater.

Since the structural capacity of the small-diameter pile is only 1013 kN, the pile capacity is hence controlled by the structural capacity, which is 1013 kN.

6.3.3 Large-diameter bored pile

The pile capacity of the large-diameter bored pile is usually derived from the end-bearing. As the concrete grades currently used are usually high, up to grade 45, giving a permissible compressive stress in the order of 12 MPa which is well in excess of the rock bearing pressure in the order of 3–10 MPa, bell-outs for enlargement of the end-bearing areas have been extensively used. The Hong Kong Foundation Code allows combined use of shaft resistance and end-bearing of rock in the determination of its geotechnical capacity under the following conditions:

(1) When bell-out is used, the determination of shaft friction is restricted to a height of one pile diameter;
(2) When bell-out is not used, the determination of shaft friction can be based on the height of two pile diameters.

A large-diameter bored pile design for a project in Hong Kong is given below for reference.

Design data:
Shaft diameter = 3000 mm
Bell out diameter = 3000 × 1.65 = 4950 mm
Thickness of the casing = 50 mm
Reduced diameter of the pile (within the socket above the bell-out) = 2900 mm
Allowable end-bearing pressure = 5000 kPa (for Category 1[c] rock)
Shaft resistance = 700 kPa (under compression or transient tension)
Shaft resistance = 350 kPa (under permanent tension)
Under compression
Effective socket length = 2.9 m (if bell-out capacity is used)
Effective socket length = 2 × 2.9 = 5.8 m (if bell-out bearing capacity is not used)
Shaft resistance of the pile (with bell-out) = $2.9 \times 2.9 \times \pi \times 700$ =18495 kN
Shaft resistance of the pile (without bell-out) = $2.9 \times 5.8 \times \pi \times 700$ =36989 kN
End-bearing capacity of the pile without bell-out = $2.9^2 \times 0.25 \times \pi \times 5000$
= 33026 kN
Capacity of the large-diameter bored pile without bell-out = end-bearing without bell-out + shaft resistance of the pile = 36989 + 33026 = 70015 kN
End-bearing capacity of the pile with bell-out = $4.95^2 \times 0.25 \times \pi \times 5000$
= 96211 kN
Under Condition (1), the geotechnical capacity of the pile is 96221 + 18495 = 114716 kN

Under Condition (2), the geotechnical capacity of the pile is 33026 + 36989 = 70015 kN. With consideration of wind effects, both capacities can be increased by 25% to become, respectively, 143395 kN and 87519 kN

6.3.3.1 *Structural design of bored pile*

Using concrete with grade 45, steel reinforcement for yield strength of 500 N/mm^2
The structural capacity of the pile using 1% reinforcements.
Reduction factor of concrete strength due to groundwater = 0.8

Area of the pile: = $0.25 \times 2900^2 \times \pi = 6605198.554\ mm^2$

f_{cu} = 45 as the cube compressive strength in N/mm², which has to be reduced by 20% to cater for concreting under water. So, effectively, $f_{cu} = 45 \times 0.8 = 36$ MPa.

The ultimate structural capacity:

= Reduction factor due to groundwater water × area of concrete × 0.45 × f_{cu} + area of steel × 0.87 × yield strength

= 6605198.554 × 0.45 × 36 + 0.87 × 6605198.554 × 0.01 × 500

= 107004 + 28733 = 135737 kN

Taking an average partial load factor of 1.4, the allowable working load is 135737/1.4 = 96955 kN.

The structural capacity with 1% steel reinforcement is less than the geotechnical capacity with bell-out.

The structural capacity is higher than the geotechnical capacity, and the allowable compression load of the pile is 79521 kN without considering wind effect and 99401 kN with considering wind effect.

Lateral resistance check (to be discussed later)

Design data:

$L = 39.32$ m $z = 0$ m

$E_p = 23.97$ kN/mm² $I_p = 2.243176$ m⁴

$T = 5.498$ m

Design deflection:

ULS: 25 mm

Check ULS:

$L/T = 7.151692$

$z/T = 0$

$F_{\delta H} = 0.95$

$H_u = F_{\delta H} \times HT^3/E_pI_p = 8513.998$ kN

The maximum lateral force: 2833.202 kN < 8513 OK

$F_{\delta M} = 1.5$

$M_u = F_{\delta M} {}^* M\ T^2/E_pI_p = 29646.31$ kNm

The maximum banding moment: 12466.09 kN < 29463 OK

6.3.4 Minipile

Minipiles are now commonly used when the loading is not high and there is difficulty in mobilizing a heavy boring machine or working with limited headroom. The most common minipile nowadays is the 219 mm diameter hollow section with four T50 bars, which will give a working capacity of 1400 kN. Larger capacity can be achieved by the use of a larger minipile with high-yield thread bars (f_y = 520 MPa). A minipile usually comprises a few (four to five, and more is not recommended) steel reinforcing bars in a drill hole (not

Table 6.2 Typical β values for skin friction estimation

Types of piles	*Type of soils*	β
Driven small displacement piles	Saprolites	0.1–0.4
	Loose to medium-density sand	0.1–0.5
Driven large displacement piles	Saprolites	0.8–1.2
	Loose to medium-density sand	0.2–1.5
Bored piles and barrettes	Saprolites	0.1–0.6
	Loose to medium-density sand	0.2–0.6
Shaft grouted bored piles or barrettes	Saprolites	0.2–1.2

exceeding 400 mm) filled with grout. Its bearing capacity is derived from the bond with rock. Due to its small size, the pile capacity is comparatively low, and it is not assumed to resist bending in the design. However, as the machinery required for construction is small and speed of construction is fast, this pile is suitable in the construction of buildings in sites where access is difficult. In addition, due to the limited load-carrying capacity (1374 kN for 4T50 pile) of the pile and the requirement of minimum spacing (the greater of 750 mm and twice the outer diameter), the pile type can normally carry buildings up to 18 storeys. The design principles of a minipile are briefly summarized as follows:

1. The loading-carrying capacity of a minipile is derived solely from the steel bars;
2. The presumed allowable bond or friction between rock and concrete is given in Table 2.2 of the Code;
3. Values for design ultimate anchorage bond stress are given in Table 8.3 of Clause 8.4.4 of Code of Practice for Structural Use for Concrete 2004;
4. The allowable buckling capacity of the minipile may be checked with considerations of lateral restraints from the grout, permanent steel casing and the surrounding soil can be included;
5. Minipiles do not take bending or lateral force in the design. Lateral load will be taken up by raking minipile. No bending moment is allowed in any cases.

Example:

Steel casing external diameter = 219 mm; steel casing thickness = 5 mm
Rock hole diameter = 190 mm;
4T50 Grade 460 high-yield deformed steel bar;
Design grout strength = 30 MPa;

Rock encountered: Category 1(c) rock, 700 kPa (presumed allowable friction between rock and concrete) is taken from Table 2.2 of the Code of practice for foundation 2004/2017;

Design of load-carrying capacity of the pile (excluding casing and grout)

(i) Due to structural strength of the reinforcing bars
$P = 4 \times 25^2 \times \pi \times 230 = 1806$ kN
(ii) To withstand the maximum load capacity of 1806 kN, the minimum socket grouted length should be $1806/(0.19\,\pi \times 700) = 4.32$ m

(iii) The bond strength of steel bars with grout also needs to be checked. As per Clause 8.4.4 and Table 8.3 of the Code of Practice for Structural Use of Concrete 2004 and 2013, the allowable bond strength is 2.739 N/mm², in excess of the allowable friction between rock and concrete. Therefore, the bond between grout and the reinforcing bars is not a controlling design criterion.

The main advantage of minipile is that only relatively small mechanical plants are required for the construction, hence they are suitable for sites in hilly terrains or with accessibility difficulties. Minipiles are hence popular for many low-rise developments in hilly terrains in Hong Kong.

A typical minipile construction method statement by a contractor is given below:

1. Drill hole with 313.9 mm internal diameter with 5 mm thick casing by Robit drilling method. The casing shall terminate at the bedrock level and not retracted.
2. Further excavate by using the bottom bit with 323.9 mm diameter to penetrate the specified rock socket length into bedrock (Grade III or better with core recovery > 85%).
3. After completion of drilling, the hole shall be flushed clean by air or water until clear water comes out from the hole.
4. Install four number T50 bars as shown in the figure into the hole together with two 20 mm diameter PVC grout tubes to the whole length of the reinforcement. Bar spacers/centralizers at 4 m spacing shall be provided.
5. Inject cement grout with a minimum strength of 40 MPa at 28 days with a maximum W/C 0.45. The grouting pressure shall not exceed overburden stress + 20 kPa.
6. Cut the reinforcement and fix the head plate at a minimum of 24 hours after the pile has been grouted.
7. During pile installation, the adjacent ground and structures should be monitored. The contractor shall prevent water drawdown during air flushing.

For illustration, a complete design of minipile for a project in Hong Kong is given below.

6.3.4.1 Typical design of minipiles

Main bar size: T50mm (high yield)
Bond stress (rock and cement grout): 700 kPa
Bond stress (rebar and cement grout): 800 kPa
Bond stress reduction factor due to tremie concreting: 0.8
Drill hole diameter: 219 mm
Socket length: 4000 mm
Length of each side about rebar and cement: 120 mm
Thickness of steel casting: 5 mm
Outer diameter of minipile: 219 mm
Thickness of the central spreader: 20 mm
Elastic modulus of steel: 200000 N/mm²
η_h: 1.3 MN/m²
Minimum pile length: 50 m
Inclined ratio (if raking pile): 1:10

6.3.4.2 Design of pile head for minipiles

Maximum working load: 1374 kN (compression)
Maximum working load: 779 kN (tension)
Steel plate size: 430 × 450
Steel plate thickness: 60 mm
Base plate p_y: 430 N/mm^2
Weld leg between rebar and stiffener plate: 230 mm
Fillet weld length between rebar and stiffener: 8 mm
Weld leg between base plate and stiffener plate: 193 mm
Fillet weld length between base plate and stiffener: 8 mm
Number of dowel bars: Six
Size of dowel bars: T32

6.3.4.3 Design of minipiles

1. **Structural capacity**
   Consider four T50 rebars only
   Allowable working load (compression): $\pi \times (50^2/4) \times 4 \times (237.5/1000)$ = 1865 kN (compression)
   Allowable working load (tension): $\pi \times (50^2/4) \times 4 \times (237.5/1000)$ = 1865 kN (tension)
2. **Bonding between rock and cement grout**
   Assume D: 219 mm diameter drill hole
   Socket length required L
   $(\pi \times 219 \times L) \times 0.7 > 1374 \times 10^3$ i.e. $L > 3567$ mm
3. **Bonding between rebar and cement grout**
   Bond stress (rebar and cement grout) = 0.8 N/mm^2
   Allowable bond strength: $0.8 \times 1 \times 1.25 = 1$ MPa
   Provide 20 mm diameter mild steel bar spreaders at 4000 cc
   Length for each side = 120 mm
   Bond perimeter = $20 \times 4 = 480$ mm
   Socket length required L
   $480 \times L \times 1.0 > 1374 \times 10^3$ $L > 286$ 3mm
   Provide L = 4000 mm socket length (OK)
4. **Buckling check of minipiles**
   According to Figure 6.24 "Foundation Design and Construction", GEO Publication No 1/2006,
   Consider the steel casting: 5 mm thick
   Outer diameter of the minipile: 219 mm
   Inner diameter of the minipile: 209 mm
   Moment inertia of casting: $\pi \times (219^4 - 209^4)/64 = 19{,}253{,}435$ mm^4
   Four 50 mm diameter steel bars
   Moment of inertia of a single bar: $I = \pi \times (50^4)/64 = 306796$ mm^4
   Section area of single bar A = $(50)^2 \times \pi/4 = 1963$ mm^2
   Thickness of the central spreader = 20 mm

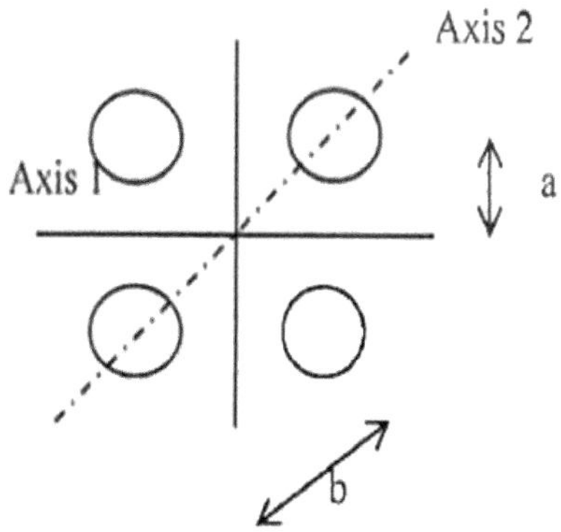

a = 35 mm, b = 35/sin45 = 49.5 mm

Axis 1 moment of inertia = $4I + 4a^2A$ = 1.08E+07 mm^4
Axis 2 moment of inertia = $4I + 2b^2A$ = 1.08E+07 mm^4
Total moment of inertia = 79253435 + 10848312 = 30,701,747 mm^4
E_p = 200000 N/mm^2
η_h = 1.3 MN/m^2
$L_c = 4(EI/\eta_h)^{0.2} = 4 \times ((200 \times 301017\ 47/(1.3/(106)))^{0.2})$ = 5435 mm
Assume free head, then critical buckling load
$P_{cr} = [\pi^2 \times EI]/[4(e+L_c/2)]^2$
$= [\pi^2 \times 200 \times 30701747]/[4(0+5435/2)]^2$ = 2012 kN > 1242 kN (OK)

5. **Design of pile head for minipile**
Maximum working load: 1242 kN (compression), 779 (tension)
Steel plate size = 430 × 450 mm
Bearing stress = (1.2 × 1242 × 1000)/(430 × 450) = 7.7 N/mm^2 < 15 N/mm^2 (G30 concrete)
Using Grade 460 steel plate, p_y = 430 N/mm^2
Steel plate thickness = 60 mm
Bending moment = 0.5(1.2 × 1242) × 0.137 = 102.1 kNm
$M_c = 1.2p_yZ = 1.2 \times 430 \times (450 \times 60^2/6)/10^6$ = 139.3 kNm > 102.1 OK
For fillet weld between rebar and stiffener
Weld length L_w = 4 × 230 = 920 mm
$P/L_w = (1.2 \times 1242 \times 10^3)/920$ = 1620 N/mm
Provide 12 mm fillet weld, weld strength = 230 × 0.7 × 12 = 1932 N/mm > 1620 N/mm OK
For fillet weld between base plate and stiffener
Weld length L_w = 12 × 193 = 2316 mm
$P/L_w = (1.2 * 1242 \times 10^3)/2316$ = 643 N/mm
Provide 8 mm fillet weld, weld strength = 230 × 0.7 × 8 = 1288 N/mm > 643 N/mm OK
Design of dowel bar
Provide six T32 bars
Tensile capacity = $175 \times (6 \times \pi \times 32^2/4)/10^3$ = 844 kN > 778.8 kN (maximum tension) OK

6.3.5 Shaft grouted piles

Extremely deep bedrock can pose a construction challenge to the conventional large-diameter bored pile on rock and the prebored H-piles socketed into rock. If friction piles are used without relying on the pile end-bearing capacity, most of the friction piles have limited pile capacities which will limit the development of superstructures. The use of shaft-grouting to enhance the shaft frictional capacity was introduced to Hong Kong in the 1990s for friction bored piles and barrettes in weak saprolites. This technique has recently been extended to areas with meta-siltstone/sandstone formation underlying marble bedrock beyond 150 m below ground. The full-scale load test for the KCRC West Rail Phase I reported by Plumbridge et al. (2000) demonstrated that the measured shaft friction capacities for shaft-grouted barrettes and bored piles in completely decomposed granite (CDG) and volcanic (CDV) could increase up to two- or three-fold when compared with those without shaft-grouting. The use of shaft-grouted barrette in MTRC Kowloon Station Mega Tower Development (Chan et al. 2004) and the private residential developments at Tung Chung Designated Area (Sze et al. 2007) also have demonstrated enhancement of shaft friction by the shaft-grouting technique in sandy alluvium, completely decomposed rhyolite (CDR), CDG and diamict deposit (DD). Lui et al. (1993) also reported a significant increase in the ultimate shaft capacity for small-diameter pressure grouting Pakt-in-Place (PIP) pile. This type of pile has been adopted in a series of projects in Yuen Long (barrettes), Hollywood Road (mini piles), Seymour Road (mini piles) and Wong Nai Chung Road (pre-bored H piles) in Hong Kong. In general, the Tube a' Manchette method is used to ensure good shaft grouting, and one of the major advantages of this grouting method is the possibility of multiple grouting.

Based on some load test results in Hong Kong, it is generally found that there is very little loading transferred to the pile base for shafted grouted piles. The ultimate shaft friction can be mobilized in colluvium or a similar type of soil or the top of the CDG zone, but it is not fully mobilized for the lower portion of the CDG zone. In general, the shaft friction capacity can be increased by several times by shaft grouting, but the percentage of increase for stiffer soil is generally lower than that for less stiff soil.

6.4 GEOTECHNICAL DESIGN OF PILE

The analysis and design of a single pile for vertical load involves the determination of its load-carrying capacity and settlement. And very often, load-carrying capacity and settlement are not independent of each other. The load-carrying capacity is often limited by predetermined settlement instead of shear failure of the soil/rock mass or the structural material of the pile. Even if shear failure of the soil/rock mass is adopted, the failure load of the pile so determined is the "ultimate load" and the "allowable load", which is the characteristic or working load the pile is allowed to carry and is the ultimate load divided by a factor of safety in the order of 2–3.

Broadly speaking, the determination of a single pile's capacity for vertical load is by "static" and "dynamic" formulae. A single pile refers to one which theoretically is an isolated one, where the effects of other piles are ignored. If effects from other piles are considered, it will be a pile group effect.

6.4.1 Static formula for a single pile

A pile's bearing capacity can be considered to be made up of two components which are friction along its pile and the resistance at its tip called "end-bearing".

$$P_p = P_s + P_b \tag{6.1}$$

where P_p is the total pile bearing capacity; P_s is the bearing due to pile friction; and P_b is the bearing capacity at end-bearing.

The shaft friction of the pile is generally determined by the alpha (α-) or beta (β-) methods. The α-method is generally applied to cohesive soil which relates the friction of the pile shaft with soil to the undrained shear strength of the soil c_u by a factor ± termed the "adhesion factor" so that the shear resistance is

$$f_s = \alpha c_u \tag{6.2}$$

And the contribution of the bearing by this friction is

$$P_s = \alpha c_u A \tag{6.3}$$

where A is the circumferential area of the pile. However, α is not a constant for a certain type of soil and some studies have shown that it may decrease with the increase of c_u. The choice of parameter should be based on local statistical results.

Nevertheless, for cohesionless soil, the behaviour is different. The shaft friction is related to the effective stress of the soil σ'_v. The relationship is through a β factor. Therefore, this approach is known as the β-approach.

$$f_s = \beta \sigma'_v \tag{6.4}$$

We may take a simple consideration by relating the normal stress of soil on the pile shaft to the vertical effective stress, say by K_0, and then the normal stress by friction coefficient expressed as $\tan\delta$ and finally set $\beta = K_0 \tan\delta$. While the determinations of K_0 and δ are empirical or experimental in nature, it is therefore convenient to combine them into a single factor of symbol β. Yet, by considering $K_0 = 1 - \sin\phi = 0.5$ based on a conservative value of $\phi = 30°$ and $\tan\delta = 0.5$ (as the coefficient of friction), it is often conservative to take $\beta = 0.5 \times 0.5 = 0.25$. Currently, β = 0.25 is commonly used by many engineers in Hong Kong in the design.

Obviously, the β value of a pile depends not only on the soil parameter, but also on the method of pile installation. Some typical values for β as measured in Hong Kong are given in Table 6.2.

Nevertheless, the reported measurement of β values can range from 0.1 to 0.6. The Code of Practice for Foundations 2017 (Hong Kong) has limited the β value to 0.25 in the absence of more accurate assessment and the factor is reduced to 0.2 for a pile in tension in the absence of a trial pile test.

In addition, due to the "arching effect" of the soil, the shaft friction does not increase indefinitely with the effective stresses. Tests by Vesic have shown that there is a critical depth beyond which the shaft friction becomes constant. The depth is usually taken as 20 pile diameters or the maximum allowable skin friction, whichever is reached earlier. The Code of Practice for Foundations 2017 (Hong Kong) has limited the ultimate skin friction to 120 kPa for tension pile.

Some engineers correlate the skin friction with the SPT. Based on limited studies in Hong Kong, the ultimate skin friction can be taken as 1.5–2.0 N in kPa, with a maximum value of 160 kPa for a driven small-diameter pile. For a large driven displacement pile, the relation can be taken as 4.0 N, with a maximum value of 250 kPa. For other countries, different correlations should be applied. For barrette formed by grab, the ultimate skin friction can be taken as 1.2 N, with a maximum value of 200 kPa. For barrette formed by cutter, the ultimate skin friction can be taken as 0.8 N, with a maximum value of 200 kPa when shaft grouting is not used, otherwise the values can be taken as 2.5 N and 200 kPa, respectively. For normal bored pile, it is similar to the barrette, except that the ultimate skin friction should be taken as 2.1 N when the shaft grouting is applied. The ultimate end-bearing capacity in soil will be taken as 10 N, with a limit of 2000 kPa.

For some types of piles, for example, small-diameter bored piles, there are also empirical approaches by relating the allowable skin friction directly to the SPT N values with limits to some values. The Code of Practice for Foundations 2017 (Hong Kong) specifies that the allowable skin friction of small-diameter bored piles formed by continuous augering to be in the order 1.0–1.6 N (in kPa) with a limit of 40 kPa.

The end-bearing component of the pile capacity often adopts a simple approach by specifying numerical values for the allowable or ultimate bearing capacity of the soil and rock as q_a or q_{ult} and the end-bearing capacity is simply the cross-sectional area multiplied by q_a or q_{ult} as

$$P_b = q_a A \tag{6.5}$$

where A is the cross-sectional area.

The q_a values are given in some codes including BS8004: 1986 and the Code of Practice for Foundations 2017 (Hong Kong) for rock and soil of different qualities. The values range from 10000 kPa for strong rock to 2000 kPa for weak rock. For soil, the values are 600 kPa to 50 kPa. In Hong Kong, the allowable end-bearing pressure is also commonly taken as 6. 7N for dry soil and 5 N (kPa) for wet soil.

However, from the principle of theoretical soil mechanics, failure mechanics similar to those of a shallow foundation apply by which the three bearing capacity factors N_q, N_c and N_γ can be applied. Therefore, the end-bearing capacity of the pile Q_b can be expressed as:

$$Q_b = \left(cN_c + qN_q + 0.5B\gamma N_\gamma\right)A_b \tag{6.6}$$

where A_b is the bearing area of the pile.

Of the three components, qN_q is the most important one as q is equal to at least the soil weight down to the pile tip which is generally of high value. The effect of the cohesive strength and γB terms are constant values which are much less than the surcharge term, hence they are commonly neglected in the design.

Similar to N_q for shallow foundations, N_q for piles also depends on the ϕ value of soil. Different researchers based on different approaches and assumptions have produced different relations as shown in Figure 6.10. The N_q by Berezantsev et al (1961) appears to be popular in many countries, and generally this factor is conservative as compared with other N_q values. This is probably the main reason for its popularity.

The work by Berezantsev (1961) appears to be commonly adopted by many engineers, and the model is shown in Figure 6.10d. The symbols in Figure 6.10 are defined as follows:

q_f = surcharge by the pile tip
q_T = the surcharge due to the overburden soil
T = skin friction acting on the soil cylinder of radius l_o
γ = unit weight of soil
φ = angle of shearing resistance of soil

$$l = \frac{\sqrt{2}ae^{\left(\frac{\pi}{4}-\frac{\varphi}{2}\right)\tan\frac{\varphi}{2}}}{\sin\left(\frac{\pi}{4}-\frac{\varphi}{2}\right)} \tag{6.7}$$

$$P_k = P_{k1} + P_{k2} \tag{6.8}$$

where P_k is the total end-bearing ultimate capacity of the pile

P_{k1} is the portion of end-bearing due to q_T which is the surcharge of the cylinder of soil of radius l_o

P_{k2} is the portion of end-bearing due to skin friction on the pile shaft.

$$P_{k1} = 2\pi a^3 \gamma\left(\frac{1}{2}A_k + B_k \alpha_T \frac{L}{d}\right) \tag{6.9}$$

$$P_{k2} = \pi a^2\left(A_T q'_T + B_T \gamma a\right) \tag{6.10}$$

$$q'_T = \frac{\tau_T(3-2v)}{4(1-v)} \tag{6.11}$$

τ_T which is the skin friction (in force per unit area) on the pile should be determined on site which may be difficult to carry out. Therefore, P_{k2} which involves τ_T may be ignored

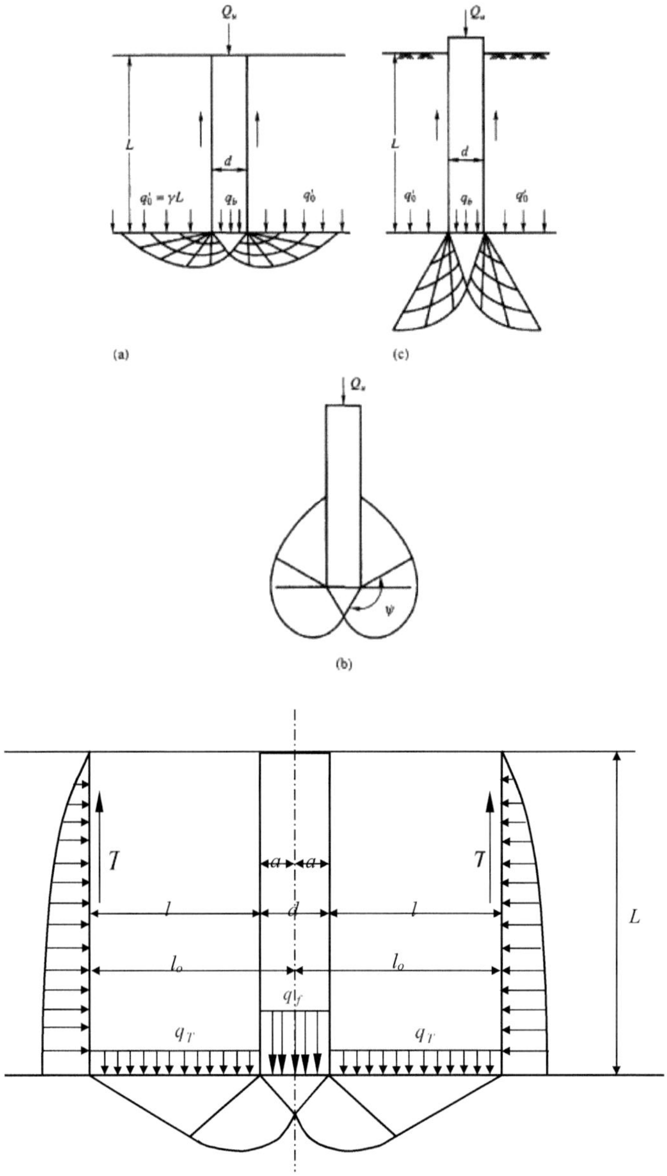

(d) failure mechanism by Berezantsev et al (1961)

Figure 6.10 Failure mechanism by N_q based on: (a) Terzaghi; (b) Meyerhof; (c) Vesic; (d) Berezantsev.

to be conservative such that the terms defined by Eqs. (6.10) and (6.11) can be dropped. Other symbols such as A_k, B_k and α_T in Eq. (6.9) are presented later.

Berezantsev et al. (1961) takes the active pressure of soil at depth z from the ground surface as follows

$$e_z = \tan\left(\frac{\pi}{4} - \frac{\varphi}{2}\right)\left\{1 - \left[1 + \frac{z}{l_o}\tan\left(\frac{\pi}{4} - \frac{\varphi}{2}\right)\right]^{1-\lambda}\right\}\frac{\gamma l_o}{\lambda - 1} \tag{6.12}$$

$$\text{where } \lambda = 2\tan\varphi\tan\left(\frac{\pi}{4} + \frac{\varphi}{2}\right) \tag{6.13}$$

The total skin friction on the cylinder of radius l_o is

$$T = \left(2\pi l_o \int_0^L e_z(z)\,dz\right)\tan\varphi \tag{6.14}$$

Since $l_o = l + a$, Eq. (6.7) can be rewritten as follows by introducing $k_1 = \frac{l_o}{d}$:

$$k_1 = \frac{l_o}{d} = \frac{1}{2}\left[1 + \frac{\sqrt{2}e^{\left(\frac{\pi}{4} - \frac{\varphi}{2}\right)\tan\frac{\varphi}{2}}}{\sin\left(\frac{\pi}{4} - \frac{\varphi}{2}\right)}\right] \tag{6.15}$$

Eq. (6.12) can then be written as:

$$e_z = \tan\left(\frac{\pi}{4} - \frac{\varphi}{2}\right)\frac{k_1}{\lambda - 1}\left\{1 - \left[1 + \frac{z}{k_1 d}\tan\left(\frac{\pi}{4} - \frac{\varphi}{2}\right)\right]^{1\ \lambda}\right\}\gamma d \tag{6.16}$$

Likewise, Eq. (6.14) can be written as

$$T = 2\pi k_1 d\tan\varphi\int_0^L e_z(z)\,dz \tag{6.17}$$

The integral in Eq. (6.14) can be worked out and becomes

$$T = 2\pi k_1 \gamma d^2 \tan\varphi\left[\left(\tan\left(\frac{\pi}{4} - \frac{\varphi}{2}\right)\frac{k_1}{\lambda - 1}\right)z + \frac{\tan^2\left(\frac{\pi}{4} - \frac{\varphi}{2}\right)\frac{1}{(\lambda - 1)d}\left(z + k_1 d\cot\left(\frac{\pi}{4} - \frac{\varphi}{2}\right)\right)^2}{\left(\tan\left(\frac{\pi}{4} - \frac{\varphi}{2}\right)\frac{z}{k_1 d} + 1\right)^{\lambda}(\lambda - 2)}\right]_0^L$$

$$= \frac{2\pi k_1 \gamma d^2 \tan\varphi}{\lambda - 1}\left[\tan\left(\frac{\pi}{4}-\frac{\varphi}{2}\right)k_1 L + \frac{\tan^2\left(\frac{\pi}{4}-\frac{\varphi}{2}\right)\left(L + k_1 d\cot\left(\frac{\pi}{4}-\frac{\varphi}{2}\right)\right)^2}{\left(\tan\left(\frac{\pi}{4}-\frac{\varphi}{2}\right)\frac{L}{k_1 d}+1\right)^{\lambda}(\lambda-2)d} - \frac{k_1^2 d}{\lambda - 2}\right] \tag{6.18}$$

Denoting the lateral soil load per metre length on the soil cylinder of radius l_o due to the active force by $E_D \frac{L}{d}\gamma d^2$, we have $T = E_D \frac{L}{d}\gamma d^2 2\pi l_o \tan\varphi = E_D \frac{L}{d}\gamma d^2 2\pi k_1 d \tan\varphi$. The quantity E_D can then be worked out as:

$$E_D \frac{L}{d}\gamma d^2 2\pi k_1 d \tan\varphi = \frac{2\pi k_1 \gamma d^2 \tan\varphi}{\lambda - 1}\left[\begin{array}{l}\tan\left(\frac{\pi}{4}-\frac{\varphi}{2}\right)k_1 L + \\ \dfrac{\tan^2\left(\frac{\pi}{4}-\frac{\varphi}{2}\right)\left(L + k_1 d\cot\left(\frac{\pi}{4}-\frac{\varphi}{2}\right)\right)^2}{\left(\tan\left(\frac{\pi}{4}-\frac{\varphi}{2}\right)\frac{L}{k_1 d}+1\right)^{\lambda}(\lambda-2)d} - \dfrac{k_1^2 d}{\lambda - 2}\end{array}\right]$$

$$\Rightarrow E_D = \frac{1}{\lambda - 1}\left[\tan\left(\frac{\pi}{4}-\frac{\varphi}{2}\right)k_1 + \frac{\tan^2\left(\frac{\pi}{4}-\frac{\varphi}{2}\right)\left(\frac{L}{d} + k_1\cot\left(\frac{\pi}{4}-\frac{\varphi}{2}\right)\right)^2}{\left(\frac{1}{k_1}\tan\left(\frac{\pi}{4}-\frac{\varphi}{2}\right)\frac{L}{d}+1\right)^{\lambda}(\lambda-2)\left(\frac{L}{d}\right)} - \frac{k_1^2}{(\lambda-2)\left(\frac{L}{d}\right)}\right] \tag{6.19}$$

E_D is a parameter depending only on φ and $\frac{L}{d}$, giving:

$$T = E_D \frac{L}{d}\gamma d^3 2\pi k_1 \tan\varphi \tag{6.20}$$

q_T in Figure 6.10 can then be worked out as the average stress due to the weight of the soil cylinder (radius l_o) minus T and then divided by the base area as

$$q_T = \frac{\pi l_o^2 \gamma L - 0.25\pi d^2 \gamma L - T}{\pi l_o^2 - 0.25\pi d^2} = \frac{\pi k_1^2 d^2 \gamma L - 0.25\pi d^2 \gamma L - E_D \frac{L}{d}\gamma d^3 2\pi k_1 \tan\varphi}{\pi k_1^2 d^2 - 0.25\pi d^2}$$

$$= \left(\frac{k_1^2 - 0.25 - E_D 2k_1 \tan\varphi}{k_1^2 - 0.25}\right)\gamma L \tag{6.21}$$

$$\text{Setting } \alpha_T = \left(\frac{k_1^2 - 0.25 - E_D 2k_1 \tan\varphi}{k_1^2 - 0.25} \right) \tag{6.22}$$

$$q_T = \alpha_T \gamma L \tag{6.23}$$

According to the failure surface assumed in Figure 6.10, Berezantsev et al. (1961) worked out the ultimate pile capacity due to surcharge q as

$$\begin{aligned} P_{k1} &= 2\pi a^3 \gamma \left(\frac{1}{2} A_k + B_k \alpha_T \frac{L}{d} \right) = 2\pi a^2 \left(\frac{1}{L/a} \right) L\gamma \left(\frac{1}{2} A_k + B_k \alpha_T \frac{L}{d} \right) \\ &= \left(\frac{1}{L/d} \right) \left(\frac{1}{2} A_k + B_k \alpha_T \frac{L}{d} \right) (\pi a^2) q \end{aligned} \tag{6.24}$$

where

$$A_k = \frac{1+\sin\varphi}{3} \left\{ \begin{aligned} &\frac{\cos(0.5\varphi) + e^{(0.5\pi - 0.5\varphi)\tan(0.5\varphi)}}{(\omega+1)\cos\varphi\cos(0.5\varphi)} \left(M^{\omega+1} - 1 \right) e^{1.5\pi\tan\varphi} \\ &-0.77\cot\varphi \left[\left(\frac{\pi}{4} - \frac{\varphi}{2} \right) e^{1.5\pi\tan\varphi} - \left(\frac{3\pi}{4} - \frac{\varphi}{2} \right) \right] \\ &+\cot\varphi (1.2 - 0.26\cot\varphi)\left(e^{1.5\pi\tan\varphi} - 1 \right) \end{aligned} \right\} \tag{6.25}$$

$$B_k = \frac{1}{3} \times \frac{1+\sin\varphi}{1-\sin\varphi} e^{\pi\tan\varphi} \left(1.5 + M^{\omega} e^{\pi\tan\varphi} \right) \tag{6.26}$$

$$M = \frac{\left[\sin\left(\frac{\pi}{4} - \frac{\varphi}{2} \right) + \sqrt{2} e^{\left(\frac{\pi}{2} - \frac{\varphi}{2} \right) \tan\left(\frac{\varphi}{2} \right)} \right] \cos\left(\frac{\varphi}{2} \right)}{\left[\cos\left(\frac{\varphi}{2} \right) + e^{\left(\frac{\pi}{2} - \frac{\varphi}{2} \right) \tan\left(\frac{\varphi}{2} \right)} \right] \sin\left(\frac{\pi}{4} - \frac{\varphi}{2} \right)} \tag{6.27}$$

$$\omega = 2\tan\varphi \tan\left(\frac{\pi}{4} - \frac{\varphi}{2} \right) \tag{6.28}$$

$$q = L\gamma \tag{6.29}$$

$$\text{Putting } N_q = \left(\frac{1}{L/d} \right) \left(\frac{1}{2} A_k + B_k \alpha_T \frac{L}{d} \right) \tag{6.30}$$

$P_{k1} = N_q(\pi a^2)q$ where πa^2 is the base area of the pile and q is the surcharge at the pile base. Therefore, $N_q = \left(\frac{1}{L/d}\right)\left(\frac{1}{2}A_k + B_k\alpha_T\frac{L}{d}\right)$ becomes the bearing capacity of the pile which depends on φ and $\frac{L}{d}$ only. Upon determination of N_q, the pile end-bearing capacity can be calculated using Eq. (6.6). If the cross section of the pile under consideration is not circular, the equivalent diameter of the pile, d, may be based on an equal base area. Summing up, the following procedures can be adopted for determination of N_q

1. The ratio L/d is first determined;
2. Calculate l/a using Eq. (6.7) and then determine $k_1 = l_o/d = l/d + 0.5$ where $l/d = 0.5l/a$;
3. Calculate λ using Eq. (6.13);
4. Calculate E_D using Eq. (6.19);
5. Calculate α_T using Eq. (6.22);
6. Calculate M using Eq. (6.27);
7. Calculate ω using Eq. (6.28);
8. With M and ω, calculate A_k using Eq. (6.25);
9. With M and ω, calculate B_k using Eq. (6.26);
10. With A_k, B_k and α_T, determine N_q using Eq. (6.30).

Figure 6.11 shows a design chart developed by the authors. It should be noted that a single relation between N_q and ϕ has been proposed by Poulos, which is commonly used by many engineers. This single relation is obtained by averaging the three curves proposed by Berezantsev et al. (1961), and the percentage error with the original relation is actually quite appreciable. A simple program N_q to replace Figure 6.11 has also been developed by Cheng, which can be obtained from natureymc@yahoo.com.hk.

6.4.2 Pile group vertical capacity

Piles are usually used in groups, and the pile capacity of a pile group is compared with a single pile. The pile capacity of a pile group is a complicated problem, and the problem greatly depends on the method of installation. For driven piles in loose to medium-density sand, the efficiency is usually greater than 1.0, as the densification and compaction effect will actually improve the soil properties. For driven pile in dense sand, it is possible that the driven process can loosen the structure of the soil, but generally this is not a critical effect. On the whole, it can be considered that the pile capacity of a pile group in sand is the sum of the individual pile capacities. On the other hand, if a pile under 1500 kN settles by 15 mm, will a pile group with two piles taking a loading of 3000 kN also settle by 15 mm? In general, it will not, as the settlement of a pile will cause soil dragdown and hence settlement to the adjacent pile. With consideration of the pile group settlement, the pile group capacity should be reduced. There are some empirical formulae in the literature for pile group capacity, but the authors; view such empirical formulae as not being

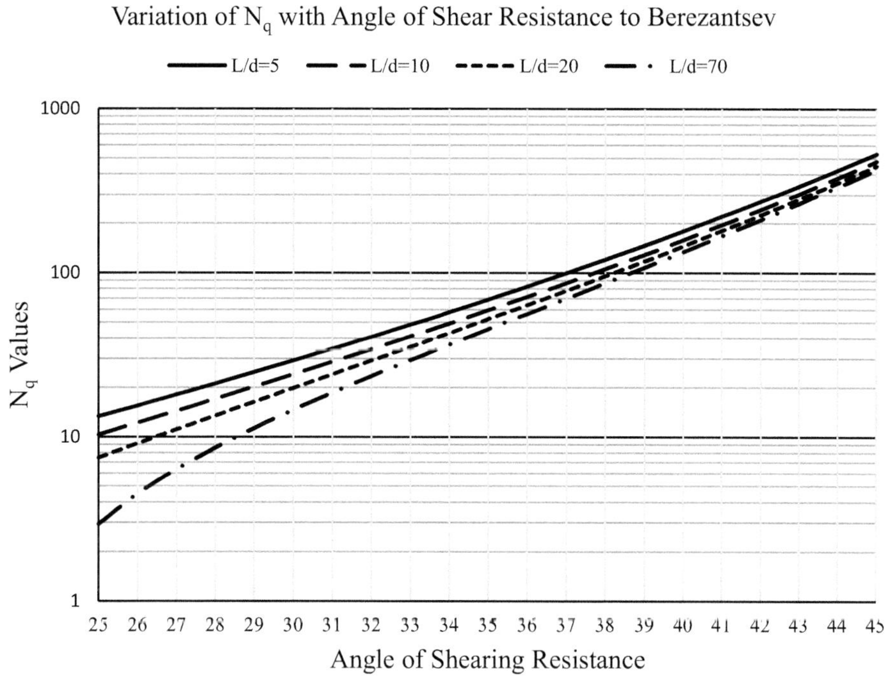

Figure 6.11 Design chart based on equations by Berezantsev et al. (1961).

credible unless they are supported by field measurements. In Hong Kong, a group reduction factor of 0.85 is commonly adopted by many engineers, but currently it is generally accepted that no reduction factor is required for pile groups in sand. For Hong Kong, the group reduction factor is not required where:

(a) The centre-to-centre spacings are of more than three times the perimeter of the piles or the circumscribed rectangles in the case of H-piles; or
(b) The load capacity of the piles is derived from the end-bearing; or
(c) The piles are rock-socketed piles; or
(d) The piles are driven to refusal with the toes resting closely onto bedrock, according to the Code of Practice for Foundation; or
(e) The piles are driven to sound bearing strata, e.g. soil with an SPT N-value of $\geqq 200$.

Pile spacing is considered to be important for a pile group. Currently, some local guidelines are as follows:

1. For driven piles and others, which derive their capacities mainly from frictional resistance, the minimum pile spacing shall be not less than the length of the perimeter of the pile or 1 m, whichever is the greater, and the piles shall be placed at not less

than half the length of the perimeter of the pile or 500 mm, whichever is the greater, from the site boundary.

2. For steel H-piles, the perimeter should be taken as $2(b + d)$, where b and d are the overall breadth and depth of the section, respectively.
3. For minipiles and socketed steel H-piles that derive their capacities mainly from bond strength of grout in rock sockets, the minimum pile spacing should be 750 mm or twice the outer diameter, whichever is the greater.
4. For bored piles and the like which derive their capacities mainly from end-bearing, the minimum clear spacing between the surfaces of adjacent piles should be based on practical considerations of positional and verticality tolerances of piles. It is recommended to provide a nominal minimum clear horizontal spacing of 500 mm between shaft surfaces or edges of bell-outs, as appropriate.

For a pile group in clay, a block failure (all the piles lumped together to form a shaft) is considered to be a possible but highly unlikely failure phenomenon. The group reduction factor is also commonly considered to be 1.0, while the settlement consideration will usually dictate the actual pile group capacity. Two programs based on the elastic approach by Poulos or the ring shear theory by Randolph with pile interactions have been developed by Cheng, which are available upon request at natureymc@yahoo.com.hk.

6.5 LATERAL LOAD ANALYSIS

In addition to the piles' primary use as axially loaded members to resist vertical loads, they are sometimes required to resist lateral loads. The lateral loads mostly come from wind and soil which often act on the pile heads. The lateral restraints are offered by the soil embedding the pile. Lateral shears and moments, and sometimes axial loads, will then be induced in the piles. The goals of designers are to determine deflections and stresses in the selected soil-pile system in order that they may be controlled within tolerable limits.

6.5.1 Ultimate analysis

The ultimate lateral load resistance of a pile can be originated from the resistance of soil. For cohesive soils, Broms (1964a) takes the ultimate resistance as $9c_u$ where c_u is the undrained shear strength of the soil. The top $1.5d$ (where d is the pile diameter) of soil resistance is ignored as illustrated for long and short piles in Figure 6.12.

In Figure 6.12(a), which shows a short pile, the equilibrium is controlled by the soil resistance under the ultimate soil pressure distribution as shown,

$$H_u = 9c_u df \tag{6.31}$$

where H_u is the ultimate shear.

As zero shear exists at $1.5d + f$ below ground, where the maximum moment in the pile will coexist, we may list

$$M_{max} = H_u(e + 1.5d + f) - H_u \times 0.5f = H_u(e + 1.5d + 0.5f) \tag{6.32}$$

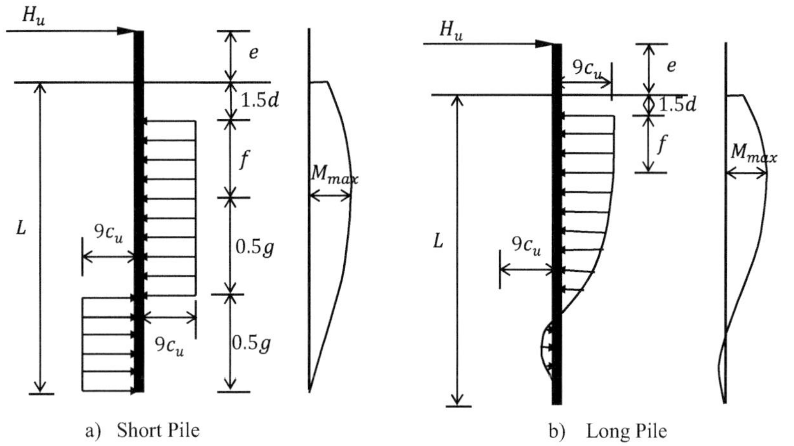

Figure 6.12 Ultimate lateral resistance of a pile in cohesive soil.

Also, taking moment at zero shear location due to push and pull soil pressure below the location, a couple constituted by the soil pressure is also equal to M_{max}, i.e.

$$M_{max} = 9c_u d \times 0.5g \times 0.5g = 2.25c_u dg^2 \tag{6.33}$$

In addition

$$L = 1.5d + f + g \tag{6.34}$$

Therefore, there are four equations [(6.31)–(6.34)] to solve H_u, M_{max}, f and g.

Equating Eqs. (6.32) and (6.33) with H_u substituting by Eq. (6.31) and simplifying

$$4f(e + 1.5d + 0.5f) = g^2$$

Substituting Eq. (6.34) to eliminate g

$$2f^2 + (4e + 6d)f = [(L - 1.5d) - f]^2$$

$$\Rightarrow 3f^2 + [(4e + 6d) + 2(L - 1.5d)]f - (L - 1.5d)^2 = 0$$

$$\Rightarrow 3f^2 + (4e + 2L + 3d)f - (L - 1.5d)^2 = 0 \tag{6.35}$$

Solving f by the above quadratic equation, H_u can be directly solved by Eq. (6.31), together with M_{max} by Eq. (6.33). If M_{max} exceeds the flexural capacity of the pile M_p at

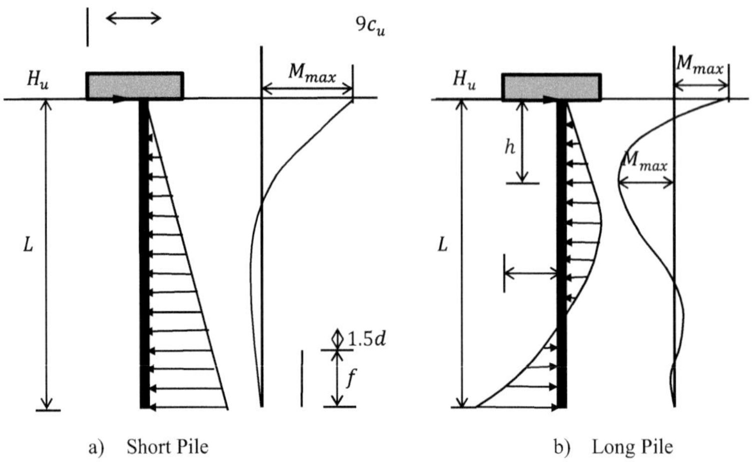

Figure 6.13 Ultimate lateral resistance of a pile in cohesionless soil.

long length, the pressure profiles below the zero shear location will change as shown in Figure 6.12(b), M_p will then be taken into account as a limiting criterion for determination of H_u. Therefore, only Eqs. (6.31) and (6.32) need be used with $M_{max} = M_p$, leaving H_u, f to be solved.

For cohesionless soil, the ultimate soil resistance is the passive pressure. Taking the arching effect of the soil into account, Broms (1964b) takes the ultimate resistance as $3K_p\sigma'$. Some more updated results actually suggest replacing 3 by K_p. The ultimate lateral resistance of the pile will be under the combined consideration of the ultimate pressure of the soil and the ultimate bending capacity of the pile, as illustrated in Figure 6.13.

For the short pile as shown in Figure 6.13(a), the full passive soil resistance is mobilized, which is equal to the ultimate shear of the pile S_{ult}

$$S_{ult} = 0.5\left(3K_p\sigma' \, dL\right) \tag{6.36}$$

The moment is the greatest at the pile head which is

$$M_{max} = 0.5\left(3K_p\sigma' \, dL\right)\left(\frac{2}{3}L\right) \tag{6.37}$$

and M_{max} is less than the flexural capacity of the pile.

However, if M_{max} is greater than the flexural capacity of the pile M_p, then the soil pressure distribution becomes that of Figure 6.12(b). Then the equation involving h which is the location of zero shear and the ultimate shear is set up as

$$S_{ult} = 0.5\left(3K_p\gamma hdL\right) \tag{6.38}$$

taking $\sigma' = \gamma h$ where γ is the unit weight of the soil with the addition of surcharge if any.

And the moment at depth h is $M_{max} = 0.5\left(3K_p\gamma hdh\right)\left(\frac{2}{3}h\right)$ and putting $M_{max} = M_p$ the flexural capacity of the pile which is a known value. Therefore,

$$M_p = 0.5\left(3K_p\gamma hdh\right)\left(\frac{2}{3}h\right) \tag{6.39}$$

Solving h using Eq. (6.39), S_{ult} can be calculated using Eq. (6.38)

If the pile head is at some depth below the ground so that σ' is not zero at the pile head level, Eqs. (6.38) and (6.39) will be modified for determination of the S_{ult}.

The foregoing describes an approach for determination of the ultimate shear of a pile. The allowable shear will then be the ultimate shear divided by a factor of safety in the order of 2–3. However, the controlling criterion for a pile resisting lateral load is often deflection, which is normally a serviceability limit state. Approaches have been developed for the estimation of the lateral deflections of piles.

6.5.2 Lateral deflection of pile

A direct method for estimating lateral deflection is to assume the embedding soil to be an elastic material and with the application of formulae such as Mindlin's equations and their integrated forms. This method involves the Young's modulus and Poisson's ratio of soil. However, the analytical methods derived are often very complicated and have to be carried out by computer. In addition, these methods will result in very high stresses at the pile heads which have to be dealt with by plastic analysis using iterative methods.

To overcome the complication as mentioned above, an approach based on the assumption of "Winkler's springs" was developed by Terzaghi (1955). The method is known as the "horizontal subgrade reaction method". By this method, the soil restraint is idealized as a series of elastic springs (independent of each other and therefore called Winkler's springs) on the pile shaft. The spring value is denoted by k_h which is called the "coefficient of horizontal subgrade reaction" and is related to the displacement as

$$p = k_h u \tag{6.40}$$

where p is the pressure and u is the horizontal displacement of the soil. k_h carries a physical meaning of stiffness which is the proportionality relating pressure and lateral displacement, i.e., the lateral pressure required to produce unit lateral displacement. In metric units it is kN/m²/m. For simulation of a pile as a structural model for solution by the stiffness method, a beam element with a series of lateral point springs as supports can be constructed as shown in Figure 6.14. In this figure, the springs are discrete point springs which carry the value of $k_h B\Delta z$ where B is the width of the pile and ΔL is the tributary

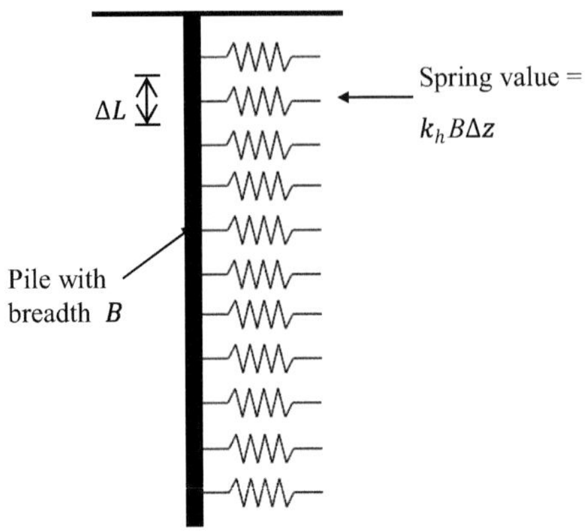

Figure 6.14 Computer model of a pile with lateral soil springs.

length of the pile the spring represents. Continuous supports can be assumed if the springs are closely spaced.

With B as the width of the pile, the load per unit length of the pile is $w = pB$. So we may list

$$w = k_h Bu \tag{6.41}$$

k_h depends on the stiffness of the soil and at the same time may depend also on the depth of soil z. A general relation is

$$k_h = k_L \left(\frac{z}{L}\right)^{\alpha} \tag{6.42}$$

where L is the level of the pile tip below ground and k_L is the k_h value at $z = L$, $\pm$ is a coefficient to be applied depending on the type of the soil. Generally, $\pm$ is taken as 0 for cohesive soil so that the k_h value is constant along the pile shaft and $\pm$ is taken as 1 for cohesionless soil so that k_h varies linearly with depth.

Therefore, the "spring values" along the pile shaft can be defined under determined values of k_h.

For cohesionless soil, k_h is expressed as

$$k_h = n_h \left(\frac{z}{B}\right) \tag{6.43}$$

where n_h is termed the "constant of horizontal subgrade reaction" and depends on the type of soil only. Terzaghi (1955) has produced values for different ranges of SPT N values. The values are reproduced with metric units in Table 6.3.

Table 6.3 Values of n_h for cohesionless soils

	Loose (SPT N value = 4–10)	*Medium (SPT N value = 11–30)*	*Dense (SPT N value = 31–50)*
n_h for dry or moist sand	2200 kN/m^2/m	6600 kN/m^2/m	17600 kN/m^2/m
n_h for submerged sand	1300 kN/m^2/m	4400 kN/m^2/m	10700 kN/m^2/m

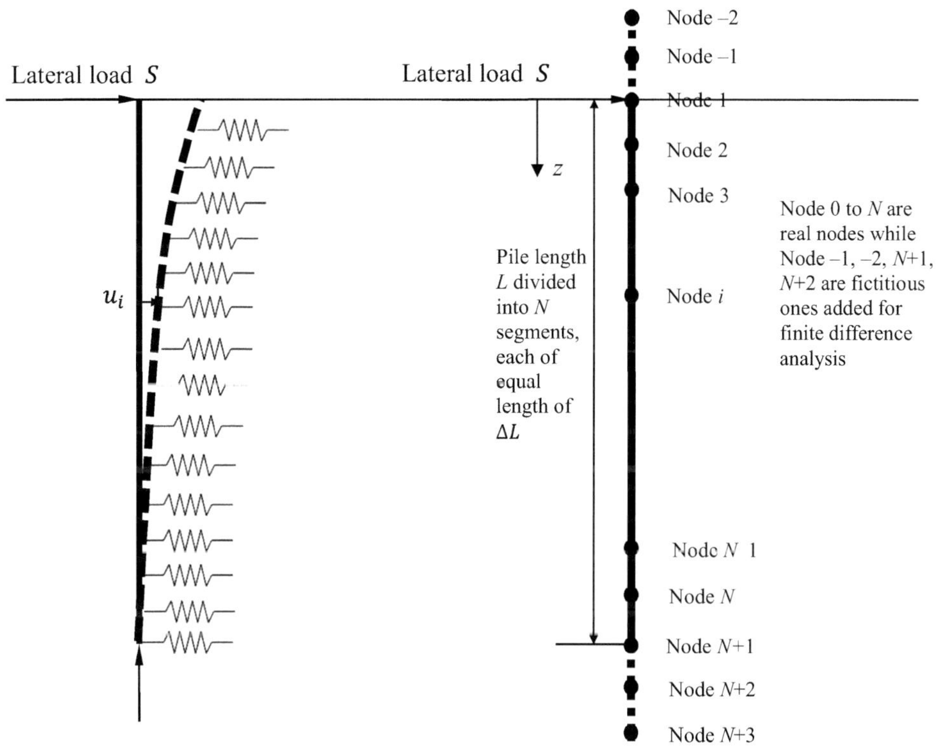

Figure 6.15 Finite difference model for analysing a pile in elastic soil.

With the k_h determined either by Eq. (6.55) or (6.56) for cohesive or cohesionless soil, simulation into the computer model as shown in Figure 6.15 can be carried out for analysis. Results comprising lateral deflections, internal forces of the pile and soil reactions can be obtained. It should be noted that the n_h is a derived soil property, and is different from the Young's modulus which can be directly measured.

Alternatively to the stiffness method, if the pile is assumed to be an elastic beam, the following relation based on basic structural analysis theory can be listed:

$$E_p I_p \frac{d^4u}{dz^4} = -w = -pd = -k_h Bu$$

$$\Rightarrow E_pI_p\frac{d^4u}{dz^4}+k_hBu=0 \tag{6.44}$$

In Eq. (6.44) z is the ordinate of the point on the pile under consideration. For cohesive soil, Eq. (6.44) becomes

$$E_pI_p\frac{d^4u}{dz^4}+k_LBu=0 \tag{6.45}$$

where k_LB is a constant. Lengthy analytical solutions exist, as found in Poulos (1980) and others.

For cohesionless soil, using Eq. (6.43) where $k_h=n_h\left(\frac{z}{B}\right)$, Eq. (6.44) becomes

$$E_pI_p\frac{d^4u}{dz^4}+n_hzu=0 \tag{6.46}$$

Both Eqs. (6.45) and (6.46) can be solved numerically by the finite difference method. Eq. (6.46) can be expressed as

$$E_pI_p\frac{u_{i-2}-4u_{i-1}+6u_i-4u_{i+1}+u_{i+2}}{\Delta L^4}+k_hBu_i=0 \tag{6.47}$$

for the ith node based on the model as shown in Figure 6.15. Figure 6.15 explains the formulation of the finite difference equations.

By the finite difference method and with reference to Figure 6.11, a pile of length L divided into N equal segments (each of length L/N) will have $N+5$ points, with $N+1$ real nodes and the end nodes $N-2, N-1, N+2, N+3$ as "fictitious nodes". $N+1$ equations can be formed based on Eq. (6.47). Four more equations are required for the complete solution. They are based on support conditions at the pile head and pile tip and the applied loads which are either shear forces or moments at the pile head. The equations are summarized in Table 6.4.

If a shear is applied at pile head, $S=E_pI_p\frac{d^3u}{dz^3}=E_pI_p\frac{u_2-2u_1+2u_{-1}-u_{-2}}{2\Delta L^3}$ shall replace the boundary condition for $S=E_pI_p\frac{d^3u}{dz^3}=0$. If a moment is applied at pile head, $M=E_pI_p\frac{d^2u}{dz^2}=E_pI_p\frac{u_2-2u_1+u_{-1}}{\Delta L^2}$ shall replace the $M=E_pI_p\frac{d^2u}{dz^2}=0$ equation.

For the condition of cohesionless soil, $k_h=n_h\left(\frac{z}{B}\right)$ is applied to Eq. (6.60) for the ith node which reads

$$E_pI_p\frac{u_{i-2}-4u_{i-1}+6u_i-4u_{i+1}+u_{i+2}}{(L/N)^4}+n_hzu_i=0$$

Table 6.4 Equations for boundary conditions for use in the finite difference method

	Conditions	*Pile head*	*Pile tip*
Restraint from lateral movement	$u = 0$	$u_1 = 0$	$u_{N+1} = 0$
Restraint from Rotation	$\theta = \frac{du}{dz} = 0$	$\frac{u_2 - u_{-1}}{2(L/N)} = 0$	$\frac{u_{N+2} - u_N}{2(L/N)} = 0$
Hinged Connections (moment = 0)	$M = E_p I_p \frac{d^2u}{dz^2} = 0$	$E_p I_p \frac{u_2 - 2u_1 + u_{-1}}{(L/N)^2} = 0$	$E_p I_p \frac{u_{N+2} - 2u_{N+1} + u_N}{(L/N)^2} = 0$
Free end (Shear of pile = 0)	$S = E_p I_p \frac{d^3u}{dz^3} = 0$	$F_p I_p \frac{u_3 - 2u_2 + 2u_{-1} - u_{-2}}{2(L/N)^3} = 0$	$E_p I_p \frac{u_{N+3} - 2u_{N+2} + 2u_N - u_{N-1}}{2(L/N)^3} = 0$

Putting $z = \frac{i}{N} L$, expanding and simplifying,

$$u_{i-2} - 4u_{i-1} + \left(6 + \frac{n_h L^5}{E_p I_p} \frac{i}{N}\right) u_i - 4u_{i+1} + u_{i+2} = 0 \tag{6.48}$$

So a single parameter $\frac{n_h L^5}{E_p I_p}$ now remains. Let $T = \sqrt[5]{\frac{E_p I_p}{n_h}}$, $\frac{n_h L^5}{E_p I_p} = (L/T)^5$. The single parameter becomes L/T which can be applied for solution in the form of:

$$u_{i-2} - 4u_{i-1} + \left(6 + \left(\frac{L}{T}\right)^5 \frac{i}{N^5}\right) u_i - 4u_{i+1} + u_{i+2} = 0 \tag{6.49}$$

Upon solution of the u_i values, coefficients such as deflection coefficient and moment coefficient can be used to multiply certain quantities to obtain the true values. Therefore, parametric studies can be conducted which result in simple coefficients for general use applications. A computer program for the lateral load analysis of piles with soil yielding (P-Y method) has been developed by the authors.

In the following graphs so derived, contained in Figures 6.16–6.18, the coefficients are to be multiplied by parameters as listed in Table 6.5 to obtain the deflections, moments and soil reaction (expressed as load per unit length) on the pile. In the vertical axis, the pile location is expressed as the coefficient multiplied by L/T.

Similar design charts can be devised for cohesive soil. For a pile group, it is commonly assumed that each pile will share the same amount of lateral load, and the lateral load analysis is carried out individually. To account for the group effect, reductions on the η_h value as functions of pile spacing in the direction of loading have been proposed in the Canadian Foundation Engineering Manual (1978) which is reproduced in Table 6.6.

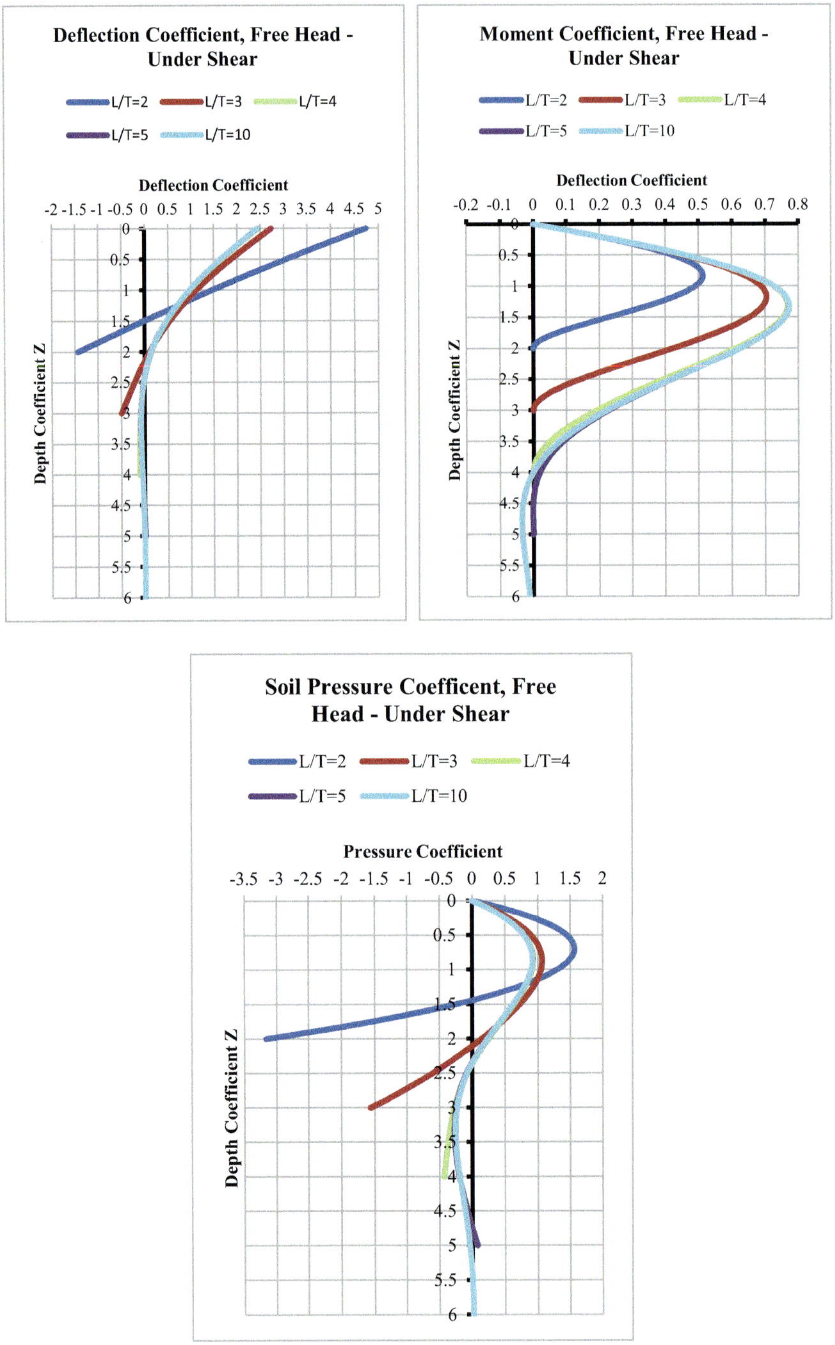

Figure 6.16 Design charts for shear at the pile head – pile head free.

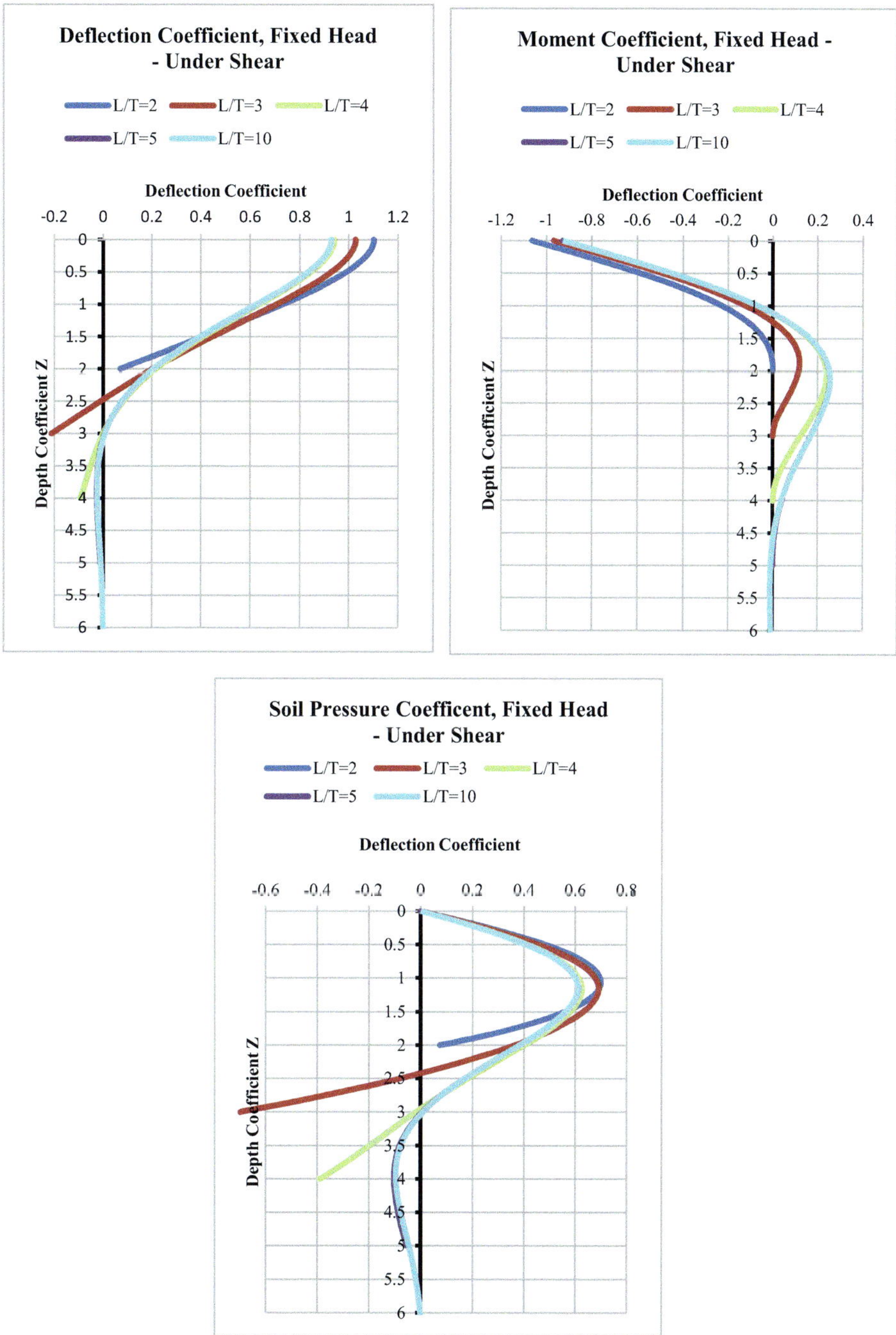

Figure 6.17 Design charts for shear at the pile head – pile head fixed.

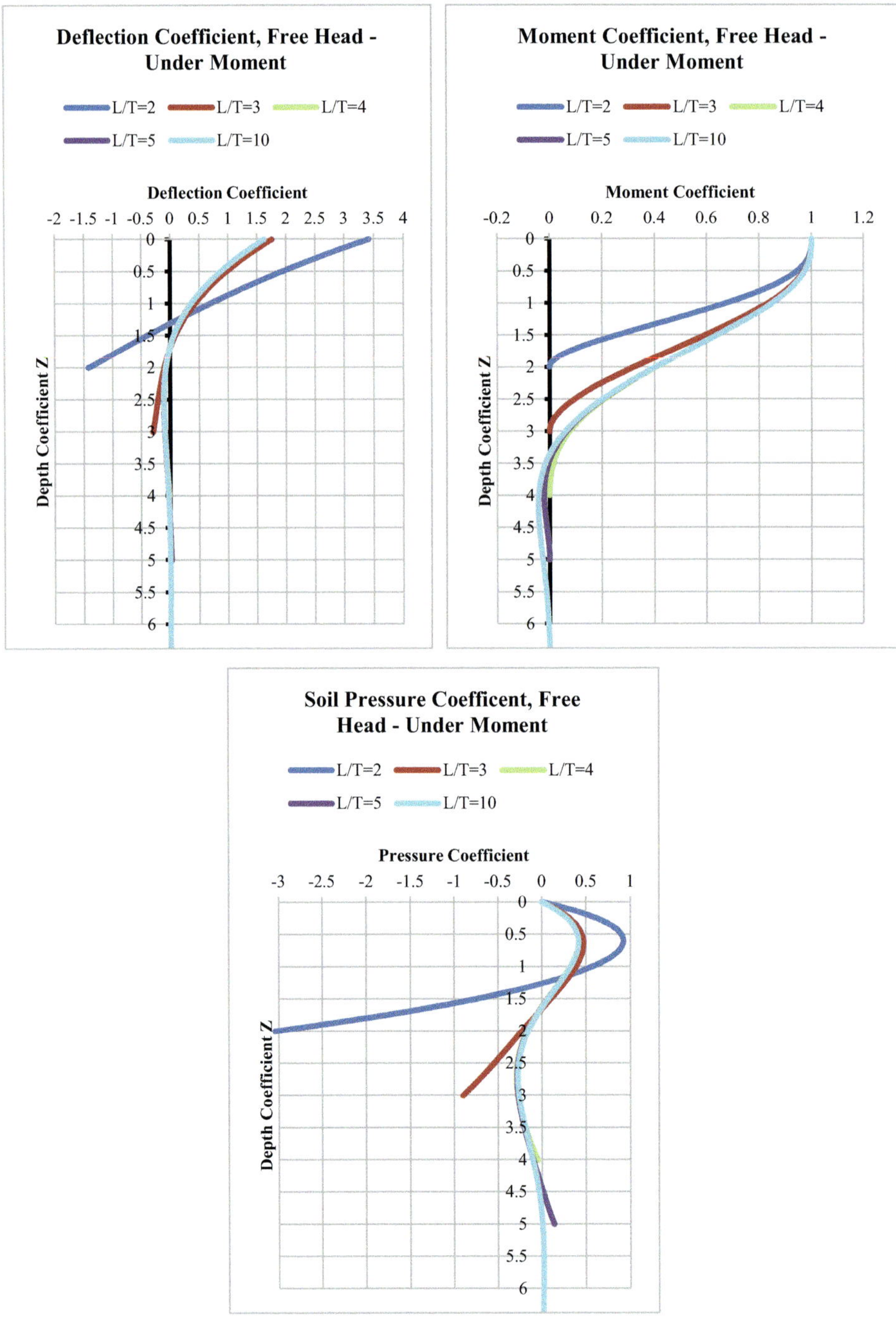

Figure 6.18 Design charts for moment at the pile head – pile head free.

Table 6.5 Application of coefficients in the finite difference method

	By shear S on pile head	*By moment M on pile head*
Deflection δ_p	$\delta_p \dfrac{ST^3}{E_p I_p}$	$\delta_p \dfrac{MT^2}{E_p I_p}$
Moment M_p	$M_p ST$	$M_p M$
Soil reaction R_p	$R_p S/T$	$R_p M/T^2$

Table 6.6 Reduction factors with spacing for lateral load analysis

Pile spacing in the direction of the applied shear (D is projected width of the pile lateral load)	3D	4D	6D	8D
Reduction factor to be applied on η_h	0.25	0.4	0.7	1.0

In the case that the lateral displacement is large and the lateral response from the soil is nonlinear, the P-Y method is commonly used by engineers. In this model, the lateral stress against displacement is either defined by a higher order function or direct stress–strain relations from the test is used as input into the program. For this case, the axial force in the pile can also be considered, which will be a stability geometric matrix as given by Eqs. (6.50) and (6.51). In general, the use of incremental displacement at the pile top is prescribed, and the pile displacement and internal forces in the pile are determined from an iteration analysis. A complete horizontal load/displacement relation can then be established for the pile. The advantage of using FEM for such conditions is that post-buckling behaviour can also be obtained easily, and this can be considered easily by the program P-Y developed by the authors or other commercial programs. Some engineers simply multiply the axial load with the lateral displacement to get an additional moment on pile, but this is not adequate if the lateral movement is large, as the additional moment will generate additional lateral displacement until the whole iteration process becomes stable.

$$\{[K_b]+[K_\sigma]+[K_s]\}[\delta]=[F] \tag{6.50}$$

$$[K_\sigma]=\frac{-P}{30L}\begin{bmatrix} 36 & 3L & -36 & 3L \\ 3L & 4L^2 & -3L & -L^2 \\ -36 & -3L & 36 & -3L \\ 3L & -L^2 & -3L & 4L^2 \end{bmatrix} \tag{6.51}$$

The use of the P-Y method with the consideration of pile axial force will be more suitable for marine piles where the lateral displacements are large. For normal building works, such fine tuning in the analysis may not be necessary.

6.6 PILE SETTLEMENT OF A SINGLE PILE AND PILE GROUP

For estimation of the pile vertical settlement for a single pile, there are several approaches. For end-bearing piles on rock, some engineers consider the pile as a column without skin friction and the pile settlement can be obtained from the axial shortening of a column (PL/AE) under an axial load. This approach is conservative but popular in Hong Kong, and it is simple enough for quick engineering calculation. To consider the skin friction of single pile, the approach by Poulos and Davis (1980) has been used in the past in Hong Kong but not currently. This approach is based on the use of the Mindlin equation where the pile is divided into a series of segments, and influence coefficients are generated. Cheng has developed a similar program using a faster numerical technique for the settlement calculation of single piles and pile groups with moment, which can be obtained from Cheng for education purposes. For pile group analysis, the assumption of rigid caps has to be used to complement the last equation for the solution using this approach. During the development of the Fortran 90 code for the pile group analysis using the approach by Poulos and Davis (1980), Cheng found problems of over-interaction if the number of piles and the number of divisions along a pile are large. This is due to the fact that the use of the Mindlin equation assumes the effect of pile extends to infinity, which is in practice not correct. From many field observations, the influence zone of a pile is usually around 5–10 pile diameters. Furthermore, many engineers comment that the design figures by Poulos and Davis (1980) are difficult to use, and many cases are not covered by those design figures.

Alternatively, the ring shear theory by Randolph (1987) can be adopted. Both the ring shear theory by Randolph (1987) and the use of the Mindlin equation by Poulos and Davis (1980) suffer from a major limitation of strain compatibility between the pile and soil, which is not possible if the pile settlement is relatively large. Randolph assumed the zone of influence to be γ_m, and the settlement of soil ω_z can be expressed as

$$\omega_s = \frac{\tau_o r_o}{G} \ln \frac{r_m}{r} \qquad r_o \le r \le r_m \tag{6.52}$$

where $\tau_0(z)$ is the skin friction adjacent to the pile, and r_0 is the radius of the pile, ω_z is the vertical displacement of the soil and r_m is the radius of influence. For the axial forces within a small pile segment

$$\frac{dP}{dz} = -2\pi r_0 \tau_0 \tag{6.53}$$

$$\text{Since } P = -A_p E_p \frac{\partial \omega_p}{\partial z} \tag{6.54}$$

$$\text{hence } A_p E_p \frac{\partial^2 \omega_p}{\partial z^2} - 2\pi r_0 \tau_0 = 0 \tag{6.55}$$

Assume computability of displacement between pile and soil, hence ω_p is equal to ω_s. Based on this, we can get

$$A_pE_p\frac{\partial^2\omega}{\partial z^2}-2\pi r_o\tau_o=0\Leftrightarrow\frac{\partial^2\omega}{\partial z^2}-\frac{2\pi G}{A_pE_p\ln\frac{r_m}{r_o}}\omega=0 \tag{6.56}$$

This is usually expressed as

$$\frac{\partial^2\omega}{\partial z^2}-u^2\omega=0 \qquad u^2=\frac{2\pi G}{A_pE_p\ln\frac{r_m}{r_o}}=\frac{K_1}{A_pE_p} \tag{6.57}$$

where $K_1 = 2\pi G/\ln(r_m/r_0)$. The general solution to Eq. (6.57) is

$$\omega(z)=C_1e^{uz}+C_2e^{-uz} \tag{6.58}$$

The axial force is given by

$$P(z)=-A_pE_p\frac{\partial\omega(z)}{\partial z}=-A_pE_pu(C_1e^{uz}-C_2e^{-uz}) \tag{6.59}$$

Put z = l, $\omega_b=C_1e^{ul}+C_2e^{-ul}$, the pile load at base is

$$P_b=-A_pE_pu(C_1e^{ul}-C_2e^{-ul}) \tag{6.60}$$

Based on Eqs. (6.59) and (6.60), we can get

$$C_1=\frac{1}{2}\left(\omega_b-\frac{P_b}{A_pE_pu}\right)e^{-ul}\quad\text{and}\quad C_2=\frac{1}{2}\left(\omega_b+\frac{P_b}{A_pE_pu}\right)e^{ul} \tag{6.61}$$

For a rigid circular plate (similar to a pile base), the displacement at the rigid pile base is given by elasticity as

$$P_b=\frac{4r_bG_b}{(1-\mu)}\omega_b=K_2\omega_b \tag{6.62}$$

where r_b can be different from r_0 and G_b can be different from G, and b represents the properties at the pile base. Finally, the pile load can be obtained as

$$P(z) = P_b\left\{\frac{K_1}{uK_2}\sinh\left[u(l-z)\right] + \cosh\left[u(l-z)\right]\right\} \tag{6.63}$$

Putting $z = 0$ in Eq. (6.63) for P_o and $z = 0$, we can obtain

$$\omega_o = P_o\left[1 + \frac{K_2}{A_pE_p}\cdot\frac{\tanh(ul)}{u}\right]\left[K_2 + K_1\frac{\tanh(ul)}{u}\right]^{-1} \tag{6.64}$$

Or $\omega_o = P_o f_{ii}$ (6.65)

Eq. (6.64) applies if the soil properties are uniform with depth. If the shear modulus varies linearly with depth, the equation is given by Randolph as

$$\frac{P_o}{G_e r_o \omega_o} = \frac{\dfrac{4\eta}{(1-\mu)\xi} + \dfrac{2\pi\rho}{\zeta}\dfrac{\tanh(ul)}{ul}\dfrac{l}{r_o}}{1 + \dfrac{4\eta}{\pi\lambda(1-\mu)\xi}\dfrac{\tanh(ul)}{ul}\dfrac{l}{r_o}} \tag{6.66}$$

where G_e = shear modulus at $z = l$, $\eta = r_b/r_o$ (r_b = radius of pile base)
$\xi = G_e/G_b$ (G_b = shear modulus for end bearing stratum)
$\rho = G_{av}/G_e$ (G_{av} = average shear modulus of soil over pile length)
μ = Poisson ratio of soil

$\lambda = E_p/G_e$ $\quad \zeta = \ln(r_m/r_0)$, $\quad ul = \left(\dfrac{2}{\zeta\lambda}\right)^{0.5}\dfrac{l}{r_o}$

If the average shear modulus is used in Eq. (6.55), the results will be close to those using Eq. (6.64), and Eq. (6.66) will be sufficiently good for general cases. The equations by Randolph have been verified to give results close to those by Poulos. Some curve fitting results with finite element results by Randolph give ζ as

If $\xi \neq 1$, $\zeta = \ln\{[0.25 + (2.5\rho(1-\mu) - 0.25)\xi]l/r_0\}$

If $\xi = 1$, $\zeta = \ln[2.5\rho(1-\mu)l/r_0]$ or $r_m = 2.5\rho(1-\mu)l$ (6.67)

Cooke has suggested that r_m can be taken as 20 r_0 while a value of 5–10 r_0 is actually commonly observed. In pile group analysis, if Eq. (6.80) is used, over-interaction may occur. Readers can try Eq. (6.64) or (6.66) for cases where the soil along the pile shaft has no stiffness, or the pile is based on a rigid mass or length of pile tends to infinity.

Example: For a 0.5 m diameter tubular pile with an internal core diameter 0.25 m, E for soil can be taken as 20000 kPa while E for the base of pile is 80000 kPa. E for concrete is 25000 MPa and the Poisson ratio of soil is 0.2. Determine the pile settlement for a vertical load of 2500 kN if length of pile = 30 and 100 m (take r_m = 10 m). What can you observe and deduce? Why?

Answer:

G_s = 8333 kPa, G_b = 33333 kPa, K_1 = 14193 kPa u = 0.0621

K_2 = 41667 kPa f_{ii} for 30 m = 4.525×10^{-6} f_{ii} for 100 m = 4.375×10^{-6}

Settlement = 11.3 mm for 30 m length and 10.94 mm for 100 m length of pile. It is noted that the use of a very long pile cannot reduce pile settlement, as the majority of the pile load is taken up by the skin friction for the first 10 m pile length.

Example:
For a 20 m length of 0.5 m diameter tubular pile with an internal core of diameter 0.25 m, E for soil and concrete can be taken as 30000 kPa and 25000 MPa. Poisson ratio can be taken as 0.2. Determine the skin friction distribution along the pile shaft (at an interval of 2 m) and the amount of pile base load for an applied load of 1500 kN. What can you observe? If a total load of 12000 kN is applied to eight piles and the piles are arranged in a 2 × 4 pattern with a spacing of 2.5 m in both directions, determine the pile settlement under this case. If an additional moment of 6000 kNm is applied to this pile group in the long direction, determine the various pile loads and settlements; assume rigid cap in your calculation (Table 6.7).

Solution:

$$L = 20 \text{ m } A_p = \pi\,(0.25^2 - 0.125^2) = 0.1473 \text{ m}^2$$

$$\therefore \quad A_p\,E_p = 3.682 \times 10^6$$

Since E_s = 30000 kPa, the base is not rigid,

$$\therefore \quad \zeta = ln\,(r_m/r_o) = ln[2.5\rho(1-\mu)\,l\,/r_o] = ln[2.5 \times 0.8 \times 20/0.25] = 5.075$$

$$G_s = E_s/[2(1+\mu)] = 12500 \text{ kPa}$$

$$\therefore \quad K_1 = 2\,\pi G/\zeta = 15476$$

$$K_2 = 4\,r_b\,G_b/(1-\mu) = 15625$$

$$u = \sqrt{[K_1/(A_pE_p)]} = 0.0648$$

$$(\tanh ul)/u = 13.282.$$

Table 6.7 Distribution of skin friction along pile shaft, based on ring shear theory

Z	0	2	4	6	8	10	12	14	16	18	20
τ	70.58	63.13	56.74	51.31	46.74	42.96	39.9	37.5	35.8	34.6	34

Put in Eq. (6.60) for $z = 0$

$P(0) = 1500 = P_b \{K_1/(u K_2) \sinh ul + \cosh ul\}$

$\rightarrow P_b = 53.95$ kN

Since $dP(z)/dz = -2\pi r_o \tau_o$

$= -P_b\{K_1/K_2 \cosh[u(l - z)] + u \sinh[u(l - z)]$

For a pile group, f_{ij} between different piles are given by Eq. (6.64) with K_1' and K_2' replacing K_1 and K_2 if $i \neq j$.

$$K_1' = \left. K_1 \ln\frac{r_m}{r_o} \middle/ \ln\frac{r_m}{R} \right. \qquad K_2' = K_2 \frac{\pi R}{2r_b} \quad \text{or} \quad K_2 \frac{\pi}{2\sin^{-1}(r_b/R)} \tag{6.68}$$

Eq. (6.68) can be proved easily by the relation:

$$\text{Pile base defelction profile } \omega_R = \frac{2r_b}{\pi R}\omega_b \tag{6.69}$$

where R is the pile centre-to-centre spacing. To use Eq. (6.68), u and K_1 are required to be computed again or can be kept constant (very small difference). For a pile group with n piles, group interaction can be considered by using the rigid cap assumption as

$$\sum_{j=1}^{n} f_{ij} P_j = \text{cap movement} \quad (i = 1 \text{ to } n) \tag{6.70}$$

$$\text{From vertical force equilibrium, } \sum P_j = P \tag{6.71}$$

Using the rigid cap assumption, the individual pile top movement is the same as the cap movement, hence Eq. (6.70) provides $n - 1$ equations as

$$\delta_1 = \delta_2, \delta_2 = \delta_3, \delta_3 = \delta_4 \,\ldots\ldots\ldots \delta_{n-1} = \delta_n, \tag{6.72}$$

Eq. (6.72) provide $n - 1$ equations for the solution of individual pile load together with Eq. (6.71). Hence, there are n equations for n unknowns, and individual pile load can be solved from Eqs. (6.71) and (6.72) (see program PILE27 by Cheng). Rigid cap assumption is not a bad assumption and has been commonly used in design in the past. For a refined analysis, flexible cap analysis by PLATE using the Randolph theory and further refinement can be used (obtainable from the author at natureymc@yahoo.com.hk).

It should be noted that in Figure 6.19, if the option coupled analysis is chosen, f_{ij} will be formed and a flexibility matrix of the pile group is formed. This flexibility matrix is inverted to form the pile group stiffness matrix which is coupled to the pile cap for the analysis. If this option is not chosen, f_{ij} will be 0, and each pile function as a vertical spring with

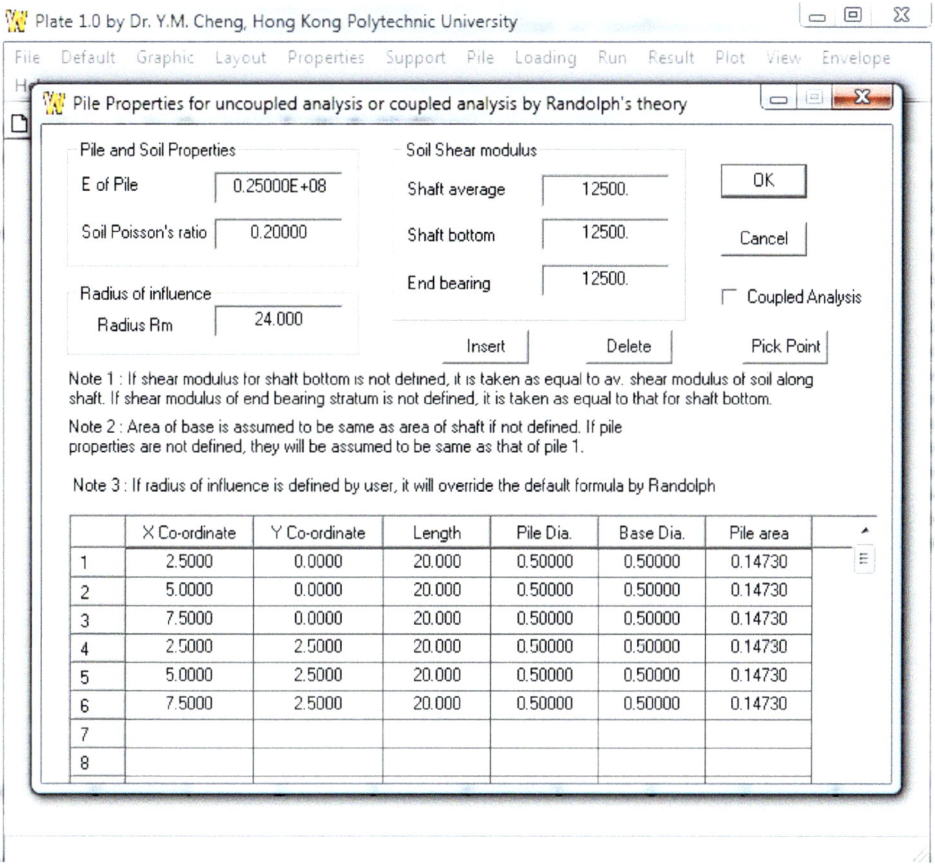

Figure 6.19 Pile options in PLATE.

a spring constant which is $1/f_{ii}$. To avoid over-interaction, the diffraction coefficients by Mylonakis and Gazetas (1998) are incorporated into PLATE to reduce the over-interaction phenomenon.

For the pile group, Randolph proposed that r_m is given by,

$$r_m = 2.5\,\rho(1-\mu)l + R_G \tag{6.73}$$

where R_G = equivalent radius of the pile group, and this term is *sometimes neglected* for simplicity. With regard to the previous example (Figure 6.20),

$$\pi R_G^2 = (2.5 \times 3 + 0.5)(2.5 + 0.5) \Rightarrow R_G = 2.764$$

$$r_m = 2.5\,l\,(1-\mu) + R_G = 42.76 \text{ m}$$

$$\zeta = ln\,(r_m/r_o) = 5.142$$

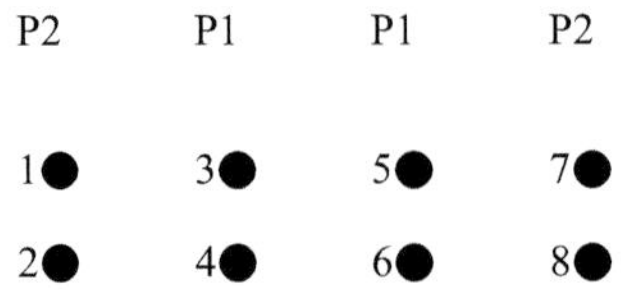

Figure 6.20 Pile group layout under vertical load for the example.

Pile 1-3	R	=	2.5	,	K_1'	=	27661	,	K_2'	=	245437
	f_{13}	=	$3.075 x 10^{-6}$								
Pile 2-3	R	=	3.54	,	K_1'	=	31506	,	K_2'	=	347097
	f_{23}	=	$2.942 x 10^{-6}$								
Pile 3-7	R	=	5	,	K_1'	=	36594	,	K_2'	=	490874
	f_{37}	=	$2.838 x 10^{-6}$								
Pile 3-8	R	=	5.59	,	K_1'	=	38500	,	K_2'	=	548797
	f_{38}	=	$2.809 x 10^{-6}$								
Pile 1-7	R	=	7.5	,	K_1'	=	45117	,	K_2'	=	736311
	f_{17}	=	$2.74 x 10^{-6}$								
Pile 1-8	R	=	7.9	,	K_1'	=	46505	,	K_2'	=	775581
	f_{18}	=	$2.729 x 10^{-6}$								

Figure 6.21 Formation of interaction coefficients f_{ij}.

$K_1 = 2\,\pi G/\zeta = 15274$ and $K_2 = 15625$

$u = \sqrt{[K_1/(A_p E_p)]} = 0.0644$

$(\tanh ul)/u = 13.33$

$f_{ii} = \{1 + (K_2/A_p E_p)[(\tanh ul)/u]\}/\{K_2 + K_1[(\tanh ul)/u]\}$

$= 4.82 \times 10^{-6}$

Interaction analysis gives the results in Figure 6.21.

Vertical force equilibrium: 12000 = 4P1 + 4P2

→ P1 + P2 = 3000

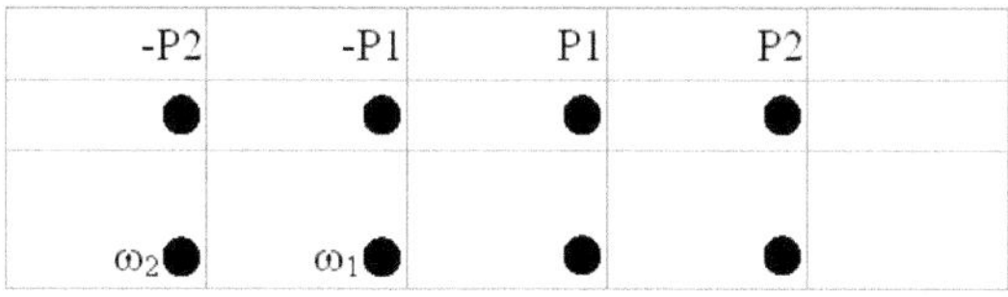

Figure 6.22 Pile group under moment.

Using rigid cap assumption: ω for pile 3 = ω for pile 1

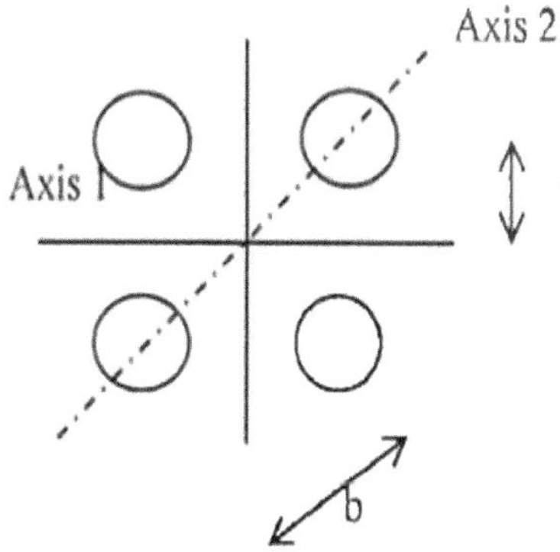

For a single pile, $\omega = 4.82 \times 1500 \times 10^{-3} = 7.2$ mm. Hence, the pile group interaction effect is very critical in the assessment of settlement. For a moment of 6000, the problem is anti-symmetric. Using moment equilibrium, the internal couple equals the external moment (Figure 6.22)

$$6000 = 2\,P_1\,(2.5) + 2\,P_2\,(7.5)$$

$$\rightarrow \quad 1200 = P_1 + 3\,P_2$$

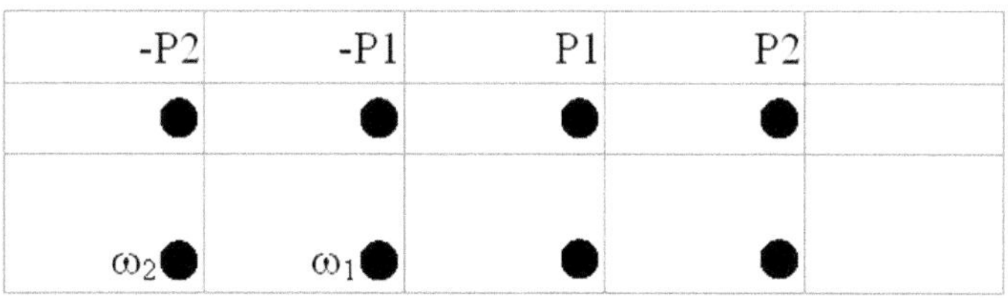

By rigid cap assumption $\omega_1/d_1 = \omega_2/d_2$, hence P1 = 92.3 kN, P2 = 369.2 kN. Hence the pile overall movements are $\omega_1 = \pm\,0.31$ mm, $\omega_3 = \pm\,0.93$ mm. If PLATE is used for this example, the results are also very close to the values above.

Example:
For a 30 m length pile with 0.5 m diameter tubular pile and an internal core diameter 0.25 m, E is 15000 kPa at the top and 35000 kPa at the bottom. E for the bearing stratum can be taken as 80000 kPa. E for concrete is 25000 MPa and Poisson ratio = 0.2. Determine the pile settlement for a vertical load of 2500 kN. If a pile group is arranged in 2 × 4 pattern with a spacing 2.2 m with a vertical load of 20000 kN and moment 10000 kN-m along the long direction determine the pile group settlement if r_m is taken as 10 m.

Solution:

K_1 = 17742.5 kPa, K_2 = 41667 kPa $f_{ii} = 4 \times 10^{-6}$ settlement = 10 mm for single pile

Assume the pile arrangement 1 3 5 7 P_1 = load at 1, 2, 7, 8
2 4 6 8 P_2 = load at 3, 4, 5, 6

$f_{12} = 2.7 \times 10^{-6}$ $f_{23} = 2.56 \times 10^{-6}$ $f_{15} = 2.37 \times 10^{-6}$
$f_{17} = 1.92 \times 10^{-6}$ $f_{38} = 2.28 \times 10^{-6}$ $f_{18} = 1.82 \times 10^{-6}$
Vertical load settlement = 51.7mm P_1 = 3968 kN P_2 = 1032 kN
Moment settlement = 2.16 mm/0.72 mm P_1 = 687.5 kN P_2 = 210.2 kN

For PLATE analysis with this problem, determine the equivalent pile spring stiffness for the pile. If the thickness of the cap is 1.2 m and Poisson ratio of concrete = 0.2, determine the pile load if the load is applied along a shear wall with 6 m length.

Answer:
$K = 1/f_{ii}$ = 25000 kN/m
Uncoupled analysis – load and settlement at P_1 = 204 kN, 0.82 mm, P_3 = 2076.8 kN, 8.36 mm, P_5 = 3606.6 kN, 14.52 mm, P_7 = 4111.8 kN, 16.56 mm

Coupled analysis – load and settlement at P_1 = 760.1 kN, 34.87 mm, P_3 = 1626.9 kN, 44.6 mm, P_5 = 2840.5 kN, 51.43 mm, P_7 = 4772.5 kN, 52.31 mm

It is noted that there are noticeable differences between the results from uncoupled and coupled analyses. The stress transfer between piles is hence not negligible, and PLATE provides a very simple solution without the use of true 3D analysis.

6.7 CLASSICAL PILE GROUP ANALYSIS

Piles are commonly used in the form of a group, as shown in Figure 6.23.

Before the availability of computers, engineers relied on the rigid cap assumption to carry out pile group analysis. The classical rigid cap method as shown in Figure 6.24 requires the following assumptions:

1. The pile cap is assumed to be a rigid plate.
2. Pile heads/cap connections are hinged connections (fixed-head pile is not allowed).
3. The pile displacement of any pile is assumed to be proportional to the offset from the neutral axis of the pile cap.
4. Length of pile is the same (varying L can be allowed)
5. Effect of soil is neglected (end-bearing and no stress transfer between piles).

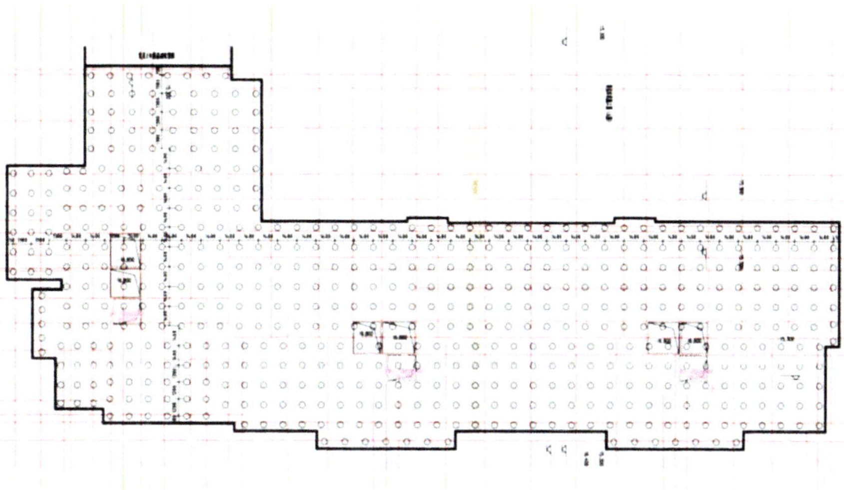

Figure 6.23 A large pile cap supported by CFG piles in Hong Kong.

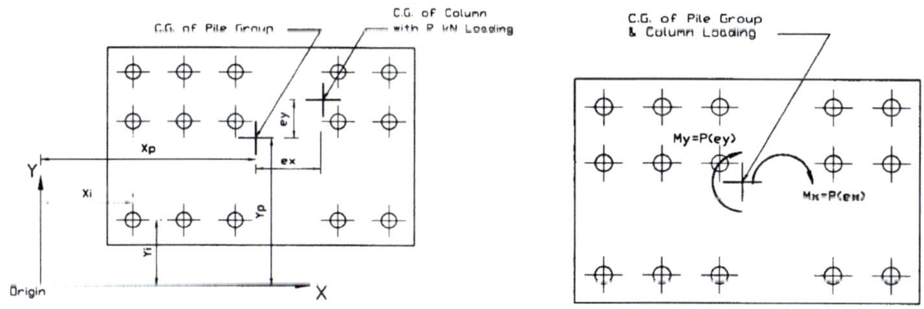

(a) Pile group under vertical load (b) pile group under moment

Figure 6.24 Pile group under vertical load and moment in rigid cap assumption.

6. The external load is taken up by the axial load in the piles, and no moment is generated in the piles.

The overall moment M_x and M_y will be equivalent to eccentricities along the x and y directions as:

$$M_x = P * e_x \text{ and } M_y = P * e_y \tag{6.74}$$

Consider the eccentric moment which will cause the cap to tilt about the pile group centroid, the displacement of each pile top will be proportional to its distance from the pile group centroid. The present problem is similar to a bolt group problem, and the classical rivet equation is used which gives the individual pile load as

$$P_i = A_i\left[\frac{P}{\Sigma A_i} + \frac{M_x - M_y\left(\frac{I_{xy}}{I_{yy}}\right)}{I_{xx} - \frac{I_{xy}^2}{I_{yy}}}(Xi - Xp) + \frac{M_y - M_x\left(\frac{I_{xy}}{I_{xx}}\right)}{I_{yy} - \frac{I_{xy}^2}{I_{xx}}}(Yi - Yp)\right] \tag{6.75}$$

where $I_{xx} = \Sigma A_i(X_i - X_p)^2$ $\quad I_{yy} = \Sigma A_i\,(Y_i - Y_p)^2$ $\quad I_{xy} = \Sigma A_i\,(X_i - X_p)\,(Y_i - Y_p)$

In most cases, the area of each pile is the same as the others within a pile group, as most engineers prefer to control more even settlement within each pile group. In this case, Eq. (6.75) simplifies to

$$P_i = \left[\frac{P}{n} + \frac{M_x - M_y\left(\frac{I_{xy}}{I_{yy}}\right)}{I_{xx} - \frac{I_{xy}^2}{I_{yy}}}(X_i - X_p) + \frac{M_y - M_x\left(\frac{I_{xy}}{I_{xx}}\right)}{I_{yy} - \frac{I_{xy}^2}{I_{xx}}}(Y_i - Y_p)\right] \tag{6.76}$$

$$\text{where } I_{xx} = \Sigma(X_i - X_p)^2 \qquad I_{yy} = \Sigma\,(Y_i - Y_p)^2 \qquad I_{xy} = \Sigma\,(X_i - X_p)\,(Y_i - Y_p) \tag{6.77}$$

For a symmetrical pile group, $I_{xy} = 0$, Eq. (6.76) will further simplify to

$$P_i = \left[\frac{P}{N} + \frac{M_x}{I_{xx}}(X_i - X_p) + \frac{M_y}{I_{yy}}(Y_i - Y_p)\right] \tag{6.78}$$

Once the pile reactions are determined, the bending moment and shear force at each point on the pile cap can be determined from one-dimensional analysis by grouping all the pile loads along a grid into a single point load (similar to simple footing design). The bending moment and shear force as determined refer to the global section and uniform reinforcement can be designed. For flexible cap analysis using the plate element method, a design strip is usually formed, and the reinforcement arrangement will not be uniform. It should be noted that for economy design of the pile cap, many engineers adopt the concept that the CG of the pile group coincides with the CG of the dead load from the superstructure for the design of pile group.

If the piles have different lengths, set $K_i = A_i\,E_i/L_i$ and replace A_i with K_i in the above formulae. The determination of the CG of the pile group should also adopt K_i instead of the area then the CG of a pile group is determined as: $\Sigma(K_iX_i)=(\Sigma K_i)X_p$, $\Sigma(K_iY_i)=(\Sigma K_i)Y_p$. Once the CG is determined, the equivalent second moment of area can be calculated based on this CG and K_i as:

$$I_{xx} = \Sigma K_i(X_i - X_p)^2 \qquad I_{yy} = \Sigma K_i(Y_i - Y_p)^2 \qquad I_{xy} = \Sigma K_i(X_i - X_p)\,(Y_i - Y_p)$$

i.e., the CG of the pile group is (0,0).

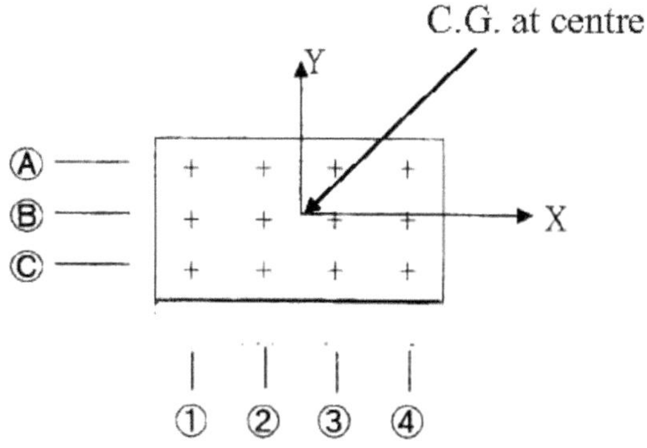

Figure 6.25 A symmetric pile group.

The pile load is then distributed according to the following modified formula

$$P_i = K_i\left[\frac{P}{\sum K_i} + \frac{M_x - M_y\left(\frac{I_{xy}}{I_{yy}}\right)}{I_{xx} - \frac{I_{xy}^2}{I_{yy}}}(X_i - X_p) + \frac{M_y - M_x\left(\frac{I_{xy}}{I_{xx}}\right)}{I_{yy} - \frac{I_{xy}^2}{I_{xx}}}(Y_i - Y_p)\right] \qquad (6.79)$$

Example: Consider the problem in Figure 6.25, determine the maximum and minimum pile loads if the vertical load, M_x and M_y are 6000 kN, 2000 kN-m and 1000 kNm, respectively. Pile spacing is 1.5 m.

Solution:

$I_{xx} = 0.75^2 \times 6 + 2.25^2 \times 6 = 33.75 \qquad I_{yy} = 1.5^2 \times 8 = 18$

For a symmetric pile group, maximum pile load at A4 = 716.7 kN
Minimum pile load at C1 = 283.3 kN

Example: For the asymmetric pile group as shown in Figure 6.26, determine the maximum and minimum pile loads if the vertical load, M_x and M_y are 8000 kN, 2000 kN-m and 1500 kNm, respectively. Pile spacing is 1.5 m.

Take the coordinate of A1 as (0,0). Taken the moment about grid 1 and A

$1.5 \times 2 + 3 \times 4 + 4.5 \times 4 = 12X \qquad \Rightarrow X = 2.75$ m from grid 1

$1.5 \times 4 + 3 \times 2 + 4.5 \times 2 = 12Y \qquad \Rightarrow Y = 1.75$ m from grid A

$I_{xx} = (-2.75)^2 \times 2 + (-1.25)^2 \times 2 + (0.25)^2 \times 4 + (1.75)^2 \times 4 = 30.75$

$I_{yy} = (-1.75)^2 \times 4 + (-0.25)^2 \times 4 + (1.25)^2 \times 2 + (2.75)^2 \times 2 = 30.75$

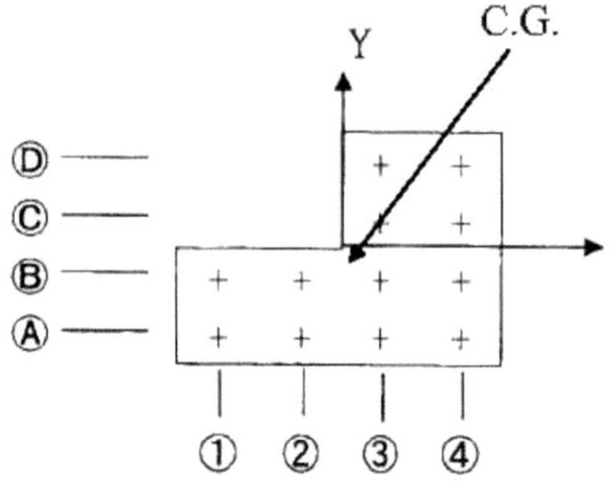

Figure 6.26 An asymmetric pile group analysis.

In fact, I_{xx} should be equal to I_{yy} as it is symmetrical about the diagonal of the pile group.

$$I_{xy} = (-2.75)(-1.75) + (-2.75)(-0.25) + (-1.25)(-1.75) + (-1.25)(-0.25) + (0.25)(-1.75) + (0.25)(-0.25) + (0.25)(1.25) + (0.25)(2.75) + (1.75)(-1.75) + (1.75)(-0.25) + (1.75)(1.25) + (1.75)(2.75) = 12$$

$$P_i = 8000/12 + [2000 - 1500(12/30.75)]/(30.75 - 12^2/30.75)X + [1500 - 2000(12/30.75)]/(30.75 - 12^2/30.75)]Y = 666.7 + 54.27X + 27.6Y$$

Maximum pile load at D4 (1.75,2.75) = 837.6 kN
Minimum pile load at A1 (–2.75,–1.75 = 469.2 kN

Additional example: Determine the maximum and minimum pile loads for a pile group with 16 piles, with the information as shown below if P = 100000 kN, M_x = 14000 kN-m, M_y = 5500 kN-m.

Coordinates of the piles

(1) (0,0)	(2) (0,2) (3)	(0,4) (4) (0,6)
(5) (2,0)	(6) (2,2) (7)	(4,0) (8) (4,2)
(9) (6,0)	(10) (6,2) (11)	(6,4) (12) (6,6)
(13) (8,0)	(14) (8,2) (15)	(8,4) (16) (8,6)

Answer:
$I_{xx} = 156$ $I_{yy} = 76$ $I_{xy} = 10$
Maximum pile load = 6795.1 kN Minimum pile load = 5721.2 kN

Additional example: Determine the pile loads for piles in the pile group as shown below using the rigid cap design method. Vertical load = 16000 kN, M_x = 3000 kN-m, M_y = 2500 kN-m (moment along axis). Note: Pile length is not constant!

Pile	x	y	Pile length (m)
1	0	0	20
2	2	0	22
3	4	0	23
4	6	0	24
5	0	2	18
6	2	2	20
7	4	2	20
8	6	2	21
9	0	4	16
10	2	4	17
11	4	4	16
12	6	4	18

Solution:
Assume AE = 1 unit, then

$$\sum K_i = \sum \frac{1}{L_i} = 0.623$$

$\Sigma K_i x_i = (\Sigma K_i) x_p$ $\Sigma K_i y_i = (\Sigma K_i) y_p$ give $x_p = 2.89$ $y_p = 2.189$
Based on $I_{xx} = \Sigma K_i (x_i - x_p)^2$ $I_{yy} = \Sigma K_i (y_i - y_p)^2$ $I_{xy} = \Sigma K_i (x_i - x_p)(y_i - y_p)$
$I_{xx} = 3.119$ $I_{yy} = 1.658$ $I_{xy} = 0.0325$

Based on Eq. (6.92), the pile loads are

P1 = 984	P2 = 980.6
P3 = 1020.3	P4 = 1056.6
P5 = 1258.8	P6 = 1227.6
P7 = 1322.2	P8 = 1349.4
P9 = 1602.4	P10 = 1619.5
P11 = 1839	P12 = 1739.8

For the pile group settlement, it is usually considered to be equal to the single pile settlement under an equivalent load, if the pile load is mainly transferred to the pile base so that pile group interaction is not important. This is generally an acceptable practice, unless the piles are closely spaced. A group reduction factor of 0.85 is applied to the pile capacity, while the pile group settlement is still commonly considered to be equal to the single pile settlement in Hong Kong due to the use of a high factor of safety on the pile capacity and the design loads. Based on many field test results on pile groups, it has been found that the pile group interaction phenomenon is small for the actual pile spacing used by the engineers. For pile group analysis, most engineers adopt the computer method as discussed in Section 6.8 in recent times. One of the important critics for the rigid cap theory is the assumption of the plane section remaining plane, which is purely an assumption which cannot be proved. The use of the computer method can also allow the consideration of pile cap stiffness and the soil support reaction between the piles, which has been demonstrated to amount to 30% in some cases.

6.8 ANALYSIS AND DESIGN OF PILE CAP

The RC design of pile caps is covered in clause 3.11.4, part 1 of BS8110. The design provisions are as follows:

1. General – Pile caps are designed either using bending theory or using the truss analogy. When the truss method is used, the truss should be of triangulated form with a node at the centre of the loaded area. The lower nodes are to lie at the intersection of the centrelines of the piles with the tensile reinforcement. Tensile forces in pile caps for some common cases are shown in Figure 6.27.
2. Truss method with widely spaced piles – Where the spacing exceeds three times the pile diameter only reinforcement within 1.5 times the diameter from the centre of the pile should be considered to form a tension member of a truss.
3. Shear forces – The shear strength of a pile cap is normally governed by the shear on a vertical section through a full width of the cap. The critical section is taken at 20% of the pile diameter inside the face of the pile, as shown in Figure 6.6. The whole of the force from the piles with centres lying outside this line should be considered.
4. Design shear resistance – The shear check may be made in accordance with provisions for the shear resistance of solid slabs given in clauses 3.5.5 and 3.5.6 of the code. The following limitations apply with regard to pile caps, as shown in Figure 6.6.
   (a) The distance a_v from the face of the column to the critical shear plane. The enhanced shear stress is $2d\ v_c/a_v$ where v_c is given in the code. The maximum

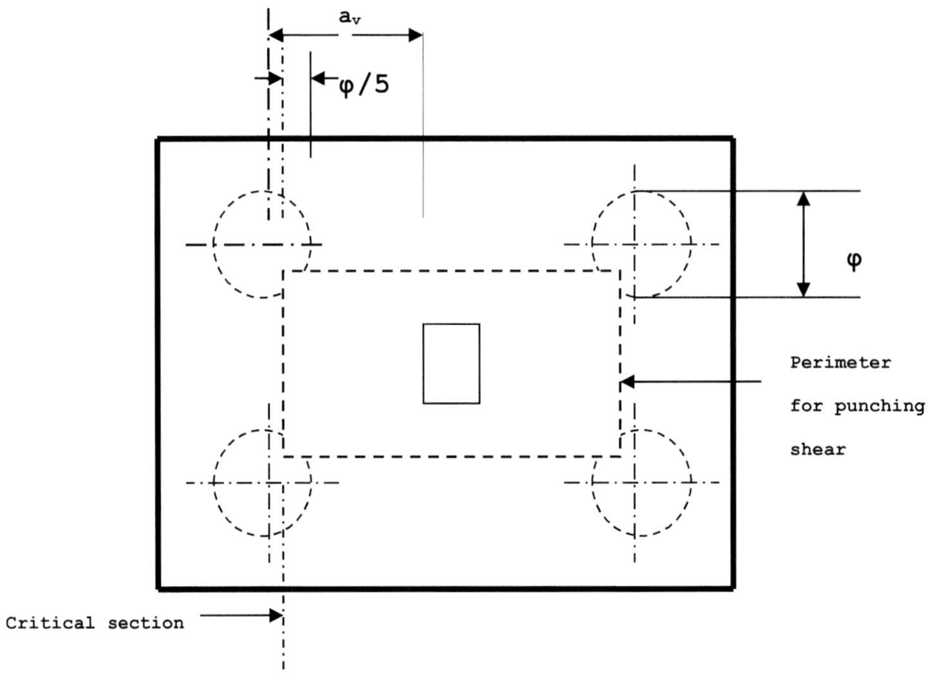

Figure 6.27 Critical section for shear design.

shear stress at the face of the column must not exceed 0.8√fcu or 5 N/mm². Similar requirements are also given in the Hong Kong Code of Practice for Structural Use of Concrete 2013.

(b) Where the pile spacing is less than or equal to three times the pile diameter φ, the enhancement can be applied over the whole of the critical section. Where the spacing is greater, the enhancement can only be applied to strips of width equal to 3 × φ centred on each pile. Minimum stirrups are not required in pile caps where $v < V_c$.

(c) The tension reinforcement should be provided with a full anchorage in accordance with section 3.12.8 of the code. The tension bars must be anchored by bending them up the sides of the pile cap.

5. Punching shear – The following two checks are required:
   1. The design shear stress on the perimeter of the column is not to exceed 0.8√fcu or 7 N/mm².
   2. If the spacing of the piles is greater than three times the pile diameter, punching shear should be checked on the perimeter shown in Figure 6.27.

From the requirements of the code, it can be realized that the shear stresses in the cap design are the major concern and the pile cap thickness is usually designed based on the shear resistance. The flexible strength of the pile cap seems not to be critical in the design, but we understand that the flexible strength of the pile cap will influence the pile load distribution significantly. The method reinforcement design for pile cap is the same as the raft foundation in Chapter 4. The twisting moment M_{xy} should be added to the moment M_x and M_y in order to form the design moment. Similarly, the combined shear V_{max} should be used instead of the design for shear V_x and V_y along the X and Y strips (added together). Some engineers need to develop an additional program (based on Excel or others) to perform such a design, as most commercial programs fail to perform the more rigorous reinforcement design approach. PLATE can however perform the moment and shear as suggested, and also design for punching shear automatically.

There are two major methods for the analysis and design of pile cap. Classically, the pile cap is treated as rigid to distribute the superstructure loadings to piles. After that, the cap can be treated as rigid beams along two different directions for the design of the reinforcement along two directions. The pile cap is then treated as rigid beams with the superstructure loadings with individual pile loads. Such a continuous beam is static in that all pile loads are known, hence the bending and shear force at different sections can be determined easily. This approach is acceptable for small pile caps with a limited number of piles.

For large pile caps with many supporting piles, the use of finite element plate analysis for the pile cap appears unavoidable. For most pile caps, the thickness of the cap is of major importance, so that the pile cap will be a thick plate where the shear deformation cannot be neglected. The problems of some commercial thick plate elements have been discussed in Chapter 1. The piles are usually modelled as spring supports, and the spring stiffness is commonly taken as a column element in Hong Kong. If the effect of soil is considered, Eqs. (6.55) and (6.56) can be adopted, but this option is absent in commercial programs (except the PLATE program by Cheng with isolated springs or coupled springs options). The pile cap is actually a raft with soil support, and some field measurements have indicated that up to 30% of the vertical load can be taken up by the raft action. To model this effect, the soil beneath can be modelled as Winkler's support or elastic half space in PLATE. Cheng has established true 3-D models for several pile caps and the results have been compared with the use of PLATE using a 2-D plate model with coupled springs and elastic half space options. The results by simple 2-D analysis by PLATE are virtually the same as those by the tedious true 3-D analysis.

For the computer modelling of a pile cap problem, there are several considerations and steps. First, a pile cap is usually thick, so that it is more appropriate to use the thick plate model for the analysis. Sometimes, ridiculous results may arise from the thick plate analysis if the mesh design is not good. Some engineers may switch the analysis to the think plate option to avoid such a problem. There are, however, also cases where the thin plate option may give poor results. The improved eight or nine nodes thick plate model proposed by Cheng can however avoid such a problem through the use of advanced formulation. For the modelling of pile, most engineers define the pile by using an elastic spring model with a spring constant defined by the strut model as $K = AE/L$. This option appears to be popular among many engineers, but it neglects the effect of skin friction on piles. Alternatively, Eq. (6.56) can be used which has been built into PLATE but no other commercial programs. With reference to Figure 6.28, PLATE offers an option of coupled analysis, where the f_{ij} will also be formed. The coupled analysis with the plate will give a pseudo 3-D analysis which is good enough for most cases.

There will be different types of loadings under load cases, and there will be different loading combinations to be considered for the analysis and design. The maximum and minimum envelopes for the internal forces with respect to different load cases will be evaluated and used for the reinforcement design. Normally, the bending and punching shear will control the design, while the shear force is usually less critical in many cases. There are also many cases in Hong Kong where the punching shear dictates the design

Figure 6.28 Coupled and uncoupled option for pile group analysis in PLATE.

so that the thickness of the pile cap needs to be increased. For the pile cap, it is normally analysed as a thick plate problem, as the thickness/span ratio will usually exceed 0.1. From the finite element analysis (through the PLATE program by Cheng or other commercial programs), design strip force will be obtained, and the reinforcement for bending and shear can be evaluated based on different concrete design codes. A sample output of the design strip forces is given in Figure 6.29. A typical reinforcement design for pile

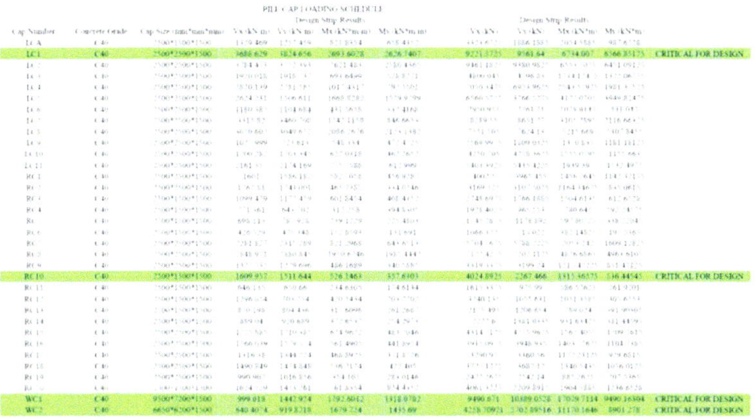

(a) Sample design strip forces output from thick plate finite element analysis for pile cap design

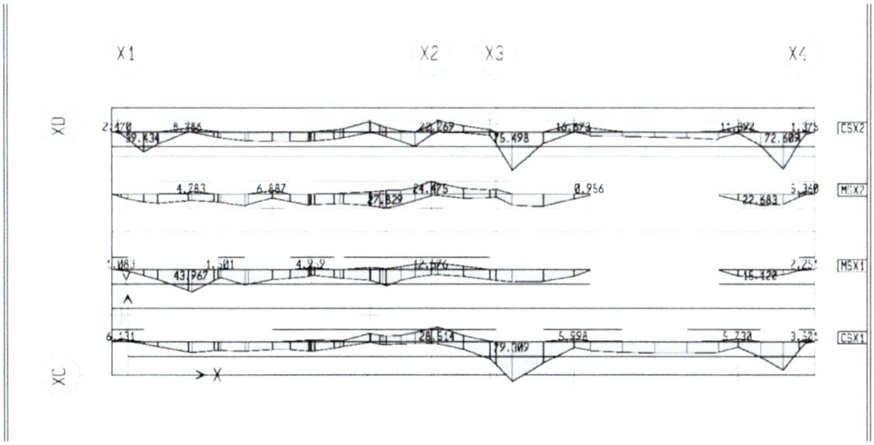

(b) Sample design strip forces from thick plate finite element analysis for pile cap design

Figure 6.29 Sample design strip forces and bending reinforcement design for a pile cap with an opening.

TABLE: Concrete Slab Design 01 - Flexural And Shear Data

Strip	Station	ConcWidth	FTopCombo	FTopMoment	FTopArea	FTopAMin	BotCombo	BotMoment	FBotArea	FBotAMin	VCombo	VForce	VArea	Status	GlobalX	GlobalY	Layer
Text	m	m	Text	kN m	mm2	mm2	Text	kN m	mm2	mm2	Text	kN	mm2/m	Text	m	m	Text
CSA4	0	2.428	1.2DL+1.2LL+1.2WIND+1.2SOIL+1UPLIFT	347.369	5759.94	0	1.4DL+1.6	9751.7865	12025.8	8526.91	1.4DL+1.6	8745.19	6407.26	OK	82.5973	59.5988	A
CSA4	1.14286	2.428	1.4DL+1.4SOIL+1.4WIND+1UPLIFT	115.843	6339.69	8526.91	1.4DL+1.4	2326.4353	4283.75	0	1.4DL+1.6	8745.19	6734.49	OK	82.5973	58.4559	A
CSA4	2.28571	2.428	1.4DL+1.6LL+1Water	5602.75	6741.85	8526.91	1.4DL+1.4	3254.7529	4477.61	8526.91	1.4DL+1.6	4642.89	3326.2	OK	82.5973	57.3131	A
CSA4	3.42857	2.428	1.4DL+1.6LL+1Water	7231.38	8463.89	8526.91	1.4DL+1.4	3276.8932	4232.76	0	1.4DL+1.6	3196.35	3326.2	OK	82.5973	56.1702	A
CSA4	4.57143	2.428	1.4DL+1.6LL+1Water	8382.75	9657.54	8526.91	1.4DL+1.4	2775.8628	3734.87	0	1.4DL+1.6	2464.88	3326.2	OK	82.5973	55.0274	A
CSA4	5.71429	2.428	1.4DL+1.6LL+1Water	9441.57	10800.9	8526.91	1.4DL+1.4	1654.9328	2716.05	0	1.2DL+1.2	3427	3326.2	OK	82.5973	53.8845	A
CSA4	6.85714	2.428	1.4DL+1.6LL+1Water	10899.5	12425.5	8526.91	1.4DL+1.4	10.7498	3387.97	0	1.2DL+1.2	7407.86	4999.59	OK	82.5973	52.7417	A
CSA4	8	2.428	1.4DL+1.6LL+1Water	13512.8	15363.1	8526.91	1.4DL+1.4	0	3103.51	0	1.2DL+1.2	7705.52	5000.28	OK	82.5973	51.5988	A
CSA4	9.14286	2.428	1.4DL+1.6LL+1Water	10750	12194	8526.91	1.4DL+1.4	136.5154	1708.79	0	1.2DL+1.2	3813.42	3326.2	OK	82.5973	50.4559	A
CSA4	10.2857	2.428	1.4DL+1.6LL+1Water	9067.53	10305.4	8526.91	1.4DL+1.4	810.2248	1170.99	0	1.2DL+1.2	2300.74	3326.2	OK	82.5973	49.3131	A
CSA4	11.4286	2.428	1.4DL+1.6LL+1Water	7659.7	8740.41	8526.91	1.4DL+1.4	1960.1441	2420.55	0	1.4DL+1.6	2277.26	3326.2	OK	82.5973	48.1702	A
CSA4	12.5714	2.428	1.4DL+1.6LL+1Water	5956.84	6895.29	8526.91	1.4DL+1.4	2655.3606	3100.45	0	1.4DL+1.6	2735.82	3326.2	OK	82.5973	47.0274	A
CSA4	13.7143	2.428	1.4DL+1.6LL+1Water	3385.96	4145.25	8526.91	1.4DL+1.4	2664.7718	3130.14	0	1.4DL+1.6	3558.49	3326.2	OK	82.5973	45.8845	A
CSA4	14.8571	2.428	1.4DL+1.6LL+1Water	691.217	1443.71	0	1.4DL+1.6	3127.353	4125.91	8526.91	1.4DL+1.6	5116.03	3326.2	OK	82.5973	44.7417	A
CSA4	16	2.428	1.4DL+1.6LL+1Water	11.9336	4435.9	0	1.4DL+1.6	9975.4052	12229.7	8526.91	1.4DL+1.6	9137.47	6819.45	OK	82.5973	43.5988	A
CSA8	0	2.428	1.4DL+1.4SOIL+1.4WIND+1UPLIFT	0	2307.85	0	1.4DL+1.6	12866.543	14742.6	8526.91	1.4DL+1.6	7569.64	4895.05	OK	98.7373	59.5988	A
CSA8	1.14286	2.428	1.4DL+1.4SOIL+1.4WIND+1UPLIFT	149.728	4971.37	8526.91	1.4DL+1.6	4523.7986	5597.27	8526.91	1.4DL+1.6	14123.5	12630.8	OK	98.7373	58.4559	A
CSA8	2.28571	2.428	1.4DL+1.6LL+1Water	7036.06	8325.62	8526.91	1.4DL+1.4	2154.9991	3145.02	0	1.4DL+1.6	7472.92	5339.67	OK	98.7373	57.3131	A
CSA8	3.42857	2.428	1.4DL+1.6LL+1Water	9477.37	10950.7	8526.91	1.2DL+1.2	60.2088	2630.02	0	1.4DL+1.6	5015.84	3326.2	OK	98.7373	56.1702	A
CSA8	4.57143	2.428	1.4DL+1.6LL+1Water	11491.9	13104.1	8526.91	1.4DL+1.4	0	1780.26	0	1.4DL+1.6	3702.86	3326.2	OK	98.7373	55.0274	A
CSA8	5.71429	2.428	1.4DL+1.6LL+1Water	13301.7	15130.5	8526.91	1.4DL+1.4	0	1265.72	0	1.2DL+1.2	3817.01	3326.2	OK	98.7373	53.8845	A
CSA8	6.85714	2.428	1.4DL+1.6LL+1Water	15300.6	17357.5	8526.91	1.4DL+1.4	0	1495.19	0	1.2DL+1.2	7520.12	4744.95	OK	98.7373	52.7417	A
CSA8	8	2.428	1.4DL+1.6LL+1Water	18116.5	20600.1	8526.91	1.4DL+1.4	0	1559.14	0	1.2DL+1.2	7716.01	4663.99	OK	98.7373	51.5988	A
CSA8	9.14286	2.428	1.4DL+1.6LL+1Water	15200.5	17248.5	8526.91	1.2DL+1.2	0	908.238	0	1.2DL+1.2	4199.25	3326.2	OK	98.7373	50.4559	A
CSA8	10.2857	2.428	1.4DL+1.6LL+1Water	12934.2	14677.4	8526.91	1.4DL+1.4	0	1188.05	0	1.2DL+1.2	2914.69	3326.2	OK	98.7373	49.3131	A
CSA8	11.4286	2.428	1.4DL+1.6LL+1Water	10699.2	12170.7	8526.91	1.4DL+1.6	0	1043.16	0	1.4DL+1.6	3326.41	3326.2	OK	98.7373	48.1702	A
CSA8	12.5714	2.428	1.4DL+1.6LL+1Water	8076.42	9278.49	8526.91	1.4DL+1.6	0	1266.78	0	1.4DL+1.6	4211.14	3326.2	OK	98.7373	47.0274	A
CSA8	13.7143	2.428	1.4DL+1.6LL+1Water	4482.63	5354.75	8526.91	1.4DL+1.4	1642.9427	1895.73	0	1.4DL+1.6	5649.95	3341.1	OK	98.7373	45.8845	A
CSA8	14.8571	2.428	1.4DL+1.6LL+1Water	920.03	1613.28	0	1.4DL+1.6	4599.8767	5673.48	8526.91	1.4DL+1.6	8323.25	6271.9	OK	98.7373	44.7417	A
CSA8	16	2.428	1.4DL+1.6LL+1Water	79.4035	3984.95	0	1.4DL+1.6	15492.196	18405.7	8526.91	1.4DL+1.6	14964.9	12741.3	OK	98.7373	43.5988	A
CSB3	0	2.428	1.2DL+1.2LL+1.2WIND+1.2SOIL+1UPLIFT	7626.93	8685.09	8526.91	1.4DL+1.6	133.3467	151.847	0	1.4DL+1.4	5832.22	3456.24	OK	71.1373	59.5988	B
CSB3	1.146	2.428	1.4DL+1.6LL+1Water	3585.61	4083.08	0	1.2DL+1.2	4406.9239	5018.34	8526.91	1.4DL+1.4	5864.02	3490.64	OK	72.2833	59.5988	B
CSB3	2.292	2.428	1.4DL+1.6LL+1Water	4352.8	4956.71	0	1.2DL+1.2	7142.0112	8132.9	8526.91	1.4DL+1.4	3401.13	3326.2	OK	73.4293	59.5988	B
CSB3	3.438	2.428	1.4DL+1.6LL+1Water	5161.84	5878	0	1.2DL+1.2	8568.1944	9756.95	8526.91	1.2DL+1.2	2363.42	3326.2	OK	74.5753	59.5988	B
CSB3	4.584	2.428	1.4DL+1.6LL+1Water	5328.9	6068.23	0	1.2DL+1.2	9343.0264	10639.3	8526.91	1.2DL+1.2	1556.2	3326.2	OK	75.7213	59.5988	B
CSB3	5.73	2.428	1.4DL+1.6LL+1Water	5195.5	5916.33	0	1.2DL+1.2	9343.3138	10639.6	8526.91	1.4DL+1.6	1318.35	3326.2	OK	76.8673	59.5988	B
CSB3	6.876	2.428	1.4DL+1.6LL+1Water	4601.63	5240.07	0	1.2DL+1.2	8754.2461	9968.81	8526.91	1.4DL+1.6	1809.85	3326.2	OK	78.0133	59.5988	B
CSB3	8.022	2.428	1.4DL+1.6LL+1Water	3369.09	3836.52	0	1.2DL+1.2	7657.9443	8720.41	8526.91	1.4DL+1.6	2507.07	3326.2	OK	79.1593	59.5988	B
CSB3	9.168	2.428	1.4DL+1.6LL+1Water	1287.42	1466.04	0	1.4DL+1.6	6300.172	7174.26	8526.91	1.4DL+1.6	3592.12	3326.2	OK	80.3053	59.5988	B
CSB3	10.314	2.428	1.4DL+1.6LL+1Water	497.296	566.291	0	1.4DL+1.6	8555.814	9742.85	8526.91	1.4DL+1.6	6680.08	4373.46	OK	81.4513	59.5988	B

(a) Reinforcement design to consider various load combination

Figure 6.30 Reinforcement layout for a typical pile cap, for a project in Hong Kong.

cap is shown in Figure 6.30. A problem which is not noticed by many engineers is about the design of the reinforcement. Many engineers consider the M_{xx} and M_{yy} for the steel reinforcement design but neglect that the principal moment is usually not along the X or Y directions. Using the principal direction of moment for steel design is, however, not practical, as the principal direction of moment will vary from point to point within a pile cap. The steel reinforcement should be adequate along any direction. To achieve this, the Wood Armer (Law and Cheng 2006) consideration for the steel reinforcement should be adopted, which is provided in Chapter 4.

For the pile/cap connection, in general, the piles will go inside the cap for some distance, depending on the pile types, hinge or rigid connection and other factors. A typical socket H-pile/cap connection is shown in Figure 6.31, and this detail applies to general steel pile/pile cap connections. It should be noted that shear studs are usually installed along the H-pile to increase the friction between the cement grout and the pile, and tests have demonstrated the effectiveness of the shear studs.

While most pile/cap connections have been designed as a pin in the past, a fixed head connection (Figure 6.32) is also commonly used currently for a reduction of lateral displacement.

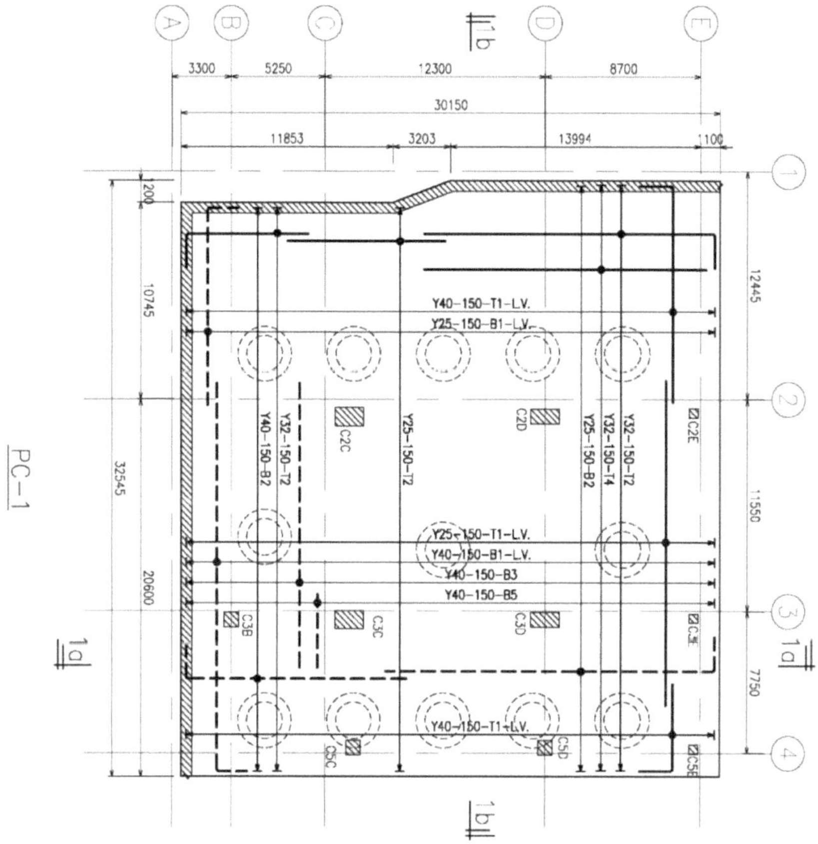

b) Plan view of reinforcement design

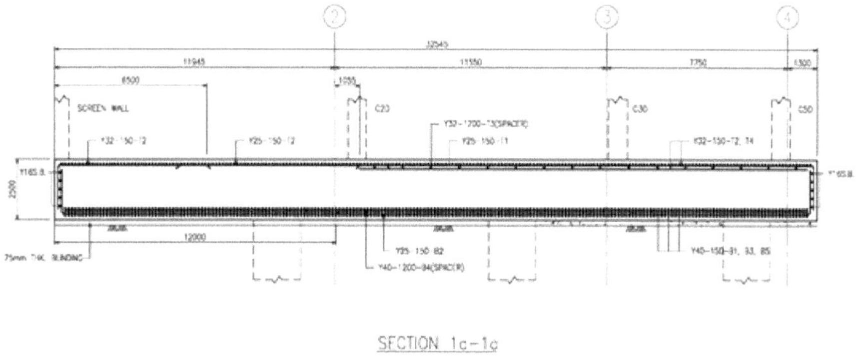

(c) Section 1c-1c for Fig.6.29b

Figure 6.30 (Continued)

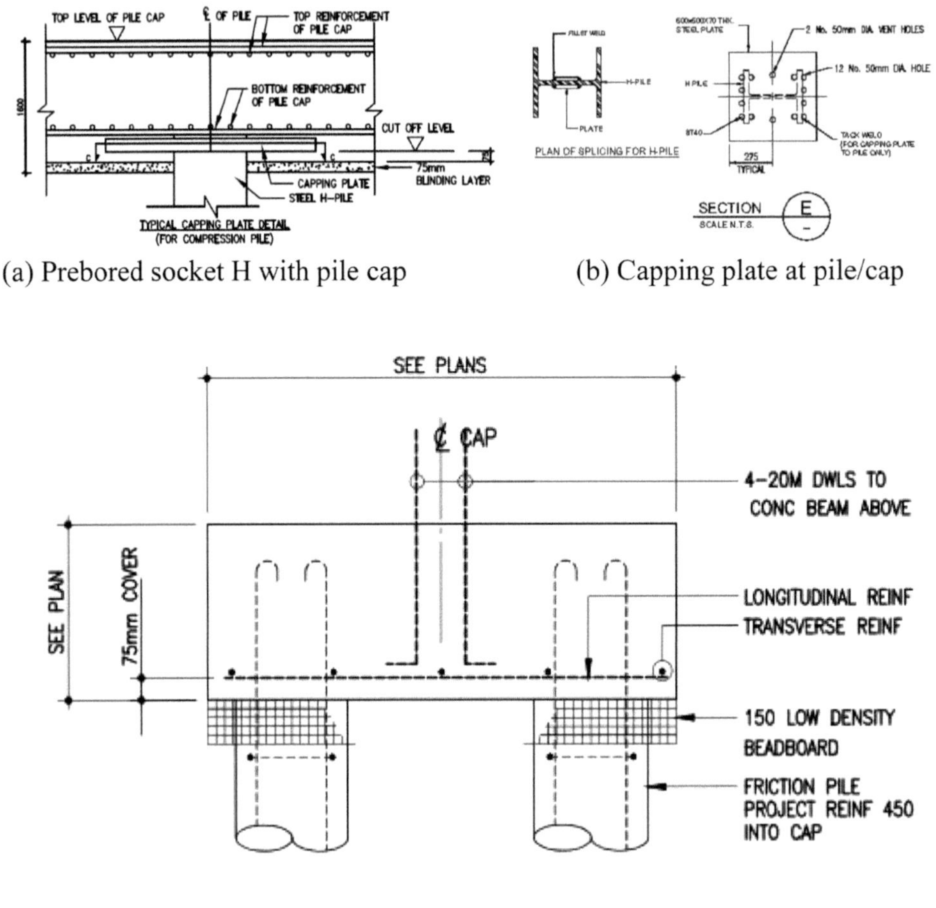

(a) Prebored socket H with pile cap

(b) Capping plate at pile/cap

c) Bored pile with pile cap (free head)

Figure 6.31 Typical pin joint details for an H-pile and a bored pile.

6.9 PILE DRIVING

6.9.1 Dynamic formula

While a pile is being installed by driving, the measured performance including temporary compressions of the pile and soil and the "set" which is the permanent downward displacement of the pile can be used to estimate its static capacity. The mathematical formula for this purpose is called a "dynamic formula". During on-site driving of pile, a guideline must be established to stop further driving of piles, otherwise the piles may be subjected to unreasonably high driving stress and get distorted or buckled. The use of Hiley's formula or a similar equation may also give a negative set during the calculation, hence some engineers adopt a heavier ram for driving to force the set value to become positive. Over-driving is hence not uncommon in Hong Kong and many other countries, and distortion of pile

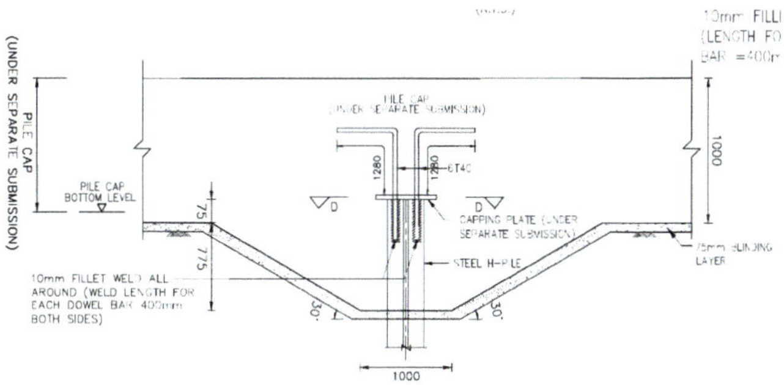

(a) Typical fixed head details for socket H pile

(b) A photo for fixed head connection with H pile

Figure 6.32 Different connection details between the pile and pile cap.

by driving as shown in Figure 6.33 has been recorded at the Kai Tak site in Hong Kong, requiring the whole pile to be extracted for examination. Extreme distortion of pile has also been recorded in Hong Kong when a pile encountered an inclined rockhead, and the pile become U-shaped upon over-driving.

The dynamic formulae are often based on the energy approach. For ease of formulation and computation, simplified assumptions are adopted. The following is a derivation of the Hiley Formula which is a popular one, with reference to Figure 6.34. The main assumptions adopted are listed as follows:

(i) The pile is a rigid body by which the movement of the pile at every portion is identical;
(ii) Energy consumed by temporary compressions of the pile, soil and the pile cushion are based on static deformation, i.e. the deformation follows a linear manner from

Figure 6.33 H-pile buckled due to hard over-driving.

zero to the final deformation of $\left(c_c + c_p + c_q\right)$ at the ultimate load of P_u where c_c, c_p and c_q are the temporary compressions of the cushion, pile and soil respectively. The energy is $0.5P_u\left(c_c + c_p + c_q\right)$;

(iii) The energy consumed during the final set measured as s is $P_u s$.

By reference to Figure 6.34, the energy of the driving hammer before impact is $E_h W_h h$ in which $W_h h$ is the loss in potential energy of the drop hammer of weight W_h dropping from height h and E_h is the hammer efficiency as a factor to discount $W_h h$ due to friction and other factors when the hammer drops.

The velocity of the hammer when it hits the pile cushion, U_h, can be found by equating $E_h W_h h$ to its kinetic energy expressed as $\frac{1}{2}\frac{W_h}{g}U_h^2$. That is

$$U_h = \sqrt{E_h 2gh} \tag{6.80}$$

Let V_h and V_p be the velocities of the hammer and the pile (and helmet) after impact, then during impact, by the Law of Conservation of Momentum (the contribution of the soil is ignored)

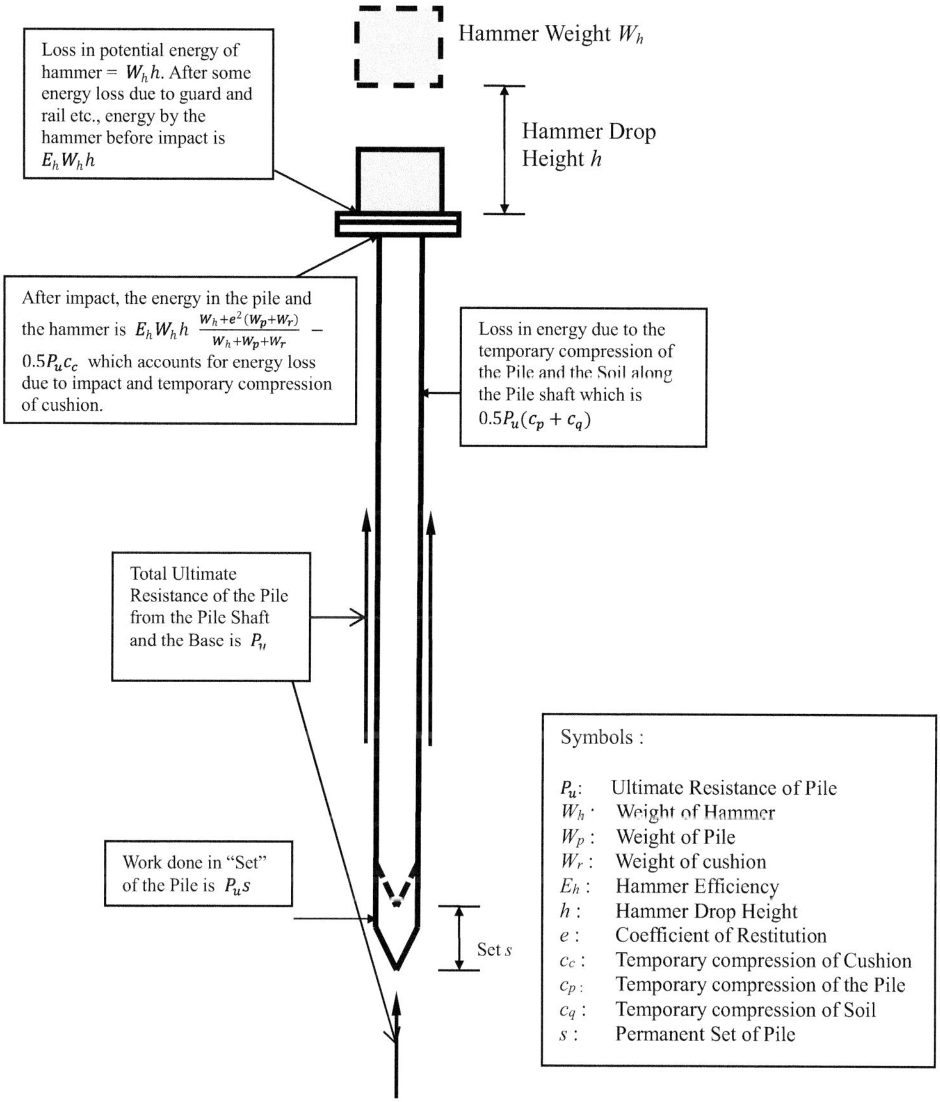

Figure 6.34 Energy transfer process as assumed by the Hiley formula.

$$\frac{W_h}{g}U_h = \frac{W_h}{g}V_h + \frac{W_p + W_r}{g}V_p \Rightarrow W_h U_h = W_h V_h + \left(W_p + W_r\right)V_p \tag{6.81}$$

Assuming the rigid body impact condition and applying Newton's Law of Restitution, we have

$$\frac{V_p - V_h}{0 - U_h} = -e \Rightarrow V_h = V_p - eU_h \tag{6.82}$$

Substituting Eq. (6.82) into Eq. (6.81) and making V_p the subject of the formula

$$V_p = \frac{(1+e)W_h}{W_h + (W_p + W_r)} U_h \tag{6.83}$$

Therefore, the kinetic energy of the pile and the cushion after impact is $\frac{(W_p + W_r)}{2g} V_p^2$.

Substituting V_p by Eqs. (6.83) and (6.80) and simplifying, the kinetic energy of the pile and the cushion after impact is

$$KE_{p+c,\ after\ impact} = E_h W_h h \left\{ \frac{(1+e)^2 W_h (W_p + W_r)}{(W_h + W_p + W_r)^2} \right\} \tag{6.84}$$

The kinetic energy of the hammer after impact is $\frac{W_h}{2g} V_h^2$. Substituting Eq. (6.83) into Eq. (6.82) and then Eq. (6.80) and simplifying,

$$V_h = \frac{W_h - e(W_p + W_r)}{W_h + W_p + W_r} \sqrt{E_h 2gh} \tag{6.85}$$

Therefore, the kinetic energy of the hammer after impact is

$$KE_{h,\ after\ impact} = E_h W_h h \left\{ \frac{W_h - e(W_p + W_r)}{W_h + W_p + W_r} \right\}^2 \tag{6.86}$$

Adding the energies of the pile with cushion and that of the hammer in Eq. (6.86), the total kinetic energy is

$$KE_{total,\ after\ impact} = E_h W_h h \left\{ \frac{W_h + e^2 (W_p + W_r)}{W_h + W_p + W_r} \right\} \tag{6.87}$$

This total energy will then be eventually transferred to the pile and dissipated through work done against soil resistances and elastic compression of the pile and cushion. Based on the assumptions (ii) and (iii) above, the total work done through dissipation in the pile and soil and set is

$$W_{p+s} = 0.5P_u\left(c_c + c_p + c_q\right) + P_u s \tag{6.88}$$

Assuming $KE_{total,\ after\ impact} = W_{p+s}$ and making P_u the subject of the formula, the Hiley formula in the following expression is arrived at:

$$P_u = \frac{E_h W_h h}{s + 0.5\left(c_c + c_p + c_q\right)}\left\{\frac{W_h + e^2\left(W_p + W_r\right)}{W_h + W_p + W_r}\right\} \tag{6.89}$$

From Eq. (6.89) we may consider that E_h caters for the discount of the pile hammer energy during falling and the factor $\frac{W_h + e^2\left(W_p + W_r\right)}{W_h + W_p + W_r}$ accounts for the discount of energy through impact. Besides the Hiley's formula, there are more than 10 similar dynamic formulae which are used by engineers.

Therefore, at the pile's "final set", it should be noted that $c_p + c_q$ and s can be measured during the final set of the pile. c_c is comparatively small. It is often assumed to be constant, ranging from 4 to 6 mm. This value is either assumed or can nowadays be measured by video camera. Other quantities can be also directly measured except E_h, which can be calibrated by a modern device as modern techniques can measure the total energy delivered to the pile. Previously, E_h was assumed to take a value of 0.7 normally. A factor of safety of 2 can be adopted as the pile capacity has been checked by the dynamic formula. Otherwise, 3.0 is adopted when the static formula using classical soil mechanics principle is used. If a hydraulic hammer is used, a hydraulic hammer factor is introduced to the upper term of Hiley's formula in Hong Kong, which is taken to be 0.6 for pile driving with a pile cushion and 0.7 for pile driving without a pile cushion. However, a pile cushion is used in practically all hydraulic hammers. If a pile cushion is not used, the factor can be taken as 0.7, and the term C_c can be omitted from Hiley's formula. The hydraulic hammer factor is usually taken to be independent of the length of pile in a typical design.

6.9.1.1 Limitations of the dynamic formula

1. In the conservation of momentum, Eq. (6.81), the whole weight of the pile cannot be taken, as it takes time for the force wave to travel to the pile base. The pile is not a rigid body upon impact, so only part of the pile experiences the force and is in motion.
2. The coefficient of restitution and the efficiency of the piling rig are taken as constants in the formula, but they are actually not constants between different blows as demonstrated by some field tests. The energy absorption of the head is also not constant and can vary with time.
3. Dynamic pile resistance is different from static resistance.

4. The rate at which the soil is sheared is not accounted for during pile driving. The high strain rates in cohesive soils during pile penetration can cause the viscous resistance of the soil to be considerably greater than the static capacity of the pile. Poskitt (1991) shows that without considering soil damping, the driving resistance can be overestimated by several times.
5. It only considers the hammer ram and the pile as concentrated masses in the transfer of energy. In fact, the driving system includes many other elements such as the anvil, helmet, and hammer cushion. Their presence also influences the magnitude and duration of peak force being delivered to the pile.
6. The energy transfer ratio is found to vary a great deal, and is largely controlled by the length, material of pile and hammer type. For a short pile, the dynamic formula tends to overestimate the energy transfer, while it will underestimate the energy transfer ratio for a long steel pile or hydraulic hammer.

In most cases, Hiley's formula will be conservative as compared with the static load test, and a factor of safety of 2 or more is possible arising from the use of Hiley's formula. If the set S as calculated is negative, some engineers apply the wave equation in the analysis, while other engineers may simply increase W_r to make S become positive. The consequence of increasing W_r is that over-driving and buckling of pile may take place, which has been observed by the authors in many projects. The use of the wave equation is a more rational approach under these circumstances.

Modern devices such as the pile driving analyser (PDA) can measure the energy delivered to the pile after impact as EMX, then Eq. (6.90) can be reduced to

$$P_u = \frac{EMX}{s + 0.5\left(c_p + c_q\right)} \tag{6.90}$$

It is interesting to note that EMX is found to be not sensitive to the length of pile. There are also some cases where the energy transfer ratio exceeds 1.0 due to various possible factors. Despite the understanding that the assumptions for the dynamic formula as listed above are too crude, the dynamic formula has been used for quite a long time due to its relative simplicity. The applicability of the formula is proved rather by a subsequent loading test and the accumulation of experience. Code requirements in relation to the applicability of the formula often involve limited range of the $c_p + c_q$ such as not less than 25 mm and not greater than 50 mm. In Hong Kong, after years of experience, the set table shall comprise the following criteria for $c_p + c_q$:

(i) Set not less than 25 mm unless it can be proven to be "driven to refusal", i.e. less than 10 mm per 10 blows which is likely on hard rock;
(ii) Not greater than 50 mm, in the case that it is greater than 50 mm by Eq. (6.89), keep to 50 mm;
(iii) $\frac{\left(c_p + c_q\right)}{L} \leq 1.15$ where L is the pile length in m, $c_p + c_q$ in mm.

An example of a set table is given in Table 6.8

Table 6.8 Set table by the Hiley formula for the Hong Kong Foundation Code

Pile Length	c_p+c_q																			
	6	*7*	*8*	*9*	*10*	*11*	*12*	*13*	*14*	*15*	*16*	*17*	*18*	*19*	*20*	*21*	*22*	*23*	*24*	*25*
15	–	–	–	–	–	50	50	50	50	50	50	50	–	–	–	–	–	–	–	–
16	–	–	–	–	50	50	50	50	50	50	50	50	50	–	–	–	–	–	–	–
17	–	–	–	–	50	50	50	50	50	50	50	50	50	50	–	–	–	–	–	–
18	–	–	–	–	50	50	50	50	50	50	50	50	50	50	46	–	–	–	–	–
19	–	–	–	50	50	50	50	50	50	50	50	50	50	50	45	40	–	–	–	–
20	–	–	–	50	50	50	50	50	50	50	50	50	50	48	43	38	33	–	–	–
21	–	–	–	50	50	50	50	50	50	50	50	50	50	47	42	37	32	27	–	–
22	–	–	–	50	50	50	50	50	50	50	50	50	50	45	40	35	30	25	–	–
23	–	–	50	50	50	50	50	50	50	50	50	50	49	44	39	34	29	–	–	–
24	–	–	50	50	50	50	50	50	50	50	50	50	47	42	37	32	27	–	–	–
25	–	–	50	50	50	50	50	50	50	50	50	50	46	41	36	31	26	–	–	–
26	–	50	50	50	50	50	50	50	50	50	50	50	45	40	35	30	–	–	–	–
27	–	50	50	50	50	50	50	50	50	50	50	49	44	39	34	29	–	–	–	–
28	–	50	50	50	50	50	50	50	50	50	50	47	42	37	32	27	–	–	–	–
29	–	50	50	50	50	50	50	50	50	50	50	46	41	36	31	26	–	–	–	–
30	50	50	50	50	50	50	50	50	50	50	50	45	40	35	30	–	–	–	–	–
31	50	50	50	50	50	50	50	50	50	50	49	44	39	34	29	–	–	–	–	–
32	50	50	50	50	50	50	50	50	50	50	48	43	38	33	28	–	–	–	–	–
33	50	50	50	50	50	50	50	50	50	50	46	41	36	31	26	–	–	–	–	–
34	50	50	50	50	50	50	50	50	50	50	45	40	35	30	25	–	–	–	–	–
35	50	50	50	50	50	50	50	50	50	49	44	39	34	29	–	–	–	–	–	–
36	50	50	50	50	50	50	50	50	50	48	43	38	33	28	–	–	–	–	–	–
37	50	50	50	50	50	50	50	50	50	47	42	37	32	27	–	–	–	–	–	–
38	50	50	50	50	50	50	50	50	50	46	41	36	31	26	–	–	–	–	–	–
39	50	50	50	50	50	50	50	50	50	45	40	35	30	–	–	–	–	–	–	–
40	50	50	50	50	50	50	50	50	49	44	39	34	29	–	–	–	–	–	–	–
41	50	50	50	50	50	50	50	50	48	43	38	33	28	–	–	–	–	–	–	–
42	50	50	50	50	50	50	50	50	47	42	37	32	27	–	–	–	–	–	–	–
43	50	50	50	50	50	50	50	50	46	41	36	31	26	–	–	–	–	–	–	–
44	50	50	50	50	50	50	50	50	45	40	35	30	–	–	–	–	–	–	–	–

Example 1: Final set calculation using a drop hammer for a precast pretensioned concrete pile (a case in Hong Kong)

The ultimate bearing capacity of pile $R = Wh^{*}H^{*}e^{*}N/(S + C/2)$

where R = ultimate pile capacity = design working load × 2 = 5400 kN;
efficiency $N = (Wh + P \times n^2)/(Wh + P)$
where Wh = weight of hammer = 120 kN; n = coefficient of restitution = 0.4
e = efficiency of free fall = 0.8; H = free fall height of hammer = 2 m
P2 = weight of helmet = 4.8 kN (no follower)
S = final set (mm) per blow
$C = C_c + C_p + C_q$ (mm)
C_c = temporary compression of pile head = 6.0 mm
Assume $C_p + C_q$ = on site temporary compression of pile and quake = 20 mm

Type of pile: 500 diameter × 125 mm thick prestressed precast concrete pile
Length of pile (L_p) = 30 m H = 2.0 m
Unit weight of pile = 3.8 kN/m P1 = 30 × 3.8 =114 kN
P = P1 + P2 = 114 + 4.8 = 118.8 kN
$N = (120 + 0.4^2 \times 118.8)/(120 + 118.8) = 0.582$
$R = 5400 = (120 \times 2 \times 0.8 \times 0.582) \times 1000/[S + 0.5\ (6 + 20)]$
s = 7.69 mm per blow or 77 mm per 10 blows

For a drop hammer, the height of free fall can be controlled easily and hence H is used. Since the pile length and temporary compression are not known in advance, different combinations of these values are considered to generate a set table for site control. Furthermore, the drop height may not be 2.0 m on site, hence different tables similar to Table 6.9 with different drop heights should be prepared.

Example 2: The ultimate bearing capacity of a steel pile by diesel hammer (a case in Hong Kong)

$$R = E * e * N/(S + C/2)$$

Table 6.9 A set table for driving of precast pretensioned concrete pile using a drop hammer for a project in Hong Kong (50 mm criterion is not applied)

Final set table for set per 10 blows without follower																	
Pile length (m)	*Temporary compression ($C_p + C_q$) mm*																
	15	*16*	*17*	*18*	*19*	*20*	*21*	*22*	*23*	*24*	*25*	*26*	*27*	*28*	*29*	*30*	*31*
20	130	125	17	115	110	105	100	95	90	85	80	75	70	65	60	55	50
21	127	122	120	112	107	102	97	92	87	82	77	72	67	62	57	52	47
22	124	119	117	109	104	99	94	89	84	79	74	69	64	59	54	49	44
23	121	116	114	106	101	96	91	86	81	76	71	66	61	56	51	46	41
24	118	113	111	103	98	93	88	83	78	73	68	63	58	53	48	43	38
25	115	110	108	100	95	90	85	80	75	70	65	60	55	50	45	40	35
26	112	107	105	97	92	87	82	77	72	67	62	57	52	47	42	37	32
27	109	104	102	94	89	84	79	74	69	64	59	54	49	44	39	34	29
28	107	102	99	92	87	82	77	72	67	62	57	52	47	42	37	32	27
29	104	99	97	89	84	79	74	69	64	59	54	49	44	39	34	29	24
30	102	97	94	87	82	77	72	67	62	57	52	47	42	37	32	27	22
31	100	95	92	85	80	75	70	65	60	55	50	45	40	35	30	25	20
32	97	92	90	82	77	72	67	62	57	52	47	42	37	32	27	22	17
33	95	90	87	80	75	70	65	60	55	50	45	40	35	30	25	20	15
34	93	88	85	78	73	68	63	58	53	48	43	38	33	28	23	18	13
35	91	86	83	76	71	66	61	56	51	46	41	36	31	26	21	16	11
36	89	84	81	74	69	64	59	54	49	44	39	34	39	34	19	14	9
37	87	82	79	72	67	62	57	52	47	42	37	32	27	22	17	12	7
38	85	80	77	70	65	60	55	50	45	40	35	30	25	20	15	10	5
39	83	78	75	68	63	58	53	48	43	38	33	28	23	18	13	8	3
40	81	76	73	66	61	56	51	46	41	36	31	26	21	16	11	6	1

where R = ultimate pile capacity = design working load × 2 = 5900 kN

$N = (W + P \times n^2)/(W + P)$
W = weight of hammer = 100 kN; n = coefficient of restitution = 0.32
e = efficiency of hammer = 0.9; P2 = weight of helmet = 4.8 kN
S = final set (mm) per blow
C_c = temporary compression of pile head = 2.5 mm
$C_p + C_q$ = on-site temporary compression of pile and quake = 32 mm
Type of pile: 305 × 305 × 180 kg/m bearing H-pile grade 55C
Length of pile (L_p) = 28 m Blow count: 38 blows/min
Energy output: = $100 \times (66/38)^2 = 301.66$ kNm
Unit weight of pile = 1.766 kN/m P1 = 28 × 1.766 = 49.44 kN
P = P1 + P2 = 49.44 + 4.8 = 54.24 kN
$N = (100 + 0.32^2 \times 54.24)/(100 + 54.24) = 0.684$
$R = 5900 = (301.66 \times 0.9 \times 0.684) \times 1000/[S + 0.5\ (2.5 + 32)]$
S = 14.22 mm per blow or 142 mm per 10 blows

The blow count is the number of blows per minute for the diesel hammer strike. Since H is not easily defined for a diesel hammer, while the blow count can be measured easily, the use of blow count is now favoured for diesel hammer strike. Again, a series of set tables similar to that in Table 6.10 should be prepared, based on a different blow count.

In Figure 6.35, the typical pile set during pile driving is taken by a labourer. Alternatively, a laser has been used in some projects, but this is usually less popular due to higher setup requirements and cost.

In Hong Kong, the following guidelines should be adopted in preparing the design of the final set table:

(a) The design final set should not be less than 25 mm per 10 blows;
(b) The design final set should not be greater than 100 mm per 10 blows;
(c) When the calculated final set is between 50–100 mm per 10 blows, the design final set should be taken as 50 mm per 10 blows;
(d) The calculated final set value should be discarded if the corresponding $(c_p+c_q)/L$ is greater than 1.15, where c_p and c_q are in mm and L is in m;
(e) For a drop hammer, the efficiency of the hammer should not be greater than 0.7 unless verified otherwise by testing (although many test results give an efficiency greater than 90%); and
(f) Where the pile is driven to bedrock, the design final set should be taken as 10 mm per 10 blows.

The temporary compression of the hammer cushion (C_c) should be taken as not less than 5 mm when a plastic cushion not exceeding 200 mm thickness is used. It is interesting to find that there are many cases where the set is never achieved despite heavy driving for a long time and depth. Upon static load tests or PDA tests, these piles however give adequate pile capacities. If a pile fails in the driving formula but complies with the pile capacity using a PDA test, should this pile be considered as acceptable? Currently, in Hong Kong,

Table 6.10 A set table for driving of precast pretensioned concrete pile using a diesel hammer for a project in Hong Kong (50 mm criterion is not applied)

Final set table for set per 10 blows									*(Blow count 38/min)*								
Pile length (m)	*Temporary compression ($C_p + C_q$) mm*																
	27	*28*	*29*	*30*	*31*	*32*	*33*	*34*	*35*	*36*	*37*	*38*	*39*	*40*	*41*	*42*	*43*
20	194	189	184	179	174	169	164	159	154	149	144	139	134	129	124	119	114
21	191	186	181	176	171	166	161	156	151	146	141	136	131	126	121	116	111
22	187	182	177	172	167	162	157	152	147	142	137	132	127	122	117	112	107
23	184	179	174	169	164	159	154	149	144	139	134	129	124	119	114	109	104
24	180	175	170	165	160	155	150	145	140	135	130	125	120	115	110	105	100
25	177	172	167	162	157	152	147	142	137	132	127	122	117	112	107	102	97
26	174	169	164	159	154	149	144	139	134	129	124	119	114	109	104	99	94
27	171	166	161	156	151	146	141	136	131	126	121	116	111	106	101	96	91
28	167	162	157	152	147	142	137	132	127	122	117	112	107	102	97	92	87
29	164	159	154	149	144	139	134	129	124	119	114	109	104	99	94	89	84
30	161	156	151	146	141	136	131	126	121	116	111	106	101	96	91	86	81
31	159	154	149	144	139	134	129	124	119	114	109	104	99	94	89	84	79
32	156	151	146	141	136	131	126	121	116	111	106	101	96	91	86	81	76
33	153	148	143	138	133	128	123	118	113	108	103	98	93	88	83	78	73
34	150	145	140	135	130	125	120	115	110	105	100	95	90	85	80	75	70
35	148	143	138	133	128	123	118	113	108	103	98	93	88	83	78	73	68
36	145	140	135	130	125	120	115	110	105	100	95	90	85	80	75	70	65
37	142	137	132	127	122	117	112	107	102	97	92	87	82	77	72	67	62
38	140	135	130	125	120	115	110	105	100	95	90	85	80	75	70	65	60
39	137	132	127	122	117	112	107	102	97	92	87	82	77	72	67	62	57
40	135	130	125	120	115	110	105	100	95	90	85	80	75	70	65	60	55

some engineers do not accept such piles, while there are also some engineers who accept such piles. Of course, the final acceptance of a pile is based on the static load test, but this is more than 50 times the cost of a PDA test so that static load tests to all piles on site would be very expensive and time consuming. Up to now, there are still many engineers who have doubts about the acceptability of PDA tests, including the authors. For the CASE method, the assumption of damping lumped at the pile tip is a highly simplified model. On the other hand, the CAPWAP or a similar approach appears to be very scientific. Since Cheng has programmed a computer code for such backward analysis, the authors' believe that the multi-solution problems and the use of the soil models are also uncertainties for which there are no perfect solutions. This is further complicated by the fact that very few engineers fully understand the theory and limitations of the wave matching process, and the authors observe that many engineers perform only a routine CAPWAP analysis which may give ridiculous soil parameters but a good signal matching in the PDA reports.

6.9.2 Wave equation method

To overcome the shortcomings of the dynamic formula as discussed above, an analytical method based on the "wave theory" on an elastic body has been developed. This analysis takes into account the phenomenon that stress waves (created in the pile by impact

Type of Hammer	: Drop Hammer	Existing Ground Level	: 5.6mPD
Hammer Mark	: KE87	Pile Length	: 42m
Wt. of Hammer	: 15T	Embedded Length	: 37.53m
Fall of Hammer	: 3000mm	No. of Blows	: 10 blows
Pile No.	: 3P52	C_p+C_q :	: 33mm
Size and Grade	: 305x305x223 (S450	Final set	: 4mm

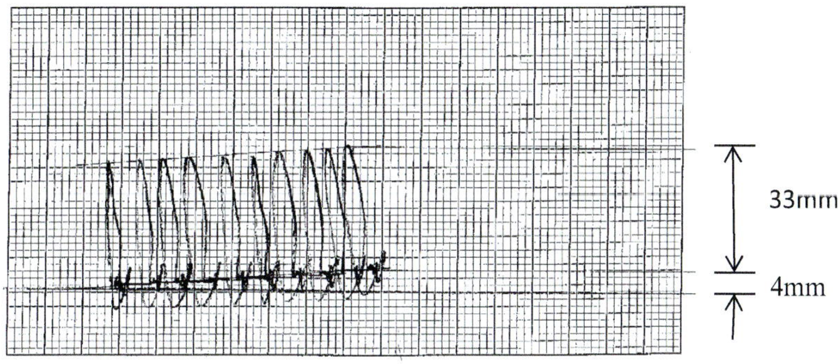

(a) Actual measurement on site for the set during driving

(b) Pile set measurement

Figure 6.35 Set measurement for driving of a steel H-pile using a diesel hammer.

of the driving hammer) of varying magnitudes move down the length of the pile at the speed of sound. The pile is not stressed equally and simultaneously along its length, as is assumed in conventional dynamic formulae. In addition, the wave equation analysis also considers individually the effects of various factors in the driving process such as pile and hammer characteristics, cushion stiffness and soil damping effects by inserting appropriate parameters. The ultimate static load capacity of the pile offered by the soil can be estimated more rationally.

The basic wave equation for pile driving follows a partial differential equation. The equation is a one-dimensional (geometrically) solution describing the displacement of any point of the pile at co-ordinate x and at time t which reads:

$$\frac{\partial^2 u}{\partial t^2} = c^2 \frac{\partial^2 u}{\partial x^2} + S(x,t) \tag{6.91}$$

where u is the displacement of a point in the pile from its original position;

x is the co-ordinate of a point on the pile, often taken as distance from the pile top;
t is time;
c is the wave speed in the pile which is equal to $\sqrt{E/\rho}$ where E and ρ are the Young's modulus and the density of the material making up the pile, respectively; and $S(x,t)$ is the term representing the soil resistance force at x and t comprising both the static and dynamic components.

The equation can be derived as follows with reference to Figure 6.36 where F is the force at the element at ordinate x and at time t. $R(x,t)$ is the resistance by the soil which can be friction along the pile shaft and end-bearing at the pile tip.

Let A be the cross-sectional area of the pile and ρ be the density of the pile. By Newton's second law, we may list

$$F + \partial F - F + R(x,t)\partial x = -\rho A \partial x \frac{\partial^2 u}{\partial t^2}$$

$$\frac{\partial F}{\partial x} = -\rho A \frac{\partial^2 u}{\partial t^2} - R(x,t) \tag{6.92}$$

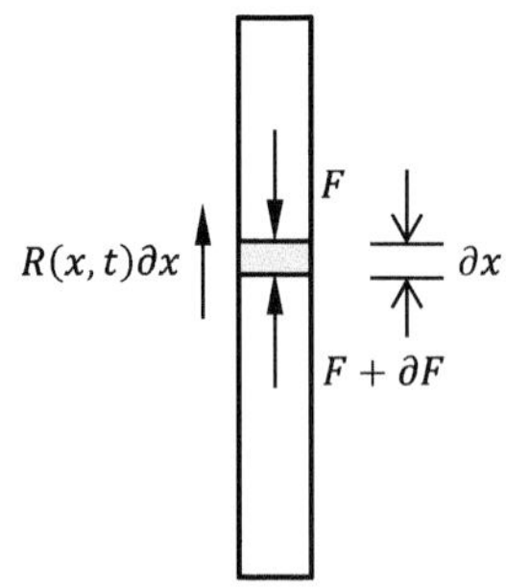

Figure 6.36 Schematic representation of a pile element for derivation of the wave equation.

However, by elasticity, F is also related to the deformation of the pile by

$$F = -AE\varepsilon = -AE\frac{\partial u}{\partial x} \Rightarrow \frac{\partial F}{\partial x} = -AE\frac{\partial^2 u}{\partial x^2} \tag{6.93}$$

$$\text{Equating} - \rho A\frac{\partial^2 u}{\partial t^2} - R(x,t) = -AE\frac{\partial^2 u}{\partial x^2} \Rightarrow \frac{\partial^2 u}{\partial t^2} = \frac{E}{\rho}\frac{\partial^2 u}{\partial t^2} + S(x,t) \tag{6.94}$$

By putting $c = \sqrt{\frac{E}{\rho}}$ which is the wave speed in the pile and $S(x,t) = -R(x,t)/\rho A$ is a term related to the resistance of the pile by soil/rock on its shaft or tip.

$$\frac{\partial^2 u}{\partial t^2} = c^2\frac{\partial^2 u}{\partial t^2} + S(x,t) \tag{6.95}$$

By Eq. (6.95), the displacement of the pile u at any ordinate x and t can be solved if all others parameters including $S(x,t)$ are known. However, in real applications, it is the other way round by which u and its derivatives, which are velocity and acceleration, are measured and by making assumptions on parameters such as the form of distribution of resistance, soil spring values, quake (elastic limit of the soil), damping constants of soil, etc., the soil resistance of the soil/rock can be calculated. An important point to note is the difference between the velocity of the force and the velocity of the particle. The velocity of the force given by C is large, ranging from 4000 to 6000 m/s typically. The actual velocity of the pile material is however very small.

For ease of analysis, Smith (1962) suggested an approach to solve the equation in which the pile is idealized into a series of masses connected by elastic springs and "dashpots", as illustrated in Figure 6.38. Each of the masses takes soil resistances, both static (dependent on displacement of soil) and dynamic (dependent on the velocity of the mass when the pile is struck by the hammer) components. A pile capacity versus the final set values can be obtained. The analysis is tedious and the iterative computation usually requires a computer program to derive the solution. A typical output by Cheng is shown in Figure 6.37, which is termed Forward by Cheng. This program is called Forward as there is a Backward program which works backward. The Backward program by Cheng accepts the PDA results and uses one of the curves to compute the other one, with some assumed soil/pile parameters which can be varied. These parameters are adjusted until the computed results match well the actual signal. Three versions of Backward programs have been developed by Cheng. The first version adopts the simplex method, while the second simulated annealing method can perform the signal matching and avoid the problem of local minimum normally, at the expense of more computer time. The third version adopts the generalized reduced gradient method, which is faster in computation, but the method is limited to local minimum close to the initial solution as defined by the engineers. It should be noted that there are actually multiple solutions in the optimization analysis, which correspond to different local minima. In general, the global minimum solution should be used, but it

1/DELTEE =	6000.	NOP(1) = 1
MP =	9	NOP(2) = 1
VELM =	6.40	NOP(3) = 1
NSTOP=	200	NOP(4) = 3
RUT =	222.4	NOP(5) = 1
FUP =	11.1	NOP(6) = 1
MO =	7	NOP(7) = 2
QSIDE =	0.0025	NOP(8) = 1
QPOINT =	0.0025	NOP(9) = 1
SIDEJ =	0.6562	NOP(10) = 1
POINTJ =	0.0328	NOP(11) = 1
AREA =	0.09	NOP(12) = 1
IPRINT =	5	

M	WAM(M)	XKAM(M)	RUM(M)	GAMMA(M)	EEM(M)	SLACK(M)	VEL(M)	Q(M)	SOILJ(M)	A(M)
1	14.6791	463385.5	0.0	579.1	0.900000	0.000000	6.401	0.002540	0.656200	0.093
2	3.6031	891220.3	0.0	0.0	0.500000	0.000000	0.000	0.002540	0.656200	0.093
3	4.4482	1134821.7	0.0	0.0	0.500000	0.000000	0.000	0.002540	0.656200	0.093
4	6.6723	630456.5	0.0	-1.0	1.000000	0.000000	0.000	0.002540	0.656200	0.093
5	6.6723	630456.5	0.0	-1.0	1.000000	0.000000	0.000	0.002540	0.656200	0.093
6	6.6723	630456.5	0.0	-1.0	1.000000	0.000000	0.000	0.002540	0.656200	0.093
7	6.6723	630456.5	70.4	-1.0	1.000000	0.000000	0.000	0.002540	0.656200	0.093
8	6.6723	630456.5	70.4	-1.0	1.000000	0.000000	0.000	0.002540	0.656200	0.093
9	6.6723	0.0	70.4	0.0	0.000000	0.000000	0.000	0.002540	0.656200	0.000
10	0.0000	0.0	11.1	0.0	0.000000	0.000000	0.000	0.002540	0.032810	0.000

T	F 1	F 2	F 3	F 5	F 7	F 8	D 1	D 2	D 3	D 5	D 7	D 8
5	21845.83	5458.69	476.39	230.32	210.05	117.97	0.006	0.001	0.001	0.001	0.001	0.001
10	28532.09	24942.18	9101.25	248.53	210.05	117.97	0.010	0.004	0.002	0.001	0.001	0.001
15	26052.45	33330.53	29773.50	783.14	210.96	117.99	0.013	0.008	0.004	0.001	0.001	0.001
20	21841.59	24616.99	22480.66	4522.51	241.65	119.41	0.015	0.010	0.007	0.001	0.001	0.001
25	13737.02	6347.00	21001.55	14705.35	586.36	148.50	0.016	0.013	0.010	0.003	0.001	0.001
30	6233.58	20334.19	19745.37	25552.52	2426.36	411.70	0.017	0.015	0.012	0.006	0.001	0.001
35	6233.58	3781.65	2071.91	27092.46	7805.84	1707.01	0.017	0.017	0.014	0.009	0.002	0.001
40	6233.58	4391.88	15527.51	19983.63	16778.56	5484.03	0.017	0.018	0.015	0.011	0.004	0.002
45	6233.58	14606.26	7327.92	12372.42	24890.38	12044.23	0.017	0.019	0.016	0.012	0.007	0.003
50	6233.58	0.00	2361.95	10170.73	25755.90	18151.15	0.017	0.020	0.017	0.013	0.010	0.006
55	6233.58	10949.51	14563.21	11375.49	17465.37	17937.76	0.016	0.021	0.018	0.014	0.011	0.009
60	6233.58	9714.84	2769.44	10259.83	5687.67	7885.13	0.016	0.022	0.019	0.015	0.012	0.012
65	5182.16	0.00	7195.06	7449.90	-2870.44	-6752.62	0.014	0.022	0.019	0.016	0.014	0.014
70	3183.26	13483.18	9447.99	5438.40	-7415.79	-14737.33	0.013	0.023	0.020	0.016	0.015	0.017
75	1184.35	255.23	0.00	2171.12	-10271.71	-10502.70	0.012	0.023	0.020	0.017	0.018	0.019

MAXIMUM COMPRESSIVE AND TENSILE FORCES OVER AREAS IN THE SEGMENTS

SEGMENT	TIME	CCMP	TIME	TENS	TIME	D MAX
1	10	28532.1	200	0.0	41	0.0173
2	15	33330.5	200	0.0	71	0.0228
3	18	35375.8	200	0.0	200	0.0232
4	24	27890.7	93	-13442.0	199	0.0465
5	33	27854.6	91	-14656.0	200	0.0454
6	41	28612.2	82	-10683.3	200	0.0436
7	48	26585.1	154	-11433.4	200	0.0426
8	52	19163.4	71	-14823.3	200	0.0420
9					187	0.0420

TOTAL SOIL RESISTANCE	=	222.41000000
PERMANENT SET OF PILE	=	0.03948796
PILE WEIGHT	=	40.03380000

Figure 6.37 Typical output from the solution of a wave equation – forward analysis.

is sometimes found that the corresponding soil parameters may not be acceptable. There should be sufficient pile penetration to ensure that the pile capacity is activated, but there are many cases where the pile penetration is small so that the test results can only reflect the activated capacity instead of the pile capacity.

There are popular methods for analysing pile capacity by a dynamic method:

(I) Use of a pile driving analyser (PDA) with the wave equation method

This method depends on the measurement of pile forces and accelerations by such electronic devices as transducers and accelerometers, as shown in Figure 6.39, which are installed at the pile head during a hammer strike. By applying wave theory analysis to these measured quantities, the load-carrying capacity, soil parameters, energy transfer, pile driving stresses and integrity of the pile can be estimated.

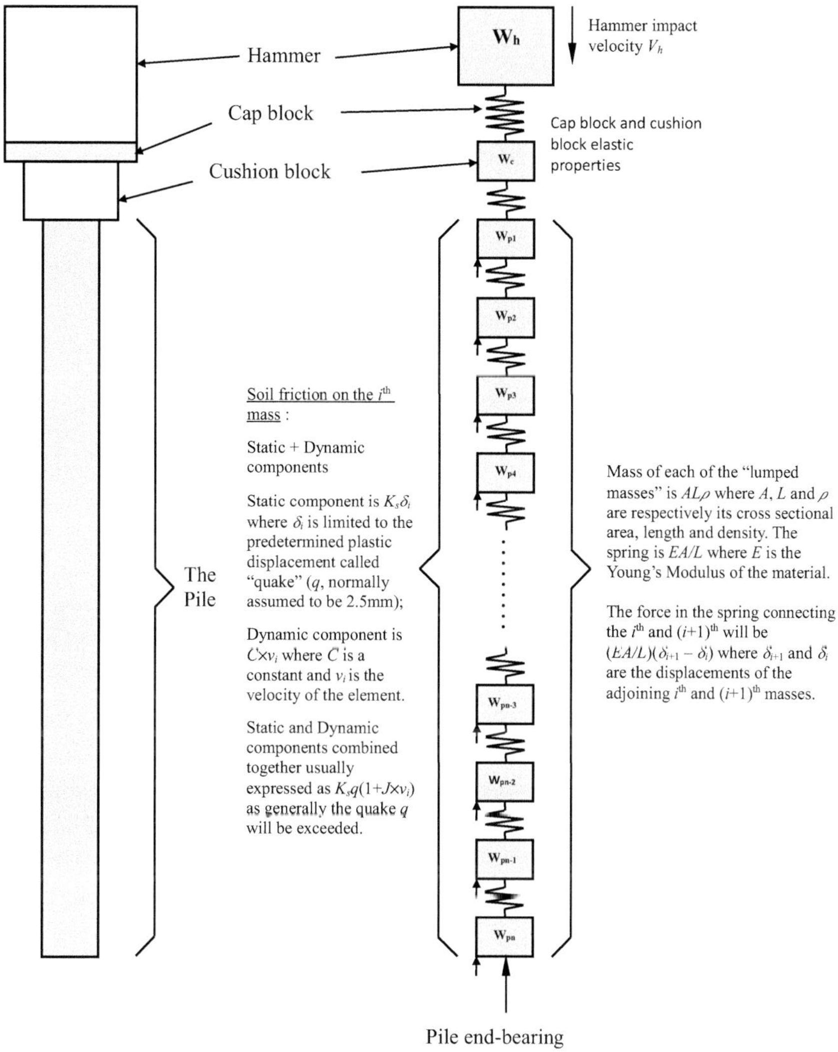

Figure 6.38 Smith's model for idealization of a pile in dynamic analysis.

Generally, the following quantities can be measured in PDA tests:

(i) Maximum energy delivered to the pile by the ram (EMX). As discussed, this energy can be used to calibrate the hammer efficiency and the impact factor used in the Hiley formula. The quantity can be found for a single hammer blow and an averaged value over 10 blows can also be calculated;

(ii) Maximum impact force (FMX) which is usually the first maximum force measured in a hammer strike. This force can be used to calculate the static capacity of the pile by the Case method commonly used in the industry as described in (II);

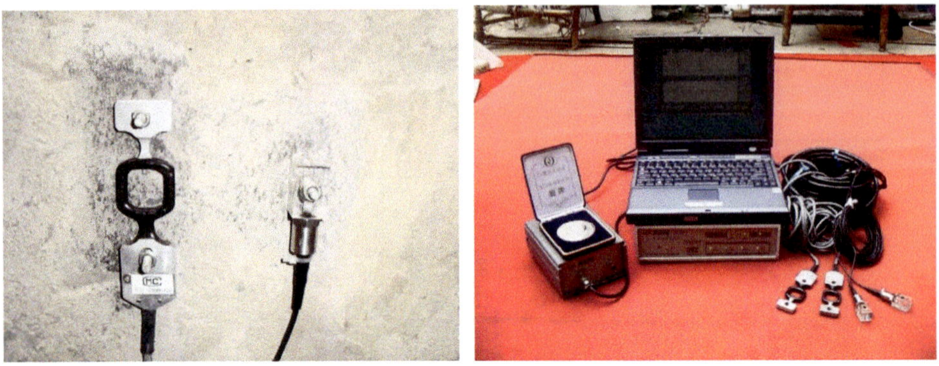

(a) Force and velocity transducer (b) PDA testing set

Figure 6.39 Equipment for a PDA test.

(iii) Forces and velocities (particle velocities) at the pile head are normally taken at 1024 (2^{10}) "time measurement points" (or multiples). (The use of the 1024 time measurement points is for facilitation of the working of "fast Fourier transform".) Thus, if the total time taken for measurement is 102.4 ms (millisecond), the readings are taken at 0.1 ms intervals;

The following quantities can then be calculated based on analysis of the measured quantities

(1) The maximum: compressive stress (CSX), which is simply the measured maximum force in the pile divided by the cross-sectional area of the pile;
(2) The ultimate static capacity of the pile (RMX) which can be calculated by the case method under various assumed J_c values (soil damping constants);
(3) The pile integrity factor (BTA) which is arrived at by examination of the wave forms;
(4) The compressive stress at the bottom of the pile (CSB);
(5) Total skin friction (SFT).

(II) Case method

The case method (Rausche et al., 1985) is a closed-form solution based on the following assumptions:

(i) The pile is uniform in section and construction material, i.e. a parameter called the impedance $Z = \dfrac{AE}{c}$ is constant where E, A and c are the Young's modulus of the pile material, cross-sectional area of the pile and wave propagation velocity of the stress-wave in the pile, respectively;
(ii) The stress wave experiences no energy loss in its transmission through the pile shaft and there are no distortions of signals;

(iii) The resistance to the dynamic component of the force is at the pile toe only, whilst that of the pile shaft is ignored;
(iv) The resistance to the dynamic component is proportional to the particle velocity.

The ultimate static capacity of the pile is given as (see later section):

$$R = \frac{(1 - J_c)}{2}(F_1 + Z_p v_1) + \frac{(1 + J_c)}{2}(F_2 - Z_p v_2) \quad (6.96)$$

where R is the ultimate static capacity of the pile;
F_1 is the pile head force measured at time t_{c1};
F_2 is the pile head force measured at time t_{c2};
v_1 is the pile head velocity measured at time t_{c1};
v_2 is the pile head velocity measured at time t_{c2};
t_{c1} is the time when the pile head force F_1 is recorded;

$t_{c2} = t_{c1} + 2L/c$

L is the pile length measured from pile head to pile toe;
c is the propagation velocity of the stress-wave in pile and can be calculated by $c = \sqrt{E/\rho}$ where E and ρ are the Young's modulus and density of the pile material, respectively;
J_c is the "lumped case damping factor" which defines the dynamic component of the pile at the pile toe. Its value depends on the type of soil and the dimensions of the pile.

With the appropriately chosen J_c value, R is determined as the maximum value that can be arrived at among various measured sets of F_1, F_2, v_1 and v_2. A detailed description of the Case method with illustration of its use by a numerical example is given in a later section. It should be noted that the case method is more accurate when the end-bearing resistance is the dominating component, as the damping factor is lumped at the pile toe only.

(III) CAPWAP (Case Pile Wave Analysis Program) or similar

CAPWAP is a computer program for pile static capacity analysis developed by the Case Institute of Technology in the late 1960s. It is based on the $F_u = -Zv_u$ model by combining the PDA data in the form of a force and velocity development against time (wave traces) with the Wave Equation Analysis of Pile (WEAP) program (Rausche et al. 1972, 1988) and computes the pile static capacity iteratively by signal matching.

Under this technique, the velocities and forces are first measured by the PDA method at the pile head when struck by a pile-driving hammer. The pile head forces (or pile head velocities) are then input into the wave equation as "input excitation" to carry out analysis under a set of initially assumed soil parameters including quake, damping constant, ultimate friction distribution along the pile shaft, etc. Normally, the back-calculated pile head velocities (or pile head forces) would be different from the measured ones. The soil parameters are then adjusted and the wave equation reanalysed until a reasonably good match between the calculated and measured values is obtained. Then, there is good reason to assume that the correct soil parameters have been ascertained and the ultimate static capacity of the pile can be calculated. However, it should be noted that the answers may not be unique, i.e., different sets of soil parameters can all result in good matching with the measured quantities. Therefore, it is a judgement as to the choice of the parameters.

(IV) Using CAPWAP to calibrate the final set table

As CAPWAP can give a fairly accurate pile load capacity, the CAPWAP capacity can be used to calibrate the parameters, including the hammer efficiency E_h and the coefficient of restitution e used in the Hiley formula by back-substitution. The calibrated formula can then be used to determine the final set table. The calibrated parameters may not carry any physical meanings by which they were derived. They may be considered simply as coefficients to fit into the Hiley formula which may only be used in the particular site and hammer with certain fall heights.

The matching mechanism of the CAPWAP method is given in a later section. In actual CAPWAP operation, an automatic matching approach is used where the matching is carried out by the optimization algorithm built in the program. However, as good matching by the algorithm can still be achieved by unreasonable soil parameters and load capacity of the pile, the results should be carefully assessed by experienced laboratories and where necessary exercise manual signal matching with soil parameters within reasonable ranges. Besides some commercial programs, Cheng has also developed the program Backward which is basically the same as the commercial programs.

Based on Hiley's formula, the driving force and capacity should be basically inversely proportional to the set. Based on the PDA tests on some test piles in Hong Kong, it has been however observed that both the actual transferred energy and the axial force in the pile are not too sensitive to the value of the set. For the stroke used for striking of pile, it is also observed that while the toe-driving force increases with the stroke, the axial force in the pile as measured from the PDA is also not too sensitive to the stroke. The toe-driving force as well as the pile axial force are also not too sensitive to the pile length during striking. All these test results illustrate that the pile-striking process is not a simple process which can be described by Hiley's formula or similar formulae used in other countries.

6.10 NEGATIVE SKIN FRICTION

Piles installed through compressible materials (e.g. fill or marine clay) can experience negative skin friction. This occurs on the part of the shaft along which the downward movement of the surrounding soil exceeds the settlement of the pile. Negative skin friction could result from consolidation of a soft deposit caused by dewatering or the placement of fill. The dissipation of excess pore water pressure arising from pile driving in soft clay can also result in consolidation of the clay. In determining the amount of negative skin friction, it is necessary to estimate the position of the neutral plane, i.e. the level where the settlement of the pile equals the settlement of the surrounding ground. For end-bearing piles, the neutral plane will be located close to the base of the compressible stratum. NSF can be estimated as $\tau = \beta\sigma_v'$, where β is taken as 0.25 in Hong Kong. Negative skin friction can sometimes amount to half of the pile capacity for some piles in the reclamation area, and is an important factor in the design of piles; a negative friction design for a pile in Hong

Table 6.11 A negative skin friction calculation for a bored pile in Hong Kong (NSF = ∫βσ'dz)

Negative skin friction of pile				*Saturated soil unit weight (kN/m³)*	21		
Pile framing mark	*Pile type*	*Perimeter of the pile (m)*	*Marine deposit top level (mPD)*	*Marine deposit bottom level*	*Marine deposit*	*Average effective stress at MD*	*NSF (kN)*
PC32a	3 m diameter bored pile	9.42	−13.2	−27	13.8	240.0	7805
PC33a	2 m diameter bored pile	6.28	−13.8	−27.1	13.3	243.9	5096
PC33b	2 m diameter bored pile	6.28	−13.6	−27	13.4	242.3	5099
PC33c	2 m diameter bored pile	6.28	−13.4	−26.9	13.5	240.6	5102
PC33d	3 m diameter bored pile	9.42	−13.2	−26.8	13.6	238.9	7656
PC33e	2 m diameter bored pile	6.28	−13	−26.7	13.7	237.2	5105
PC33f	2 m diameter bored pile	6.28	−14.2	−27.7	13.5	249.5	5292
PC33g	3 m diameter bored pile	9.42	−14	−27.5	13.5	247.3	7866
PC33h	2 m diameter bored pile	6.28	−13.8	−27.3	13.5	245.1	5197
PC33i	2 m diameter bored pile	6.28	−13.6	−27.2	13.6	243.4	5199
PC33j	2 m diameter bored pile	6.28	−13.3	−27.1	13.8	241.1	5227
PC33k	2 m diameter bored pile	6.28	−14.4	−28.2	13.8	253.5	5494
PC33p	3 m diameter bored pile	9.42	−14.2	−28.1	13.9	251.8	8246
PC33q	2 m diameter bored pile	6.28	−14	−27.9	13.9	249.5	5448
PC33r	2 m diameter bored pile	6.28	−13.8	−27.7	13.9	247.3	5400
PC33s	3 m diameter bored pile	9.42	−13.6	−27.6	14	245.6	8102
PC33t	2 m diameter bored pile	6.28	−14.1	−27.5	13.4	247.9	5217

Kong is given in Table 6.11. Some engineers argue that due to the effect of creep, the skin friction will reduce gradually with time. Also, with the application of the vertical load, the pile will settle and release some of the negative skin friction. Based on these factors, some engineers do not believe in the need to consider negative skin friction in the design of piles. The authors tend to agree with this view in general.

6.11 PILE TESTS

Testing of foundations may be required for one or more of the following purposes:

(a) To ascertain the performance of the foundation under load;
(b) To establish or justify the design parameters; and
(c) To verify the structural integrity of the foundation units.

There are many methods of testing piles. These methods can be grouped into three categories:

(a) The static load principle,
(b) The dynamic load principle, and, and
(c) Other tests.

For tests based on the static load principle, the load applied to the test pile is static, that is, not moving. Two types of tests based on the static load principle are:

(a) The maintained load test (ML), and
(b) The constant rate of penetration (CRP) test.

For tests based on the dynamic load principle, the load applied is dynamic, that is, moving or vibrating. Three types of tests based on the dynamic load principle are:

(a) The driving formula test,
(b) The Case-Goble test, and
(c) the steady-state dynamic response test.

Other tests include:

(a) Sonic logging,
(b) Sonic echo tests,
(c) Lateral load test,
(d) Coring test,
(e) Vibration test,
(f) Ultrasonic echo sounder test.

The type of test selected and the number of tests required should be appropriate to the type of foundation and suitable for the purpose required. In this unit, some of the methods of testing piles are illustrated and described. The tests described include both the static and dynamic methods.

6.11.1 Static load test on pile

To ensure the stability of the test assembly, careful consideration should be given to the provision of a suitable reaction system. The geometry of the arrangement should also aim to minimize any interactions between the test pile, reaction system and reference beam supports. It is advisable to have, say, a 10–20% margin on the capacity of the reaction against maximum test load. Traditionally, a static loading test is carried out by jacking a pile against a kentledge (Figure 6.40) or a reaction frame supported by tension piles or ground anchors (Figure 6.41). In recent years, the Osterberg load cell (O-cell) has been widely adopted for static loading tests for large-diameter cast-in-place concrete piles. It can also be used in driven steel piles.

Besides the static load test on the vertical compression and tension capacity of pile, lateral load tests are sometimes carried out on piles taking up a very large lateral load.

The typical procedures for the static load tests are:

1. The test load is applied in two equal increments up to the design pile capacity under working load, which is then released. Loading is reapplied in four equal increments up to twice the design pile capacity and maintained for at least 72 hours before removal.
2. The load at each load incremental stage will be maintained for 10 minutes or longer until the rate of settlement is less than 0.05 mm in 10 minutes.
3. The pile is considered to fail the static load test if the maximum pile head settlement exceeds ($2WL/A_pE_p + D/120 + 4$ mm), where D is the diameter or the least lateral dimension of the pile.

(a) Kentledge test used concrete block and load cell

(b) arrangement of hydraulic jack

Figure 6.40 A typical static load test for a pile using a kentledge.

(a) Use of tension pile

(b) use of anchors

Figure 6.41 Static load test using a reaction frame and tension piles/anchor.

4. In some building codes, the residual pile head settlement is also considered as the acceptance criteria in the load test, but the authors view that such criteria are not meaningful. According to the Hong Kong code, the residual settlement at the pile head cannot exceed $D/120 + 4$ mm and 25% of the maximum pile head settlement during the test.

For interpretation of the pile capacity static load test results, there are many methods available which include: single tangent method, double tangent method, Vander Vee's method, De Beer's method, Brinch Hansen's 80% criterion, Brinch Hansen's 90% criterion, Fuller and Hoy's method, Butler and Hoy's method, Davisson's method (Hong Kong method), Carroll's method, Mazurkiewicz's method, Chin's method and others. In general, the basic concepts behind these methods are the pile capacity and settlement

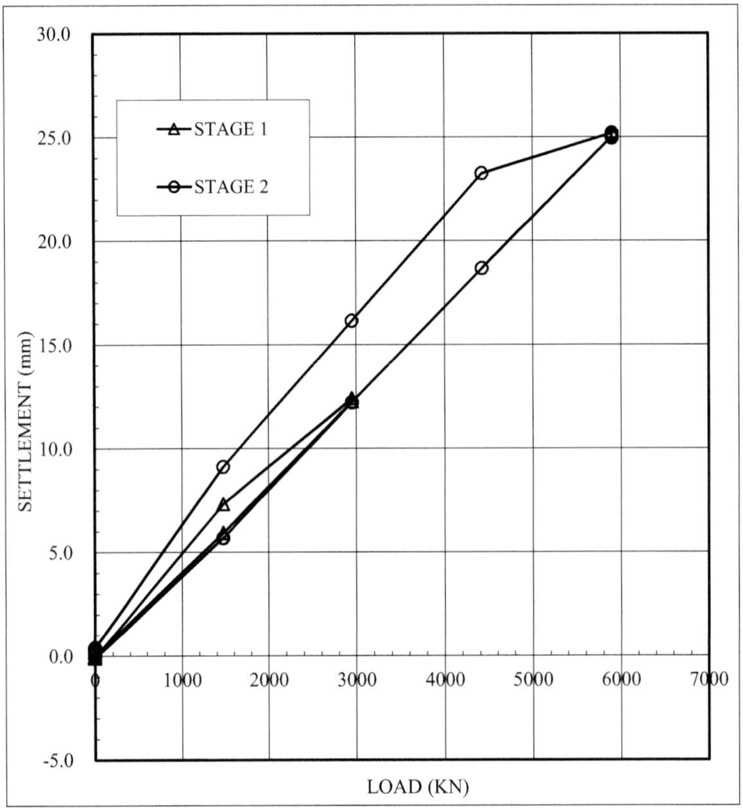

Figure 6.42 A typical static load test result.

considerations. The pile capacity obtained from these methods can differ greatly, and the engineer must be clear in mind about the meaning of pile capacity. More commonly, pile settlement is the critical issue, and the authors' belief that settlement consideration can be a more meaningful consideration in the pile capacity evaluation. A typical static load test result is shown in Figure 6.42 for illustration.

From Table 6.12, it is clear that there is a wide spread of pile capacities from different methods of interpretation. The Davisson's method, which is a settlement-based criterion adopted in Hong Kong, appears to be on the low side as compared with those capacity-controlled criteria. Many engineers assume axial shortening of pile PL/AE as the pile settlement. Based on some field tests in Hong Kong, as shown in Table 6.13, it is found that this approach is not always on the safe side.

In the literature, there are different methods for interpreting the pile capacity from static load tests. Davisson's method, which is based on the consideration of pile settlement, is the approach adopted in Hong Kong. In Table 6.14, which is based on series static load tests in Hong Kong, it is noted that there are wide differences between different interpretations, and Davisson's method gives almost the lowest pile capacity.

Table 6.12 Results for different methods of interpretation for a static pile load test in Hong Kong

Ultimate load predicted method	*PCC5-2*	*7P2*	*CPW7-P4*	*CPW11-P3A*	*PCB3-P2*	*E4-H5*	*KL5a-H4*	*C5-H8*
Single tangent method	10274	12848	9779	9000	7526	12360	8818	7407
Double tangent method	10842	13454	10221	9455	7777	12730	9455	8000
Brinch Hansen's 90 percent criterion	11200	14100	10675	10000	8000	13000	10136	8527
Fuller and Hoy's method	11295	14000	10675	9925	7984	13125	10136	8527
Butler and Hoy's method	11053	13697	10432	9749	7808	12730	9756	8370
Davisson's method	11150	14000	10000	10000	8000	10000	9182	7950
Carroll's method	9390	11624	8789	8000	6777	10850	8273	6950
Mazurkiewicz's method	12000	14727	11243	10667	8731	13750	10615	9256
Chin's method	14300	20000	12500	12500	10000	14286	12500	10000
Mean ultimate load	**11278**	**14272**	**10479**	**9922**	**8067**	**12537**	**9875**	**8332**

Table 6.13 Comparisons between the measured and calculated settlements from series of field test results

Pile no.	*Pile length (m)*	*Test load (kN)*	*Measured settlement (mm)*	*Calculated settlement (mm)* $\left(\frac{PL}{AE}\right)$
A	23.58	6163	23.860	30.92
B	33.60	7090	28.493	40.77
C	21.41	5900	24.210	26.87
D	26.65	7080	23.710	32.29
E	30.76	5900	28.430	38.61
F	20.73	7096	27.903	25.18
G	16.91	7096	25.185	20.57
H	20.51	5920	28.190	25.83
I	22.90	5920	32.775	28.84
J	20.76	5950	25.190	26.28

Table 6.14 Different methods of load interpretation for series of static load tests in Hong Kong

Ultimate load predicted method	*PCC5-2*	*7P2*	*CPW7-P4*	*CPW11-P3A*	*PCB3-P2*	*E4-H5*	*KL5a-H4*	*C5-H8*
Single tangent method	10274	12848	9779	9000	7526	12360	8818	7407
Double tangent method	10842	13454	10221	9455	7777	12730	9455	8000
Brinch Hansen's 90 percent criterion	11200	14100	10675	10000	8000	13000	10136	8527
Fuller and Hoy's method	11295	14000	10675	9925	7984	13125	10136	8527
Butler and Hoy's method	11053	13697	10432	9749	7808	12730	9756	8370
Davisson's method	11150	14000	10000	10000	8000	10000	9182	7950
Carroll's method	9390	11624	8789	8000	6777	10850	8273	6950
Mazurkiewicz's method	12000	14727	11243	10667	8731	13750	10615	9256
Chin's method	14300	20000	12500	12500	10000	14286	12500	10000
Mean ultimate load	**11278**	**14272**	**10479**	**9922**	**8067**	**12537**	**9875**	**8332**

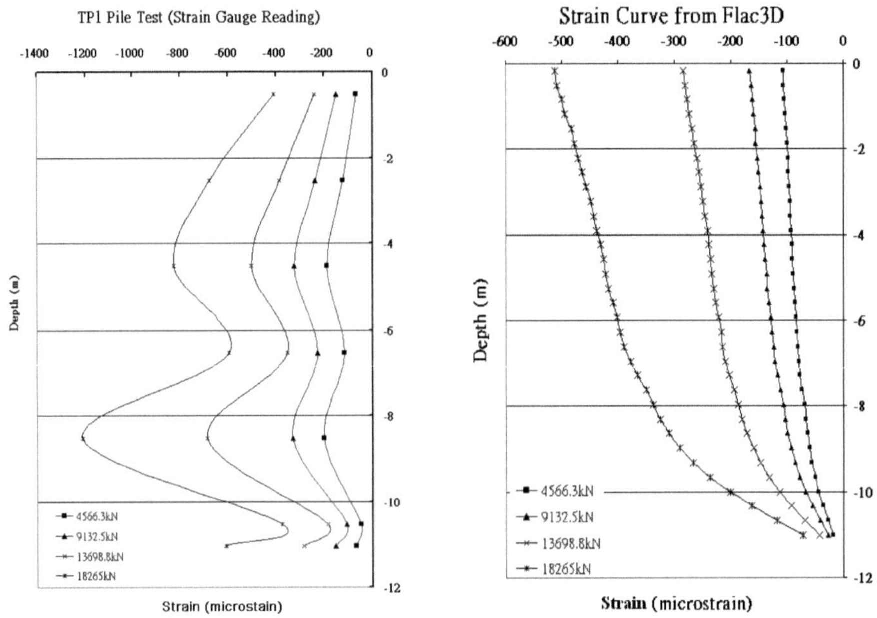

Strain Reading from field test

Strain Reading from program FLAC

Figure 6.43 Test results for Test Pile 1.

It is also found that numerical studies on the pile axial load distribution from static load tests may not match well with the field test results, no matter how the material parameters are tuned. Three test piles, TP1 to TP3, with detailed instrumentations are analysed as shown in Figures 6.43–6.45.

The field tests for TP2, which is a socket H-pile, are generally more in line with numerical studies, and more details about the results are given below. The axial displacement and the axial load along TP2 are shown in Figure 6.46. Such a good prediction appears to be uncommon, according to the authors' experience.

Another interesting pile load test is shown in Figure 6.47. Since the elastic parameters for soil are based on the standard penetration result N, the statistical relation is adjusted till a good match between the computational results and the test results are obtained in Figure 6.47. A good match is given by E = 5.42N MPa, which is much greater than the E = 2N MPa which is commonly used.

A series of static load tests in sandy soil assessed by the static formula, Davisson's method and Hiley's method are shown in Table 6.15. From the results in this table, if the pile capacity by Davisson's method (settlement control) is taken as the reference, it appears no particular approach gives good predictions under all circumstances. Davisson's method, however, does not give the true pile capacity, hence the capacity by this method is underestimated.

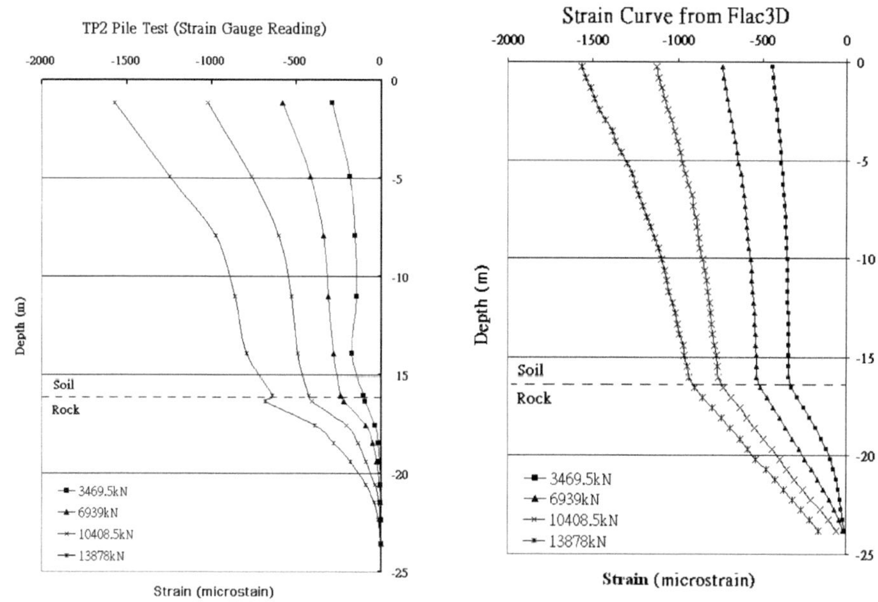

Strain Reading from field test

Strain Reading from FLAC

Figure 6.44 Test results for Test Pile 2.

6.11.2 Pile integrity tests

The integrity of a pile can be assessed by various means, including the low strain and large strain method. Some of the more common techniques are described here.

6.11.2.1 Down-hole techniques

The use of down-hole techniques is common for the assessment of the pile integrity for a bored pile. There are three major down-hole test techniques:

1. Cross-hole sonic logging,
2. Gamma-gamma logging,
3. Parallel seismic testing.

In general, these methods examine the quality of the concrete but not the pile base condition. Of these three tests, cross-hole sonic logging is most commonly used by engineers.

6.11.2.2 Low-strain echo test

The sonic-echo test as shown in Figure 6.48 is performed by striking the pile head with a light hammer and measuring the response of the pile with a sensor (accelerometer or

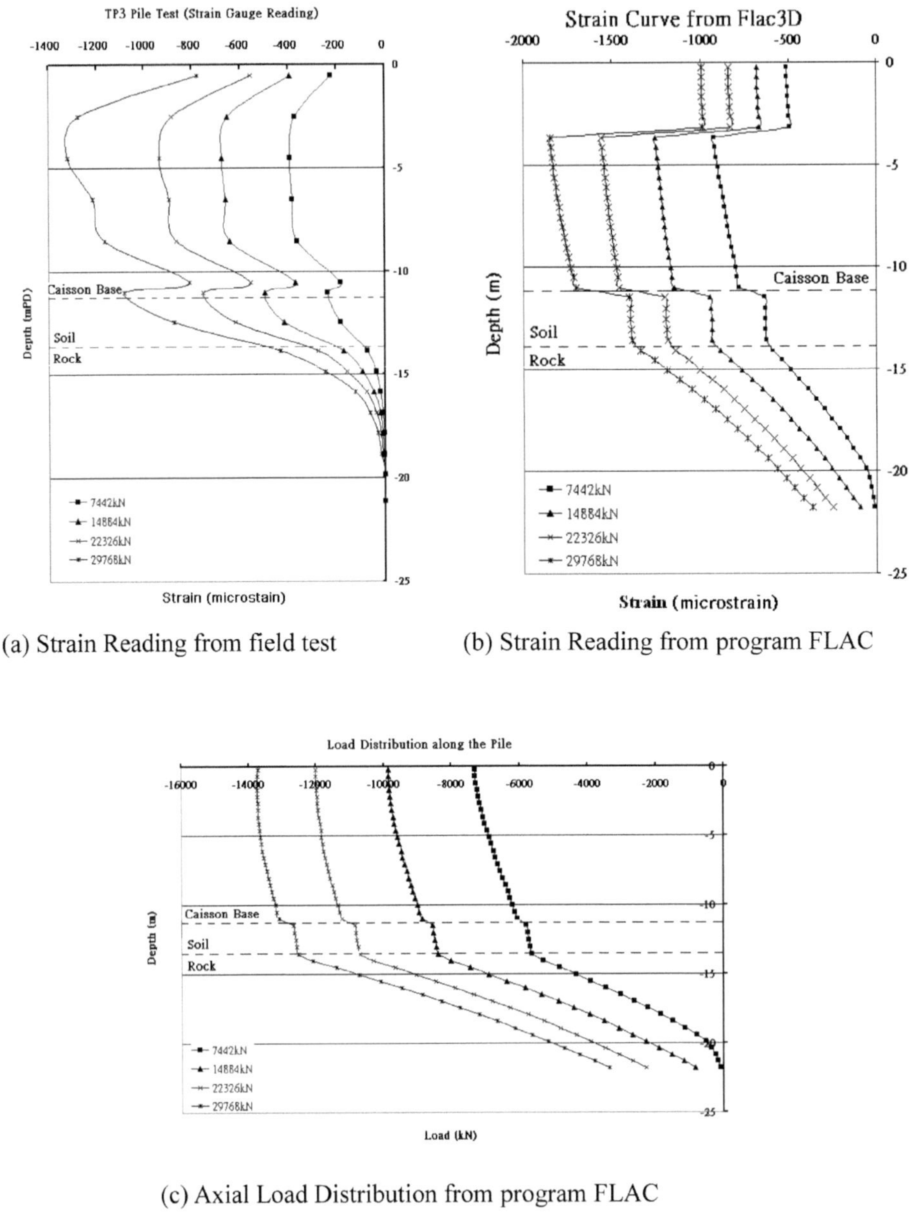

Figure 6.45 Test results for Test Pile 3.

geophone velocity transducer) coupled to the pile head as shown in Figure 6.49. In general, the sensor should be placed close to the centre of the pile, with the pile head prepared and levelled. The sensor should be firmly attached to the pile head to pick up the signal. The hammer should be struck at a distance of about 300 mm from the sensor.

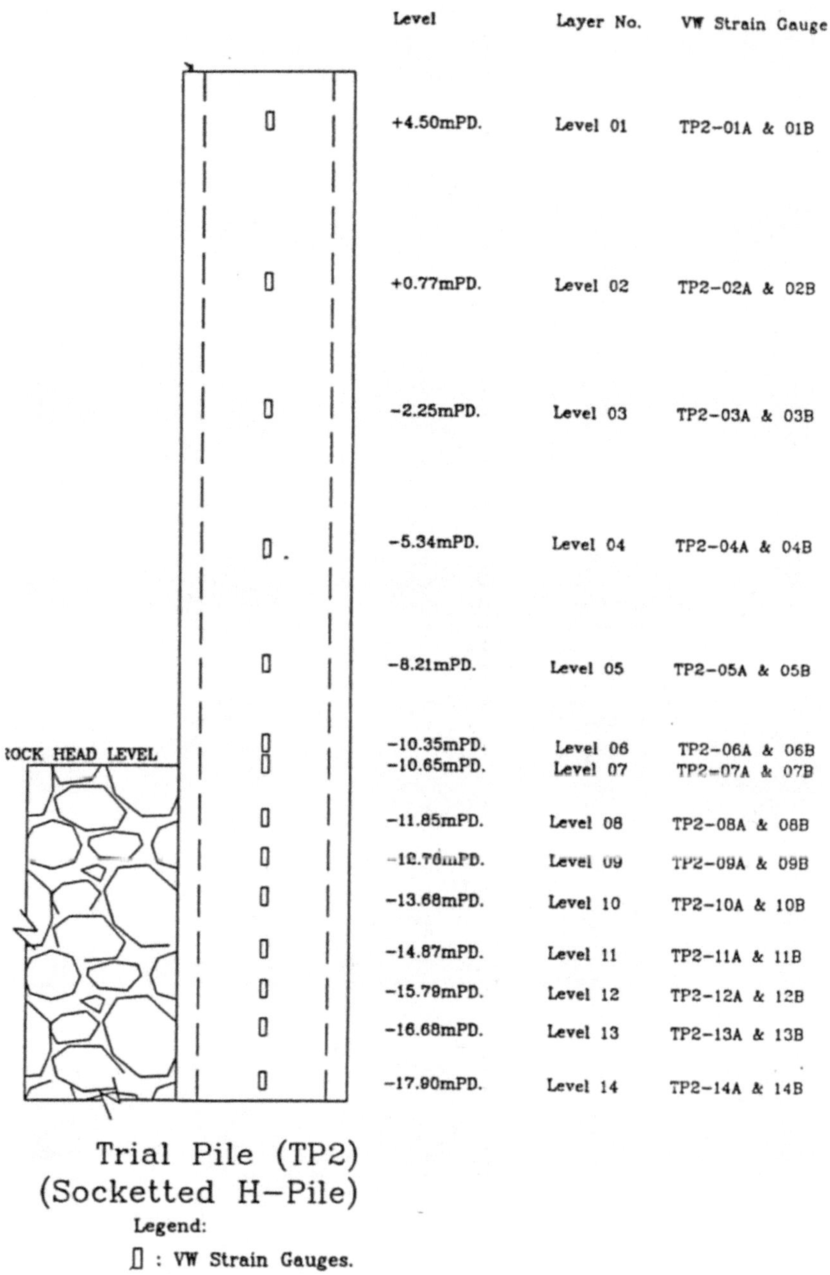

(a) Strain gauges location in Test Pile 2

Figure 6.46 Skin friction vs. applied load for Test Pile 2.

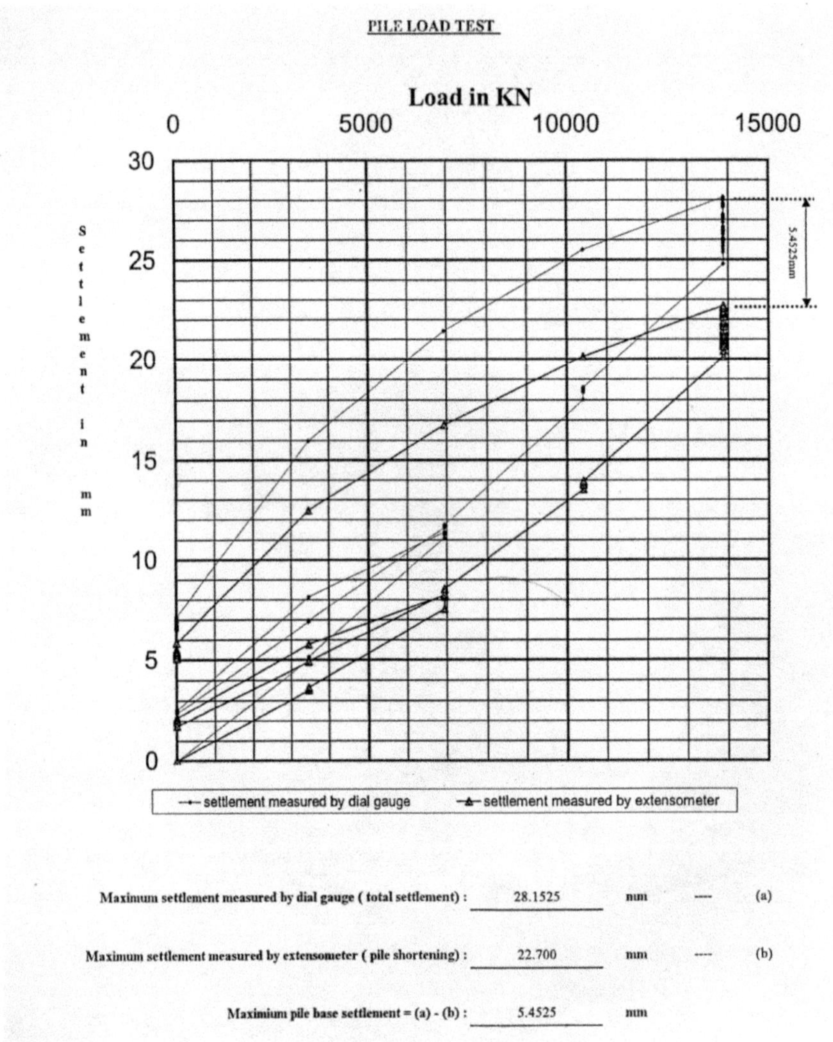

(b) Load-Settlement Curve of TP2 (measured by Dial Gauge and Extensometer in Field Test)

Figure 6.46 (Continued)

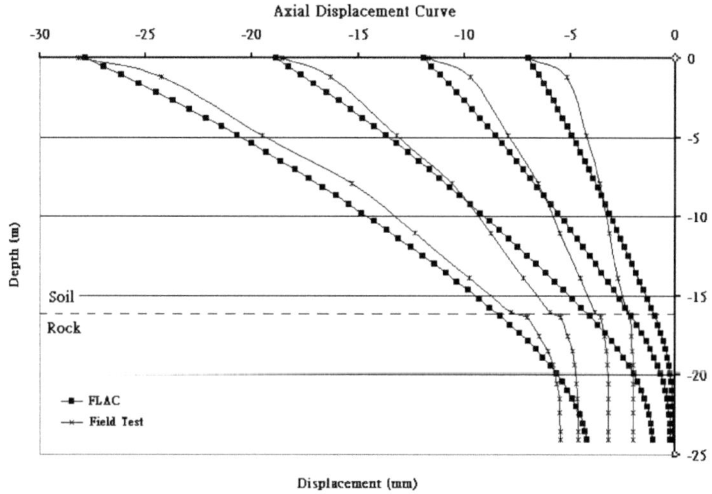

(c) Axial Displacement

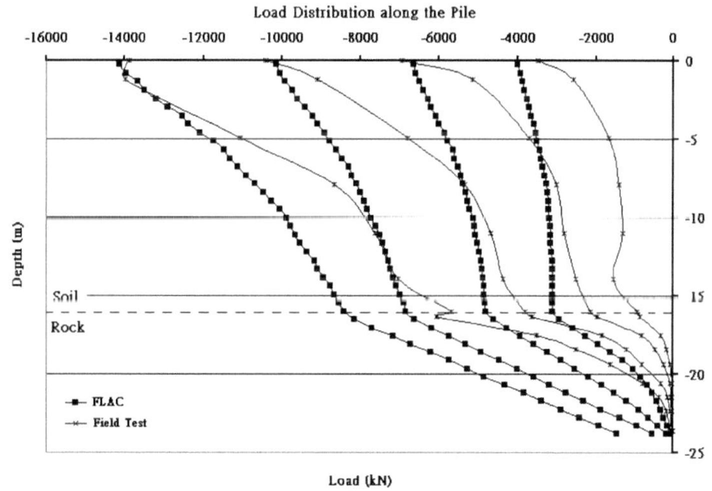

(d) Axial Load Distribution

Figure 6.46 (Continued)

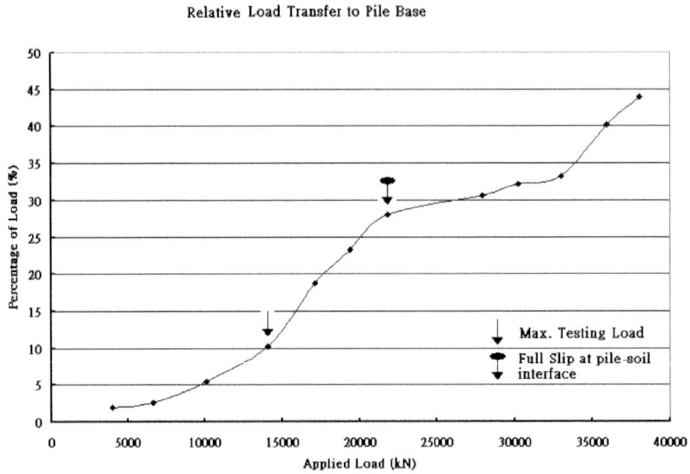

(e) Relative Load Transfer to Pile Base against applied load

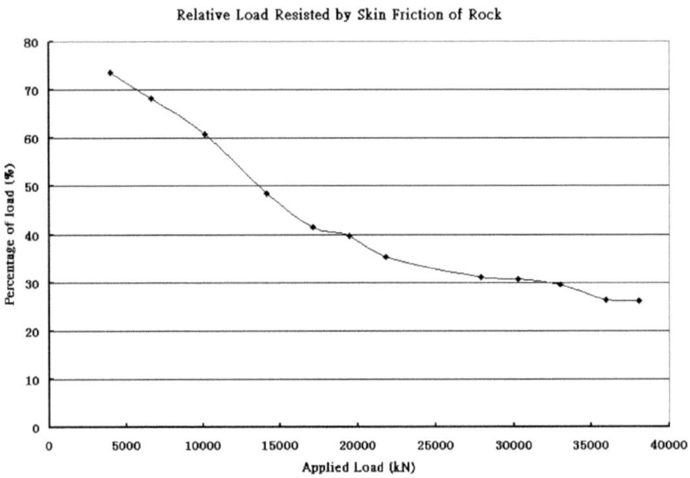

(f) Relative Load on Skin Friction of Pile

Figure 6.46 (Continued)

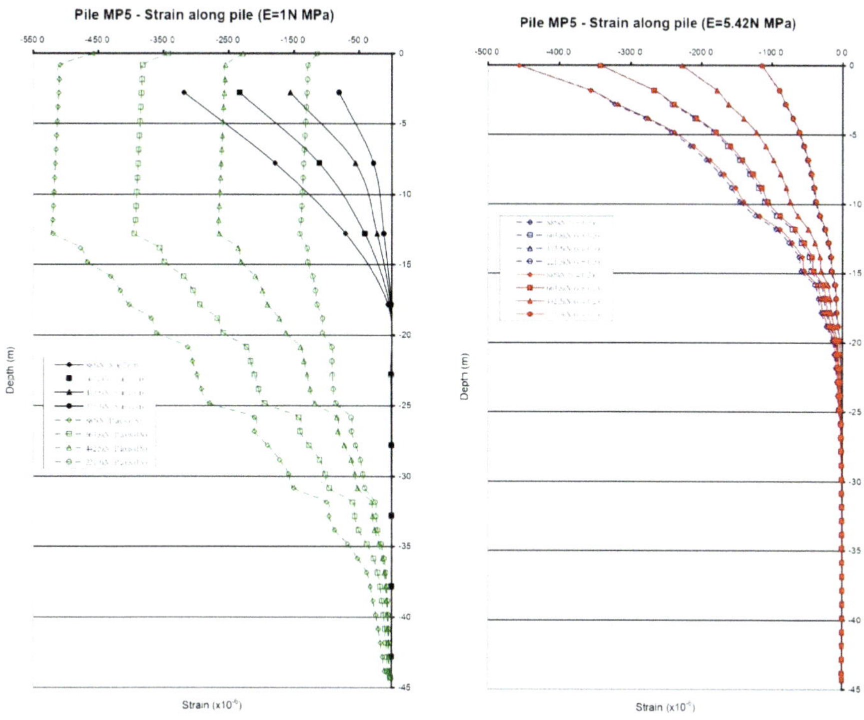

(a) E=1N MPa and hardening model (b) E=5.42N and elastic model

Figure 6.47 Pile load test and different interpretations for a pile load test in Hong Kong.

Table 6.15 Eleven test pile load capacity comparisons

Pile no.	*Static formula (kN)*	*Tomlinson method*	*Load test (Davisson's method)*	*Hiley's formula*
1	5430	7892	6500	6919
2	6829	10736	8000	5835
3	6120	8158	5500	6425
4	4751	4873	4700	9157
5	5471	5233	7200	6462
6	5932	8518	6500	5523
7	5299	6703	5500	6100
8	5959	3861	5700	7091
9	5086	7118	5800	6600
10	5260	5306	6200	7617
11	5764	7286	8800	6461

Figure 6.48 Low-strain echo test on site.

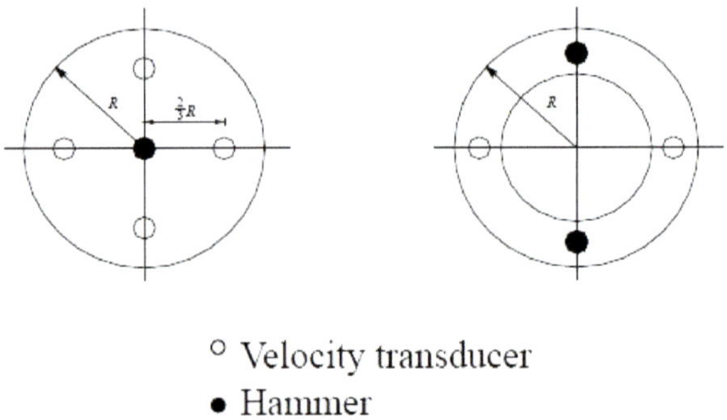

Figure 6.49 Arrangement of the hammer and sensor for a solid and a hollow pile.

The hammer blow generates a compressive stress wave which is channelled down the pile shaft as a "bar-wave". The latter is partly reflected back towards the pile head by any change in impedance within the pile, and typical results are shown in Figure 6.50. These impedance changes can be a result of changes in pile section, concrete density or shaft-soil properties. The stress wave is transmitted through the pile at a velocity, v_b (where v_b is the bar-wave velocity of propagation through the pile material) and the time lapse, t, between the hammer impulse and the arrival of the reflected waves at the pile head from the pile tip is a measure of the distance travelled by the stress wave, such that:

$$t = \frac{2L}{v_b} \tag{6.97}$$

where L represents the distance to the reflecting surface (pile tip in this case).

If the value of v_b is known, or can be estimated within reasonable limits, then t will give an estimate of the pile length or the depth to any other reflecting surface within the pile.

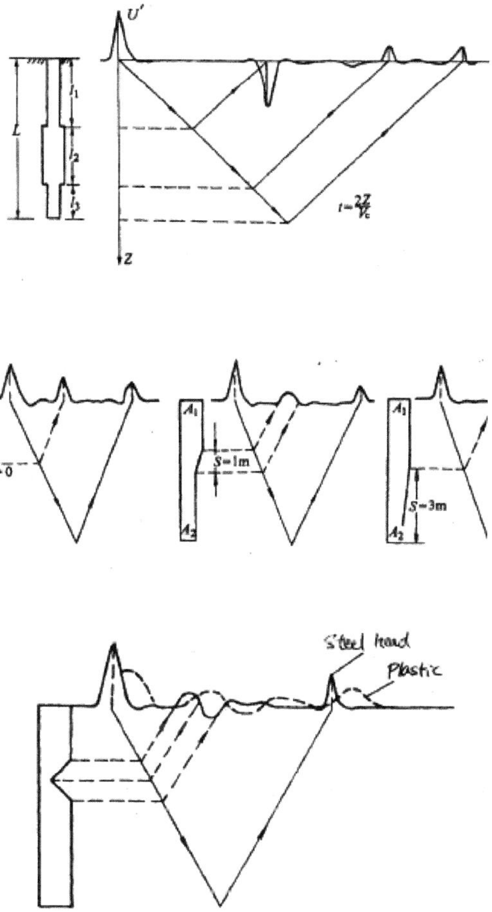

Figure 6.50 Some typical responses from an echo test.

If the pile length is known, then a comparison can be made between the length calculated from the test result and the known length, in order to verify that the depth to the reflecting surface is correct. Typical results for bulging and necking are shown in Figure 6.51.

Typical test procedure
The material in the pile head must be prepared such that no delamination or "micro-cracking" is present, to ensure a clean transmission of the stress wave down the pile. The sensor is coupled to the pile head, usually with a grease- or gel-based couplant, and the pile is struck with the hammer at or near the pile axis. Normally, a hammer weighing less than 1 kg with a plastic impact tip is used. Heavier hammers have sometimes been found to give better results for large piles greater than 1 m in diameter. Typical sensors used for the echo test are shown in Figure 6.52.

The test is repeated several (at least three) times in order to obtain representative samples by averaging of these individual results. The more hammer blows recorded, the greater the reduction in the effects of random signals (noise) from other site activities or system noise. As the effects of this extraneous noise are reduced, so the repeatable parts of the signal are

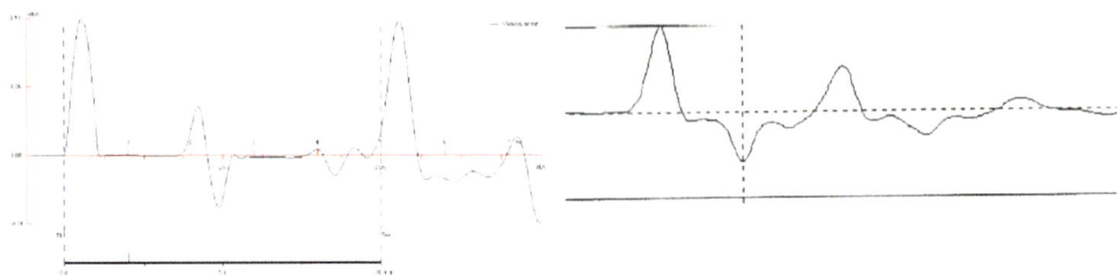

Figure 6.51 Typical results for necking and bulging.

Figure 6.52 Sensors used in a low-strain test (velocity transducer at left, accelerometer at right).

enhanced. As a general rule, background noise can be reduced by a factor of $\sqrt{n}$, where n is the number of superimposed signals from tests on the same pile.

6.11.2.3 Vibration test

In the vibration test, a vibrator with a regulator is placed on top of the pile, typically a small-diameter bored pile. A sinusoidal force is applied with a frequency which can be adjusted. The frequency keeps on increasing during the test, and the response of the pile head velocity against frequency is monitored. It is found that when the frequency of the force activates the resonance of the pile, a peak value will be obtained. Based on the solution of the wave equation, it can be demonstrated that the mobility Z which is given by V/F is:

$$Z = \frac{1}{V_c A_c \gamma_c} \frac{\frac{EA_c f}{kV_c} + \tan\frac{fL}{V_c}}{\frac{EA_c f}{kV_c}\tan\left(\frac{fL}{V_c}\right) - 1} \tag{6.98}$$

Where f is the angular frequency of the applied force, k is the elastic spring constant at the pile base from the end-bearing, and V_c is the speed of the wave force. For resonance, with k tends to infinity

$$Z = \frac{1}{V_c A_c \gamma_c} \tan\frac{fL}{V_c} \tag{6.99}$$

When k tends to 0, the resonance condition is given by

$$Z = \frac{1}{V_c A_c \gamma_c} \cot\frac{fL}{V_c} \tag{6.100}$$

No matter the value of k, the resonance will be a repeating function, and the difference between the resonance frequency Δ will give the resonance length which is

$$L = \frac{V_c}{2\Delta} \tag{6.101}$$

The mean of the F/V relation with the frequency is called the characteristic mobility, which should be theoretically equal to $1/V_c A_c \gamma_c$. Based on this test, the resonance length and the quality of concrete can be estimated. If the resonance length is smaller than the actual constructed length, it can be concluded that there are some problems with the quality of the concrete somewhere below the ground, and the location can actually be determined from the resonance length. The limitations of the vibration include:

(a) The pile is not suitable for long pile, as the signal is easily damped for pile with a length to diameter ratio of about 20 in stiff and dense soil and 30 in loose soil;
(b) Small but structurally significant variations in the wave velocity through weak concrete zone may not be detected;
(c) The test is sensitive to abrupt changes, but not gradual changes, in the pile cross section, hence it is not applicable to the tapered pile;
(d) Unable to quantify the vertical extent of section changes or the lateral position of the defects;
(e) Vertical cracks cannot be detected.

The use of a vibration test is not simple, and its applications are now largely replaced by the use of large strain test.

6.11.3 Koden test (ultrasonic echo sounder test)

For checking the size and shape of the bored pile and bell-out, the Koden test is commonly adopted. The Koden test adopts the use of ultrasonic wave and the reflection to measure the size and shape of the bored pile. The sensor is lowered at the centre of the bored pile, and the results at two perpendicular axes are measured immediately on site, as shown in Figure 6.53.

6.11.3.1 Borehole video inspection test

The borehole video inspection test (Figure 6.54) offers an indirect and inexpensive way to inspect and investigate the condition of the borehole directly. It can provide a continuous inspection of the bored pile with a 360° view before installation of the reinforcement cage and concreting.

6.11.4 Large strain test

A hammer is used to generate sufficient penetration at the pile tip and to activate the pile capacity. Based on the force and velocity measured at the top, the skin friction, base capacity and soil parameters are back-computed. The integrity of the pile and defects can be estimated from the PDA test, as shown in Figure 6.55. The cost of large-strain PDA tests is about 10% that of a classical kentledge test, hence these tests are becoming more popular currently. Pile integrity can be assessed from the PDA result which is claimed by the developer, and some recommendations are provided by the PDA supplier for the pile integrity assessment. The locations of the defects can be determined from the time of wave reflection, while the quality of the pile shaft can be estimated from the β value as obtained from the PDA tests. The authors tend to be conservative in using such a criterion, and minor defects are generally not detected by the PDA. There are some inherent limitations to the large strain test interpretations, which appear to be open-ended questions which cannot be answered with satisfaction in all cases. The pile capacity can also be based on CASE or CAPWAP.

The arrangement for the strain gauge and velocity/acceleration transducer for the PDA test is shown in Figure 6.55. In general, two pairs of sensors are installed along the opposite diameter of a pile, and the mean of the results is used to eliminate the effect of the eccentricity during pile driving. The typical output from a PDA test for a pretensioned segmental pile is shown in Figure 6.56, where $Z=AE/c$, and c is the velocity of the wave inside the pile which can be determined from E = (specific weight in kN/m^3/g)c^2.

Normally, the velocity of the wave can be taken as 5123 m/s for a steel pile and can be assessed on site if possible for a concrete pile. In general, many PDA tests are performed on re-strike piles, and it is recommended that a minimum of 10 re-strikes be performed before carrying out the PDA test. A typical PDA test report shall cover the following information:

1. Hammer used for testing;
2. Blow selected for analysis;
3. Final set, temporary compression of the pile and soil during the test;
4. Static capacity of the pile (mobilized), based on assumed J value;
5. Pile stresses at the time of testing;

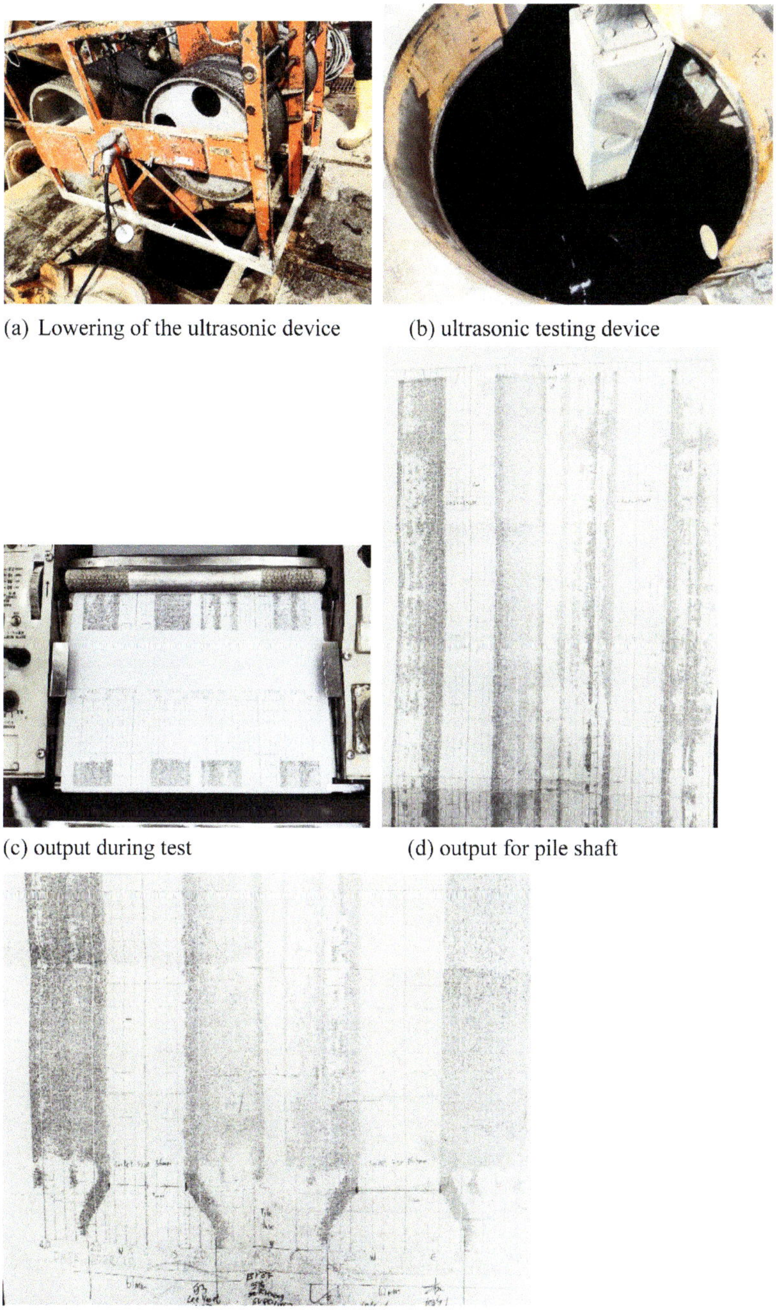

(a) Lowering of the ultrasonic device

(b) ultrasonic testing device

(c) output during test

(d) output for pile shaft

(e) output for bellout

Figure 6.53 Koden test arrangement and typical results for pile shaft and bell-out.

Figure 6.54 Borehole video inspection.

6. Pile hammer performance;
7. Extent and locations of structural damage in the pile, if any;
8. Discussions on anomalies, if any.

Some of the limitations of the PDA test include:

1. Vertical crack – usually not shown up in the test result;
2. Fine horizontal crack – the impact stress wave can pass through the fine crack easily with little loss in energy;
3. Some subsurface conditions – debris at pile base, cover to reinforcement, non-homogeneous condition.

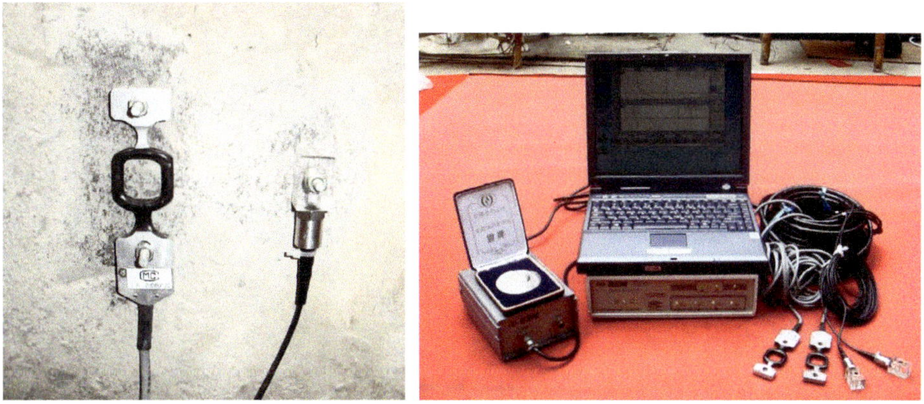

Figure 6.55 Velocity and strain transducers for a large-strain test.

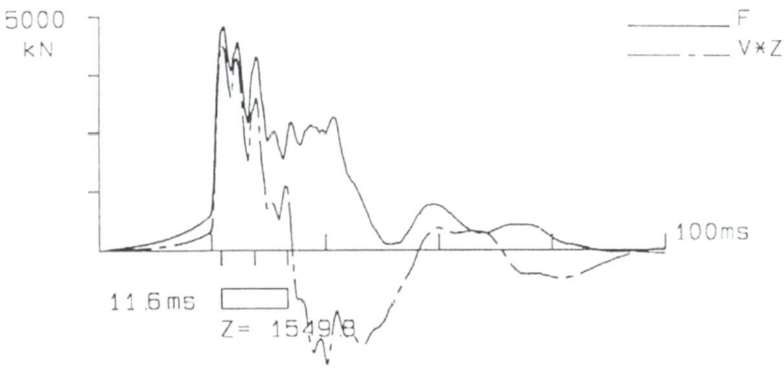

Figure 6.56 A PDA test result for a precast pretensioned pile in Hong Kong.

Some of the more important outputs from the PDA tests are given below:

CSX/FMX – the maximum compression stress/force measured at the transducer location;
VMX/AMX – maximum downward velocity/acceleration measured at the transducer location;
EMX – maximum energy transmitted past the transducer;
FT1/VT1 – force/velocity at impact or "time 1";
BTA – pile integrity factor for damage analysis;
RSP – static Case-Goble capacity;
RMX – maximum Case-Goble capacity (using J_c) occurs at time TMX (time increment delay after time 1 for RMX) during the blow.

Table 6.16 Calibration efficiency of a diesel hammer using EMX

Pile no.	*Calibrated efficiency (%)*
1	75.5
2	76.6
3	90
4	86.5
5	83
6	86
7	87
8	74
9	73
10	64.5

Based on the actual energy delivered to the pile, the efficiency of the driving (based on EMX) can be calibrated. The results of the analysis for 10 pile tests in Hong Kong using a diesel hammer is given in Table 6.16. Since this efficiency refers to the actual energy delivered to a pile, the efficiency of the diesel hammer must be greater than the values in Table 6.16. It appears that an efficiency of 80% for a diesel hammer in Hiley's formula which is adopted in Hong Kong is conservative in general.

From the PDA test result as shown in Figure 6.56, the downward and upward wave in the pile can also be computed with Eqs. (6.102)–(6.105) (*M* means measured results in these equations), and the results for the downward and upward waves are also usually given the PDA test report.

$$v_d = \frac{1}{2}\left(v_M + \frac{F_M}{Z}\right) \tag{6.102}$$

$$v_u = \frac{1}{2}\left(v_M - \frac{F_M}{Z}\right) \tag{6.103}$$

$$F_d = \frac{1}{2}\left(F_M + Zv_M\right) \tag{6.104}$$

$$F_u = \frac{1}{2}\left(F_M - Zv_M\right) \tag{6.105}$$

The digitized results for force, acceleration and velocity are recorded on a hard disk for later post-processing. A sample form for the PDA test report is shown below.

The beta value β is the ratio for the change of impedance *Z* from one position to another position (Z_1/Z_2), which can be determined from $(F_{d,1} + F_{u,1})/(F_{d,1} - F_{u,1})$ and provides a

Table 6.17 Interpretation of β values

Beta	Pile condition
1.0	Uniform
0.8–1.0	Slight damage
0.6–0.8	Damaged
Below	Broken

measure for the reduction of cross-sectional area at the damage location. Rausche and Goble (1972) suggested the guidelines for the β values in Table 6.17.

A typical CAPWAP summary result is given below for reference.

Soil segment	*Location*	*Ru*	*Sum of Ru*	*Unit resistance (kN/m)*	*Unit resistance per area (kN/ m²)*	*Quake*
1						
2						

Toe resistance = 4623 kN Total shaft friction = 2417 kN

Soil model parameters:

Smith damping factor	0.1 (shaft)	0.63 (toc)
Case damping factor	0.62 (shaft)	2.12 (toe)
Damping type	Viscous (shaft)	Smith+viscous (toe)
Reloading level	100 (shaft)	100 (toe)
Soil plug weight		4.0 kN (Toe)

CAPWAP match quality = 4.51
Observed: final set = 0.1mm Blow count = 10000 b/m
Computed: final set = 0.3mm Blow count = 3237 b/m
Maximum top compressive stress = 310 MPa (T = 31.6 ms)
Maximum compressive stress = 325.4 MPa (z = 10.0 m, T = 34.8 ms)
Maximum tensile stress = –53.42 MPa (z = 8.0 m, T = 51.3 ms)
Maximum energy transmitted EMX = 95.8 J
Maximum measured top displacement (TMX) = 23.5 mm

Although the use of β can be useful, there are several precautions in using this value:

1. Incorrect wave speed or modulus are used;
2. Significant bending is present in pile during driving;
3. For a very short pile, this value may not be useful.

Ten test piles in Hong Kong are assessed by both the Broms approach and the CAPWAP approach. Again, there are notable differences between the two approaches, and the authors do not believe that CAPWAP is more superior than the EMX approach. These methods can only be viewed as approximate assessments of the pile capacity (Table 6.18).

Table 6.18 Comparisons between the use of EMX and CAPWAP approaches for pile capacity assessment

Pile no.	*EMX approach by Broms*	*CAPWAP result (kN)*
1	4592	5608
2	5700	7437
3	4750	5706
4	4854	4574
5	6520	5615
6	5395	6000
7	7475	7500
8	5500	7040
9	4165	5600
10	5658	7020

6.11.5 Coring and sonic tests to bored pile

For large-diameter bored pile, it is common to core concrete to examine the quality. One of the critical points in the coring test is the interface condition (Figure 6.57) between the pile and bedrock. Without careful control, the interface may be missed easily during the coring. In addition, steel pipe may be left in the piles, and sonic tests will be carried out to examine the quality of the in situ concrete. In general, the denser the concrete, the faster will be the ultrasonic wave. A typical sonic test is shown in Figure 6.57.

The quality of the sonic test may be affected by

1. Actual density contrast in concrete;
2. Reinforcement amount in the vicinity;
3. Distance between emitter and receiver;
4. Distance of travelling path from the edge;
5. Condition of sonic tube – bent or detached from concrete due to shrinkage or other reasons can affect the result;
6. Calibration range of the equipment;
7. The actual density contrast may out of range;
8. Size of tube – in general, a smaller size may give better results.

For the sonic test, the arrangement of the sonic tubes is usually controlled by the length and area of the bored pile. Some of the typical arrangements used in Hong Kong are given in Figures 6.58 and 6.59.

A coring test can locate the interface problem (Figure 6.60) easily and clearly. The limitations of coring tests include:

1. Minor anomalies are not uncommon, even in sound concrete;
2. Signal anomalies may not imply any defects;
3. Only a significant density contrast will give a noticeable response – minor anomalies cannot be easily justified by practical means.

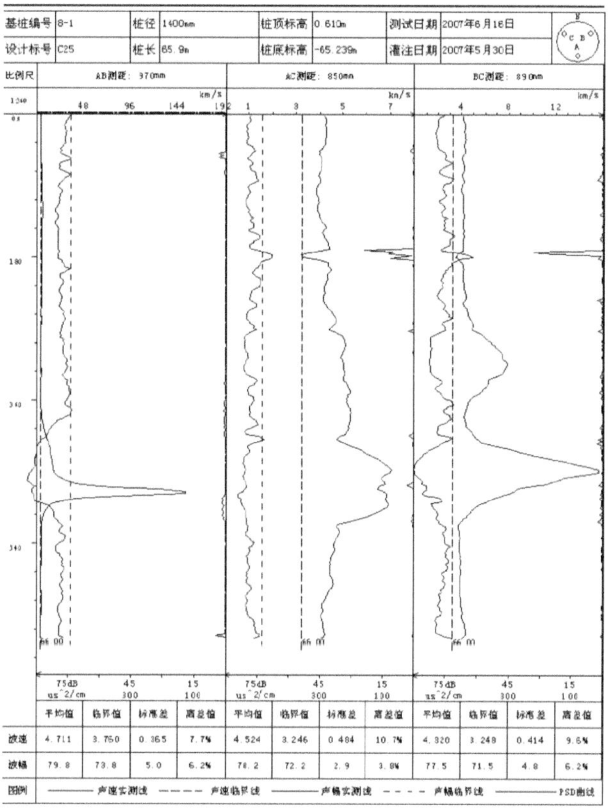

Figure 6.57 Typical sonic test results.

6.12 CASE AND CAPWAP ANALYSIS

For the large-strain PDA tests, the Case method and CAPWAP can be used for the interpretation. In most cases, the wave propagation in pile can be considered as a one-dimensional problem, as a PDA test is more suitable to those small-diameter bored piles or driven piles. For large-diameter bored piles, the three-dimensional effect can be important, but this is not likely in practical applications.

6.12.1 Case method

The case method is based on the basic wave equation developed by Smith as follows:

$$\frac{\partial^2 u}{\partial t^2} = c^2 \frac{\partial^2 u}{\partial x^2} + S(x,t) \tag{6.106}$$

where u is the particle displacement of the pile at location x and time t;
c is the compressive wave velocity in the pile and is equal $\sqrt{E/\rho}$ where E is the Young's modulus and ρ is the density of the pile material; $S(x,t)$ is the soil/rock resistance at location x and time t.

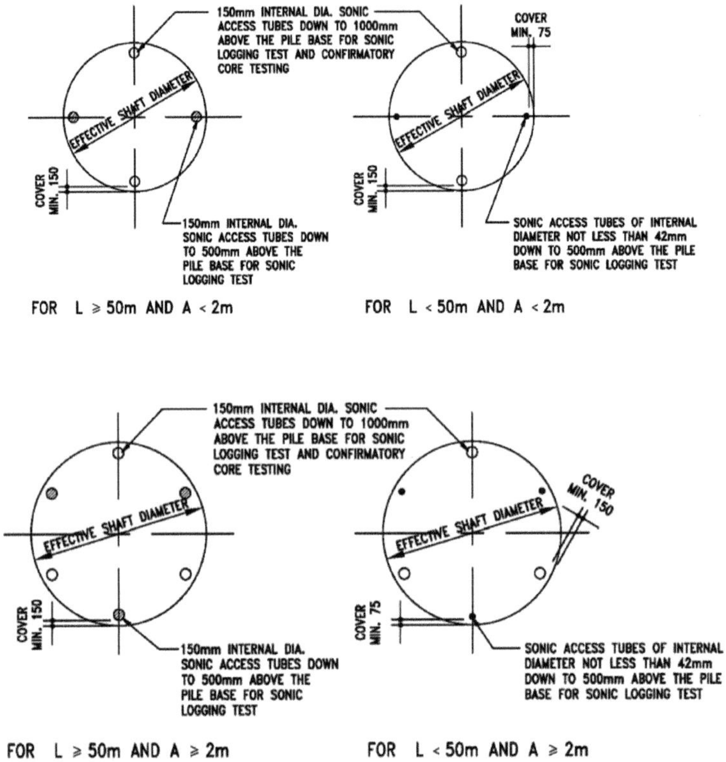

Figure 6.58 Layout of sonic access tubes inside a bored pile, Hong Kong practice.

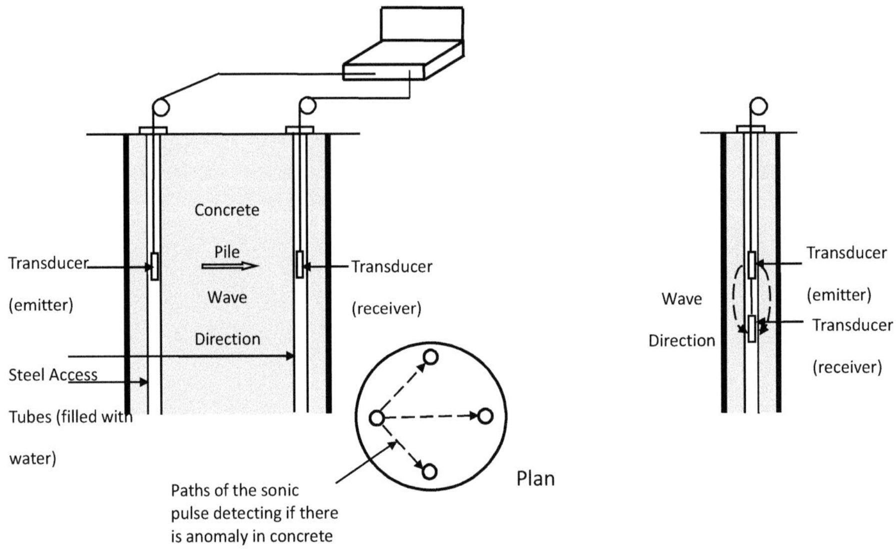

Figure 6.59 Typical twin tube and single tube sonic test.

Figure 6.60 Interface as obtained from a coring test.

The basic assumptions are as follows:

(i) The pile is uniform in section and construction material, i.e. the impedance $Z = AE/c$ is constant;
(ii) The stress wave experiences no energy loss in its transmission through the pile shaft and there are no distortions of signals;
(iii) The resistance to the dynamic component of the force is at the pile toe only, whilst that of the pile shaft is ignored;
(iv) The resistance to the dynamic component is proportional to the particle velocity;
(v) The problem is a one-dimensional wave propagation problem.

6.12.1.1 Derivation of the formula for the case method

Eq. (6.106) describes the particle displacement of the pile at any position x and time t, the general solution of the wave equation is given mathematically as:

$$u(x,t) = h(x - ct) + f(x + ct) \tag{6.107}$$

where h and f are arbitrary functions determined by the initial boundary conditions.

Physically h and f may respectively be regarded as the downward and upward waves travelling in opposite directions but with the same speed c. h and f cannot be measured directly, but the sum of these two waves, which is the actual wave in the pile, can be measured by the strain gauges and accelerometers in such tests as PDA tests. The shapes of the two waves remain unchanged in their course of transmission. However, at any level of the pile the side resistance R will degenerate into two resistance waves (both of magnitude $0.5R$), one of which is an upward compressive wave and the other is a downward tension wave. The resistance wave serves to increase the resistance and decrease the velocity of the pile at its top. As the total resistance of the pile is the sum of the static and dynamic resistance generated by the striking hammer, the case method is used to determine the static resistance, which is the true resistance of the pile in the course of its service life by wave analysis with a number of assumptions. The following sets out the detailed derivation of the formula to be used.

For the downward particle velocity (the actual moving velocity of the section)

$$v_d = \frac{\partial h(x-ct)}{\partial t} = \frac{\partial h(x-ct)}{\partial x} \cdot \frac{\partial x}{\partial t} = \frac{\partial h(x-ct)}{\partial x} \cdot (-c)$$

As strain $\varepsilon_d = \frac{-\partial h(x-ct)}{\partial x} = \frac{v_d}{c}$

the downward force $F_d = AE\varepsilon_d = AE\left(\frac{v_d}{c}\right) = \frac{AE}{c} v_d = Zv_d$, (6.108)

by putting $Z = AE/c$, called the impedance as a mechanical property of the pile. Similarly, it can be deduced that the upward force is

$$F_u = -Zv_u \tag{6.109}$$

where v_u is the upward particle velocity. Therefore, generally, it can be described that the velocity and force at any section of the pile shaft are made up of the upward and downward components as illustrated in Figure 6.61(a). That is:

$$v = v_u + v_d \tag{6.110}$$

$$F = F_u + F_d \tag{6.111}$$

The individual upward or downward components of the wave cannot be measured directly. Instead, the total sum v and F are measured by the PDA. If the measured velocity and force at any section M are, respectively, v_M and F_M, by replacing v and F by v_M and F_M and re-arranging Eq. (6.108) to Eq. (6.111), we can write:

$$v_d = \frac{1}{2}\left(v_M + \frac{F_M}{Z}\right) \tag{6.112}$$

$$v_u = \frac{1}{2}\left(v_M - \frac{F_M}{Z}\right) \tag{6.113}$$

$$F_d = \frac{1}{2}(F_M + Zv_M) \tag{6.114}$$

$$F_u = \frac{1}{2}(F_M - Zv_M) \tag{6.115}$$

At the pile base, if it is free without any base resistance as illustrated in Figure 6.61(b),

$$F = 0 \quad \text{i.e.} \quad F_u = -F_d \tag{6.116}$$

$$v_u = v_d; \quad v = v_u + v_d = 2v_d \tag{6.117}$$

At the pile base, if it is restrained from movement as illustrated in Figure 6.61c

$$v = 0 \quad \text{i.e.} \quad v_u + v_d = 0 \Rightarrow v_u = -v_d \tag{6.118}$$

$$F_u = F_d; \quad F = F_u + F_d = 2F_d \tag{6.119}$$

Eq. (6.119) is important as it shows by theory that the pile tip penetration force is at most twice the pile axial force during driving. Without this dynamic effect, driving of the pile will require a much higher driving force which may damage it.

The actual pile base condition will be between "free" and "restrained", depending on the stiffness of the pile base subgrade. Reference is now made to Figure 6.61(d) for analysis of the side friction. Above and below the section i, the forces and velocities can be formulated as:

$$\text{Above the section: } F_1 = F_{1u} + F_{1d}; \ v_1 = v_{1u} + v_{1d}; \quad F_1 = -Zv_1 \tag{6.120}$$

$$\text{Below the section: } F_2 = F_{2u} + F_{2d}; \ v_2 = v_{2u} + v_{2d}; \quad F_2 = Zv_2 \tag{6.121}$$

$$\text{For equilibrium and compatibility } R_i = F_1 - F_2; \quad v_1 = v_2 \tag{6.122}$$

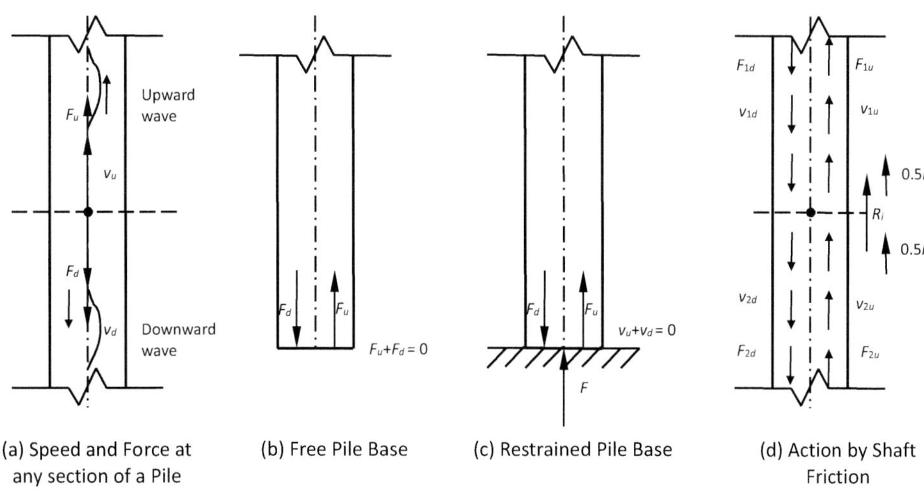

Figure 6.61 Pile forces and velocities.

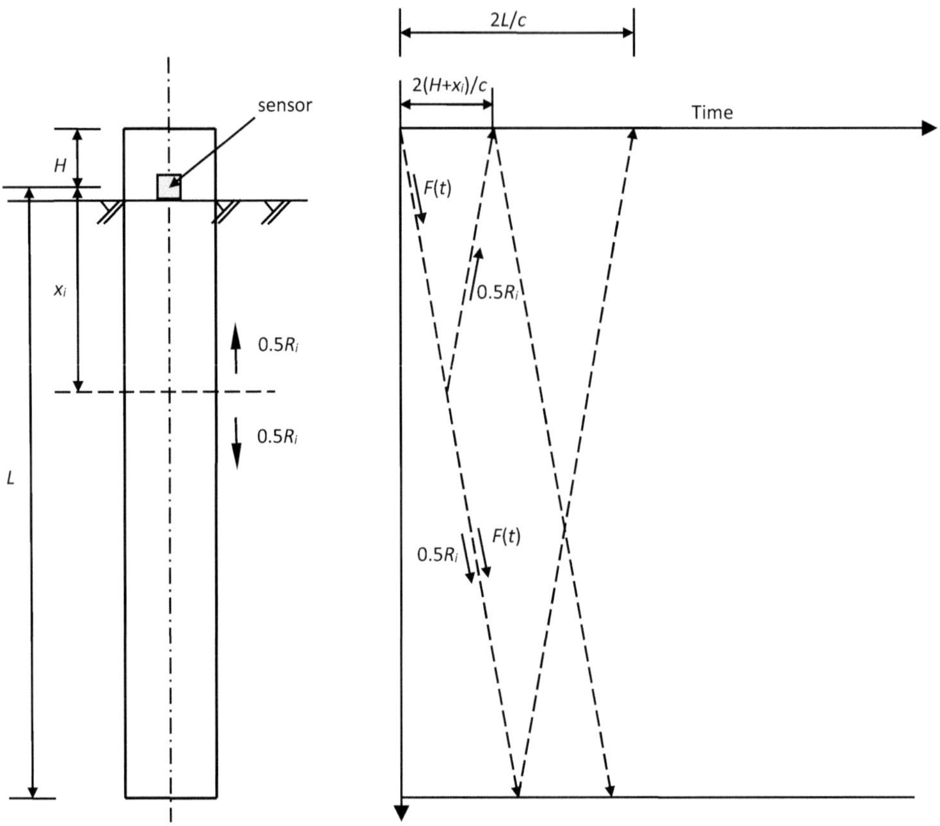

Figure 6.62 Wave transmission in a pile shaft.

As $R_i = F_1 - F_2 = (F_{1u} + F_{1d}) - (F_{2u} + F_{2d}) = (F_{1u} - F_{2u}) + (F_{1d} - F_{2d})$, and solving Eqs. (6.120)–(6.121),

$$\left(F_{1u} - F_{2u}\right) = 0.5R_i \text{ and } \left(F_{1d} - F_{2d}\right) = 0.5R_i \tag{6.123}$$

implying that R_i is split into two equal parts, each of $0.5R_i$. One generates an upward compressive wave and the other a downward tension wave.

Consider Figure 6.62, illustrating the transmission of waves in the pile shaft with particular reference to a section at x_i below ground. The pile is of length L below ground, and its head at height H above ground, cross-sectional area A, Young's modulus E and density ρ (the last two are for determination of wave speed c by $\sqrt{E/\rho}$). A sensor is installed at ground level.

When the hammer strikes the pile at its top, only a compressive downward wave is created with particle velocity $v_M(t) = F_M(t)/Z$ where $v_M(t)$ and $F_M(t)$ are recorded by the sensor.

When the compressive wave is transmitted to the pile tip, which is assumed to be free after time L/c, a reflective wave (tension wave by which $F_u = -F_d = -F$ and $v_u = v_d = v$) is generated and reaches the sensor after time $2L/c$. The wave then subsequently reaches the top level of the pile at which it is reflected back to the sensor at $2(L+H)/c$. So the velocity and force detected by the sensor at time t is due to the wave reaching the sensor at time t and all the previous waves (n numbers) reaching the sensor after reflections at top level and tip of the pile as listed in Eqs. (6.124) and (6.125),

$$\begin{aligned} v_{MF}(t) &= \frac{1}{Z}\left[F(t) + F\left(t - \frac{2L}{c}\right) + F\left(t - \frac{2L}{c} - \frac{2H}{c}\right) + F\left(t - \frac{4L}{c} - \frac{2H}{c}\right)\right. \\ &\quad \left. + F\left(t - \frac{4L}{c} - \frac{4H}{c}\right) + \ldots..\right] \\ &= \frac{1}{Z}\left[F(t) + \sum_{j=1}^{n} F\left[t - \frac{2jL}{c} - \frac{2jH}{c}(j-1)\right] + \sum_{j=1}^{n} F\left[t - \frac{2jL}{c} - \frac{2jH}{c}\right]\right] \end{aligned} \tag{6.124}$$

$$F_{MF}(t) = F(t) - \sum_{j=1}^{n} F\left[t - \frac{2jL}{c} - \frac{2jH}{c}(j-1)\right] + \sum_{j=1}^{n} F\left[t - \frac{2jL}{c} - \frac{2jH}{c}\right] \tag{6.125}$$

As for the side friction R at x_i as indicated in Figure 6.62, similar to the generated upward waves, we can put down

$$\begin{aligned} v_{MRu}(t) &= \frac{1}{Z}\left[\left(-\frac{1}{2}\right)\sum_{j=0}^{n} R_i\left[t - \frac{2x_i}{c} - \frac{2jL}{c} - \frac{2jH}{c}\right]\right. \\ &\quad \left. + \left(-\frac{1}{2}\right)\sum_{i=0}^{n} R_i\left[t - \frac{2x_i}{c} - \frac{2jL}{c} - \frac{2jH}{c}(j-1)\right]\right] \end{aligned} \tag{6.126}$$

$$\begin{aligned} F_{MRu}(t) &= \frac{1}{2}\left[\sum_{j=0}^{n} R_i\left[t - \frac{2x_i}{c} - \frac{2jL}{c} - \frac{2jH}{c}\right]\right. \\ &\quad \left. - \sum_{j=0}^{n} R_i\left[t - \frac{2x_i}{c} - \frac{2jL}{c} - \frac{2jH}{c}(j-1)\right]\right] \end{aligned} \tag{6.127}$$

If there are N segments each giving friction forces R_1, R_2 ...... R_N, then

$$\begin{aligned} v_{MRu}(t) &= \frac{-1}{2Z}\left[\sum_{i=1}^{N}\sum_{j=0}^{n} R_i\left[t - \frac{2x_i}{c} - \frac{2jL}{c} - \frac{2jH}{c}\right]\right. \\ &\quad \left. + \sum_{i=1}^{N}\sum_{j=0}^{n} R_i\left[t - \frac{2x_i}{c} - \frac{2jL}{c} - \frac{2jH}{c}(j-1)\right]\right] \end{aligned} \tag{6.128}$$

$$F_{MRu}(t)=\frac{1}{2}\left[\sum_{i=1}^{N}\sum_{j=0}^{n}R_i\left[t-\frac{2x_i}{c}-\frac{2jL}{c}-\frac{2jH}{c}\right]\right.$$
$$\left.-\sum_{i=1}^{N}\sum_{j=0}^{n}R_i\left[t-\frac{2x_i}{c}-\frac{2jL}{c}-\frac{2jH}{c}(j-1)\right]\right] \tag{6.129}$$

Similarly for the downward waves,

$$v_{MRd}(t)=\frac{-1}{2Z}\left[\sum_{i=1}^{N}\sum_{j=0}^{n}R_i\left[t-\frac{2x_i}{c}-\frac{2jL}{c}-\frac{2jH}{c}(j-1)\right]\right.$$
$$\left.+\sum_{i=1}^{N}\sum_{j=0}^{n}R_i\left[t-\frac{2x_i}{c}-\frac{2jL}{c}-\frac{2jH}{c}\right]\right] \tag{6.130}$$

$$F_{MRd}(t)=\frac{1}{2}\left[\sum_{i=1}^{N}\sum_{j=0}^{n}R_i\left[t-\frac{2x_i}{c}-\frac{2jL}{c}-\frac{2jH}{c}(j-1)\right]\right.$$
$$\left.-\sum_{i=1}^{N}\sum_{j=0}^{n}R_i\left[t-\frac{2x_i}{c}-\frac{2jL}{c}-\frac{2jH}{c}\right]\right] \tag{6.131}$$

The signal v_M and F_M as detected by the sensor is the sum of the above.

$$v_M = v_{MF} + v_{MRu} + v_{MRd} \tag{6.132}$$

$$F_M = F_{MF} + F_{MRu} + F_{MRd} \tag{6.133}$$

Energy loss in the form of radiating damping to the surrounding has been ignored in the above derivation. Therefore, error may be more significant if the testing time is long and involves many time steps.

Consider only times at t_1 and $t_1+\frac{2L}{c}$ where the signals $v_M(t_1), F_M(t_1), v_M\left(t_1+\frac{2L}{c}\right)$, $F_M\left(t_1+\frac{2L}{c}\right)$ are taken and added up (with the v_M multiplied by Z for matching of units) and by Eqs. (6.124) to (6.131), we can arrive at

$$F_M(t_1)+F_M\left(t_1+\frac{2L}{c}\right)+Zv_M(t_1)-Zv_M\left(t_1+\frac{2L}{c}\right)$$
$$=\sum_{i=1}^{N}R_i(t_1)+\sum_{i=1}^{N}R_i\left(t_1+\frac{2L}{c}-\frac{2x_i}{c}\right) \tag{6.134}$$

Assuming R_i is constant, i.e. $R_i(t_1) = R_i\left(t_1 + \frac{2L}{c} - \frac{2x_i}{c}\right)$

So the total pile shaft resistance $R_T = \frac{1}{2}\left[\sum_{i=1}^{N} R_i(t_1) + \sum_{i=1}^{N} R_i\left(t_1 + \frac{2L}{c} - \frac{2x_i}{c}\right)\right]$

$$\Rightarrow R_T = \frac{1}{2}\left[F_M(t_1) + F_M\left(t_1 + \frac{2L}{c}\right)\right] + \frac{Z}{2}\left[v_M(t_1) - v_M\left(t_1 + \frac{2L}{c}\right)\right] \quad (6.135)$$

Eq. (6.135) represents the basic equation for the case method. However, R_T comprises both the static and dynamic components, i.e.

$$R_T = R_{static} + R_{dynamic} \Rightarrow R_{static} = R_T - R_{dynamic} \quad (6.136)$$

R_{static} is to be found, which is the capacity of the pile during its service life.

Assuming $R_{dynamic}$ originated from the pile tip and that $R_{dynamic}$ is directly proportional to the velocity where the constant of proportionality is a "damping coefficient", J_p. That is

$$R_{dynamic}(t) = J_p v_{toe}(t) \quad (6.137)$$

When the wave generated by the blow at the top reaches the pile tip, we can write

$$F_{toe} = F_M(t) - \frac{1}{2}\sum_{i=1}^{N} R_i(t) = F_M(t) - \frac{1}{2}R_T(t) \quad (6.138)$$

$$v_{toe} = 2v_d = 2\frac{F_{toe}}{Z} = \frac{2}{Z}\left[F_M(t) - \frac{1}{2}R_T(t)\right] = \frac{1}{Z}\left[2F_M(t) - R_T(t)\right] \quad (6.139)$$

Putting $J_c = J_p/Z$ which is called the "lumped case damping factor" and substituting Eqs. (6.135), (6.137) and (6.139) into Eq. (6.136) gives

$$\begin{aligned} R_{static}(t_1) = {} & \frac{1}{2}\left[F_M(t_1) + F_M\left(t_1 + \frac{2L}{c}\right)\right] + \frac{Z}{2}\left[v_M(t_1) - v_M\left(t_1 + \frac{2L}{c}\right)\right] \\ & - J_c Z \times \frac{1}{Z}\left[2F_M(t) - \frac{1}{2}\left[F_M(t_1) + F_M\left(t_1 + \frac{2L}{c}\right)\right]\right. \\ & \left.\left. - \frac{Z}{2}\left[v_M(t_1) - v_M\left(t_1 + \frac{2L}{c}\right)\right]\right]\right] \end{aligned}$$

After simplifying, the static component of resistance of the pile at time t_1 is

$$R_{static}(t_1) = \frac{(1-J_c)}{2}\left[F_M(t_1) + Zv_M(t_1)\right] + \frac{(1+J_c)}{2}$$
$$\left[F_M\left(t_1 + \frac{2L}{c}\right) - Zv_M\left(t_1 + \frac{2L}{c}\right)\right] \tag{6.140}$$

By Eq. (6.140), the ultimate static resistance of the pile at time t_1 can be estimated if the pile driving analyser (PDA) can produce readings of forces and velocities at time t_1 (any chosen time) and $t_1 + \frac{2L}{c}$ which are $F_M(t_1), v_M(t_1); F_M\left(t_1 + \frac{2L}{c}\right),\ v_M\left(t_1 + \frac{2L}{c}\right)$ and with the assumed values of J_c. In the equation, the following symbols are reiterated:

R_{static}	is the static resistance of the pile at pile at time t_1;
J_c	is the "lumped case damping factor" which needs to be assumed or calibrated;
L	is the length of the pile below ground;
c	is the wave velocity in the pile shaft and can be determined by $c = \sqrt{E/\rho}$ where E and ρ are, respectively, the Young's modulus and density (mass, not weight per unit volume) of the pile material;
Z	is the impedance of the pile determined by $Z = AE/c$ where A is the cross-sectional area of the pile;
$F_M(t_1)$	is the force reading detected by the PDA sensor at time t_1;
$v_M(t_1)$	is the velocity reading detected by the PDA sensor at time t_1;
$F_M\left(t_1 + \frac{2L}{c}\right)$	is the force reading detected by the PDA sensor at time $t_1 + \frac{2L}{c}$;
$v_M\left(t_1 + \frac{2L}{c}\right)$	is the velocity reading detected by the PDA sensor at time $t_1 + \frac{2L}{c}$

6.12.2 Large strain pile driving wave equation back analysis

Based on the theory of characteristics, the integral along the positive and negative directions of the characteristics line gives the following iterative form:

$$\text{Along } \frac{dx}{dt} = C, Z_{i-1}V_{i,j} + F_{i,j} = Z_{i-1}V_{i-1j} + F_{i-1j} - S^{+}_{i-1} \tag{6.141}$$

$$\text{Along } \frac{dx}{dt} = -C, Z_iV_{i,j} - F_{i,j} = Z_iV_{i+1,j-1} + F_{i+1,j-1} - S^{-}_{i} \tag{6.142}$$

$$\text{where } S_{i-1}^{+} = \frac{S_u(i-1)}{O_k(i-1)}\left(u_{i-1j} - u_{pi-1j}\right)\left(1 + J_{si-1}V_{i-1,j}\right) \tag{6.143}$$

$$S_i^{-} = \frac{S_u(i)}{Q_k(i)}\left(u_{i+1,j-1} - u_{pi+1,j-1}\right)\left(1 + J_{si}V_{i+1,j-1}\right) \tag{6.144}$$

Where Z, V, F, Q, S_u, Q and J are the impedance, velocity of wave, force in pile, elastic limit of skin friction, ultimate skin resistance, quake and damping constant of the Smith model, respectively. By using the boundary conditions on the top of the pile, the above-mentioned problem will be satisfactorily answered. The pile is divided into n elastic bar units for analysis. For each bar unit, it is assumed that all the soil resistances act at the bottom of the unit, and the impedance of the pile changes only at the interface between two adjoining units, and wave distortion does not occur inside the units. After the time interval Δt, upward travelling wave $P_u(i-1,j)$ and downward travelling wave $P_d(i-1,j-1)$ become $P_{(i,j)}$ at the top of the unit. When $t = j\Delta t$, considering the transmission and reflection of the upward and downward travelling waves, and the upward and downward travelling waves of the skin resistance $S(i,t)$, the upward and downward travelling waves at section i are, respectively, given as follows:

$$P_u(i,j) = 2\cdot\frac{Z_i}{Z_{i+1}+Z_i}P_u(i+1,j) + \frac{Z_{i+1}-Z_i}{Z_{i+1}+Z_i}P_d(i,j-1) + \frac{Z_i}{Z_{i-1}+Z_i}S(i,j) \tag{6.145}$$

$$P_d(i,j) = 2\cdot\frac{Z_i}{Z_i+Z_{i-1}}P_u(i-1,j-1) + \frac{Z_{i-1}-Z_i}{Z_i+Z_{i-1}}P_u(i,j-1) - \frac{Z_{i+1}}{Z_i+Z_{i+1}}S(i,j) \tag{6.146}$$

At the pile tip, the upward and downward travelling waves will be:

$$P_d\left(N_{p,j}\right) = P_d\left(N_{p-1,j-1}\right) \tag{6.147}$$

$$P_u\left(N_{p,j}\right) = -P_d\left(N_{p-1,j-1}\right) + S\left(N_{s,j}\right) + S\left(N_{s+1,j}\right) \tag{6.148}$$

At time t, the measured force and speed on the top of pile are $P_m(j)$ and $V_m(j)$. If we take the location of the sensor as the boundary and the measure velocity as the boundary condition, then we have

$$P_u(1,j) = P_u(2,j-1) \tag{6.149}$$

$$P_d(1,j) = ZV_m(j) + P_u(2,j-1) \tag{6.150}$$

Therefore, the force–time curve obtained by calculation will be given by

$$P_c(j) = P_d(i,j) + P_u(i,j) = ZV_m(j) + 2P_u(2,j-1) \tag{6.151}$$

The particle velocity and displacement $V(i,j)$ and $d(i,j)$ could be given by the equations as follows:

$$V(i,j)=\frac{P_d(i,j)}{Z_i}-\frac{P_u(i,j)}{Z_{i+1}} \tag{6.152}$$

$$d(i,j)=d(i,j-1)+\frac{\Delta t}{2}\left[V(i,j-1)+V(i,j)\right] \tag{6.153}$$

If the Smith soil model is used, three parameters will be defined: the maximum static resistance S_u, the largest elastic deformation Q and the damping coefficient J_s. Soil resistance $S(i,j)$ is divided into static resistance $S_s(i,j)$ and dynamic resistance $S_d(i,j)$ as given by

$$S(i)=S_s(i,j)+S_d(i,j) \tag{6.154}$$

in which the static soil resistance of the Smith model is

$$S_s(i,j)=\frac{S_u(i)}{Q(i)}\left[S(i,j)-DE(i,j)\right] \tag{6.155}$$

$$DE(i,j)=\begin{cases} d(i,j)-Q(i) & \text{when } d(i,j)-DE(i,j-1)\geq Q(i) \\ DE(i,j-1) & \text{when } -Q(i)\leq d(i,j)-DE(i,j-1)\leq Q(i) \\ d(i,j)+Q(i) & \text{when } -Q(i)>d(i,j)-DE(i,j-1) \end{cases} \tag{6.156}$$

where $DE(i,j)$ is the plastic soil displacement. The dynamic soil resistance of the Smith model is

$$R_d(i,j)=\left|R_s(i,j)\right|J_s(i)V(i,j) \tag{6.157}$$

where $J_s(i)$ is the soil Smith damping coefficient with a dimension s/m. In order to fine-tune the calculated force as compared with the measured force, a soil particle can be added at the pile tip so that at any time t, there is a soil-particle inertia force $S_{SM}(j)$ on the pile tip given by,

$$R_{SM}(j)=W_s\left[V(mp,j)-V(mp,j-1)\right]/(g\Delta t) \tag{6.158}$$

where W_s is the soil mass defined by the engineer, $V(mp,j)$ is the velocity of the particle on the pile tip, g is acceleration due to gravity and Δt is the time step.

A damper with a damping coefficient J_{ms} could be added to the pile tip, which is used to simulate the energy dissipation near to the pile tip. The damping force $R_{dm}(j)$ is given by

$$R_{dm}(j)=J_{ms}V(mp,j) \tag{6.159}$$

Therefore, on the pile tip,

$$R(N_p,j) = R_s(N_p,j) + R_d(N_p,j) + R_{sm}(j) + R_{dm}(j) \tag{6.160}$$

According to the above-mentioned formulas, the force–wave curve $F_c(t)$ can be calculated according to the measured $V_{m(t)}$ for different times and segments by successive cycling of the above processes.

Generally, the error between the calculated and measured force values can be represented by an objective function *F(x)* as

$$F(x) = \sum_{j=1}^{N_{time}} ABS\left(F_c(j) - F_m(j)\right) \tag{6.161}$$

$$\text{where } x = \left\{\left(R_u(i), Q(i), J_s(i), R_t, Q_t, ff, W_s, J_{ms}\right), i = 1, N_{pile}\right\} \tag{6.162}$$

and N_{time} is usually taken as 1024 points so that fast Fourier transform can be performed easily for processing of the signal. The minimization of Eq. (6.161) is well-known to have many local minima during the analysis, and great care and experience are required for the proper analysis. According to the CAPWAP or similar procedures, the soil parameters and skin friction and pile capacity are varied until the calculated force–wave curve matches well with that from the measurement. This is actually an inverse analysis of the pile driving wave equation, which is a difficult process.

Actually, the authors have noticed some design calculations by engineers where the matching is good but the soil parameters are not realistic. In some commercial programs, the users are allowed to choose the option of fully automatic matching or manual tuning of the soil parameters in the matching. Some inexperienced engineers simply adopt the automatic matching option without considering the acceptability of the soil parameters. To deal with Eq. (6.161), it is better to use a constrained optimization method so that unrealistic soil parameters will not be tried during the analysis. This is also the approach used by Cheng, and more details can be found in the works by Cheng et al. (2020). Sample programs for the forward analysis (similar to WEAP and others) or the backward analysis (similar to CAPWAP or others) can be obtained from Cheng at natureymc@yahoo.com.hk

6.12.3 Simplex method for signal matching solution

Eq. (6.161) is adopted for analysis of the signals obtained from a PDA test so that the pile capacity and soil parameters can be back-analysed. In program CAPWAP/C, the numerical method is not given, and the user can only use the program as a black box. Cheng has developed similar programs using three numerical methods, and the computer coding can be obtained from Cheng for education purposes. The first one is based on the simplex method through which the objective function values of the simplex vertices constituted by the n +1 vertices of the n-dimensional solution space are compared. The worst point is then removed and substituted by a new and better point which will then gradually approach the minimum point. For a two-dimensional problem, the simplex is a triangle, while for a three-dimensional problem, it will be a tetrahedron.

The initial simplex is usually taken as a regular domain with all sides equal, which is known as the regular simplex. The method to constitute the initial simplex is to define an initial vertex as

$$X^{(0)} = \left[a_1, a_2, \cdots a_n\right]^T \tag{6.163}$$

The remaining n vertices are as follows:

$$\begin{aligned} X^{(1)} &= \left[a_1 + p, a_2 + q, \cdots a_n + q\right]^T \\ X^{(2)} &= \left[a_1 + q, a_2 + p, \cdots a_n + q\right]^T \\ &\cdots\cdots\cdots\cdots \\ X^{(n)} &= \left[a_1 + q, a_2 + q, \cdots a_n + p\right]^T \end{aligned} \tag{6.164}$$

The rule to form $X^{(1)}$ to $X^{(n)}$ is: for $X^{(i)}$, its ith component will be $a_i + p$, and all the remaining components are defined as $a_j + q$. In the formula, p and q can be easily decided by the sides. If we assume that the length of the side is A, the length connecting $X^{(1)}$ and $X^{(0)}$ will be

$$\begin{aligned} \left\|X^{(1)} - X^{(0)}\right\| &= \sqrt{\left(a_1 + p - a_1\right)^2 + \left(a_2 + p - a_2\right)^2 + \cdots + \left(a_n + p - a_n\right)^2} \\ &= \sqrt{p^2 + (n-1)q^2} = A \end{aligned} \tag{6.165}$$

For the side connecting $X^{(2)}$ and $X^{(1)}$, the length will be

$$\begin{aligned} \left\|X^{(2)} - X^{(1)}\right\| &= \sqrt{\left(a_1 + q - a_1 - p\right)^2 + \left(a_2 + p - a_2 - q\right)^2 + \cdots + \left(a_n + q - a_n - q\right)^2} \\ &= \sqrt{(q-p)^2 + (p-q)^2} = \sqrt{2(p-q)^2} = A \end{aligned} \tag{6.166}$$

The distance between any other two points is given by either Eq. (6.165) or Eq. (6.166). The simultaneous solution of the two equations will give

$$\begin{cases} p = \dfrac{\sqrt{n+1} + n - 1}{n\sqrt{2}} A \\[2ex] q = \dfrac{\sqrt{n+1} - 1}{n\sqrt{2}} A \end{cases} \tag{6.167}$$

The search process for the simplex method consists of four steps, including reflection, expansion, contraction and reducing the side length, which are introduced as follows.

1. Reflection

Calculate the function values of the initial simplex of n +1 vertices and find the worst point $X^{(h)}$ and the best point $X^{(e)}$, that is

$$\begin{cases} f\left(X^{(h)}\right)=\max\left\{f\left(X^{(0)}\right),f\left(X^{(1)}\right),\cdots,f\left(X^{(n)}\right)\right\} \\ f\left(X^{(e)}\right)=\min\left\{f\left(X^{(0)}\right),f\left(X^{(1)}\right),\cdots,f\left(X^{(n)}\right)\right\} \end{cases} \tag{6.168}$$

Remove $X^{(h)}$ and calculate the centre $X^{(n+1)}$ of the remaining n vertices, that is,

$$X^{(n+1)} = \frac{1}{n}\left(\sum_{i=0}^{n} X^{(i)} - X^{(h)}\right) \quad (i \neq h) \tag{6.169}$$

Calculate the reflection point $X^{(n+2)}$ as

$$X^{(n+2)} = X^{(n+1)} + \alpha\left(X^{(n+1)} - X^{(h)}\right) \tag{6.170}$$

where α is a reflection coefficient and is generally taken as 1, then

$$X^{(n+2)} = 2X^{(n+1)} - X^{(h)} \tag{6.171}$$

2. Expansion

The new simplex is formed after removing $X^{(h)}$ and adding $X^{(n+2)}$. If $X^{(n+2)}$ in the new simplex is not the worst point [that is, $f\,(X^{(n+2)}) < f\,(X\,(h))$], but not the best point, the key is to find $X^{(h)}$ and $X^{(e)}$ based on Eq. (6.171), and reflect again. If $X^{(n+2)}$ in the new simplex is the best point, that is $f\left(X^{(n+2)}\right) \leq f\left(X^{(e)}\right)$, carry on the expansion and find the next expansion point.

$$X^{(n+3)} = X^{(n+1)} + \beta\left(X^{(n+2)} - X^{(n+1)}\right) \tag{6.172}$$

where β is the expansion coefficient and β > 1. If $f\left(X^{(n+3)}\right) \leq f\left(X^{(n+2)}\right)$, $X^{(n+2)}$ is substituted by $X^{(n+3)}$.

3. Contraction

If $X^{(n+2)}$ in the new simplex from which $X^{(h)}$ has been removed is the worst point, that is,

$$f\left(X^{(h)}\right) > f\left(X^{(n+2)}\right) > f\left(X^{(i)}\right) \quad (i = 1,2,\cdots,n) \tag{6.173}$$

Give up $X^{(n+2)}$ and find the contraction point $X^{(n+4)}$ between $X^{(n+1)}$ and $X^{(n+2)}$ as

$$X^{(n+4)} = X^{(n+1)} + \gamma\left(X^{(n+2)} - X^{(n+1)}\right) \tag{6.174}$$

where γ is the contraction coefficient and $0 < \gamma < 1$. If $f\left(X^{(n+4)}\right) < f\left(X^{(h)}\right)$, $X^{(h)}$ is substituted by $X^{(n+4)}$.

4. Reducing the side length

If $f\left(X^{(n+4)}\right) \geq f\left(X^{(h)}\right)$, keep the best point $X^{(e)}$ fixed, and the other n points approaching $X^{(e)}$, so that the length of the simplex could reduce by a half, that is

$$X^{(i)} = X^{(e)} + \frac{1}{2}\left(X^{(i)} - X^{(e)}\right) \quad (i = 0, 1, \cdots, n, \text{but } i \neq \text{e}) \tag{6.175}$$

In this way we can get a new simplex, and then process the above four steps again. For each of the new simplexes, the termination criterion as given below is examined

$$\left\{\frac{1}{n+1}\sum_{i=0}^{n}\left[f\left(X^{(i)} - f\left(\overline{X}^{(n+1)}\right)\right)\right]^2\right\}^{\frac{1}{2}} \leq \varepsilon \tag{6.176}$$

If the termination criterion is satisfied, then stop the search. The centre point $\overline{X}^{n+1}$ of the $(n+1)$ vertices is the X^* associated with the minimum value of the objective function. Other termination criteria have also been used with success. The simplex method has been found to be highly effective and efficient for many problems, but the algorithm can also become highly inefficient for some other problems, particularly highly nonlinear problems. Cheng has also adopted the simulated annealing method and the generalized reduced gradient (GRG) method for the present problem with satisfaction, and the resulting codes can perform similarly to the CAPWAP program which is popular in Hong Kong and many other countries. The performance of the simplex method as compared with the simulated annealing method and the GRG method (Lasdon et al. 1974) is difficult to assess as it appears to be case dependent, but the simulated annealing method appears to be the most stable method over a variety of problems, while the GRG method is the fastest method in general.

For the large strain pile test results in Figure 6.63, the length and diameter of the precast pretensioned concrete pile are 26 m and 450 mm, respectively. Based on the time for the reflected signal to travel to the pile top, the Young's modulus of the concrete pile can be back-calculated as 45250 MPa. The maximum impact force from the diesel hammer is 4787 kN. With the simplex method, the end-bearing, skin friction and static pile capacity are found to be 4103 kN, 1942 kN and 5853 kN, respectively, with a prescribed damping factor of 0.1 s/m and a quake of 2.5 mm using manual signal matching. If completely automatic signal matching is used (signal matching approach 2), an even better signal matching is obtained with a damping factor of 0.5 s/m and a quake of 1.3 mm (which are outside the normal ranges in Hong Kong), and the static pile capacity is estimated to be 5415 kN with some major fluctuation in the skin friction at the middle of the pile. Multiple solutions

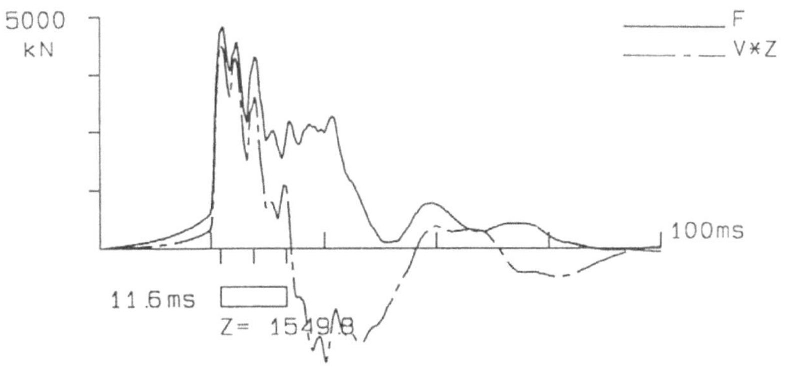

Figure 6.63 A PDA test in Hong Kong, for a precast pretensioned pile.

in signal matching are commonly encountered by many engineers, and major differences between soil parameters and pile capacities obtained from different sets of matching are not uncommon. The static pile capacity by the CASE analysis (case damping factor = 0.15) is 5765 kN, while the pile capacity is found to be 6231 kN from a static load test. Based on the harmony/particle search optimization, Eq. (6.151) is minimized by assigning an upper and lower bound to each variable (quake, damping, soil parameters, skin resistance and base capacity) based on the experience of the engineer, and a static pile capacity of 5946 kN is obtained by the harmony search method (different parameters for different segments of pile). Since the upper and lower bounds are established by soil mechanics principles and experience, unrealistic values can be avoided in the signal matching process. This approach possesses the advantages of manual control as well as automatic signal matching, and has been found to be efficient based on some projects in Hong Kong. It should also be noted that the multiple maxima after the first maximum are the results of the wave reflection at the connections between individual pile segments, for which butt welding to the steel end plate is used which will create a zone of higher stiffness.

6.13 MONITORING PLAN

Some monitoring plans have been discussed in Chapters 3 and 5, and the monitoring for piling works are discussed in this section. In Hong Kong, the alarm level is divided into three levels – Alert, Alarm and Action – according to the monitored values and the previously agreed values or values specified in the government codes or particular specification. For the typical monitoring plan of the prebored driven H-pile, ground movement, differential settlement and vibration are monitored (Figure 6.64) to safeguard the adjacent buildings with raft foundations only. The frequency of monitoring will depend on the item and various factors. In general, settlement monitoring at a frequency of twice each week is adequate. For the vibration, the frequency of monitoring will be two to three times each day using manual measurement, or once every half hour for IoT monitoring.

The locations for the measurements in Figure 6.65 are generally determined from the locations of works, adjacent buildings and ground conditions. To minimize the vibrations and sound during piling, a monitoring plan as shown in Figures 6.66 and 6.67 is proposed for this project. For such work, IoT monitoring without human intervention is the best solution.

The monitoring is based on Alert-Alarm-and-Action (AAA) mechanism.

Instrument	Alert		Alarm		Action		Checkpoint		Frequency (per day)	
	PNAP	Proposed	PNAP	Proposed	PNAP	Proposed	PNAP	Proposed	PNAP	Proposed
Differential Settlement	1:600	1:800	1:450	1:500	1:300	1:300	-	-	-	1
Ground Settlement Monitoring Marker	12	12	18	18	25	25	-	SM1-SM9	-	1
PPV for village house (Tai Kiu Village), mm/s (Continuous/Transient)	-	1.5	-	2	3	3	-	V5,V6,V9, V10,V11,V12	-	3
PPV for low-rise building, mm/s (Continuous/Transient)	-	3.7	-	5.6	7.5	7.5	-	V2,V3,V4	-	1
PPV for nullah, mm/s (Continuous/Transient)	-	7.5	-	11.2	15 (PNAP)	15	-	V1,V7,V8	-	1

Figure 6.64 Monitoring plan for a drive H-pile with preboring in Hong Kong.

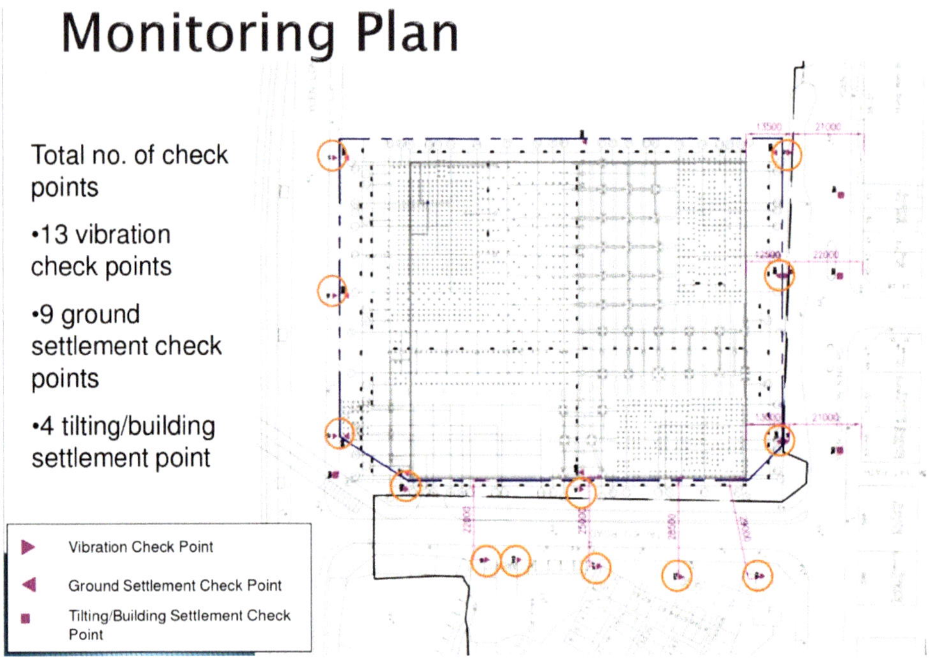

Figure 6.65 Settlement and vibration check.

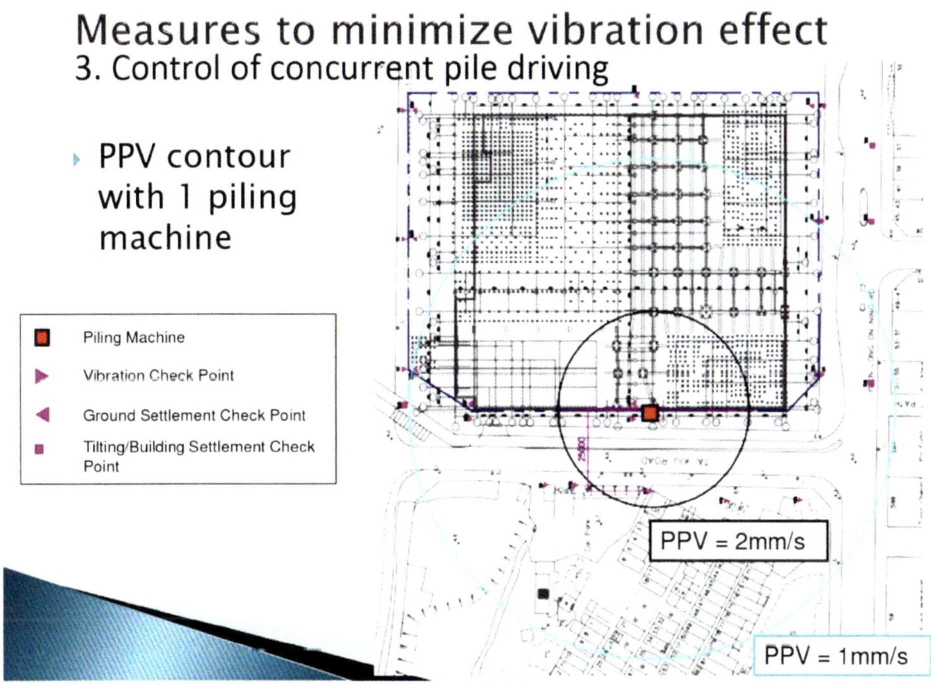

Figure 6.66 Vibration control for one piling machine on site.

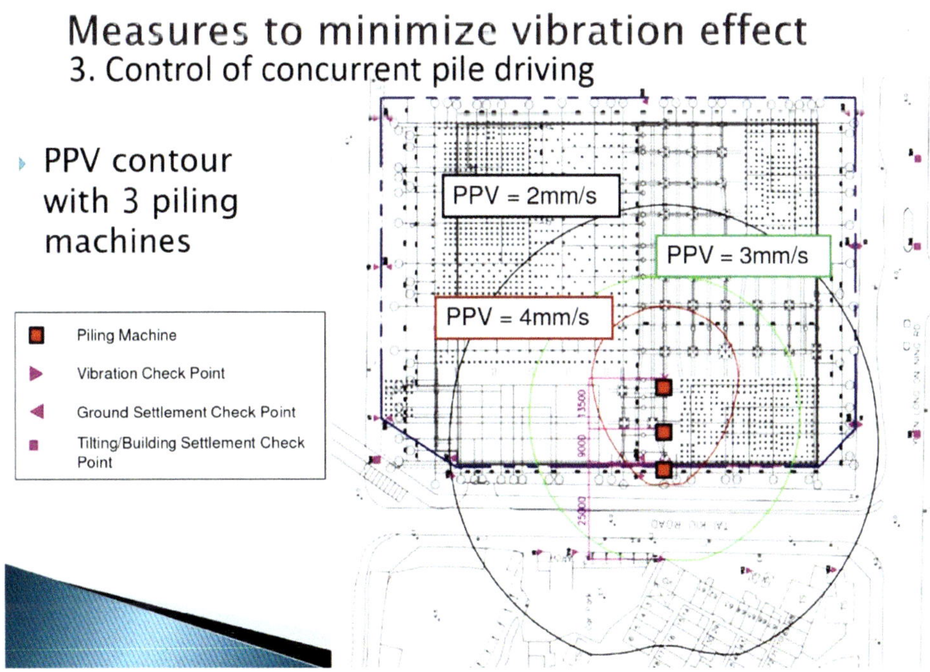

Figure 6.67 Vibration control for three piling machines on site.

References

Berezantsev VG, Khristoforov VS and Golubkov VN (1961), Load bearing capacity and deformation of piled foundation, *Proceedings of the 5th International Conference on Soil Mechanics*, II, 11–15.

Canadian Geotechnical Society (2006), *Canadian Foundation Engineering Manual*, BiTech Publisher Ltd.

Chan, G, Lui, JYH, Lam, K, Yin, KK, Law, CW, Lau, R, Chan, A, & Hasle, R (2004), Shaft grouted friction barrette piles for a super high-rise building, *Proceedings, New Perspectives in the Design and Construction of Foundation Structures*, HKIE, 83–98.

Cheng YM, Law CW and Liu LL (2021), *Foundation analysis and design*, CRC Press.

Lasdon L.S., Fox R.L., Ratner M.W. (1974), Nonlinear optimization using the generalized reduced gradient method. Revue française d'automatique, informatique, recherche opérationnelle. *Recherche opérationnelle*, 8(V3), 73–103. www.numdam.org/item/RO_1974__8_3_73_0/

Law CW and Cheng YM (2006), *Improved thick plate analysis and design*, HKIE Transaction, pp. 49–58.

Lui, SPY, Cheung, SPY and Chan, AKC (1993), Pressure grouted minipiles for a 12-storey residential building at the mid-levels scheduled area in Hong Kong, *Proceedings second international conference on soft soil engineering*, Guangzhou, pp. 419–424.

Mylonakis G and Gazetas G (1998), Settlement and additional internal forces of grouped piles in layered soil, *Geotechnique*, 48(1), 55–72.

Plumbridge, GD, Littlechild, BD, Hill, SJ and Pratt, M (2000). Full scale shaft grouted piles and barrettes in Hong Kong–A first. *Proceedings of the nineteen annual seminar of the geotechnical division of the HKIE*, Hong Kong, pp. 157–166.

Poskitt TJ (1991), Energy lossess in pile-driving due to soil rate effects and hammer misalignment, *Proceedings of the Institution of Civil Engineers*, 91(4), 823–851.

Poulos HG and Davis EH (1980), *Pile foundation analysis and design*, John Wiley & Sons.

Randolph, MF (1987). *PIGLET, a computer program for the analysis and design of pile groups*. Report Geo 87036. Department of Civil Engineering, University of Western Australia, Nedlands.

Rausche F, Goble GG, Garland E and Likins Jr. (1985), Discussion of "*Dynamic determination of pile capacity*" March, ASCE, *Journal of Soil Mechanics and Foundations*, 1985, 111(3), 367–383.

Rausche F, Likins GE and Hussein M (1988), *Pile integrity evaluation by impact methods, Third Intl. Conference on the application of stress wave theory on piles*, BiTech Publisher Ltd., Vancouver, pp. 44–55.

Rausche F, Moses F and Goble GG (1972), Soil resistance predictions from pile dynamics, *ASCE, Journal of Soil Mechanics and Foundations*, SM9, September, 917–937.

Smith EA (1962), Pile driving analysis by the wave equation, *Transactions on ASCE*, 127(part 1), 1145–1193.

Smith EAL (1960), Pile driving analysis by the wave equation. *Journal of the Soil Mechanics and Foundations Division*, ASCE, 86, 35–64.

Sze, JWC, Lam, AKM, Pappin, JW and Chan, KM (2007), Design and construction of shaft-grouted friction barrette in Tung Chung designated area, *Proceedings of the HKIE geotechnical division annual seminar 2007*, HKIE, pp. 299–304.

Terzaghi K (1955), Evaluation of coefficient of subgrade reaction, *Geotechnique*, 5(4), 297–326.

Further Reading

Bolton M, Kitamura A, KusaKabe O and Terashi M (2021), *New horizons in piling, development and application of press-in piling*, CRC Press.

Bowles JE (1996), *Foundation analysis and design*, 5th ed., The McGraw-Hill Companies, Inc.
Broms BB (1964a), Lateral resistance of piles in cohesive soils, *J.S.M.F.D., ASCE* 90(SM2), 27–63.
Broms BB (1964b), Lateral resistance of piles in cohesionless soils, *J.S.M.F.D., ASCE* 90(SM3), 123–156.
Cheng YM (2004), Nq factor for pile foundation by Berezantzev, *Geotechnique*, LIV, 149–150.
Cheng YM, Wang JH, Li L and Fung WH (2020), *Frontier in civil engineering vol. 4: Numerical methods in geotechnical engineering part II*, Benjamin Press.
Fleming K, Weltman A, Randolph M and Elson K (2009), *Piling engineering*, 3rd ed., Taylors and Francis.
Guo WD (2013), *Theory and practice of pile foundations*, CRC Press.
Hertlein B and Davis A (2006), *Nondestructive testing of deep foundations*, John Wiley.
Kaynia AM (2022), *Analysis of pile foundations subject to static and dynamic loading*, CRC Press.
Mays T (2015), *Design guide for pile caps*, Concrete Reinforcing Steel Institute.
Rajapakse R (2016), *Pile design and construction rules of thumb*, 2nd edition, Elsevier.
Reese LC and Impe WV (2011), *Single piles and pile groups under lateral loading*, CRC Press.
Tomlinson M and Woodward J (2008), *Pile design and construction practice*, 5th ed., Taylors and Francis.
Viggiani C, Mandolini A and Russo G (2012), *Piles and pile foundations*, Spon Press.

Chapter 7

Slope stability analysis and stabilization

7.1 GENERAL INTRODUCTION

Landslides have been a problem in many developed cities all over the world. In Hong Kong, landslides have caused loss of life and property damage. In the 50 years after 1947, more than 470 people were killed by failures of mostly cut slopes, fill slopes and some retaining walls. Since 1977, the government has spent lots of money and effort in slope stabilization, however, there is still an average of about 300 incidents each year. On 18 June 1972, a 4 0m high fill slope embankment, as shown in Figure 7.1, at Sau Mau Ping Estate collapsed and killed 71 people. This accident was followed by the collapse of the hillside above a steep temporary excavation at Po Shan Road which triggered a landslide that destroyed a 12-storey residential building and killed 67 people. Some more famous slope failure cases from Hong Kong and China are given in Figures 7.1–7.5. There have been many slope failure cases in China in the last 10 years, and these cases clearly illustrate the importance of proper slope stability analysis and stabilization. There are also active research works in slope stability analysis and stabilization in Hong Kong, due to the needs of the society.

Many methods of analysis have been developed for slope stability analysis over the last 50 years, and there are still many new methods that have evolved in the recent years. Broadly speaking, the methods can be grouped under the following categories:

1. Limit equilibrium method (LEM) – most popular up to the present;
2. Limit analysis method (LAM), including the discontinuity layout optimization method (DLO);
3. Plasticity formulation;
4. Finite element/difference method in form of strength reduction method (SRM) – another option among the engineers;
5. Distinct element method (DEM);
6. Spectral element method (SEM);
7. Meshless method – smoothed particle hydrodynamics method (SPH) and material point method (MPM).

Under these seven major categories, there are also many variations. It is interesting to note that the classical finite element approach is seldom adopted for slope stability analysis, as it is not easy to define the initial stresses, constitutive model as well as the load path precisely. The strength-reduction approach is hence adopted instead of the classical

DOI: 10.1201/9781032650586-7

Figure 7.1 Fill slope failures at Sau Mau Ping, Hong Kong, 1972.

Figure 7.2 Fill slope failure at Po Shan Road, Hong Kong, 1972.

Figure 7.3 Natural slope failure at Sum Wan, Hong Kong 1995.

Figure 7.4 Fill slope failure at Shenzhen, China, 2015.

finite element approach in practice. Most engineers and researchers hence still consider the ultimate limit state of the system for the assessment of many slope stability, bearing capacity and lateral earth pressure problems, and LEM is the most popular approach used by engineers at present. Currently, there are more than 30 limit equilibrium formulations in the literature, but only about five of them are commonly used in routine design by

Figure 7.5 A large-scale three-dimensional slope failure at Sai Kung, Hong Kong in 2016.

engineers. Many commercial programs are also available for LEM analysis. The finite element method in the form of the strength reduction approach (SRM) is the second most popular approach used for practical works, and there are also several commercial programs available for SRM analysis. For the other methods, they are not commonly adopted for routine engineering design, and only limited computer programs are available for engineers and researchers.

Although all slope failures are three-dimensional (3D) in nature, two-dimensional (2D) analyses are considered in most practical cases. The reasons behind such a phenomenon are: (1) insufficient site investigations to define a good 3D ground profile in most projects; (2) very limited computer programs are available for 3D analysis; (3) tedious to define a 3D model and to perform the analysis; (4) difficulty in the analysis of results and visualization; (5) 3D analysis will tend to produce a slightly larger factor of safety, if the same soil parameters are used for 2D and 3D analysis; and (6) there are drawback in all 3D analyses due to the effect of the intermediate stress and the stress dependency of the soil parameters on the locations and existing stresses.

In this chapter, the basic formulation of the 2D slope stability methods will be discussed. A brief discussion of 3D analysis and the uses of some more advanced methods will also be covered. The demonstration version of the program SLOPE 2000 can be obtained from Cheng at natureymc@yahoo.com.hk. Readers should note that there are both English and Chinese versions of SLOPE 2000, and the program is also a solution module in the large-scale geotechnical analysis/design package GeoSuite (from versions 1 to 4.0). SLOPE 2000 has overcome many technical problems in other commercial programs, in particular the convergence and the search for critical solution problems. Currently, there are 12 two-dimensional and 7 three-dimensional methods of analysis available in version 2.5 of SLOPE 2000, and some methods of analysis appear to be absent in most of the other commercial programs (Janbu's rigorous, Sarma, China load factor methods).

There are two major approaches in carrying out slope stability analyses. The first approach is the *total stress approach*, which applies to clayey slopes or slopes with saturated sandy soils under short-term loadings where the pore pressure cannot dissipate within a short time. Currently, this approach is not commonly used, but most commercial programs still retain such an option for engineers. The second approach is the *effective stress approach* where drained conditions prevail so that the effective soil parameters will be more appropriate. Some engineers perform both total stress and effective stress analyses for a slope, with different accepted factors of safety.

7.2 DEFINITION OF THE FACTOR OF SAFETY

The factor of safety, F, for slope stability analysis is commonly defined as the ratio of the ultimate shear strength divided by the mobilized shear stress at the verge of failure, though other definitions also exists. F is usually assumed to be constant along the slip surface, and it is commonly defined with respect to the force or moment equilibrium as follows:

1. Moment equilibrium: generally used for the analysis of rotational landslides, and the factor of safety F_m defined with respect to moment is given by:

$$F_m = \frac{M_r}{M_d} \tag{7.1}$$

Where M_r is the sum of the resisting moments and M_d is the sum of the driving moment about a prescribed moment point. For circular failure surface, the centre of the circle is usually taken as the moment point for convenience. For a non-circular failure surface, any arbitrary moment point can be adopted. It should be noted that for the Bishop method, which does not satisfy horizontal force equilibrium, the factor of safety will depend on the choice of the moment point. Nevertheless, some computer programs allow Bishop method to be used for a non-circular slip surface. It should also be noted that in most stability formulations, the global moment equilibrium is usually considered, while the local moment equilibrium is not enforced, except for the Janbu rigorous method.

2. *Force equilibrium*: generally applied to translational failures. The factor of safety F_f defined with respect to force is given by:

$$F_f = \frac{F_r}{F_d} \tag{7.2}$$

where F_r is the sum of the resisting forces and F_d is the sum of the driving forces. In the finite element method or similar, the factor of safety is applied on the material parameters so as to bring the system to failure, which is equivalent to the rigorous method in the limit equilibrium method.

In all the stability formulations, the vertical force equilibrium must always be enforced, while the horizontal force or moment equilibrium may be sacrificed. Surprisingly, the results from such deficiencies are usually not critical as compared with the rigorous formulations.

In Eqs. (7.1) and (7.2), if external loads/moments in the form of soil nails, pile or others are considered, the factor of safety can further be defined in two different ways. For example, if the resisting moment from soil nails is denoted as M_e, the factor of safety by moment can further be defined as

$$F_{m1} = \frac{M_r + M_e}{M_d} \quad \text{or} \quad F_{m2} = \frac{M_r}{M_d - M_e} \tag{7.3}$$

If M_e (or F_e) is small, the two definitions of the factor of safety in Eq. (7.3) are close to each other. If M_e is large, the difference between the two definitions can be significant, and F_{m2} can be much great than F_{m1}. Actually, the later equation in Eq. (7.3) may even be negative by definition, while the first equation in Eq. (7.3) always remains positive. In using any computer program, the user should check carefully for the definition of the factor of safety. Actually, F_{m2} is more commonly adopted by many computer programs, but some commercial programs also include an option to use F_{m1}.

Besides the limit equilibrium method which has used by many engineers, there are some other methods such as the finite element method, limit analysis, distinct element method and others. These methods are discussed in this chapter.

7.3 SLOPE STABILITY ANALYSIS – LIMIT EQUILIBRIUM METHOD

For "simplified methods", force and moment equilibrium cannot be fulfilled simultaneously, while the "rigorous methods" try to satisfy both equilibrium at the expense of some additional assumptions. Different methods give slightly different factors of safety (sometimes great differences, but not commonly) in general, but it appears that all design codes do not have a clear requirement on these two factors of safety, and a single factor of safety is actually specified in many design codes. A slope may hence actually possess several factors of safety according to the methods of analysis, and only the extremum method by Cheng et al. (2010) gives a unique factor of safety satisfying both force and moment equilibrium (global and local) without the use of assumptions which cannot be proved. Under the extremum principle, there is only one factor of safety which can satisfy all the force and moment equilibrium (locally and globally) and the maximum strength of the system is fully mobilized. Based on this principle, any system will possess only one factor of safety, and this definition can avoid the difficult dilemma of multiple factors of safety using different method of analysis.

Under all the common classical approaches, slope stability analysis is a statically indeterminate problem. Different assumptions have to be adopted so that the problem can be solved. Before the invention of the computer, the early limit equilibrium methods were simple enough to be computed by hand calculation; for example, the infinite slope analysis (Haefeli, 1948) and the $\phi_u = 0$ undrained analysis (Fellenius, 1918). With the advent of computers, more advanced methods have been developed, in particular, the limit equilibrium methods which are based on the techniques of slices which can be vertical, horizontal or inclined. The first slice method was by Fellenius (1927), and there was rapid development of the slice methods in the 1950s and 1960s by Bishop (1955), Janbu et al. (1956), Janbu (1957), Lowe and Karafiath (1960), Morgenstern and Price (1965), Spencer (1967) and Janbu (1973). The common features of the methods of slices have been summarized by Zhu et al. (2003):

(a) The sliding body over the failure surface is divided into a finite number of slices which are usually cut vertically, but horizontal as well as inclined cuts have also been used by various researchers. The vertical cut is a more reasonable assumption, as gravity acts only in the vertical direction. In general, the differences between different methods of cutting are not major. Recently, some researchers have put forward formulations where the whole soil mass without any slice is considered using two moment points. The author, however, do not favour such an approach, as the internal yielding along any section is not considered in the analysis. The line of thrust outside the soil mass can happen easily using such a formulation.
(b) The shear strength along the slip surface is mobilized to the same degree at the limit state. That means that there is only a single factor of safety along the slip surface. There are some formulations where the factor of safety can vary along the failure surface (close to a progressive failure consideration), but such methods usually require additional assumptions and are not adopted for practical purposes.
(c) Some assumptions about inter-slice force relations are required so that the problem can become statically determinate. The no-slice formulation can avoid this limitation at the expense of the internal consistency problem. The extremum principle by Cheng et al. (2010) can, however, avoid all these limitations at the expense of intensive computation.
(d) The factor of safety is computed from force, moment or both force and moment equilibrium equations. No formulation is rigorous in that there is no way to ensure the complete satisfaction of force and local/global moment equilibrium automatically.

The shear strength mobilized along a slip surface depends on the effective normal stress σ' acting on the failure surface. Frohlich (1953) analysed the influence of the σ' distribution on the slip surface on the calculated *F*. He suggested an upper and lower bound for the possible *F* values. Similar approaches have also been proposed by Baker (1980) and Cheng et al. (2013). Cheng et al. (2013) have discussed the stability formulation based on the basal force distribution. This will be discussed in more depth in the next chapter of this book.

For slope stability analysis, the mobilized shear strength τ_m along the failure surface is given by:

$$\tau_m = \tau_f / F \tag{7.4}$$

where F is the factor of safety. The ultimate shear strength τ_f is given by the Mohr-Coulomb relation as

$$\tau_f = c' + \sigma_n' \tan\phi' \quad \text{or} \quad c_u \tag{7.5}$$

where c' = cohesion, σ_n' = effective normal stress, ϕ' = angle of internal friction and c_u = undrained shear strength.

The limit equilibrium method is the most popular approach in slope stability analysis, which can be traced back to Coulomb's method in retaining wall earth pressure determination. For a slope stability problem, this method is well known to be a statically indeterminate problem (as there are more than one slice in the problem as against the Coulomb's method), and assumptions on the interslice shear forces are required to render the problem statically determinate. Consider the static equilibrium conditions and the concept of limit equilibrium; the number of equations and unknown variables are summarized in Tables 7.1 and 7.2.

From the above tables, it is clear that the classical slope stability problem is statically indeterminate in the order of $6n - 2 - 4n = 2n - 2$. In other words, additional $(2n - 2)$ assumptions have to be introduced to solve this problem. If the width of the slice is not sufficient, the locations of the base normal forces are usually assumed to be at the middle of each slice, hence $n - 2$ equations have to be introduced. The most common additional assumptions are either the location of the interslice normal forces or the relation between the interslice normal and shear forces (or equivalently the inclination of the internal forces). That will introduce $n - 1$ additional equations (n slice has only $n - 1$ interfaces), so the problem will now become over-specified by 1 after the introduction of such assumptions. There is no LEM method which can avoid this limitation and satisfy all the global and local force/moment

Table 7.1 Summary of system of equation (n – number of slices)

Equations	*Condition*
n	Moment equilibrium for each slice
$2n$	Force equilibrium in the X and Y directions for each slice
n	Mohr-Coulomb failure criterion
$4n$	Total number of equations

Table 7.2 Summary of unknowns

Unknowns	*Description*
1	Safety factor
n	Normal force at the base of slice
n	Location of normal force at base of slice
n	Shear force at base of slice
n-1	Interslice horizontal force
n-1	Interslice tangential force
n-1	Location of interslice force (line of thrust)
6n-2	Total number of unknowns

equilibrium automatically. The extremum principle by Cheng et al. (2010, 2011b), which is practically equivalent to the lower bound method, is an advancement over the classical LEM in that all the local and global force/moment equilibrium can be satisfied without introducing the interslice force function concept, however the number of computations required for extremum analysis far exceeds that based on the classical LEM methods. Extremum is hence more suitable for academic research than for practical applications, except for some special cases where the results are sensitive to the interslice force function.

The limit equilibrium method can be classified into two main categories: "simplified" methods and "rigorous" methods. For the simplified methods, either force or moment equilibrium is satisfied, but not both at the same time. For the rigorous methods, both force and moment equilibrium are satisfied, but usually the analysis is more complicated and tedious, and non-convergence in analysis is quite commonly encountered due to the introduction of the additional assumptions. The author has to stress that there is no limit equilibrium method which can guarantee convergence automatically. The author has come across many cases where very strange results can arise from the "rigorous" methods (which should be eliminated because the internal forces are unacceptable). In this respect, no method is particularly better than others under all cases, although methods which have more careful consideration on the internal forces will usually perform better in general. Morgenstern (1992), Cheng as well as many other researchers have found that most of the commonly used methods of analysis give results which are similar to each other. In this respect, there is not a strong need to fine-tune the "rigorous" slope stability formulations except for isolated cases, as the interslice shear forces have only small effects on the factor of safety in general.

7.3.1 Rigorous limit equilibrium formulation

For rigorous formulations which consider both the force and moment equilibrium, consider the equilibrium of force and moment of the slope as shown in Figure 7.6.

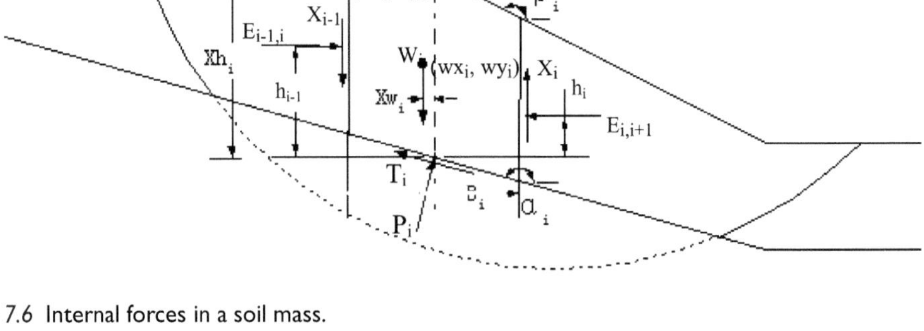

Figure 7.6 Internal forces in a soil mass.

With respect to Figure 7.6, the following assumptions are used in the analysis:

1. φ for the first slice ($\varphi_{0,1}$) and the last slice ($\varphi_{n,n+1}$) and R for the first slice ($R_{0,1}$) and the last slice ($R_{n,n+1}$) are set to 0 as the internal forces at the two ends are 0.
2. X for the first slice ($X_{R0,1}$) and the last slice ($X_{Rn,n+1}$) are also 0.
3. Based on *force* equilibrium, the following equations can be derived easily.

$$R_{12} = \frac{F*\left[-(W_1+Q_1)\sin\alpha_1 + H_1\cos\alpha_1\right] + f_1\left[(W_1+Q_1)\cos\alpha_1 + H_1\sin\alpha_1 - U_1\right] + c_1L_1}{F*\sin(\varphi_{12}+\beta_1-\alpha_1) + f_1\cos(\varphi_{12}+\beta_1-\alpha_1)} \tag{7.6}$$

$$R_{23} = \frac{F*\left[-(W_2+Q_2)\sin\alpha_2 + H_2\cos\alpha_2\right] + f_2\left[(W_2+Q_2)\cos\alpha_2 + H_2\sin\alpha_2 - U_2\right] + c_2L_2}{F*\sin(\varphi_{23}+\beta_2-\alpha_2) + f_2\cos(\varphi_{23}+\beta_2-\alpha_2)} + \frac{R_{12}\left[F*\sin(\varphi_{12}+\beta_1-\alpha_2) + f_2\cos(\varphi_{12}+\beta_1-\alpha_2)\right]}{F*\sin(\varphi_{23}+\beta_2-\alpha_2) + f_2\cos(\varphi_{23}+\beta_2-\alpha_2)} \tag{7.7}$$

$$R_{i,i+1} = \frac{F*\left[-(W_i+Q_i)\sin\alpha_i + H_i\cos\alpha_i\right] + f_i\left[(W_i+Q_i)\cos\alpha_i + H_i\sin\alpha_i - U_i\right] + c_iL_i}{F*\sin(\varphi_{i,i+1}+\beta_i-\alpha_i) + f_i\cos(\varphi_{i,i+1}+\beta_i-\alpha_i)} + \frac{R_{i-1,i}\left[F*\sin(\varphi_{i-1,i}+\beta_{i-1}-\alpha_i) + f_i\cos(\varphi_{i-1,i}+\beta_{i-1}-\alpha_i)\right]}{F*\sin(\varphi_{i,i+1}+\beta_i-\alpha_i) + f_i\cos(\varphi_{i,i+1}+\beta_i-\alpha_i)} \tag{7.8}$$

$$R_{n-1,n} = \frac{F*\left[-(W_n+Q_n)\sin\alpha_n + H_n\cos\alpha_n\right] + f_n\left[(W_n+Q_n)\cos\alpha_n + H_n\sin\alpha_n - U_n\right] + c_nL_n}{F*\sin(\varphi_{n-1,n}+\beta_{n-1}-\alpha_n) + f_n\cos(\varphi_{n-1,n}+\beta_{n-1}-\alpha_n)} \tag{7.9}$$

where
F = factor of safety, W_i = weight of slice i, U_i = pore pressure at base i,
Q_i = vertical load at slice i, H_i = horizontal load at slice i, α_i = base angle at slice i,
f_i = tanϕ at base i, $R_{i,j}$ = interslice internal force,
β_i = angle between interslice direction and horizontal, set to 90° in present study,
$\varphi_{i,j}$ = angle of inclination of interslice force with horizontal and tanφ = $\lambda f(x)$, or
$X = \lambda f(x)E$,
$c_i = c$ at base i, L_i = base length at slice i,
$R_{i,i+1}$ denote the resultant of E and X along any interface i.

If we substitute Eq. (7.6) into Eq. (7.7), Eq. (7.7) into Eq. (7.8), and so on until $R_{n,n+1}$, and using the fact that $R_{n,n+1} = 0$, a polynomial of order n for K will be established. n factors of safety based on force equilibrium can be obtained from this polynomial equation. Moment equilibrium (global) is then back-checked, and the solution will continue by adjusting some of the assumptions (usually λ) until moment equilibrium is also achieved. Any internal force system which satisfies force equilibrium or both force and moment equilibrium will have the same equations as above. For a method which does not satisfy horizontal force equilibrium (like Bishop's method), the above method cannot be used in the determination of the factor of safety and the classical iteration method must be used.

For *moment* equilibrium (take moment about the midpoint of each slice base):

$$W_1 * X_{w1} + Q_1 * X_{Q1} + H_1 * X_{H1} - R_{12} * X_{R12} = 0 \tag{7.10}$$

$$W_2 * X_{w2} + Q_2 * X_{Q2} + H_2 * X_{H2} - R_{23} * X_{R23} + R_{12} * \left[X_{R12} + L_1 * \cos(\varphi_{12} + \beta_1 - \gamma_1)\right] = 0 \tag{7.11}$$

$$W_i * X_{wi} + Q_i * X_{Qi} + H_i * X_{Hi} - R_{i,i+1} * X_{Ri,i+1} + R_{i-1,i} * \left[X_{Ri-1,i} + L_{i-1} * \cos\left(\varphi_{i-1,i} + \beta_{i-1} - \gamma_{i-1}\right)\right] = 0 \tag{7.12}$$

$$W_n * X_{wn} + Q_n * X_{Qn} + H_n * X_{Hn} + R_{n-1,n} * \left[X_{Rn-1,n} + L_{n-1} * \cos\left(\varphi_{n-1,n} + \beta_{n-1} - \gamma_{n-1}\right)\right] = 0 \tag{7.13}$$

Slope stability methods, such as the Janbu simplified method, Corps of Engineers method, Lowe and Karafiath method, Janbu rigorous method, Spencer method and Sarma method, have various assumptions on φ which are:

1. The angle β of each interface with the horizontal direction. Except Sarma's method for non-vertical slice, all the other methods assume β to be equal to 90° or vertical slices are considered.
2. The interslice force angle (φ), where tanφ is equal to $\lambda f(x)$. For example, φ is assumed to be 0 in the Janbu simplified method and is assumed to be constant in the Spencer method.

It should be noted that most of the "rigorous" methods adopt the overall moment equilibrium instead of the local moment equilibrium in the formulation, except for the Janbu rigorous method (1973) and the extremum method by Cheng et al. (2010). The line of thrust can be back-computed from the internal forces after the stability analysis. Since the local moment equilibrium equation is not adopted explicitly, the line of thrust may fall outside the slice by using the assumption $X = \lambda f(x)E$, which is clearly unacceptable, and it can be considered that the local moment equilibrium is not satisfied at a certain slice. The effects of the local moment equilibrium are however usually not critical towards the factor of safety, as the effect of the interslice shear force is usually small in most cases (although there are exceptions). The engineers should, however, check the location of the thrust line as a good practice after performing those "rigorous" analyses, but it appears that most commercial programs fail to do so.

7.3.1.1 Solution procedure

First, a prescribed $f(x)$ will be specified, and λ will vary from 0 (or a small negative value) gradually, and the maximum is controlled by $\tan\varphi = \lambda^* f(x)$. The factor of safety ($F$) is then computed using Eqs. (7.6)–(7.9). The solution of all the positive factors of safety can be obtained by a double QR method for the Hessenberg matrix arising from Eqs. (7.6)–(7.9). Next, the interslice force (using force equilibrium) and its location (using moment equilibrium) can be obtained from the above equations directly without using iterative analysis. The factor of safety will then be examined with respect to the following conditions:

1. R will resist slippage of the slope, hence it must be positive.
2. The base and interface normal forces computed must be greater than 0.
3. Moment equilibrium is achieved for a particular λ.

For those methods which do not satisfy moment equilibrium, the requirements on moment equilibrium need not be considered. If the above requirements are satisfied, the computed F value is acceptable and the analysis is complete. If there is no solution, then the analysis fails to converge under the given $f(x)$, but this does not mean that there is no factor of safety. Factor of safety always exists, but the prescribed $f(x)$ may not be appropriate under this case.

The advantage of the present method is that the factor of safety and internal forces with respect to force equilibrium are obtained directly without any iteration analysis. This is important in analysis as there are many cases where the use of an iterative method fails to give a converged result or an incorrect converged result is obtained. The time required for double QR computation is not excessively long as interslice normal and shear forces are not required to be determined in obtaining a factor of safety. If the above double QR method cannot give a solution, then the problem has no solution under the given $f(x)$, by the nature of the problem. None of the iteration methods can assess the nature of the problem, as failure to converge may be simply a problem of the iteration analysis.

7.3.2 Interslice force function and Morgenstern-Price method

The interslice shear force X is assumed to be related to the interslice normal force E by the relation $X = \lambda f(x)E$. Some researchers have used another format as $X = f(x)[E\tan\phi' + C]$ so that C can be taken as the average cohesive strength multiplied by the interface length. This approach is, however, not commonly adopted, as the relation between E and X is not simple and will create many difficulties in the numerical computation. There is no theoretical basis to determine $f(x)$ for a general problem, as the slope stability problem is statically indeterminate by nature. More detailed discussion is provided about $f(x)$ by the lower bound method. There are seven types of $f(x)$ commonly in use:

Type 1: $f(x) = 1.0$. This case is equivalent to the Spencer method and is commonly adopted by many engineers and reported in the literature. Consider the case of a sandy soil with

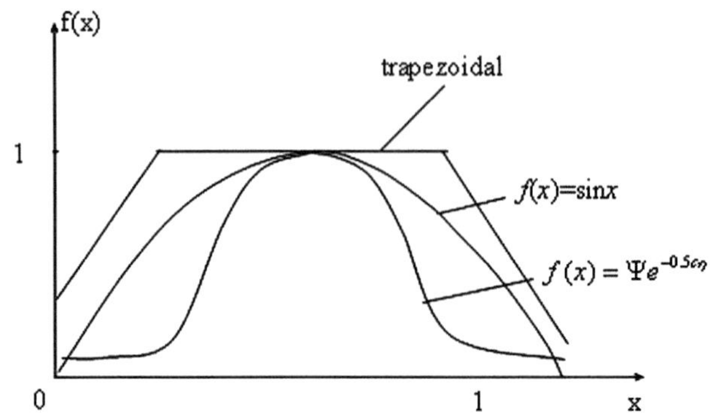

Figure 7.7 Shape of some interslice shear force function.

$c'=0$. If the Mohr-Coulomb relation is applied to the interslice force relation, $X = E\ tan\phi'$, then $f(x) = 1.0$ and $\lambda = tan\phi$. Since there is not a strong requirement to apply the Mohr-Coulomb relation for the interslice forces, $f(x)$ should be different from 1.0 in general. It will be demonstrated in a later section that $f(x) = 1.0$ is actually not a realistic or good choice for some special cases. The convergence of this interslice force function is generally not as good as that by the type 2 function.

Type 2: $f(x) = \sin(x)$. This is a relatively popular alternative to $f(x) = 1.0$. This function is adopted purely because of its simplicity and good convergence.

Type 3: $f(x)$ = trapezoidal shape shown in Figure 7.7. Type 3 $f(x)$ will reduce to type 1 as a special case, but it is seldom adopted in practice as there is no physical ground to define this function.

Type 4: $f(x)$ = error function or the Fredlund-Wilson-Fan force function (1986), which is in the form of $f(x) = \Psi\exp(-0.5c^n\eta^n)$, where Ψ, c and n have to be defined by the user. η is a normalized dimensional factor which has a value of –0.5 at the left exit end and 0.5 at the right exit end of the failure surface. η varies linearly with the x-ordinates of the failure surface. This error function is actually based on an elastic finite element stress analysis by Fan et al. (1986). Since the stress state in the limit equilibrium analysis is the ultimate condition and is different from the elastic stress analysis by Fan et al. (1986), the suitability of this interslice force function cannot be justified simply by the elastic analysis. It is also difficult to define the suitable parameters for a general problem with soil nails, the water table and external loads. Some engineers have reported that the convergence for this function is good, but there is not such a common belief among all engineers.

For the first four types of functions shown above, they are commonly adopted in the Morgenstern-Price and GLE methods, and both the moment (globally) and force equilibrium (locally and globally) can be satisfied simultaneously. A completely arbitrary interslice force function is theoretically possible, but there is no simple way or theoretical background in defining this function, hence the arbitrary interslice force function is seldom considered in practice.

Type 5: Corps of Engineers interslice force function. $\lambda f(x)$ is assumed to be constant and is equal to the slope angle defined by the two extreme ends of the failure surface. Moment equilibrium cannot be achieved by this function as $\lambda f(x)$ is prescribed to be known. The function appears to be not popular nowadays.

Type 6: Lowe-Karafiath interslice force function. $\lambda f(x)$ is assumed to be the average of the slope angle of the ground profile and the failure surface at the section under consideration. Similar to the Corps of Engineers method, moment equilibrium cannot be achieved by this function as $\lambda f(x)$ is prescribed to be known. The function also appears to be not popular nowadays.

Type 7: $f(x)$ is defined as the tangent of the base slope angle at the section under consideration, and this assumption is used in the load factor method in China. Similar to the previous two cases, this method cannot satisfy moment equilibrium. This approach appears to be used solely within China, and is not used by engineers outside China.

For types 5–7 interslice force functions, only force equilibrium is enforced in the formulation. The factors of safety from these methods are, however, usually very close to those by the "rigorous" methods, and are usually better than the results by the Janbu simplified method (1957). In fact, the Janbu simplified method (1957) is given by the case of $\lambda = 0$ for the Corps of Engineers method, Lowe-Karafiath method (1960) and the load factor method, and results from the Janbu analysis (1957) can also be taken as the first approximation trial in the Morgenstern-Price analysis (1965).

The Morgenstern-Price method is usually based on the interslice force function types 1–4, though the use of arbitrary function is possible and is occasionally used. If type 1 interslice force function is used, this method will reduce to the Spencer method. The Morgenstern-Price method satisfies the force and global moment equilibrium. Since the local moment equilibrium equation is not used in the formulation, the internal forces of individual slices may not be acceptable. For example, the line of thrust (centroid of the interslice normal force) may fall outside the soil mass from the Morgenstern-Price analysis. The GLE method is basically similar to the Morgenstern-Price method, except that the line of thrust is determined and is closed at the last slice. The acceptability of the line of thrust for any intermediate slice may still be unacceptable from the GLE analysis. In general, the results from these two methods of analysis are very close.

Based on a Mohr circle transformation analysis, Chen and Morgenstern (1983) have established that $\lambda f(x)$ for the two ends of a slip surface should be equal to the ground slope angle (if there is no external load). Other than this requirement, there is no simple way to establish $f(x)$ for a general problem. Since the requirement by Chen and Morgenstern (1983) applies only under an infinitesimal condition, it is seldom adopted in practice. Even though there is no simple way to define $f(x)$, Morgenstern (1992), among others, has however pointed out that for normal problems, F from different methods of analyses are similar, so that the assumptions on the internal force distributions are not major issues for practical use except for some particular cases. In view of the difficulty in prescribing a suitable $f(x)$ for general problem, most engineers adopt $f(x) = 1.0$ which is satisfactory for most cases. Cheng et al. (2010) have however established the upper and lower bounds of the factor of safety and the corresponding $f(x)$ based on the extremum principle which will be discussed later. Cheng et al. (2010) have also demonstrated that the extremum principle can satisfy the requirement of $\lambda f(x)$ for the two ends of a slip surface as stipulated by Chen and Morgenstern (1983).

7.3.3 Janbu's rigorous method

Janbu's rigorous formulation is different from the previous formulation in that the line of thrust is assumed to be known. The line of thrust is the locus for the centroid of the internal forces, and it can be assumed to be 0.33–0.4 times the interface length for normal practice. Janbu (1957) described a method of analysis which may be applied to both circular and non-circular slip surfaces (see Figure 7.6). The original formulation was extended to cover the analysis of bearing capacity and earth pressure problems by Janbu (1957, 1973). This was actually the first method of slices in which overall force equilibrium and overall moment equilibrium are satisfied.

Janbu formulated the general equations of equilibrium by resolving vertically and parallel to the base of each slice. By considering the overall force equilibrium an expression for the factor of safety F_f was obtained. To render the problem statically determinate, the position of the line of thrust of the interslice forces is assumed. By taking moments about the centre of the base of each slice, overall moment equilibrium is implicitly satisfied, and the interslice shear forces can be calculated. These are then inserted into the expression for the factor of safety, and both overall force equilibrium and overall moment equilibrium are satisfied.

For any slice: at base–normal stress σ, shear stress τ, pore pressure u

$$\text{Failure criterion: ultimate shear strength } \tau_f = c' + (\sigma - u)\tan\phi' \tag{7.14}$$

Mobilized shear strength $\tau_m = \tau_f / F$ where F is the factor of safety.

$$\text{Now } P = \sigma l \quad T = \tau_m l \quad \text{so } T = \frac{1}{F}\left(c'l + (P - ul)\tan\phi'\right) \tag{7.15}$$

$$\text{Resolve vertically: } P\cos\alpha + T\sin\alpha = W - \left(X_R - X_L\right) \tag{7.16}$$

Rearranging, and substituting for T gives

$$P = [W - (X_R - X_L) - \frac{1}{F}(c'l\sin\alpha - ul\tan\phi'\sin\alpha)]/m_\alpha \tag{7.17}$$

$$\text{where } m_\alpha = \cos\alpha\left(1 + \tan\alpha\frac{\tan\phi'}{F}\right) \tag{7.18}$$

$$\text{Resolve parallel to base of slice: } T + (E_R - E_L)\cos\alpha = (W - (X_R - X_L))\sin\alpha \tag{7.19}$$

$$E_R - E_L = (W - (X_R - X_L))\tan\alpha - \frac{1}{F}(c'l + (P - ul)\tan\phi')\sec\alpha \tag{7.20}$$

Take moments about centre of base of slice (for thin slice):

$$E_R b \tan\alpha_t - X_R b - (E_R - E_L)h_t = 0 \tag{7.21}$$

$$\text{or } X_R = E_R \tan\alpha_t - (E_R - E_L)\frac{h_t}{b} \tag{7.22}$$

Overall FORCE equilibrium, in the absence of surface loading

$$\Sigma(E_R - E_L) = 0 \quad \Sigma(X_R - X_L) = 0 \tag{7.23}$$

$$\text{From Eq. (7.23)}, \Sigma(E_R - E_L) = \Sigma(W - (X_R - X_L))\tan\alpha - \frac{1}{F_f}\Sigma(c'l + (P - ul)\tan\phi')\sec\alpha = 0 \tag{7.24}$$

Factor of safety F is sometimes sensitive to the line of thrust. X obtained should be checked with E to ensure the correct factor of safety. If violated, adjust the line of thrust.

$$\text{Hence, } F_f = \frac{\Sigma(c'l + (P - ul)\tan\phi')\sec\alpha}{\Sigma(W - (X_R - X_L))\tan\alpha} \tag{7.25}$$

The solution is reached iteratively. First it is assumed that $X_R - X_L = 0$. Then values of E and X are calculated using Eqs. (7.20) and (7.22) above; the values of the shear force X lag by one iteration. As moment equilibrium is satisfied by Eq. (7.21), $F_f = F_m$. In Janbu's original formulation, P is eliminated and the following expression for F is obtained:

$$F = \frac{\Sigma[c'b + (W - (X_R - X_L) - ub)\tan\phi']/n_\alpha}{\Sigma(W - (X_R - X_L))\tan\alpha} \quad \text{in which } \eta_\alpha = \text{cosa } m_\alpha \tag{7.26}$$

Solution of Janbu's rigorous analysis

1. Perform Janbu's simplified method first to get an estimate of F as the initial trial.
2. Assume $X = 0$ in the first trial, determine base normal P from Eq. (7.17) and E from Eq. (7.24).
3. Determine X from Eq. (7.22) and hence ΔX.
4. Put ΔX in Eq. (7.17) and get new P, and hence E from Eqs. (7.17) and (7.24).
5. Repeat steps (3) and (4) until convergence. In case convergence is not achieved, it is recommended to vary the location of the thrust line. There is, however, no guideline to vary the thrust line to achieve automatic convergence in a general case.

Janbu's rigorous method (1973) appears to be appealing in that the local moment equilibrium is used in the intermediate computation. The internal forces and local moment equilibrium will hence be acceptable if the analysis can converge. As suggested by Janbu (1973), the line of thrust ratio is usually taken as one-third of the interslice height, which is basically compatible with the classical lateral earth pressure distribution. It should be noted that the equilibrium of the last slice is not checked in the Janbu rigorous method (1973), therefore the local moment equilibrium from the Janbu rigorous method (1973) is not strictly rigorous, but appears to not have been noticed by many engineers in the past. As mentioned earlier, once assumptions are introduced, the problem becomes over-specified by one equation, so the moment equilibrium of the last slice cannot be enforced in the Janbu rigorous method. A practical limitation of this method is the relatively poor convergence in the analysis, particularly when the failure surface is highly irregular or there are external loads. This is due to the fact that the line of thrust ratio is predetermined in the analysis, while the internal force computation is greatly affected by this thrust line location. The previous formulation where $f(x)$ is prescribed can achieve convergence better than the Janbu rigorous method, but the back-computer thrust line location may lie outside the soil mass, which means that the local moment equilibrium is not achieved. The constraints in the Janbu rigorous method (1973) are greater than those in the other methods, hence convergence is usually poorer. If the method is slightly modified by assuming $ht/h = \lambda f(x)$, where ht = height of line of thrust above slice base and h = length of the vertical interslice, the convergence of this method may be improved. There is, however, difficulty in defining $f(x)$ for the line of thrust, and hence this approach is seldom considered. Cheng has developed another version of the Janbu rigorous method with slightly better convergence which is implemented in the program SLOPE 2000.

7.3.4 Sarma's method

Sarma (1979) proposed a completely different approach to compute the factor of safety based on the critical acceleration that is required to bring a soil mass to a state of limiting equilibrium. He also assumed the slope of internal tangential force distribution $X(x)$ is known and is given by:

$$X(x) = \lambda f_3(x)\left[c_{avg}H(x) + (K_{avg} - R_u)\frac{\gamma_{avg}H^2}{2}\tan\phi_{avg}\right] \tag{7.27}$$

where: $X(x)$ = interslice forces parallel to slice interface
λ = scaling factor determined as given later in Eq. (7.42)
$f_3(x)$ = scaling function determined by the user, usually chosen as 1.0
$\gamma_{avg}, c_{avg}, K_{avg}, \tan\phi_{avg}$ = average soil strength parameter along interface of the slice (see Eqs. 7.45–7.49)

H = height of slice
Ru_i = pore pressure parameter equal to the ratio of pore water pressure to the total overburden pressure

Considering the vertical and horizontal equilibrium of the slice *i* with horizontal and vertical surcharge *Q* and *V*, the vertical and horizontal equilibrium equations are obtained as:

$$P_i \cos\alpha + T_i \sin\alpha_i = W_i + V_i - \Delta X_i \tag{7.28}$$

$$T_i \cos\alpha - P_i \sin\alpha_i = KW_i + \Delta E_i + \Delta Q_i \tag{7.29}$$

where T = shear force at slice base. It is assumed that under the action of KW, the full shear strength of soil is mobilized. Hence

$$S = c' + (\sigma - u)\tan\phi' \Rightarrow T_i = (P_i - U_i)\tan\phi' + c'b_i \sec\alpha \tag{7.30}$$

From Eqs. (7.28) and (7.30):

$$P_i \cos\alpha_i + (c'b_i \sec\alpha_i + P_i \tan\phi' - U_i \tan\phi_i')\sin\alpha_i = W_i + \Delta V_i - \Delta X_i$$

$$\Rightarrow P_i = [W_i + \Delta V_i - \Delta X_i - c_i' \, b_i \tan\alpha_i + U_i \tan\phi_i' \sin\alpha_i]\cos\phi / [(\cos(\phi_i' - \alpha_i))]$$

$$\Rightarrow P_i = [W_i + \Delta V_i - \Delta X_i - c_i' \, b_i \tan\alpha_i + Ru_i W_i \tan\phi_i' \sin\alpha_i]\cos\phi \,/\, [(\cos(\phi_i' - \alpha_i))]$$

$$\text{and } T_i = (W_i + \Delta V_i - \Delta X_i - P_i \cos\alpha_i) / \sin\alpha_i \tag{7.31}$$

Substituting P_i by Eq. (7.31) gives:

$$T_i = [(W_i + \Delta V_i - \Delta X_i)\sin\phi_i' + c_i' b_i \cos\phi_i' - Ru_i W_i \sin\phi_i'] \,/\, [\cos(\phi_i' - \alpha_i)] \tag{7.32}$$

$$\text{Putting Eq.(7.32) into Eq.(7.29)}, \Delta X_i \tan(\phi_i' - \alpha_i) + \Delta E_i + \Delta Q_i = D_i - KW_i \tag{7.33}$$

$$D_i - \Delta Q_i = (W_i + \Delta V_i)\tan(\phi_i' - \alpha_i) + \frac{c_i' b_i \cos\phi_i' \sec\alpha_i - U_i \sin\phi_i'}{\cos(\phi_i' - \alpha_i)}$$

$$D_i - \Delta Q_i = (W_i + \Delta V_i)\tan(\phi_i' - \alpha_i) + \frac{(c_i' b_i \cos\phi_i' - Ru_i W_i \sin\phi_i')\sec\alpha_i}{\cos(\phi_i' - \alpha_i)} \tag{7.34}$$

Consider the horizontal equilibrium of the whole mass where $\Sigma\Delta E_i$ cancel out:

$$\Sigma\Delta Q_i + \Sigma\Delta X_i \tan(\phi_i' - \alpha_i) + \Sigma K W_i = \Sigma D_i \tag{7.35}$$

For the moment equilibrium, we can take the moment about the centre of gravity of the whole sliding soil mass (including vertical surcharge) to eliminate the weight term W_i and the vertical surcharge terms ΔV_i.

$$\Sigma\Delta Q_i(y_Q - y_g) + \Sigma(T_i \cos\alpha_i - P_i \sin\alpha_i)(y_g - y_i) \\ - \Sigma(P_i \cos\alpha_i + T_i \sin\alpha_i)(x_g - x_i) = 0 \tag{7.36}$$

where (x_g, y_g) = coordinates of the centre of gravity of the whole soil mass

(x_i, y_i) = coordinates of the mid-point of the slice base

Eqs. (7.28), (7.29), (7.33) and (7.36) give

$$\Sigma\Delta Q_i(y_Q - y_g) + \Sigma(KW_i + \Delta E_i + \Delta Q_i)(y_g - y_i) - \Sigma(W_i + \Delta V_i - \Delta X_i)(x_g - x_i) = 0$$

$$\Sigma\Delta Q_i(y_Q - y_g) + \Sigma[D_i - \Delta X_i \tan(\phi_i' - \alpha_i)](y_g - y_i) - \Sigma(W_i + \Delta V_i - \Delta X_i)(x_g - x_i) = 0$$

$$\Sigma\Delta X_i[(y_i - y_g)\tan(\phi_i' - \alpha_i) + (x_g - x_i)] = \Sigma W_i(x_g - x_i) \\ + \Sigma D_i(y_i - y_g) - \Sigma\Delta Q_i(y_Q - y_g) \tag{7.37}$$

Sarma (1979) then assumed that the resultant shear force can be expressed as Eq. (7.27) and f_3 is assumed to be known. This is equivalent to assuming the shape of the distribution of the interslice shear force. However, the magnitude of the interslice shear force is not assumed to be known. Therefore, substituting Eq. (7.27) into Eqs. (7.35), (7.36) and (7.37), gives

$$\Sigma\Delta Q_i + \lambda\Sigma F_3 \tan(\phi_i' - \alpha_i) + K\Sigma W_i = \Sigma D_i \tag{7.38}$$

$$\lambda\Sigma f_3[(y_i - y_g)\tan(\phi_i' - \alpha_i) + (x_g - x_i)] = \Sigma W_i(x_g - x_i) \\ + \Sigma D_i(y_i - y_g) - \Sigma\Delta Q_i(y_Q - y_g) \tag{7.39}$$

With an estimated value of f_3, Eqs. (7.38) and (7.39) can be solved to obtain λ and K which can be expressed as follows:

$$\lambda = S_2/S_3 \tag{7.40}$$

$$K = (S_1 - \lambda S_4)/\Sigma W_i \tag{7.41}$$

where: $S_1 = \Sigma D_i - \Sigma\Delta Q_i$ (7.42)

$$S_2 = \Sigma W_i(x_g - x_i) + \Sigma D_i(y_i - y_g) - \Sigma\Delta Q_i(y_Q - y_g) \tag{7.43}$$

$$S_3 = \Sigma f_3[(y_i - y_g)\tan(\phi_i' - \alpha_i) + (x_g - x_i)] \tag{7.44}$$

$$S_4 = \Sigma f_3 \tan(\phi_i' - \alpha_i) \tag{7.45}$$

The lever arm of the interslice normal force is given by

$$h_{i+1} = [E_i h_i - 0.5_i \tan\alpha_i (E_i + E_{i+1}) - 0.5_i (X_i + X_{i+1}) \\ - \Delta Q_i (y_Q - y_i) + \Delta V_i (X_V - X_i)]/E_{i+1} \tag{7.46}$$

This K value will give the critical acceleration for the soil mass under analysis, while the corresponding λ gives the change of the interslice shear force by Eq. (7.27). In order to determine the factor of safety for the soil mass, an adjusted factor of safety is applied to the properties of the soil in the calculation of the mobilized shear force at the slice base by Eq. (7.42) until the value of K is zero (i.e. acceleration = 0).

For the choice of the assumed function $f_3(x)$, Sarma (1979) carried out a detailed study and found that a value of 1.0 is applicable for most cases. If there are some local sections where X and E violate the failure criterion, $f_3(x)$ can be slightly reduced. Again, there is no simple way to define $f_3(x)$ so that a value of 1.0 is commonly adopted.

For non-homogeneous soil, the normal force acting on the interslice surface i in soil layer j which can be used as the start for estimating E in the iteration analysis which is given by

$$E_{i,j} = a_{i,j} W_{i,j} (h_{i,j} - h_{i,j+1}) + \frac{a_{i,j}\gamma_{i,j}(h_{i,j} - h_{i,j+1})^2}{2} + b_{i,j}(h_{i,j} - h_{i,j+1}) + d_{i,j} P_{wij} \tag{7.47}$$

where:

$$a_{i,j} = \frac{1 - \sin(\beta_{i,j})\sin\phi'_{i,j}}{1 + \sin(\beta_{i,j})\sin\phi'_{i,j}} \tag{7.48}$$

$$b_{i,j} - \frac{-2c'_{i,j}\cos\phi'_{i,j}\sin(\beta_{i,j})}{1 + \sin(\beta_{i,j})\sin\phi'_{i,j}} \tag{7.49}$$

$$d_{i,j} = \frac{2\sin\phi'_{i,j}\sin(\beta_{i,j})}{1 + \sin(\beta_{i,j})\sin\phi'_{i,j}} \tag{7.50}$$

$$\beta_{i,j} = 2\alpha_{i,j} - \phi'_{i,j} \tag{7.51}$$

$W_{i,j}$ = weight per unit area at the level j along the i-th interslice surface

$$= \sum_{k=1}^{j-1} \gamma_{i,k}(h_{i,k} - h_{i,k+1}) \tag{7.52}$$

$V_{i.j}$ = the piezometric height at level j along interslice surface i

$$P_{wij} = \frac{1}{2}\gamma_w(V_{i,j} + V_{i,j+1})(h_{i,j+1} - h_{i,j}) \tag{7.53}$$

$h_{i,j}$ = the y-coordinate of j-th soil boundary at the i-th slice

$\gamma_{i,j}$ = density of the soil layer j at slice i

γ_w = density of water.

The shear force acting on the interslice surface i for the non-homogeneous case is given by:

$$X_i = \lambda f_3(x)[(K_{i,ave} - Ru_{i,ave})\left(\frac{\gamma_{i,ave}H_i^2}{2}\right)\tan\varphi'_{i,ave} + c_{i,ave}H_i] \tag{7.54}$$

where:

$$\gamma_{i,ave} = \frac{\Sigma\gamma_j(h_{i,j} - h_{i,j+1})}{H_i} \tag{7.55}$$

$$Ru_{i,ave} = \frac{\Sigma P_{wij}}{\gamma_{i,ave}H_i^2/2} \tag{7.56}$$

$$K_{i,ave} = \frac{\Sigma E_{i,j}}{\gamma_{i,ave}H_i^2/2} \tag{7.57}$$

$$\tan\phi'_{i,ave} = \frac{\Sigma(E_{i,j} - P_{wij})\tan\phi'_j}{\Sigma(E_{i,j} - P_{wji})} \tag{7.58}$$

$$c_{i,ave} = \frac{\Sigma c'_j(h_{i,j} - h_{i,j+1})}{H_i} \tag{7.59}$$

H_i = Height of interslice i.

For the actual solution required in the Sarma (1979) method, Cheng has adopted a bracket method where an upper bound and lower bound of the factor of safety are defined. K will then be determined from Eq. (7.41). If there is a change in the sign of K based on the upper and lower bounds of the factor of safety, then the chosen bound will be adequate and the bound will be narrowed. If there is no change in the sign of K based on the upper and lower bounds of the factor of safety, another range will be chosen. It is interesting to note that very few commercial programs implement Sarma's method, but the author is not sure of the exact reason for this phenomenon.

7.4 SIMPLIFIED METHOD OF ANALYSIS

Although the rigorous limit equilibrium formulation has been well received by engineers, the simplified formulations which comply with either force or moment equilibrium only are still well favoured by engineers because the factor of safety from the simplified methods is usually close to the results from the rigorous methods (usually a slightly smaller value). In view of the various uncertainties in the soil parameters, some engineers do not bother to adopt those rigorous formulations for which a computer program is required. In Hong Kong, there is, however, a very special consideration which is almost not present in other countries. The cost of land is very expensive in Hong Kong, hence developers prefer to acquire as much usable area as possible within the lot boundary. To achieve this, the slope angle must be as steep as possible. A rigorous formulation is hence more popular in Hong Kong as that several percent increase in the factor of safety is still treasured by engineers. However, in other countries, the simplified methods are also well received by the engineers.

There are two major simplified methods which are commonly used by engineers. Actually, for the interslice force function types 5–7 in Section 7.3.2, they are also referred to as the simplified method. The solutions of these cases are, however, special cases in the rigorous formulation, and the computer code for these simplified methods can actually be modified from the rigorous formulation with only several additional lines of program statements, hence these special cases are not grouped under this section.

7.4.1 Bishop and Janbu simplified methods

Bishop's method

Bishop's method is one of the most popular slope stability analysis methods and is used worldwide. This method satisfies only the moment equilibrium but not the horizontal force equilibrium, and applies only for a circular failure surface. The centre of the circle is taken as the moment point in the moment equilibrium equation for convenience, and the factor of safety will be different if a different moment point is adopted. Bishop's method has been used for some non-circular failure surfaces, but Fredlund et al. (1992) have demonstrated that the factor of safety will be dependent on the choice of the moment point because there is a net unbalanced horizontal force in the system. The use of the Bishop method for a non-circular failure surface is generally not recommended (though it is allowed by some commercial programs) because of the unbalanced horizontal force problem, and this can be important for problems with loads from earthquakes or soil reinforcements. This method is simple for hand calculation and the convergence is fast. It is also virtually free from convergence problems, and the results are very close to those obtained by the "rigorous" methods. If a circular failure surface is sufficient for the design and analysis, this method can be a very good solution for engineers. When applied to an undrained problem with $\phi = 0$, the Bishop method and the Swedish method are identical.

In Bishop's formulation, $X_L = X_R = 0$ is assumed, hence Eq. (7.17) simplifies to

$$P = \left[W - \frac{1}{F}(c'l\sin\alpha - ul\tan\phi'\sin\alpha)\right]/m_\alpha \tag{7.60}$$

Overall MOMENT equilibrium about the centre of circle: $\Sigma WR\sin\alpha = \Sigma TR$ (7.61)
where R is the radius of the circular slip surface. Rearranging and substituting for T gives:

$$F_m = \frac{\Sigma(c'l + (P - ul)\tan\phi')}{\Sigma W\sin\alpha} = \frac{\sum \dfrac{\left[c'b + (W - ub)\tan\phi'\right]\sec\alpha}{\left[1 + \dfrac{\tan\alpha\tan\phi'}{F}\right]}}{\sum W\sin\alpha} \tag{7.62}$$

As this equation contains F on both sides it has to be solved iteratively. Convergence is usually quick and so the method is suitable for hand calculation, although it is time consuming. Once F is found, P and T can be determined from Eqs. (7.60) and (7.15), respectively.

Janbu's simplified method

For the Janbu simplified method (1957), force equilibrium is completely satisfied while moment equilibrium is not satisfied. This method is also popular worldwide as it is rapid in computation with only a few convergence problems. This method can be used for a non-circular failure surface which is commonly observed in sandy-type soils. Janbu (1973) later proposed a "rigorous" formulation which is more tedious in computation. By examining the overall horizontal force equilibrium, a value of the factor of safety F_o is obtained. Resolve parallel to the base of the slice:

$$T + (E_R - E_L)\cos\alpha = W\sin\alpha \tag{7.63}$$

$$\text{Hence } E_R - E_L = W\tan\alpha - \frac{1}{F}(c'l + (P - ul)\tan\phi')\sec\alpha \tag{7.64}$$

For the overall horizontal force equilibrium, in the absence of surface loading:

$$\Sigma(E_R - E_L) = 0 \tag{7.65}$$

$$\text{And } \Sigma(E_R - E_L) = \Sigma W\tan\alpha - \frac{1}{F}\Sigma(c'l + (P - ul)\tan\phi')\sec\alpha = 0 \tag{7.66}$$

$$\text{Hence } F_o = \frac{\Sigma(c'l + (P - ul)\tan\phi')\sec\alpha}{\Sigma W\tan\alpha} = \frac{\sum\left[c'b + (W - ub)\tan\phi\right]/\eta_\alpha}{\sum W\tan\alpha} \tag{7.67}$$

$$\text{where } \eta_\alpha = \left(1 + \frac{\tan\alpha \tan\phi'}{F}\right)\cos^2\alpha \tag{7.68}$$

Based on the ratio of the factors of safety from the "rigorous" and "simplified" analyses, Janbu (1973) proposed a correction factor F_0 given by Eq. (7.69) (converted to equation form by the author) for the Janbu simplified method (1957). When the factor of safety from the simplified method is multiplied with this correction factor, the result will be close to that from the "rigorous' analysis.

$$\text{For } c, \phi > 0, \quad f_0 = 1 + 0.5\left[\frac{D}{l} - 1.4\left(\frac{D}{l}\right)^2\right]$$

$$\text{For } c = 0, \quad f_0 = 1 + 0.3\left[\frac{D}{l} - 1.4\left(\frac{D}{l}\right)^2\right]$$

$$\text{For } \phi = 0, \quad f_0 = 1 + 0.6\left[\frac{D}{l} - 1.4\left(\frac{D}{l}\right)^2\right] \tag{7.69}$$

For the correction factor shown above and Figure 7.8, l is the length joining the left and right exit points, while D is the maximum thickness of the failure zone with reference to this line. D is tedious to determine by drawing the figures to take the maximum value. It can be obtained easily by determining the maximum depth from the slip surface to the base line by forming the normal line equation through the base nodal points to the base line. Since the correction factors by Janbu (1973) are based on limited case studies and are purely empirical in nature, the uses of these factors for complicated non-homogeneous slopes have been questioned by some engineers. Based on the author's experience, it is found that the factor of safety for the simplified Janbu method is usually close to that by

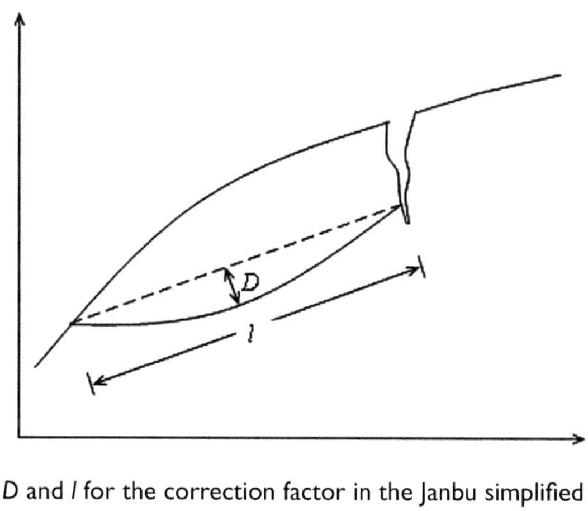

Figure 7.8 Definitions of D and l for the correction factor in the Janbu simplified method.

the Morgenstern-Price method, if this correction factor is used. Since the interslice shear force can sometimes generate a high factor of safety for some complicated cases which may occur in dam and hydropower projects, the use of the Janbu simplified method (1957) is preferred over other methods in these kinds of projects in China.

The Lowe and Karafiath method (1960) and the Corps of Engineers method are based on the interslice force function types 5 and 6. These two methods satisfy only force equilibrium but not moment equilibrium. In general, the Lowe and Karafiath method (1960) will give results close to those from the "rigorous" method, even though the moment equilibrium is not satisfied. For the Corps of Engineers method, it may lead to a high factor of safety in some cases, and some engineers adopt a lower interslice force angle to account for this problem (Duncan and Wright 2005); this practice is also adopted by some engineers in China. Similarly, the load transfer in China satisfies only the force equilibrium as well.

7.4.2 Davis and German method

Davis' method, first developed at the University of California by Shen et al. (1978, 1981a,b,c), was adopted in the USA, therefore it is also called the United States method. This method, which is shown in Figure 7.9, is a limit equilibrium method where the assumptions of the analysis are:

i) The soil is homogeneous;
ii) The face of the excavation is vertical (modified to be an inclined surface by Cheng);

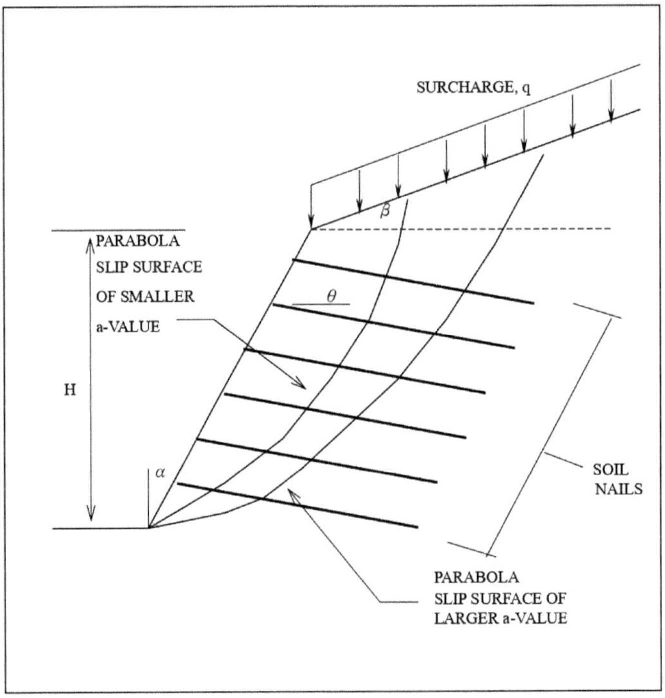

Figure 7.9 Modified Davis method for soil nailed system by Cheng.

iii) The ground surface is horizontal;
iv) A satisfactory drainage system exists giving zero pore pressure;
v) Horizontal spacing and vertical spacing of nails are equal;
vi) The nail length is uniform throughout the soil-nailed mass;
vii) The effective length of the nail is the length of nail beyond the potential failure surface;
viii) The pullout resistance is mobilized by the friction and/or adhesion at the soil re-bar interfaces within the effective zone;
ix) Only tensile strength of the re-bar is considered (i.e. shear and bending strength are ignored).

The German method, developed by Stocker et al. (1979) and Gassler and Gudehus (1981), is an analysis based on a two-part wedge mechanism as shown in Figure 7.10. The two-part wedge mechanism was adopted following experimental and full-scale tests conducted on soil-nailed walls (reported in Stocker et al. 1979 and Gassler and Gudehus 1981). This method of analysis depends on the limit equilibrium approach in which force polygons are constructed to satisfy the force equilibrium. The assumptions of the analysis are as follows:

i) The soil is homogeneous;
ii) The slip surface is bilinear;
iii) The shearing resistance of soil is entirely mobilized along the potential slip surface;
iv) Full drainage system is implemented; and
v) The layout of soil nails is uniform, i.e. uniform length, vertical/horizontal spacing of soil nails.

As shown in Figure 7.10, the two wedges above the sliding surface are separated by a vertical line. The position of this vertical line and in turn the configuration of the upper and lower wedges depend on the slope of the lower wedge. The angle of the assumed sliding surface below the lower wedge, θ_B, is varied to give the most critical condition. When the assumed sliding surface touches the end of the soil nail zone, the lower wedge is generated by drawing a vertical line from the intersecting point to the slope crest. For the assumed sliding surface below the upper wedge, the angle, θ_A, is set to $(\pi/4 + \phi'/2)$.

The shearing resistance of the soil, as defined by Mohr-Coulomb's failure criterion, is assumed to be entirely mobilized along the potential sliding surface. The total design forces in the nails (as represented by resultant ΣTp in Figure 7.10) are determined from pull-out tests which are used for design unless they exceed the elastic limit of the re-bar itself. No contribution of bending stiffness is taken into account. The global factor of safety, *FS*, is given as the ratio of the sum of the available resisting limit nail forces, ΣTp, to the total force required to maintain limit equilibrium, ΣT. The total force required to maintain limit equilibrium, ΣT, can be obtained by considering the force polygon, as shown in Figure 7.10.

Fortran codes for the modified Davis' and German methods are available from Cheng for education purposes. In general, the use of the classical slope stability method will be adequate for the analysis, hence these methods are less popular now.

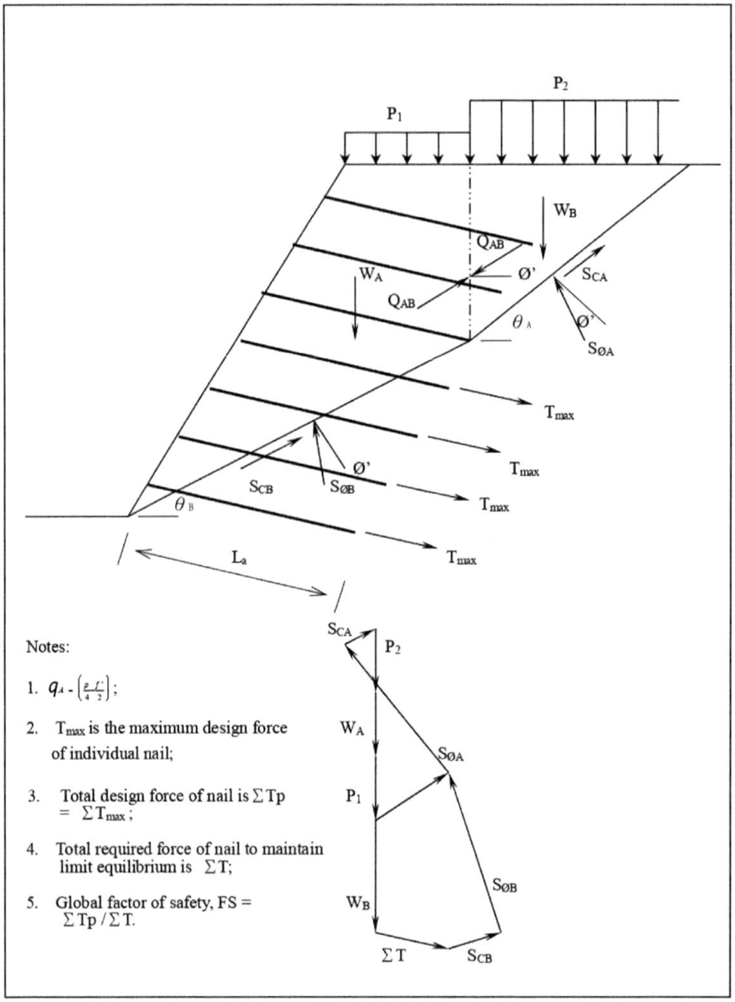

Figure 7.10 The German method.

7.5 NUMERICAL EXAMPLES ON SLOPE STABILITY ANALYSIS

For illustration of the numerical process, example 1, which a simple slope as shown in Figure 7.11, is studied. The slope is given by coordinates (4,0), (5,0), (10,5) and (12,5), while the water table is given by (4,0), (5,0), (10,4) and (12,4). The soil parameters are: unit weight = 19 kN/m^3, c'=5 kPa and ϕ'=36°. To define a circular failure surface, the coordinates of the centre of rotation and the radius should be defined. Alternatively, a better method is to define the x-ordinates of the left and right exit ends and the radius of the circular arc. The latter approach is better as the left and right exit ends can usually be estimated easily from engineering judgement. In the present example, the x-ordinates of

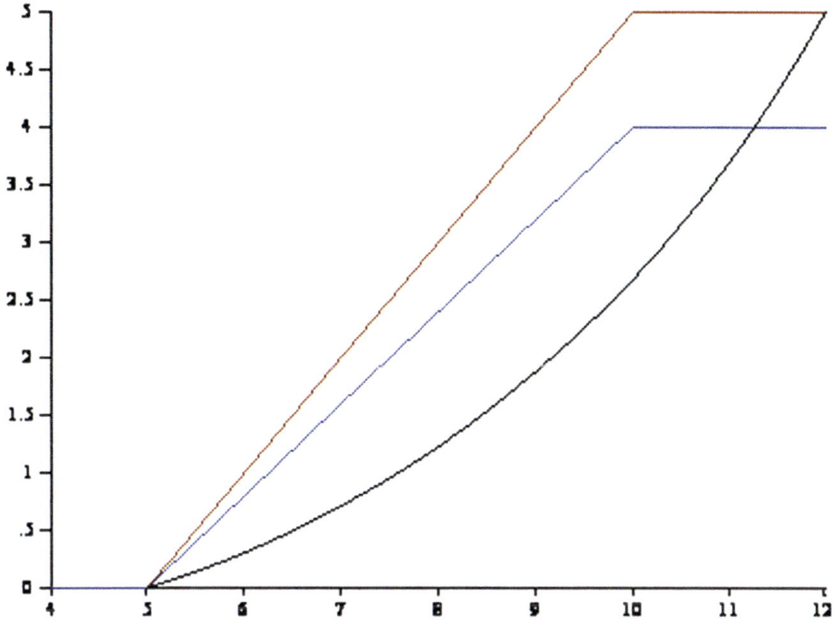

Figure 7.11 Numerical examples for a simple slope.

the left and right exit ends are defined as 5.0 m and 12.0 m, while the radius is defined as 12 m. The soil mass is divided into 10 slices for analysis and the details are given below.

Slice	*Weight (kN)*	*Base angle (°)*	*Base length (m)*	*Base pore pressure (kPa)*
1	2.50	16.09	0.650	1.57
2	7.29	19.22	0.662	4.52
3	11.65	22.41	0.676	7.09
4	15.54	25.69	0.694	9.26
5	18.93	29.05	0.715	10.99
6	21.76	32.52	0.741	12.23
7	23.99	36.14	0.774	12.94
8	25.51	39.94	0.815	13.04
9	32.64	45.28	1.421	7.98
10	11.77	52.61	1.647	0.36

The results of analyses for the problem in Figure 7.11 are given in Table 7.3. For the Swedish method or the ordinary method of slices, where only the moment equilibrium is considered while the interslice shear force is neglected, the factor of safety from the global moment equilibrium takes the simple form of:

$$F_m = \frac{\Sigma\left(c'l + \left(W\cos\alpha - ul\right)\tan\phi'\right)}{\Sigma W\sin\alpha} \tag{7.70}$$

Table 7.3 Factors of safety for the failure surface shown in Figure 7.11

F	*Bishop*	*Janbu simplified*	*Janbu rigorous*	*Swedish*	*Load factor*	*Sarma*	*Morgenstern Price*
	1.023	1.037	1.024	0.991	1.027	1.026	1.028

Note: The correction factor is applied to the Janbu simplified method. The results for the Morgenstern-Price method using $f(x) = 1.0$ and $f(x) = \sin(x)$ are the same. Tolerance in the iteration analysis for the simplified method is 0.0005.

A factor of safety of 0.991 is obtained directly from the Swedish method for this example without any iteration. Based on an initial factor of safety of 1.0, the successive factors of safety during the Bishop iteration analysis are 1.0150, 1.0201, 1.0219, 1.0225 and 1.0226. For the Janbu simplified method (1957), the successive factors of safety during the iteration analysis using the Janbu simplified method are 0.9980, 0.9974 and 0.9971. Based on a correction factor of 1.0402, the final factor of safety from Janbu's simplified analysis is 1.0372. If the double QR method is used for the Janbu simplified method (1957), a value of 0.9971 is obtained *directly* from the first positive solution of the Hessenberg matrix without using any iteration analysis. For the Janbu rigorous method (1973), the successive factors of safety based on iteration analysis are 0.9980, 0.9974, 0.9971, 1.0102, 1.0148, 1.0164, 1.0170, 1.0213, 1.0228, 1.0233 and 1.0235. For the Morgenstern-Price method (1965), a factor of safety of 1.0282 and the internal forces are obtained directly from the double QR method without any iteration analysis. The variation of F_f and F_m with respect to λ using the iteration analysis for this example is shown in Figure 7.12. It should be noted that F_f is generally more sensitive to λ than F_m, and the two lines may not meet for some cases, which can be considered as no solution to the problem. There are cases where the lines are very close but actually do not intersect. If a tolerance large enough is defined, then the two lines can be considered as having an intersection and the solution converge. This type of "false" convergence is experienced by many engineers in Hong Kong. These two lines may be affected by the choice of the moment point, and convergence can sometimes be achieved by adjusting the choice of the moment point. The results as shown in Figure 7.12 assume the interslice shear forces to be zero in the first solution step, and this solution procedure appears to be adopted in many commercial programs. Cheng et al. (2008) have, however, found that the results shown in Figure 7.12 may not be the true result for some special cases, and this will be discussed further in a later section.

From Table 7.3, it is clear that the Swedish method is a very conservative method as first suggested by Whitman and Bailey (1967). In addition, Janbu's simplified method (1957) will also give a smaller factor of safety if the correction factor is not used. After the application of the correction factor, Cheng found that the results from Janbu's simplified method (1957) are usually close to those "rigorous" methods. In general, the factors of safety from different methods of analysis are usually close to each other, as pointed out by Morgenstern (1992).

Example 2 is a simple slope with a water table, as shown in Figure 7.13, which will be used to illustrate the detailed calculation procedures for various limit equilibrium methods.

For this problem, 10 slices have been chosen for illustration of the intermediate calculation process. In general, for any point with a major change in the geometry of the ground

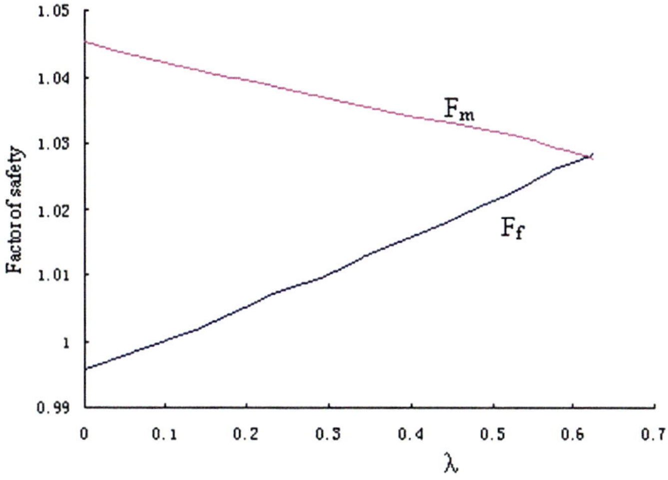

Figure 7.12 Variations of F_f and F_m with respect to λ for the example in Figure 7.9.

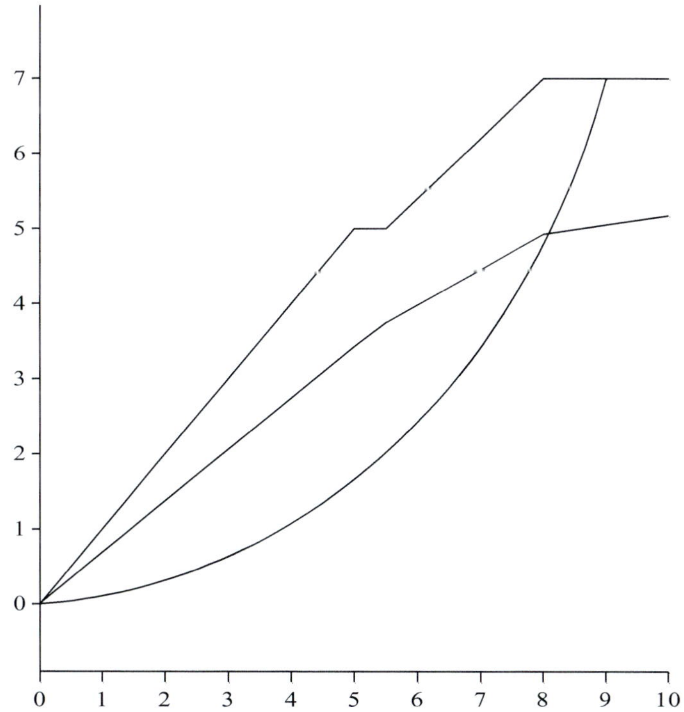

Figure 7.13 Example 2 for illustration.

Table 7.4 Coordinates of the circular slip surface and corresponding ground surface level for Figure 7.13

Node no.	*X (m)*	*Y (m)*	*Ground level (m)*
1	0.0	0.0	0.0
2	0.625	0.054	0.625
3	1.25	0.147	1.25
4	1.875	0.282	1.875
5	2.5	0.46	2.5
6	3.75	0.954	3.75
7	5.0	1.663	5.0
8	5.5	2.018	5.0
9	6.75	3.144	6.0
10	8.0	4.789	7.0
11	9.0	7.0	7.0

Table 7.5 Basic slice details (V = vertical surcharge, H = horizontal load, inward as positive, α = base inclination of slip surface, *l* = base length of slice, u = pore water pressure, Xc/Yc = coordinates of the centroid of slice) for Figure 7.13

No.	*weight (kN)*	*V (kN)*	*H (kN)*	*α (deg)*	*l (m)*	*u (kPa)*	*wsinα*	*Xc*	*Yc*
1	3.48	0.00	0.00	4.92	0.627	1.87	0.298	0.416	0.223
2	10.19	0.00	0.00	8.52	0.632	5.37	1.510	0.907	0.511
3	16.39	0.00	0.00	12.17	0.639	8.45	3.454	1.581	0.856
4	22.07	0.00	0.00	15.86	0.650	1.11	6.030	2.2	1.226
5	58.61	0.00	0.00	21.58	1.344	14.11	21.56	3.157	1.849
6	74.09	0.00	0.00	29.55	1.437	16.60	36.54	4.393	2.734
7	30.54	0.00	0.00	35.43	0.614	17.14	17.70	5.245	3.272
8	70.23	0.00	0.00	42.01	1.682	14.34	47.01	6.119	3.879
9	59.83	0.00	0.00	52.77	2.066	6.53	47.634	7.346	5.097
10	20.46	0.00	0.00	65.66	2.427	0.68	18.640	8.333	6.261

profile, a division of slice should be added. Other than these turning points, the slices can be divided evenly for simplicity. For the problem as shown in Figure 7.13, the basic slice details are generated in Tables 7.4 and 7.5.

For Bishop's method (1955), the denominator $\Sigma w_i \sin\alpha_i$ is a constant and is evaluated as 200.366. For the numerator $\Sigma[c'b_i+(w_i - u_ib_i)\tan\phi]m_\alpha$, where $m_\alpha = \sec\alpha/(1 + \tan\alpha\tan\phi/F)$, using an initial factor of safety 1.0, the results of iteration analysis are given in Table 7.6.

For the present problem, if the seventh iteration is evaluated, the factor of safety will remain stationary at 0.845. It should be noted that the factor of safety during iteration analysis usually changes in a monotonic way. A large fluctuation in the temporary factor of safety is usually associated with a non-convergence case, which is however highly uncommon for a circular slip surface with the simplified methods.

For the same problem, if the Janbu simplified method (1957) is used, the denominator $\Sigma w_i \tan\alpha_i$ is a constant and is evaluated as 285.749. For the numerator $\Sigma[c'b_i + (w_i - u_ib_i)\tan\phi]\eta_\alpha$, where $\eta_\alpha = \sec^2\alpha/(1 + \tan\alpha\tan\phi/F)$, using an initial factor of safety 1.0; the results of the iteration analysis are shown in Table 7.7.

Table 7.6 Numerators in iteration analysis using the Bishop method for Figure 7.13

Slice	*F = 1.0*	*F = 0.897*	*F = 0.863*	*F = 0.852*	*F = 0.847*	*F = 0.846*
1	2.720	2.702	2.695	2.693	2.692	2.692
2	5.522	5.462	5.440	5.432	5.429	5.428
3	8.025	7.906	7.862	7.846	7.840	7.838
4	10.265	10.072	10.002	9.976	9.967	9.964
5	26.271	25.631	25.400	25.316	25.285	25.273
6	32.792	31.753	31.382	31.246	31.197	31.179
7	13.418	12.923	12.748	12.684	12.660	12.652
8	32.295	30.918	30.433	30.258	30.193	30.170
9	33.270	31.529	30.925	30.707	30.627	30.600
10	15.095	14.107	13.771	13.651	13.607	13.5916
	Σ = 179.672 F = 0.897	Σ = 173.003 F = 0.863	Σ = 170.658 F = 0.852	Σ = 169.809 F = 0.847	Σ = 169.498 F = 0.846	Σ = 169.384 F = 0.845

Table 7.7 Numerators in iteration analysis using the Janbu simplified method for Figure 7.13

Slice	*F = 1.0*	*F = 0.856*	*F = 0.806*	*F = 0.787*	*F = 0.779*	*F = 0.775*	*F = 0.774*
1	2.730	2.704	2.693	2.688	2.686	2.686	2.685
2	5.583	5.496	5.459	5.443	5.437	5.434	5.433
3	8.210	8.033	7.959	7.928	7.915	7.909	7.907
4	10.671	10.382	10.262	10.211	10.190	10.181	10.178
5	28.251	27.259	26.852	26.682	26.610	26.580	26.567
6	37.695	35.979	35.286	34.998	34.878	34.827	34.806
7	16.467	15.597	15.249	15.105	15.045	15.020	15.009
8	43.465	40.815	39.769	39.339	39.159	39.084	39.052
9	54.984	50.889	49.303	48.657	48.388	48.275	48.227
10	36.628	33.238	31.962	31.447	31.233	31.144	31.106
	Σ = 244.684 F = 0.856	Σ = 230.392 F = 0.806	Σ = 224.794 F = 0.787	Σ = 222.500 F = 0.779	Σ = 221.542 F = 0.775	Σ = 221.140 F = 0.774	Σ = 220.970 F = 0.773

It is interesting to note that Janbu's method (1957) usually requires more iterations to achieve convergence than Bishop's method, and the factor of safety from Janbu's method (excluding f_0) appears to be always on the low side as compared with other methods (except under the action of nails). If the correction factor which is 1.061 is adopted for the present problem, Janbu's method (1957) will give an overall factor of safety of 0.82 which is compared to 0.845, 0.859, 0.854 and 0.854 for Bishop's method (1955), the load factor method, Sarma's method (1973) and Spencer's method (1967), respectively.

To determine the internal forces, the base normal and shear force, P and T, can be determined by Eqs. (7.50) and (7.15), using the converged factor of safety.

Slice	*P*	*T*	*For Bishop's method*
1	3.224	3.185	
2	9.341	6.422	
3	14.767	9.274	
4	19.589	11.788	

Slice	*P*	*T*	*For Bishop's method*
5	51.207	29.900	
6	64.255	36.883	
7	26.827	14.966	
8	62.381	35.685	
9	51.263	36.188	
10	14.108	16.072	
Slice	*P*	*T*	*For Janbu's method*
1	3.200	3.461	
2	9.262	6.950	
3	14.611	9.998	
4	19.341	12.663	
5	50.395	31.953	
6	62.966	39.156	
7	26.223	15.815	
8	60.729	37.519	
9	49.236	37.728	
10	13.004	16.571	

It should be noted that the base normal and shear forces are not sensitive to the use of different methods of analysis, which is also supported by the results shown above. On the other hand, there are greater differences in the interslice shear and normal forces among different methods of analysis.

For the interslice shear force, it is simply assumed as zero for the Janbu and Bishop methods. For the interslice normal force under the Bishop formulation (1955), some commercial programs produce the output for the users. Since Bishop's method (1955) cannot satisfy horizontal force equilibrium, the use of horizontal force equilibrium from left to right or from right to left will produce different internal normal forces. The author hence does not give the interslice normal force for Bishop's method (1955) in SLOPE 2000 to avoid the users making mistakes in the interpretation. For the interslice normal force under Janbu's simplified formulation (1957), it can be determined using Eq. (7.64).

Once E_R for slice 1 is determined using the fact that E_L is zero for this slice, E_R from slice 1 will become E_L for slice 2 and Eq. (7.64) can hence be used to determine E_R for slice 2. Furthermore, since the number of internal force is 1 less than the number of slices, the interslice normal force is hence given by

Section I	*E*	*For Janbu's method*
1	3.17	
2	8.67	
3	15.37	
4	22.26	
5	33.45	
6	36.45	
7	34.14	
8	21.37	
9	5.0	

Table 7.8 Overturning and restoring moment from earthquakes and soil nails

Slice	*Overturning moment*	*Restoring moment*
1	3.401	0
2	9.654	0
3	14.963	0
4	19.328	0
5	47.687	49.576
6	57.323	15.275
7	20.499	0
8	42.883	0
9	29.248	85.013
10	7.619	0

In the presence of earthquakes and soil nail forces, the problem will be more complicated. Consider Figure 7.13, with the application of a horizontal earthquake coefficient 0.1, a horizontal inward pressure of 10 kPa is applied on the ground surface from $x = 3$ to $x = 4$ and a soil nail is applied at $x = 4$, with a length of 5.8 m and an angle of inclination of 10°.

Consider the soil nail with a nail head coordinate (4,4) and an angle of inclination of 10°, the soil nail intersects with the slip surface at (6.998,3.471). Behind the slip surface, the effective nail length is 2.755 m. The effective bond load using Hong Kong practice is given by 6.04 kPa with a bond load factor of safety of 2.0. The effective nail load is hence given by 6.04 × 2.755 = 16.65 kN. Using the coordinate (6.998,3.471), the nail load should be applied at slice 9, and the nail load is decomposed to a vertical load of 16.4 kN and an inward horizontal load of 2.89 kN. Given the coordinate of the centre of rotation and radius of rotation as (−0.544,9.985) and 10 m, the restoring moment arising from the horizontal load and the soil nail and the over-turning moment arising from the earthquake load can be determined; these are shown in Table 7.8.

The overall moment is summed up to be 102.741, and this term is added to the denominator of Bishop's equation (7.62) in the form of 102.741/R, where R = 10 m for the present circular slip surface. The final value will be 200.366 + 102.741/10 = 210.64. On the other hand, the vertical component of the nail load (16.4 kN) will be added to W under the vertical force equilibrium consideration. The horizontal force due to earthquake and nail load are not added in Eq. (7.62) directly, as horizontal force equilibrium is not considered in the Bishop method. These two terms enter indirectly into Bishop's equation through the overall moment term in the denominator of Eq. (7.62). Based on these results, the factor of safety of this slope will be 0.794 for the Bishop method, which is shown in Table 7.9. For Janbu's method, the denominator will become $\Sigma(\tan\alpha + H)$, and the results for the horizontal load on each slice are given in Table 7.10.

From Table 7.10, the net horizontal loading will be 10.189 kN. The vertical component of the soil nail load will be added to W for vertical force equilibrium, and the denominator will be 299.743 (consider also the vertical component of the soil nail in slice 9). The iteration analysis using the Janbu simplified method will be given by Table 7.11.

There are some simplified methods which rely on the assumption of interslice force function but fulfil only force but not moment equilibrium. For example, in the Corps of Engineers method, *f(x)* is assumed to be constant and is equal to the slope angle defined by

Table 7.9 Iteration analysis for the Bishop method with earthquake, external load and soil nails using the over-shooting method (note: *is changed by soil nail compared with Table 7.6)

Slice	*F = 1.0*	*F = 0.794*
1	2.720	2.680
2	5.522	5.389
3	8.025	7.761
4	10.265	9.841
5	26.271	24.873
6	32.792	30.542
7	13.418	12.353
8	32.295	29.351
9	35.011*	31.139
10	15.095	13.040

Σ = 181.414 Σ = 166.97

F = 0.863 F = 0.794 converged

Table 7.10 Horizontal load from earthquake and soil nail and external load

Slice	*Earthquake*	*Soil nail/external load*
1	0.348	0
2	1.019	0
3	1.639	0
4	2.207	0
5	5.861	7.5
6	7.409	2.5
7	3.054	0
8	7.023	0
9	5.983	16.4
10	2.046	0

Table 7.11 Iteration analysis for Figure 7.13 using the Janbu simplified method

Slice	*F = 1.0*	*F = 0.739*	*F = 0.727*	*F = 0.725*
1	2.730	2.676	2.672	2.672
2	5.583	5.402	5.391	5.390
3	8.210	7.846	7.825	7.821
4	10.671	10.080	10.045	10.040
5	28.251	26.240	26.124	26.107
6	37.695	34.255	34.061	34.032
7	16.467	14.735	14.639	14.625
8	43.465	38.238	37.953	37.910
9	57.862	49.477	49.033	48.967
10	36.628	30.153	29.824	29.775

Σ = 247.561 Σ = 219.102 Σ = 217.568 Σ = 217.339

F = 0.826 F = 0.731 F = 0.726 F = 0.725 converged

Table 7.12 $f(x)$ for the Lowe-Karafiath method

Section	$\lambda f(x)$
1	0.0856
2	0.148
3	0.209
4	0.270
5	0.360
6	0.476
7	0.554
8	0.633
9	0.744

the two extreme ends of the failure surface. In the Lowe-Karafiath method (1960), $\lambda f(x)$ is assumed to be the average of the slope angle of the ground profile and the failure surface at the section under consideration. For the problem in Figure 7.13, $f(x)$ is given in Table 7.12.

Based on this $\lambda f(x)$, the factor of safety is determined to be 0.867 directly from the double QR method. For this problem, if the China type load factor method is used, the factor of safety will be 0.859, while the Corps of Engineers method gives a value of 0.881. In general, the author finds that the results from the Lowe-Karafiath method (1960) are comparable to those from the other rigorous methods and appear to be more reliable in general. On the other hand, the Swedish method gives a value of 0.756, which is much smaller than the other methods. The Swedish method is well known to be conservative and is not recommended generally to be used.

While the simplified method for a single slip surface is simple enough for hand or spreadsheet calculation, the calculation required for the rigorous method is so complicated that a computer program is required for the analysis. Very few books try to outline the detailed procedures required in the rigorous analysis, and this book will try to help readers to understand the procedures required in individual rigorous methods. The readers can implement their own programs based on the procedures as provided in this section.

7.5.1 Morgenstern-Price (Spencer) method

In this section, the Spencer method will be considered, which is the most popular version used by engineers for normal routine design work. The extension of the Spencer method to the Morgenstern-Price method will also be discussed in this section. Consider the example in Figure 7.11 again.

Resolve horizontally: $T\cos\alpha - P\sin\alpha + E_R - E_L = 0$ (7.71)

Rearranging, and substituting for T gives

$$E_R - E_L = P\sin\alpha - \frac{1}{F}\left[c'l + (P - ul)\tan\phi'\right]\cos\alpha \qquad (7.72)$$

Since $\Delta X = \lambda\Delta E$, hence putting Eq. (7.71) into Eq. (7.72) can yield an expression for P which depends on λ and F only. ΔE is then determined by Eq. (7.72) and hence E is

Table 7.13 Base normal and shear forces in first step which is actually Janbu simplified analysis (using initial factor of safety 1.0 and $\lambda = 0$)

Slice	*Base normal*	*Base shear*
1	3.200	3.460
2	9.262	6.949
3	14.611	10.000
4	19.341	12.661
5	50.397	31.950
6	62.968	39.153
7	26.224	15.814
8	60.732	37.516
9	49.239	37.726
10	13.006	16.570

Table 7.14 Base normal, shear and ΔE in the first iteration (Fos with respect to moment = 0.853, $\lambda = 0$)

Slice	*P*	*T*	ΔE
1	2.029	2.675	3.174
2	5.867	5.372	5.499
3	9.210	7.728	6.693
4	12.123	9.788	6.893
5	31.435	24.699	11.179
6	39.122	30.267	3.006
7	15.706	12.225	-2.316
8	36.613	29.002	-12.771
9	35.750	29.164	-16.375
10	11.363	12.810	-5.021

formed successively. X is then formed by using $X = \lambda E$. In the first step of the solution, Janbu simplified analysis is used to determine the first trial factor of safety (Bishop factor of safety can be used for a circular surface), with λ being zero in the first step in the double QR analysis. The base normal and shear forces can then be determined accordingly. For the present example, the factor of safety will be 0.773 as given previously. The base normal and shear forces as shown in Table 7.13 will be stored for use when a new λ value is tried.

Using an initial $\lambda = 0$, base normal and shear forces, P and T, can be determined accordingly, which are given in Table 7.14. Based on a moment point at (4.91,8.585), P, T and W can take the moment about the moment point to evaluate the factor of safety with respect to the moment.

The factor of safety of 0.853 has not converged with respect to moment consideration, and in the next step, this value will be used for next iteration analysis. The results of analysis for the second step of iteration are shown in Table 7.15.

The factor of safety has changed from 0.853 to 0.862 in the second step, and the process continues until the value becomes stable at step 4. Based on the converged base normal

Table 7.15 Base normal, shear and ΔE in the second iteration (Fos with respect to moment = 0.862, λ = 0)

Slice	*P*	*T*	ΔE
1	2.055	2.693	2.871
2	5.954	5.433	4.914
3	9.382	7.848	5.880
4	12.396	9.979	5.895
5	32.327	25.324	8.752
6	40.539	31.260	0.132
7	16.371	12.690	−3.462
8	38.431	30.274	−15.488
9	37.983	30.728	−19.180
10	12.581	13.662	−6.358

Table 7.16 Base normal, shear and ΔE in the first step (Fos with respect to moment = 0.848, λ = 0.635)

Slice	*P*	*T*	ΔE
1	5.048	4.789	5.640
2	10.356	8.515	8.855
3	13.961	11.054	9.900
4	16.395	12.780	9.448
5	36.925	28.543	13.784
6	39.136	30.280	3.010
7	13.970	11.009	−2.591
8	29.500	24.021	−12.800
9	27.762	23.571	−14.393
10	8.874	11.067	−3.682

force, the factor of safety with respect to force as given by Eq. (7.73) is used which gives F_f = 0.773. This factor of safety is different from the moment factor of safety F_m which is 0.862. $F_m - F_f$ is hence a positive value and λ must be assigned until this value becomes zero. The user can assigned a value of 0.1 (or any other value) and increase gradually until the term $F_m - F_f$ has a change in sign, and a more precise value of λ can then be interpolated. The author has chosen λ to be 0.635 (converged value) for illustration purposes. The results of iteration analysis are shown in Tables 7.16 and 7.17.

$$F_f = \frac{\Sigma(c'l + (P - ul)\tan\phi')\cos\alpha}{\Sigma P \sin\alpha} \tag{7.73}$$

Based on the results in Table 7.17, the factor of safety with respect to force is 0.857, and hence the factor of safety of this system is given as 0.857 with λ = 0.635.

If the double QR method is used, a λ value will be assumed and the factor of safety with respect to force is determined *directly* without any iteration analysis. The net moment (in Table 7.18) based on the corresponding λ and F_f will then be used to evaluate the net

Table 7.17 Base normal, shear and ΔE in the fourth step (Fos with respect to moment = 0.857, λ = 0.635)

Slice	*P*	*T*	ΔE
1	5.097	4.823	5.074
2	10.500	8.616	7.887
3	14.217	11.233	8.684
4	16.767	13.040	8.088
5	38.001	29.297	10.857
6	40.616	31.313	0.001
7	14.595	11.447	−3.669
8	31.072	25.122	−15.148
9	29.648	24.892	−16.762
10	9.974	11.837	−4.890

Table 7.18 Temporary results during double QR analysis

Trial	λ	*Net moment*
1	0	10.167
2	0.05	9.471
3	0.1	8.732
4	0.15	7.957
5	0.2	7.150
6	025	6.319
7	0.3	5.470
8	0.35	4.608
9	0.4	3.740
10	0.45	2.871
11	0.5	2.004
12	0.55	1.144
13	0.6	0.293
14	0.65	−0.545
15	0.617	−0.001

moment on the system, and the factor of safety will be 0.854 with λ = 0.617 for the present problem. These results differ slightly from the results using iteration analysis because of the truncation involved in the tremendous calculations required in both analyses. Such a small difference is, however, acceptable from a practical point of view.

The above procedures can be extended to the Morgenstern-Price method easily by using the fact that $X_i = \lambda f(x_i)E$. Other than this small change, there is virtually no difference between the Spencer method and the Morgenstern-Price method in the calculation procedures.

7.5.2 Janbu's rigorous method

The Janbu rigorous method differs from other rigorous methods in that the thrust line is assumed to be known, and the resultant interslice forces E_i were assumed to be acting on a "line of thrust" and β= inclination of line of thrust to horizontal at the interface. The numerical procedures as outlined in this section are based on the author's implementation instead of the original Janbu rigorous approach. By taking moments about the base centre of each slice and assuming width b is small, the following equations can be obtained:

$$E_{i+1} b \tan\beta + \frac{(X_i + X_{i+1})b}{2} + (E_{i+1} - E_i)\left(h_i - \frac{b\tan\alpha}{2}\right) = 0 \quad (7.74)$$

$$\frac{X_i + X_{i+1}}{2} = -E_{i+1}\tan\beta - \frac{E_{i+1}h_i}{b} + \frac{E_{i+1}\tan\alpha}{2} + \frac{E_i h_i}{b} - \frac{E_i \tan\alpha}{2} \quad (7.75)$$

Where $b = l\cos\alpha$, h = height of the line of thrust and β is the slope of the line of thrust. If the width of a slice is small, $X_i \approx X_{i+1}$ and $E_i \approx E_{i+1}$, X_i can be approximated by Eq. (7.75). In the original formulation by Janbu, the width of the slice should be small enough for it to be valid. The author has relaxed this requirement and has found that this will slightly improve the convergence. In the first step of the Janbu rigorous method, the Janbu simplified method is actually performed first. The previous results for the Janbu simplified method will hence be adopted as the initial values in the Janbu rigorous analysis. Based on the initial internal forces and a thrust ratio of 0.333, from Eq. (7.75), P is determined and is used in Eq. (7.74) to give ΔE. Using the fact that E_1 is zero for the first slice, E can then be determined successively and the results are given in Table 7.19. X is then determined directly from Eq. (7.75) (Table 7.20).

Starting from an initial factor of 1.0, the Janbu simplified method gives the first trial factor of safety of 0.773. In the Janbu rigorous analysis using the original Janbu formulation, the factor of safety starts from 0.826 and converges at 0.849. Using the author's modification where the width of each slice is taken as finite, the factor of safety varies from 0.845 to 0.848 and converges at 0.849, and the results of analysis are shown in Tables 7.21–7.23.

Table 7.19 E from successive iterations in Janbu rigorous analysis

Slice	*Iteration 1*	*Iteration 2*	*Iteration 3*
1	3.174	3.839	3.957
2	8.674	11.625	12.656
3	15.367	20.791	22.592
4	22.261	29.454	31.329
5	33.443	41.364	42.814
6	36.454	42.880	43.491
7	34.141	38.132	38.458
8	21.376	23.490	23.308
9	5.013	7.473	6.772

Table 7.20 X from successive iterations in Janbu rigorous analysis

Slice	*Iteration 1*	*Iteration 2*	*Iteration 3*
1	1.231	1.489	1.534
2	5.572	7.560	8.258
3	11.246	15.054	16.154
4	16.651	21.094	21.971
5	23.719	28.225	28.748
6	26.244	28.734	28.575
7	6.784	5.973	5.629
8	7.080	8.312	7.749
9	3.211	−1.277	−2.005

Table 7.21 E from successive iterations in Janbu rigorous analysis using Cheng's formulation

Slice	*Iteration 1*	*Iteration 2*	*Iteration 3*	*Iteration 4*
1	3.174	4.418	4.993	5.252
2	8.674	11.685	12.819	13.122
3	15.367	19.964	21.384	21.663
4	22.261	28.002	29.578	29.885
5	33.443	39.713	41.070	41.142
6	36.454	39.774	40.148	39.961
7	34.141	36.284	36.902	36.931
8	21.376	21.407	22.122	22.267
9	5.013	4.044	3.878	3.443

Table 7.22 X from successive iterations in Janbu rigorous analysis using Cheng's formulation

Slice	*Iteration 1*	*Iteration 2*	*Iteration 3*	*Iteration 4*
1	2.189	3.046	3.442	3.621
2	5.990	7.829	8.340	8.297
3	10.451	12.920	13.441	13.568
4	15.094	18.027	18.692	18.793
5	23.441	26.040	26.211	25.912
6	21.490	19.130	18.207	17.987
7	19.627	21.243	23.830	24.786
8	11.135	8.291	7.051	6.428
9	4.485	4.929	5.861	5.428

Table 7.23 Factor of safety from original and modified Janbu rigorous analysis

Iteration	*FS for original formulation*	*FS for modified formulation*
1	0.826	0.845
2	0.838	0.848
3	0.845	0.849
4	0.849	0.849

Table 7.24 Bounds for Sarma analysis

Trial	*Lower bound*	*Upper bound*	λ
1	0.8	1	0.000718
2	0.8	0.9	0.000784
3	0.85	0.9	0.000750
4	0.85	0.875	0.000766
5	0.85	0.863	0.000775
6	0.85	0.856	0.000780
7	0.853	0.856	0.000777
8	0.853	0.854	0.000779
9	0.853	0.855	0.000778
10	0.854	0.854	0.000778

Table 7.25 R_u, K_i and E during iteration analysis for Sarma's method

Interface	R_u	K_i	E
1	0.335	1.560	4.710
2	0.329	1.251	14.068
3	0.322	1.067	25.031
4	0.314	0.857	33.002
5	0.301	0.646	46.712
6	0.280	0.525	54.097
7	0.281	0.441	36.255
8	0.255	0.287	21.658
9	0.136	0.174	7.863

7.5.3 Sarma's method

For the example in Figure 7.13, a bounding method is used by the author, and the bounds during the trial analysis are given in Table 7.24.

The critical situation (when factor of safety = 1) is obtained as –0.0777, while the corresponding λ is given by 0.000611. During the initial analysis, R_u and K_i from Eqs. (7.56) and (7.57) are given in Table 7.25, while E is determined from Eq. (7.57).

When the factor of safety is determined, X can be determined from Eq. (7.54). Base normal and shear force will then be given by Eqs. (7.31) and (7.32).

Although the Sarma method (1979) can converge well for many cases, the thrust line back-calculated from this method may sometimes lie outside the soil mass, which is not acceptable. Implementation of the Sarma method (1979) by Cheng does not require iteration analysis, and using an initial factor of safety from the Janbu or Bishop method, the results can be bracketed within a narrow range easily.

7.6 MISCELLANEOUS CONSIDERATIONS ON SLOPE STABILITY ANALYSIS

7.6.1 Acceptability of the failure surfaces and results of analysis

Based on an arbitrary interslice force function, the internal forces which satisfy both the force and moment equilibrium may not be kinematically acceptable, but this issue is seldom considered by engineers. The acceptability conditions of the internal forces include:

1. Since the Mohr-Coulomb relation is not used along the vertical interfaces between different slices, it is possible though not common that the interslice shear forces and normal forces may violate the Mohr-Coulomb relation.
2. Except for the Janbu rigorous method and the extremum method for which the resultant of the interslice normal force must be acceptable, the line of thrust from other "rigorous" methods which are based on overall moment equilibrium may lie outside the failure mass which is not possible. In fact, all the other methods of analysis cannot guarantee that the thrust line location must be acceptable, as the thrust line location is not used in most rigorous formulations. The extremum approach by Cheng et al. (2010) can overcome this problem at the expense of intensive computation.
3. The interslice normal forces should not be in tension. For the interslice normal forces near to the crest of the slope where the base inclination angles are usually high, if *c*' is high, it is highly likely that the interslice normal forces will be in tension in order to maintain the equilibrium. This situation can be eliminated by the use of tension crack. Alternatively, the factor of safety with tensile interslice normal forces for the last few slices may be accepted (which is the practice as adopted in most commercial programs), as the factor of safety is usually not sensitive to these tensile forces. On the other hand, tensile interslice normal forces near the slope toe are usually associated with special shape failure surfaces with kinks, steep upward slope at the slope toe or an unreasonably high/low factor of safety. The factors of safety associated with these special failure surfaces need special care in the assessment and should be rejected if the internal forces are unacceptable. Such a failure surface should also be eliminated during the location of the critical failure surfaces.
4. The base normal forces may be negative near the toe and crest of the slope. For negative base normal forces near to the crest of the slope, the situation is similar to the tensile interslice normal forces and may be tolerable (as adopted in most commercial programs). For negative base normal forces near to the toe of the slope which is physically unacceptable, it is usually associated with a deep-seated failure with a high upward base inclination. Since a very steep exit angle is not likely to occur, it is possible to limit the exit angle during automatic location of the critical failure surface.

If the above criteria are strictly enforced for all the slices of a failure surface, many slip surfaces will fail to converge. One of the reasons for this is the effect of the last slice when the base angle is large. Based on the force equilibrium, tensile interslice normal force will be created easily if *c*' is high. This result can propagate so that the results for the last few slices will be in conflict with the criteria above. If the last few slices are not strictly enforced, the factor of safety will be acceptable when compared with other methods of analysis. A suggested procedure is that if the number of slices is 20, only the first 15 slices are checked against the criteria above. It should also be noted that most commercial programs do not check the acceptability of the internal forces, and an unacceptable factor of safety associated with ridiculous internal forces may occasionally appear. Cheng has talked with the developer of a famous slope stability program, and was disappointed that the developer leaves this to the engineer's decision instead of internal checking during the computation.

7.6.2 Tension crack

As the condition of limiting equilibrium develops with a factor of safety close to 1, a tension crack shown in Figure 7.10 may form near the top of the slope through which no tensile strength can develop. If the tension crack is filled with water, a horizontal hydrostatic force P_w will generate additional driving moment and driving force which will reduce the factor of safety. The depth of a tension crack z_c can be estimated as:

$$z_c = \frac{2c\sqrt{K_a}}{\gamma} \tag{7.76}$$

where K_a is the Rankine active pressure coefficient. The presence of a tension crack will tend to reduce the factor of safety of a slope, but the precise location of a tension crack is difficult to be estimated for a general problem. It is suggested that if a tension crack is required to be considered, it should be specified at different locations and the critical results can then be determined. Sometimes, the critical failure surface with and without a tension crack can differ appreciably, and the location of the tension crack needs to be assessed carefully. In SLOPE 2000 by Cheng or some other commercial programs, the location of the tension crack can be varied automatically during the search for the critical failure surface.

7.6.3 Earthquake

Earthquake loadings are commonly modelled as vertical and horizontal loads applied at the centroid of the sliding mass, and the values are given by the earthquake acceleration factors k_v/k_h (vertical and horizontal) multiplied by the weight of the soil mass. This quasi-static simulation of earthquake loading is simple in implementation but should be sufficient for most design purposes, unless the strength of soil can be reduced by more than 15% due to the earthquake action. Beyond that, a more rigorous dynamic analysis may be necessary which will be more complicated, and more detailed information about the earthquake acceleration as well as the soil constitutive behaviour are required. Usually, a single earthquake coefficient may be sufficient for the design, but a more refined earth dam earthquake code is specified in DL5073-2000 in China. The design earthquake coefficients will vary according to the height under consideration, which will be different for different slices. Though this approach appears to be more reasonable, most of the design codes and existing commercial programs do not adopt this approach. The program SLOPE 2000 by Cheng can, however, accept this special earthquake code.

7.6.4 Water and seepage

An increase in pore water pressure is one of the main factors for slope failure. Pore water pressure can be defined in several ways. The classical pore pressure ratio r_u is defined as $u/\gamma h$, and an average pore pressure for the whole failure mass is usually specified for the analysis. Several different types of stability design charts are also designed using an average pore pressure definition. The use of a constant averaged pore pressure coefficient is obviously a highly simplified approximation. With the advancements in computer hardware and software, the uses of these stability design charts are now mainly limited to the preliminary designs only. The pore pressure coefficient is also defined as a percentage of the vertical surcharge applied on the ground surface in some countries. This definition of the pore pressure coefficient is, however, not commonly used.

If pore pressure is controlled by the ground water table, u is commonly taken as $\gamma_w h_w$, where h_w is the height of the water table above the base of the slice. This is the most commonly used method to define the pore pressure, which assumes that there is no seepage and the pore pressure is hydrostatic. Readers should note that there are other definitions of pore pressure coefficients which are not commonly used, and the use of pore pressure coefficients is now scarce. Alternatively, a seepage analysis can be conducted and the pore pressure can be determined from the flow-net or the finite element analysis. This approach is more reasonable but is less commonly adopted in practice due to the extra effort to perform a seepage analysis. More importantly, it is not easy to construct a realistic and accurate hydrogeological model to perform the seepage analysis.

Pore pressure can also be generated from the presence of a perched water table. In a multi-layered soil system, a perched water table may exist together with the presence of a water table if there are great differences in the permeability of the soil. This situation is rather common for the slopes in Hong Kong. For example, slopes at mid-levels in Hong Kong Island are commonly composed of fill at the top, which is underlain by colluvium and completely decomposed granite. Since the permeability of completely decomposed granite is one to two orders of magnitude less than that for colluvium and fill, a perched water table can easily be established within the colluvium/fill zone during heavy rainfall while the permanent water table may be within the completely decomposed granite zone. Consider Figure 7.14, a perched water table may be present in soil layer 1 with respect to the interface between soils 1 and 2 due to the permeability of soil 2 being 10 times less than that of soil 1. For a slice base between A and B, it is subjected to the perched water table effect and pore pressure should be included in the calculation. For a slice base between B and C, no water pressure is required in the calculation while the water pressure at the slice base between C and D is calculated using the ground water table only.

For the problem shown in Figure 7.15, if EFG which is below the ground surface is defined as the ground water table, the pore water pressure will be determined by EFG directly. If the ground water table is above the ground surface and undrained analysis is adopted, ground surface CDB is impermeable and the water pressure arising from AB will become external load on surface CDB. For drained analysis, the water table given by AB should be used, but vertical and horizontal pressure corresponding to the hydrostatic pressure should be applied on surfaces CD and DB. Thus a trapezoidal horizontal and vertical pressure will be applied to surfaces CD and DB, while the water table AB will be used to determine the pore pressure.

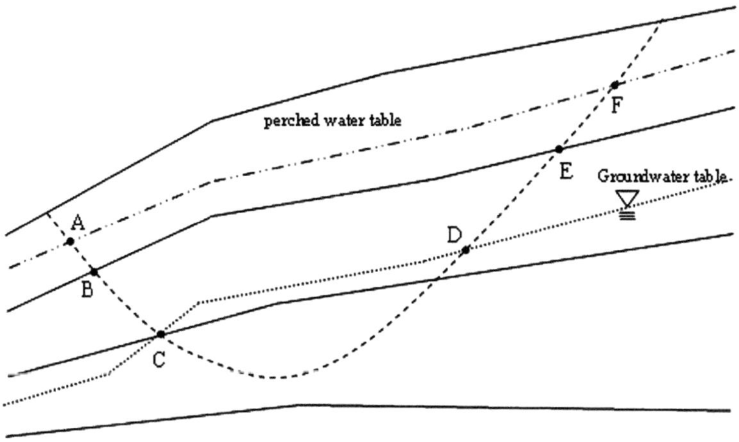

Figure 7.14 Perched water table in a slope.

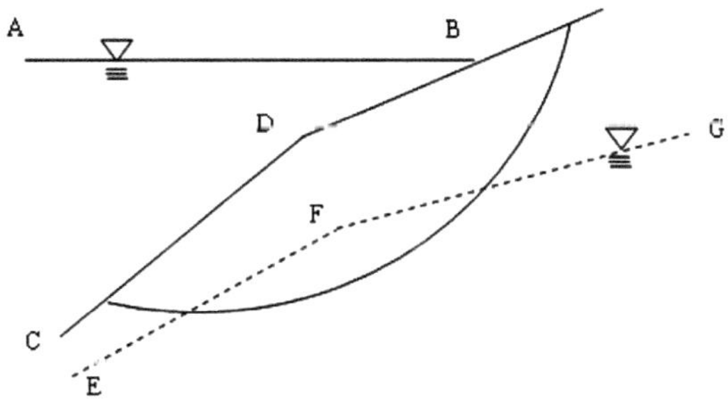

Figure 7.15 Modelling of ponded water.

The use of hydrostatic water pressure is usually on the safe side and is commonly adopted by engineers. Some programs (SLOPE 2000) accept the definition of the seepage field from a seepage analysis. The actual water pressure from the seepage analysis can be used to define the water pressure, which is then superimposed by the forces induced from the hydraulic gradient. The average horizontal/vertical component of the hydraulic gradient multiplied by the unit weight of water will be applied at the centroid of the wetted zone of each slice for the analysis.

For the treatment of the interslice forces, usually the total stresses instead of the effective stresses are used. This approach, though slightly less rigorous in the formulation, can greatly simplify the analysis and is adopted in virtually all commercial programs. Greenwood (1987) and Morrison and Greenwood (1989) have reported that this error is particularly significant where the slices have high base angles with a high water table. King (1989) and Morrison and Greenwood (1989) have also proposed revisions to the classical effective stress limit equilibrium method. Duncan and Wright (2005) have also reported that some "simplified" methods can be sensitive to the assumption of the total or effective interslice normal forces in the analysis.

7.6.5 Saturated density of soil

The unit weights of soil above and below the water table are not the same and may differ by 1–2 kN/m^3. For computer programs which cannot accept the input of saturated density, this can be modelled by the use of two different types of soil for a soil which is partly submerged. Alternatively, some engineers assume the two unit weights to be equal in view of the small differences between them.

7.6.6 Moment point

For simplified methods which satisfy only the force or moment equilibrium, the Janbu (1957) and Bishop methods (1955) are the most popular ones adopted by engineers. There is a perception among some engineers that the factor of safety from the moment equilibrium is more stable and is more important than the force equilibrium in stability formulation (Abramson et al. 2002). However, true moment equilibrium depends on the satisfaction of force equilibrium. Without force equilibrium, there is no moment equilibrium. Force equilibrium is, however, totally independent of the moment equilibrium. For methods which satisfy only the moment equilibrium, the factor of safety depends on the choice of moment point. For a circular failure surface, it is natural to choose the centre of the circle as the moment point, and it is also well known that the Bishop method can yield very good results even when the force equilibrium is not satisfied. Fredlund et al. (1992) have discussed the importance of the moment point on the factor of safety for the Bishop method, and the Bishop method cannot be applied to a general slip surface because the unbalanced horizontal force will create a different moment contribution for a different moment point. Baker (1980) pointed out that for the "rigorous" methods, the factor of safety is independent of the choice of the moment point. Cheng et al. (2008a) have, however, found that the mathematical procedures to evaluate the factor of safety may be affected by the choice of moment point. In fact, many commercial programs allow the user to choose the moment point for analysis. The double QR method by Cheng (2003) is not affected by the choice of the moment point in the analysis and is a very stable solution algorithm.

7.6.7 Use of soil nail/reinforcement

Soil nailing is a slope stabilization method by introducing a series of thin elements called nails to resist tension, bending and shear forces in the slope. The reinforcing elements are usually made of round cross-section steel bars. Nails are installed sub-horizontally into the soil mass in a prebored hole, which is fully grouted. Occasionally, the initial portions of some nails are not grouted, but this practice is not commonly adopted. Nails can also be driven into the slope, but this method of installation is uncommon in practice. A typical soil nail arrangement is provided in Figure 7.16.

The fundamental principle of the soil nail is the development of tensile force in the soil mass which renders the soil mass stable. Although only tensile force is considered in the analysis and design, soil nail function by a combination of tensile force, shear force and bending action is difficult to analyse. The use of the finite element by Wei and Cheng (2009b) demonstrated that the bending and shear contribution to the factor of safety is generally not significant, and the current practice in soil nail design should be sufficient for

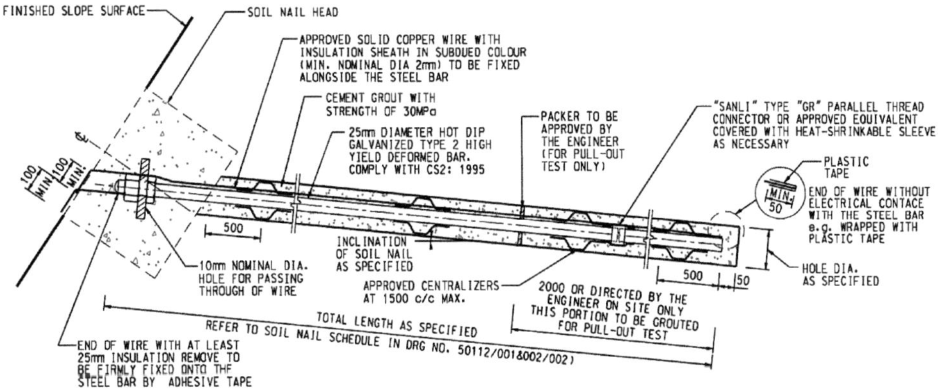

Figure 7.16 Typical soil nail details used in Hong Kong.

most cases. Nails are usually constructed at an angle of inclination from 10° to 20°. For an ordinary steel bar soil nail, a thickness of 2 mm is assumed as the corrosion zone so that the design bar diameter is in total 4 mm less than the actual diameter of the bar according to Hong Kong practice. The nail is usually protected by galvanization, paint, epoxy and cement grout. For a critical location, protection by expensive sleeving similar to that in a rock anchor may be adopted. Alternatively, FRP and CFRP may be used for soil nails which are currently under consideration by the author.

The practical imitations of soil nails include:

1. Lateral and vertical movement may be induced from excavation, and the passive action of soil nail is not as effective as the active action of anchor.
2. Difficulty in installation under some groundwater conditions.
3. Suitability of soil nail in loose fill is doubted by some engineers – stress transfer between the nail and soil is difficult to be established.
4. Collapse of drill hole before the nail is installed can happen easily in some ground conditions.
5. For a very long nail hole, it is not easy to maintain the alignment of the drill hole.

There are several practices in the design of soil nails. One of the precautions in the adoption of soil nails is that the factor of safety of a slope without a soil nail must be greater than 1.0 if a soil nail is going to be used. This is due to the fact that a soil nail is a passive element, and the strength of the soil nails cannot be mobilized until soil tends to deform. The effective nail load is usually taken as the minimum of:

(a) The bond strength between the cement grout and soil;
(b) The tensile strength of the nail, which is limited to 55% of the yield stress in Hong Kong, and 2 mm sacrificial thickness of bar surface is allowed for corrosion protection;
(c) The bond stress between the grout and the nail.

In general, only factors (a) and (b) are controlled in the actual design. The bond strength between the cement grout and soil is usually based on one of the following criteria:

(a) The effective overburden stress between grout and soil controls the unit bond stress on the soil nail, and it is estimated from the formula ($\pi c'D+2D\sigma_v'\tan\phi'$) for Hong Kong practice, while the Davis method (Shen et al. 1981a) allows inclusion of the angle of inclination, and D is the diameter of the grout hole. A safety factor of 2.0 is commonly applied to this bond strength in Hong Kong. During the calculation of the bond stress, only the portion behind the failure surface (effective bond length) is taken into consideration in the calculation.
(b) Some laboratory tests suggest that the effective bond stress between the nail and soil is relatively independent of the vertical overburden stress. This is based on the stress redistribution after the nail hole is drilled and the surface of the drill hole should be stress free. The effective bond load will then be controlled by the dilation angle of the soil. Some of the laboratory tests in Hong Kong have shown that the effective overburden stress is not important for the bond strength. On the other hand, some field tests in Hong Kong have shown that the nail bond strength depends on the depth of embedment of the soil nail. It appears that the bond strength between cement grout and soil may be governed by the type of soil, method of installation and other factors, and the bond strength may be dependent on the overburden height and time in some cases but this is not a universal behaviour.
(c) If the bond load is independent of the depth of embedment, the effective nail load will then be determined in a proportional approach, as shown in Figure 7.17

For a soil nail of length L, bonded length L_b and total bond load T_{sw}, L_e for each soil nail and T_{mob} for each soil nail are determined from the formula below:

$$\text{For slip 1: } T_{mob} = T_{sw} \tag{7.77}$$

In this case the slip passes in front of bonded length and the full magnitude is mobilized to stabilize the slip

$$\text{For slip 2: } T_{mob} = T_{sw} \times (L_e/L_b) \tag{7.78}$$

In this case the slip intersects the bonded length and only a proportion of the full magnitude provided by the nail length behind the slip is mobilized to stabilize the slip.

The effective nail load is usually applied as a point load on the failure surface in the analysis. Some engineers, however, model the soil nail load as a point load at the nail head or

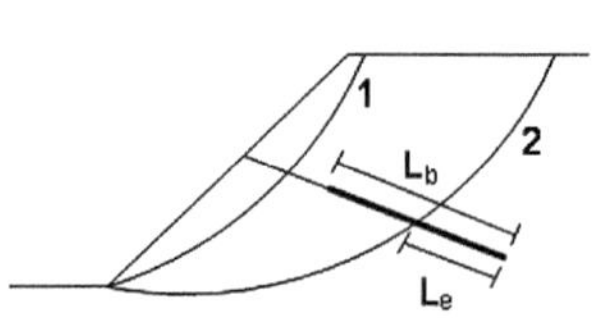

Figure 7.17 Definition of effective nail length in the bond load determination.

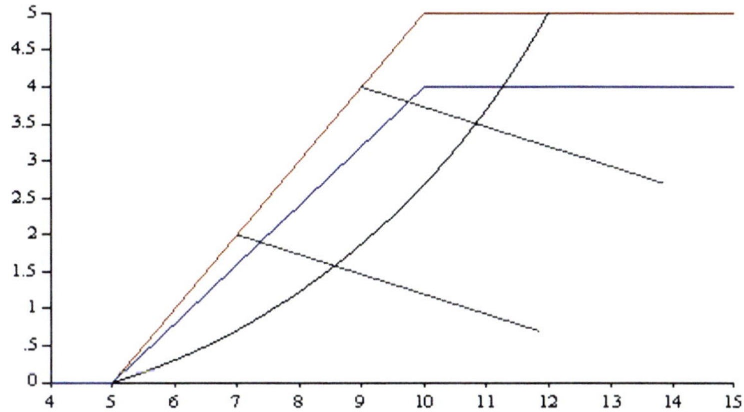

Figure 7.18 Two rows of soil nails are added to the problem in Figure 7.11.

Table 7.26 Factor of safety for the failure surface shown in Figure 7.11 (correction factor is applied in the Janbu simplified method)

F	*Bishop*	*Janbu simplified*	*Janbu rigorous*	*Swedish*	*Load factor*	*Sarma*	*Morgenstern Price*
	1.807	1.882	fail	1.489	1.841	1.851	1.810

as a distributed load applied on the ground surface. In general, there is no major difference in the factors of safety from these minor variations in treating the soil nail forces.

The effectiveness of the soil nail can be illustrated by adding two rows of 5 m length soil nails inclined at an angle of 15° to the problem shown in Figure 7.11, which is shown in Figure 7.18. The x-ordinates of the nail heads are 7.0 and 9.0. The total bond load is 40 kN for each nail which is taken to be independent of the depth of embedment, while the effective nail loads are obtained as 27.1 kN and 24.9 kN considered by a simple proportion as given in Figure 7.17. The results of analysis shown in Table 7.26 have illustrated that: (1) the Swedish method is a conservative method in most cases; and (2) the Janbu rigorous method (1973) is more difficult to converge as compared with other methods. It is also noticed that when an external load is present, there are greater differences between the results from different methods of analysis.

7.6.8 Failure to converge

Fail to converge during the solution of the factor of safety is sometimes found for "rigorous" methods which satisfy both force and moment equilibrium, and is quite serious for non-circular failure surfaces. If this situation is found, the initial trial factor of safety can be varied and convergence is occasionally achieved (but not guaranteed). Alternatively, the double QR method by Cheng (2003) can be used, as this is the ultimate method in the solution of the factor of safety. If no physically acceptable answer can be determined from

Table 7.27 Comparisons of the results from program A with SLOPE 2000 for 30 non-circular failure surfaces with soil nails, using Spencer analysis (detailed information can be obtained from the author)

Case	*Program A*	*Slope 2000*
1	Fail	1.2401
2	Fail	1.439
3	Fail	1.1308
4	Fail	1.3295
5	Fail	1.494
6	Fail	1.4826
7	Fail	1.2726
8	Fail	1.2446
9	1.586	1.5722
10	Fail	1.1457
11	Fail	1.8187
12	Fail	2.396
13	Fail	1.9984
14	Fail	1.5873
15	Fail	1.4557
16	1.7584	1.7584
17	Fail	1.3991
18	Fail	1.5732
19	Fail	1.495
20	Fail	1.4074
21	Fail	1.3438
22	Fail	1.5323
23	Fail	1.7703
24	Fail	1.9606
25	Fail	1.2796
26	1.266	1.264
27	1.471	1.473
28	1.232	1.2367
29	1.493	1.4761
30	1.357	1.3221

the double QR method, then there is no physical result for the specific method of analysis with the given assumptions ($f(x)$ or thrust line). Under such conditions, the simplified methods can be used to estimate the factor of safety, or the extremum principle by Cheng et al. (2010) may be adopted to determine the factor of safety. The convergence problem of the "rigorous" method will be studied in more details in later sections, and there are more case studies which are provided in the user guide of SLOPE 2000. The seriousness of the convergence problem can be illustrated by the use of the double QR method in SLOPE 2000 as compared with the iteration analysis in two famous commercial programs. As shown in Tables 7.27 and 7.28, 30 prescribed non-circular failure surfaces with soil nails are considered using the Spencer–Price analysis ($f(x) = 1$). SLOPE 2000 by the author can provide satisfactory answers for all 30 cases (no convergence requirement in double QR analysis), but the two commercial programs fail to give answers in many cases. From the author's experience, if the failure surfaces are relatively non-smooth with heavy external

Table 7.28 Comparisons of the results from program A with SLOPE 2000 for 30 non-circular failure surfaces with soil nail, using Spencer analysis (detailed information can be obtained from the author)

Case	*Program B*	*Slope 2000*
1	Fail	1.2401
2	1.407	1.439
3	1.1	1.1308
4	1.382	1.3295
5	Fail	1.494
6	Fail	1.4826
7	1.303	1.2726
8	Fail	1.2446
9	1.575	1.5722
10	1.103	1.1457
11	1.753	1.8187
12	Fail	2.396
13	Fail	1.9984
14	Fail	1.5873
15	Fail	1.4557
16	1.542	1.7584
17	1.381	1.3991
18	1.527	1.5732
19	Fail	1.495
20	Fail	1.4074
21	1.362	1.3438
22	1.572	1.5323
23	1.771	1.7703
24	1.829	1.9606
25	Fail	1.2796
26	1.275	1.264
27	1.483	1.473
28	1.230	1.2367
29	1.483	1.4761
30	1.373	1.3221

loads, many commercial programs fail to converge in the analysis easily. Such a failure to converge is, however, purely a false alarm, as SLOPE 2000 using the double QR method is able to give satisfactory results in all 30 test cases as well as other problems. If the double QR method cannot provide an answer to a problem, then by the nature of the problem, there is no answer to the problem with the given interslice force function $f(x)$. Engineers can, however, vary $f(x)$ and get an answer in most cases, but such tuning will not be automatic by nature, as the form of $f(x)$ for which an answer exists cannot be auto-evaluated by nature. The results as shown have clearly illustrated the limitations of the classical limit equilibrium formulations. The author has received many cases from different engineers about serious failures to converge during the MP analysis, particularly for problems with many nails. In most cases, results are actually available from SLOPE 2000, while answers are available for all cases if the extremum principle is used.

For the test cases as given Tables 7.27 and 7.28 as well as other test cases not shown in these two tables, interested readers can obtain the input files for the two commercial programs and for SLOPE 2000 for study from Cheng at natureymc@yahoo.com.hk. These test cases can illustrate the difficulties in ensuring convergence as well as numerical problems that may come up from the computation in many commercial programs using iteration analysis.

7.6.9 Location of the critical failure surface

The minimum factor of safety as well as the location of the critical failure surface are required for the proper design of a slope. For a homogeneous slope with a simple geometry and no external load, the log-spiral failure surface will be a good solution for the critical failure surface. In general, the critical failure surface for a sandy soil with a small c' value and high ϕ' will be close to the ground surface, while the critical failure surface will be a deep-seated one for a soil with a high c' value and small ϕ'. With the presence of an external vertical load or soil nail, the critical failure surface will generally drive the critical failure surface deeper into the soil mass. For a simple slope with a heavy vertical surcharge on top of the slope (typical abutment problem), the critical failure surface will be approximately a two-wedge failure from the non-circular search. This failure mode is also specified by the German code for abutment design. For a simple slope without any external load or soil nail, the critical failure surface will usually pass through the toe. Based on the above characteristics of the critical failure surface, engineers can manually locate the critical failure surface with ease for a simple problem. The use of the factor of safety from the critical circular or log-spiral failure surface (Frohlich 1953; Chen 1975), which will be slightly higher than that from the non-circular failure surface, is also adequate for simple problems.

For complicated problems, the above guidelines may not be applicable, and it will be tedious to carry out a manual trial and error in locating the critical failure surface. Automatic search for the critical circular failure surface is available in nearly all of the commercial slope stability programs. A few commercial programs also offer an automatic search for non-circular critical failure surface with some limitations. Since the automatic determination of the effective nail load (controlled by the overburden stress) appears to be not available in most commercial programs, engineers often have to perform the search for the critical failure surface by manual trial and error and the effective nail load is separately determined for each trial failure surface. To save time, only limited failure surfaces will be considered in the routine design. Cheng has found that reliance only on manual trial and error in locating the critical failure surface may not be adequate, with the adoption of modern optimization methods to overcome this problem. This important topic is discussed in a later section of this chapter.

7.6.10 Design of loose fill slope

There have been several major failures of loose fill slopes in Hong Kong in the past, with many deaths and loss of properties, and so the Hong Kong government and the Hong Kong Institution of Engineers have carried out a study on the design method for loose fill slope. The design methodology is now provided in Geoguide 7 (2008). Loose fill slope is critical because there are: (1) lower soil parameters; and (2) contraction under shear and strain softening under undrained conditions. The use of soil nails may not be sufficient to prevent local failures, as the development of bond and the zone of stabilization for soil nail in loose fill are somewhat questionable. In view of that, grillage is commonly used on loose fill slopes. Reinforced concrete grillage anchored by a soil nail (usually at the intersection of the grillage beams) resists the earth pressure, and this facing system spread the nail forces to the retained ground instead relying purely on the stabilization effect. The required stabilizing pressure on the slope facing, which is assumed to be triangular in shape, is determined by limit equilibrium analyses. In Hong Kong, a conservative lower bound value of steady-state strength equal to 0.2 times the mean effective stress at the onset of liquefaction is usually assumed (most test values are actually far greater than 0.2 for granular soil in Hong Kong). Using the conservative steady-state shear strength as the design shear strength and the determination of shear strength parameters from remoulded fill specimens, a relatively low factor of safety of 1.1 is recommended for use in Hong Kong. A typical analysis for a loose fill slope in Hong Kong is provided in Figure 7.19.

Many engineers determine the required nail forces by using the peak shear strength of the loose fill with a higher factor of safety. The peak mean effective stress is commonly taken as $(\sigma_x' + \sigma_y' + \sigma_z')/3$. If we assume $\sigma_x' = \sigma_x' = K_0 x \sigma_z'$, then peak mean effective stress $p_{peak}' = (1 + 2\ K_0)x\ \sigma_z'/3$. The steady-state undrained shear strength c_{zss} is then taken as 0.2^* p_{peak}' or $0.2x(1 + 2\ K_0)x\ \sigma_z'/3$. This value is then defined as the loose soil parameter in a suitable computer program for analysis.

7.6.11 3D analysis

All slope failure mechanisms are 3D in nature, but 2D analysis is performed for most practical cases. The difficulties associated with true 3D analysis are: (1) sliding direction, (2) satisfaction of 3D force and moment equilibrium, (3) relating the factor of safety to the previous two factors, and (4) a great amount of computational geometrical calculations are required. At present, there are still many practical limitations in the adoption of 3D analysis, and there are only a few 3D slope stability programs which are suitable for general use. Simplified 3D analysis for a symmetric slope is available in SLOPE 2000 by Cheng, and true 3D analysis for a general slope has been developed which is named SLOPE3D. 3D slope stability analysis is discussed in detail in a later section.

7.7 LIMIT ANALYSIS METHOD – DLO

7.7.1 Introduction to limit analysis

The limit analysis adopts the concept of an idealized stress–strain relation, i.e., the soil is assumed to be a rigid perfectly plastic material with an associated flow rule. Without carrying out the step-by-step elasto-plastic analysis, the limit analysis can provide solutions to many problems. Limit analysis is based on the bound theorems of classical plasticity theory (Drucker et al., 1951; Drucker and Prager, 1952). The general procedure of limit analysis is to assume a kinematically admissible failure mechanism for an upper bound solution or a statically admissible stress field for a lower bound solution, and the objective function will be optimized with respect to the control variables. Early efforts at limit analysis were mainly made on using the direct algebraic method or analytical method to obtain the solutions for slope stability problems with simple geometry and soil profile (Chen, 1975). Since closed-form solutions for most practical problems are not available,

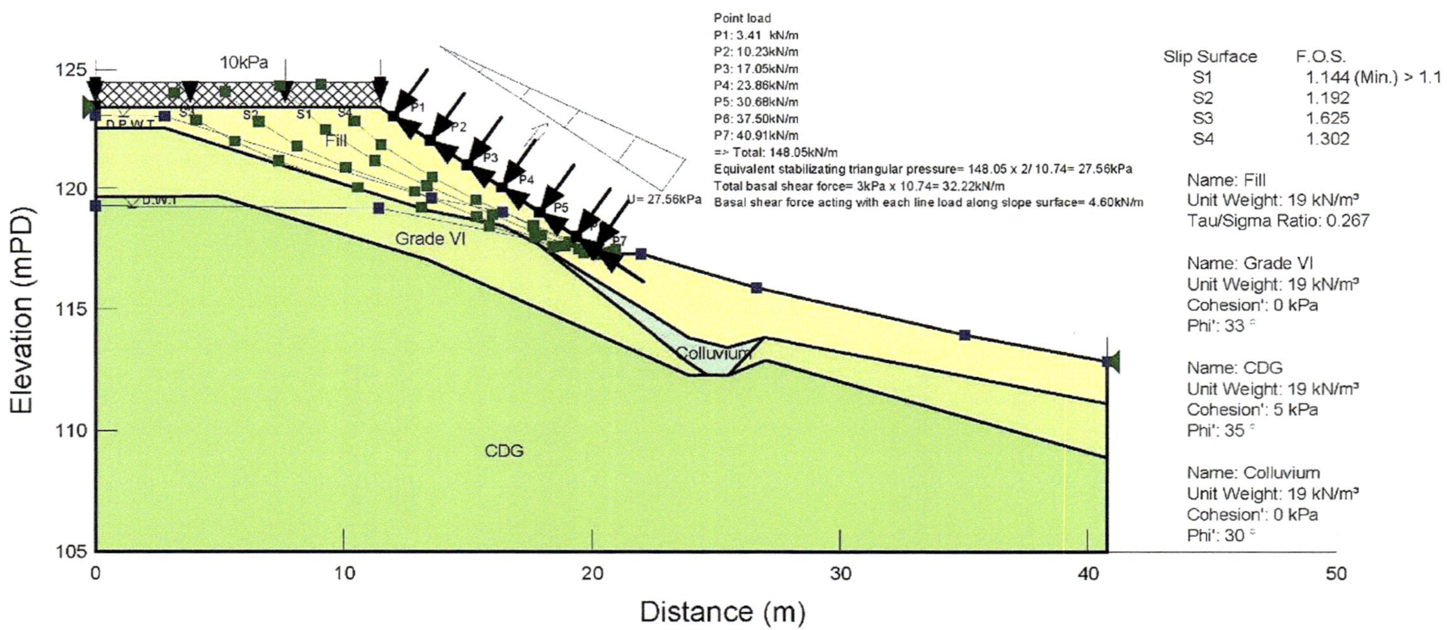

Figure 7.19 Typical design example for loose fill slope design.

latter attention has been shifted to employing the slice techniques in traditional limit equilibrium to the upper bound limit analysis (Michalowski, 1995; Donald and Chen, 1997).

Limit analysis is based on two theorems: (a) the lower bound theorem, which states that any statically admissible stress field will provide a lower bound estimate of the true collapse load; and (b) the upper bound theorem, which states that when the power dissipated by any kinematically admissible velocity field is equated with the power dissipated by the external loads, then the external loads are upper bounds on the true collapse load (Drucker and Prager, 1952).

A statically admissible stress field is one that satisfies the equilibrium equations, stress boundary conditions, and yield criterion. A kinematically admissible velocity field is one that satisfies strain and velocity compatibility equations, velocity boundary conditions, and the flow rule. When combined, the two theorems provide a rigorous bound on the true collapse load. Application of the lower bound theorem usually proceeds as described next. (a) First, a statically admissible stress field is constructed. Often it will be a discontinuous field in the sense that we have a patchwork of regions of constant stress that together cover the whole soil mass. There will be one or more particular values of stress that are not fully specified by the conditions of equilibrium. (b) These unknown stresses are then adjusted so that the load on the soil is maximized but the soil remains unyielded. The resulting load becomes the lower bound estimate for the actual collapse load.

Stress fields used in lower bound approaches are often constructed without a clear relation to the real stress fields. Thus, the lower bound solutions for practical geotechnical problems are often difficult to find. Collapse mechanisms used in the upper bound calculations, however, have a distinct physical interpretation associated with actual failure patterns and thus have been extensively used in practice.

7.7.2 Discontinuity layout optimization

Discontinuity layout optimization (DLO) is a novel limit analysis-based slope stability analysis method which was proposed by Smith and Gilbert (2007). DLO can identify a critical layout of lines of discontinuity, which form a failure, by using the rigorous mathematical optimization techniques. These lines of discontinuity are virtually the slip lines in two-dimensional geotechnical stability problems. DLO can be used to identify critical translational sliding block failure mechanisms which are similar to the method of slices solution for slope stability problems. This traditional method, however, works only with failure soil mass in a few sliding blocks. DLO can overcome this limitation by the lines of discontinuity which define the boundaries between the rigid moving blocks of material being divided into a large number of sliding blocks. The failure mechanisms are actually the upper bound theory corresponding to a load factor or strength factor. The load factor or strength factor obtained from the trial process for any arbitrarily assumed mechanism is increased if necessary in order to find the exact solution which brings a slope to collapse. Based on this theory, the DLO procedure, in essence, can be applied to a wide range of geotechnical stability problems by trial-and-error processes.

In the DLO procedure, the discontinuities are assumed to be variables that consider the relative displacement along the discontinuities. By using the relative displacement, compatibility can be checked at each node by applying the simple linear equation involving these variables. Finally, an objective function can be defined in accordance with the energy

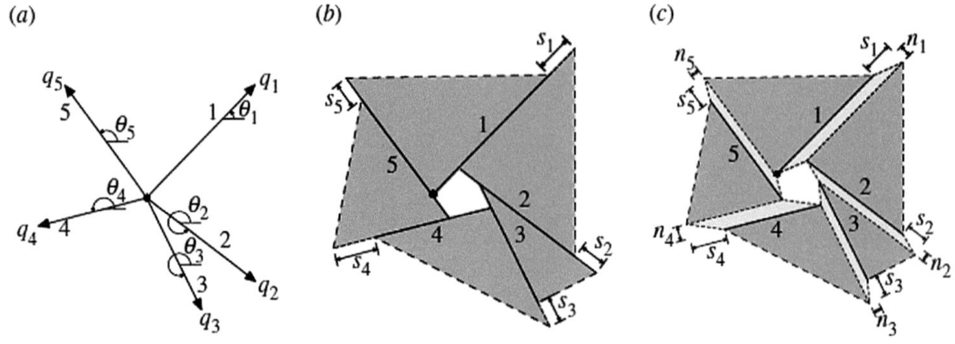

Figure 7.20 Equilibrium and compatibility conditions at a node: (a) equilibrium at (unloaded node in a truss); (b) compatibility at node in slip mechanism (translation only); (c) compatibility at node in slip mechanism (translation and dilation) (from Smith and Gilbert, 2007).

dissipation along all discontinuities when the failure occurs. A linear optimization problem can then be determined by a linear function of the slip displacement variables.

The DLO procedure expresses the limit analysis problem entirely in terms of lines of discontinuity instead of elements as in the classical continuum problem (Smith and Gilbert 2007). Using DLO, a large number of potential discontinuities are set up at different orientations; while the continuum-based element formulations, discontinuities are typically restricted to lie only at the edges of elements. With the use of modern optimization algorithms, an optimized solution can be achieved easily. After the initial success by Smith and Gilbert (2007), there have been different works in DLO by Clarke et al. (2013), Smith and Gilbert (2013), Bauer and Lackner (2015), Al-Defae and Knappett (2015), Leshchinsky (2015), Vahedifard et al. (2014) and Leshchinsky and Ambauen (2015). The original DLO formulation suffers from the limitation that only the translation mechanism can be considered. In view of such a limitation, Gilbert et al. (2010) and later Smithy and Gilbert (2013) extended the DLO formulation to cover the rotational formulation. Since DLO is actually a numerical form of limit analysis, the basic limitation of limit analysis is similar to that for DLO.

As discussed above, the compatibility of the displacement of the nodes is important to achieve the potential critical mechanisms. There are three cases to demonstrate the equilibrium and compatibility conditions at a node in Figure 7.20. Figure 7.20a considers an unloaded node in a truss:

$$\sum_{i=1}^{5} \alpha_i q_i = 0 \tag{7.79}$$

$$\sum_{i=1}^{5} \beta_i q_i = 0 \tag{7.80}$$

where $\alpha_i = \cos\theta_i$ and $\beta_i = \sin\theta_i$. Compatibility at the node in the slip is considered for translation only in Figure 7.20b

$$\sum_{i=1}^{5} \alpha_i s_i = 0 \tag{7.81}$$

$$\sum_{i=1}^{5} \beta_i s_i = 0 \tag{7.82}$$

where s_i is the shear displacement for i discontinuity. Compatibility at the node in the slip considers both the translation and dilation in Figure 7.20c:

$$\sum_{i=1}^{5} \alpha_i s_i - \beta_i n_i = 0 \tag{7.83}$$

$$\sum_{i=1}^{5} \beta_i s_i + \alpha_i n_i = 0 \tag{7.84}$$

where n is normal displacement for i discontinuity.

7.7.3 Some results from discontinuity layout optimization

The details of the DLO are covered by Smith and Gilbert, and the method is currently available in the program LimitState GEO. Interested readers should consult these commercial programs about the operations and underlying principles. It should be noted that there is only one commercial program for DLO at present, and 3D DLO is still not available for practical purposes, due to various technical problems and limitations. A typical output from DLO is given in Figure 7.21 for reference. It should be noted that both the factor of safety and critical slip surface from the DLO match very well with the results from LEM

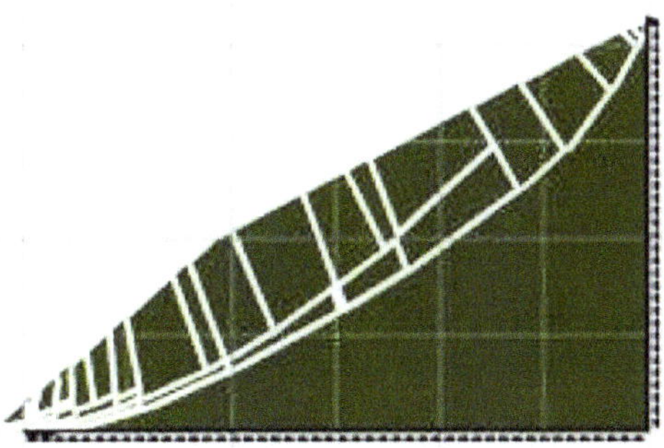

Figure 7.21 An output from DLO analysis.

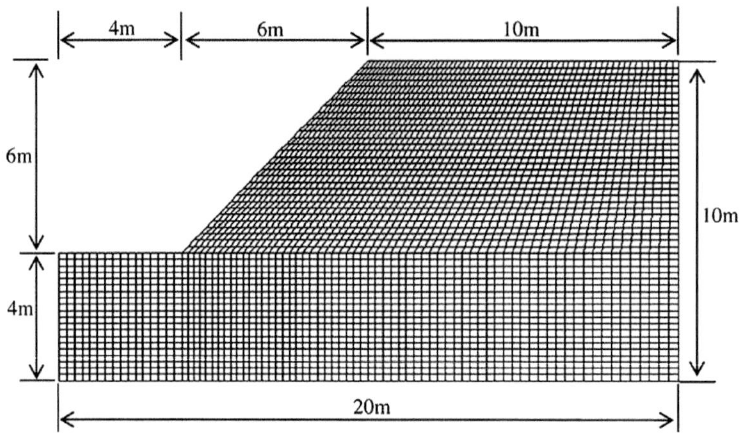

Figure 7.22 Geometry of a standard problem by Cheng et al. (2007).

or SRM for this case, but the author has also found that there are many cases for which noticeable differences between the results from DLO and LEM/SRM are found, particularly when some special soil parameters are defined.

With reference to the standard slope stability problem by Cheng et al. (2007) as shown in Figure 7.22 a total of 23 cases with different soil parameters are studied. It is noted that the results from DLO and LEM match very well except for a few cases (cases 6, 11, 16, 21, 22, 23). On the other hand, the results from SRM1 (zero dilation angle) and SRM2 (dilation angle = friction angle) match even better with the results using LEM. Actually, Cheng has carried out many case studies with the DLO and SRM and has found that the DLO at present has more problem cases than SRM, particularly for some special combinations of soil parameters. For normal cases, the differences between DLO and other methods are small (Table 7.29).

Yu et al. (1998) have provided a very detailed comparison between the use of limit analysis and LEM, and it was found that the results from the two methods are similar and comparable in most cases for relatively simple problems. Recently, DLO has been adopted for slope stability analysis by Leshchinsky and Ambauen (2015), and it was found that the results by DLO and LEM are generally comparable. Leshchinsky and Ambauen (2015) have, however, found some cases for which there are notable differences between the DLO and LEM, and they have concluded that DLO requires less assumption on the location of collapse, and therefore may be more preferable than LEM, especially for complex, yet realistic, geotechnical problems. After reviewing the examples by Leshchinsky and Ambauen (2015), the author tends to disagree with their results and comments. There are some limitations in the works by Leshchinsky and Ambauen (2015) which include: (1) the use of a classical LEM method which is greatly affected by the convergence problem (Cheng et al. 2008, Cheng et al. 2010); (2) the critical failure surface has not been determined [Fig. 12 from Leshchinsky and Ambauen (2015) has only considered 151 surfaces]; and (3) the interslice force function can be critical in complex problems. As discussed by Cheng (2003), Cheng et al. (2010) and Cheng et al. (2013), these three problems can lead to

Table 7.29 Comparisons between the results from LEM, SRM and DLO

Case	*Cohesion c' (kPa)*	*Friction angle Φ' (degree)*	*FOS (LEM)*	*FOS (DLO)*	*FOS by SRM (SRM1)*	*FOS by SRM (SRM2)*
1	2	5	0.25	0.26	0.25	0.25
2	2	15	0.50	0.51	0.52	0.50
3	2	25	0.74	0.76	0.77	0.75
4	2	35	1.01	1.04	1.07	1.03
5	2	45	1.35	1.40	1.42	1.36
6	5	5	0.41	0.44	0.41	0.41
7	5	15	0.70	0.73	0.72	0.71
8	5	25	0.98	1.00	1.02	0.99
9	5	35	1.28	1.31	1.33	1.30
10	5	45	1.65	1.70	1.71	1.69
11	10	5	0.65	0.71	0.66	0.65
12	10	15	0.98	1.04	1.00	0.995
13	10	25	1.30	1.35	1.34	1.33
14	10	35	1.63	1.70	1.69	1.65
15	10	45	2.04	2.09	2.09	2.07
16	20	5	1.06	1.22	1.12	1.12
17	20	15	1.48	1.61	1.51	1.50
18	20	25	1.85	1.96	1.87	1.87
19	20	35	2.24	2.34	2.26	2.27
20	20	45	2.69	2.79	2.72	2.74
21	5	0	0.20	0.24	0.22	0.22
22	10	0	0.40	0.47	0.44	0.44
23	20	0	0.80	0.95	0.88	0.88

relatively poor solutions in some cases by classical LEM, and the extremum principle by Cheng et al. (2010), Cheng et al. (2011b) and Cheng et al. (2013) has overcome these problems and can provide solutions similar to some classical plasticity problems which are not possible with the classical LEM. A fair comparison and commentary on these methods must be based on reliable and robust analyses that identify the differences between DLO and LEM.

In Figure 7.23 (Fig. 12 by Leshchinsky and Ambauen 2015), there are great differences between the critical results by the author and Leshchinsky and Ambauen (2015). The soil parameters are unit weight = 20 kN/m^3, c' = 0 kPa and ϕ' = 30° for soil layer 1; unit weight = 19 kN/m^3, c' = 0 kPa and ϕ' = 45° for soil layer 2; and unit weight = 19 kN/m^3, c' = 10 kPa and ϕ' = 0° for soil layer 3. On the left-hand side of the critical slip surface by Leshchinsky and Ambauen (2015), there is a very sudden change in the slope of the critical failure surface which seems unlikely to happen. At the right-hand side of the critical slip surface by Leshchinsky and Ambauen (2015), the critical slip surface is nearly vertical, which is also highly unlikely, as the friction angle of soil layers 1 and 2 are 30° and 40°, respectively, with zero cohesive strength. When this same slip surface by Leshchinsky and Ambauen (2015) is considered with the M-P method using $f(x)$ = sin(x), the author actually gets a factor of safety of 1.05, which is significantly greater than the result of 0.95 by Leshchinsky and Ambauen (2015).

This problem was then reanalysed by Cheng using LEM to locate the critical slip surface. For this problem, the use of $f(x)$ = 1 is poor in convergence, and the author gets a

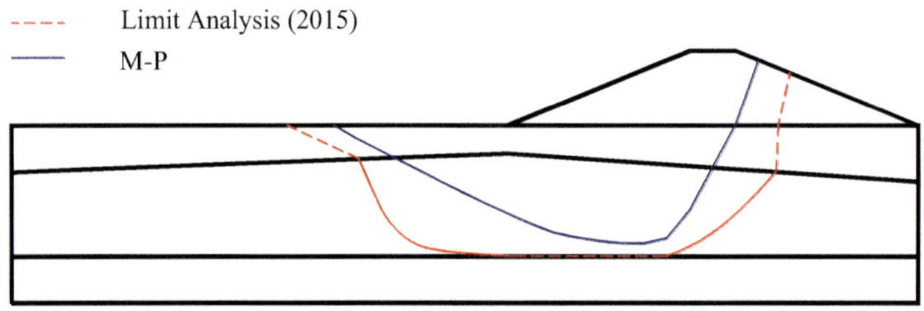

Figure 7.23 Comparisons of DLO and LEM for Fig. 12 by Leshchinsky and Ambauen (2015), FOS = 0.95 by DLO and 1.0 by Spencer method by Leshchinsky and Ambauen (2015) using 151 trial surface, 0.92 by M-P method using *f(x)* = sin(x) and 0.97 by *f(x)* = 1.0 by the author using about 20000 trials with a simulated annealing optimization method.

slightly different critical slip surface and a factor of safety of 0.97 by using $f(x) = 1.0$. As mentioned by Cheng et al. (2008, 2010), $f(x)$ can be critical in some cases, which will affect the optimized solution. In this respect, Cheng has also adopted the extremum principle (Cheng et al. 2010, 2011, 2013) and obtained a critical solution of 0.915, which is close to that by using $f(x) = \sin(x)$. Since the extremum principle, which is practically equivalent to the lower bound approach, satisfies all the force and moment equilibrium with acceptable internal forces and is free from convergence problem, this result can be considered to be a very good estimation of the critical result for the present problem. Cheng believes that the critical result by Leshchinsky and Ambauen (2015) is possibly a local minimum instead of being the global minimum.

Cheng has adopted an accuracy of 0.001 in all the global optimization searches in the present study, and the global minima of each example have been tested with different optimization algorithms for confirmation. Based on the above case studies, it can be concluded that some of the past reported results in the literature which are not optimized with the modern optimization algorithms may not be reliable enough for comparisons. In particular, for the presence of a soft band which is a difficult problem, the present study and the works by Cheng and Li (2007) and Cheng et al. (2012) have demonstrated that great care must be taken in order to obtain a good result. Furthermore, as a relatively new computational method, DLO has been demonstrated to be affected by the soft band or local minima problem. Overall, the author views that the problems presented in this section are not fundamental deficiencies of the DLO. Instead, they highlight the limitations of the numerical technique in implementing the DLO up to the present moment. With a refined and improved numerical technique coupled with DLO, the author expects that better results will be produced in the future. On the other hand, it is dangerous to compare the advantages and limitations of different stability methods based on old results or computer programs with limitations. Some of the comments in earlier literature are possibly distorted by the limitations of the computational technique in computer programs instead of being the actual comparisons of different stability analysis methods.

For the problem in Figure 7.24 with a soft band at soil layer 2 as discussed by Cheng et al. (2007b), surprising results are again obtained by DLO. The unit weight of the soils are 19 kN/m^3, and c'=20 kPa and ϕ'=35° for soil layer 1; c'=0 kPa and ϕ'=25° for soil layer 2; and c'=10 kPa and ϕ'=35° for soil layer 3. As discussed by Cheng et al. (2007a), it appears that some SRM programs are affected by the size of the solution domain. The factor of

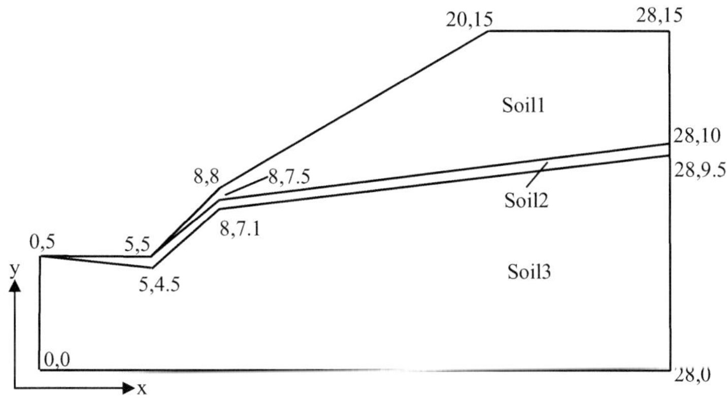

Figure 7.24 A slope problem with a soft band as discussed by Cheng et al. (2007).

safety for LEM was obtained as 0.927 by the Spencer method by Cheng et al. (2007b), and this value lies within the SRM1 and SRM2 results by Plaxis and the new version of Phase (8.0). On the other hand, the factor of safety appears to be highly dependent on the nodal number adopted in the analysis. Even if 2000 nodal number is adopted, the factor of safety from DLO still appears to be unsatisfactory, which is shown in Table 7.30. The results by DLO are higher than those by LEM or SRM under all cases in Table 7.30, and the differences are not minor. Surprisingly, the critical failure surface from DLO as shown in Figure 7.25 is similar to that by LEM or SRM (Cheng et al. 2007b). From Table 7.30, it can be concluded that the most influential factor in a proper DLO analysis is the nodal number.

7.8 FINITE ELEMENT ANALYSIS OF SLOPE STABILITY

In the classical limit equilibrium and limit analysis methods, the progressive failure phenomenon cannot be estimated. Some researchers propose using the finite element method to overcome some of the basic limitations in the traditional methods of analysis. At present, there are two major applications of the finite element in slope stability analysis.

The first approach is to perform an elastic (or elasto-plastic) stress analysis by applying the body force (weight) due to soil to the slope system. Once the stresses are determined, the local factors of safety can be determined easily from the stresses and the Mohr-Coulomb criterion. The global factor of safety can also be defined in a similar way by determining the ultimate shear force and the actual driving force along the failure surface. Pham and Fredlund (2003) have adopted the dynamic programming method to perform this optimization search, and suggested that this approach can overcome the limitations of the classical limit equilibrium method. The author, however, believes that the elastic stress analysis is not a realistic picture of the slope at the ultimate limit state. In view of these limitations, the author does not think that this approach is really better than the classical approach. It is also interesting to note that both the factor of safety and the location of the critical failure surface from such analysis are usually close to that by the limit equilibrium method. To adopt the elasto-plastic finite element slope stability analysis, one precaution should be

Table 7.30 FOS for different nodal numbers and tolerances of analysis ($c' = 0$, $\phi' = 25°$ for the soft band) for the problem in Figure 7.22

28 m domain size, solution tolerance 0.01, different nodal density

Case	*nodal No.*	*FOS by DLO*	*FOS by LEM*	*FOS difference with LEM (DLO %)*
1	250	1.356	0.927	–46.28
2	500	1.069	0.927	–15.32
3	1000	1.082	0.927	–16.72
4	2000	1.055	0.927	–13.81

28 m domain size, nodal density 500, different solution tolerance

Case	*solution tolerance*	*FOS by DLO*	*FOSby LEM*	*FOS difference with LEM (DLO %)*
1	0.01	1.069	0.927	–15.32
2	0.001	1.069	0.927	–15.32
3	0.004	1.069	0.927	–15.32
4	0.005	1.069	0.927	–15.32

Solution tolerance 0.01, nodal density 500, different domain size

Case	*Domain size (m)*	*FOS by DLO*	*FOSby LEM*	*FOS difference with LEM (DLO %)*
1	28	1.069	0.927	–15.32
2	20	1.093	0.927	–17.91
3	12	1.025	0.927	–10.57

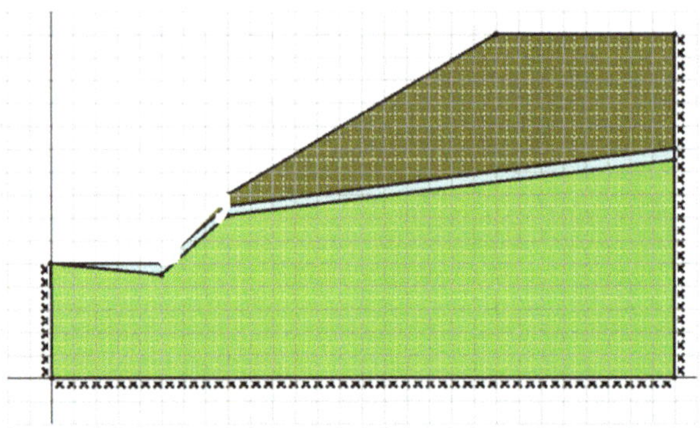

Figure 7.25 Critical failure surface from DLO for the problem in Figure 7.24.

noted. If the deformation is so large that the finite element mesh is greatly modified, the geometric non-linear effect may induce a major effect on the results.

The second finite element slope stability approach is the strength reduction method (SRM). The shear strength reduction technique was proposed as early as 1975 by

Zienkiewicz et al. (1975) and has since been used by Naylor (1982) and Matsui and San (1992), among others; its application has been mostly in the past decade due to the increasing speed of computers. In the SRM, the gravity load vector for a material with unit weight γ_s is determined from Eq. (7.85) as:

$$\{f\} = \int [N]^T \{\gamma_s\} dv \tag{7.85}$$

where $\{f\}$ is the equivalent body force vector and $[N]$ is the shape factor matrix. The constitutive model adopted in the non-linear element is usually the Mohr-Coulomb criterion, but other constitutive models are also possible, though they are seldom adopted in practice. The material parameters c' and ϕ' are reduced according to

$$c_f = c'/F \; ; \; \phi_f = \tan^{-1}\{\tan(\phi'/F)\} \tag{7.86}$$

The factor of safety F keeps on changing until the ultimate state of the system is attained, and the corresponding factor of safety will be the factor of safety of the slope. The termination criterion is usually based on one of the following:

1. The nonlinear equation solver cannot achieved convergence after a pre-set maximum number of iteration;
2. There is a sudden increase in the rate of change of displacement in the system;
3. A failure mechanism has developed.

The location of the critical failure surface is usually determined from the contour of the maximum shear strain or the maximum shear strain rate. There is no simple way to determine a precise slip surface in SRM, as compared with that in LEM. In general, a relatively clear slip surface region can be obtained, and the slip surface from SRM is usually close to that by the LEM. There are, however, also many cases (particularly cases with soil nail) where a clear slip surface region is not obtained from the SRM, or the slip surface is relatively large in size (Figure 7.26b). This is one of the practical limitations in the use of SRM for routine slope stability analysis.

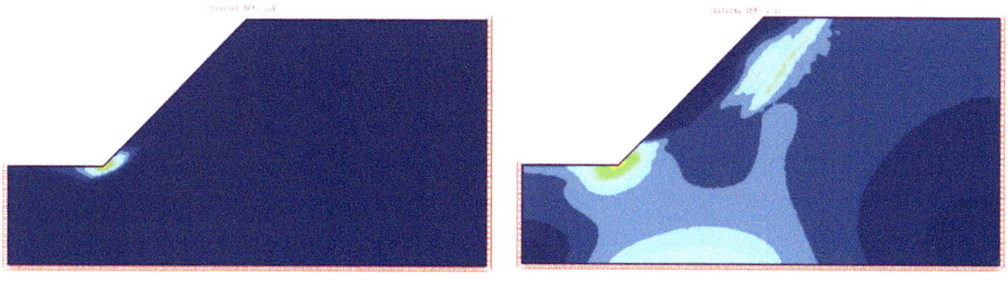

(a) SRM1 (b) SRM2

Figure 7.26 SRM analysis of the problem as shown in Figure 7.22: (a) $c' = 2$ kPa, $\phi' = 45°$, $\psi' = 0$; (b) $c' = 2$ kPa, $\phi' = 45°$, $\psi' = 45°$.

Among a number of commonly used linear and non-linear constitutive models, the elastic-perfectly plastic model with Mohr-Coulomb failure criterion is mostly adopted. It has been shown that the sudden transition of soil behaviour from elastic state to plastic state in the elastic-perfectly plastic model helps identify the factor of safety when applying a strength-reduction technique (Dawson et al. 1999). Others have used more complicated constitutive models in the strength reduction method. For example, Matsui and San (1992) used Duncan and Chang's hyperbolic model (Duncan and Chang 1970), and Zhang et al. (2011) used a revised Duncan and Chang's model. However, identifying the failure becomes more difficult. In addition, the factor of safety by strength reduction is found insensitive to the constitutive model used, even though more complicated constitutive models have advantages in predicting the loading–displacement behaviour. This is because the stress-dependent stiffness behaviour and hardening effects are excluded when using the strength-reduction technique (Plaxis 2014; Alkasawneh et al. 2008). Zhang et al. (2011)'s study also revealed that the factors of safety by using Duncan and Chang's model and using the revised model are almost the same.

One difficulty in the strength-reduction stability analysis lies in the definition of slope failure criteria. So far, there is no uniform definition in the literature. Matsui and San (1992) proposed to use the contour of shear strain as a failure indicator; for example, their study of an embankment slope showed that a well-defined failure zone develops from the toe to top surface when the contour of shear strain exceeds 15%. In contrast, Griffiths and Lane (1999) and Dawson et al. (1999) took non-convergence (out-of-balance force) as an indicator for failure in a finite element method and finite difference method, respectively. Others considered slopes to have failed when there are contiguous plastic zones through the toe to the crest (e.g. Zheng et al. 2009).

The main advantages of the SRM are as follows: (i) the critical failure surface is found automatically from the localized shear strain arising from the application of gravity loads and the reduction of shear strength; (ii) it requires no assumption on the interslice shear force distribution; and (iii) it is applicable to many complex conditions and can give information such as stresses, movements and pore pressures which are not possible with the LEM. Griffiths and Lane (1999) pointed out that the widespread use of the SRM should be seriously considered by geotechnical practitioners as a powerful alternative to the traditional limit equilibrium methods. One of the important criticisms of SRM is the relatively poor performance of the finite element method in capturing the localized shear band formation. While the determination of the factor of safety is relatively easy and consistent, many engineers find that it is not easy to determine the critical failure surfaces in some cases as the yield zone spreads over a wide domain instead of localizing within a soft band. Other limitations of the SRM include the choice of an appropriate constitutive model and parameters, boundary conditions and the definition of the failure condition/failure surface; a detailed comparison between the SRM and LEM is given in a later section.

Although SRM appears to have many advantages, it is not easy to identify the critical slip surface from the analysis, and the results can be sensitive to many factors, so that users may need to carry out many trial-and-error procedures to obtain a reasonable result. As shown in Figure 7.26, a change in the dilation angle has a very small effect on the factor of safety, but a very major change in the critical slip surface. In fact, it is not easy to identify a clear slip surface for the results in Figure 7.26. On the other hand,

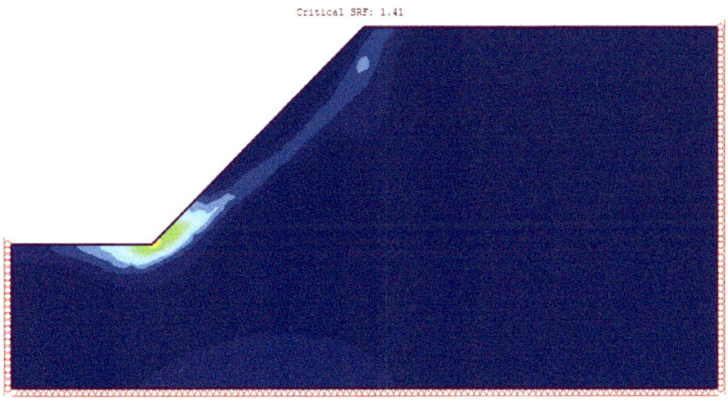

Figure 7.27 SRM analysis of the problem as shown in Figure 7.22, c' = 2 kPa, ϕ' = 45°, ψ' = 35°.

if the dilation angle is revised to 35°, a clear slip surface can be identified, as shown in Figure 7.27. In this respect, engineers may need to use LEM to assess the acceptability of the results from SRM!

Cheng has carried out many studies on the SRM analysis of slopes, for both 2D and 3D problems. Many interesting and useful results have been found using SRM, but Cheng has also found many practical problems for the SRM. The analyses are limited by the ability of the nonlinear finite element equilibrium equation solver near to the failure state and the identification of the failure condition. Many surprising results can be found from every SRM program that Cheng has tested. Though these limitations are not actually the fault of the SRM, they also reflect the difficulties in the nonlinear analysis around the limit state, where the stiffness matrix approaches 0.

7.9 DISTINCT ELEMENT ANALYSIS

When a soil slope reaches its critical failure state, the critical failure surface is developed and the FOS of the slope is 1.0. The continuum methods such as FEM or LEM will fail to characterize the post-failure mechanisms in the subsequent failure process. For a rock slope comprised of multiple joint sets which control the mechanism of failure, a discontinuum modelling approach is actually more appropriate. Discontinuum methods treat the problem domain as an assemblage of distinct, interacting bodies or blocks which are subjected to external loads and are expected to undergo significant motion with time. This methodology is collectively named the discrete element method (DEM). The DEM will be a more suitable tool for the study of progressive failure and the flow of the soil mass after initiation of slope failure, though this method is not efficient for the analysis of a stable slope. The development of DEM represents an important step in the modelling and understanding of the mechanical behaviour of joint rock masses or soil slope with FOS < 1.0. Although continuum codes can be modified to accommodate discontinuities, this procedure is often difficult and time consuming. In addition, any modelled inelastic displacements are further limited to elastic orders of magnitude by the analytical principles exploited in developing the solution procedures. In contrast, discontinuum analysis permits sliding along and opening/closure between particles or blocks. Nowadays, discontinuum modelling is starting to constitute the commonly applied numerical approach to rock slope analysis.

Particle flow code is employed by Cheng for slope stability studies, and it employs the distinct-element method with the following assumptions: (1) particles are circular in PFC2D/spherical in PFC3D and elastic; (2) contacts between particles occur over a vanishingly small area (i.e. at a point); (3) the magnitude of elastic particle "overlap" of nominal radii is related to the contact force by a force–displacement law similar to linear springs that mimic the physics of Hertz elastic spheres (Cundall, 1988); and (4) tensional and shear bonds can exist between particles. With these codes, it is possible to model granular flow, translational movement of blocks, fracture of intact rock and dynamic response to blasting or seismicity. The breaking of bonds between circular particles roughly simulates intact rock fracture and failure (although not fracture propagation). Deformation between particles due to shear or tensile forces can also be included, where slip between adjacent particles is prescribed in terms of the frictional coefficients that limit the contact shear force. Particle flow codes are thus able to simulate material from the macro level of fault- or joint-bounded blocks to the micro scale of grain-to-grain contact, with the main limiting factors being the computing time and memory requirements. In this sense, it becomes possible to model a number of soil/rock slope failure processes, and subsequently, the runout of the failed material down the slope. At present, these codes are predominantly a research tool, but its potential is being widely recognized in mining, petroleum and civil engineering.

To assess the complete failure mechanism, the distinct element method (in terms of particle flow approach) can provide a qualitative assessment. Two problem generation approaches have been used by Cheng. A slope can be formed by an assembly of particles or triangular rigid blocks. To avoid the use of an excessive number of particles or rigid blocks which requires extensive computation time for analysis, a limited number in the range of 10000–100000 is used by Cheng. Initially, the initial stress state of the system is generated from the known soil mechanism principle. Model generation is quite complicated and crucial in PFC. Circle particles as balls in PFC2D are used to simulate the soil particles. The boundaries of the model are defined by a series of walls that contain the particles. In general, the usual objective in creating an irregular packing is to fill some given space with particles at a given porosity and to ensure that the assembly is at equilibrium. In this study, particles are falling under self-weight gravity at random layer by layer, lying on the bottom of the model, which represents the slope generation in nature. This kind of specimen generation technique is called the "rainy method". Radius expansion, which is called the "expand method", is also used in the particle generation, achieving in a trial-and-error manner to satisfy the requirement of porosity ratio. Thus two ways of specimen generation are combined and incorporated in this simulation model. Once the initial state is established, the change of the water table/pore pressure or the application of an external load will be applied to the system. The complete displacement history of the system from initial movement to a complete collapse can be qualitatively assessed. While the use of the distinct element method is difficult to provide a factor of safety for design, the collapse mechanism can be assessed which is not possible with all the classical methods discussed.

For the parameter determination, the friction coefficient μ at the contact for each ball in PFC is taken as the value of $\tan\varphi$ (φ is the macro friction angle of soil). For the bond strength, the definition in PFC is relatively complicated. The particle flow code is based on contact constitutive law in theory. To simulate a medium with cohesion, optional bonding is defined in PFC. Unlike the friction coefficient, the bond cohesion in PFC cannot be compared with the macro cohesive strength of soil. It is specified that bonded contacts may carry tension, and the bonds have finite strengths in tension and shear. The contact bonds which reproduce the effect of an adhesion acting over the vanishingly small area of the contact point are chosen to be used in this study, and the contact bond constitutes the tensile normal bond strength (n_bond) and shear bond strength (s_bond). Both bonds can be envisioned as a kind of glue joining the two particles. The contact-bond glue is of a vanishingly small size that acts only at the contact point. For a particle flow analysis, it is vital to choose the appropriate material parameters for the model in the analysis. Because of the discrete, particle-based nature of the model, specification of material properties and boundary conditions is more difficult than with available continuum methods. The material parameters for the particle flow analysis are, however, different from the macro-material properties as they are controlled by the micro-properties such as the particle-size distribution, packing, particle and bond stiffnesses, particle friction coefficients, and bond strengths. The macro-behaviour reflects the average behaviour and is controlled by the micro-material parameters, grading of the materials and state of packing. It is extremely difficult to determine the micro-material parameters directly, so the micro-material parameters will be varied over wide ranges until the macro-behaviours are reproduced.

As an illustration, a 5 m high 45° slope is modelled with the distinct element method by imposing the initial condition in the first step. In the initialization of the model, soil slopes with different angles were generated by the rainy method, particles fall under self-weight gravity at random lying on the bottom of the model layer by layer, which represent the slope generation in nature. The vertical stress is basically equal to the overburden stress, while an at-rest pressure coefficient of 0.5 is employed in the present example. The unit weight of the particle is 17 kN/m^3, while the friction factor is 0.5. Due to raining, a 4 m water table is established which is equivalent to a body force of –9.81 kN/m^3 applied to the particle system. The slope finally collapses, which is shown in Figure 2.28. The results of the intermediate analysis shown in Figure 7.28 are interesting. When the number of time steps is small, no distinct failure zone can be observed. Starting from 3000 time steps, a failure zone is observed from the displacement vector plot, and this failure zone stops to expand at a time step of about 13000. The failure domain is relatively stable over the remaining analysis and keeps moving until the slope finally collapses at a time step of 51000. It should be noted that the failure mass moves above the stable zone, which is basically constant after a time step of 13000. The power of the distinct element is that while the ultimate limit state can be estimated from the limit equilibrium and finite element method, the final collapse mechanism or the flow of the failure mass can be estimated from the distinct element, which is not possible with the classical methods. The results shown in Figure 7.28 are, however, qualitative, and precise results for design are difficult to determine from the distinct element method at present.

To stabilize the slope, two soil nails are added to the system shown in Figure 7.29a. The soil nails are modelled by a collection of particles connected together, and 4 m water is then applied to the system. The final displacement field is shown in Figure 7.29b, which indicates that the soil nails have effectively inhibited the collapse of the slope.

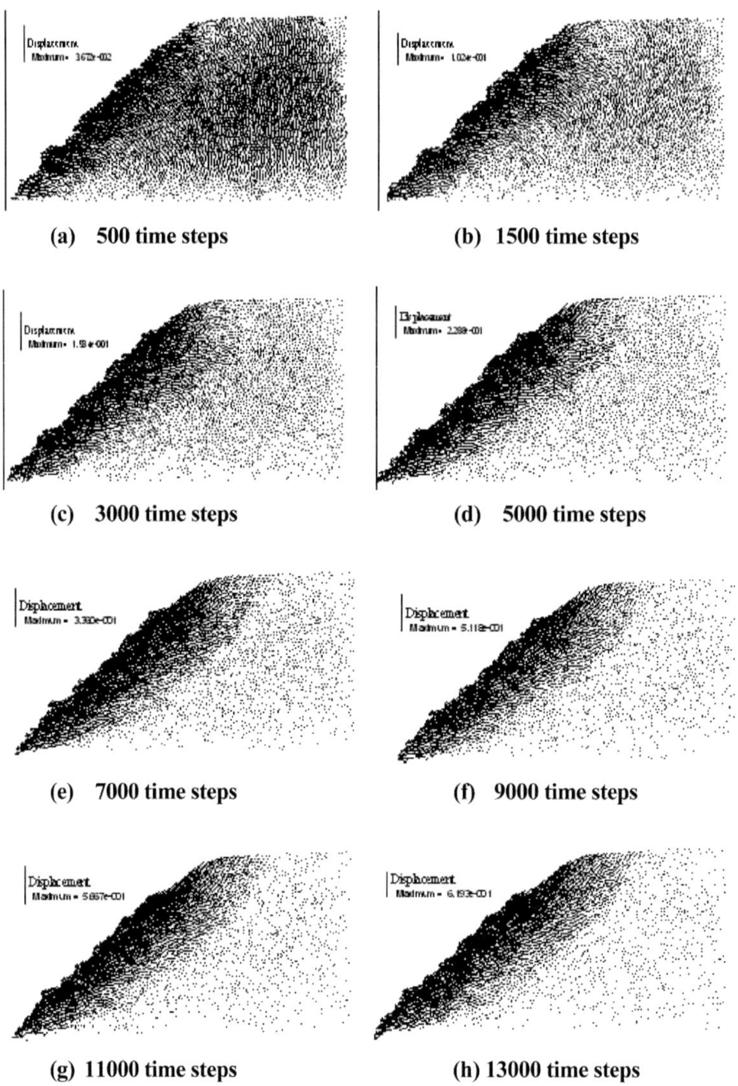

Figure 7.28 Displacement of the slope at different time steps when a 4 m water level is imposed.

It is also interesting to note that a failure mechanism is usually formed by the distinct element analysis, as compared with the difficulty in the SRM. As shown in Figure 7.30, a well-defined failure mechanism is not found by the SRM.

7.10 LOCATION OF CRITICAL FAILURE SURFACE

The critical/minimum factor of safety of a slope must be determined before the slope stabilization works can be designed. One of the important slope design works is determination of the minimum factor of safety and the critical slip surface for the slope stabilization

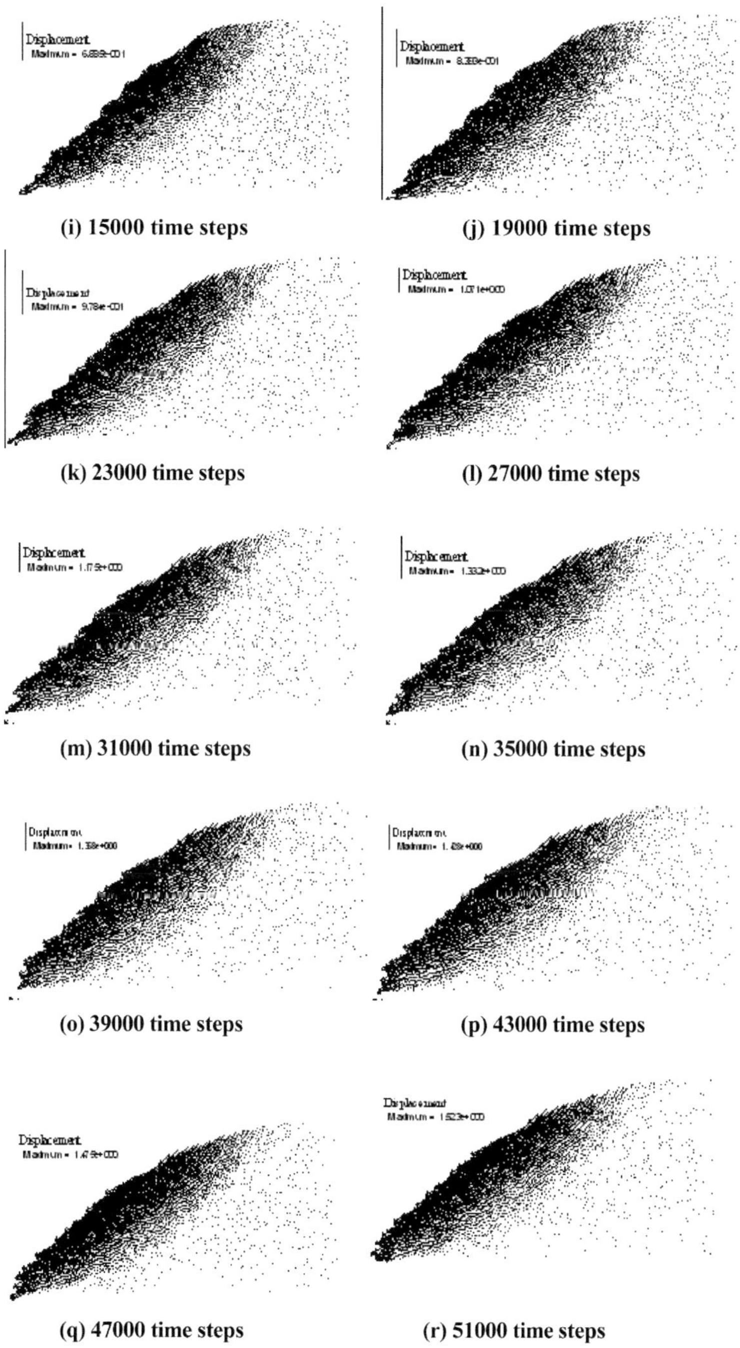

Figure 7.28 (Continued)

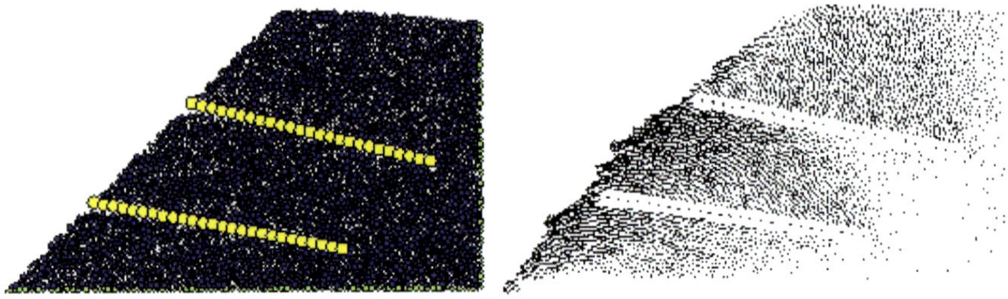

(a) 2 soil nails inclined at 10° installed imposed

(b) Displacement field after 4m water is

Figure 7.29 Effect of soil nails installation.

Figure 7.30 DEM analysis of slope with two layers of soil.

works. According to the upper bound theory, any prescribed failure surface will be an upper bound to the true solution. For the critical failure surface which corresponds to the global minimum, some of the difficulties and interesting phenomena in locating the critical failure surface will be discussed. Consider a one-dimensional function $y = f(x)$ defined over a solution domain as shown in Figure 7.31. The local minima where the gradients of the function are equal to 0 ($f'(x) = 0$) are given by points c, e, i, k, m and o, C and D, while the global minimum is defined by point g. The sharp change of y within the region e to i is practically possible for a slope stability problem. For example, for the Fei Tsui Road failure in Hong Kong in 1995, the failure was initiated within a very narrow band of soft material which was not detected in the site investigation process. For multi-variable optimization analysis required by the slope stability problem, the factor of safety objective function is highly complicated, and the problem will be a complicated N-P hard type which has attracted the attention of many researchers.

Another special feature about the critical failure surface for a simple slope is shown in Figure 7.32. For most practical problems, there are only minor changes in the factors of safety if the trial failure surfaces fall within the shaded region shown in Figure 7.32. In this respect, there is not a strong need to determine the precise location of the critical failure surface if the geometry and ground conditions for a slope are simple. For complicated

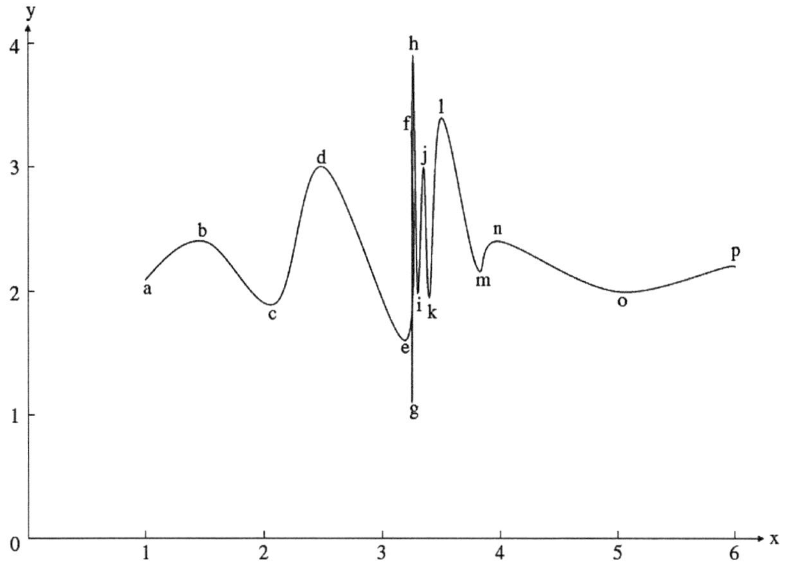

Figure 7.31 A simple one-dimensional function illustrating the local minima and the global minimum.

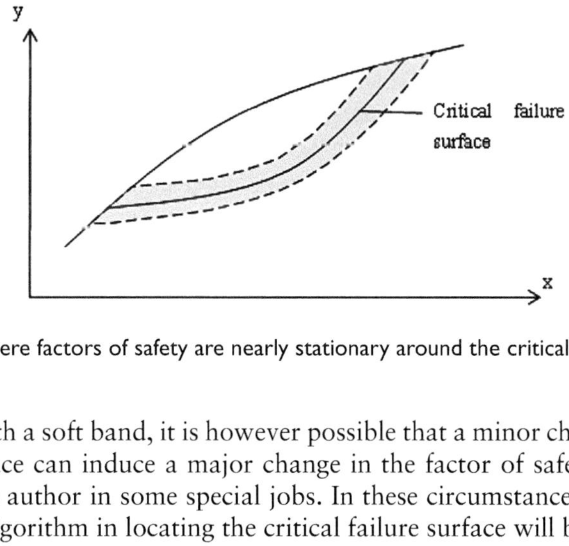

Figure 7.32 Region where factors of safety are nearly stationary around the critical failure surface.

slopes or slopes with a soft band, it is however possible that a minor change in the location of the failure surface can induce a major change in the factor of safety, which has been experienced by the author in some special jobs. In these circumstances, the robustness of the optimization algorithm in locating the critical failure surface will be important for the analysis and design.

The failure surface can be divided into the circular and non-circular failure surfaces (some programs offer an ellipse option). A circular (or ellipse) failure surface is actually a sub-set of the non-circular failure surface, but it is useful because of: (1) some stability formulations apply to the circular failure surface only; and (2) the critical circular failure surface is a good approximation of the critical solution for some simple problems and is simple to be evaluated. Classically, for a circular failure surface, the location of the critical failure surface is usually determined by the method of grids shown in Figure 7.33. There are only three control variables in this case: x and y ordinates of the centre of rotation and the radius of the failure surface. Each grid point is used as the centre of rotation, while different radii are considered (10–20 cases) for the circular failure surface, and the minimum factor of safety for different radii is assigned to this grid point. Different factors of safety are hence assigned to different grid points, and the trend of the global minimum can be assessed by drawing the contour of the factor of safety from the grid points. This method is robust and simple to operate, but the accuracy will depend on the spacing between the grid points. The specified grid must also be large enough to embrace all possible local minima and the global minimum in order to obtain a clear picture of the distribution of the factor of safety. The grid method is simple to implement and is available in most commercial slope stability programs (but not in SLOPE 2000). Visually, it is easy to view the local minimum and global minimum, but the extension of such a geometrical method to non-circular cases is not possible.

For a general non-circular failure surface, the number of control variables which are controlled by the number of points for the failure surface is usually much more than three. To locate the critical failure surface which may take any shape and location, the geometric method will not be possible. Some programs offer an option where each vertex is bounded within a rectangle, but such an algorithm is inefficient in application and requires lots of engineer effort in defining the solution domain for each control variable. Some of the special features of the objective function of the safety factor *F* for this case include:

1. The objective function of the safety factor *F* is usually non-smooth, non-convex and discontinuous over the solution domain. Discontinuity of the objective function can be generated by: generation of an unacceptable failure surface; "failure to converge" of the objective function; presence of obstructions in the form of sheet pile, retaining wall, large boulders, tension crack or others. Gradient type optimization methods are applicable only for continuous function and will break down if there are discontinuities in the objective function.
2. Chen and Shao (1983) have demonstrated that multiple minima similar to that shown in Figure 7.33 will exist in general. Duncan and Wright (2005) have also shown the existence of multiple local minima even for a simple homogeneous slope, which was also illustrated by Cheng et al. (2007b). The local minimum close to the initial trial will be obtained by the classical gradient type optimization methods. If an initial trial close to the global minimum is used, the global minimum can usually be found by classical methods, but a good initial trial is difficult to establish for a general multi-variable problem. The success of a global optimization algorithm to escape from the local minima for an initial solution far from the global minimum is crucial in the slope analysis problem.
3. A good optimization algorithm should be effective and efficient over different topography, soil parameters and loadings. The analysis should also be insensitive to the optimization parameters.

Various classical optimization methods for the non-circular failure surface have been proposed and used in the past. SLOPE 2000 appears to be the first program to adopt the heuristic optimization method in the location of the critical slip surface, and some

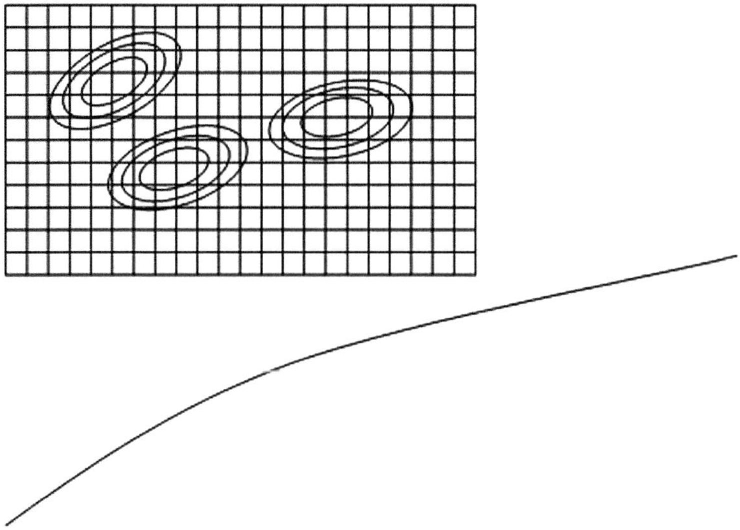

Figure 7.33 Grid method and presence of multiple local minima for circular search.

commercial programs later also adopted a similar technique in their calculation engine. Baker and Gaber (1977, 1978) have proposed the use of variational principle, but this method is complicated even for a simple slope and is not adopted for practical problems. Moreover, if the gradient of the global minimum is not zero, the variational principle will miss the critical solution. Chen and Shao (1983) and Nguyen (1985) have suggested the use of the simplex method for this problem, which is actually suitable only for linear problems. The simplex method has been adopted by the program EMU developed by Chen, and it works fairly well for simple problems. The simplex method fails to work well for some complicated cases, and also fails to work automatically for cases where the local minimum and global minimum differ by a very small value but the critical slip surface differs significantly in the location. Celestino and Duncan (1981) have adopted the alternating variable method, while Arai and Tagyo (1985) and Yamagami and Jiang (1997) have adopted the conjugate-gradient method and dynamic programming, respectively. These classical methods are applicable mainly for continuous functions, but they are limited by the presence of the local minimum, as the local minimum close to the initial trial will be obtained in the analysis. There is also a possibility that the global minimum within the solution domain is not given by the condition that the gradient of the objective function $\nabla f = 0$, and a good example has been illustrated by Cheng (2003). The presence of the other local minima or the global minimum will not be obtained by the classical methods unless a good initial trial is adopted, however a good initial trial is difficult to establish for a general problem.

Greco (1996) and Malkawi et al. (2001) adopted the Monte Carlo technique for locating the critical slip surface with success for some cases, but there is no precision control on the accuracy of the global minimum. Zolfaghari et al. (2005) adopted the genetic algorithm, while Bolton et al. (2003) used the leap-frog optimization technique to evaluate the minimum factor of safety. All of the above methods are based on the use of static bounds to the control variables, which means that the solution domain for each control variable is fixed and pre-determined by engineering experience. Cheng (2003) developed a procedure which transformed the various constraints and the requirement of a kinematically acceptable failure mechanism to the evaluation of upper and lower bounds of the control variables, and the simulated annealing algorithm is used to determine the critical slip surface. The control variables are defined with dynamic domains which change during the solution, and the bounds are controlled by the requirement for a kinematically acceptable failure mechanism. Through such an approach, there is no need to define the pre-determined static solution domain for each control variable based on engineering experience, and precision control during the search for the critical solution will be possible.

In view of the limitations of the classical optimization methods, the current approach in recent commercial programs to locate the critical failure surface is the use of the heuristic global optimization methods. Within the last 10 years, many modern heuristic optimization methods have been proposed and used for slope stability problems and other types of engineering problems, and the author believes that there are actually no great differences among the individual optimization methods. The author also finds that no single optimization method can outperform the other methods in all cases. The term heuristic is used for algorithms which find solutions among all the possible ones, but they do not guarantee that the best will be found, therefore, they may be considered as approximate and not accurate algorithms. These algorithms usually find a solution close to the best one, and they find it fast and easily. Another important feature is that the requirement for human judgement or interaction should be minimized or even eliminated if possible, and Cheng has come across some hydropower projects in China where there are several weak zones (strong local minima) for which nearly all existing methods fail to work well.

There are two major issues in the location of the critical failure surface, which are the generation of the trial failure surface and the global optimization search for the critical failure surface. These two aspects are discussed in the following sections.

7.10.1 Generation of trial failure surface

For a classical gradient-type optimization method, once an initial trial is defined, the refinement of the critical failure surface will be given by the gradient of the objective function (which can be obtained by simple finite difference operation if the problem can converge). For a slope stability problem, particularly for "rigorous' formulations, failure to converge is common, so that the objective function will not be a continuous differentiable function. For a simplified method, the use of numerical differentiation may work if the initial choice is well chosen, which is, however, usually not possible. The use of the heuristic optimization method is currently the most popular and effective method for these problems. For the heuristic global optimization methods, trial failure surfaces are required to be generated which are controlled by the bounds for each control variable. Different methods in generating the failure surfaces have been proposed by Greco (1996), Cheng et al. (2007a, 2007b, 2007c, 2007e, 2008b, 2008c, 2009 2012), Bolton (2003), Li et al. (2005) and Zolfaghari (2005). In general, these methods are very similar in their basic operations. The coordinates of the points defining the failure surface are taken as the control variables, and

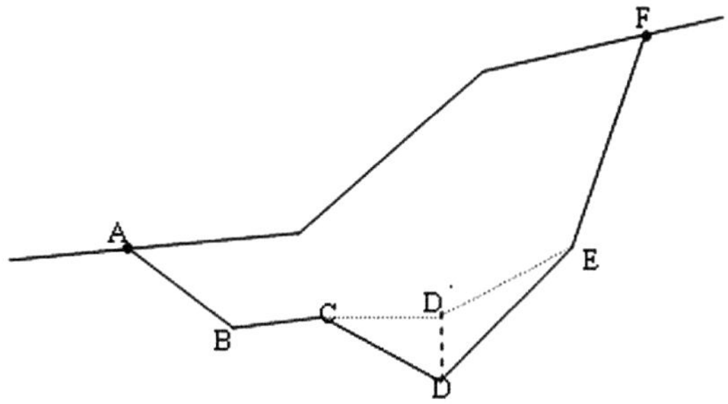

Figure 7.34 A failure surface with a kink or non-convex portion.

lower and upper bounds are assigned to each control variable. Consider the failure defined by ABCDEF shown in Figure 7.34. If each control variable is defined over static lower and upper bounds, point D, which is unlikely to be acceptable for a normal problem, can be generated by the random number generator. Since segment CD will be a kink which hinders the development of the failure, D is highly unlikely to be acceptable except for some special cases which are discussed later.

To generate a convex surface (this requirement can be relaxed in SLOPE 2000 if necessary) by the method proposed by Cheng (2003), consider a typical failure surface ACDEFB shown in Figure 7.34. The x-ordinates of the two exit ends A and B are taken as the control variables of the objective function and the upper and lower bounds of these two variables are specified by the engineer (bounds for the first two control variables are fixed). The static bounds for the first two control variables can be defined easily for the present problem with engineers' experience. Once the two exit ends A and B of the failure surface are defined, the requirements of the kinematically acceptable mechanism can be implemented as:

1. The x-ordinates of the interior points C, D, E and F of the failure surface can be obtained by uniform division of the horizontal distance between A and B which is Xright-Xleft. The x-ordinates of C, D, E and F are hence not control variables. Alternatively, the division can be made to follow the slope profile and the x-ordinates of the interior points are also not control variables.
2. Points A and B are connected and C1 is determined as a point located vertically above C. The y-ordinate of C1 is the lower value of either: (1) the y-ordinate of the ground profile as determined by x-ordinate of C, (2) the y-ordinate of the point lying along the line joining points A and B and determined by the x-ordinate of C. C1 is the upper bound to the y-ordinate of the first inter-slice. The lower bound of the y-ordinate of C (third control variable) is set by Cheng (2003) as C1-AB/4. In fact, such a lower bound can allow for a deep-seated failure surface and is adequate for all the cases that Cheng has encountered. The lower bound of the y-ordinate of C can be set to C1-AB/5 (instead of C1-AB/4 which is a conservative estimation of the lower bound) in most situations without affecting the solution. The y-ordinate of point C is a control variable of the objective function and is confined within the upper and lower bounds as determined in step 2.

3. Once a y-ordinate of C is chosen in the simulated annealing analysis, it connects A and C and extrapolates the line to G which is defined by the x-ordinate of point D. The lower bound of the y-ordinate of point D will be point G in order to maintain a concave failure shape. The upper bound of D, which is D1, is determined in the same way as for point C1. If part of the ground profile lies below the line joining B and C and affects the determination of D1 (for example point J in Figure 7.35), it connects C and J instead of B and C and determines the upper bound as D2 instead of D1.
4. Perform step 3 for the remaining points until all upper and lower bounds of the control variables are defined.
5. To allow for a non-concave failure surface which is unlikely to occur in reality, an option where the lower bound of point E will be set to the lower of the value as determined in step 3 above or the y-ordinate of point D is allowed. The y-ordinate of point E cannot be lower than that of D or else there will be a kink in the failure surface which prevents failure from occurring. The lower bound to the y-ordinate is sometimes totally eliminated, which is required for problems with a soft band. A non-convex failure surface can hence be generated from the present proposal by removing the lower bound requirement as required in the present method.

In Figure 7.35, the control variables are the x-ordinates of A and B and the y-ordinates of points C, D, E and F. A control variable vector X is used to store these control variables and the order of the control variables must be X_A, X_B, Y_C, Y_D, Y_E, Y_F. For the location of the global minimum of the objective function, engineers need to define only the upper and lower bounds of the first two control variables. An initial trial will be determined in a way

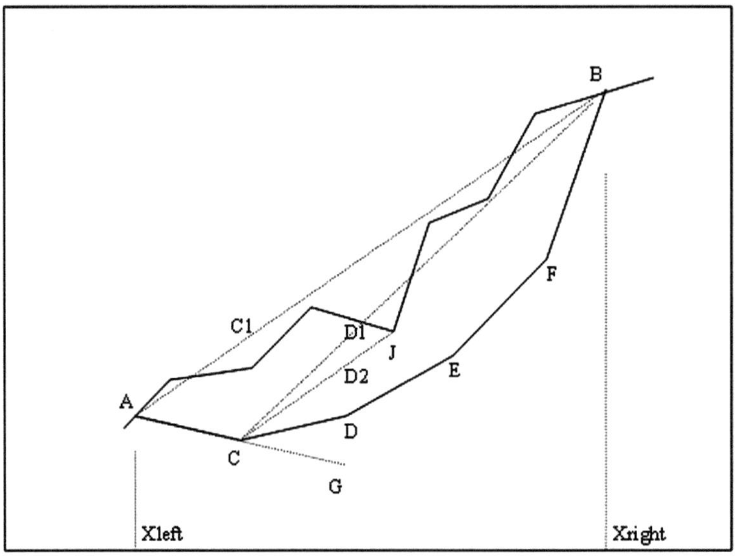

Figure 7.35 Generation of dynamic bounds for non-circular surface.

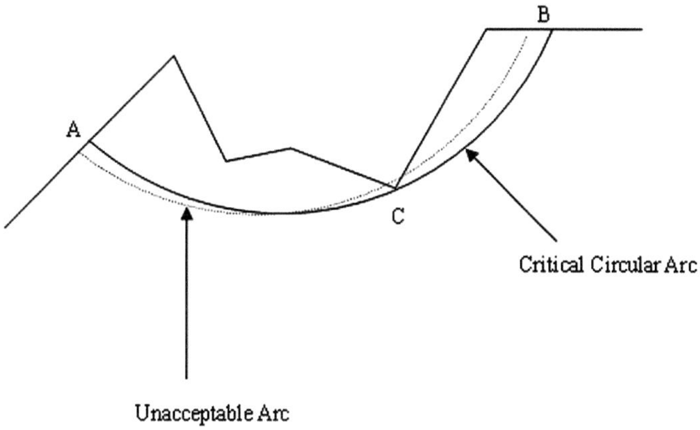

Figure 7.36 Dynamic bounds to acceptable circular surface.

similar to the approaches shown above. The upper and lower bounds of the other control variables will then be calculated according to steps 2 and 3 above. If the number of slices is n, then the number of control variables will be $n + 1$. If rock is encountered in the problem, the lower bound determination shown above has to be modified slightly. In steps 2 and 3 above, the lower bound will either be the y-ordinate of point G or the y-ordinate of the rock profile as determined by the x-ordinate of D.

For a circular failure surface, there are only three control variables, which are the x and y coordinates of the centre of rotation and the radius of the failure surface. Cheng (2003), however, adopts the x-ordinates of the two exit ends and the radius of the failure surface as the three control variables in analysis as it is easier to define the upper and lower bounds for the two exit ends (see Figure 7.36). This approach is also used by many commercial programs. The control variable vector X will be X_A, X_B, r. For the lower and upper bounds of the radius, the lower bound is set to half of the length of line AB which is the minimum possible radius. The upper bound of the radius is set to 50 times AB (any value which is not too small will be acceptable). An unacceptable failure surface will not be generated in the analysis and the constraints will control the lower and upper bounds of the radius when the two exit ends are defined. The constraints include:

1. The failure surface cannot cut the ground profile at more than two points within the two exit ends. As seen in Figure 7.36, point C will control the upper bound of the radius.
2. The failure surface cannot cut into rock stratum, which will control the lower bound of the radius.
3. The y-ordinate of the centre of rotation is higher than the y-ordinate of the right exit end. For this case, the last slice cannot be defined. This constraint will also control the lower bound of the radius.

In the present method, the first two variables, which are the x ordinates for the left and right ends, are varied within the user-defined lower and upper bounds, which are constant

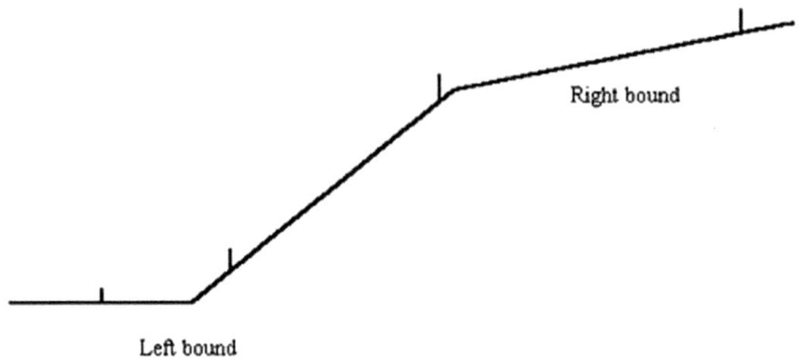

Figure 7.37 Domains for the left and right ends decided by engineers to define a search for the global minimum.

during the analysis. These two bounds are usually obvious in an engineering sense, or else the engineers can specify a large domain for the sake of safety. Besides these two variables, the bounds for the remaining variables (y ordinates of failure surface) are computed sequentially according to the guidelines shown above for circular and non-circular failure surfaces. The bounds from the present method are dynamic and different from classical simulated annealing methods or other global optimization methods where the bounds remain unchanged during the analysis. The generation of trial failure surfaces and the search direction will then proceed in accordance with the normal simulated annealing procedure, and the global minimum can be located easily with very high accuracy under the present proposal. The minimization process in the present formulation will depend on the lower and upper bounds of the left and right exit ends shown in Figure 7.37, which can be decided easily with experience and following engineering principles. For inexperienced engineers, wider ranges can be defined for the lower and upper bounds and the number of trials required for analysis will only increase slightly with the increase in the left and right ranges, which is another major advantage of the approach by Cheng (2003). For example, Cheng (2003) finds that when the ranges for the left and right exit ends are increased two-fold, the number of trials required will remain unchanged in many cases and may increase by less than 15% in some rare cases.

In the present algorithm, the x-ordinates are not considered as the control variables in order to reduce the number of control variables. This is usually satisfactory, as Cheng (2003) finds that the y-ordinates are more important than the x-ordinate in the factor of safety. Cheng (2003, 2007) also proposed that the x-ordinates can be adopted as the control variables. This approach will approximately double the number of control variables, and is considered to be useful only for those problems controlled by a soft band where the factor of safety is highly sensitive to the x-ordinates as well.

7.10.2 Global optimization methods

Global optimization problems are typically difficult to solve, and in the context of combinatorial problems, they are often N-P hard types for slope stability problems. The difficulties in performing the global optimization analysis and the requirement for a robust optimization algorithm have been discussed in Section 7.1. With the developments in computer software and hardware, many artificial intelligence-based algorithms based on natural selection and the mechanisms of population genetics have been developed. These algorithms are commonly applied in pattern recognition, electronic, production/control engineering or signal processing systems. These new heuristic optimization algorithms have been successfully applied to many different disciplines for both continuous and discrete optimization problems, but there are only limited uses of these methods in slope stability problems.

Many global optimization methods can be used for the present problem, and readers are recommended to read the relevant materials for which many books and papers have come out in the last 10 years. The author has implemented different algorithms in SLOPE 2000 for research purposes, and some of the options include:

1. Simulated annealing algorithm (SA);
2. Harmony search and different modified/improved versions (HM);
3. Particle swarm optimization search and different modified/improved versions (PSO);
4. Any-colony method;
5. Tabu search;
6. Genetic algorithm (GA);
7. Coupling methods, for example, PSO-HM, HM-complex search, etc.

Based on the various trial tests, the author has found that virtually all these methods can work well except under isolated cases. Each method may encounter problem cases, and so far, the author has not been able to find a single method which can pass through all the test cases, particularly for some cases with special soil parameters. Although these special cases are seldom encountered, it is preferable to have a solution algorithm which fails to work with only an extremely small probability. In this respect, the use of a coupled method or even double optimization method may be adopted. In the double optimization method, based on the optimized result from the first optimization algorithm, the bounds to each control variables can be narrowed so that the second optimization method will generate refined results of the analysis. Through this approach, the probability that the critical slip surface can be found is greatly enhanced.

SA, SHM, MHM, PSO, Tabu and ant-colony methods were first used by Cheng (2003, 2007, 2008), with various modifications to suit the slope stability problems. For the first demonstration of the applicability of these modern optimization methods, eight test problems are used to illustrate the effectiveness of Cheng's proposal on the modified SA algorithm (2003), with problems 4 and 8 shown in Figures 7.38 and 7.39. Problems 1–3 are similar to problem 4 except for the external load. For problems 1–4, which are shown in Figure 7.38, there are two types of soils with a water table. In problem 1, there is no external load while a horizontal load is applied in problem 2. Vertical load is applied in problem 3, while both vertical load and horizontal loads are applied in problem 4. Problems 5–7 are also similar to problem 8 except for the external load. For problems 5–8, which are shown in Figure 7.39, there are three types of soils, a water table and a perched water table. In problem 5, there is no external load, while horizontal load is applied in problem 6. Vertical load is applied in problem 7, while both vertical load and horizontal loads are applied in problem 8. The cohesive strengths of soils for problems 1–4 are 5 kPa and 2 kPa, respectively, for soil 1 and 2, while the corresponding cohesive strengths for

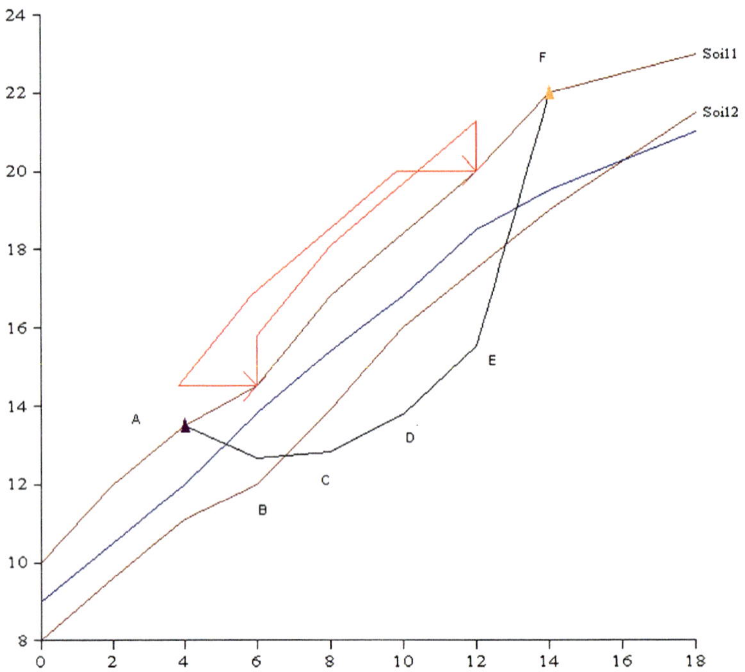

Figure 7.38 Problem 4 with horizontal and vertical load (critical failure surface is shown by ABCDEF).

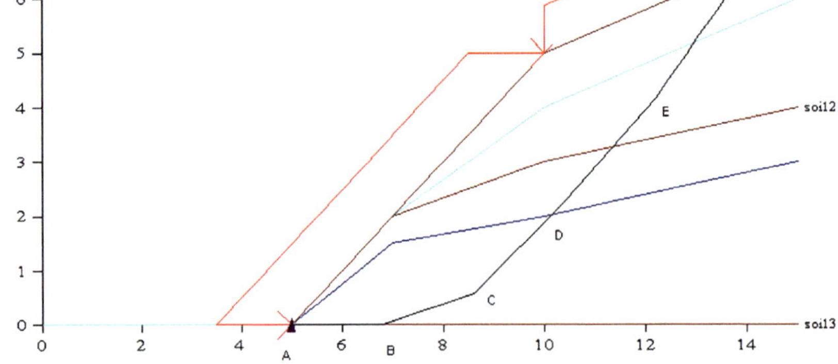

Figure 7.39 Problem 8 with horizontal and vertical load (critical failure surface is shown by ABCDEF).

problems 5–8 are 5 kPa, 2 kPa and 5 kPa. The friction angles of soils for problems 1–4 are 35° and 32°, respectively, for soil 1 and 2, while the corresponding friction angles for problems 5–8 are 32°, 30° and 35°. The unit weight of soil is kept constant at 19 kN/m^3 in all these cases. It is not easy to minimize the factor of safety for these problems by manual trial and error as the precise location of the failure surface will greatly influence the factor of safety. The minimum reference factors of safety are determined by an inefficient but robust pattern search approach. To limit the amount of computer time used, the number of slices is limited to five in these studies and the slices are divided evenly.

The critical solutions for these test cases are shown in Figures 7.38 and 7.39 by ABCDEF. The x-ordinates of the left exit end A (4.0 for problems 1–4 and 5.0 for problems 5–8) and right exit end F (14.0 for all problems) of the failure surfaces are fixed so that only the y-ordinates of B, C, D, E are variables (x-ordinates of B, C, D, E are obtained by even division). There are hence four control variables in the present study. Based on the critical result, BCDE obtained from the minimization analysis (round up to two decimal places), a grid is set up 0.5 m directly above and below B, C, D, E as obtained by the simulated annealing analysis. The spacings between the upper and lower bounds are hence 1.0 m for all four control variables. The grid spacing for each control variable is 0.01 m, so that each control variable can take 101 possible locations. The present grid spacing is fine enough for pattern search minimization and all the possible combinations of failure surfaces are tried which are 101 × 101 × 101 × 101 or 10406041 combinations.

The factors of safety for the test cases as shown in Table 7.31 have clearly illustrated that the combined use of the proposed failure surface generation and the simulated annealing method is able to minimize the factors of safety with high precision, and the results are similar to those obtained by pattern search based on 10406041 trials. The location of the critical failure surface obtained from simulated annealing analysis for problem 4 shown in Table 7.32 is very close to that obtained by pattern search, and similar results are also obtained for all the other problems. The results in Tables 7.31 and 7.32 clearly illustrate the capability of the proposed modified SA algorithm in minimizing the factors of safety, so that the burden on the engineers can be relieved by the adoption of modern global optimization techniques. Besides the simulated annealing method, the other global optimization as modified by Cheng methods (2007b) can also work with satisfaction for these eight problems.

Table 7.31 Comparison between minimization search and pattern search for eight test problems using simulated annealing method (tolerance in minimization search = 0.0001)

Case	*Trials required in SA*	*FS from SA*	*FS from pattern search*
1	10081	0.7279	0.7279
2	10585	0.8872	0.8872
3	9577	0.7684	0.7685
4	10585	0.9243	0.9243
5	12097	0.7727	0.7726
6	13105	1.1072	1.1072
7	11593	0.7494	0.7492
8	12601	1.0327	1.0327

(FS = factor of safety, SA=simulated annealing analysis.)

Table 7.32 Coordinates of the failure surface with minimum factor of safety from SA and from pattern search for Figure 7.39 (values with * are fixed and are not control variables)

Point	*x-ordinate*	*y-ordinate from SA*	*y-ordinate from pattern search*
A	4	13.5*	13.5*
B	6	12.677	12.67
C	8	12.831	12.82
D	10	13.784	13.78
E	12	15.539	15.54
F	14	22.0*	22.0*

Table 7.33 Comparisons between the number of trials required for dynamic bounds and static bounds in simulated annealing minimization

Case	*Trials in DB*	*Trials from SB*
1	10081	19823
2	10585	21023
3	9577	17234
4	10585	22131
5	12097	23968
6	13105	25369
7	11593	23652
8	12601	25104

(SB = static bounds, DB = dynamic bounds.)

To illustrate the advantages of the present dynamic bound technique as compared with the classical static bounds to the control variables, the same problems are considered with static bounds analysis. The static bounds are defined as 0.5 m above and below the critical failure surface BCDE, and the results are shown in Table 7.33 (the same minimum values are obtained from the two analyses). It is clear that the present proposal can greatly reduce the time of computation as compared with the classical simulated annealing technique which is highly beneficial for real problems. This advantage is particular important when the number of control variables is great.

7.10.3 Presence of soft band/Dirac function

If is a very thin soft band where the soil parameters are particularly low is present in the problem, the critical failure surface will be greatly controlled by this soft band. This type of problem poses great difficulty as normal random number generation (uniform probability) is used within the solution domain, and this feature is difficult to capture automatically. The thickness of the soft band can be so small that it can be considered as a Dirac function within the solution domain. Such failures have been reported in Hong Kong and the slope failure at Fei Tsui Road is one of the famous examples in Hong Kong where failure was controlled by a thin band of soil. To overcome the limitation of the normal random number generation, the probability of the variables falling within that

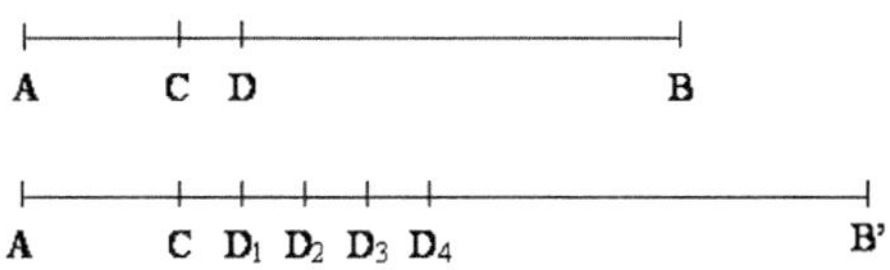

Figure 7.40 Transformation of domain to create special random number with weighting.

special region must be increased, which is equivalent to a non-uniform random number generation.

For a thin soft band, the probability of the control variables falling within this region will be small with the use of the classical random number. From principles of engineering, the probability of the control variables falling within this soft band region should, however, be greater than that falling within other regions. For this difficult case, Cheng et al. (2007) proposes to increase the probability of the search within the soft band. Since the location of the soft-band region is not uniform within the solution domain, it is difficult to construct a random function with increased probability within the soft-band region at different locations. This problem can be solved by a simple transformation proposed by Cheng, shown in Figure 7.40, where a classical random function is used in simulated annealing analysis. In Figure 7.40, the actual domain for a control variable x_i ($N + 1 > i > 2$ in present method) is represented by a segment AB with a soft band CD in between AB. For control variables x_j where $I \neq j$, the location of the soft band CD and the solution bound AB for control variable x_i will be different from that for control variable x_j. For segment AB, several virtual domains with a width of CD for each domain are added adjacent to CD, as shown in Figure 7.40. The transformed domain AB' is used as the control domain of variable x_i. Every point generated within the virtual domain D1-D2, D2-D3, D3-D4 is mapped to the corresponding point in segment CD1. This technique is effectively equivalent to giving more chances to those control variables within the soft band. The weighting to the variables within the soft band zone can be controlled easily by the simple transformation suggested in Figure 7.40. To the best of the author's knowledge, the search for the critical failure surface with a 1 mm thick soft band has never been minimized successfully, but this has been solved effectively by the proposed domain transformation by Cheng (2007). The transformation technique is coded into SLOPE 2000 and has been used to overcome a very difficult hydropower project in China where there are several layers of highly irregular soft band. For that project, several commercial programs were used to locate the critical failure surface without satisfaction.

7.11 DETERMINATION OF THE BOUNDS ON THE FACTOR OF SAFETY AND *F*(*X*)

To determine the bounds of the factor of safety and $f(x)$, the slope as shown in Figure 7.41 can be considered. For a failure surface with n slices, there are $n - 1$ interfaces and hence $n - 1$ $f(x_i)$. $f(x)$ will lie within the range of 0–1.0, while the mobilization factor λ and the objective function FOS based on Morgenstern-Price's method (1965) will be determined for each set of $f(x_i)$. The maximum and minimum factors of safety of a prescribed failure surface satisfying force and moment equilibrium will then be given by the various possible $f(x_i)$ satisfying Eq. (7.87).

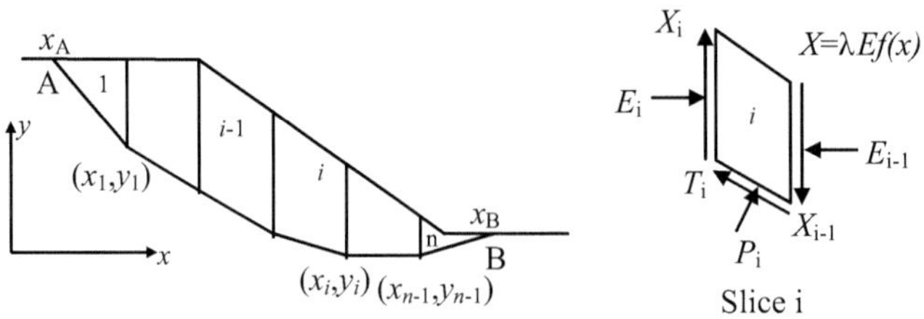

Figure 7.41 Slip surface and slices.

Maximize (or minimize) FOS subject to $0 \le f(x_i) \le 1.0$ for all i (7.87)

In carrying out the optimization analysis as given by Eq. (7.87), the constraints from the Mohr-Coulomb relation along the interfaces between slices as given by Eq. (7.88) should be considered.

$X \le E \tan(\phi') + c'L$ and line of thrust within soil mass (7.88)

where L is the vertical length of the interface between slices. The constraint given by Eq. (7.88) should also satisfy the requirement that the line of thrust of the internal forces lies within the soil mass, and Eq. (7.88) can have a major impact on the FOS in some cases, which will be illustrated by numerical examples in the following section. Since other than the $f(x)$, the Morgenstern-Price's method (1965) is totally governed by force and moment equilibrium, the maximum and minimum factors of safety found from varying $f(x)$ will provide the upper and lower bounds to the factor of safety of the slope that are useful for some difficult problems.

Since the objective function is a higher order function, the factor of safety is obtained by the double QR method by Cheng (2003). The simulated annealing method, which is more stable but less efficient, is used to determine the extrema (maximum or minimum factor of safety) with any given slip surface in Eq. (7.87). To evaluate the global minimum factor of safety of a slope, another global optimization analysis should be carried out for the factor of safety, which is an outer loop of the global optimization analysis. To ensure that "false" failure-to-converge due to iteration analysis (Cheng et al. 2008a) is not encountered so as to reduce the discontinuity of the objective function, the factor of safety is determined by the more time-consuming but robust double QR method. Since the factor of safety is available for practically all of the failure surfaces, the more efficient harmony search method can be used for locating the critical failure surface. The complete process is computationally intensive, but the use of modern global optimization processes can make this process a reality on a personal computer within an acceptable computation time.

Pan (1980) stated that the slope stability problem is actually a dual optimization problem which is not well known outside China. This is equivalent to the use of lower and upper bounds in conjunction for the ultimate state of a system. On one hand, the soil mass should redistribute the internal forces to resist the failure, which will result in a maximum factor of safety for any given slip surface, and this is called the maximum extremum principle. On the other hand, the slip surface with the minimum factor of safety is the most probable failure surface, which is called the minimum extremum principle. The maximum and minimum extremum principles are actually equivalent to the lower and upper bound methods, which are well known and will not be discussed here. Mathematically, the solution from the use of variational principles is an extremum of a function, and this is also equal to the global maximum/minimum of the function, which can also be determined from an optimization process. The "present proposal" can be viewed as a form of the discretized variational principle.

The maximum extremum principle is not new in engineering, and the ultimate limit state of a reinforced concrete beam is actually the maximum extremum state where the compressive zone of the concrete beam will propagate until a failure mechanism is formed. The ultimate limit state design of reinforced concrete or steel is equivalent to the maximum extremum principle, and this principle has been neglected in slope stability analysis for about 40 years! For any prescribed failure surface, the maximum "strength" of the system will be mobilized when a continuous yield zone is formed which is similar to a concrete beam. Pan's extremum principle (1980) can provide a practical guideline for slope stability analysis, and it is equivalent to the calculus of variation method used by Baker and Garber (1978), Baker (1980) and Revilla and Castillo (1977). This dual extremum principle has been proved by Chen (1998) based on lower and upper bound analyses. The maximum extremum is actually the lower bound solution, and the present approach is actually a lower bound approach as well as a variational principle approach.

Cheng et al. (2010) carried out the analysis in the following ways:

1. The factor of safety of a given failure surface with a given $f(x)$ will be determined by the use of the double QR method.
2. $f(x)$ will be taken as the control variable in the optimization process by the simulated annealing method, and the maximum factor of safety for a given failure surface and the associated $f(x)$ will be determined from an optimization analysis. The maximum extremum will be taken as the factor of safety for a prescribed failure surface, $f(x)$ can hence be directly obtained from the use of the optimization principle.
3. In addition to the maximum extremum factor of safety, Cheng et al. (2010) also investigated $f(x)$ for the minimum extremum factor of safety, which will provide a lower bound of the factor of safety for comparison.
4. The problem of convergence is investigated through the "present proposal".
5. the relation between failure to converge and the shape of $f(x)$ is also investigated.

Cheng et al. (2010) applied the simulated annealing method (Cheng 2003) complying with Eqs. (7.87) and (7.88) to evaluate the two extrema of the factor of safety, and the method has been coded into a general purpose program (SLOPE 2000). To determine the maximum and minimum extrema by the simulated annealing method, a tolerance of 0.0001 is used to control the optimization search and the factor of safety determination. This tolerance will terminate the search in a particular solution path during the optimization process (see also Cheng 2007a for details of the heuristic optimization methods). Since Cheng et al. (2010) adopted 15 slices in the computation, there are, in total, 14 $f(x_i)$ unknowns in the analysis, and the number of trials required to evaluate the two extrema ranges from 25000 to 32000, which is controlled by the tolerance during the

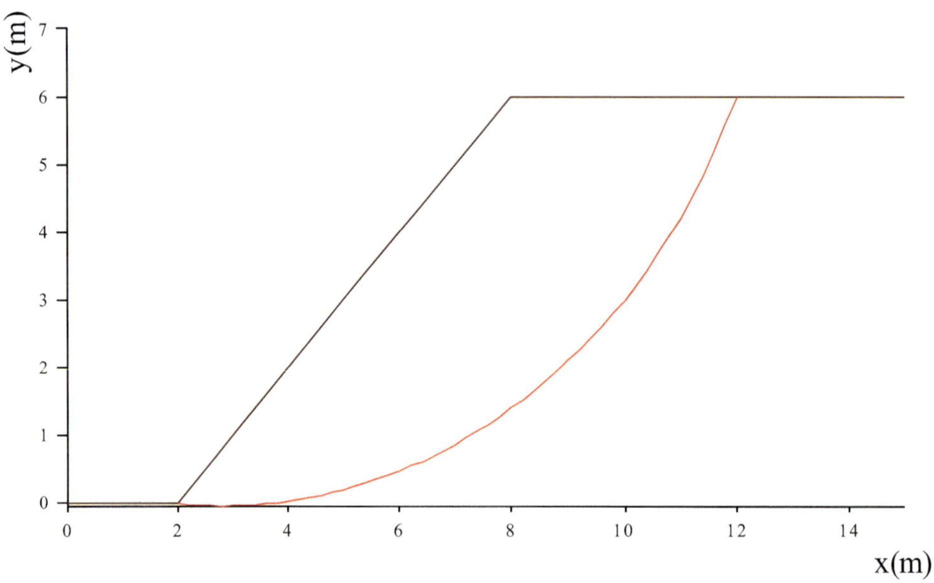

Figure 7.42 A simple slope with a circular failure surface.

optimization search. Based on this study, it was found that about 30–80% of the trials can converge when $f(x)$ is varied, and those trials that fail to converge are controlled by either Eq. (7.87) or no physically acceptable answer can be found from the double QR method. The number of trials which fail to comply with Eq. (7.88) is about 3–5 times that where no physically acceptable answer can be found by the double QR method, so the compliance with Eq. (7.88) (together with the requirement on the line of thrust), which has been neglected in the Morgenstern-Price method (1965), is actually important if an arbitrary $f(x)$ is defined.

Consider a very simple 45° 6 m high slope with a circular failure surface as shown in Figure 7.42. The unit weight of the soil is 19 kN/m^3, while c' and ϕ' vary as shown in Tables 7.34a and 7.34b. The differences between the two extrema are less than 2% of the results given by Spencer's method (1967), which clearly indicates that the factor of safety is not sensitive to $f(x)$ and that Spencer's method (1967) gives a good result for this example (see Table 7.34a). It is also interesting to find that while the FOS is not sensitive to Eq. (7.88), λ is quite sensitive to the Mohr-Coulomb relation along the interfaces as the mobilization of interslice shear force to achieve maximum and minimum resistance will involve a higher λ value. It is also observed that values of λ from the two extrema are generally greater than those from Spencer's analysis (1967), and these observations also apply to many other examples.

For the slope as shown in Figure 7.43 with the soil parameters given in Table 7.35, the various factors of safety are given in Tables 7.36a and 7.36b. With only the lower nail present, the differences between the two extrema as compared with Spencer's result (1967) are about 5.9% and 4.4% when Eq. (7.88) is used or not used, respectively. The corresponding results/differences when the two soil nails are present are 10.5% and 4.1%.

Table 7.34a Factors of safety from lower bound and Spencer's analysis for Figure 7.42 [* Spencer's result violates Eq. (7.88)]

FOS	*Max. FOS no Eq. (7.88)*	*Max FOS with Eq. (7.88)*	*Min. FOS no Eq. (7.88)*	*Min. FOS with Eq. (7.88)*	*Spencer*
c' = 0 kPa, ϕ' = 20°	0.759	0.749	0.738	0.743	0.745*
c' = 0 kPa, ϕ' = 40°	1.753	1.733	1.702	1.708	1.718
c' = 5 kPa, ϕ' = 20°	1.017	1.012	1.002	1.003	1.007
c' = 5 kPa, ϕ' = 40°	2.008	1.998	1.966	1.965	1.98
c' = 10 kPa, ϕ' = 20°	1.280	1.277	1.268	1.267	1.272
c' = 10 kPa, ϕ' = 40°	2.263	2.261	2.230	2.229	2.242

Table 7.34b λ from lower bound and Spencer's analysis for Figure 7.42 [* Spencer's result violates Eq. (7.88)]

FOS	*Max. FOS no Eq. (7.88)*	*Max FOS with Eq. (7.88)*	*Min. FOS no Eq. (7.88)*	*Min. FOS with Eq. (7.88)*	*Spencer*
c' = 0 kPa, ϕ' = 20°	1.867	1.0	1.888	1.0	0.522*
c' = 0 kPa, ϕ' = 40°	1.82	1.151	1.886	0.901	0.522
c' = 5 kPa, ϕ' = 20°	1.89	0.892	1.873	1.758	0.457
c' = 5 kPa, ϕ' = 40°	1.857	1.208	1.896	1.903	0.491
c' = 10 kPa, ϕ' = 20°	1.711	1.024	1.889	1.893	0.407
c' = 10 kPa, ϕ' = 40°	1.855	1.432	1.846	1.907	0.464

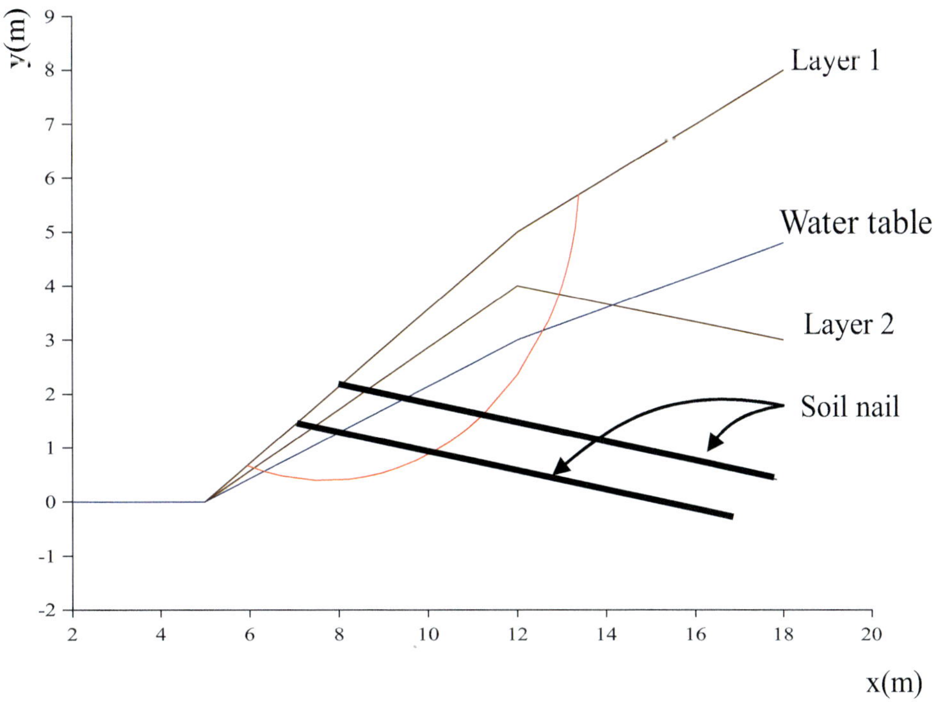

Figure 7.43 A problem with two soils, two soil nails, and a water table.

Table 7.35 Soil parameters for Figure 7.43

Soil	*Unit weight (kN/m³)*	*Saturated unit weight (kN/m³)*	*c' (kPa)*	*φ' (°)*
Top	18	20	5	36
Second layer	15	17	3	30

Table 7.36a Factors of safety from lower bound approach and Spencer's analysis for Figure 7.43

Case	*Max. FOS no Eq. (7.88)*	*Max FOS with Eq. (7.88)*	*Min. FOS no Eq. (7.88)*	*Min. FOS with Eq. (7.88)*	*Spencer*
Bottom nail	1.856	1.841	1.750	1.763	1.790
Two nails	2.661	2.600	2.398	2.498	2.515

Table 7.36b λ from lower bound approach and Spencer's analysis for Figure 7.43

Case	*Max. FOS no Eq. (7.88)*	*Max FOS with Eq. (7.88)*	*Min. FOS no Eq. (7.88)*	*Min. FOS with Eq. (7.88)*	*Spencer*
Bottom nail	1.149	0.944	0.924	1.902	0.488
Two nails	1.435	1.281	1.149	2.011	0.547

It can be observed that when the soil nail or external load is present, the choice of $f(x)$ has a noticeable impact on the results, and the compliance with Eq. (7.88) is also a critical issue which should be considered in the determination of extrema. The $f(x)$ for this problem is shown in Figure 7.44, and this interslice force function is clearly not similar to the commonly used functions. Based on the concept of extrema, the statically indeterminate nature of $f(x)$ is actually statically determinate for any complicated problem, and the procedure to determine $f(x)$ is completely independent of the difficulty of the problem.

7.12 THREE-DIMENSIONAL SLOPE STABILITY ANALYSIS

All slope failures are practically three-dimensional (3D) in nature, and near two-dimensional failures are scarce. Two-dimensional analysis is, however, usually carried out in practice, as this will greatly simplify the analysis with the limited site investigations that are available for normal engineering projects. There are many drawbacks in most existing 3D limit equilibrium slope stability methods which include:

1. Direction of slide is not considered in most of the existing slope stability formulation, so that the problems under consideration must be symmetrical in geometry and loading. Some formulations vary the direction of slide using a symmetric formulation until the minimum factor of safety is found for a given prescribed failure surface, but such an approach is highly inefficient due to the various geometry determinations required for the axes rotation.

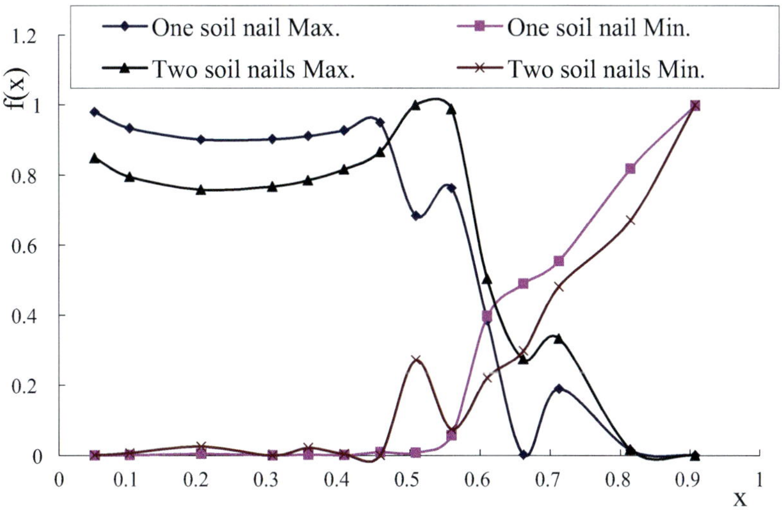

Figure 7.44 *f(x)* for the problem in Figure 7.43.

2. Location of the critical non-spherical 3D failure surface under general conditions is a difficult N-P hard type global optimization problem which has not been solved effectively and efficiently. For 3D problems, this is particularly important, as many more variables are required to define a general 3D surface than a 2D surface.
3. Existing methods of analyses are numerically unstable under transverse horizontal forces.

Because of the above-mentioned limitations, 3D analysis based on limit equilibrium is actually less convenient as compared with 3D SRM analysis, which is contrary to the 2D situation. Nevertheless, many engineers still prefer 3D limit equilibrium analysis, as they can get a feeling of the problem easily from their past 2D experience. Cavounidis (1987) demonstrated that the factor of safety for a normal slope under three-dimensional analysis is greater than that under two-dimensional analysis, but this result is based on the same soil parameters for 3D and plane strain cases, which is actually not correct. The 3D factor of safety is usually several times higher than the corresponding 2D analysis, however, after the adjustment of the soil parameters for plane strain condition, which is about 10% higher than the 3D condition, the factor of safety for 3D analysis may actually be lower than the corresponding 2D analysis, and this is one of the reasons for the 3D failures observed in practice. Another critical factor may be the different ground conditions for 2D and 3D considerations, which can be considered only from a 3D basis.

Most of the existing 3D methods rely on the assumption of a plane of symmetry in the analysis. For complicated ground conditions, this assumption is no longer valid, and the failure mass will fail along a direction with *least resistance* so that the sliding direction will also control the factor of safety of a slope. Stark and Eid (1998) have also demonstrated that the factor of safety of a 3D slope is controlled by the direction of slide and a symmetric failure may not be suitable for a general slope. Yamagami and Jiang (1996, 1997) and Jiang and Yamagami (1999) developed the first method for asymmetric problems where the classical stability equations (without direction of slide/direction of slide is zero) are used, while the direction of slide is considered by a minimization of the factor of safety with respect to rotation of axes. Yamagami and Jiang's formulation (1996, 1997) can be very time consuming even for a single failure surface, as the formation of columns and the determination of geometry information with respect to rotation of axes is the most time-consuming computation in the stability analysis. Huang et al. (2002) proposed the first method for the 3D asymmetrical Bishop method where the sliding direction enters directly into the determination of the safety factor. The generalized 3D slope stability method by Huang et al. (2002) is practically equivalent to the Janbu rigorous method with some simplifications on the transverse shear forces. Since it is difficult to satisfy completely the line of thrust constraints in the Janbu rigorous method, which is well known in 2D analysis, the generalized 3D method by Huang et al. (2002) will also face a converge problem so that this method is less useful for practical problems.

At the verge of failure, the soil mass can be considered as a rigid body. The direction of slide has three possibilities:

1. Soil columns are moving in the same direction with a unique sliding direction – adopted by Cheng and Yip (2007) and many other researchers in the present formulation.
2. Soil columns are moving towards each other – violates the assumption of rigid failure mass and is not considered.
3. Soil columns are moving away from each other – adopted by Huang et al. (2002). Such a condition can also be studied with the use of DEM.

For three-dimensional analysis, the potential failure mass of a slope is divided into a number of columns, which is a simple extension of the corresponding 2D analysis. At the equilibrium condition, the internal and external forces acting on each soil column are shown in Figure 7.45. The weight of the soil and vertical load are assumed to act at the centre of each column for simplicity. This assumption is not exactly true, but is good enough if the width of each column is sufficiently small, and the resulting governing equations will be greatly simplified and should be adequate for practical purposes. For practical purposes, Cheng uses more than 10000 columns for many 3D slope analyses. The assumptions required in the 3D formulation by Cheng are:

1. Mohr-Coulomb failure criterion is valid.
2. For the Morgenstern-Price method, the factor of safety is determined based on the sliding angle where factors of safety with respect to force and moment are equal.
3. Sliding angle is the same for all soil columns (Figure 7.46).

By Mohr-Coulomb criteria, the global factor of safety, F, is defined as

$$F = \frac{S_{fi}}{S_i} = \frac{C_i + N_i' \cdot \tan\phi_i}{S_i} \tag{7.89}$$

where F is the factor of safety, S_{fi} is the ultimate resultant shear force available at the base of column i, N'_i is the effective base normal force, C_i is $c'A_i$ and A_i is the base area

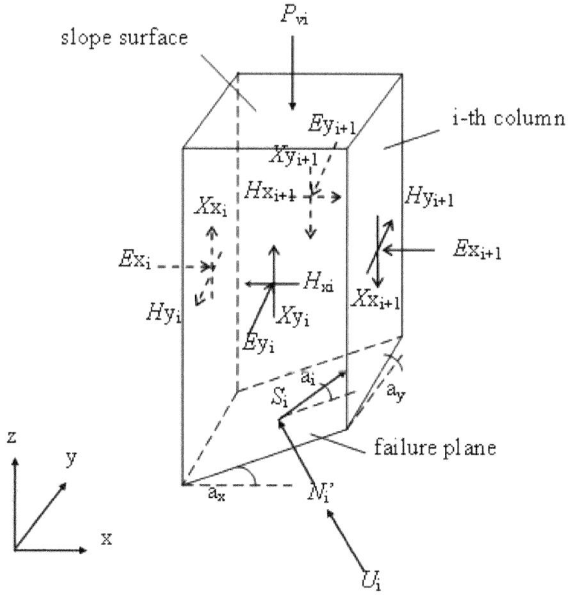

Figure 7.45 External and internal forces acting on a typical soil column.

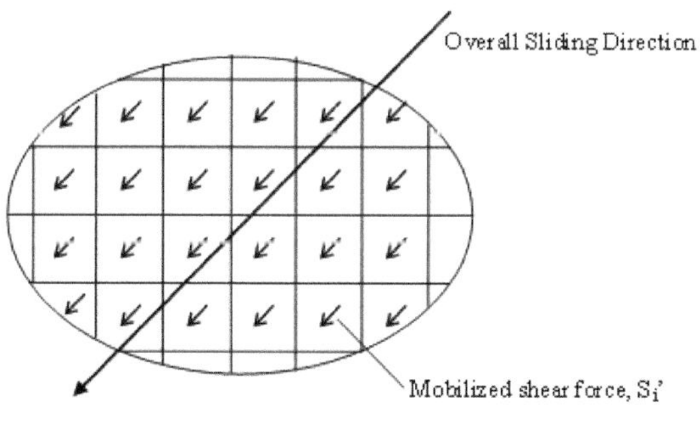

Figure 7.46 Unique sliding direction for all columns (on plan view).

of the column. The base shear force *S* and base normal force *N* with respect to the x, y and z directions for column *i* are expressed as the components of forces by Huang et al. (2002):

$$S_{xi} = f_1 \cdot S_i;\ S_{yi} = f_2 \cdot S_i;\ \text{and}\ S_{zi} = f_3 \cdot S_i$$

$$N_{xi} = g_1 \cdot N_i;\ N_{yi} = g_2 \cdot N_i;\ \text{and}\ N_{zi} = g_3 \cdot N_i \tag{7.90}$$

in which $\{f_1 \cdot f_2 \cdot f_3\}$ and $\{g_1 \cdot g_2 \cdot g_3\}$ are unit vectors for S_i and N_i (see Figure 7.45). The projected shear angle *a*' (individual sliding direction) is the same for all the columns in the

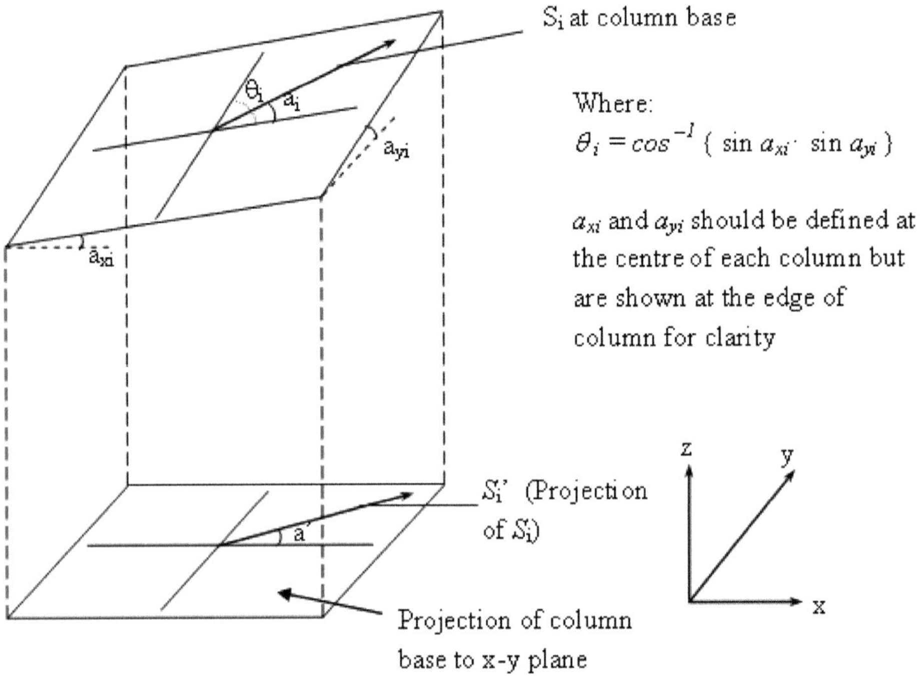

Figure 7.47 Relationship between projected and space shear angle for the base of column *i*.

x–y plane in the present formulation, and by using this angle, the space shear angle a_i (see Figure 7.47) can be found for each column and is given by Huang et al. (2002) as Eq. (7.91):

$$a_i = \tan^{-1}\{\sin\theta_i/[\cos\theta_i + (\cos a_{yi}/\tan a' \cdot \cos a_{xi})]\} \tag{7.91}$$

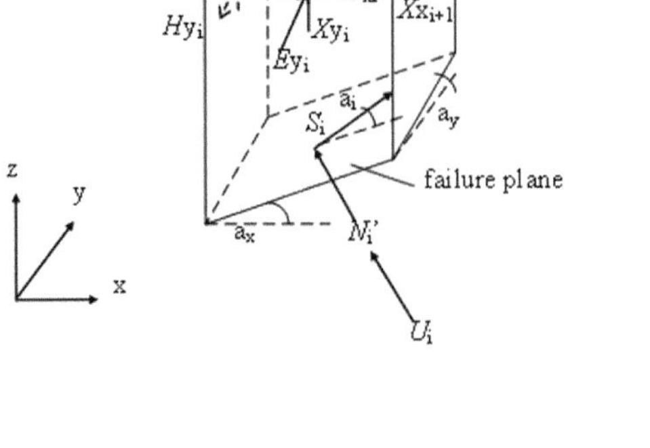

Where:

a_i is the space sliding angle for sliding direction with respect to the direction of slide projected to the x–y plane [see also a' in Figure 7.46 and Eq. (7.91)]
a_x, a_y are the base inclination along the x and y directions measured at the centre of each column (shown at the edge of the column for clarity)
Ex_i, Ey_i are the intercolumn normal forces in the x and y directions, respectively
Hx_i, Hy_i are the lateral intercolumn shear forces in the x and y directions, respectively
N_i', U_i are the effective normal force and base pore water force, respectively
P_{vi}, S_i are the vertical external force and base mobilized shear force, respectively
Xx_i, Xy_i are the vertical intercolumn shear force in the plane perpendicular to the x and y directions

$$n_i = \left\{ \frac{\pm\tan a_{xi}}{J}, \frac{\pm\tan a_{yi}}{J}, \frac{1}{J} \right\} = \{g_1, g_2, g_3\}$$

[negative adopted by Huang et al. (2002) and positive adopted by Cheng and Yip (2007)]

$$s_i = \left\{ \frac{\sin(\theta_i - a_i)\cdot\cos a_{xi}}{\sin\theta_i}, \frac{\sin a_i\cdot\cos a_{yi}}{\sin\theta_i}, \frac{\sin(\theta_i - a_i)\cdot\sin a_{xi} + \sin a_i\cdot\sin a_{yi}}{\sin\theta_i} \right\} = \{f_1, f_2, f_3\}$$

in which $J = \sqrt{\tan^2 a_{xi} + \tan^2 a_{yi} + 1}$

An arbitrary intercolumn shear force function $f(x,y)$ is assumed in the present analysis, and the relationships between the intercolumn shear and normal forces in the x and y directions are given as:

$$Xx_i = Ex_i\cdot f(x,y)\cdot\lambda_x;\quad Xy_i = Ey_i\cdot f(x,y)\cdot\lambda_y \tag{7.92}$$

$$Hx_i = Ey_i\cdot f(x,y)\cdot\lambda_{xy};\quad Hy_i = Ex_i\cdot f(x,y)\cdot\lambda_{yx} \tag{7.93}$$

where λ_x and λ_y are the intercolumn shear force X mobilization factors in the x and y directions, respectively, λ_{xy} and λ_{yx} are the intercolumn shear force H mobilization factors in the x–y and y–x planes, respectively.

Taking the moment about the z-axis at the centre of the i-th column, the relations between lateral–intercolumn shear forces can be expressed as:

$$\Delta y_i\cdot(Hx_{i+1} + Hx_i) = \Delta x_i\cdot(Hy_{i+1} + Hy_i) \tag{7.94}$$

From Eq. (7.94), $$Hx_{i+1} = \frac{\Delta x_i}{\Delta y_i}\cdot(Hy_{i+1} + Hy_i) - Hx_i \tag{7.95}$$

From Eq. (7.94), $$Hy_{i+1} = \frac{\Delta y_i}{\Delta x_i} \cdot (Hx_{i+1} + Hx_i) - Hy_i \quad (7.96)$$

where Δx_i and Δy_i are the widths of the column defined in Figure 7.48. Hx_i and Hy_i for the exterior columns should be zero in most cases or equal to the applied horizontal forces if defined. By using the property of complementary shear (or moment equilibrium in the x–y plane), Hy_{i+1} or Hx_{i+1} can then be determined from Eqs. (7.93), (7.95) and (7.96) accordingly, so only λ_{xy} or λ_{yx} is required to be determined but not both. The important concept of complementary shear force which is similar to the complementary shear stress ($\tau_{xy} = \tau_{yx}$) in elasticity has not been used in any 3D slope stability analysis method in the past but is crucial in the present formulation. It should be noted that Huang et al. (2002) have assumed Hy_i to be 0 for an asymmetric problem in order to render the problem determinate, which is valid for symmetric failure only. Although the concept of complimentary shear is applicable only in an infinitesimal sense, if the size of the column is not great, this assumption will greatly simplify the equations. More importantly, Cheng and Yip (2007) have demonstrated that the effect of λ_{xy} or λ_{yx} is small, and the error in this assumption is actually not important.

Force equilibrium in X, Y and Z directions
Considering the vertical and horizontal forces equilibrium for the *i*-th column (Figures 7.49 and 7.50) in the z, x and y directions give:

$$\Sigma F_z = 0 \rightarrow N_i \cdot g_{3i} + S_i \cdot f_{3i} - (W_i + P_{vi}) = (Xx_{i+1} - Xx_i) + (Xy_{i+1} - Xy_i) \quad (7.97)$$

$$\Sigma F_x = 0 \rightarrow S_i \cdot f_{1i} - N_i \cdot g_{1i} + P_{hxi} - Hx_i + Hx_{i+1} = Ex_{i+1} - Ex_i \quad (7.98)$$

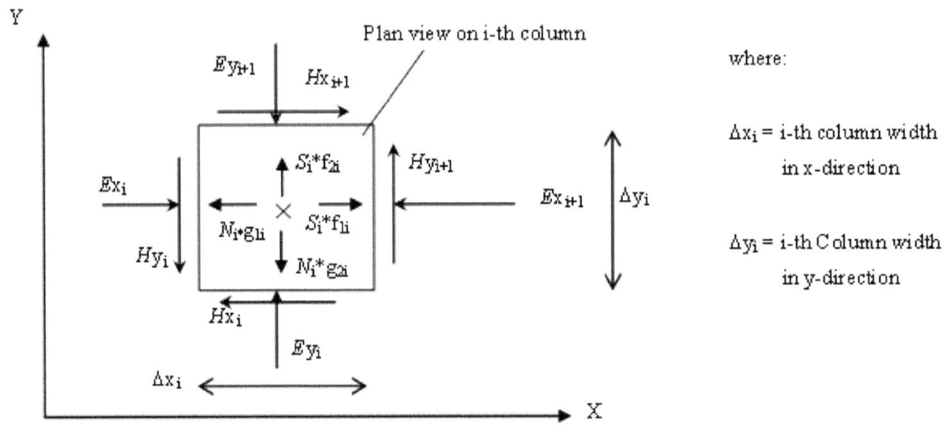

Figure 7.48 Force equilibrium in the x–y plane.

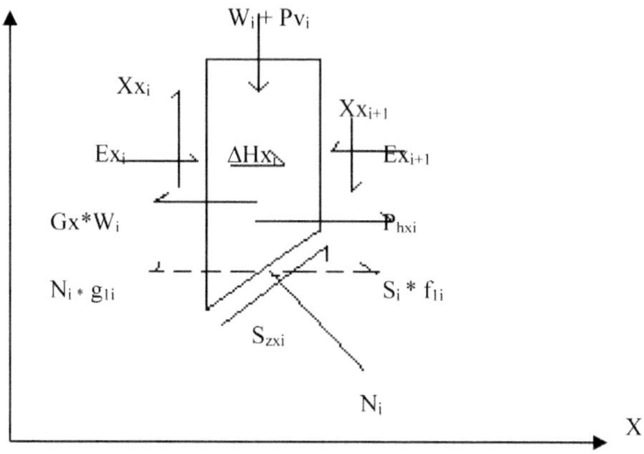

Figure 7.49 Horizontal force equilibrium in the x direction for a typical column (ΔHx_i = net lateral intercolumn shear force).

$$\Sigma F_y = 0 \rightarrow S_i \cdot f_{2i} - N_i \cdot g_{2i} + P_{hyi} - Hy_i + Hy_{i+1} = Ey_{i+1} - Ey_i \tag{7.99}$$

Solving Eqs. (7.95), (7.98) and (7.97), the base normal and shear forces can be expressed as:

$$N_i = A_i + B_i \cdot S_i; \quad S_i = \frac{C_i + (A_i - U_i) \cdot \tan\phi_i}{F\left(1 - \dfrac{B_i \cdot \tan\phi_i}{F}\right)} \tag{7.100}$$

$A_i = \dfrac{W_i + P_{vi} + \Delta Ex_i \cdot \lambda_x + \Delta Ey_i \cdot \lambda_y}{g_{3i}}$; $B_i = -\dfrac{f_{3i}}{g_{3i}}$; $U_i = u_i A_i$ (u_i = average pore pressure at the *i*-th column)

7.12.1 Overall force and moment equilibrium in the x and y directions

Considering the overall force equilibrium in the x direction:

$$-\sum Hx_i + \sum N_i \cdot g_{1i} - \sum S_i \cdot f_{1i} = 0 \tag{7.101}$$

Let $F_x = F$ in Eq. (7.89), using Eq. (7.101) and rearranging Eq. (7.101), the directional safety factor F_x can be determined as:

$$F_x = \frac{\Sigma[(N_i - U_i) \cdot \tan\phi_i + C_i)] \cdot f_{1i}}{\Sigma N_i \cdot g_{1i} - \Sigma Hx_i}, \quad 0 < F_x < \infty \tag{7.102}$$

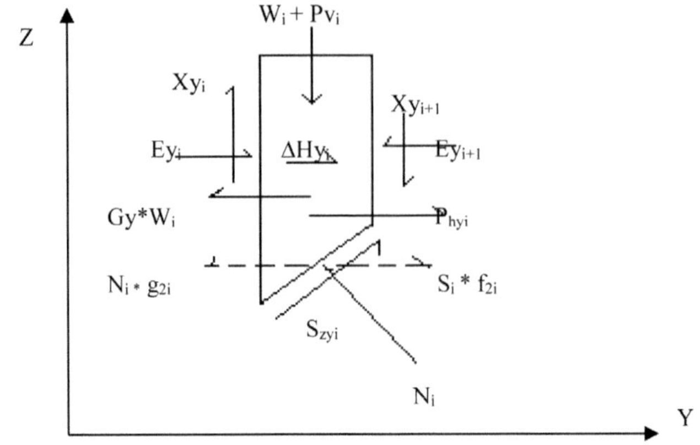

Figure 7.50 Horizontal force equilibrium in the y direction for a typical column (ΔHy_i = net lateral intercolumn shear force).

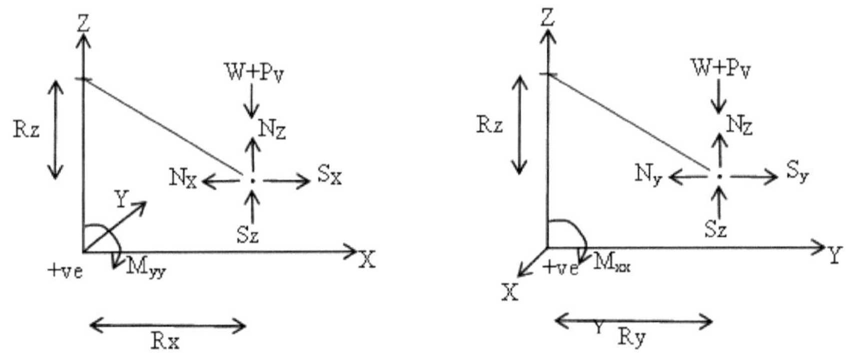

Figure 7.51 Moment equilibrium in the x and y directions.

(Earthquake loads and net external moments are not shown for clarity.)

From the overall moment equilibrium in the x direction (Figure 7.51):

$$\sum(W_i + P_{vi} - N_i \cdot g_{3i} - S_i \cdot f_{3i}) \cdot RX + \sum(N_i \cdot g_{1i} - S_i \cdot f_{1i}) \cdot RZ = 0 \qquad (7.103)$$

RX, RY and RZ are the lever arms to the moment point. Similarly, considering the overall force equilibrium in the y direction:

$$-\sum Hy_i + \sum N_i \cdot g_{2i} - \sum S_i \cdot f_{2i} = 0 \qquad (7.104)$$

Let $F_y = F$ in Eq. (7.89), using Eq. (7.100) and rearranging Eq. (7.101), the directional safety factor (F_y) can be determined as:

$$F_y = \frac{\Sigma[(N_i - U_i)\cdot \tan\phi_i + C_i)]\cdot f_{2i}}{\Sigma N_i \cdot g_{2i} - \Sigma Hy_i}, \quad 0 < F_y < \infty \tag{7.105}$$

Overall moment equilibrium in the y direction (Figure 7.51):

$$\sum(W_i + P_{vi} - N_i \cdot g_{3i} - S_i \cdot f_{3i})\cdot RY + \sum(N_i \cdot g_{2i} - S_i \cdot f_{2i})\cdot RZ = 0 \tag{7.106}$$

Based on a trial sliding angle, λ_x is kept changing with a specified interval in Eq. (7.102), until the calculated F_x satisfies the overall moment equilibrium Eq. (7.103) in the x direction. A similar procedure is applied to λ_y until the calculated F_y also satisfies the overall moment equilibrium Eq. (7.106) in the y direction. If F_x is not equal to F_y, the sliding angle will be varied until $F_x = F_y$ and then force as well as moment equilibrium will be achieved. Since all the equilibrium equations have been used in the formulation, there is no equation to determine λ_{xy} unless additional assumptions are specified. In the present formulation, Cheng and Yip (2007) suggest that λ_{xy} can be specified by the user or can be determined from the minimization of the factor of safety with respect to λ_{xy}. The problem associated with λ_{xy} and the importance of this parameter are further discussed in a later section.

7.12.2 Reduction to 3D Bishop and Janbu simplified methods

The 3D asymmetric Morgenstern-Price method takes a relatively long time for solution and the convergence is less satisfactory as compared with the simplified method. The initial solutions from 3D Janbu or Bishop analysis can be adopted to accelerate the Morgenstern-Price solution, and many engineers may still prefer to use the simplified method for routine design. The proposed Morgenstern-Price formulation will be simplified by considering only force or moment equilibrium equations and neglect all the intercolumn vertical and horizontal shear forces. Considering *overall moment equilibrium* in the x direction and about an axis passing through (x_o,y_o,z_o) (centre of rotation of the spherical failure surface) and parallel to the y-axis. Let $F_{my} = F$ in Eq. (7.89), and rearrange Eq. (7.103) to give:

$$F_{my} = \frac{\sum\{[K_{yi}\cdot[f_{1i}RZ_i + f_{3i}RX_i]\}}{\sum(W_i + P_{vi})\cdot RX_i + \sum N_i \cdot (g_{1i}\cdot RZ_i - g_{3i}\cdot RX_i)} \tag{7.107}$$

The corresponding F_{mx} is obtained from Eq. (7.108) as

$$F_{mx} = \frac{\sum\{K_{xi}\cdot[f_{2i}RZ_i + f_{3i}RY_i]\}}{\sum(W_i + P_{vi})\cdot RY_i + \sum N_i \cdot (g_{2i}\cdot RZ_i - g_{3i}\cdot RY_i)} \tag{7.108}$$

in which: $K_{yi} = \dfrac{\left\{C_i + \left[\dfrac{(W_i + P_{vi})}{g_{3i}} - U_i\right]\tan\phi_i\right\}}{1 + \dfrac{f_{3i}\cdot\tan\phi_i}{g_{3i}\cdot F_{my}}}$; $K_{xi} = \dfrac{\left\{C_i + \left[\dfrac{(W_i + P_{vi})}{g_{3i}} - U_i\right]\tan\phi_i\right\}}{1 + \dfrac{f_{3i}\cdot\tan\phi_i}{g_{3i}\cdot F_{mx}}}$

Considering *overall moment equilibrium* about an axis passing through (x_o, y_o, z_o) and parallel to the z-axis gives:

$$\sum(-N_i \cdot g_{1i} + S_i \cdot f_{1i}) \cdot RY + \sum(N_i \cdot g_{2i} - S_i \cdot f_{2i}) \cdot RX = 0 \tag{7.109}$$

Let $F_{mz} = F$ in Eq. (7.89), and rearrange Eq. (7.109) to give:

$$F_{mz} = \frac{\sum[K_{zi} \cdot (f_{2i} \cdot RX_i - f_{3i} \cdot RY_i)]}{\sum N(g_{2i} \cdot RX_i - g_{1i} \cdot RY_i)}; \quad K_{zi} = \frac{\left\{C_i + \left[\frac{(W_i + P_{vi})}{g_{3i}} - U_i\right]\tan\phi_i\right\}}{1 + \frac{f_{3i} \cdot \tan\phi_i}{g_{3i} \cdot F_{mz}}} \tag{7.110}$$

For the 3D asymmetric Bishop method, at the moment equilibrium point, the directional factors of safety, F_{mx}, F_{my} and F_{mz}, are equal to each other. Under this condition, the global factor of safety F_m based on moment can be determined as:

$$F_m = F_{mx} = F_{my} = F_{mz} \tag{7.111}$$

The sliding direction can be found by changing the *projected shear direction* at a *specified angular interval*, until F_{mx}, F_{my} and F_{mz} are equal to each other. In reality, there is no way to ensure complete 3D moment equilibrium in the Bishop method as Eq. (7.109) is redundant and is not used in the present method or the method by Huang et al. (2002) as Eqs. (7.107) and (7.108) are already sufficient for the solution of the factor of safety. The left-hand side of Eq. (7.108) can hence be viewed as an *unbalanced moment term*. For a completely symmetric slope, this term is exactly zero and 3D moment equilibrium is automatically achieved. In general, this term is usually small if the asymmetrical loading or sliding direction is not great. Cheng and Yip (2007) hence adopt Eqs. (7.107) and (7.108) in the formulation, which is equivalent to assigning $F_{mx} = F_{my}$. This is a limitation of the present 3D asymmetric Bishop simplified method as well as all the other existing 3D Bishop methods for general asymmetric problems as Eq. (7.109) is a redundant equation.

By neglecting the inter-column shear forces for Janbu analysis, Eqs. (7.102) and (7.105) simplify to

$$F_{sx} = \frac{\sum A_{xi}\left(f_{1i} + \frac{f_{3i} \cdot g_{1i}}{g_{3i}}\right)}{\sum \frac{g_{1i}}{g_{3i}} \cdot (W_i + P_{vi})}; \quad A_{xi} = \frac{\left\{C_i + \left[\frac{(W_i + P_{vi})}{g_{3i}} - U_i\right]\tan\phi_i\right\}}{1 + \frac{f_{3i} \cdot \tan\phi_i}{g_{3i} \cdot F_{sx}}} \tag{7.112}$$

$$F_{sy} = \frac{\sum A_{yi}\left(f_{2i} + \frac{f_{3i} \cdot g_{2i}}{g_{3i}}\right)}{\sum \frac{g_{2i}}{g_{3i}} \cdot (W_i + P_{vi})}; \quad A_{yi} = \frac{\left\{C_i + \left[\frac{(W_i + P_{vi})}{g_{3i}} - U_i\right]\tan\phi_i\right\}}{1 + \frac{f_{3i} \cdot \tan\phi_i}{g_{3i} \cdot F_{sy}}} \tag{7.113}$$

For the 3D asymmetric Janbu method, at force equilibrium point, the directional factors of safety, F_{sx} and F_{sy}, are equal to each other. Under this condition, the global factor of safety F_f based on force can be determined as

$$F_f = F_{sx} = F_{sy} \tag{7.114}$$

Since the factor of safety is also used in vertical force equilibrium, 3D force equilibrium is completely achieved in the 3D Janbu simplified method.

Based on the theory and the program SLOPE3D developed by Cheng, the large-scale 3D slope failure as shown in Figure 7.52 has been analysed by Cheng. The 3D profile was supplied by engineers in form of GIS files. These files were then transformed to simple text files, which were further processed by Cheng by developing two small programs. The final data were then read in by the program SLOPE3D as shown in Figure 7.52. Columns are then generated for this 3D profile, and more than 10000 columns have been used by Cheng in the analysis. The stability analysis used by Cheng is the 3D Morgenstern-Price method as outlined in this section, and $f(x,y)$ is taken to be 1.0 for simplicity. After the analysis, the factor of safety of the post-failure condition was evaluated as acceptable, hence the road was opened for public after the analysis. It is interesting to note that the formulation by Cheng and Yip (2007) with a unique sliding direction can avoid the lateral force numerical instability problem as compared with the varying sliding direction approach by Huang et al. (2002). Cheng's formulation is also adopted by commercial program SLIDE3D. SLOPE3D can locate the critical 3D failure surface by spherical, ellipsoidal and NURBS surfaces, while the general shape NURBS surface appears to be unique in the 3D slope stability program.

The importance of 3D slope stability analysis can be further illustrated by the effect of curvature on the factor of safety for a highway slope in Figure 7.53. This effect is common for many highway slopes and this effect is actually noticeable in Figure 7.53, but is seldom

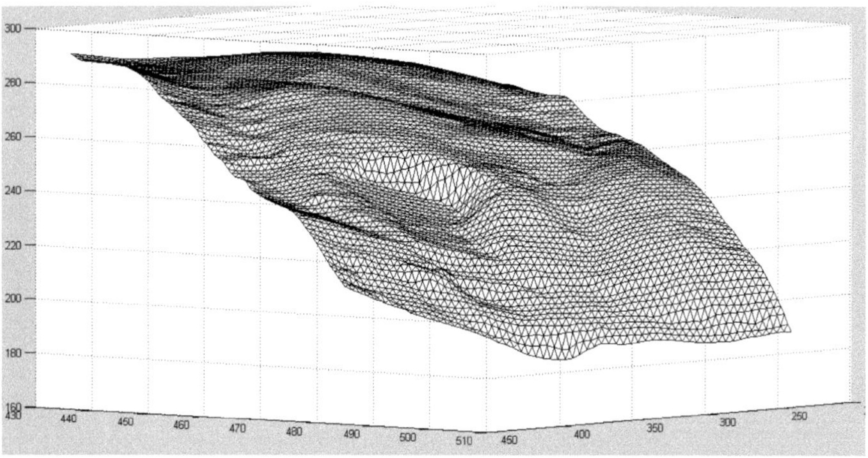

Figure 7.52 3D analysis of the problem in Figure 7.5, the geometry is generated from the GIS file for this project.

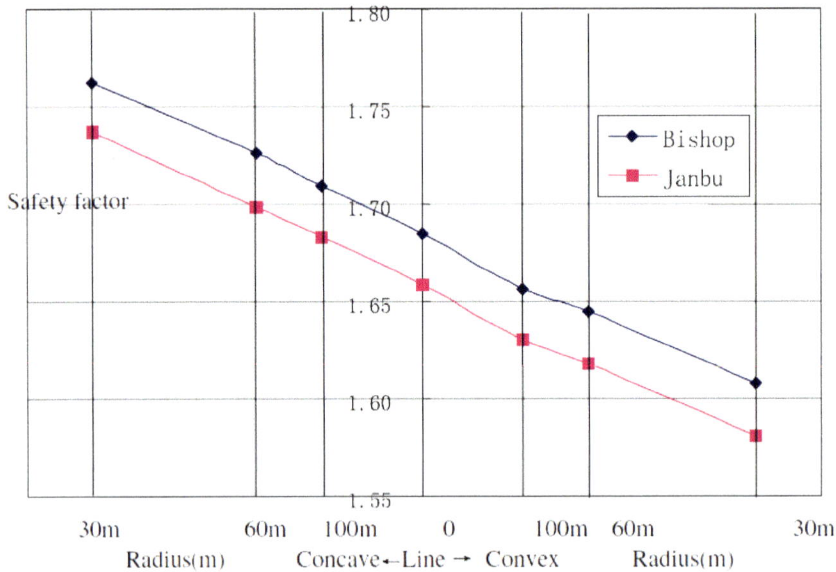

Figure 7.53 Effect of curvature on the factor of safety.

considered in practice for routine design. SLOPE3D has many interesting features which are shown in Figure 7.54. In Figure 7.54a, the geometry of the problem is defined by the 3D information of boreholes, from ground surface to different layers of soil. The users only need to define borehole information, which can greatly relieve the burden in 3D data preparation. 3D nails and loadings can be defined in SLOPE3D, and a general non-spherical or non-ellipsoidal search is achieved through the use of the NURBS function with an arbitrary number of control points.

7.13 SPECTRAL ELEMENT METHOD

The spectral element method was developed by Patera in 1984. This method is based on the weak/variational formulation of the differential problem, and the number of collocation points inside each element can be increased, also, it is conceptually similar to the p-FEM. The spectral method is a form of discretization method for the approximate solution of weak form partial-differential equations. Spectral element methods rely on the use of higher order polynomial approximation with great accuracy, but are less capable of dealing with complex geometries. The spectral method can be classified as the Galerkin method, however, it has a strong resemblance to collocation methods. The spectral method is different from other approximate methods in terms of the "globality" of its basis functions. Most other numerical methods converge by decomposition of the solution domain into smaller subdomains using low-order approximation. In contrast, the spectral method does not decompose the computational domain and converges to solution simply by increasing the order of approximation. While the domain decomposition brings basis functions with support over only a few neighbouring subdomains, the support stays

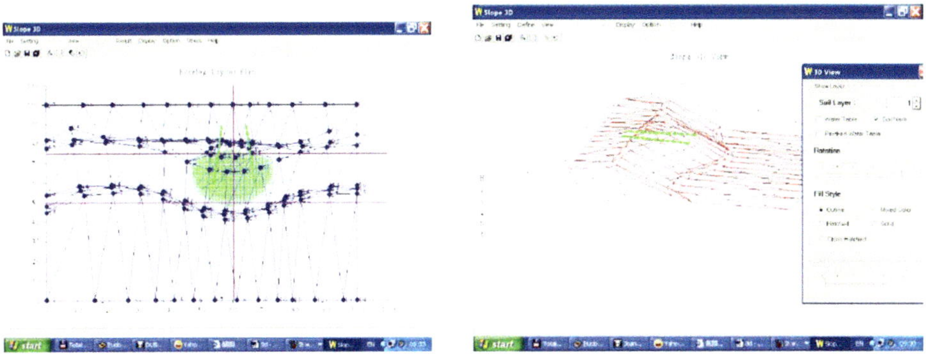

(a) Use of triangulation to define 3D geometry (b) 3D soil nails

Figure 7.54 Use of SLOPE3D.

with the whole computational domain in the spectral method. The spectral method takes advantage of the fast convergence of coefficients in expansion of the solution to the trial functions. Fast convergence of the coefficients results in high accuracy, even though the expansion series is truncated. Detailed treatment of the SEM and sample codes have been provided and applications of the SEM are given by Tiwari (2015). SEM can be viewed as a special form of SRM for slope stability analysis.

FEM uses separate nodes/points for the numerical integration and interpolation, and produces a relatively dense matrix and increased computing time. The integrals in SEM are evaluated based on the Gauss–Lobatto–Legendre (GLL) quadrature points. Since both the integration point and interpolation point are the same, it will lead to a diagonal matrix and therefore simplifies the analysis. The method adopts tensor-product Lagrange interpolants within each element, where the nodes of these shape functions are placed at the zeros of Legendre polynomials (Gauss-Lobatto points), mapped from the reference domain $[-1, 1] \times [-1, 1]$ to each element. For smooth functions, it can be shown that the resulting interpolants converge exponentially faster as the order of the interpolant is increased. The efficiency in numerical computation is achieved by using the Gauss–Lobatto quadrature for evaluating the elemental integrals. The quadrature points reside at the nodal points, which enables fast tensor-product techniques to be used for iterative matrix solution methods. Gauss–Lobatto quadrature hence naturally results in diagonal mass matrices, which is computationally light as compared with the classical finite element method. Depending on the size of the resulting matrix, two approaches are commonly used to solve the matrix equation. When the matrix size is large but the degree of polynomial is small, the use of a preconditioned conjugate gradient method or similar is popular. For the reverse condition, the Schur complement is commonly adopted. The spectral element method was originally developed for solving the computational fluid dynamic problems, but was later extended to the solution of many problems in solid mechanics.

So far, there are only very limited applications of the SEM in geotechnical problems, except for some slope stability problems. Gharti et al. (2012, 2017) considered elasto-plastic 3D SEM analysis with parallel implementation. Similar to SRM, the factor of safety is determined by plotting the maximum displacement against the trial factor of safety, for which the author has found it to be less reliable when there is soil reinforcement in a slope. Gharti et al. (2012) also found that the results from SRM and SEM agree very well in general. Tiwari et al. (2014, 2015) carried out a series of works in applying SEM in slope stability problems. For interested readers, the open source program Specfem3D-Slope (version 1.2 at present) by Gahrti et al. (2017) is a good reference for trying the SEM slope stability analysis. Another open source program is the Nektar++ 4.4.1 for fluid dynamics problems. Detailed documentation is provided so that users can operate the program to solve many slope stability problems. Comparing the SEM and SPH, which will be discussed later, the most distinct difference is the ability of SPH to handle very large deformations, which is not possible with the SEM.

The basis functions in the finite element method apply only locally, involving incorporating only the adjacent elements containing the node x_i belonging to the function ϕ_i. On the other hand, the basis functions in the spectral method are "global", which means that the support of all basis functions is for the whole solution domain. Instead of a low degree of polynomials in FEM, the spectral method adopts a higher degree of polynomials. In SEM, the basis functions are infinitely differentiable, which is different from that in SEM. Furthermore, for fluid flow by SEM, the polynomial approximation for the velocity is usually two degrees higher than that for the pressure (to avoid spurious mode).

Based on the study by Pech and others, it was found that if SEM can be applied for a problem, it will generally require fewer nodes and higher accuracy as compared with the classical FEM. So far, SEM is still far from mature, and a general-purpose program which can consider a variety of problems is not still available. For the application of SEM in slope stability analysis, it is possible that SEM offers some enhancement in the computations, but one of the critical problems with SRM is the difficulty in determining the ultimate state and the critical failure surface, which are demonstrated in Chapter 1. The author believes that these limitations will apply simultaneously to SEM and FEM, and SEM is not better than classical FEM in this respect.

7.14 GOVERNMENT REQUIREMENTS

An acceptable factor of safety should be based on the consideration of the recurrent period of heavy rainfall, the consequences of slope failures, knowledge about the long-term behaviour of the geological materials and the accuracy of the design model. The requirements adopted in Hong Kong are given in Tables 7.37 and 7.38, and these values have been found to be satisfactory in Hong Kong. For the slopes at the Three Gorges Project in China, the slopes are very high and steep, and there is a lack of previous experience as well

Table 7.37 Recommended factors of safety *F* (GEO, Hong Kong, 1984)

		Risk of human losses		
		Negligible	*Average*	*High*
Risk of economic losses	Negligible	1.1	1.2	1.4
	Average	1.2	1.3	1.4
	High	1.4	1.4	1.5

Table 7.38 Recommended factors of safety for rehabilitation of failed slopes (GEO, Hong Kong, 1984)

Risk of human losses		
Negligible	*Average*	*High*
$F > 1.1$	$F > 1.2$	$F > 1.3$
F for recurrency period of 10 years		

long-term behaviour knowledge of the geological materials, a higher factor of safety is hence adopted for the design. In this respect, an acceptable factor of safety shall fulfil the basic requirements from soil mechanics principles as well as the long-term performance of the slope. The author would like to emphasize that the definition of the factor of safety, method of analysis and the accuracy in the optimization search are never clearly specified in either the Hong Kong code, Euro code or any other engineering codes. The precise values of the factor of safety are very critical in Hong Kong, as many slopes are formed with an angle as steep as possible to provide more space for development.

Geotechnical engineers should consider the current slope conditions as well as the future changes, such as the possibility of cuts at the slope toe, deforestation, surcharges and excessive infiltration. For very important slopes, there may be a need to monitor the pore pressure and suction using a tensiometer and piezometer, and the displacement can be monitored by inclinometers, GPS or microwave reflection. Use of strain gauges or optical fibres in soil nails to monitor the strain and nail loads may also be considered if necessary. For large-scale projects, use of the classical monitoring methods is expensive and time-consuming, and the use of GPS has become increasingly popular in recent years. Cheng is currently working on the IoT monitoring of slopes and retaining walls, and such measures can help to provide continuous monitoring of the safety of these works without significant human intervention.

7.15 SLOPE PROTECTION AND STABILIZATION

To protect and stabilize a slope, there are several major groups of methods, which include surface protection, surface drainage, subsurface drainage and stabilization by inclusions. There are various methods for slope stabilization which are discussed below. More detailed discussions have been provided by Ortigao and Sayao (2004).

7.15.1 Surface protection

Chunam plastering and shotcrete are commonly used for slope surface protection. Chunam is a surface protection formed by using a clay and cement mixed plaster. The typical thickness of the plaster is around 40–50 mm for permanent works. Since the application of chunam is labour intensive, it has been largely replaced by the use of shotcrete which can be sprayed. To avoid cracking of the shotcrete, wire mesh may be placed on the surface of the slope. In addition, the use of masonry and stone pitching are also common, as they look better than shotcrete and chunam. There are also cases where a concrete surface is formed on the surface of slopes.

The surface of a slope can be improved by some gentler means, such as:

1. Hydroseeding – this is the application of grass seed mixed with fertilizer and nutrients in an aqueous solution by a spraying method. Different types of grasses can be specified for aesthetic purposes. The root of the grass will also be an organic reinforcing fibre and hold the surface soil.
2. Turfing – This is the direct application of grass with developed roots onto the slope surface. The relatively mature grass will grow easier and strengthen the surface.
3. Planting of trees and shrubs – this will provide better visual results and a strengthening effect to the slope from the deep roots.

The Hong Kong Government is committed to maintaining the highest standards of slope safety, while at the same time ensuring that the appearance of the slopes should not be visually intrusive. In this regard, the current policy for man-made slopes is to preserve existing vegetation as far as is possible, and to use vegetation as a slope surface cover on both new and upgraded slopes. As a result, man-made slopes are being made to blend with their surroundings, minimizing adverse visual impacts. Similarly, stabilization works over large areas of natural hillsides are avoided, as far as possible, on the grounds that these would generally be both impractical, largely because of access difficulties, and environmentally damaging, in terms of both the disturbance to the surrounding area and the visual eyesore produced. Consequently, risk is usually mitigated by adjusting development layouts, incorporating buffer zones, and erecting downslope measures such as debris-resisting barriers.

In situations where the long-term repair of natural terrain landslide scars is necessary, soil bioengineering measures offer a low-cost, potentially effective, sustainable, largely maintenance-free, and environmentally acceptable alternative to conventional slope works. Importantly, the use of soil bioengineering measures on natural hillsides in Hong Kong is entirely compatible with current environmental concerns, including greening initiatives, ecological compatibility, and sustainability. Practically, vegetation cover is less heat and light reflective than a hard cover, the plants extract water from the soil, and their roots bind the soil particles.

Until 2003, experience with the remedial use of vegetation in eroded areas of natural terrain in Hong Kong was largely restricted to direct planting, particularly in areas of severe soil erosion. Most notably, planting was carried out in country parks (by AFCD), and close to new towns (DevOs, CEDD). Soil bioengineering techniques had not previously been applied in Hong Kong, although they are used successfully elsewhere.

In April 2003, the GEO of CEDD began a pilot project to assess the suitability of soil bioengineering measures for minimizing the deterioration of natural slopes in areas of natural terrain affected by recent, shallow landsliding and related gully erosion. The project had two main objectives, to identify measures that are capable of reinforcing the soil mass, and thereby increase the resistance of the slope to further debris movement, and to identify means of accelerating the natural revegetation of deteriorating slopes, which in turn would enhance the local ecosystems.

A range of potentially suitable measures was selected, and field trials carried out on several natural terrain slopes across Hong Kong. The findings and recommendations are presented in GEO Report No. 227 (July 2008). Currently, applied field trials are being carried out on selected man-made slopes to investigate the suitability of use of a range of bioengineering techniques in different urban and rural settings.

Soil bioengineering combines biological elements with engineering design principles. For example, engineering requirements may dictate highly compacted soil for fill slopes, while plants prefer relatively loose soil. Using a sheep's foot roller for compaction is a solution that would integrate biological and engineering requirements because it compacts the soil, but also allows plant establishment in the resulting depressions in the slope. Differing needs can generally be integrated through creative approaches and occasional compromises in planning and design.

7.15.2 Surface drainage

Surface drainage including U-channel, step channel, sand-trap and others are commonly used in combination for surface protection. The surface drainage system will help to reduce the amount of infiltration. The application of no-fine concrete facing and toe drain can also help to divert rain water quickly, and these methods are also commonly used in Hong Kong. Typical surface drainage is shown in Figure 7.55. It must also be noted that the channels and catch-pit can be easily blocked by leaves, rubbish and soil and the blockage can be particularly serious after a typhoon.

7.15.3 Subsurface drainage

There are many forms of subsurface drainage, which include raking drains, weep holes, wells and drainage tunnels. For raking drains, as shown in Figure 7.56b, geotextile is usually used to wrap over the drain to prevent clogging. Weep holes, as shown in Figure 7.56a, are usually short (around 1 m) and are commonly used to divert ground water that is close to the surface. The photograph in Figure 7.56a was taken 12 hours after heavy rainfall, yet there is still significant underground water coming out from the weep hole through the fractures in the rock slopes which has been protected with shotcrete. A raking drain is longer than a weep hole, with a length of more than 10 m in many slopes in Hong Kong. It can help to divert the ground water and lower the ground water table.

(a) U channel and catch-pit on berm (b) steep channel on slope

Figure 7.55 Typical details for surface drainage on slope.

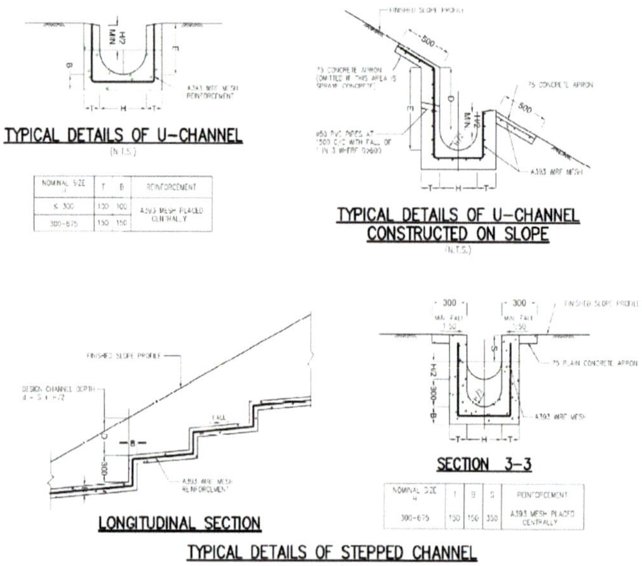

c) Typical details of U and step channels in Hong Kong

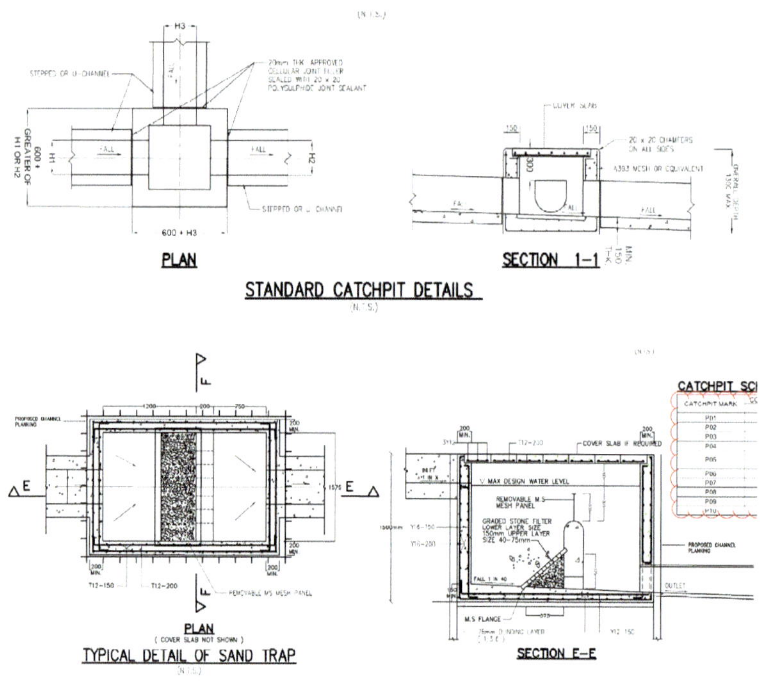

d) Typical details for catchpit

Figure 7.55 (Continued)

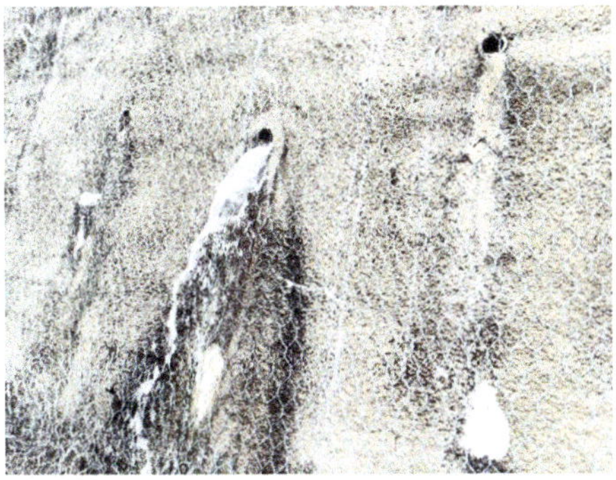

(a) Weep hole

(b) raking drain

Figure 7.56 Weep hole with geotextile to prevent clogging.

7.15.4 Inclusions and stabilization

The most commonly used inclusions and stabilizations include dentition, rock dowel and bolts, soil nail, anchor, buttress, retaining wall, pile and sprayed concrete. Dentition is the filling up of cavities in rocks which helps to strength the boulders. A rock dowel is commonly used in conjunction with dentition to help stabilize boulders that may have some potential for failure, but where removal of such boulders may create other problems so that they are not removed. Since a dowel which is not grouted functions simply by shear, the capacity to stabilize a boulder is passive and limited. The use of a rock bolt with the application of some bolting force will be more active towards the stabilization of larger boulders or unstable zones. At the head of the rock bolt, as shown in Figure 7.57, there is a base plate applying force to the rock surface, and the bolt head is commonly protected by a concrete block.

A soil nail is commonly used to stabilize slopes in Hong Kong and many other countries, and the procedures for installation are illustrated in Figure 7.58. A reinforcement steel bar is the most common material for soil nails, as it is cheap and readily available. Although a soil nail is protected by cement grout, there may be potential long-term corrosion, hence some soil nails are epoxy grouted in corrosive environments. Currently, glass fibre-reinforced polymer (GFRP) and carbon fibre-reinforced polymer (CFRP) soil nails are also used to some extent, but the costs of these materials prohibit their widespread use (Cheng et al. 2015). After installation of soil nails, three or more soil nails are commonly selected for pullout tests (Figure 7.59) in accordance with the testing procedures as illustrated in Figure 7.60 for the quality control of the soil nails. More details on the classical and FRP soil nails can be found in Cheng (2015). Some typical pullout tests results for steel, GFRP and CFRP nails by Cheng are given in Figures 7.61 and 7.62.

It must be noted that the size of the nail head cannot be too small, otherwise it will not restrain the surface soil from failure, which has happened in Hong Kong. The nail head is hence designed as a footing/retaining structure in Hong Kong, which is considered to be reasonable for the high-angled slopes in Hong Kong. In fact, all of the slope stability programs assume that effective nail length provides friction to restrain the soil from failure.

Figure 7.57 Typical concrete block at head of rock bolt/soil nail.

If the nail head is too small, it is possible that the friction in front of the effective soil nail zone is mobilized so that the soil mass will fail by debonding from the front of the soil nail (termed "front failure" by Cheng). Such a failure mode is possible and has happened in Hong Kong, but is not incorporated into any commercial slope stability programs to the best of the author's knowledge, except for SLOPE 2000 by Cheng. In the soil nail option menu, this failure mode can be chosen easily by a simple click of a check box.

In the case where the extent of fracture in rock is large, more active stabilization is enabled with the use of a rock anchor. Unlike soil nails which are commonly fully grouted (although there are many soil nails in Hong Kong where the front 2 m is not grouted to avoid transferring stresses to the slope surface), a rock anchor is grouted only for the effective zone, and the anchor force is typically large. The problems with rock anchor installations include: (1) long-term monitoring of the anchor load is required; and (2) drilling to good competent rock to form the grouting zone may be difficult in some highly fractured zones.

The use of a buttress and retaining wall to stabilize slopes is common in Hong Kong. There are many rock slopes where boulders are protected by the construction of an adjoining buttress. For some soil slopes, it may not be possible to cut and carry out any stabilization work, in such a case, the use of a buttress will be a very good solution. A typical buttress is shown in Figure 7.63; a buttress is chosen because there is no need to reduce the space for construction of the classical retaining wall.

7.15.5 Use of grillage

When the soil is loose, the use of grillage as shown in Figures 7.64 and 7.65 is common in many countries. In general, the use of soil nails or even anchors is required to provide the support to the grillage system. The grillage facing functions somewhat like a retaining

Figure 7.58 Typical procedures for the installation of soil nails.

system to ensure that the loose fill is effectively retained. Vertical nails are usually provided at the slope toe to absorb any unbalanced forces arising from the possible construction deviation in the alignment of the soil nails. A minimum grillage embedment of 0.5 m below the slope surface is provided in Hong Kong to ensure good contact of the grillage with the fill. To prevent local instability involving the outflow of the saturated loose fill through the grillage openings, a minimum coverage of 50% of the slope surface is commonly adopted in Hong Kong.

(a) Set up of hydraulic jack, load cell

(b) GFRP nail, Tube a' Manchette grouted

Figure 7.59 Pullout test of GFRP soil nails by Cheng.

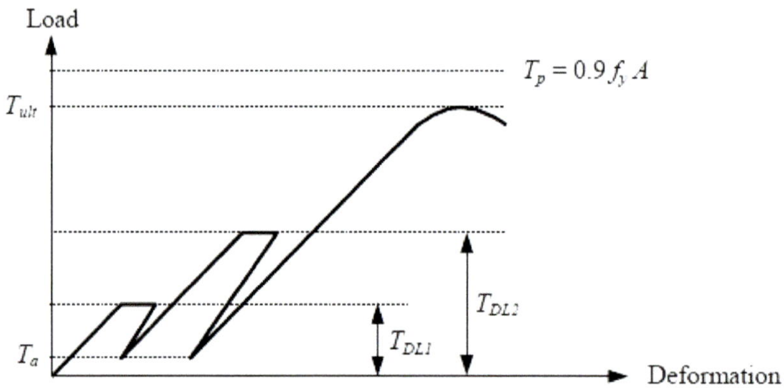

Figure 7.60 Typical pull out test procedures for soil nail in Hong Kong.

7.15.6 Flexible and rigid barrier

To safeguard against the debris flow problem, the uses of both rigid and flexibility barriers are adopted in Hong Kong. In general, if the potentially dangerous zone is close to premises, the use of a rigid barrier is more common. The flexibility barrier has the advantages of being fast in construction, economical and easy to protect a large area, and the applications of both types of barrier are clearly illustrated in Figures 7.66a and 7.66b.

A typical flexible barrier system as shown in Figure 7.67 consists of:

(1) Energy dissipaters;
(2) Bracing cables;
(3) Interception panel;
(4) Post
(5) Net – the mesh size should correspond to D_{90} of the incoming particles;

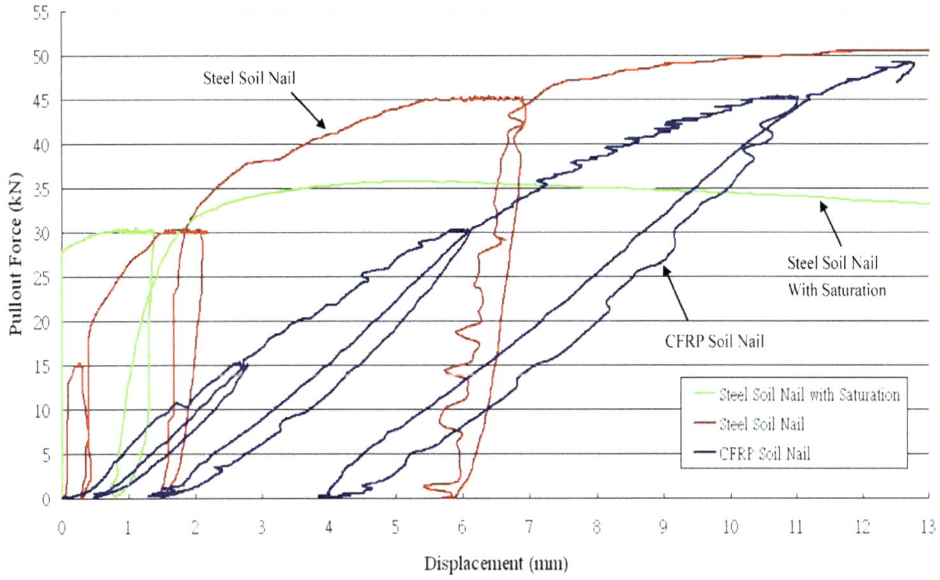

Figure 7.61 Pull out test for steel, GFRP and CFRP soil nails by Cheng et al. (2015).

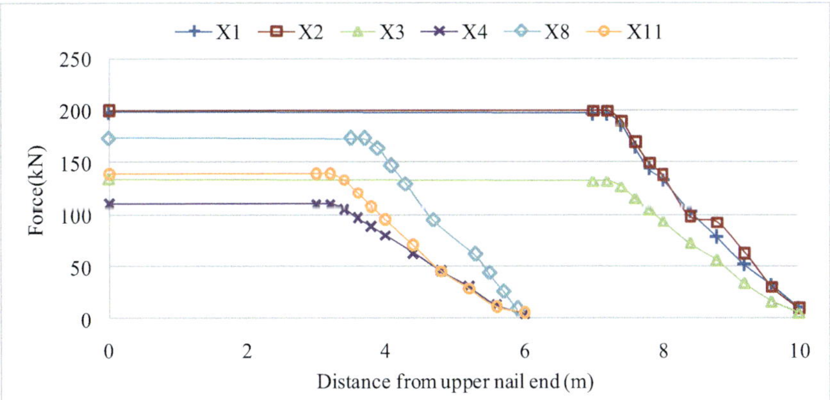

Figure 7.62 Axial force in some soil nails.

(6) Brake element – large plastic deformation capacity to reduce peak load;
(7) Lateral anchors – cable anchors with a flexible anchor head to suit the layout of the barrier.

The protection net is spanned with supporting cables fixed along the area of the embankment with edge cables and the cables are anchored in the embankments. If the span exceeds 15–20 m, intermediate supports will be required for the supporting cables. Brake elements

Figure 7.63 Typical buttress wall.

as shown in Figure 7.67c in the supporting cables will stretch when loaded so that the cables will align with respect to the load action. For overflow, the top supporting cable must be provided with abrasion protection. The flexible barrier requires simple and quick in installation. The structure also looks transparent and is well integrated into the landscape. In general, the corrosion protection of the barrier components should maintain a minimum event-free service life of between 30–50 years.

For rigid barriers, many such barriers are constructed in accordance with Figure 7.68. The main wall is usually supported on minipiles, while soil nails (which may be keyed into rock) are provided at the bottom of the base slab.

7.16 DESIGN OF MONITORING SYSTEM FOR A RIGID BARRIER

In this section, a rigid barrier monitoring project as shown in Figure 7.70 by Cheng is used for illustration. The site is located at the hillside about 100 m to the northeast of Hang Tsuen, new Tung Chung, Hong Kong special administrative region. The rigid retaining wall facade is 19.5 m long with an exposed height of 4.5 m, and the two sides of the rigid barrier are approximately 6 m long and 7 m high. The rigid barrier is approximately trapezoid in shape. The monitored retaining wall is a reinforced concrete wall with minipiles and soil nails. Since there is no detailed relevant guidance for the monitoring schemes in Hong Kong, some of the specifications from China are adopted in the design proposal:

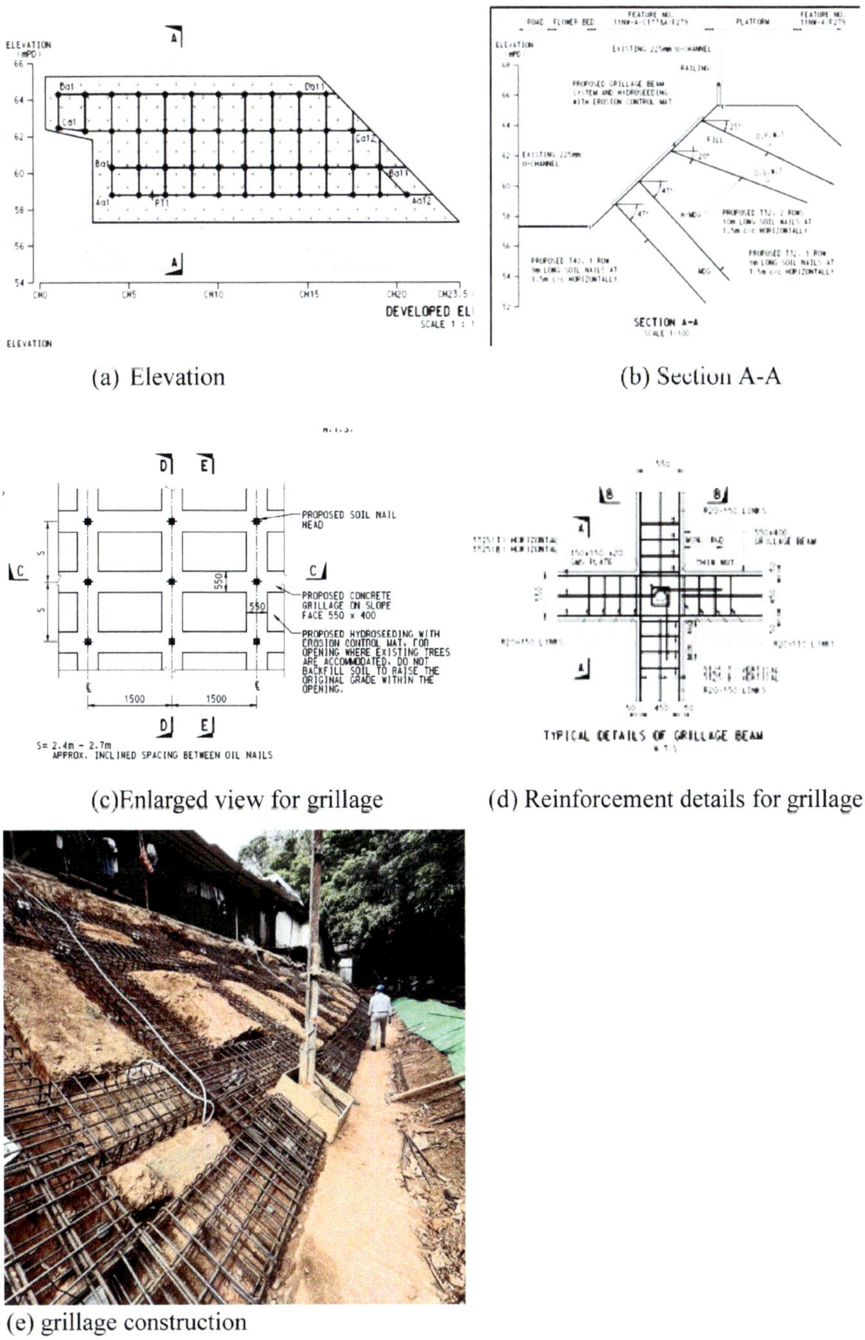

(a) Elevation

(b) Section A-A

(c)Enlarged view for grillage

(d) Reinforcement details for grillage

(e) grillage construction

Figure 7.64 A typical grillage system for loose fill slope in Hong Kong.

Figure 7.65 A grillage system with vegetation.

1. Regulations on prevention and control of geological disasters
2. Code for risk assessment of geological hazards (DZ/T 0286-2015)
3. Code for monitoring landslides, landslides and debris flows (DZ/T 0221-2006)
4. Code for geotechnical engineering investigation (gb50021-2001) (2009 edition)
5. Technical specification for construction slope engineering (GB 50330-2013)
6. Engineering survey specification (GB 50026-2016)
7. Technical specification for monitoring of construction foundation pit engineering (GB 50497-2009)
8. Building deformation measurement code (JGJ 8-2016)
9. Design code for lightning protection of buildings (GB 50057-2010)
10. Verification regulations of ultrasonic rangefinder (JJG 928-1998)
11. Regulations for precipitation observation (SL 21-2015)

The monitoring system of this project includes two parts. The first part is the acquisition end and the second part is the cloud data processing system. The design framework of this system is shown in Figure 7.69.

(1) Acquisition end system

The acquisition end system is supported by debris flow pressure sensors, distance sensor, image acquisition sensor integration and data transmission module, the debris flow

(a) A typical rigid barrier with houses in-front

(b) A typical flexible barrier

Figure 7.66 Two major types of debris flow barrier in Hong Kong.

pressure sensor or any distance sensor collected beyond the limits of the data, the boot image sensor to collect the image and the collected data through the data transmission module and image real-time transmission to the cloud data processing centre.

(2) Cloud data processing system

After receiving the data transmitted by the collecting end, the cloud data processing system conducts real-time processing on the measured data and images and prepares the monitoring report, which is sent to the specified user.

In this project, IoT pressure sensors, distance sensors and a camera are installed at the rigid barrier, as shown in Figure 7.70, to monitor the debris flow. For the system, three **detection modes** are implemented:

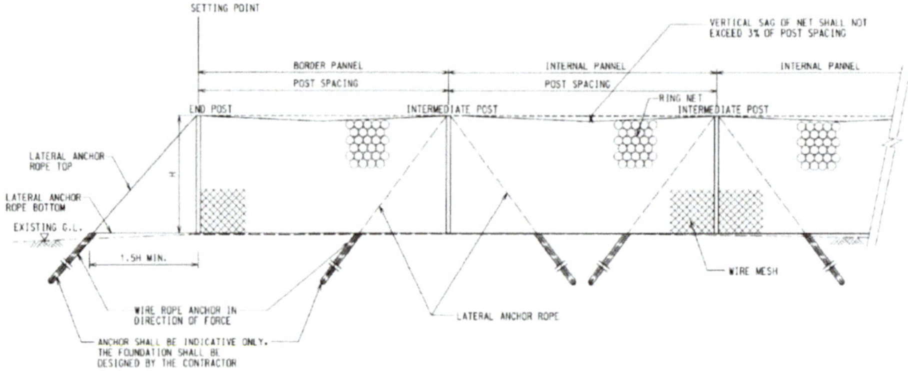

(a) Typical elevation of flexible barrier

(b) Intermediate post and rope net

(c) Brake ring (when activated, dissipate energy from the rope net)

Figure 7.67 Some components for a flexible barrier.

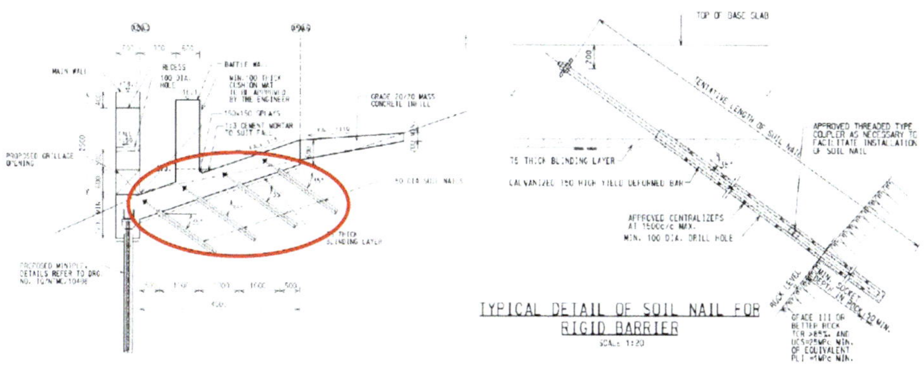

(a) Section through a typical rigid barrier

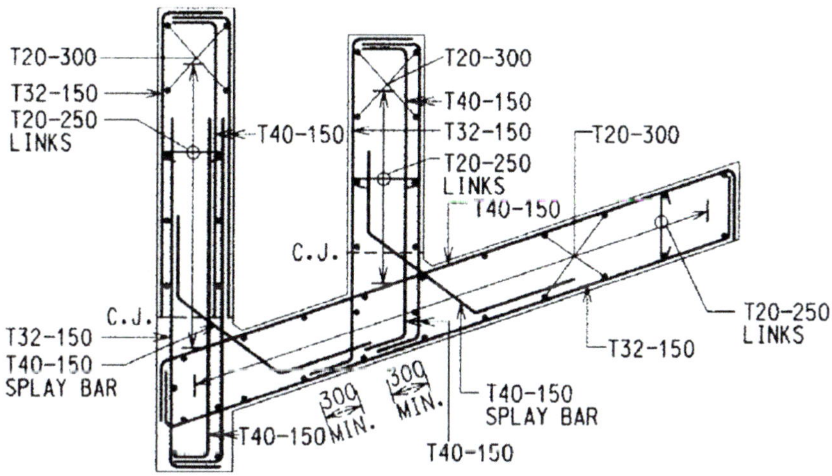

(b) Typical reinforcement details for rigid barrier

Figure 7.68 Some details for a typical rigid barrier.

(1) Surveillance Mode

The surveillance mode relates to those periods when the debris impact sensors have not been triggered for unscheduled readings by a debris impact incident. Under such conditions, the system shall collect readings (e.g. debris thickness, stress level, images) as well as data on basic sensor conditions (i.e. power level, temperature, etc.) every 4 hours. Sensors and industrial cameras are able to detect and capture impact, debris thickness and site photos at any time within the 4-hour data transmission intervals.

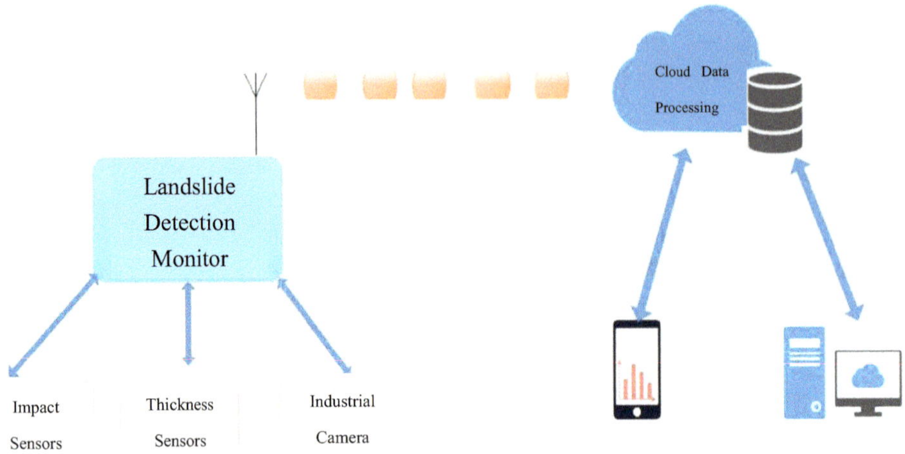

Figure 7.69 Monitoring design frame diagram.

(a) Instrumentation (b) Locations of sensors at rigid barrier

Figure 7.70 Pressure sensors, ultrasonic distance sensors and camera IoT monitoring of a rigid barrier by the author at Tung Chung.

(2) Detection mode

The detection mode relates to those periods when the debris impact sensors have been triggered for unscheduled readings by a debris impact incident. Under such conditions, the system collects time-stamped readings of stress level and debris thickness, and stress level and debris thickness at the time of impact, as well as data on basic sensor conditions (i.e. power level, temperature, etc.). Additional readings are then collected at a maximum of 2-minute intervals for a period of 1 hour after the detection mode has been activated, after which the system reverts to the surveillance mode for continued monitoring. This mode is initiated whenever the sensor detects a debris impact event. The activation of the detection mode triggers the collection of photographic images by the impact validation camera(s).

(3) Instruction Mode

This mode includes the collection of one sensor reading as well as data on basic sensor conditions (i.e. power level, temperature, etc.) when remotely activated through the off-site database management system.

7.16.1 Hardware design scheme

(1) Debris flow impact sensor: select at least three stress and pressure sensors, place them according to measurement requirements, and connect the measured data with wire. There is a sub-transmission terminal for each set of pressure transducers, and the battery is connected through a set of wires to the sub-terminal. The sub-terminal is then connected to the sensors through another set of wires which allow both electric supply and data transmission.
(2) Distance measurement sensor: at least three waterproof ultrasonic distance measurement sensor are, required. Measurement data are transmitted through wired output. There is a sub-transmission terminal for each set of ultrasonic ranging sensors, and the battery is connected through a set of wires to the sub-terminal. The sub-terminal is then connected to the sensors through another set of wires which allow both electric supply and data transmission.
(3) Image sensor: industrial type camera, output image data. An industrial camera is able to take images every 4 hours even when there is no collision to trigger the detection mode. Also, it allows remote control to change the monitoring frequency. The industrial camera is connected to the data acquisition and transmission terminal gateway to capture and upload the site photos to the on-line system.
(4) Data acquisition and transmission terminal gateway: integrated into the industrial camera. This collects the data of the above sensors through sub-transmission terminals for each set of sensors and the industrial camera and uploads them to the cloud system through 4G wireless transmission. This transmission terminal has the capacity to store and process data through a wireless network (4G) for transmission. This product has the function of local data storage in an SD card which is built inside the terminal, and sends the collected data to the cloud according to

the requirements. If the data fail to be sent due to data interruption, the function of the continued transmission can be realized. A 16GB SD card is inserted into the terminal.

(5) Power supply mode: the solar power supply mode is adopted. The basic configuration is: solar panel (130 W/18 V), colloidal energy storage battery (150 Ah/12 V), solar controller (15 A/12 V). This can guarantee the normal operation of sensors and the gateway system for at least 3 years. In the absence of a power supply, the colloidal energy storage battery can maintain normal operation of the whole system for at least 1 month when the power is fully charged.
(6) Waterproof: each piece of equipment is made waterproof, with waterproof function.
(7) The equipment management system monitors the running status of the equipment at all times. When the equipment fails, it promptly sends the equipment failure information to relevant personnel and reminds them to repair and replace the equipment. Relevant personnel regularly inspect the monitoring site and carry out performance tests on the monitoring equipment to ensure the normal operation of the system.

The details and specifications of the individual sensors are covered by Cheng (2019) and will not be repeated here. Besides the hardware sensors, the system is also an important component for IoT monitoring. Three necessary environments (network environment, software environment and hardware environment) are required to support the stable operation of the rigid barrier IoT real-time monitoring and warning system. A team of IT engineers is required to support the development and maintenance of the IoT monitoring system. A typical automatic output of the pressure sensor readings from the system is given in Figure 7.71.

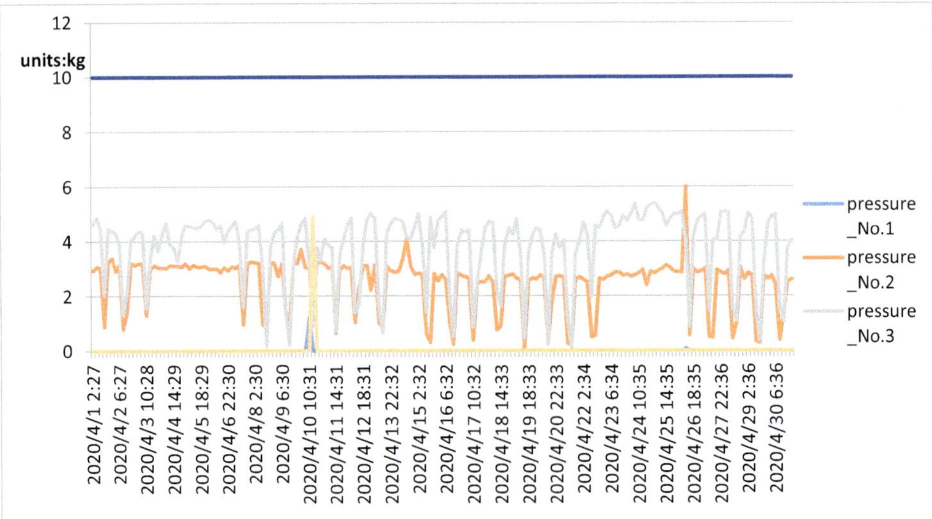

Figure 7.71 Pressure sensor results sent to cloud through 4G and processed in the server.

Appendix: Unification of bearing capacity, lateral earth pressure and slope stability problems

Lateral earth pressure, ultimate bearing capacity and slope stability issues are three important and classical problems which are well considered by the use of the limit equilibrium method, limit analysis method and method of characteristics in the past. It is interesting to note that these three topics are usually considered separately in most books or research studies, and different methods of analyses have been proposed for individual problems, even though they are governed by the same requirements for the ultimate conditions. Since the governing equations and boundary conditions for these problems are actually the same, the author believes that each problem can be viewed as the inverse of the other problems which will be demonstrated in this section.

For bearing capacity and lateral earth pressure problems, use of the slip line method is more common, as the geometry under consideration is usually more regular in nature. For a slope stability problem where the geometry is usually irregular with complicated soil reinforcements and external loads, the use of the slip line method is not practically possible. Engineers usually adopt the limit equilibrium method (an approximate slip line form) with different assumptions on the internal force distribution (*f(x)*) for the solution of the problems. Morgenstern (1992) commented that for most practical problems, the uses of different assumptions on the internal forces are not important. Cheng et al. (2010) and Cheng et al. (2011b) have developed the extremum principle and demonstrated the equivalence of the maximum extremum from the limit equilibrium method and the classical plasticity solution by a simple footing on clay. Cheng et al. (2010) treat *f(x)* as a variable to be determined, and complete equilibrium is enforced during the search for the maximum extremum of a system. Cheng et al. (2010) also pointed out that as long as an *f(x)* is prescribed, the limit equilibrium solution will be a lower bound to the ultimate limit state, which is the lower bound theorem. Under the ultimate condition where the strength of a system is fully mobilized, *f(x)* is determinate by this requirement, which is a boundary condition which has not been used in the past. Cheng et al. (2010) pointed out that every kinematically acceptable failure surface should have a factor of safety. Failure to converge in the classical limit equilibrium analysis is caused by the use of an inappropriate *f(x)* in the analysis.

In this section, the author first demonstrates the equivalence between the classical lateral earth pressure and bearing capacity problem by the slip line method. The equivalence between the lateral earth pressure problem and slope stability problem will then be illustrated by the use of the extremum principle. Based on these results, it can be concluded that the three classical problems are equivalent in the basic principles, and each problems can be viewed as the inverse of the other problems.

Bearing capacity for a shallow foundation has been studied by many researchers using different methods. Except for the term N_γ (weight term), N_c and N_q terms for the cohesive strength and surcharge are the same among different researchers (except Terzaghi 1943) by using different methods of analysis. Sokolovskii (1960), Booker and Zheng (2000) and Cheng and Au (2005) have solved the bearing capacity problem under various conditions by the method of characteristics. Martin (2005) pointed out that the method of characteristics can be used to establish the exact plastic collapse load for any combination of the parameters c, φ, γ, B and q for the exact bearing capacity calculations, and a powerful program ABC has been produced for such analysis. For lateral earth pressure,

Chen (1975), Kerisel and Absi (1990), Cheng (2003), Subra and Choudhury (2006), Liu et al. (2009a, 2009b), Peng and Chen (2013), Liu (2014), Vo and Russell (2014, 2016) and many others have also obtained three active/passive earth pressure coefficients for a plane strain problem based on the limit analysis and the method of characteristic. Cheng et al. (2007d) also obtained the three lateral earth pressure coefficients for an axi-symmetric problem based on the method of characteristic. In general, the basic numerical method is adopted by all authors with various modifications or adaptations applied to individual cases.

A1: Basic slip line theory

The slip line method (or method of characteristic) considers the yield and equilibrium of a soil mass controlled by Mohr-Coulomb's criterion, and it is a typical lower bound method. A typical slip line system is shown in Figure A1, and the equations are governed by the α and β characteristic equations given by Sokolovskii (1965) in Eqs. (2.37) and (2.38).

If the self-weight of soil is neglected (i.e. $\gamma = 0$), Eqs. (2.37) and (2.38) can be simplified as follows:

$$-\sin 2\mu \frac{\partial p}{\partial S_\alpha} + 2R\frac{\partial \theta}{\partial S_\alpha} = 0 \tag{A1}$$

$$\sin 2\mu \frac{\partial p}{\partial S_\beta} + 2R\frac{\partial \theta}{\partial S_\beta} = 0 \tag{A2}$$

Integrating the above equations along the α and β lines gives:

$$\ln\bar{p} - 2\theta\cot 2\mu = C_\alpha \Rightarrow \ln\bar{p} - 2\theta\tan\phi = C_\alpha \text{ (along } \alpha \text{ line)} \tag{A3}$$

$$\ln\bar{p} + 2\theta\cot 2\mu = C_\beta \Rightarrow \ln\bar{p} + 2\theta\tan\phi = C_\beta \text{ (along } \beta \text{ line)} \tag{A4}$$

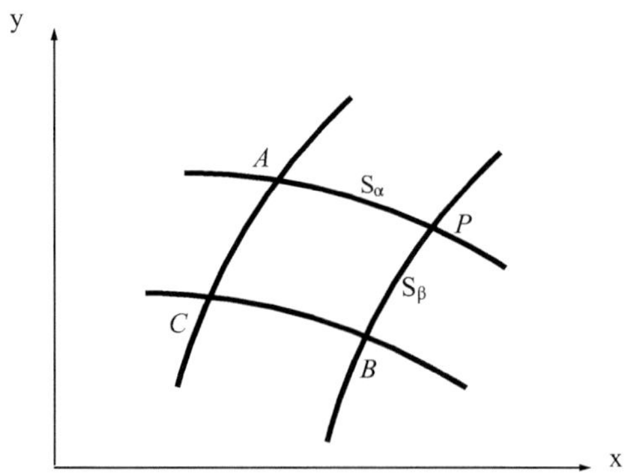

Figure A1 α and β lines in slip line solution.

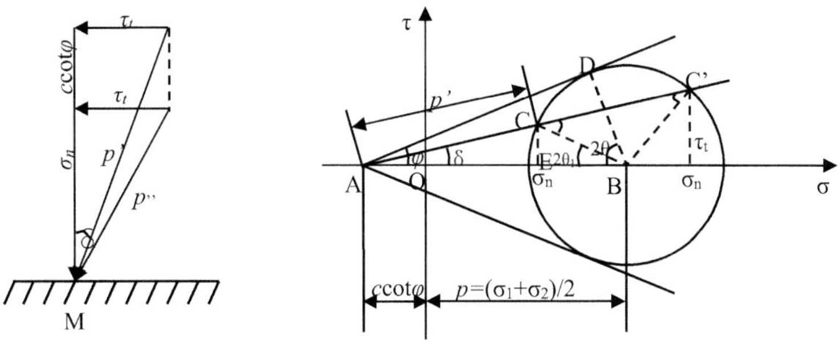

(a) stresses at boundary (b) Mohr circle for stress at boundary

Figure A2 The stress state of point M on a rough surface.

where $\bar{p} = p + c\cot\phi$, c is cohesive strength of soil and C_α and C_β are constants. The sign convention used in this book is compressive stress being positive, while the θ angles and others are taken as positive if measured in a counter-clockwise direction. For a more general condition, the friction along the interface between the soil and the footing or the retaining wall is considered. If a pressure p' acts on one point M on the soil surface and the angle between p' and the normal line to boundary is δ (i.e. friction angle, Figure A2), p' can be described as:

$$p' = \sqrt{(\sigma_n + c\cot\phi)^2 + \tau_t^2} \tag{A5}$$

where σ_n and τ_t are normal and shear stress acting on the point M, respectively.

The stress state at the point M can be represented by the Mohr circle as shown in Figure A2b. AD is the Mohr-Coulomb line and the angle between AD and the horizontal line AB is φ. AC represents stress p', and the angle between AC and the horizontal line is δ (∠CAB), and δ is taken as positive in Figure A2a in a clockwise direction, which corresponds to an active pressure condition. The line AC intersects the Mohr circle at point C, and the abscissa and ordinate of the point C are σ_n and τ_t, respectively. Line OB denotes the characteristic stress p, ∠ABD = 2θ, ∠ABC = $2\theta_1$, and angle θ_1 is the inclination angle between the major principal stress and boundary. It can easily be seen from the geometrical relationship in the Figure A2b that:

$$\tau_t = p'\sin\delta \tag{A6}$$

$$\overline{\sigma_n} = \sigma_n + c\cot\phi = p'\cos\delta \tag{A7}$$

Therefore, Eq. (A6) divided by Eq. (A7) gives,

$$\frac{\tau_t}{\sigma_n + c\cot\phi} = \frac{\sin\delta}{\cos\delta} = \tan\delta \tag{A8}$$

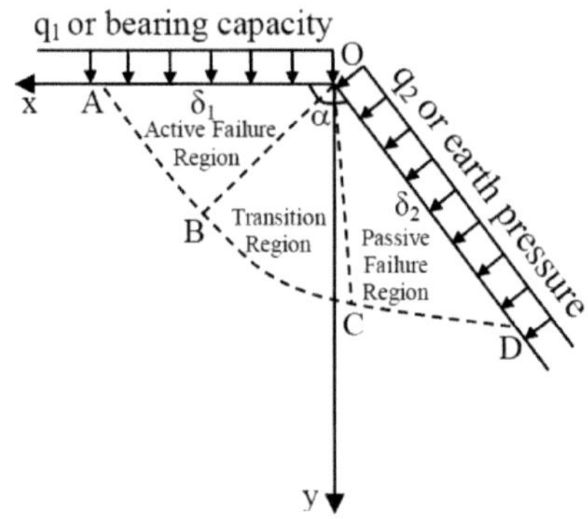

Figure A3 Unified model of bearing capacity and lateral earth pressure problem.

For convenience, δ in Eq. (A7) is a boundary condition which is defined in terms of τ_f, σ_n, c and φ instead of purely τ_f, σ_n for simplicity. This boundary condition is usually the footing or the back of a retaining wall (surface AO in Figure A3) where σ_n is constant (γ is neglected here). The true friction can be transformed to the apparent friction δ easily. According to the geometrical relationship in Figure A2b,

$$\overline{\sigma_n} = \sigma_n + c\cot\phi = \overline{AB} - \overline{BE} = \overline{AB} - \overline{BC}\cos 2\theta_1 = \overline{AB} - \overline{BD}\cos 2\theta_1 \tag{A9}$$

$$\tau_t = \overline{CE} = \overline{BC}\cos 2\theta_1 = \overline{BD}\sin 2\theta_1 \tag{A10}$$

From triangle ABD, it is known that,

$$\overline{AB} = p + c\cot\phi \tag{A11}$$

$$\overline{BD} = \overline{AB}\sin\phi = (p + c\cot\phi)\sin\phi \tag{A12}$$

Substitute the above equations, in Eqs. (A9) and (A10),

$$\overline{\sigma_n} = \sigma_n + c\cot\phi == (p + c\cot\phi) - (p + c\cot\phi)\sin\phi\cos 2\theta_1 \tag{A13}$$

$$\tau_t == (p + c\cot\phi)\sin\phi\sin 2\theta_1 \tag{A14}$$

Therefore, Eq. (A14) divided by Eq. (A13) gives:

$$\frac{\tau_t}{\sigma_n + c\cot\phi} = \frac{(p + c\cot\phi)\sin\phi\sin 2\theta_1}{(p + c\cot\phi) - (p + c\cot\phi)\sin\phi\cos 2\theta_1} = \frac{\sin\phi\sin 2\theta_1}{1 - \sin\phi\cos 2\theta_1} \tag{A15}$$

Solving Eqs. (A8) and (A15) simultaneously,

$$\frac{\sin\delta}{\cos\delta}=\frac{\sin\phi\sin 2\theta_1}{1-\sin\phi\cos 2\theta_1} \tag{A16}$$

Rearranging Eq. (A16) gives:

$$\frac{\sin\delta}{\sin\phi}=\sin\left(2\theta_1+\delta\right) \tag{A17}$$

The general solution to Eq. (A17) is

$$2\theta_1+\delta=2m\pi+\Delta \tag{A18}$$

$$\text{and } 2\theta_1+\delta=\left(2m+1\right)\pi-\Delta \tag{A19}$$

where $\Delta=\arcsin\dfrac{\sin\delta}{\sin\phi}$. Combining the above two solutions,

$$2\theta_1+\delta=2m\pi+k\Delta+\left(1-k\right)\frac{\pi}{2} \tag{A20}$$

$$\text{or } \theta_1=m\pi+\frac{1}{2}\left(k\Delta-\delta\right)+\left(1-k\right)\frac{\pi}{4} \tag{A21}$$

where $k = \pm 1$; m is an integer and generally $m = 0$ or $m = \pm 1$. According to the triangle ABC in Figure A2b,

$$p+c\cot\phi=p'\frac{\sin\Delta}{\sin\left(\Delta-k\delta\right)}=\frac{\left(\sigma_n+c\cot\phi\right)}{\cos\delta}\frac{\sin\Delta}{\sin\left(\Delta-k\delta\right)} \tag{A22}$$

Based on Eqs. (A20) and (A21), the pressure and friction angle on the soil boundary can be converted into the characteristic pressure p and θ_1. For the angle θ between the major principal stress σ_1 and the boundary, if the boundary is horizontal, $\theta = \theta_1$. If the boundary inclines to the horizon at α, $\theta = \theta_1 + \alpha$. It is noted here that the sign ($\pm$) of k is defined based on the direction of displacement of the boundary when the soil attains the limit equilibrium state. When the boundary of soil moves in the direction of p', $k = -1$; when the boundary of soil moves in the opposite direction of p', $k = 1$. In other words, if the soil is in active failure state, $k = -1$; otherwise, the soil is in the passive failure state and $k = 1$.

For a plane strain case, the ultimate bearing capacity of a shallow foundation can be determined from the "superposition" approach which has been shown to be an approximate but good assumption for normal problems by Michalowski (1997) and Cheng

(2002). For a strip footing loaded vertically in the plane of symmetry, the ultimate bearing capacity pressure q_u is given by the bearing capacity factors N_γ, N_c and N_q as:

$$q_u = q_{u\gamma} + q_{uc} + q_{uq} = \frac{1}{2}\gamma B N_r + cN_c + qN_q \tag{A23}$$

Similarly, the total active earth pressure is considered to be a combination of the effects due to the weight of the soil ($p_{a\gamma}$), the cohesive strength of soil (p_{ac}) and the surcharge loading (p_{aq}). The lateral active earth pressure coefficients $K_{a\gamma}$, K_{ac} and K_{aq} and the total active earth pressure can be expressed as

$$p_a = p_{a\gamma} + p_{ac} + p_{aq} = \gamma h K_{ar} - cK_{ac} + qK_{aq} \tag{A24}$$

N_q is derived by Cheng as

$$N_q = \frac{\cos\delta_1}{\cos\delta_2}\frac{\sin(\Delta_1+\delta_1)\sin\Delta_2}{\sin(\Delta_2-\delta_2)\sin\Delta_1}\exp\left[(2\alpha-\pi+\Delta_1+\delta_1+\Delta_2-\delta_2)\tan\phi\right] \tag{A25}$$

A similar procedure can be used to derive N_c as:

$$N_c = \left\{\frac{\cos\delta_1}{\cos\delta_2}\frac{\sin(\Delta_1+\delta_1)\sin\Delta_2}{\sin(\Delta_2-\delta_2)\sin\Delta_1}\exp\left[(2\alpha-\pi+\Delta_1+\delta_1+\Delta_2-\delta_2)\tan\phi\right]-1\right\}\cot\phi \tag{A26}$$

For the lateral earth pressure problem, K_{aq} is derived as:

$$K_{aq} = \frac{\cos\delta_2}{\cos\delta_1}\frac{\sin(\Delta_2-\delta_2)\sin\Delta_1}{\sin(\Delta_1+\delta_1)\sin\Delta_2}\exp\left[(\pi-2\alpha-\Delta_1-\delta_1-\Delta_2+\delta_2)\tan\phi\right] \tag{A27}$$

K_{ac}, is given by:

$$K_{ac} = \left\{1-\frac{\cos\delta_2}{\cos\delta_1}\frac{\sin(\Delta_2-\delta_2)\sin\Delta_1}{\sin(\Delta_1+\delta_1)\sin\Delta_2}\exp\left[(\pi-2\alpha-\Delta_1-\delta_1-\Delta_2+\delta_2)\tan\phi\right]\right\}\cot\phi \tag{A28}$$

It is clear that the relation between N_q and K_{aq} for a general rough interface is given as:

$$N_q \cdot K_{aq} = 1 \tag{A29}$$

It is also noted that the following relation among N_c, K_{aq} and K_{ac} can be developed

$$N_c = K_{ac}/K_{aq} \tag{A30}$$

Eq. (A30) can also be viewed from another point of view. The active pressure along OD induced from the uniform surcharge q_1 along AO will be negative for the cohesive strength. If q_1 is chosen such that the lateral earth pressure on OD is exactly 0, this condition will correspond to the ultimate bearing capacity induced by cohesive strength. Using this concept, Eq. (A30) is also determined from Eqs. (A26), (A27) and (A28). The bearing capacity factors N_c and N_q are hence demonstrated to be related to the active lateral earth pressure coefficients K_{ac} and K_{aq} by Eqs. (A27) and (A28). For the terms N_γ and $K_{a\gamma}$ related to the self weight of soil, analytical expression is not possible and numerical computation will be adopted for the comparisons which will be explained in a later section.

For N_γ, Sokolovskii (1965), Cheng and Au (2005) and others determined it by solving Eqs. (2.37) and (2.38) from right to left. For the ultimate pressure at the bottom of the footing on a level ground, the slip line program SLIP by Cheng and Au (2005) starts with a uniform distributed load (very small and tends to zero, similar to that by Martin [2004]) along OD (which is horizontal now) and the construction of the slip-line field begins from the right-hand side to the left-hand side. Since N_γ from SLIP is defined by the average value of the bearing stress along OA from bearing capacity determination (after Sokolovskii 1966) from the Prandtl's mechanism (Prandtl, 1920), numerically, it is the same as the passive pressure $K_{p\gamma}$ from KP, or

$$N_\gamma = K_{p\gamma} \tag{A31}$$

N_γ can be considered as the corresponding passive earth pressure coefficient, and the classical bearing capacity equation $\gamma BN_\gamma/2$ can be considered as equivalent to $K_p\gamma H^2/2$. The classical definition of bearing capacity $\gamma BN_\gamma/2$ is hence a better definition as compared with using γBN_γ, as it actually reflects the fact that bearing capacity and passive pressure problems are just image problems of each other.

The author should point out a major issue for the N_γ results as given by Martin (2004a,b) and some other researchers, for which the failure mechanism is based on an asymmetric failure mode. Although most of the available solutions based on the method of characteristic use an asymmetric failure mechanism, Graham et al. (1988) have adopted a near-symmetric failure mechanism for problems with a sloping ground with the method of characteristic. Cheng and Au (2005) pointed out that the results by Graham et al. (1988) are not good as compared with experimental results due to the use of symmetric failure mechanism. However, the results by Sokolovskii (1965) and other famous classical results are actually based on a symmetric failure mechanism. If the author adopts asymmetric failure similar to that by Martin (2005), the results from program SLIP by Cheng will be equal to those by ABC. Cheng and Li (2017) clearly demonstrate that N_γ can also be determined from the lateral earth pressure program KP by Cheng.

Based on the results above, it can be concluded that the bearing capacity problem and lateral earth pressure problem are inverse image problems which can be unified under the same formulation. It is true that it is more convenient to obtain the three bearing capacity factors in the traditional way by solving Eqs. (2.37) and (2.38) from right to left, but it is also very possible to solve the bearing capacity problem by solving the problem from left to right as a lateral earth pressure problem at the expense of slightly more effort in the computation.

To determine the N_c factor, a surcharge is applied underneath a foundation with a value equal to cN_c without any external surcharge while the self weight is maintained at zero.

For factor N_q, the self weight and cohesive strength are maintained at zero while a surcharge of 1 unit is applied outside the foundation and the foundation load is maintained at N_q. If the slip surface based on the classical Prandtl's or Hill's mechanism (Chen 1975) is used, the factors of safety as determined for different φ are exactly 1.0 if the extremum principle is used and *f(x)* is kept on varying until the maximum factor of safety is found. For the term N_r, a triangular pressure is applied underneath the foundation, and the maximum pressure is adjusted until the critical factor of safety from the extremum limit equilibrium method is 1.0. The critical slip surface is allowed to be automatically determined from the extremum principle in this analysis, and the resulting critical slip surface with a factor of safety 1.0 from the extremum principle will be very close to that by slip line analysis. The failure zone at the right of Figure A4 is a typical triangular zone underneath the footing with a curved narrow transition zone, and this slip surface is far from the Hill's or Prandtl's mechanism as adopted by Chen (1975). The slip surface from the extremum principle is close to that shown in Figure A4, and the N_r from the extremum principle limit equilibrium slope stability analysis is also very close to that by the slip line method. For N_c and N_q, if the slip surfaces are not specified but are searched from the extremum principle in the way suggested by Cheng et al. (2007a, 2010), the critical slip surfaces from the extremum limit equilibrium solution are exactly equal to that from the slip line solution as shown by Cheng et al. (2013).

It is also interesting to note that the lower bound principle has been clearly illustrated in the present analysis. For the case of factor N_c using Prandtl's mechanism, the author has analysed the intermediate results during the extremum analysis. *f(x)* is kept on changing during the analysis, with an initial value of 1.0 for all x_i during the simulated annealing optimization analysis. With *f(x)* = 1.0, the factor of safety is actually 0.925, which is far from 1.0. As *f(x)* is changed, the factor of safety converges towards 1.0 during the simulated annealing analysis, *but no factor of safety greater than 1.0 can be obtained*. The same results are also obtained for the case of N_q. The results as shown in Figure A5 are a good illustration of the lower bound principle.

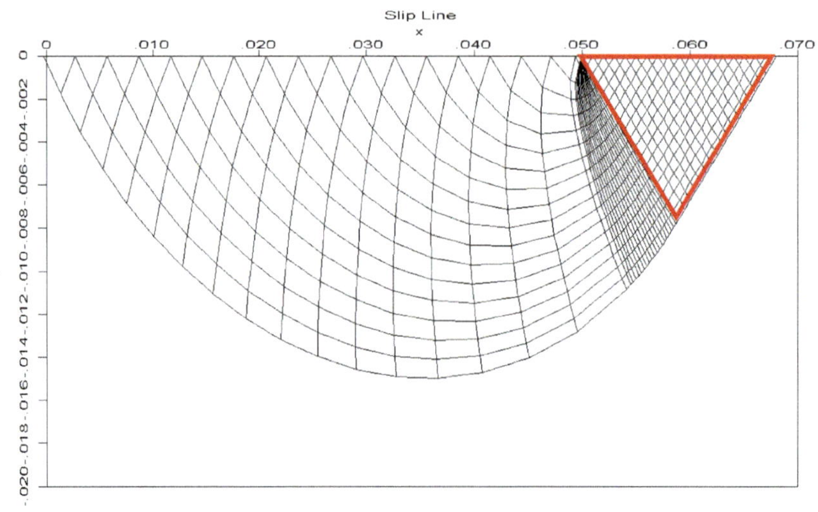

Figure A4 Slip line for the bearing capacity problem N_r when φ = 10°.

In previous sections, the author has applied the method of characteristics to study two classical and important geotechnical problems: the active lateral earth pressure problem and the ultimate bearing capacity problem. The active lateral earth pressure and bearing capacity problems are demonstrated to be equivalent, except for the ease of mathematical manipulation. Based on the slip line theory, the coefficients N_q, N_c, K_{aq} and K_{ac} are derived and the relations among them are established. Two finite difference programs, ABC and KA, are further used to confirm the validity of the results as derived in the present study. Cheng and Au (2005) have found that iterative analysis is required for a normal bearing capacity problem but not for lateral active earth pressure (Cheng 2003). When the retaining wall gradually becomes horizontal and behaves as a bearing capacity problem, it is found that iterative analysis will be necessary for the active pressure program to achieve a good result (similar to slip line analysis of a bearing capability problem). Other than the ways to solve Eqs. (3.5) and (3.6), there are no practical differences between the two classical geotechnical problems, and the two problems can actually be viewed as equivalent problems under the ultimate condition.

It has been demonstrated that classical bearing capacity and active lateral earth pressure problems are actually the same, as they are controlled by both the yield and equilibrium equations. Based on the present study, for a normal shallow foundation problem, it can be viewed as a lateral earth pressure problem and vice versa. The surcharge (ultimate bearing capacity) behind an imaginary retaining wall which generates net upward stresses outside the foundation (or equivalently the retaining wall) will be the ultimate limit state of the system. A bearing capacity problem can hence be viewed as a lateral earth pressure problem in this manner.

Cheng has demonstrated the lower bound principle in Figure A5 using the extremum analysis. Based on the minimum of the extremum from the limit equilibrium method, the author has also demonstrated that there are no practical differences between the slope stability problem and the lateral earth pressure and bearing capacity problems, provided that the extremum from the slope stability analysis is used in the comparisons. Based on the present study, it can be concluded that the three geotechnical problems are practically the same problem – the ultimate condition of the system where the maximum resistance of the system is fully mobilized. In this respect, the three geotechnical problems can be considered as equivalent under the ultimate conditions.

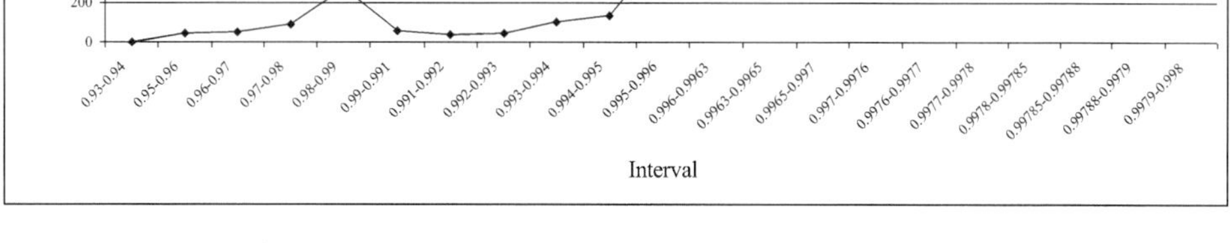

Figure A5 Distribution of acceptable factor of safety during simulated annealing analysis for N_c (same results for N_q) with $\varphi = 30°$ using $f(x)$ as the variable.

Based on these previous works, it is clear that there is not a real difference between the slope stability and bearing capacity problem in the literature, which is also illustrated clearly in the present work. Overall, it can be said that there is no physical difference between slope stability, lateral earth pressure or bearing capacity problems. Each problem can be viewed as the other problem from a slightly different point of view. Such unification is not surprising, as all three problems are controlled by the same yield and ultimate condition criterion. The author's program can also be used for all three problems, with only minor changes in the procedures of operation, but no change to the computer coding is needed. Hence, these three important geotechnical problems can be considered as unified if the ultimate condition is considered. Actually, the slope stability problem is also a determinate problem under this condition.

References

Abramson LW, Lee TS, Sharma S and Boyce GM (2002), *Slope stability and stabilization methods*, 2nd ed., John Wiley.

Al-Defae AH and Knappett JA (2015), Newmark sliding block model for pile reinforced slopes under earthquake loading, *Soil Dynamics and Earthquake Engineering*, 75, 265–278.

Alkasawneh W, Malkawi AH and Albataineh JNM (2008), A comparative study of various commercially available programs in slope stability analysis, *Computers and Geotechnics*, 35(3), 428–435.

Arai K and Tagyo K (1985), Determination of noncircular slip surfaces giving the minimum factor of safety in slope stability analysis, *Soils and Foundations*, 25, 43–51.

Baker R (1980), Determination of the critical slip surface in slope stability computations, *International Journal of Numerical and Analytical Methods in Geomechanics*, 4, 333–359.

Baker R and Garber M (1977), Variational approach to slope stability, In: *Proceedings of the ninth international conference on soil mechanics and foundations engineering*, Vol. 2, Tokyo, Japan, pp. 9–12.

Baker R and Garber M (1978), Theoretical analysis of the stability of slopes, *Geotechnique*, 28, 395–411.

Bauer S and Lackner R (2015), Gradient-based adaptive discontinuity layout optimization for the prediction of strength properties in matrix–inclusion materials, *International Journal of Solids and Structures*, 63, 82–98.

Bentley (2014), *Plaxis user manual*, Bentley.

Berezantsev VG, Khristoforov VS and Golubkov VN (1961), Load bearing capacity and deformation of piled foundation, *Proceedingsof the 5th International Conference on Soil Mechanics*, II, 11–15.

Bishop AW (1955), The use of the slip circle in the stability analysis of earth slopes, *Geotechnique*, 5(1), 7–17.

Bolton HPJ, Heymann G and Groenwold A (2003), Global search for critical failure surface in slope stability analysis, *Engineering Optimization*, 35(1), 51–65.

Celestino TB and Duncan JM (1981), Simplified search for non-circular slip surface, *Proceedings of 10th international conference on soil mechanics and foundation engineering*, Stockholm, Sweden, pp. 391–394.

Chan PS (1999), *Distinct element slope stability analysis, Master thesis*, Hong Kong Polytechnic University, Hong Kong.

Chang CS (1992), Discrete element method for slope stability analysis, *Journal of Geotechnical Engineering, ASCE*, 118(12), 1889–1905.

Chen WF (1975), *Limit analysis and soil plasticity*, Elsevier, Burlington, VT.

Chen ZY and Morgenstern NR (1983), Extensions to generalized method of slices for stability analysis, *Canadian Geotechnical Journal*, 20(1), 104–109.
Chen Z and Shao C (1983), Evaluation of minimum factor of safety in slope stability analysis, *Canadian Geotechnical Journal*, 25(4), 735–748.
Cheng YM (1998), Advancement and improvements in discontinuous deformation analysis, *Computers and Geotechnics*, 22(2), 153–163.
Cheng YM (2002), Slip line solution and limit analysis for lateral earth pressure problem. In: *The ninth conference on computing in civil and building engineering*, April 3–5, Taipei, Taiwan, pp. 311–314.
Cheng YM (2003), Locations of critical failure surface and some further studies on slope stability analysis, *Computers and Geotechnics*, 30, 255–267.
Cheng YM (2007), Global optimization analysis of slope stability by simulated annealing with dynamic bounds and Dirac function, *Engineering Optimization*, 39(1), 17–32.
Cheng YM, Albert Yeung TY, Tham LG, Au SK, So TC, Choi YK and Chen J (2009), New soil nail material-pilot study of grouted GFRP pipe nails in Korea and Hong Kong, *Journal of Civil Engineering Materials, ASCE*, 21(3), 93–102.
Cheng YM and Au SK (2005), Slip line solution of bearing capacity problems with inclined ground, *Canadian Geotechnical Journal*, 42, 1232–1241.
Cheng YM, Au SK and Yeung T (2015), Laboratory and field evaluation of several types of soil nails for different geological conditions, *Canadian Geotechnical Journal*, 53, 634–645.
Cheng YM, Choi YK, Woo SB and Chen J (2005), Pilot tests on pressure-grouted GFRP soil nails in Hong Kong, *Korean Geotechnical Society National Conference*, June 26, Seoul, Korea, 3–12.
Cheng YM, Hu YY, Wei WB (2007d), General axisymmetric active earth pressure by method of characteristic – theory and numerical formulation, *Journal of Geomechanics, ASCE*, 7(1), 1–15.
Cheng YM, Lansivaara T, Baker R and Li N (2013), The use of internal and external variables and extremum principle in limit equilibrium formulations with application to bearing capacity and slope stability problems, *Soils and Foundation*, 53(1), 130–143.
Cheng YM, Lansivaara T and Siu J (2008), Impact of convergence on slope stability analysis and design, *Computers and Geotechnics*, 35(1), 105–115.
Cheng YM, Lansivaara T and Wei WB (2007a), Two dimensional slope stability analysis by limit equilibrium and strength reduction methods, *Computers and Geotechnics*, 34, 137–150.
Cheng YM and Lau CK (2014), *Soil slope stability analysis and stabilization–new methods and insights*, 2nd ed., Spon Press.
Cheng YM, Li DZ, Li L, Sun YJ, Baker R and Yang Y (2011), Limit equilibrium method based on approximate lower bound method with variable factor of safety that can consider residual strength, *Computers and Geotechnics*, 38, 628–637.
Cheng YM and Li L (2007), Particle swarm optimization algorithm for non-circular failure surface in two dimensional lope stability analysis, *Computers and Geotechnics*, 34(2), 92–103.
Cheng YM, Li L and Chi SC (2007b), Studies on six heuristic global optimization methods in the location of critical slip surface for soil slopes, *Computers and Geotechnics*, 34, 462–484.
Cheng YM, Li L Chi SC and Wei WB (2008), Determination of critical slip surface using artificial fish swarms algorithm, *Journal of Geotechnical and Geoenvironmental Engineering, ASCE*, 134(2), 244–251.
Cheng YM, Li L, Lansivaara T, Chi SC and Sun YJ (2008), Minimization of factor of safety using different slip surface generation methods and an improved harmony search minimization algorithm, *Engineering Optimization*, 40(2), 95–115.
Cheng YM, Li L, Sun YJ and Au SK (2012), A coupled particle swarm and harmony search optimization algorithm for difficult geotechnical problems, *Structural and Multidisciplinary Optimization*, 45, 489–501.

Cheng YM and Li N (2017), Equivalence between bearing capacity, lateral earth pressure and slope stability problems by slip-line and extremum limit equilibrium methods, *International Journal of Geomechanics*, 17(12), ASCE, 04017113.

Cheng YM, Li N and Fung WH, and Li L. (2019), Laboratory and field test and distinct element analysis of Debris flow, *Natural Hazards and Earth System Sciences*, 19, 181–199.

Cheng YM, Liu HT, Wei WB and Au SK (2005), Location of critical three-dimensional non-spherical failure surface by NURBS functions and ellipsoid with applications to highway slopes, *Computers and Geotechnics*, 32(6), 387–399.

Cheng YM and Yip CJ (2007), Three-dimensional asymmetrical slope stability analysis–Extension of Bishop's, Janbu's, and Morgenstern–Price's techniques, *Journal of Geotechnical and Geoenvironmental Engineering*, 133(12), 1544–1555.

Cheng YM, Zhao ZH and Sun YJ (2010), Evaluation of interslice force function and discussion on convergence in slope stability analysis by the lower bound method, *Journal of Geotechnical and Geoenvironmental Engineering,ASCE*, 136(8), 1103–1113.

Cheng YM, Lansivaara T, Baker R and Li N (2013), The use of internal and external variables and extremum principle in limit equilibrium formulations with application to bearing capacity and slope stability problems, *Soils and Foundation*, 53(1), 130–143.

Cheng YM and Zhu LJ (2004), Unified formulation for two dimensional slope stability analysis and limitations in factor of safety determination, *Soils and Foundations*, 44(6), 121–128.

Chowdhury R (2010), *Geotechnical slope analysis*, CRC Press.

Clarke SD, Smith CC and Gilbert M (2013), Modelling discrete soil reinforcement in numerical limit analysis, *Canadian Geotechnical Journal*, 50, 705–715.

Cundall PA (1988), Formulation of a three-dimensional distinct element model – Part I: A scheme to detect and research contacts in a system composed of many polyhedral blocks, International Journal of Rock Mechanics Mining Sciences and Geomechanics, 25(3), 107–116.

Dawson EM, Roth WH and Drescher A. (1999), Slope stability analysis by strength reduction, *Geotechnique*, 49(6), 835–840.

Donald I and Chen ZY (1997), Slope stability analysis by the upper bound approach: Fundamentals and methods, *Canadian Geotechnical Journal*, 34(6), 853–862.

Drucker DC, Greenberg W and Prager W (1951), The safety factor of an elastic plastic body in plane strain, *Transactions of the ASME, Journal of Applied Mechanics*, 73, 371.

Drucker DC and Prager W (1952), Soil mechanics and plastic analysis or limit design, *Quarterly of Applied Mathematics*, 10, 157–165.

Duncan JM and Chang CM (1970), Nonlinear analysis of stress and strain in soils, *Journal of Soil Mechanics and Foundations Division, ASCE*, 96(SM5), 1629–1653.

Duncan JM and Wright SG (2005), *Soil strength and slope stability*, John Wiley, Hoboken, NJ.

Fan K, Fredlund DG and Wilson GW (1986), An interslice force function for limit equilibrium slope stability analysis, *Canadian Geotechnical Journal*, 23, 287–296.

Fellenius W (1918), Kaj-och jordrasen i Goteborg, *Teknisk Tidskrift V.U.*, 48, 17–19.

Fellenius W (1927), *Erdstatische Berechnungen mit Reibung und Kohasion* (in German), Ernst, Berlin, Germany.

Fredlund DG, Zhang ZM and Lam L (1992), Effect of axis on moment equilibrium in slope stability analysis, *Canadian Geotechnical Journal*, 29(3), 456–465.

Frohlich OK (1953), The factor of safety with respect to sliding of a mass of soil along the arc of a logarithmic spiral, In: *Proceedings of the Third International Conference on Soil Mechanics and Foundation Engineering*, Vol. 2, Zurich, Switzerland, pp. 230–233.

Gassler G and Gudehus G (1981). Soil nailing–some aspects of a new technique. *Proceedings of the Xth ICSMFE*, Vol. 3, Stockholm, pp. 665–670.

Geotechnical Control Office (1984), *Geotechnical manual for slopes*, 2nd ed., HKSAR Government.

Geotechnical Engineering Office (1976), *Report on the slope failures at Sau Mau Ping, August*, Hong Kong Government.

Geotechnical Engineering Office (1994), *Report on the Kwun Lung Lau landslide of 23 July 1994*, HKSAR Government.
Geotechnical Engineering Office (1995), *Guide to slope maintenance (geoguide 5), CEDD*, HKSAR Government.
Geotechnical Engineering Office (1996a), *Report on the Fei Tsui road landslide of 13 August 1995*, HKSAR Government.
Geotechnical Engineering Office (1996b), *Report on the Shum Wan road landslide of 13 August 1995*, HKSAR Government.
Geotechnical Engineering Office (1996c), *GEO report no. 52: investigation of some major slope failures between 1992 and 1995*, HKSAR Government.
Geotechnical Engineering Office (1998), *GEO report no. 78: report on the Ching Cheung road landslide of 3 August 1997*, HKSAR Government.
Geotechnical Engineering Office (2001), *GEO report no. 115, regional variation in extreme rainfall values*, HKSAR Government.
Geotechnical Engineering Office (2002), *GEO report no. 56, application of prescriptive measures to slopes and retaining walls*, 2nd ed., HKSAR Government.
Geotechnical Engineering Office (2003), *Technical guidance note no. 11: enhancing the reliability and resolution of engineered soil cut slope*, HKSAR Government.
Geotechnical Engineering Office (2004a), *GEO technical guidance note no. 15–guidelines for classification of consequence-to-life category for slope features*, HKSAR Government.
Geotechnical Engineering Office (2004b), *GEO technical note TB1/2004–review of cases with problems encountered during soil nail construction (1993–2003)*, HKSAR Government.
Geotechnical Engineering Office (2007), *GEO publication no. 1/2007–engineering geological practice in Hong Kong*, HKSAR Government.
Geotechnical Engineering Office (2007), *Thirty years of slope safety practice in Hong Kong*, GEO, HKSAR Government.
Geotechnical Engineering Office (2008), *Guide to soil nail design and construction (Geoguide 7)*, HKSAR Government.
Geotechnical Engineering Office and Hong Kong Institution of Engineers (2011a), *Design of soil nails for upgrading loose fill slopes*, HKSAR Government.
Geotechnical Engineering Office (2011b), *Technical guidelines on landscape treatment for slopes–GEO publication no. 1/2011*, HKSAR Government.
Gharti HN, Komatitsch D, Oye V, Martin R and Tromp J (2012), Application of an elastoplastic spectral-element method to 3D slope stability analysis, *Interantional Journal for Numerical Methods in Engineering*, 91, 1–26.
Gharti HN, Komatitsch D, Langer L, Martin R, Oye V, Tromp J, Vaaland U, Yan ZZ (2017), SPECFEM3D_GEOTECH v1.2.0, An open-source, parallel and cross-platform geotechnical engineering application.
Gilbert M, Smith C and Pritchard T (2010), Masonry arch analysis using discontinuity layout optimisation. *Proceedings of the institution of civil engineers: Engineering and computational mechanics* 163(3), 155–166.
Griffiths DV and Lane PA (1999), Slope stability analysis by finite elements, *Geotechnique*, 49(3), 387–403.
Graham J, Andrew M. and Shields DH (1988), Stress Characteristics for Shallow Footing in Cohesionless Slopes, *Canadian Geotechnical Journal*, 25(2), 238–249.
Greco VR (1996), Efficient Monte Carlo technique for locating critical slip surface, *Journal of Geotechnical Engineering, ASCE*, 122, 517–525.
Greenwood JR (1987), Effective stress stability analysis. Discussion in *Ninth european conference on soil mechanics and foundations*, Dublin, Ireland, September 1987, Vol. 3, Post Conference Proceedings, Balkema, 1989, pp. 1082–1108.
Haefeli R (1948), The stability of slopes acted upon by parallel seepage, *Proceedings of the 2nd International Conference on Soil Mechanics and Foundation Engineering*, 1, 57–62.

Hazzard JF, Maxwell SC and Young RP (1998), Micromechanical modelling of acoustic emissions, In: *Proceedings of ISRM/SPE Rock Mechanics in Petroleum Engineering*, Eurock 98, Trondheim, Norway, pp. 519–526, SPE 47320.

Huang CC, Tsai CC, and Chen YH (2002). Generalized method for three-dimensional slope stability analysis. *Journal of Geotechnical and Geoenvironmental Engineering*, 128(10), 836–848.

Huang YH (2014), *Slope stability analysis by the limit equilibrium method*, ASCE Press.

Hungr O (1987), An extension of Bishop's simplified method of slope stability analysis to three dimensions, *Geotechnique*, 37(1), 113–117.

Janbu N (1957), Earth pressure and bearing capacity by generalized procedure of slices, in *Proceedings of 4th International Conference on Soil Mechanics*, 207–212.

Janbu N (1973), Slope stability computations, In RC Hirschfield and SJ Poulos (eds), *Embankment-dam engineering*, John Wiley, pp. 47–86.

Janbu N, Bjerrum L and Kjaernsli B (1956), *Soil mechanics applied to some engineering problems*, Norwegian Geotechnical Institute, Publ. No. 16, Norway.

Jiang JC and Yamagami T (1999), Determination of the sliding direction in three dimensional slope stability analysis, in *Proceedings of 44th Symposium on Geotechnical Engineering*, 193–200.

Kerisel J. and Absi E. (1990), Active and Passive Earth Pressure tables, A. A. Balkema

King GJW (1989), Revision of effective stress method of slices, *Geotechnique*, 39(3), 497–502.

Lam L and Fredlund DG (1993), A general limit equilibrium model for three dimensional slope stability analysis, *Canadian Geotechnical Journal*, 30(6), 905–919.

Leshchinsky B (2015), Bearing capacity of footings placed adjacent to c'-ϕ' slopes, *Journal of Geotechnical and Geoenvironmental Engineering*, ASCE, 141(6), 04015022.

Leshchinsky B and Ambauen S (2015), Limit equilibrium and limit analysis: comparison of benchmark slope stability problems, *Journal of Geotechnical and Geoenvironmental Engineering, ASCE*, 141(10), 04015043.

Leshchinsky D and Huang CC (1992), Generalized three-dimensional slope stability analysis, *Journal of Geotechnical Engineering, ASCE*, 118, 1559–1576.

Liu FQ (2014), Lateral earth pressures acting on circular retaining walls, *International Journal of Geomechanics,* 14(3), 04014002.

Liu FQ, Wang JH and Zhang LL (2009a), Axi-symmetric active earth pressure obtained by the slip line method with a general tangential stress coefficient, *Computers and Geotechnics*, 36(1–2), 352–358.

Liu FQ, Wang JH and Zhang LL (2009b). Analytical solution of general axisymmetric active earth pressure. *International Journal for Numerical and Analytical Methods in Geomechanics*, 33(4), 551–565.

Lowe J and Karafiath L (1960), Stability of Earth dams upon drawdown, in *Proceedings of the first Pan-American conference on soil mechanics and foundation engineering*, Vol. 2, Mexico City, pp. 537–552.

Martin CM (2004), User Guide for ABC – Analysis of Bearing Capacity (v1.0). Department of Engineering Science, University of Oxford.

Malkawi AIH, Hassan WF, and Sarma SK (2001). Global search method for locating general slip surface using Monte Carlo techniques. *Journal of Geotechnical and Geoenvironmental Engineering*, 127, 688–698.

Martin CM (2005), Exact bearing capacity calculations using the method of characteristics, Proceedings of the 11th International Conference on IACMAG, Turin, Vol. 4, Italy, pp. 441–450.

Matsui T and San KC (1992), Finite element slope stability analysis by shear strength reduction technique, *Soils and Foundations*, 32(1), 59–70.

Michalowski RL (1995), Slope stability analysis: A kinematical approach, Geotechnique, 45(2), 283–293.

Michalowski RL (1997), An estimate of the influence of soil weight on bearing capacity using limit analysis, *Soils and Foundations*, 37(4), 57–64.

Morgenstern NR (1992), The evaluation of slope stability–a 25-year perspective, stability and performance of slopes and embankments–II, *Geotechnical Special Publication*, 31, ASCE.

Morgenstern NR and Price VE (1965), The analysis of stability of general slip surface, *Geotechnique*, 15(1), 79–93.

Morrison IM and Greenwood JR (1989), Assumptions in simplified slope stability analysis by the method of slices, *Geotechnique*, 39, 503–509.

Naylor DJ (1982), Finite elements and slope stability, In *Numerical Methods in Geomechanics, Proceedings of the NATO* Advanced Study Institute, Lisbon, Portugal, 1981, pp. 229–244.

Nguyen VU (1985), Determination of critical slope failure surfaces, *Journal of Geotechnical Engineering, ASCE*, 111(2), 238–250.

Ortigao JAR and Sayao AS (2004), *Handbook of slope stabilization*, Springer Verlag.

Peng MX and Chen J (2013), Slip-line solution to active earth pressure on retaining walls, *Geotechnique*, 63(12), 1008–1019.

Pham HTV and Frelund DD (2003), The application of dynamic programming to slope stability analysis, *Canadian Geotechnical Journal*, 40(4), 830–847.

Plaxis (2014), User manual, Bentley Systems.

Prandtl L (1920), Uber die Harte plastischer Koroer, Nachr. Kgl. Ges. Wiss. Gottingen, *Math. Phys. KL*: 74–85.

Revilla J and Castillo E (1977), The calculus of variations applied to stability of slopes, Geotechnique, 27, 1–11.

Sarma SK (1973), Stability analysis of embankments and slopes, *Geotechnique* 23(3), 423–433.

Sarma SK (1979), Stability analysis of embankments and slopes, *Journal of the Soil Mechanics and Foundations Division*, 105(GT12), 1511–1522.

Sarma SK (1987), A note on the stability of slopes, *Geotechnique*, 37(1), 107–111.

Sarma SK and Tan D (2006), Determination of critical slip surface in slope analysis, *Geotechnique*, 56(8), 539–550.

Shen CK, Bang S, Hermann LR and Romstad KL (1978). A reinforced lateral earth support system. *Proceedings of ASCE Symposium on Earth Reinforcement*, Pittsburgh, Pennsylvania, pp. 764–793.

Shen CK, Bang S, Hermann LR (1981a). Ground movement analysis of earth support system. *Journal of Geotechnical Engineering Division, Proceedings of ASCE*, 107(GT12), pp. 1609–1624.

Shen CK, Bang S, Romstad KM, Kulchin L and De Natale JS (1981b). Field measurements of an earth support system. *Journal of Geotechnical Engineering Division, Proceedings of ASCE*, 107(GT12), 1625–1642.

Shen CK, Hermann LR, Romstad KM, Bang S, Kim YS and De Natale JS (1981c). An insitu earth reinforcement lateral support system. *Department of Civil Engineering, University of California*. Report No. 81-03, Mar. 130p. NSF Report No. NSF/CEE-81059.

Smith C and Gilbert M (2007), Application of discontinuity layout optimization to plane plasticity problems, *Proceedings of the Royal Society A*, 463(2086), 2461–2484.

Smith CC and Gilbert M (2013), Identification of rotational failure mechanisms in cohesive media using discontinuity layout optimization, *Geotechnique*, 63(14), 1194–1208.

Sokolovskii VV (1960), *Statics of soil media* (translated by DH Jones and AN Scholfield), Butterworths Scientific, London.

Sokolovskii VV (1965), *Statics of Granular Media* (translated by Lusher J.K. and Daniel A.W.T.), Pergamon Press.

Spencer E (1967), A method of analysis of the stability of embankments assuming parallel inter-slice forces, *Geotechnique*, 17, 11–26.

Stark TD and Eid HT (1998), Performance of three-dimensional slope stability methods in practice, Journal of Geotechnical and Geoenvironmental Engineering, ASCE, 124(11), 1049–1060.

Stocker MF, Korber GW, Gassler G and Gudehus G (1979). Soil nailing. *Proceedings of international conference on soil reinforcement*, Paris. pp. 469–474.

Subra Rao KS, Choudhury D (2005), Seismic passive earth pressures in soils, *Journal of Geotechnical and Geoenvironmental Engineering*, 131(1), 131–135.

Terzaghi K (1943), Theoretical soil mechanics, New York: Willey, US.
Thornton C (1997), Coefficient of restitution for collinear collisions of elastic perfectly plastic spheres, *Journal of Applied Mechanics*, 64, 383–386.
Tiwari RC, Bhandary NP and Yatabe R (2014), Spectral element analysis to evaluate the stability of long and steep slopes, *Acta Geotechnica*, 9, 753–770.
Tiwari RC, Bhandary NP and Yatabe R (2015), 3-D elasto-plastic spectral element application to evaluate the stability of large-scale landslides, *Geomechanics and Geoengineering*, 10(4), 271–289.
U.S. Corps of Engineers (2003), Slope stability.
Vahedifard F, Leshchinsky B, Sehat S and Leshchinsky D (2014), Impact of cohesion on seismic design of geosynthetic-reinforced earth structures, *Journal of Geotechnical and Geoenvironmental Engineering*, ASCE, 140(6), 04014016.
Vo T and Russell AR (2014), Slip line theory applied to a retaining wall–unsaturated soil interaction problem, *Computers and Geotechnics*, 55(1), 416–428.
Vo T and Russell AR (2016), Bearing capacity of strip footings on unsaturated soils by the slip line theory, *Computers and Geotechnics*, 74(1), 122–13.
Wei WB and Cheng YM (2009a), Soil nailed slope by strength reduction and limit equilibrium methods, *Computers and Geotechnics*, 37, 602–618.
Wei WB and Cheng YM (2009b), Strength reduction analysis for slope reinforced with one row of piles, *Computers and Geotechnics*, 36, 1176–1185.
Wei WB and Cheng YM (2010), Stability analysis of slope with water flow by strength reduction method, *Soil and Foundations*, 50(1), 83–92.
Wei WB, Cheng YM and Li L (2009a), Three-dimensional slope failure by strength reduction and limit equilibrium methods, *Computers and Geotechnics*, 36, 70–80.
Whitman RV and Bailey WA (1967), Use of computers for slope stability analyses, *Journal of the Soil Mechanics and Foundation Division*, ASCE, 93(4), 475–498.
Yamagami T and Jiang JC (1996), Determination of the sliding direction in three dimensional slope stability analysis, *Proceedings of the 2nd International Conference on Soft Soil Engineering*, 1, 567–572, Nanjing, May.
Yamagami T and Jiang JC (1997), A search for the critical slip surface in three dimensional slope stability analysis, *Soils and Foundations*, 37(3), 1–16.
Yu HS, Salgado R, Sloan SW and Kim JM (1998), Limit analysis versus limit equilibrium for slope stability, Journal of Geotechnical and Geoenvironmental Engineering, ASCE, 124(1), 1–11.
Zhang J, Zhang LM and Tang WH (2011), New methods for system reliability analysis of soil slopes, *Canadian Geotechnical Journal*, 48(7), 1138–1148.
Zienkiewicz OC, Humpheson C and Lewis RW (1975), Associated and nonassociated visco-plasticity and plasticity in soil mechanics. *Geotechnique*, 25(4), 671–689.
Zheng H (2012), A three-dimensional rigorous method for stability analysis of landslides, *Engineering Geology*, 145, 30–40.
Zheng H, Zhou CB and Liu DF (2009), A robust solution procedure for the rigorous methods of slices, *Soils and Foundations*, 49(4), 537–544.
Zhu DY, Lee CF and Jiang HD (2003), Generalised framework of limit equilibrium methods for slope stability analysis, *Geotechnique*, 53(4), 377–395.
Zhu DY, Lee CF, Qian QH and Chen GR (2005), A concise algorithm for computing the factor of safety using the Morgenstern–Price method, *Canadian Geotechnical Journal*, 42(1), 272–278.
Zhu DY, Lee CF, Qian QH, Zou ZS and Sun F (2001), A new procedure for computing the factor of safety using the Morgenstern–Price's method, *Canadian Geotechnical Journal*, 38(4), 882–888.
Zolfaghari AR, Heath AC and McCombie PF (2005), Simple genetic algorithm search for critical non-circular failure surface in slope stability analysis, Computers and Geotechnics, 32(3), 139–152.

Chapter 8

Numerical methods in geotechnical engineering

8.1 GENERAL INTRODUCTION

Except for simple cases, analytical solutions are usually not available for most practical geotechnical problems, and good examples are shown in Chapters 1 and 5 where staged constructions are critical in the analysis. In many cases, the geometry, loadings, construction sequences, material behaviour, ground water conditions as well as other factors have to be understood to perform a realistic assessment of construction works.

Many engineering problems are governed by different types of differential/integral equations. A simple example is the beam on elastic ground problem, Eq. (8.1):

$$EI\frac{d^4y}{dx^4} = q = -k_s'y \tag{8.1}$$

where q is the distributed load, EI is the bending stiffness of the beam, k_s' is the Winkler's spring constant of soil multiply with the beam width and y is the vertical displacement of the beam. If the problem is uniform with constant EI and k_s', Eq. (8.1) can be solved with analytical solutions. As discussed in Chapter 4, a numerical method is unavoidable for more complicated cases. For a raft foundation modelled by a thin plate, the governing equation is given as Eq. (8.2):

$$D\nabla^4\omega = q(x,y) - k_s\omega \tag{8.2}$$

where D is $Eh^3/12(1 - \upsilon^2)$, and υ, q, h and ω are the Poisson ratio of the plate material, distributed load, thickness of plate and vertical displacement, respectively. Solutions to this biharmonic equation for simple geometry and loading cases are usually expressed in the form of Fourier series, Navier or Levy solutions. For a thick plate, which is the normal case for most foundation problems, the shear deformation is significant in the analysis. The governing equation for thick plate is given by Eq. (8.3).

$$\frac{-K_sEh}{2(1+v)}\left(\nabla^2\omega + \frac{M}{D}\right) = q(x,y)$$

DOI: 10.1201/9781032650586-8

$$-D(1-v)\nabla^2\varphi_x-(1+v)\frac{\partial M}{\partial x}+\frac{K_s Eh}{(1+v)}\left(\frac{\partial \omega}{\partial x}+\varphi_x\right)=0$$

$$-D(1-v)\nabla^2\varphi_y-(1+v)\frac{\partial M}{\partial y}+\frac{K_s Eh}{(1+v)}\left(\frac{\partial \omega}{\partial y}+\varphi_y\right)=0 \tag{8.3}$$

where φ_x and φ_x are the rotation along the x and y axes, K_s is the shear correction factor and $M = (Mxx + Myy)/(1 + \upsilon)$. The solution to Eq. (8.3) is possible only with a numerical method.

For the stress equilibrium of a soil medium which forms the governing equation for many geotechnical programs, it is given by Eq. (8.4) as:

$$\frac{\partial \sigma_{xx}}{\partial x}+\frac{\partial \tau_{xy}}{\partial y}+\frac{\partial \tau_{xz}}{\partial z}=fx$$

$$\frac{\partial \tau_{xy}}{\partial x}+\frac{\partial \sigma_{yy}}{\partial y}+\frac{\partial \tau_{yz}}{\partial z}=fy \tag{8.4}$$

$$\frac{\partial \tau_{xz}}{\partial x}+\frac{\partial \tau_{yz}}{\partial y}+\frac{\partial \sigma_{zz}}{\partial z}=fz$$

where σ and τ are the normal and shear stresses, while f is the body forces in different directions. Lengthy analytical solutions to Eq. (8.4) are available for some simple cases, which have been given by Poulos and Davis (1974) and many others.

For a seepage problem, the two-dimensional problem is governed by the Laplace equation [Eq. (8.5)].

$$\frac{\partial^2 \phi}{\partial x^2}+\frac{\partial^2 \phi}{\partial y^2}=0 \qquad \text{for isotropic soil}$$

$$k_x\frac{\partial^2 \phi}{\partial x^2}+k_y\frac{\partial^2 \phi}{\partial y^2}=0 \qquad \text{for anisotropic soil} \tag{8.5}$$

where k_x and k_y are the permeabilities of the soil along the x and y directions and ϕ is the total head. The use of complex analysis to solve Eq. (8.5) is possible for some simple cases, but is generally not used now. For 1D and axisymmetric consolidation in clay, it is governed by the dispersion equation [Eq. (8.6)] as

$$\frac{\partial u}{\partial t}=c_v\frac{\partial^2 u}{\partial z^2} \qquad \text{one-dimensional consolidation}$$

$$\frac{\partial u}{\partial t}=c_v\frac{\partial^2 u}{\partial z^2}+c_r\left(\frac{\partial^2 u}{\partial r^2}+\frac{1}{r}\frac{\partial u}{\partial r}\right) \qquad \text{axisymmetric consolidation} \tag{8.6}$$

where c_v and u are the coefficient of consolidation and pore water pressure, respectively. For 3D condition, the governing equation Biot consolidation is given by Eq. (8.7).

$$-G\nabla^2 u_x + \frac{G}{1-2\nu}\frac{\partial}{\partial x}\left(\frac{\partial u_x}{\partial x}+\frac{\partial u_y}{\partial y}+\frac{\partial u_z}{\partial z}\right)+\frac{\partial u}{\partial x}=0$$

$$-G\nabla^2 u_y + \frac{G}{1-2\nu}\frac{\partial}{\partial y}\left(\frac{\partial u_x}{\partial x}+\frac{\partial u_y}{\partial y}+\frac{\partial u_z}{\partial z}\right)+\frac{\partial u}{\partial y}=0 \tag{8.7}$$

$$-G\nabla^2 u_z + \frac{G}{1-2\nu}\frac{\partial}{\partial z}\left(\frac{\partial u_x}{\partial x}+\frac{\partial u_y}{\partial y}+\frac{\partial u_z}{\partial z}\right)+\frac{\partial u}{\partial z}=0$$

where G and υ are the shear modulus and Poisson ratio of the soil. For ultimate limit state analysis, the hyperbolic differential slip-line equations [Eq. (8.8)] (see Chapter 2) have no analytical solution unless the weight of the soil is negligible.

$$\alpha \text{ characteristic: } -\frac{\partial p}{\partial S_\alpha}\sin 2\mu + 2R\frac{\partial \theta}{\partial S_\alpha} + \gamma[\sin(\alpha+2\mu)\frac{\partial y}{\partial S_\alpha} + \cos(\alpha+2\mu)\frac{\partial x}{\partial S_\alpha}] = 0$$

$$\beta \text{ characteristic: } \frac{\partial p}{\partial S_\beta}\sin 2\mu + 2R\frac{\partial \theta}{\partial S_\beta} + \gamma[\sin(\alpha-2\mu)\frac{\partial y}{\partial S_\beta} + \cos(\alpha-2\mu)\frac{\partial x}{\partial S_\beta}] = 0 \tag{8.8}$$

In Eq. (8.8), γ, c and φ are the unit weight of soil, cohesive strength and friction angle of soil, $\mu = (\pi/4 - \phi/2)$, θ is the angle between the direction of major principal stress and the y-axis, S_α and S_β are the arcs along the α and β lines, α is the angle between the direction of the body force and the vertical direction (y-axis). p and R are the mean stress and radius of the Mohr-circle at failure (see Chapter 2 for details):

$$p = \frac{\sigma_1+\sigma_3}{2};\quad R = \frac{\sigma_1-\sigma_3}{2} = p\sin\phi + c\cos\phi \tag{8.9}$$

The pile-driving equation is given by Eq. (8.10), where u, c, f, l, ρ and A are the axial displacement of pile, velocity of wave force, skin friction, perimeter of pile, density and sectional area of pile, respectively.

$$\frac{\partial^2 u}{\partial t^2} - c^2\frac{\partial^2 u}{\partial z^2} + \frac{fl}{\rho A} = 0 \tag{8.10}$$

The differential equations as discussed above can also be cast into an integral form in many cases. There are also other governing differential equations for other types of problems, however, only those commonly encountered in geotechnical problems will be discussed in this chapter. Many numerical methods have been developed to deal with different cases, and only those related to geotechnical problems will be discussed in this chapter.

8.2 FINITE ELEMENT METHOD (FEM)

The geometries of many geotechnical problems are complicated with non-homogeneous ground conditions, the presence of water, different types of internal and external loadings and complicated soil–structure interactions. Towards this, the use of numerical methods is inevitable. The finite element method is the most popular numerical method which forms the basis of many commercial programs; two-dimensional problems will be discussed here for illustration. The extension of two-dimensional finite element analysis to three-dimensional or axisymmetric analysis is straight-forward, but this is covered here.

Generally, there are two kinds of elastic plane problems (Zienkiewicz 2011): plane strain and plane stress. For geotechnical problems, there is practically no plane stress problem.

8.2.1 Plane strain

Figure 8.1 shows a 2D cross-section of a structure with surface force or constraint. The internal and external forces must be parallel to the cross-section for a 2D problem.

The section as shown in Figure 8.1 is perpendicular to the z-axis so that all the field quantities (stress, strain, displacement and others) depend only on the two coordinates (x, y). For plane strain condition, Eq. (8.11) applies.

$$\gamma_{zx} = \gamma_{zy} = \varepsilon_z = 0 \tag{8.11}$$

Based on Eq. (8.11), we can deduce that

$$\tau_{zx} = \tau_{xz} = 0, \tau_{zy} = \tau_{yz} = 0, \gamma_{xz} = \gamma_{yz} = 0 \tag{8.12}$$

While the normal strain is 0, the normal stress σ in the z direction is not 0, which is given by the stress–strain relations as:

$$0 = \varepsilon_z = \frac{1}{E}\left[\sigma_z - v\left(\sigma_x + \sigma_y\right)\right]$$

$$\sigma_z = v\left(\sigma_x + \sigma_y\right) \tag{8.13}$$

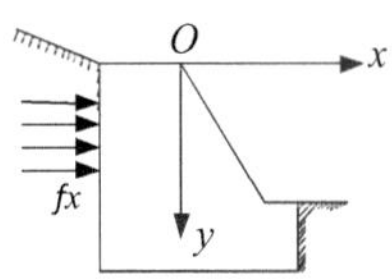

Figure 8.1 Cross-section of a 2D structure.

By substituting σ_{zz} into the stress–strain relations, we have

$$\mu_x = \frac{1-\nu^2}{E}\left(\sigma_x - \frac{\nu}{1-\nu}\sigma_y\right) \tag{8.14}$$

$$\mu_y = \frac{1-\nu^2}{E}\left(\sigma_y - \frac{\nu}{1-\nu}\sigma_x\right) \tag{8.15}$$

$$\gamma_{xy} = \frac{2(1+\nu)}{E}\tau_{xy} \tag{8.16}$$

The stress–strain relations for a 2D plane strain problem are usually expressed in a matrix form as:

$$\{\sigma\} = [\boldsymbol{D}]\{\varepsilon\} \tag{8.17}$$

$$[\boldsymbol{D}] = \frac{E_0}{1-\nu_0^2}\begin{bmatrix} 1 & \nu_0 & 0 \\ \nu_0 & 1 & 0 \\ 0 & 0 & \frac{1-\nu_0}{2} \end{bmatrix} \tag{8.18}$$

where $[\boldsymbol{D}]$ is the elastic matrix with $E_0 = \frac{E}{1-\nu^2}$ and $\nu_0 = \frac{\nu}{1-\nu}$.

8.2.2 Plane stress

The plane strain condition is a special case of three-dimensional problems, while plane stress is only approached in the limit as the thickness of the loaded body tends to zero. The plane stress problem in Figure 8.2 is a deep beam plane stress problem with thickness h.

The stresses on all parallel z-planes for the plane stress problem will be small and negligible, so we can write

$$\tau_{zx} = \tau_{zy} = \sigma_z = 0 \tag{8.19}$$

Obviously, we can deduce that

$$\tau_{xz} = \tau_{yz} = 0 \tag{8.20}$$

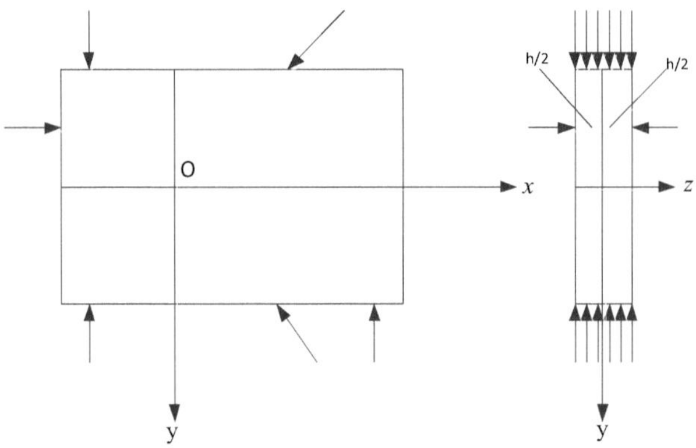

Figure 8.2 Deep beam with identical thickness.

The stress–strain relations can be expressed as

$$\varepsilon_x = \frac{1}{E}\left(\sigma_x - \nu\sigma_y\right) \tag{8.21}$$

$$\varepsilon_y = \frac{1}{E}\left(\sigma_y - \nu\sigma_x\right) \tag{8.22}$$

$$\gamma_{xy} = \frac{2(1+\nu)}{E}\tau_{xy} \tag{8.23}$$

For simplicity, the stress–strain relations for a 2D plane stress problem are usually expressed in matrix form as:

$$\{\sigma\} = [\boldsymbol{D}]\{\varepsilon\} \tag{8.24}$$

$$[\boldsymbol{D}] = \frac{E_0}{1-\nu_0^2}\begin{bmatrix} 1 & \nu_0 & 0 \\ \nu_0 & 1 & 0 \\ 0 & 0 & \dfrac{1-\nu_0}{2} \end{bmatrix} \tag{8.25}$$

where $[\boldsymbol{D}]$is the elastic matrix with $E_0 = E$ and $\nu_0 = \nu$.

8.2.3 Fundamentals of FEM

There are two major and popular approaches to FEM, which are the variational principle and weighted residual method (WRM), however there are also other less popular principles which may be more effective under certain special circumstances. The principle of virtual work (including both principle of virtual displacement and virtual stress) which is the most popular approach for FEM formulation in elasticity problems is a weak form of the equivalent integration for the governing equilibrium equations. The principles of minimum potential energy and complementary energy are two variational approaches which are equivalent to the elasticity equations.

8.2.3.1 Principle of virtual displacement

The principle of virtual displacement is the weak form of the equivalent integration for the equilibrium equations and force boundary conditions. For the equilibrium equations and force boundary conditions *f* in index notation

$$\sigma_{ij,j} + f_i = 0, \quad (\text{in domain } V) \tag{8.26}$$

$$\sigma_{ij} n_j - T_i = 0, \quad (\text{on domain boundary } S_\sigma) \tag{8.27}$$

where n and T are the normal vector and traction along the boundary. In WRM, the variation of true displacement δu_i and its boundary value (i.e., $-\delta u_i$) can be selected as the weight functions in the equivalent integration

$$\int_V \delta u_i \left(\sigma_{ij,j} + f_i\right) \mathrm{d}V - \int_{S_\sigma} \delta u_i \left(\sigma_{ij} n_j - T_i\right) \mathrm{d}S = 0 \tag{8.28}$$

The weak form of equation is given by Eq. (8.29) as:

$$\int_V \left(-\delta\varepsilon_{ij}\sigma_{ij} + \delta u_i f_i\right) \mathrm{d}V + \int_{S_\sigma} \delta u_i T_i \mathrm{d}S = 0 \tag{8.29}$$

In Eq. (8.29), the first item represents the work done by the stresses under the virtual strain, while the remaining items represent the work done by the body force and surface tractions under the virtual displacement. The summation of the internal and external virtual work is equal to 0, which is the principle of virtual displacement, and the force system will then satisfy equilibrium equations.

8.2.3.2 Principle of minimum potential energy (PMPE)

Based on Eq. (8.29), by substituting Eq. (8.14) for σ_{ij}, we can deduce that

$$\int_V \left(\delta\varepsilon_{ij} D_{ijkl} \varepsilon_{kl} - \delta u_i f_i\right) \mathrm{d}V + \int_{S_\sigma} \delta u_i T_i \mathrm{d}S = 0 \tag{8.30}$$

Using the symmetry of the constitutive matrix D_{ijkl}, we obtain

$$\left(\delta\varepsilon_{ij}\right)D_{ijkl}\varepsilon_{kl} = \delta\left(\frac{1}{2}D_{ijkl}\varepsilon_{ij}\varepsilon_{kl}\right) = \delta U\left(\varepsilon_{mn}\right) \tag{8.31}$$

where $U\left(\varepsilon_{mn}\right)$ is the unit volume strain energy. For linear elasticity

$$-\delta\phi\left(u_i\right) = f_i\delta u_i, -\delta\psi\left(u_i\right) = T_i\delta u_i \tag{8.32}$$

Eq. (8.30) is further simplified to

$$\delta\Pi_P = 0 \tag{8.33}$$

Π_p is the total potential energy of the system expressed as

$$\Pi_P = \Pi_P\left(u_i\right) = \int_V \left(\frac{1}{2}D_{ijkl}\varepsilon_{ij}\varepsilon_{kl} - f_i u_i\right)\mathrm{dV} - \int_{S_\sigma} T_i u_i \mathrm{d}S \tag{8.34}$$

Eq. (8.33) shows that, among all the potential displacements, the total potential energy of the system will take the stationary value at the real displacement which is the principle of minimum potential energy.

8.2.4 General expressions and implementation procedure of FEM

The FEM procedures always follow an orderly step-by-step process which is easy to be programmed and used by engineers. For illustration, a three-node triangular element for plane problems is considered.

8.2.4.1 Discretization of domain

The first step in FEM is to divide the solution domain into subdivisions or elements, and the more elements and degrees of freedom for a problem, the more accurate the solution will be. The number, type, size and arrangement of the elements are critical to the

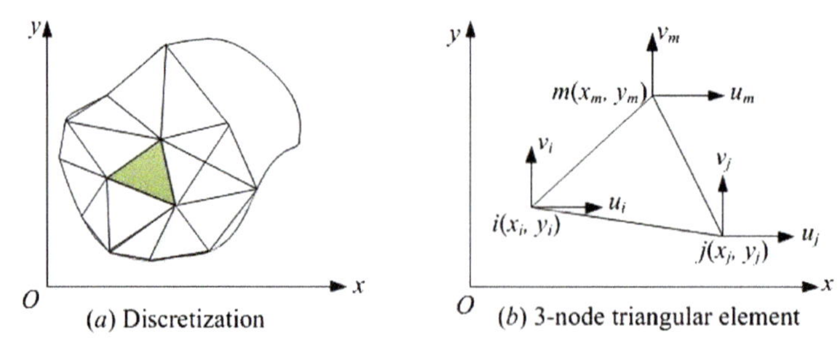

Figure 8.3 Discretization of a 2D domain by a three-node triangular element.

numerical analysis. A typical discretization with three-node triangular element is shown schematically in Figure 8.3.

Mesh generation can be a difficult process for a general irregular domain, particularly when the geometry is irregular with openings and locations of loadings and connections (columns and others). There are some high-end programs that target towards mesh generation, which are considered necessary for highly complicated problems.

8.2.4.2 Interpolation or displacement model

As can be seen from Figure 8.3(b), each node has two degrees of freedom (DOFs) or two displacement components which are stored in a column vector as:

$$\boldsymbol{a}_i = \begin{bmatrix} u_i \\ v_i \end{bmatrix} \quad (i,j,m) \tag{8.35}$$

In total, each triangular element has six nodal displacements. Putting all the displacements a in a column vector, we can obtain the element nodal displacement column matrix as

$$\boldsymbol{a}^e = \begin{bmatrix} \boldsymbol{a}_i \\ \boldsymbol{a}_j \\ \boldsymbol{a}_m \end{bmatrix} = \begin{bmatrix} u_i & v_i & u_j & v_j & u_m & v_m \end{bmatrix}^{\mathrm{T}} \tag{8.36}$$

In displacement-based FEM, nodal displacement is chosen as the basic unknowns. The displacement at any arbitrary point is based on interpolation on the three nodal displacements within each element. For a three-node triangular element with linear interpolation, the displacements in both the x and y directions within each element are

$$u = \beta_1 + \beta_2 x + \beta_3 y \tag{8.37}$$

$$v = \beta_4 + \beta_5 x + \beta_6 y \tag{8.38}$$

where β are coefficients to be determined. By substituting the six nodal displacement components into these equations, the displacement mode is expressed as

$$u = \mathrm{N}_i u_i + \mathrm{N}_j u_j + \mathrm{N}_m u_m \tag{8.39}$$

$$v = \mathrm{N}_i v_i + \mathrm{N}_j v_j + \mathrm{N}_m v_m \tag{8.40}$$

where

$$\mathrm{N}_i = \frac{1}{2A}\left(a_i + b_i x + c_i y\right) \qquad (i,j,m) \tag{8.41}$$

In Eq. (8.39), N_i, N_j and N_m are called the interpolation function or shape function, A is the area of the triangular element, and $a_i, b_i, c_i \cdots, c_m$ are constants related to the coordinates of the three nodes. Eqs. (8.39) and (8.40) are expressed in the form of a matrix as

$$\boldsymbol{u} = \begin{bmatrix} u \\ v \end{bmatrix} = \begin{bmatrix} N_i & 0 & N_j & 0 & N_m & 0 \\ 0 & N_i & 0 & N_j & 0 & N_m \end{bmatrix} \begin{bmatrix} u_i \\ v_i \\ u_j \\ v_j \\ u_m \\ v_m \end{bmatrix} = \boldsymbol{N}\boldsymbol{a}^e \tag{8.42}$$

where $\boldsymbol{N}$ is the shape function matrix and $\boldsymbol{a}^e$ is the element nodal displacement vector. The element strains $\boldsymbol{\varepsilon}$ are

$$\boldsymbol{\mu} = \begin{bmatrix} \varepsilon_x \\ \varepsilon_y \\ \gamma_{xy} \end{bmatrix} = \mathbf{L}\boldsymbol{u} = \mathbf{L}\boldsymbol{N}\boldsymbol{a}^e = \mathbf{L}\begin{bmatrix} \boldsymbol{N}_i & \boldsymbol{N}_j & \boldsymbol{N}_m \end{bmatrix}\boldsymbol{a}^e$$

$$= \begin{bmatrix} \boldsymbol{B}_i & \boldsymbol{B}_j & \boldsymbol{B}_m \end{bmatrix}\boldsymbol{a}^e = \boldsymbol{B}\boldsymbol{a}^e \tag{8.43}$$

where $\mathbf{L}$ is the differential operator and $\boldsymbol{B}$ is the element strain displacement matrix given as

$$\boldsymbol{B}_i = \mathbf{L}\boldsymbol{N}_i = \begin{bmatrix} \frac{\partial}{\partial x} & 0 \\ 0 & \frac{\partial}{\partial y} \\ \frac{\partial}{\partial y} & \frac{\partial}{\partial x} \end{bmatrix} \begin{bmatrix} N_i & 0 \\ 0 & N_i \end{bmatrix} = \begin{bmatrix} \frac{\partial N_i}{\partial x} & 0 \\ 0 & \frac{\partial N_i}{\partial y} \\ \frac{\partial N_i}{\partial y} & \frac{\partial N_i}{\partial x} \end{bmatrix} \quad (i, j, m) \tag{8.44}$$

Substitute Eq. (8.43) in Eq. (8.7),

$$\boldsymbol{\sigma} = \begin{bmatrix} \sigma_x \\ \sigma_y \\ \tau_{xy} \end{bmatrix} = \boldsymbol{D}\boldsymbol{\varepsilon} = \boldsymbol{D}\boldsymbol{B}\boldsymbol{a}^e = \boldsymbol{S}\boldsymbol{a}^e \tag{8.45}$$

where

$$S = DB = D\begin{bmatrix} B_i & B_j & B_m \end{bmatrix} = \begin{bmatrix} S_i & S_j & S_m \end{bmatrix} \tag{8.46}$$

S is the element stress matrix. The strain and stress matrices are constant for the triangular element because the displacement mode is a first-order function, and differentiating this function will be constant function.

8.2.4.3 Stiffness equilibrium equation (SEE) of FEM derived from PMPE

The total potential energy Π_p in Eq. (8.34) can be expressed in a matrix formulation as follows:

$$\Pi_p = \int_\Omega \frac{1}{2}\boldsymbol{\varepsilon}^{\mathrm{T}} \boldsymbol{D\varepsilon} t\mathrm{d}x\mathrm{d}y - \int_\Omega \boldsymbol{u}^{\mathrm{T}} \boldsymbol{f} t\mathrm{d}x\mathrm{d}y - \int_{S_\sigma} \boldsymbol{u}^{\mathrm{T}} \boldsymbol{T} t\mathrm{d}S \tag{8.47}$$

where t, f and T denote the thickness, body force and surface force, respectively. The total potential energy is the summation of the contributions from all the elements. Therefore, substituting Eqs. (8.42) and (8.43) into Eq. (8.47) gives

$$\Pi_p = \sum_e \Pi_p^e = \sum_e \left(\boldsymbol{a}^{e\mathrm{T}} \int_{\Omega_e} \frac{1}{2} \boldsymbol{B}^{\mathrm{T}} \boldsymbol{DB} t\mathrm{d}x\mathrm{d}y \boldsymbol{a}^e \right) -$$

$$\sum_e \left(\boldsymbol{a}^{e T} \int_{\Omega_e} \boldsymbol{N}^T \boldsymbol{f} t dx dy \right) - \sum_e \left(\boldsymbol{a}^{e\mathrm{T}} \int_{S_\sigma^e} \boldsymbol{N}^{\mathrm{T}} \boldsymbol{T} t\mathrm{d}x\mathrm{d}y \right) \tag{8.48}$$

Eq. (8.48) can also be expressed as

$$\boldsymbol{K}^e = \int_{\Omega_e} \boldsymbol{B}^{\mathrm{T}} \boldsymbol{DB} t\mathrm{d}x\mathrm{d}y, \boldsymbol{P}_f^e = \int_{\Omega_e} \boldsymbol{N}^{\mathrm{T}} \boldsymbol{f} t\mathrm{d}x\mathrm{d}y$$

$$\boldsymbol{P}_S^e = \int_{S_\sigma^e} \boldsymbol{N}^{\mathrm{T}} \boldsymbol{T} t\mathrm{d}x\mathrm{d}y, \ \boldsymbol{P}^e = \boldsymbol{P}_f^e + \boldsymbol{P}_S^e \tag{8.49}$$

where $\boldsymbol{K}^e$ and $\boldsymbol{P}^e$ are the element stiffness matrix and equivalent element nodal load matrix, respectively. Substituting Eq. (8.49) into Eq. (8.48), the total potential energy can be simplified as

$$\Pi_p = \boldsymbol{a}^T \frac{1}{2} \sum_e \left(\boldsymbol{K}^e \right) \boldsymbol{a} - \boldsymbol{a}^T \sum_e \left(\boldsymbol{P}^e \right) \tag{8.50}$$

Given

$$K = \sum_e K^e, P = \sum_e P^e \tag{8.51}$$

Eq. (8.50) is further simplified as

$$\Pi_P = \frac{1}{2} a^T K\ a - a^T P\ a \tag{8.52}$$

where K and P are global stiffness matrix and global nodal load matrix, respectively.
For PMPE, the variation of Π_P is equal to 0 and the unknown variable is a or

$$\frac{\partial \Pi_P}{\partial a} = 0 \tag{8.53}$$

which finally comes to the SEE of FEM as

$$K\ a = P \tag{8.54}$$

Without the provision of adequate boundary condition, the system as represented by Eq. (8.54) is actually singular, as rigid body movement will not induce any stress in the system.

8.2.4.4 Derivation of element stiffness matrices (ESM) for a three-node triangle

For a three-node triangular element, the element strain matrix B is constant, thus Eq. (8.49) gives

$$K^e = B^T D B t A = \begin{bmatrix} K_{ii} & K_{ij} & K_{im} \\ K_{ji} & K_{jj} & K_{jm} \\ K_{mi} & K_{mj} & K_{mm} \end{bmatrix} \tag{8.55}$$

of which the sub-matrix

$$K_{ij} = \begin{bmatrix} k_{ij}^{xx} & k_{ij}^{xy} \\ k_{ij}^{yx} & k_{ij}^{yy} \end{bmatrix} \tag{8.56}$$

It should be noted that the element stiffness matrix in Eq. (8.56) is symmetric, and the computational memory required in a FEM program can be reduced by using this property.

8.2.4.5 Assembling of ESMs and ENLMs

Most of the elements in the matrix $\boldsymbol{K}$ are 0 because each node is only connected with a few surrounding elements. To reduce the computer memory requirement, a rectangular matrix can be used to store the global stiffness matrix (which is a square matrix), and the half band width *bandwidth* is

$$bandwidth = (1 + NDIF) \times NDOF \tag{8.57}$$

where *NDIF* is the largest absolute difference between the element node numbers among all the elements in the mesh. Besides the bandwidth storage, the use of profile or skyline storage is also popular with slightly less memory requirement. Eq. (8.54) is usually solved by the direct Gaussian method, while the use of an iterative solver can be useful and effective in some very large problems.

In summary, the global stiffness matrix is symmetric, banded distribution and singular. Singularity will vanish if appropriate boundary conditions are imposed to Eq. (8.54) to eliminate the rigid body motion. However, a truncation error can introduce a false boundary condition for a singular matrix which is difficult to be avoided.

8.2.5 Isoparametric element and numerical integration

Although the triangular element can be formed easily and can always fit well with a complicated boundary, the accuracy of this element is low. Despite such a limitation, the triangular element is still popular among many commercial programs due to the necessity of simple mesh generation. To achieve better accuracy, the use of a higher order element and isoparametric formulation is common adopted. Consider a four-node quadrilateral element in Figure 8.4. We can transform it from Figures 8.4(a) and 8.4(b) for numerical

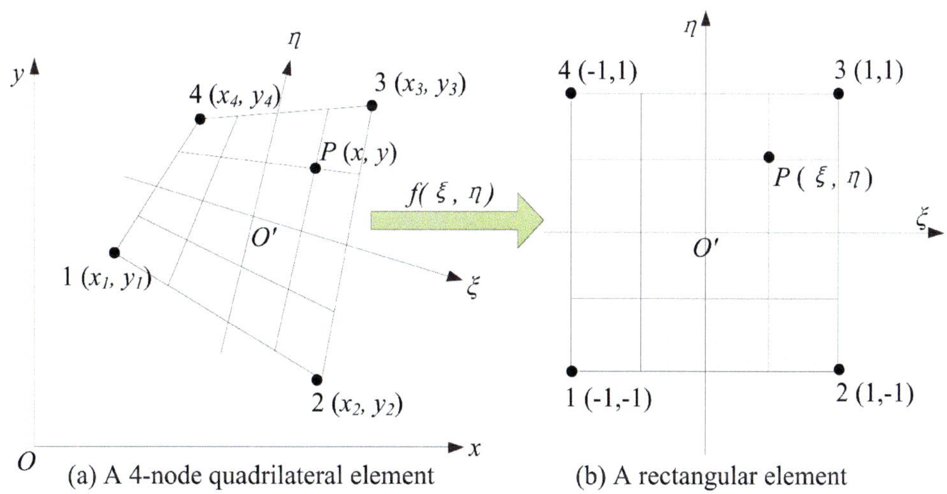

Figure 8.4 Isoparametric transition.

integration. Figure 8.4(a) is defined in the Cartesian coordinate system, while Figure 8.4(b) is defined in the local coordinate system within a specific domain (i.e., $\xi, \eta \in (-1,1)$). The relation between these two kinds of coordinate system can be described as:

$$\begin{Bmatrix} x \\ y \end{Bmatrix} = f \begin{Bmatrix} \xi \\ \eta \end{Bmatrix} \tag{8.58}$$

The displacement at any point (x,y) is:

$$x = \sum_{i=1}^{m} N_i' x_i \qquad y = \sum_{i=1}^{m} N_i' y_i \tag{8.59}$$

where (x_i, y_i) are coordinates in the Cartesian i-th node in the local coordinate system, N_i' is interpolation function of the i-th node in the local coordinate system, and m is the number of nodes chosen to transform the coordinates. Eq. (8.59) can be further expanded as

$$\begin{Bmatrix} x \\ y \end{Bmatrix} = \begin{bmatrix} N_1 & 0 & N_2 & 0 & N_3 & 0 & N_4 & 0 \\ 0 & N_1 & 0 & N_2 & 0 & N_3 & 0 & N_4 \end{bmatrix} \begin{Bmatrix} x_1 \\ y_1 \\ x_2 \\ y_2 \\ x_3 \\ y_3 \\ x_4 \\ y_4 \end{Bmatrix} \tag{8.60}$$

The element displacement model can be written as

$$\begin{Bmatrix} u \\ v \end{Bmatrix} = \begin{bmatrix} N_1 & 0 & N_2 & 0 & N_3 & 0 & N_4 & 0 \\ 0 & N_1 & 0 & N_2 & 0 & N_3 & 0 & N_4 \end{bmatrix} \begin{Bmatrix} u_1 \\ v_1 \\ u_2 \\ v_2 \\ u_3 \\ v_3 \\ u_4 \\ v_4 \end{Bmatrix} \tag{8.61}$$

where $\boldsymbol{J}$ is the Jacobi matrix while the interpolation functions are given by

$$N_1 = \frac{1}{4}(1-\xi)(1-\eta), \quad N_2 = \frac{1}{4}(1+\xi)(1-\eta)$$

$$N_3 = \frac{1}{4}(1+\xi)(1+\eta), \quad N_2 = \frac{1}{4}(1-\xi)(1+\eta) \tag{8.62}$$

Since the shape functions are expressed in the natural coordinates, the derivative and integration transformation relationships are required when an isoparametric element is used.

8.2.5.1 Derivative and integral transformation

By the law of partial differential,

$$\frac{\partial N_i}{\partial \xi} = \frac{\partial N_i}{\partial x}\frac{\partial x}{\partial \xi} + \frac{\partial N_i}{\partial y}\frac{\partial y}{\partial \xi},$$

$$\frac{\partial N_i}{\partial \eta} = \frac{\partial N_i}{\partial x}\frac{\partial x}{\partial \eta} + \frac{\partial N_i}{\partial y}\frac{\partial y}{\partial \eta} \tag{8.63}$$

or in matrix form

$$\begin{Bmatrix} \dfrac{\partial N_i}{\partial \xi} \\ \dfrac{\partial N_i}{\partial \eta} \end{Bmatrix} = \begin{bmatrix} \dfrac{\partial x}{\partial \xi} & \dfrac{\partial y}{\partial \xi} \\ \dfrac{\partial x}{\partial \eta} & \dfrac{\partial y}{\partial \eta} \end{bmatrix} \begin{Bmatrix} \dfrac{\partial N_i}{\partial x} \\ \dfrac{\partial N_i}{\partial y} \end{Bmatrix} = J \begin{Bmatrix} \dfrac{\partial N_i}{\partial x} \\ \dfrac{\partial N_i}{\partial y} \end{Bmatrix} \tag{8.64}$$

The inverse of Eq. (8.64) gives

$$\begin{Bmatrix} \dfrac{\partial N_i}{\partial x} \\ \dfrac{\partial N_i}{\partial y} \end{Bmatrix} = J^{-1} \begin{Bmatrix} \dfrac{\partial N_i}{\partial \xi} \\ \dfrac{\partial N_i}{\partial \eta} \end{Bmatrix} \tag{8.65}$$

where

$$J = \begin{bmatrix} \dfrac{\partial x}{\partial \xi} & \dfrac{\partial y}{\partial \xi} \\ \dfrac{\partial x}{\partial \eta} & \dfrac{\partial y}{\partial \eta} \end{bmatrix} = \begin{bmatrix} \sum_{i=1}^{4} \dfrac{\partial N_i}{\partial \xi} x_i & \sum_{i=1}^{4} \dfrac{\partial N_i}{\partial \xi} y_i \\ \sum_{i=1}^{4} \dfrac{\partial N_i}{\partial \eta} x_i & \sum_{i=1}^{4} \dfrac{\partial N_i}{\partial \eta} y_i \end{bmatrix}$$

$$= \begin{bmatrix} \dfrac{\partial N_1}{\partial \xi} & \dfrac{\partial N_2}{\partial \xi} & \dfrac{\partial N_3}{\partial \xi} & \dfrac{\partial N_4}{\partial \xi} \\ \dfrac{\partial N_1}{\partial \eta} & \dfrac{\partial N_2}{\partial \eta} & \dfrac{\partial N_3}{\partial \eta} & \dfrac{\partial N_4}{\partial \eta} \end{bmatrix} \begin{bmatrix} x_1 & y_1 \\ x_2 & y_2 \\ x_3 & y_3 \\ x_4 & y_4 \end{bmatrix} \tag{8.66}$$

The areas in the Cartesian coordinate system and the natural coordinate system are related by

$$ds = dxdy = |J|d\xi d\eta \tag{8.67}$$

where $|J|$ is the determinant of the Jacobian matrix J. Therefore, element stiffness matrix and equivalent nodal load matrix in Eq. (8.45) can be transformed to the global coordinates using

$$K^e = \int_{\Omega_e} B^T D B t |J| d\xi d\eta, P_f^e = \int_{\Omega_e} N^T f |J| d\xi d\eta$$

$$P_S^e = \int_{S_A^e} N^T T |J| d\xi d\eta \tag{8.68}$$

For the evaluation of this integral equation, an analytical solution may be possible for some cases, however the Gaussian integration method is employed in most cases. In the actual integration, two and three integration points along each direction of integration are commonly used, but one integration point is also adopted in some elements. Since the discretized system is usually over-stiff, the use of two integration points along each direction of integration will slightly reduce the stiffness of the matrix with better results and so is more commonly used.

8.2.6 Solution of equation and post-processing

The system of equation is singular by nature, and boundary conditions have to be imposed to eliminate the rigid body movement as well as complying with the requirements of the problems. For the solution of the system of equation, the Gaussian method is adopted in most computer programs. It should be noted that due to extensive amounts of computation during the solution, the loss of accuracy due to truncation can be significant. The use of double precision (or even more) may be necessary for some cases. Out-core solver has been popular in the past when computer memory was limited, but it appears to be not popular now. Instead, the iterative solver by the method of relaxation, conjugate gradient or other methods can be used. Cheng has developed a finite element using the conjugate gradient method and compared it with the Gaussian method. It is found that the conjugate gradient method can be efficient for very large problems, but is less efficient for smaller problems. During the solution, yielding or tension at the base of a raft foundation may occur which requires stress redistribution, hence the solution process can be nonlinear, depending on the type of problem.

After the solution, post-processing of the output to give stress/strain, moment/shear or other values is necessary. It is interesting to find that many problems in the post-processing modules in commercial programs were discovered by Cheng and Law. Also, different methods of post-processing can also be important to recover the stress/strain or moment/shear.

Some important problems in FEM which may not recognized by many engineers include:

1. Constant stresses or varying stresses within the element which is used by the engineer.
2. Nodal stresses for the same node from different elements are not the same. Average values of different nodal stresses usually give better results, but some engineers simply pick up the largest nodal value of a node from several surrounding elements for design in the past!
3. For four-node or higher order elements, there will be points with superconvergence. Better accuracy can be achieved by taking this advantage (PLATE by Cheng) which may result in better accuracy.
4. Without the imposition of BC, the global stiffness matrix is singular. Yet the equilibrium equation can still be solved because the computer truncation error generates the false boundary condition.

Cheng has developed many finite element programs for teaching and research purposes which can be obtained at natureymc@yahoo.com.hk. The available programs include plane stress/strain problems, 2D/3D truss and frames, thin/thick plate bending problems, consolidation in 1D and 2D (Biot), seepage problems (confined or free surface flow), slope stability problems, pile-driving problems and others. Mesh generation, band-width profile optimizers have also been developed by Cheng for education purposes.

8.3 FINITE DIFFERENCE METHOD (FDM)

FEM is a powerful tool for space discretization, but not for time discretization. For many time-related problems, the time component is advanced using the finite difference method, while the FEM is used for space discretization. The finite difference method is a direct approximation of the differentials without using variational principle or similar, and is actually a more direct method as compared with the finite element method. Ordinary differential equations or partial differential equations are converted to a system of linear equations for the solution.

For Taylor's expansion:

$$f(x_0 + h) = f(x_0) + \frac{f'(x_0)}{1!}h + \frac{f''(x_0)}{2!}h^2 + ... + R_n(x) \tag{8.69}$$

where R_n is a residual term. If h is small enough, R_n will be very small and negligible. The differential can hence be approximated as

$$f'(x_0) \approx \frac{f(x_0 + h) - f(x_0)}{h} \tag{8.70}$$

A better approximation for the differential can be obtained by the mean of the forward and backward difference as

$$f'(x_0) \approx \frac{f(x_0 + h) - f(x_0 - h)}{2h} \tag{8.71}$$

Using this approximation, the second or higher order differentials can be obtained as

$$f''(x_0) \approx \frac{f(x_0+h) - 2f(x_0) + f(x_0-h)}{h^2} \tag{8.72}$$

The use of the finite difference method has been illustrated in Section 6.5.2. For the solution of the Laplace seepage equation [Eq. (8.73)], it can be solved by the finite difference method. The two-dimensional soil medium is first divided into a regular rectangular mesh with "nodes" at which the total heads are to be determined.

$$K_x \frac{\partial^2 h}{\partial x^2} + K_y \frac{\partial^2 h}{\partial y^2} = 0 \tag{8.73}$$

Eq. (8.73) can be approximated as

$$K_x \frac{h_{i,j-1} - 2h_{i,j} + h_{i,j+1}}{\Delta x^2} + K_y \frac{h_{i-1,j} - 2h_{i,j} + h_{i+1,j}}{\Delta y^2} = 0 \tag{8.74}$$

where Δx and Δy are the intervals between adjacent points in the X and Y directions along the rectangular mesh, as shown in Figure 8.5. If setting $K_x = K_y$, $\Delta x = \Delta y$, Eq. (8.74) can be simplified to

$$h_{i-1,j} + h_{i,j-1} - 4h_{i,j} + h_{i,j+1} + h_{i+1,j} = 0 \tag{8.75}$$

N number of equations can be formulated in accordance with Eq. (8.75) for N number of nodes, together with the use of "fictitious nodes" outside the soil medium, which is the usual assumption in finite difference analysis. To ensure an adequate number of equations for solution for the N real nodes and the fictitious nodes, additional "boundary conditions"

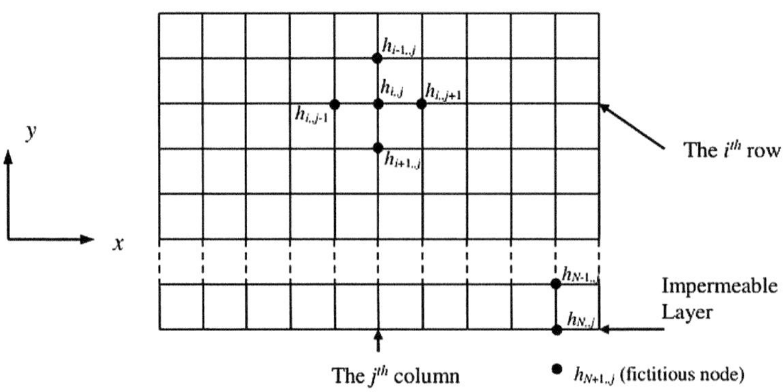

Figure 8.5 Finite difference solution for seepage problem.

equations have to be included. For the example in Figure 8.5 where $\Delta x = \Delta y = 0.5$ m, these boundary conditions include:

(i) Total head h remains constant at the nodes along the top "free surface" of the upstream and downstream sides which are, respectively, 9 m and 6 m;
(ii) At the nodes along a vertical edge considerably far from the wall, the hydrostatic conditions are considered normal without flow, i.e. the total head h remains constant. The total head at a node at z below the top surface along the far vertical edge at the upstream side, the elevation head is $9 - z$ and the pressure head is z so that the total head remains at $h = (9 - z) + z = 9$. Similarly, for the downstream side the total head of the nodes along the far edge remains at $h = 6$. The far edges are taken as 10 m from the wall;
(iii) The speeds of flow in (1) the vertical direction (Y direction) at the impermeable layer and (2) in the horizontal direction at the wall are 0, i.e., (1) $v_y = K_y \frac{\partial h}{\partial y} = 0 \Rightarrow \frac{h_{N+1,j} - h_{N-1,j}}{2\Delta y} = 0 \Rightarrow h_{N+1,j} = h_{N-1,j}$ where $h_{N-1,j}$ and $h_{N-1,j}$ are, respectively, the heads of the nodes just above the impermeable layer and the "fictitious nodes" beneath the impermeable layer. Similar boundary conditions can be formulated for (2).

The solution yields the total head contours (or equi-potential lines) in Figure 8.6. A set of "flow-lines" perpendicular to the total head contours are superimposed onto the total head contours to form the "flow net". Even though finite diffference is simple to use for a seepage problem, the authors prefer the use of the finite element method using the Galerkin formulation, for the sake of irregular geometry and non-homogeneous soil conditions.

The authors have also carried out a study on a thin plate problem using the finite difference analysis of the biharmonic plate bending equation. A rectangular plate of length 9 m

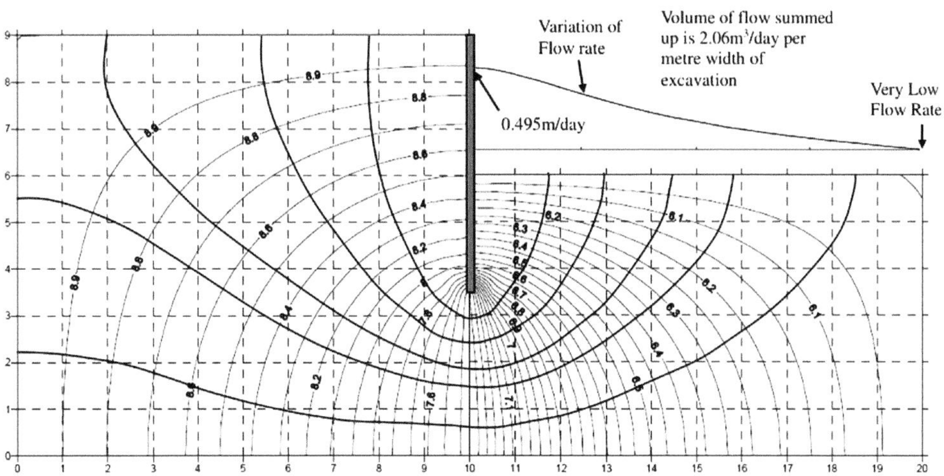

Figure 8.6 Flow-net from finite difference solution.

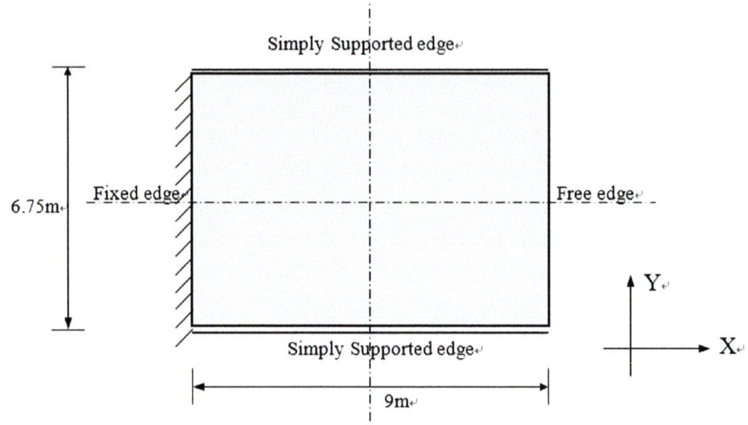

Figure 8.7 A simple thin plate under study.

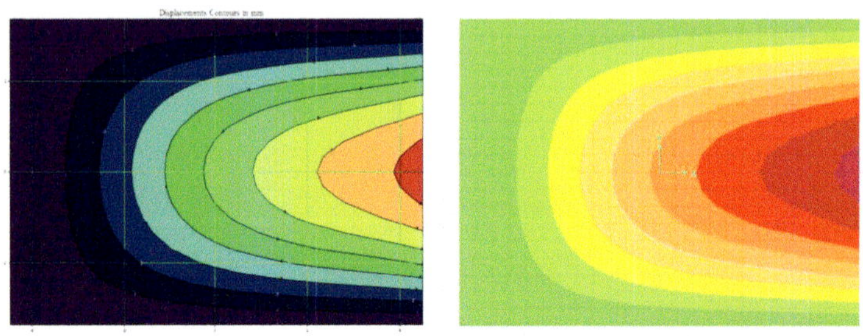

(a) Finite difference analysis of displacement (b) finite element solution

Figure 8.8 Comparisons between finite difference and finite element analysis for a thin plate.

in the X direction and width 6.75 in the Y-direction, as shown in Figure 8.7, is used for comparison. The plate is a 300 mm thick plate with Young's modulus E = 26400000 kN/m^2. In the spreadsheet, the plate is divided into 24 intervals in the X direction and 18 intervals in the Y direction. The three types of edge supports allowed by the spreadsheet comprise: (i) fixed (restrained from rotation and settlement); (ii) simply supported (restrained from settlement only); and (iii) entirely free are used on the four edges of the plate. Referring to the diagram below, the top and bottom edges are designed as simply supported; the left edge is fixed and the right ledge is free. The results of this analysis are shown in Figure 8.8, in which the maximum displacement is 8.47 mm for finite difference analysis and 8.44 mm for finite element analysis.

Despite the good match in the displacement from both analyses, the internal forces from the two methods have some notable differences in a local sense. The moments are lumped across a number of nodes to form a section for design. In this case, if the results of five adjacent nodes (with the node under the point load in the centre) are lumped to form a section, the design moments for both approaches can be calculated as in Figure 8.9.

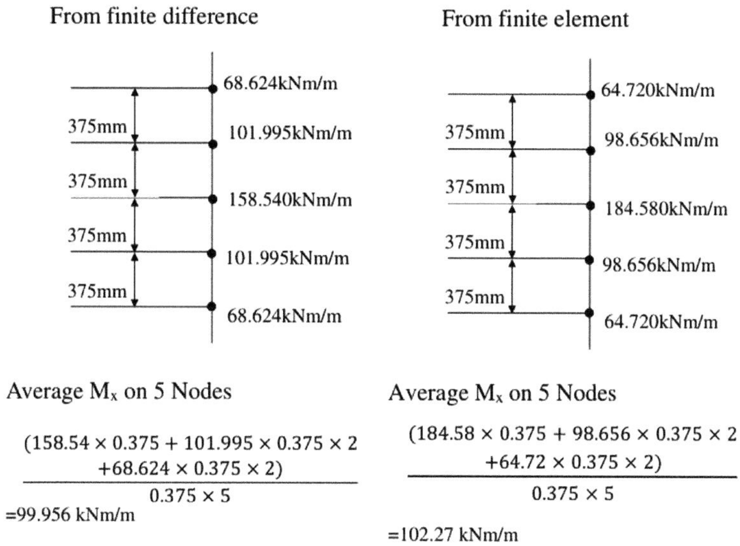

Figure 8.9 Differences between finite element and finite differences in the post-processing.

Of course, the use of a finer mesh can reduce such a discrepancy, but the results here can also illustrate some minor differences between the results from FDM and FEM, in particular, the derivatives of the displacements are used to determine the internal forces.

8.4 DISTINCT ELEMENT METHOD

The most commonly used numerical methods for continuous systems are the FEM and FDM (BEM is not commonly used due to various limitations), where the materials and displacements concerned are continuous without breakage or separation. The neighbourhood of all points will remain unchanged throughout the whole physical process. Some special algorithms have been developed for material fractures in continuum-based numerical methods, such as the displacement discontinuity technique in BEM by Crouch and Starfield (1983) and special joint elements by Goodman (1976). However, the limitations of these methods include:

(1) Large slip and opening of elements are not allowed in order to maintain the macroscopic material continuity;
(2) The amount of special fracture elements must be low so that the global stiffness matrix can be maintained well-posed without severe numerical instabilities; and
(3) Complete breakage and rotation of blocks/elements are either not allowed or treated with special algorithms.

The authors have adopted the zero thickness interface element by Goodman (1976) in developing a soil–structure interaction program. The practical problems in using this joint element include the difficulty of mesh generation. Currently, most engineers will specify a thin layer of material to simulate the interface instead of using the joint element directly in finite element analysis. Rock masses are always dissected by joints, faults, cracks or other discontinuities which control the failure and sliding of the masses. The complicated structure of rock masses and the discontinuities lead to a complicated non-linear mechanical response. Research work in the deformation of jointed rock masses is very important in modelling complicated important geotechnical problems. For such a system, the use of the classical continuity-based numerical methods will face great difficulty if the deformation is large. The use of the discontinuity-based approach will be more appropriate under such conditions.

The commonly used discontinuous numerical methods include: finite element method (FEM) with joints, distinct/discrete element method (DEM), rigid-block spring method (RBSM), contact-spring model, discontinuous deformation analysis (DDA) and numerical manifold method (NMM). The joint element method with zero thickness by Goodman (1976) is effective for only a few discontinuities and small deformations. The "no" penetration and no tension requirement at the interface cannot be strictly satisfied. If there are many discontinuities in a rock mass, the use of a joint element is very difficult to be adopted for general cases. Element separation is also possible for the joint element. There are two major DEM approaches: blocky structure and assembly of grains. There are hence two major discontinuity-based numerical approaches: blocky and granular. Actually, the granular approach is a special case of the blocky type problem, but due to the shape of the grains, contact detection can be a much easier task, and special techniques are available for the solution which is not possible for the blocky type problem.

For DEM, the micro-parameters are difficult to measure directly, hence DEM is still mainly used for qualitative but not quantitative assessment. The commonly used DEM is an explicit method based on the finite difference principles by Cundall (1971). Later, the works by Cundall were extended to UDEC and 3DEC (Cundall, 1980; Cundall and Hart, 1985, 1992). The granular approach has been developed for simulating the mechanical behaviour of granular materials (Cundall and Strack, 1979) with the development of early code BALL (Cundall, 1978) and Trubal (Strack and Cundall, 1984) which were later enhanced to PFC2D and PFC3D (Itasca, 1995). The main trend in the development and application of the method in rock engineering is represented by the history and results of the code groups UDEC/3DEC, but there are some new codes that have been released recently (EDEM, ESSS Rocky and some open source codes).

In DEM, the models can be soft or hard models, which means that penetration can be allowed or suppressed in the analysis. The block or grain can also be rigid or deformable, which is independent of the joint model. Rigid block formulation is simpler in the analysis and is sufficiently good if the confining stress or deformability of the block is small, and while the rigid particle approach is usually assumed as the confining stress is usually not high. For the penetration, Cheng (1998) found that if joint penetration is completely eliminated, the time for analysis can be extremely long with oscillation in the computation.

For DEM or DDA, block or granular formulation, the use of artificial damping to dissipate the kinetic energy is necessary, otherwise the computation will keep on fluctuating. There are three major types of damping that are used in commercial programs: viscous damping, self-adaptive damping and Coulomb damping. For viscous damping, the damping force is proportional to the velocity of the block/particle, which is also somewhat similar to the damping for pile-driving analysis. For maximum efficiency in the computation, the damping coefficient should be slightly smaller than the critical damping value. For a complicated system with different block sizes, it is difficult to choose a suitable damping, and trial-and-error analysis may be required in some cases. To overcome this problem, Cundall proposed the use of self-adaptive damping, which is used in UDEC. There are two major types of self-adaptive damping: viscous and local self-adaptive types. Self-adaptive damping varies at different locations in a system, but usually the system will be over-damped. Coulomb type damping always points opposite to the direction of movement in the analysis.

For the time-step required in the analysis, it must be controlled to an acceptable level for convergence in the analysis, which means that a typical analysis may requires thousands or even millions of cycles of computations. DEM is hence a very time-consuming analysis, and the use of GPU and parallel processing are under active research in order to overcome this major limitation. A rough guideline for the time-step is given by Eq. (8.76).

$$\Delta t = 2\sqrt{\frac{m}{k}} = \frac{2}{\omega_n} \tag{8.76}$$

Since $\omega_n = 2\pi/T$, hence

$$\Delta t < \frac{T}{\pi} \tag{8.77}$$

where k and m are the stiffness and mass of the contact spring, respectively. In actual numerical computations, Δt is usually chosen as $T_{min}/10$ to ensure convergence in the computation.

In particle DEM, two types of bonds are usually defined: contact and parallel bonds (Itasca, 1995). The particles are only partially connected at the contact points which can change at different time steps. Under low confining stresses, the tangential bonds will be weak and the particles can flow like a fluid under very small shear stresses. Particle DEM runs according to a time-difference scheme with a force–displacement law to each contact and a contact updating scheme. There are two types of contact, which are ball–wall contact and ball–ball contact. In each cycle, the contact is updated from the known particles and known wall positions. Force–displacement law is applied on each contact, and a new contact force is then calculated according to the relative motion and constitutive relation. Based on the resultant force, the moment and contact acting on the particles, the law of motion is applied to each particle to update the velocity and direction of travel. Although the particle is rigid, the soft contact with finite normal stiffness allows a small overlap between the particles for the determination of stresses.

Discontinuous soil mass can be modelled with the help of a distinct-element method in the form of particle flow code, e.g. program PFC2D/3D (Itasca, 2004) or other codes. Open source particle flow codes are available which are useful for some relatively simple problems. Spherical particles interact through frictional sliding contacts. Simulation of joint bounded blocks may be realized through specified bond strengths. The law of motion is repeatedly applied to each particle and force–displacement law to each contact. The particle flow method enables modelling of granular flow, fracture of intact rock, transitional block movements, dynamic response to blasting or seismicity, and deformation between particles caused by shear or tensile forces. These codes also allow modelling of subsequent failure processes of rock slope, e.g. simulation of rock fracture (Eberhardt, 2003). The method of particle flow code employs a time-stepping, explicit calculation scheme (Cundall and Strack, 1979), which has advantages over traditional implicit calculation schemes in that it can handle a large number of particles with modest memory requirements because there are no matrices to be inverted because the solution scheme solves the full dynamic equations of motion by dynamic propagation of waves through the material, and the velocities of the waves are dependent on the stiffness, density and packing of particles (Hazzard et al., 1998). This also allows physical instabilities such as shear-band formation to be modelled without numerical difficulty because kinetic energy that accompanies shear-band formation is released and dissipated in a physically realistic way (Itasca, 1999). On the other hand, the use of dynamic relaxation to solve the system equation in DEM requires extensive cycle time to reach the equilibrium state. If the time-step chosen is excessive, the results obtained from the analysis can be meaningless. In practical application, greater care, knowledge and judgement are required for DEM as compared with FEM, and a large problem using DEM may require days or even weeks for the solution.

8.4.1 The force–displacement law and law of motion

The force–displacement law is described for both ball–ball and ball–wall contacts. The contact arises from contact occurring at a point. For ball–ball contact, the normal vector is directed along the line between the ball centres. For ball–wall contact, the normal vector is directed along the line defining the shortest distance between the ball centre and the wall. The contact force vector F_i is composed of normal and shear components in a single plane surface as

$$F_i = F_{ij}^n(t) + F_{ij}^s(t+\Delta t) \tag{8.78}$$

The force acting on particle i in contact with particle j at time t is given by

$$F_{ij}^n(t) = K_n(r_i + r_j - l_j(t)) \tag{8.79}$$

where r_j and r_i stand for particle i and particle j radii, $l_{ij}(t)$ is the vector joining both centres of the particles and k_n represents the normal stiffness at the contact. The shear force acting on particle i during a contact with particle j is determined by:

$$F_{ij}^s(t+\Delta t) = \pm\min(F_{ij}^s(t) + K_s\Delta s_{ij}, f\left|F_{ij}^n(t+\Delta t)\right|) \tag{8.80}$$

where f is the particle friction coefficient and k_s represent the tangent shear stiffness at the contact. The new shear contact force is found by summing the old shear force [min $F_{ij}(t)$]

with the shear elastic force. Δs_{ij} stands for shear contact displacement increment occurring over a time-step Δt.

$$\Delta s_{ij} = V_{ij}^{s} \Delta t \tag{8.81}$$

where V_{ij}^{s} is the shear component of the relative velocity at contact between particles i and j over the time-step Δt. The displacement and rotation of each block or particle are updated at every time-step according to the following

$$u^{(t+\Delta t/2)} = u^{(t-\Delta t)} + \left(\frac{\sum F_i^{(t)}}{m} + g_i \right) \Delta t \tag{8.82}$$

$$\theta^{(t+\Delta t/2)} = \theta^{(t-\Delta t/2)} + \left(\frac{\sum M_i^{(t)}}{I} \right) \Delta t \tag{8.83}$$

$$x_i^{(t+\Delta t)} = x_i^{(t)} + x_i^{(t+\Delta t/2)} \Delta t \tag{8.84}$$

$$\theta^{(t+\Delta t)} = \theta^{(t)} + \theta^{(t+\Delta t/2)} \Delta t \tag{8.85}$$

The motion of a particle is determined by the resultant force and moment acting on it. The motion induced by resultant force is called translational motion. The motion induced by resulting moment is rotational motion. The equations of motion are written in vector form as follows:

Translational motion:

$$\sum_j F_{ij} + m_i g + F_i^d = m_i x_i'' \tag{8.86}$$

Rotational motion:

$$\sum_j r_i F_{ij} + M_i^d = I_r \theta_i'' \tag{8.87}$$

where x_i" and θ_i" stand for the translational acceleration and rotational acceleration of particles i. I_i stands for moment of inertia. F_i^d and M_i^d are the damping force and damping moment, respectively. Unlike finite element formulation, there are now three degrees of freedom for 2D problems and six degrees of freedom for 3D problems. In Cundall and Strack's explicit integration distinct element approach, solution of the global system of equations is avoided by considering the dynamic equilibrium of the individual particles rather than solving the entire system simultaneously. That means that Newton's law of motion is applied directly. This approach also avoids the generation and storage of the large global stiffness matrix that will appear in finite element analysis. On the other hand, the implicit DDA approach that will be discussed later will generate a global stiffness matrix which is even larger than that in the finite element analysis. Actually, an implicit DEM formulation for DEM is possible, but this will also result in a large matrix so that many DEM codes do not adopt such a formulation.

For particle-based DEM, it has been found that there are large fluctuations in the results of analysis between particles, which reflect the influence of the local phenomenon. Such results are not surprising, as the results are highly sensitive to the interaction between particles and the time-step under which the results are monitored. Such local results can be meaningless unless the results are monitored over a long time and larger region. A number of quantities in a DEM model are hence defined with respect to a specified measurement circle. Since a measurement circle is used, the stress in the particle is described as the two in-plane forces acting on each particle per volume of the particle. Average stress is defined as the total stresses in the particles divided by the volume of measurement circle. Thus, the shape of a particle is not important, as the reported stress is easily scaled by volume unit. The average stress tensor σ_{ij} of the material is defined by

$$\sigma_{ij} = \frac{1}{V}\sum_{N}\sigma_{ij}^{p}V_{p} = \frac{1}{V}\int\sigma_{ij}^{p}dV \tag{8.88}$$

where N is the number of particles contained within V, σ_{ij}^{p} is the average stress tensor in the particle (p), and V_p is the volume of the particles.

A contact surface constructed among two block edges represents a rock joint. The data elements are formed to demonstrate point contacts for each pair of blocks that are in touch. This situation includes blocks touching along their common edge, a corner meets a corner or a corner contacts another corner, which are valued. Blocks may be hung-up by the sharp corner, and stress concentration would also be induced by a sharp corner. However, it is able to achieve a realistic representation by rounding the corners in order that blocks can slide past one another smoothly during such a situation. The blocks/particles can either be deformable or undeformable, and this will affect the model in generating the contact model. If the block/particle is deformable, overlapping is allowed, and it is advisable to define the contact in terms of the normal and shear contact relations. If the blocks/particles are undeformable, a directional spring may be sufficient to model the contact behaviour.

There are two major types of contacts for a block approach, namely corner-to-corner contacts and edge-to-corner contacts. The former is more critical because there is a rock joint closed along its whole length. It is assumed that the joints extending the joint among the two contacts separate it in half, with each mid-length supporting its own contact stress. In each associated length and point contact, it is required to compute the incremental normal and shear displacements. For the normal direction, it is assumed that the stress–displacement relationship is linear and is controlled by the normal stiffness k_n

$$\Delta\sigma_n = -k_n\Delta u_n \tag{8.89}$$

where

$\Delta\sigma_n$ = increment of effective normal stress

Δu_n = increment of normal displacement

Considering a joint, if the tensile strength T is exceeded or equivalently less than the effective normal stress, effective normal stress is equal to zero. Similarly, a constant shear stiffness (k_s) controls the shear strength. The normal and shear stiffnesses are difficult to measure directly, which is well-known for everyone working in DEM. The authors adopt the parameter-matching method, where the micro-parameters are tuned so that the macro-behaviour can be reflected in the computation. Such a technique has been used by the authors and many other researchers. The shear stress (τ_s) is governed by both cohesive strength (c) and frictional angle (ϕ).

$$\left|\tau_s\right| \leq c + \sigma_n tan\phi = \tau_{max} \tag{8.90}$$

$$\Delta\tau_s = -k_s \Delta u_s^\theta \tag{8.91}$$

or, if

$$\left|\tau_s\right| > \tau_{max} \tag{8.92}$$

$$\tau_s = sign\left(\Delta u_s\right)\tau_{max} \tag{8.93}$$

where
Δu_s^θ = the incremental shear displacement in elastic component
Δu_s = the total incremental shear displacement.

For a blocky system, the joint/fracture and spacing are usually specified for a solution domain. More than one set of discontinuity can be defined for a system, and generally, a system will be less stable when the number of discontinuity sets increases. For the problem in Figure 8.10 and Table 8.1, if there is only one joint set in the problem, the tunnel is safe for different type of shapes. On the other hand, two joint sets can lead to failure and falling of blocks easily. The authors have developed a simple block generation scheme based on the joint set information as given by the users. More advanced options for block generation are available in some commercial programs. With reference to Figure 8.11, two set of joints are used to define the blocks for a system by Cheng and Zhang (2008), and some triangular blocks will be formed due to the formation of the horseshoe tunnel. Due to the excavation, the self-weight of the blocks gradually induces failure of the system, and the progressive failure and falling of blocks are seen in Figure 8.11. The formation of a realistic blocky system can be very important for the proper analysis of many underground construction works.

8.4.2 Limitations of the distinct element method

A stiffness model, a slip model and a bonding model. The stiffness model provides an elastic relationship between the contact force and relative displacement. The slip model enforces a relationship between the shear and normal contact forces such that the two contacting

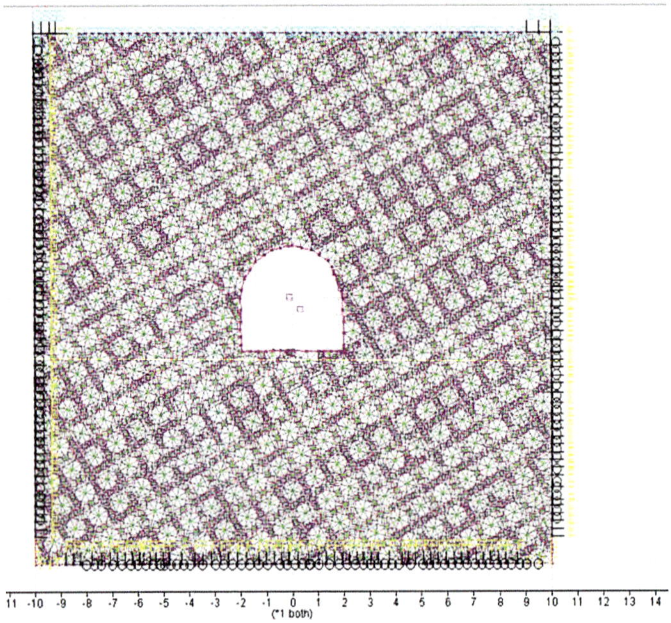

Figure 8.10 Two set Joint 30 and120 (deg) with 1 m (x direction) spacing for a horseshoe tunnel.

balls may slip relative to one another. The bonding model serves to limit the total normal and shear forces that the contact can carry by enforcing the bond-strength limits.

A number of quantities in a PFC model are defined with respect to a specified measurement circle. These quantities include coordinate number, porosity, sliding fraction, stress and strain rate. The coordination number and stress are defined as the average number of contacts per particle. Only particles with centroids that are contained within the measurement circle are considered in computation. In order to account for the additional area of particles that is being neglected, a correction factor based on the porosity is applied to the computed value of stress.

Since a measurement circle is used, stress in a particle is described as the two in-plane forces acting on each particle per volume of particle. Average stress is defined as the total stress in a particle divided by the volume of the measurement circle. Thus, the shape of a particle is regardless of the average stress measurement because the reported stress is easily scaled by volume unity. The reported stress is interpreted as the stress per volume of the measurement circle.

While distinct element methods are powerful tools for numerical analysis of discontinuous materials in geotechnical problems and geological mechanics, there are also some fundamental limitations to be considered. The granular materials are considered to be packed assemblies of particles in DEM simulation, and the mechanical interaction between particles is assumed to be simple. However, in reality, the contact between particles is highly complex and it is difficult to be detected explicitly. Also it is difficult to generate complicated model geometry. In addition, Thornton (2015) mentioned the development

Table 8.1 The DEM analysis results under different number sets of joints and tunnel shape

Shape	*Number sets of joint*	*Joint dip angle*	*Cycle*	*Time*	*Maximum displacement*	*Maximum principal stress*	*Minimum principal stress*	*Maximum joint shear displacement*	*Remark*
Unit	–	*Deg*	–	*s*	*m*	*Pa*	*Pa*	*m*	–
Circular	1	30	20000	1.982E+00	2.804E-02	3.407E+06	–2.683E+07	2.579E-03	–
	2	30&120	10000	1.172E+00	6.339E-01	1.754E+06	–3.301E+07	6.122E-01	Block failure
	1	60	10000	1.810E+00	3.081E-02	2.448E+06	–2.143E+07	1.111E-02	–
	2	60&150	20000	2.120E+00	1.872E+00	8.879E+06	–3.677E+07	9.279E-01	Block failure
Rectangle	1	30	5000	1.160E+00	4.345E-02	2.119E+06	–3.107E+07	2.449E-02	–
	2	30&120	5000	6.850E-01	1.804E-01	3.523E+06	–4.433E+07	1.120E-01	Block failure
	1	60	10000	1.791E+00	4.346E-02	3.446E+06	–3.324E+07	2.818E-02	–
	2	60&150	10000	1.155E+00	6.353E-01	6.437E+06	–4.448E+07	3.049E-01	Block failure
Horseshoe	1	30	10000	1.984E+00	3.118E-02	2.635E+06	–2.586E+07	2.347E-02	–
	2	30&120	10000	1.168E+00	6.246E-01	3.443E+06	–3.072E+07	5.922E-01	Block failure
	1	60	10000	1.796E+00	3.789E-02	3.581E+06	–3.132E+07	2.256E-02	–
	2	60&150	20000	2.124E+00	8.899E-01	5.763E+06	–3.548E+07	7.203E-01	Block failure

Figure 8.11 Generation of blocks based on two discontinuity sets for a horseshoe tunnel, with the progressive failure and free falling of blocks.

of accurate contact constitutive models for use in discrete element analysis as being non-trivial. The effective stress in a soil will govern its response and including pore water pressures in a discrete element framework is non-trivial. The relationship between the micro-parameters and the macro-parameters appears to be only partially considered, and most engineers determine the micro-parameters by a back-analysis with some basic test results. Empirical relationships between the micro- and macro-parameters are available for some problems, but the applicability of such empirical relations to general conditions is questionable.

8.5 DISCONTINUOUS DEFORMATION ANALYSIS (DDA)

The discontinuous deformation analysis (DDA) method is an alternative to the DEM by Cundall, which was proposed by Shi and Goodman (1985), Shi (1988) and Shi (1993). The global equilibrium equations are established by minimizing the total potential energy, which is similar to FEM. Cheng (1998) found that if the no-penetration criteria as originally proposed by Shi and Goodman (1985) are adopted, oscillation in computation is common. Cheng has relaxed this requirement with the introduction of normal and tangential springs instead of the penalty as in the original formulation to achieve better modelling of the joint condition.

Shi (1988, 1993) did not adopt the concept of joint model in DDA but used a penalty vector (large value) to remove the penetration between blocks. This penalty vector follows the direction of the displacement and is just a conventional elastic spring. This approach is greatly different from a realistic joint model where the normal and shear stiffnesses usually differ by a great amount. Cheng (1998) proposed to use a simple but realistic joint model incorporating normal and shear stiffness instead of a penalty vector. Consider Figure 8.12, where a corner moves from position a to b so that the corner is located inside a block (penetration allowed). If we separate the normal and tangential effects by using the concept of normal stiffness k_n and shear stiffness k_s, the work done, II, by the normal displacement d_n and tangential displacement d_s is given by

$$\text{II} = \frac{1}{2}(k_n d_n^2 + k_s d_s^2) \tag{8.94}$$

where the normal displacement d_n and tangential displacement d_s are given by $d\cos\phi$ and $d\sin\phi$, and ϕ is the angle between the displacement vector and the face of a block. If a,b

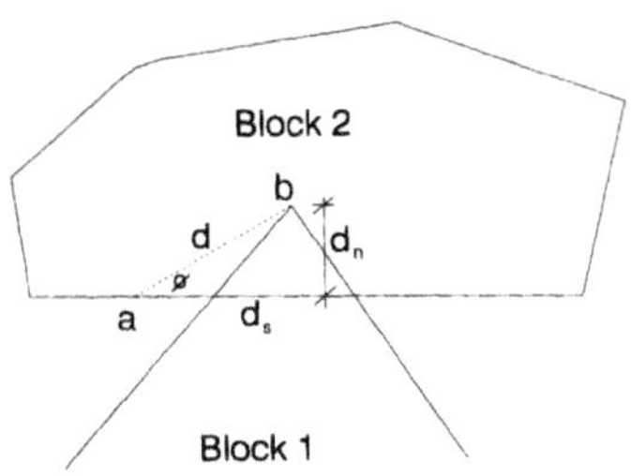

Figure 8.12 Penetration of a corner into a block.

represent the point of intersection of the displacement vector with the face of a block and the final position of vertex, and let the coordinates of a and b be denoted by (x_1,y_1) and (x_2,y_2), while the displacements caused by the internal forces arising from the penetration are given by (u_1,v_1) and (u_2,v_2) for points a,b. Eq. (8.94) can be represented by

$$\Pi = \frac{1}{2}[k_n \sin^2\phi + k_s \cos^2\phi][u_1, v_1]\cdot\begin{bmatrix}u_1\\v_1\end{bmatrix} - 2[u_1, v_1]\cdot\begin{bmatrix}u_2\\v_2\end{bmatrix} + [u_2, v_2]\cdot\begin{bmatrix}u_2\\v_2\end{bmatrix}$$

$$+2(x_1 - x_2, y_1 - y_2)\cdot\begin{bmatrix}u_1\\v_1\end{bmatrix} - 2(x_1 - x_2, y_1 - y_2)\cdot\begin{bmatrix}u_2\\v_2\end{bmatrix} + (x_1 - x_2, y_1 - y_2)\cdot\begin{bmatrix}x_1 - x_2\\y_1 - y_2\end{bmatrix} \quad (8.95)$$

Using the deformation matrix [D] as given by Shi

$$\begin{bmatrix}u_1\\v_1\end{bmatrix} = [T_1][D_1] \quad \text{and} \quad \begin{bmatrix}u_2\\v_2\end{bmatrix} = [T_2][D_2] \quad (8.96)$$

Eq. (8.95) can be simplified by expressing using [T] and [D] as

$$\begin{aligned}\Pi = \frac{1}{2}[k_n \sin^2\phi + k_s \cos^2\phi]\{[D_1]^T[T_1]^T[T_1]\,[D_1] - 2[D_1]^T[T_1]^T[T_2]\,[D_2]\\ + [D_2]^T[T_2]^T[T_2]\,[D_2] + 2[D_1]^T[T_1]^T\begin{bmatrix}x_1 - x_2\\y_1 - y_2\end{bmatrix}\\ - 2[D_2]^T[T_2]^T\begin{bmatrix}x_1 - x_2\\y_1 - y_2\end{bmatrix} + (x_1 - x_2, y_1 - y_2)\begin{bmatrix}x_1 - x_2\\y_1 - y_2\end{bmatrix}\}\end{aligned} \quad (8.97)$$

The contribution of the joint normal and shear stiffness to the equilibrium coefficient matrix [k] is given by the minimization of the energy with respect to the displacement as

$$[k_{11}] = [k_n \sin^2\phi + k_s \cos^2\phi][T_1]^T[T_1]$$

$$[k_{12}] = -[k_n \sin^2\phi + k_s \cos^2\phi][T_1]^T[T_2]$$

$$[k_{21}] = -[k_n \sin^2\phi + k_s \cos^2\phi][T_2]^T[T_1]$$

$$[k_{22}] = [k_n \sin^2\phi + k_s \cos^2\phi][T_2]^T[T_2] \quad (8.98)$$

The equivalent forces are given by

$$[F_1] = -[k_n \sin^2\phi + k_s \cos^2\phi][T_1]^T\begin{bmatrix}x_1 - x_2\\y_1 - y_2\end{bmatrix} \quad (8.99)$$

$$[F_1] = -[k_n \sin^2\phi + k_s \cos^2\phi][T_2]^T\begin{bmatrix}x_1 - x_2\\y_1 - y_2\end{bmatrix} \quad (8.100)$$

In using Eqs. (8.94)–(8.100), the concept of simple joint model is incorporated into DDA by Cheng (1998), which is better than the penalty vector by Shi.

Although DDA appears to be close to DEM in many respects, to be more accurate, it is actually closer to the FEM because:

1. Minimization of the total potential energy is used to establish the equilibrium equations.
2. The displacements of the blocks are used for the simultaneous equilibrium equations.
3. Submatrices/vectors for the stiffness, mass and loading are formed and added to the equilibrium equations, which is very similar to the FEM in many respects.
4. Displacement locking of contacting blocks is used in DDA, which is similar to the bar elements in FEM.

8.6 NUMERICAL MANIFOLD METHOD (NMM)

The numerical manifold method (NMM) was also proposed by Shi in 1993. It is a very flexible numerical analysis method which contains and combines widely used FEM and joint- or block-oriented DDA in a unified form. Based on block system theory, this method combines the advantages of DDA, FEM and analytical methods together. In this method, the concept of numerical manifold in modern mathematics is applied, the continuous and discontinuous cover functions are used to make a unified form to solve both continuous and discontinuous problems. It can be reduced to the classical finite element method when no discontinuity occurs or the DDA method when a mathematical mesh is the same as a physical mesh. NMM can be used for general conditions which are found in many geotechnical problems. NMM is attracting increasing numbers of researchers worldwide, especially from geotechnical areas. Since 1991, when this method was first proposed by Shi, many developments and extensions to NMM have been accomplished.

Lin and Mo (1994) proposed a manifold method with the element free Galerkin method. Zhou et al. (1997) modelled bending of beam and crack propagation (1999) by the element free manifold method. Koo and Chern (1996) completed the extension of displacement cover functions from one order to high order for achieving higher accuracy in analysis. Shyu and Salami (1995) introduced a four-node isoparametric finite element in the manifold method. Ohtsubo (1997) compared the results with different covers or different nodes in NMM. Wang et al. (1999) studied the manifold method with four physical covers forming an element. Although NMM is attracting the efforts of many researchers, there are still many limitations to it at present which are discussed later. Qiu (1996) proposed the manifold method without the use of penalty springs. Wang et al. (1995) studied the approximation theories for the manifold method. There are many research works in applying the manifold method to fracture mechanics. Zhang (1997, 1999) simulated crack propagation and thermal fracture by NMM and modelled crack propagation by coupling the boundary element method with the manifold method. Chiou (1997, 2000, 2002) and Tsay (1999) studied the prediction of crack growth and mixed-mode fracture propagation by the manifold method, while Zhou et al. (1999) also studied crack propagation problems using a slightly different approach.

The basics of DDA and NMM are well covered by the works by Shi, and their works will not be repeated here. Cheng has some works in three-dimensional DDA and manifold methods, and these works are covered here.

Based on the finite cover systems, the manifold method has the potential to meet more engineering requirements. The term "manifold" here comes from the topological manifold and differential manifold, which is the main subject of differential geometry, algebraic topology, differential topology and modern algebra of mathematics. The difference between the "numerical manifold" here and the traditional differential manifold is as follows: the global functions of the differential manifold are highly differentiable and entirely defined without reference to the covers; the global functions of the numerical manifold here are defined based on the covers and are only piecewise differentiable, mostly discontinuous on the contact interfaces.

The manifold connects many individual folded domains together to cover the entire material volume. Then, the global behaviour can be computed by functions defined in local covers. The new method has separated and independent mathematical covers and physical mesh: the mathematical covers define only the fine or rough approximations such as the real material boundary, while the physical mesh defines the integration fields.

The mathematical covers are chosen by engineers and may consist of finite overlapping covers which occupy the whole material volume. The conventional meshes and regions, such as regular grids or finite element meshes, can be transferred to finite mathematical covers. Based on finite covers, the manifold method is flexible enough to contain and combine well-developed analytical methods, the widely used FEM method and block-oriented DDA method in a unified form.

The physical mesh includes the boundaries of material volume, joints, blocks and the interfaces of different material zones. The constantly changing water surfaces are also part of the physical meshes. The physical mesh represents material conditions which cannot be chosen artificially. The physical cover system is formed by both mathematical covers and physical meshes. If the joints or block boundary divide a mathematical cover to two or more completely disconnected domains, those domains are defined as physical covers. Therefore, physical covers are a subdivision of mathematical covers by discontinuities. The manifold method is more suitable for computing large deformations, moving boundaries of both continuums and jointed or blocky materials.

The manifold method can perform the computations of a finite element method for continuous material. After transferring the finite element mesh to finite covers of the manifold method, the joint lines can be defined as the new physical mesh. The same finite element mesh can then compute joints in the same material volume. The finite element meshes can be used to define finite covers for the manifold method. Considering any node, all elements having this node form a mathematical cover (called a "star" in algebraic topology).

In Figure 8.13, the mathematical covers are as follows:

Node	*Element*	*Element*	*Element*
1	1,2,3		
2	1,2,3	2,4,5	2,5,3
3	1,2,3	2,5,3	3,5,6
4	2,4,5		
5	2,4,5	2,5,3	3,5,6
6	3,5,6		

Any original finite element is the common area of the mathematical cover of its nodes. The physical covers are defined as follows:

(1) The region of physical cover is the materials contained in the mathematical cover, or mathematically speaking the intersection of the mathematical cover and the material field.
(2) If the material boundaries, block boundaries or fractures divide the mathematical cover to totally isolated regions, each region is a physical cover. Therefore the physical covers are the subdivision of the mathematical covers.

The mathematical covers and physical covers of Figure 8.13 are as follows:

Math	*Physical cover*	*Physical cover*
V_1	1_1	
V_2	2_1	2_2
V_3	3_1	3_2
V_4	4_1	4_2
V_5	5_1	5_2
V_6	6_1	6_2

The displacement functions are independent of the material boundary. If the material occupies only part of the element, the displacement functions are still the same. the manifold element is the common area of several physical covers, the global displacement functions are taken as a percentage from the displacement functions of these covers. The meaning of weight functions $w_{e(i)}(x,y)$ is the weighted average, which is taken from a percentage from each node displacement $u_{e(i)}, v_{e(i)}$. The way to define a weight function is very

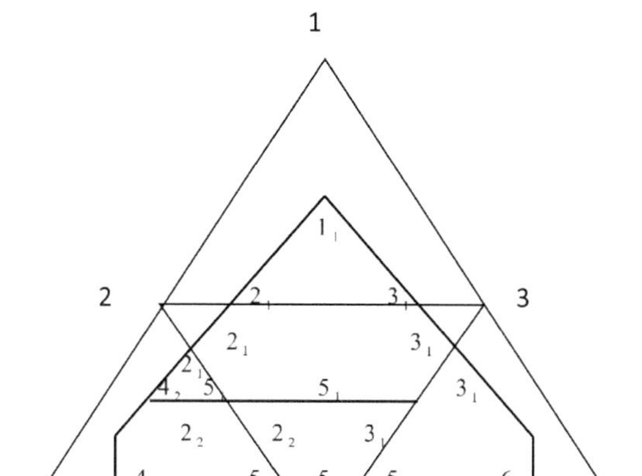

Figure 8.13 Finite element covers of two blocks.

important in the analysis. Suppose element E is the common area of covers $(C_i, i = 1,2,..m)$, the displacement functions can be expressed as:

$$u(x,y) = \sum_{i=1}^{m} w_i(x,y) u_i(x,y)$$
$$v(x,y) = \sum_{i=1}^{m} w_i(x,y) v_i(x,y) \tag{8.101}$$

where $w_i(x,y)$ is the weight function of cover i and satisfies the following equations:

$$0 \le w_i(x,y) \le 1, \quad (x,y) \in C_i; w_i(x,y) = 0, \quad (x,y) \notin C_i \tag{8.102}$$

$$\sum_{i=1}^{m} w_i(x,y) \equiv 1, \quad (x,y) \in E \tag{8.103}$$

Eq. (8.101) can be expressed as:

$$\begin{Bmatrix} u(x,y) \\ v(x,y) \end{Bmatrix} = \begin{bmatrix} w_i(x,y) f_{ij}(x,y) & 0 \\ 0 & w_i(x,y) f_{ij}(x,y) \end{bmatrix} \begin{Bmatrix} d_{i,2j-1} \\ d_{i,2j} \end{Bmatrix} \tag{8.104}$$

where d_{ij} are the unknowns and f_{ij} is the basic function $x^k y^l$.

For a triangular finite element cover, the manifold element E is the common area of covers i, j, k. The shape functions of the element are

$$\begin{pmatrix} N_i(x,y) \\ N_j(x,y) \\ N_k(x,y) \end{pmatrix} = \frac{1}{\Delta} \begin{pmatrix} x_j y_k - x_k y_j & y_j - y_k & x_k - x_j \\ x_k y_i - x_i y_k & y_k - y_i & x_i - x_k \\ x_i y_j - x_j y_i & y_i - y_j & x_j - x_i \end{pmatrix} \begin{pmatrix} 1 \\ x \\ y \end{pmatrix} \tag{8.105}$$

where $\Delta = (x_j - x_i)(y_k - y_i) - (x_k - x_i)(y_j - y_i)$. If we take this, it is easy to prove that $w_i(x,y)$ fulfils Eqs. (8.103) and (8.104). The shape functions of a triangular finite element can hence be taken as weight functions of the manifold element. In quadrilateral finite element covers, the shape functions of a quadrilateral isoparametric element can be expressed as:

$$N_i = \frac{1}{4}(1 + \xi_0)(1 + \eta_0) \quad (i = 1,2,3,4) \tag{8.105}$$

where $\xi_0 = \xi_i \xi$, $\eta_0 = \eta_i \eta$. It is easy to prove $\sum_{i=1}^{4} N_i(\xi, \eta) = 1$ and $0 \le N_i(\xi,\eta) \le 1$. Similarly we can take $w_i(x, y) = N_i(\xi, \eta)$ as the weight functions of a manifold element.

8.6.1 Equilibrium equations

The total potential energy Π is the summation of all the potential energy sources: individual stress and forces. In the following, the potential energy of each force or stress and their differentiations are computed separately.

(1) The strain potential energy Π_e produces the stiffness matrix;
(2) The potential energy Π_σ of initial stresses produces the initial stress matrix;
(3) The potential energy Π_p of the point load produces the point load matrix;
(4) The potential energy Π_w of the body load produces the body load matrix;
(5) The potential energy Π_i of inertia produces mass matrix;
(6) The strain potential energy Π_s of contact springs produces contact matrices;
(7) The potential energy Π_f of friction forces.

The details of these submatrices are given by Shi (1993) and extension works by Cheng (1998) which will not be repeated here.

In real geotechnical problems, it is commonly found that a domain under investigation may be composed of zones of continuous and discontinuous media. The coupling method of DDA and FEM has been developed by the authors previously for this kind of problem. DDA is applied to discontinuous zones, while FEM is applied to continuous zones. On the interface, the displacement and stress are transferred from one zone to another step by step to keep the continuity and compatibility. In theory, the condition of compatibility may be reached if the time step is small enough (close to zero). In practice, it is impossible to take a time step close to zero because DDA is a dynamic analysis based on time step. A very small time step means tremendous computation time, which may be intolerable.

The manifold method applies finite covers to cover the entire material volume. On each cover, the manifold method defines an independent cover displacement function. The cover displacement functions on individual covers are connected together to form a global displacement function on the entire material volume. The global displacement functions are the weighted averages of local independent cover functions on the common part of several covers. Using the finite cover systems, continuous, jointed or blocky materials can be computed in a mathematically consistent manner. The condition of compatibility is satisfied automatically.

A manifold analysis mesh of tunnelling excavation is shown in Figure 8.14 for illustration purposes. The area of computation is 100 m × 100 m with two joints. A tunnel with a 5 m half-circle on top is excavated in the centre of the computing area. Triangular finite element covers are applied to generate 208 manifold elements. The material properties used for rock material are: $E = 3 \times 10^{10}$ Pa, $\mu = 0.24$, $\rho = 2 \times 10^3$ kg/m^3, $c = 2 \times 10^6$ Pa,

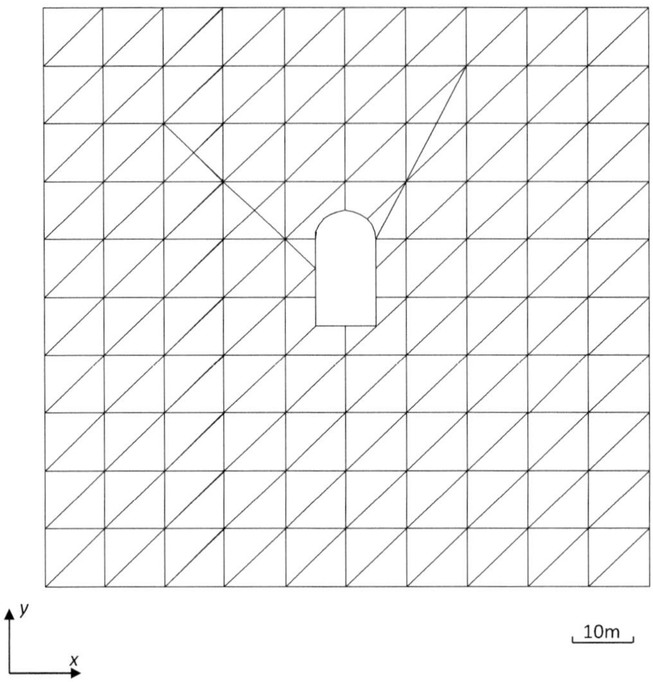

Figure 8.14 Manifold element mesh based on triangular finite cover.

$\phi = 42°$, $K_n = 8 \times 10^9$ Pa/m, $K_s = 2.7 \times 10^9$ Pa/m. The normal stiffness and shear stiffness for the joints are 200 MPa and 100 MPa. the cohesive strength and friction angle of the joint are 0.2 Mpa and 20°, respectively.

Initial stresses are calculated by the gravity of rock material. The time step is taken as 0.001 s. After calculation for 200 time steps, the displacement distribution around the tunnel is shown in Figure 8.15a. The deformations at the joint nodes A and B are discontinuous, while the deformations at other nodes are continuous.

The displacement distribution around the tunnel by the DDA method is shown in Figure 8.15(b). It is very clear that all the point displacements around the tunnel are discontinuous and larger than those using the manifold method. This result is reasonable as the DDA method applies discontinuous block theory while the manifold method applies continuous mathematical covers. For the manifold method, discontinuous deformation only occurs at the physical boundary, joints and interfaces, while continuous deformation will occur for all continuous areas, which is actually correct for the present problem.

Figure 8.16 shows the displacements and deformation in the computing area by the manifold method. The whole system is found to be stable with only a small deformation at the two joints from the manifold calculation. The whole system is unstable and destroyed with large deformation and slippage around the tunnel.

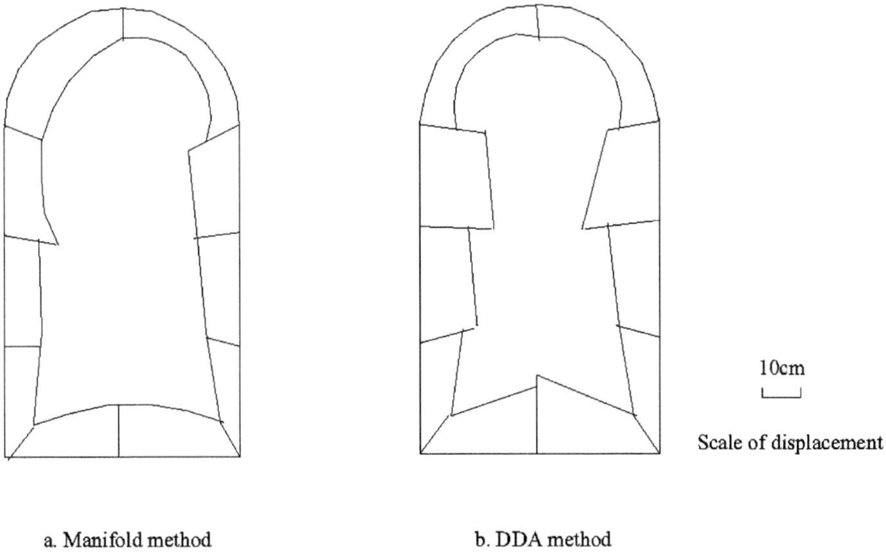

Figure 8.15 Displacement distribution around the tunnel.

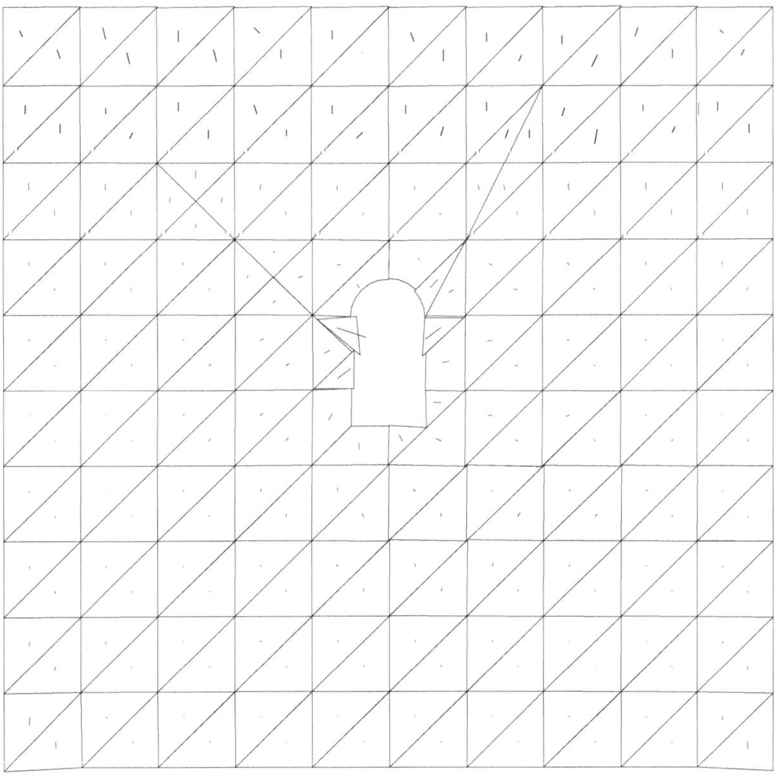

Figure 8.16 Displacement and deformation by manifold method.

8.6.2 Coal extraction – compared with coupling method of DDA and FEM

The second example is shown in Figure 8.17. The whole region under the study is 700 m × 550 m. The two vertical sides are fixed along the X direction, while the lower boundary is fixed along the X and Y directions. The inside rectangular region with a dimension of 200 m × 60m is analysed using the DDA method, while the outer region is analysed by FEM. The four seams of coat with a height of 2.5 m for each seam were dug in 10 steps, with the length of each step being 10 m. The material properties and parameters used for the calculation are: $E = 1 \times 10^{10}$ Pa, $\mu = 0.34$, $t = 0.001$ s, $\rho = 2 \times 10^3$ kg/m^3, $c = 5 \times 10^3$ Pa, $\phi = 25°$, $K_n = 8 \times 10^9$ Pa/m, $K_s = 2.7 \times 10^9$ Pa/m.

The rectangular isoparametric cover of the manifold method is used to simulate the same example. The manifold element mesh is shown in Figure 8.18. The computing grid for the inter-region adjacent to excavation is the same as by the DDA method. In this region, each block is a physical cover, hence there is no overlap between any two covers, and the manifold method is automatically transformed to DDA in this case. The results from excavation using manifold analysis after 10 steps are shown in Figure 8.19. It is found that the local displacements and deformations by the manifold method are slightly less than those by the coupling method and the trend in movement is very similar. That means that the coupling method and manifold method basically perform similarly, which also demonstrates the capability of the manifold method in replacing the coupling method in analysis. There are, however, some small differences in the results from the two different

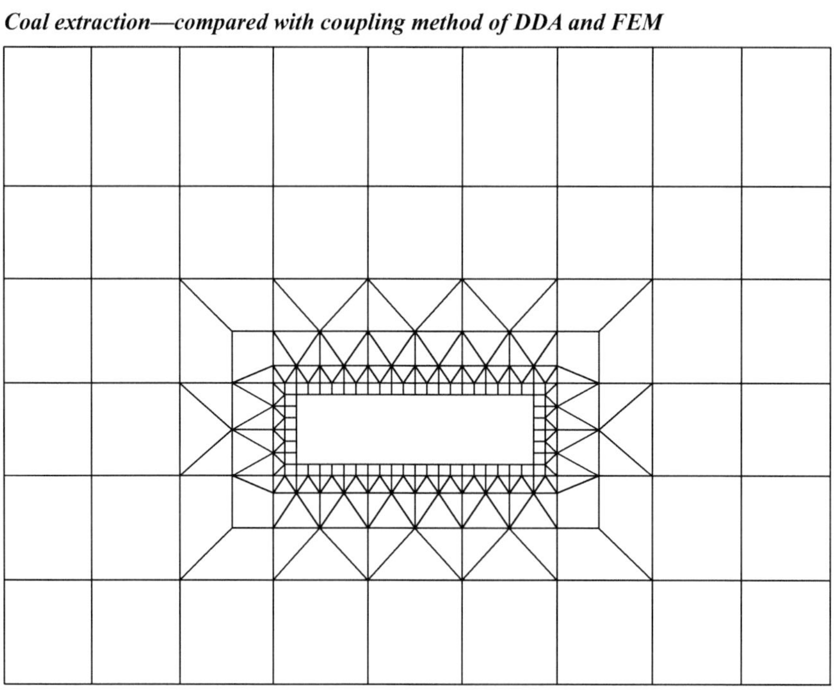

Figure 8.17 Coal extraction.

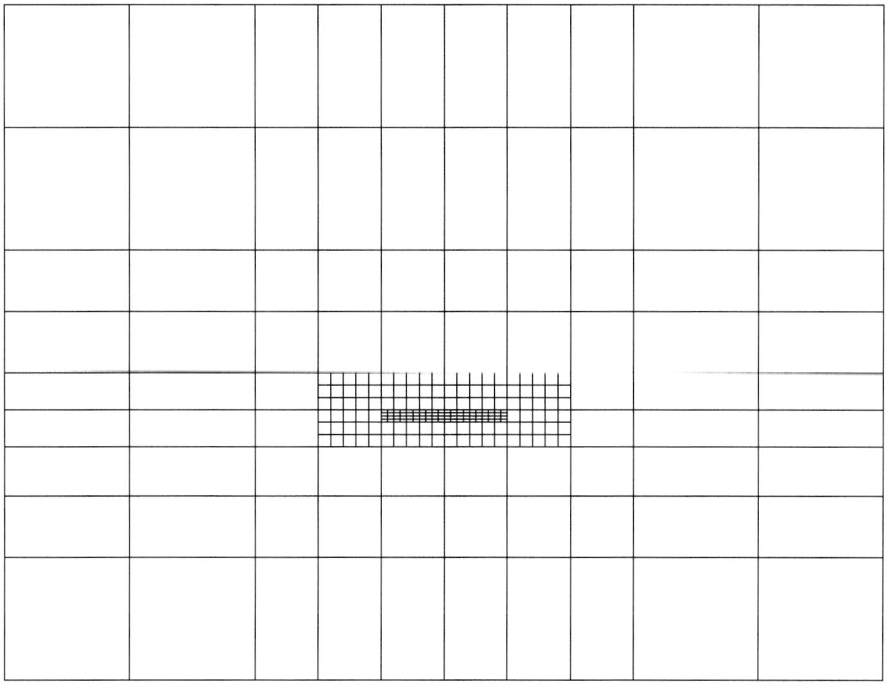

Figure 8.18 Manifold element mesh using rectangular finite element cover.

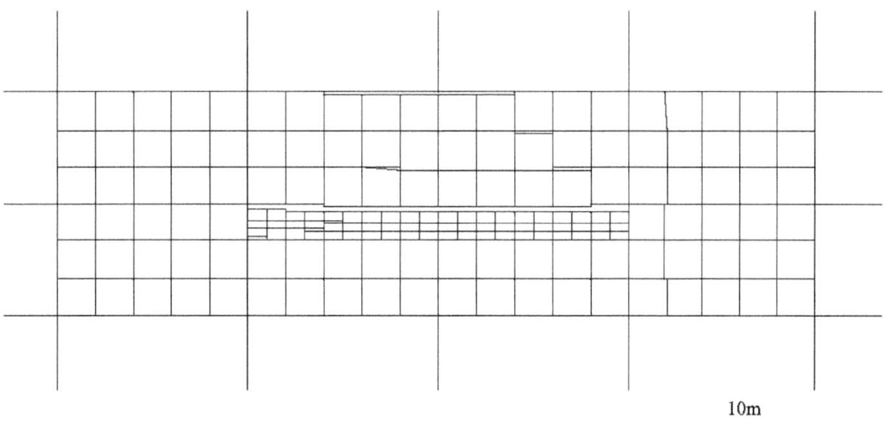

Figure 8.19 Deformation of excavation region after 10 steps of digging by manifold method.

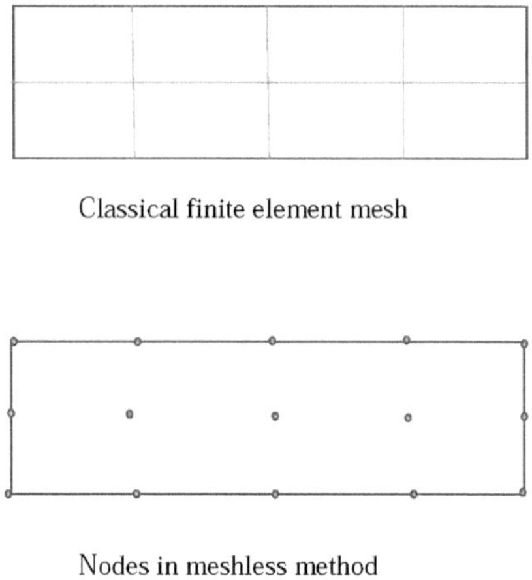

Figure 8.20 Comparisons between Finite element method and meshless method.

analyses. Cheng et al. (2002), Cheng and Zhang (2003) and Cheng and Zhang (2004) applied DEM and manifold methods to a series of underground excavation problems, and similar works have been carried out by many other researchers.

8.7 MESHLESS METHODS

The major limitations of the classical numerical methods (FEM, FDM, BEM) include the difficulty in dealing with interfaces, large deformation, separation of materials, moving domains and other issues. There are actually many engineering problems which are not well modelled by the classical continuity-based methods. The authors have considered a pile penetration problem where the pile tip was moved by 10 m using the program Abaqus. Continuous remeshing and mapping of the stresses from the old geometry to new geometry have to be carried out, which is an extremely time-consuming and tedious process.

For many process like debris flow, sand flow in silos, mixing of the geomaterials and separation of blocks are common, the continuity-based methods are totally useless in these cases. One of the critical reasons for the development of the meshless method is that mesh-free and mesh-adaptive discretizations are often well suited to cope with the geometric changes of the domain of interest, e.g. free surfaces and large deformations. Meshfree methods are based only on a set of independent points without a mesh, as shown in Figure 8.20. The nodal distribution in Figure 8.20 does not really form a mesh, as the relation between each node is not required to construct the approximation (or interpolation) functions of the unknown variable field functions. The only information required by truly meshless methods is the spatial location of each node discretizing the problem domain. The nodes can be spaced regularly or irregularly, and to a certain extent this is similar to the graded mesh generation in the classical FEM. More nodes can be placed at regions with rapid changes in the stresses and strains, and fewer nodes can be placed at other regions. The classical FEM relies on the use of local approximation polynomials, and may fail to work properly when there is a high local oscillatory solution or high deformation locally, which often appears when very high-order interpolation functions are used. The meshless method provides a good solution for this p-refinement problem of FEM with the help of kernel functions without increasing the cost of the solution, but keeping a reasonable degree of accuracy. Meshless method shape functions have greater smoothness, which means that one can go to higher order derivatives as well. Many traditional numerical methods (finite differences, finite elements or finite volumes) have trouble with high-dimensional problems, but the uses of the meshfree method for three-dimensional methods appear to be natural.

Different meshfree methods have been proposed and evolved after the development of the first meshfree methods (the smoothed particle hydrodynamics [SPH] by Gingold and Monaghan 1977 and Lucy 1977). These include the diffuse element method (DEM) by Nayrole et al. (1992), the element-free Galerkin method (EFG) by Belytschko et al. (1994), the reproducing kernel particle method (RKPM) by Liu et al. (1995), the partition of unity finite element method (PUFEM) by Babuska and Melenk (1997), the h-p Clouds by Duarte and Oden (1996), the moving least-square reproducing kernel method (MLSRK) by Liu et al. (1997), the meshless local boundary integral equation method (LBIE) by Zhu and Atluri (1998), the meshless local Petrov Galerkin method (MLPG) by Atluri and Shen (2002), the meshless point collocation methods by Aluru (2000), the meshless finite point method by Oñate and Idelsohn (1998) and more. Broadly speaking, a meshless method can be grouped by the definition of the shape functions and/or the minimization method of the approximation. The minimization may be via a strong form as in the point collocation approach (with no associated mesh) or a weak form as in the Galerkin methods, which actually require an auxiliary mesh or cell structure.

Meshfree methods have been under active research recently, particularly for three-dimensional problems. One of the main reasons for this is the difficulty in generating good mesh for problems with complicated geometry.

One of the first meshless methods proposed was the smooth particle hydrodynamics (SPH) method, which was the basis for a more general method known as the reproducing kernel particle method (RPKM). Starting from a completely different and original idea, the moving least squares shape function (MLS) has become very popular in the meshless community. More recently, the equivalence between MLS and RPKM for polynomial basis has been proven, so that both methods may now be considered to be based on the same shape functions. The MLS shape function has been successfully used in a weak form (Galerkin) with a background grid for the integration domain by Nayroles et al. (1992) and, in a more accurate way, by Belytschko and his co-workers. Oñate and Idelsohn (1988) used MLS in a strong form (point collocation) avoiding the background grid. Liu et al. (1997) used the RPKM in a weak form, while Aluru (2000) used it in a strong form. Other authors use different integration rules or weighting functions with the same shape functions.

Some critical weaknesses of meshless methods are:

1) In some cases, it is difficult to introduce the essential boundary conditions.
2) For some methods, it is laborious to evaluate the shape function derivatives.
3) Often, too many Gauss points are needed to evaluate the weak form.
4) The shape functions usually have a continuity order higher than C0. This decreases the convergence of the approximation and makes it more difficult to introduce discontinuities such as those due to heterogeneous material distributions.

5) Some methods do not work for irregular point distributions, or need complicated node connectivity to give accurate results.

It is probable that some of the drawbacks just mentioned are the reasons why some meshless methods have not been successfully used in 3D problems.

In the FEM, the nodal connectivity is predefined by the use of elements, hence there is no difficulty in carrying out the integration to determine the stiffness matrix. In the meshfree method, there is no element, hence, for each interest point x_i of the problem, and domain concentric areas or volumes are defined, and the nodes inside these areas or volumes belong to the influence-domain of node x_i. In general, the interest points are equivalent to the integration points from the background integration mesh. For methods using the collocation point methods or the nodal integration scheme, the interest points are the nodes of the nodal discretization. The shape and size of the influence-domain, which depends on the relative position of the interest point, affect the quality of the results. It is recommended that all the influence-domains possess approximately the same number of nodes inside, and the size of the influence-domain should be dependent on the nodal density around the interest point. For irregularly spaced nodal points, the use of a fitted Gaussian integration mesh to achieve each influence-domain that possesses an equal amount of nodes is better than a regular influence-domain. A new approach to establish the influence-domains in meshless methods was proposed by Belinha et al. (2009). Instead of using the classical fixed influence-domains, a spatial collocation of the nodes discretizing the problem domain to determine directly the influence-domains is used, using mathematical concepts such as Voronoï diagrams and Delaunay triangulation to determine the nodal connectivity of each node belonging to the global nodal set. Since these influence-domains are determined based on the geometric and spatial relations between the Voronoï cells obtained from the Voronoï diagram of the nodal distribution, these influence-domains are called "influence-cells".

8.7.1 Smoothed particle hydrodynamics method (SPH)

Smoothed particle hydrodynamics is a Lagrangian method, which means that an object is represented by a set of particles, and each particle has its own material properties and relation with the nearby particles at each time step (Liu and Liu, 2003). The representations of field function *f(x)* at particle *i* could be written in discretized particle approximation form as

$$f(x_i) = \sum_{j=1}^{n} \frac{m_j}{\rho_j} f(x_j) W(x_i - x_j, \mathrm{h})\mathrm{dx} \tag{8.106}$$

where the support domain is perimeter with radius k_h, m_j and ρ_j are the mass and density of an arbitrary particle j within the support domain, respectively, W is the smoothing function, and h is the smoothing length to control the width of the smoothing function and number of neighbouring particles within the smoothing function, which governs the accuracy of simulation by limiting total number of particles as shown in Figure 8.21.

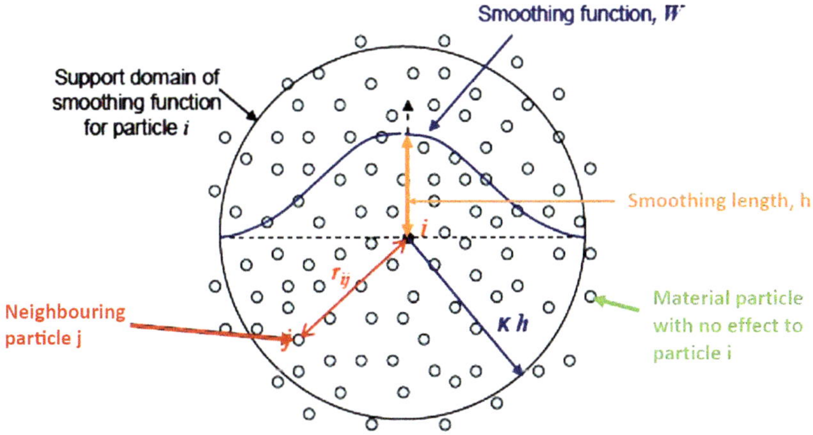

Figure 8.21 The affected neighbouring particles to particle *i* within the support domain.

A series of smoothing functions is introduced and applied in SPH for various problems. The cubic spline kernel function, which has been proposed by Monaghan and Lattanzio (1985), is chosen for the following example, as it has a narrower compact support compared with Gaussian kernel function. Therefore, the smoothing function in two dimensions could be expressed as:

$$
\begin{aligned}
W &= \frac{1}{\pi h^2}\left(\frac{15}{10} - \frac{15}{7}s^2 + \frac{15}{14}s^3\right) && \text{when } 0 \le s < 1 \\
&= \frac{1}{\pi h^2}\left(\frac{5}{14}(2 - s^2)\right) && \text{when } 1 \le s < 2 \\
&= 0 && \text{when } s \ge 2
\end{aligned}
\tag{8.106}
$$

where $s = r_{ij}/h$, and r_{ij} is the distance between particles *i* and *j*. Based on Figure 4.54 and Eq. (8.107), the smoothing function tends to zero when the radius of the support domain is equal to $2h$. Chen and Qiu (2011) indicated that a larger value of smoothing length h would enhance the accuracy of the result by increasing the interaction between the given particle and neighbouring particles within the support domain, and vice versa. The continuity equation and momentum equation considered in this work can be expressed as follows:

$$\frac{D\rho}{Dt} = -\rho\frac{\partial u^\beta}{\partial x^\beta} \tag{8.107}$$

$$\frac{Du^\alpha}{Dt} = -\frac{1}{\rho}\frac{\partial \sigma^{\alpha\beta}}{\partial x^\beta} + g^\alpha \tag{8.108}$$

where u is the velocity vector, ρ is the density, g is the acceleration, $\sigma^{\alpha\beta}$ is the stress tensor and α and β denote the directions of the coordinates. The stress tensor in Eq. (8.108) could be expressed as:

$$\sigma^{\alpha\beta} = -p\delta^{\alpha\beta} + \tau^{\alpha\beta} \tag{8.109}$$

where p is the isotropic pressure. According to the works by Bui et al. (2008), the isotropic pressure p could be expressed as

$$\mathrm{p} = -K\varepsilon_v = K\frac{\rho}{\rho_{init}} - K \tag{8.110}$$

where K is the bulk modulus, ε_v is the volumetric strain and ρ_{init} is the initial density. The shear stress tensor $\tau^{\alpha\beta}$ in Eq. (8.109) is proportional to the shear strain rate and viscosity given by

$$\tau^{\alpha\beta} = \mu\varepsilon^{\alpha\beta} \tag{8.111}$$

where μ is the viscosity of water and $\varepsilon^{\alpha\beta}$ is the shear strain rate. The shear strain rate in Eq. (8.111) could be expressed as the resultant of the elastic and plastic shear strain rate as

$$\varepsilon^{\alpha\beta} = \varepsilon_e^{\alpha\beta} + \varepsilon_p^{\alpha\beta} \tag{8.112}$$

Using the SPH formulation, an in-house Fortran 95 program has been developed for a trial test, and a slope failure in Hong Kong was chosen for study. GEO (1996) conducted an investigation and prepared a report, named GEO report No. 178, for the Shum Wan Road landslide in 1995 through a series of geological surveying, ground investigation works and laboratory tests. The results indicated that this landslide was induced by the local shear strength reduction in the clay seam layer in a concave scar area, since about 380 mm rainfall was recorded in the 30 hours before the slope sliding, and the post-failure slope is shown in Figure 8.22. The SPH model for this problem is shown in Figure 8.23.

Since the SPH method relies on the interaction between particles, the accuracy of analysis would decrease due to an insufficient number of particles at the edge. Libersky et al. (1993) suggested the use of ghost particles for reflecting the boundary condition. During the analysis, it is unavoidable that some particles of T7 and T11 will move backwards, since there are no fixed particles to restraint them, which will affect the accuracy of the analysis. In order to solve this problem, Adami (2014) suggested additional pressure on the fixed particles in order to prevent the leakage of particles during the simulation, whereas Yim (2007) recommended to use ghost particles to surround the edge of model. In this problem, ghost particles are adopted to form a symmetry plane to prevent undesirable backward movement of particles. The symmetry plane is formed according to the coordinate of particle 3751, which is the highest particle before simulation in the model which is at the right edge of domains T3, T7 and T11 in Figure 8.23.

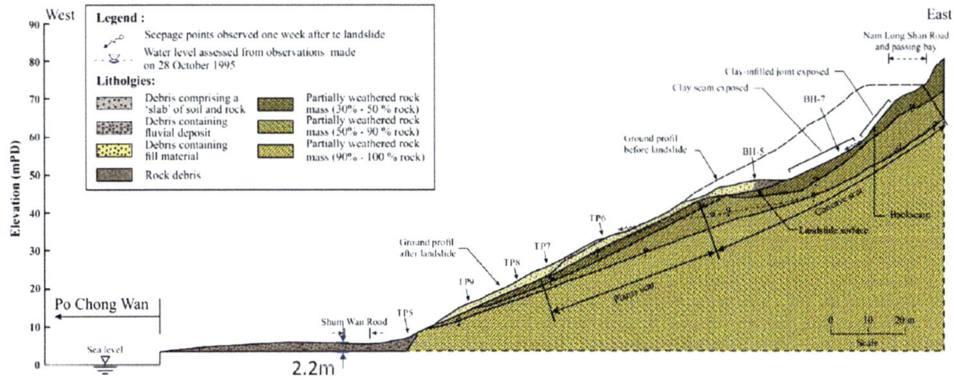

Figure 8.22 Section view of post-failure profile in Shum Wan Road landslide (GEO, 1996).

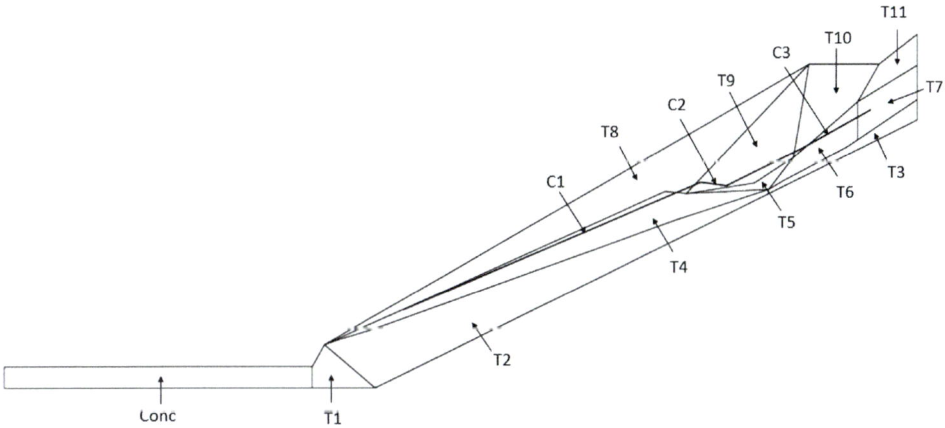

Figure 8.23 SPH analysis of the Shum Wan Road slope failure.

In order to simulate the self-weights of the tuff and clay seam in the analysis, body force is prescribed in the z direction for the SPH particles. This load is basically affected by the acceleration, load factor and time. The load factor will automatically multiply the defined acceleration and generate a constant gravity load through the analytical process. Time step is used to maintain the stability of numerical simulation, and more accurate solutions will be obtained with a shorter time step. The time required for computation ,however, also increases due to the increase in the solution steps. In this model, the time step is taken as 0.9 s, while the termination time is considered as 30 s. Several cases with different soil parameters have been considered for this problem, and the results for case 4 appear to match better with the field measurement results. The results of the analysis are shown in Figure 8.24. In general, the results from SPH can match well with the post-failure condition of the slope, and such results are not possible with the use of LEM, DLO or SRM.

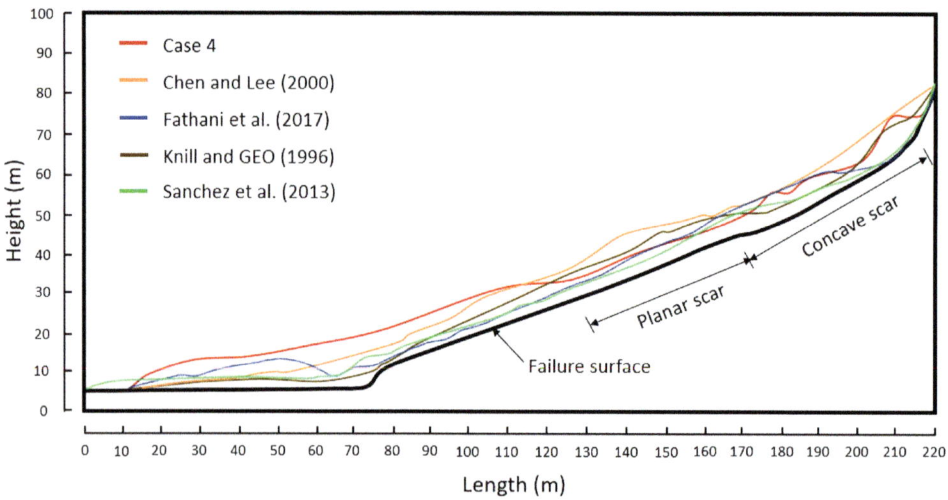

Figure 8.24 SPH simulated post-failure profiles compared with other results and field measurement results.

8.7.2 Materials point method (MPM)

The material point method also belongs to the group of meshless methods. The MPM was first published by Sulsky et al. (1994) for simulating the behaviour of continuum material and is an extension of the particle-in-cell (PIC) method developed by Harlow at Los Alamos National Laboratory in 1957, which is an ancestor of MPM and has adopted both Lagrangian and Eulerian descriptions to solve the problems in fluid dynamics. Sulsky et al. (1994) reformulated the PIC and developed the MPM. In MPM, the material body is discretized into a finite number of particles called "material points" in Lagrangian description and a fixed Eulerian background grid provides a medium for determining the spatial gradient of the particles, and for solving the equations for the conservation of mass, momentum and energy and mapping back the results to the particles (Sulsky et al., 1994). Ma et al. (2009) compared the performance and features of MPM with another meshless method called smoothed particle hydrodynamics (SPH) and found MPM to be prominent in the aspects of formulation, neighbour searching, consistency of the shape function, ability to deal with the tensile instability, application of boundary conditions and the time consumption of each time step. MPM has shown its capability and efficiency on solving the engineering problems with large deformation. MPM is a kind of meshless method which discretizes the material into a group of material points moving through a Eulerian background grid. All material properties such as mass, velocity, stress, strain and state variables will be stored in the material points, which allows the method to track material interfaces easily and to establish history-dependent constitutive models. No mesh distortion or element entanglement will be generated in the MPM as the Eulerian background grid is employed to solve the equations of motion of the material points (Figure 8.25). This makes the MPM a better technique to simulate the large deformation of material.

The Lagrangian description as shown in Figure 8.25 is one of the fundamental approaches for describing the motion of material. The material properties such as mass

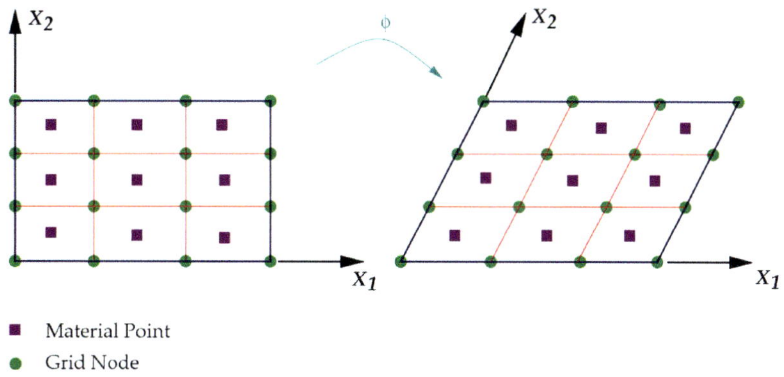

Figure 8.25 Lagrangian description or material description.

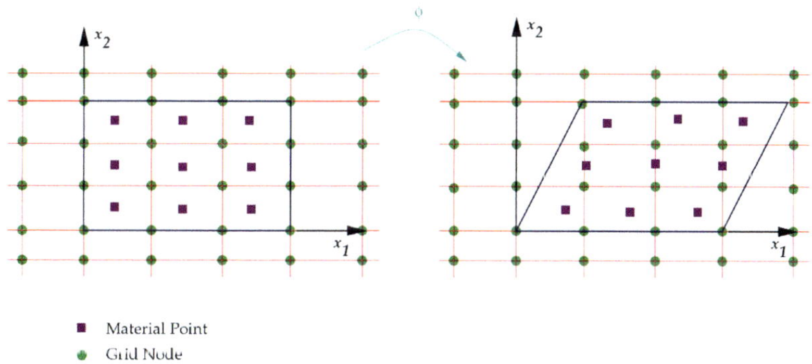

Figure 8.26 Eulerian description or spatial description.

and velocity can be determined by following each individual particle as it travels through space and time. In FEM, a mesh is embedded into the targeted material and each grid will deform with the material (Zhang et al., 2017). Since there is no convective effect, the boundary condition can be automatically imposed. It is easy to track the material interface in Lagrangian description and it is suitable for constructing a history-dependent model. However, serious element distortion, as shown in Figure 8.25, may cause significant numerical errors, or more seriously, a negative element area may lead to the termination of the computation. Although the problem of element distortion can be solved by remeshing and remapping the results to the new grid, a further numerical error may be induced during the interpolation process (Zhang et al., 2017).

In contrast with the Lagrangian description, the Eulerian description, as shown in Figure 8.26, provides a background grid fixed in space, and the grid will not deform when the material flows through it (Zhang et al., 2017). Since there is no element distortion, it is appropriate to model the material with a large fracture or deformation. However, the Eulerian approach only calculates the material quantities of the particles within the material but cannot accurately determine the location of the material interface or moving boundaries (Zhang et al., 2017). The particles are difficult to track in a Eulerian approach and it is not suitable for constructing the history-dependent model. The background grid will not deform with the material, hence it can be used to locate the material point in space.

According to Zhang et al. (2017), the area of the solid body is represented by a material domain Ω. The material domain Ω will be discretized into a set of material points moving through the Eulerian background grid. Each material point represents a subdomain Ωp and the material properties including the mass, momentum, energy, strain, stress and internal state variables will be stored in the material point for history-dependent modelling. The MPM computational cycle can be separated into two phases, which are the Lagrangian phase and the Eulerian/convective phase. In the Lagrangian phase, the grid acts as the mesh in a finite element method where the material points are embedded on the plane of the mesh. Meanwhile, the mass and momentum will be mapped to the grid nodes using shape function, and the standard finite element formulation can be applied onto the mesh and solves the momentum equation of the material domain. In this stage, the grid acts as an updated Lagrangian frame and the material points will move following the deformed shape of the grid. Followed by the Eulerian phase, the grid will be reset to the original location while the material points will remain in the updated location within a time step and prepare for the next computational cycle.

There are a total of four steps in the computational cycle of MPM, and these steps are repeated in each cycle. Steps (a)–(c) belong to the Lagrangian phase and step (d) belongs to the Eulerian phase. The details in each step are as follows:

Step (a) – The particle mass and momentum will be mapped to the grid nodes of the grid where the particle stays by using shape function.
Step (b) – The velocity and acceleration at the grid nodes will be calculated.
Step (c) – The position and velocity of the particle will be mapped from the grid nodes by using the shape function.
Step (d) – The deformed grid will be reset to the original position.

The shape function is employed to interpolate the material properties and variables (e.g. mass and momentum) between the grid nodes and the material point. The particles are attached to the grid in MPM and the spatial coordinates x_{ip} of a particle p can be interpolated from the grid nodal spatial coordinates x_{iI} as

$$x_{ip} = N_{Ip} x_{iI}, \quad \mathrm{i} = 1,2,3 \ldots \mathrm{n}_g \tag{8.113}$$

where the subscript I denotes the variables associated with the grid nodal point I, the subscript p denotes the variables associated with the particle p and n_g is the total number of grid nodal points related to the particle p. The governing equations will be:

Conservation of mass

$$\rho J = J_0 \tag{8.114}$$

Conservation of momentum

$$\sigma \bullet \nabla + \rho b = \rho \dot{v} \tag{8.115}$$

Conservation of energy

$$\dot{\rho e} = D : \sigma + \rho s + \nabla(k\nabla T) \tag{8.116}$$

Conservation of equation

$$\sigma^{v} = \sigma^{v}(D, \sigma, ...) \tag{8.117}$$

Rate of deformation

$$dv = v(x + dx, t) - v(x, t) = \text{L}\cdot\text{dx} \quad \text{and } \text{L} = \frac{\partial v}{\partial x} \tag{8.118}$$

$$\text{D} = (\text{L} + \text{L}^{\text{T}})/2$$

Boundary conditions: prescribed traction or prescribed velocity
Initial conditions: prescribed velocity or prescribed displacement

A Fortran 95 program using MPM has been developed by Cheng, and this program has been used to study the landslide on Shum Wan Road. The reliability and validity of the MPM on analysing the large deformation of the progressive slope failure will be assessed by comparing the final failure configuration generated by computer analysis based on MPM to the profile of the slope on Shum Wan Road after progressive failure. In the MPM simulations, the targeted object is discretized into a finite number of particles. The conservation of mass and momentum will then be interpolated between the particles and the grid nodes. Therefore, the slope should first be discretized into particles for executing the computer simulation. The ID of the first particle, the x-, y- and z-coordinates of the four corners of a trapezoid, the number of portions needed to be divided along the x-, y- and z-directions, the material number and the mass of the particles have to be defined. Since the number of particles on each pair of opposing edges in a trapezoid is equal, the slope profile was divided into several trapezoids for particle generation to ensure the uniformity of the distribution of the particles and to prevent the particles from being over-concentrated or over-dispersed on the short or long edge, respectively (Figure 8.27).

For slope failure simulation with soil only, the results of the MPM analysis are shown in Figure 8.28. By comparing the final failure profile of the slope in the two cases above, a relatively smooth and distinct circular failure zone can be observed in the slope with more particles. On the other hand, the particles in the slope with less particles tend to have a vertical collapse during the failure process and the height of the failure profile of the slope with less particles is comparatively low in contrast with the one with more particles. The reason for the vertical collapse may be the loosely spaced particles, with large voids between the particles. In addition, there is a separation between the soil particles at the bottom of the slope due to the unbalanced and non-uniform distribution of load distribution at the top of the slope. These phenomena demonstrate that the accuracy of the slope stability simulation increases by increasing the number of particles.

For slope failure simulations with tuff underneath the soil (case 2), the results of the analysis are shown in Figure 8.29. The final profile at the end of the analysis for case 3 is

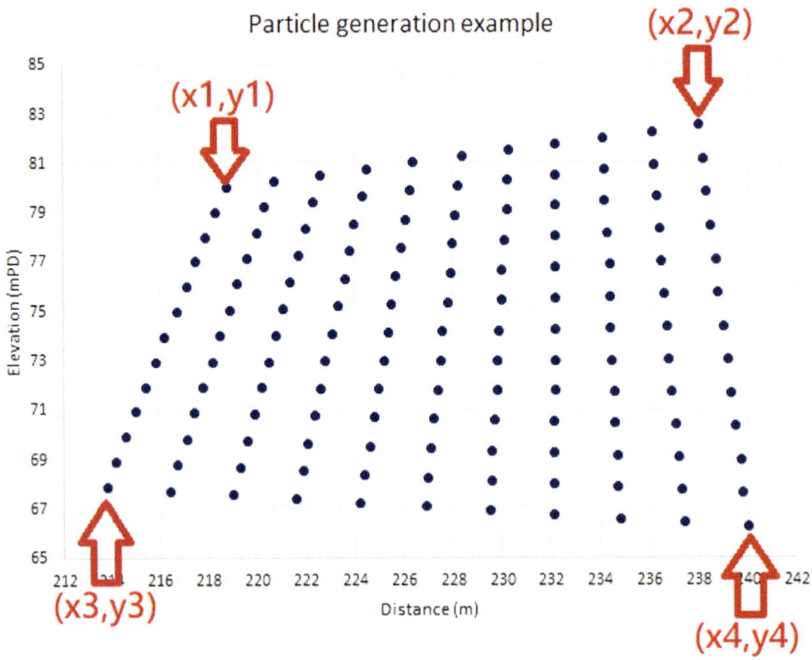

Figure 8.27 An example of particle generation.

shown in Figure 8.30. The authors have carried out different combinations of geological conditions for these problems, and the results for other cases will not be shown here.

The authors would like to conclude that while the meshless method can be a useful tool for the assessment of the slope failure process, there are actually many limitations to these methods for real engineering. From the debris flow tests by Cheng et al. (2019), it is observed that segregation will occur during the flow process. Such a segregation phenomenon keeps developing through the whole flow process, and the authors do not believe that any macro-constitutive model can model such a phenomenon for a general case. Such segregation has been seen for many debris flow cases in Hong Kong and other places, and the examples in applying meshless methods to real problems as shown in this chapter can only be viewed as approximate/qualitative analyses of the actual flow process. In addition, for a slope with a soft band which is only 2–5 mm in thickness, the use of meshless methods encounters many practical and technical difficulties, and the situation can be even worse than that with the FEM. On the other hand, a soft band of even 1 mm in thickness is not a problem in SLOPE2000 by Cheng.

8.8 OPTIMIZATION ANALYSIS IN GEOTECHNICAL ENGINEERING

In the ultimate limit state analysis of many geotechnical problems, the final solution will be given by the critical solution of the governing equations. Basically, there are two major types of optimization methods: continuous smooth function optimization by gradient

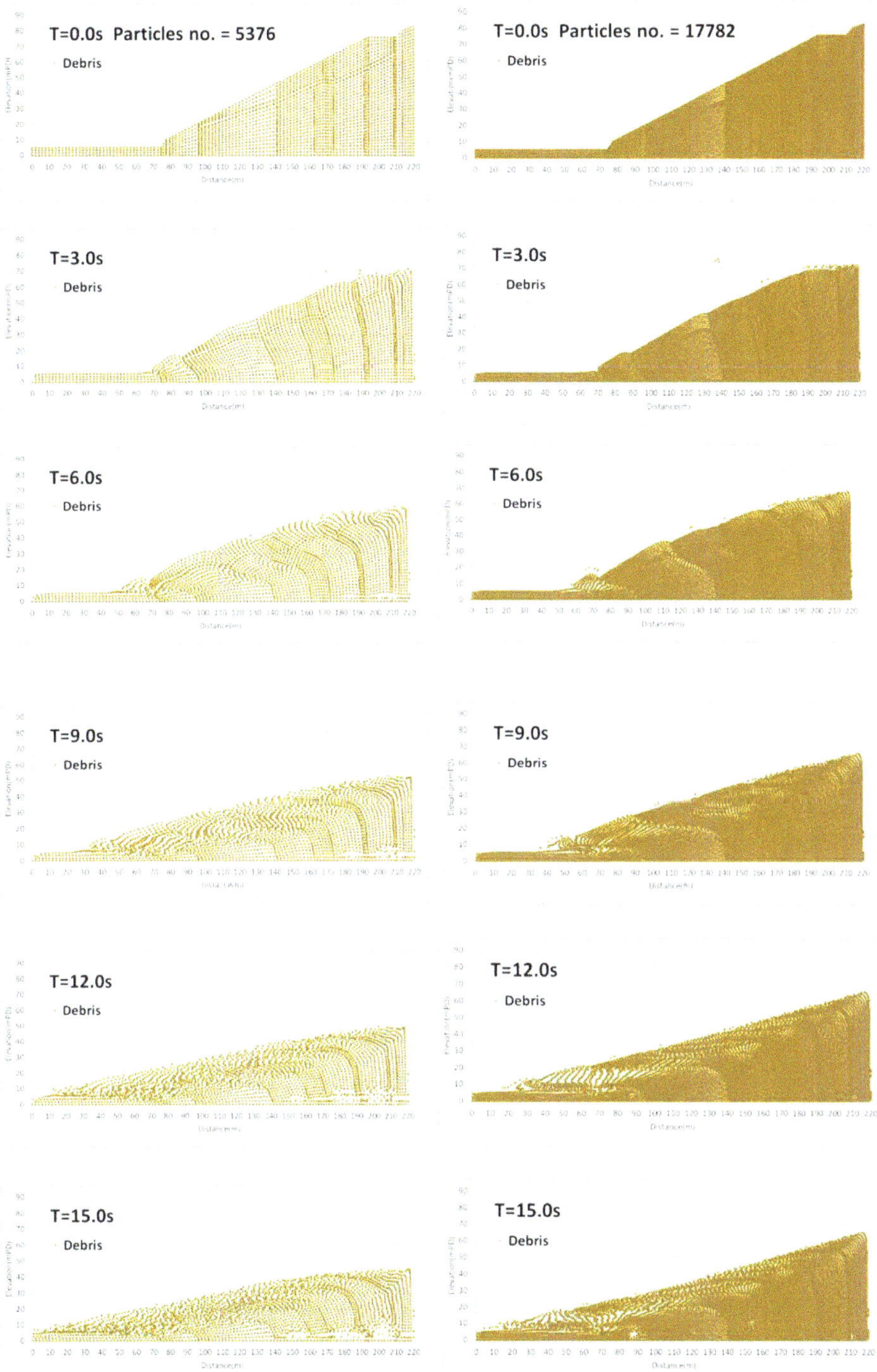

Figure 8.28 Sequence of slope failure for slope with 5376 (Case 1-1) (left) and 17782 (Case 1-2) (right) particles.

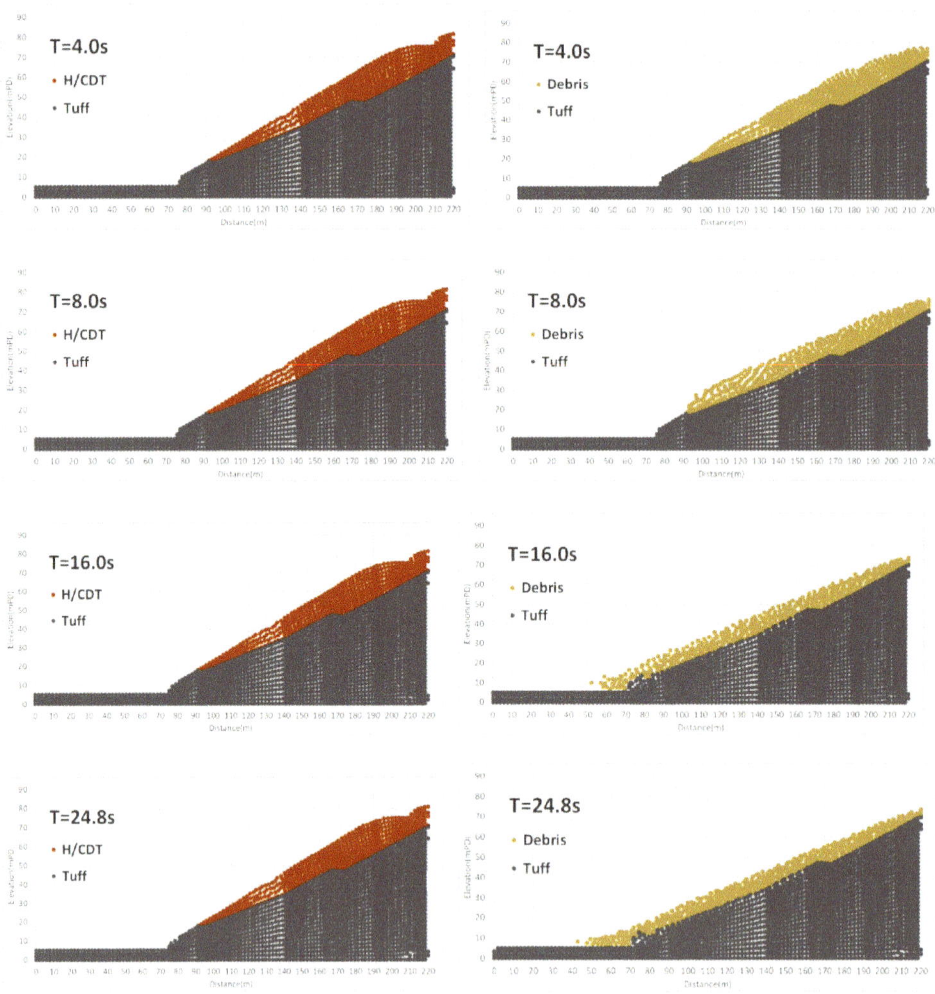

Figure 8.29 Slopes failure simulations by using H/CDT (Case 2-1) (left) and debris (Case 2-2) (right) with E = 12.5 MPa and v = 0.35.

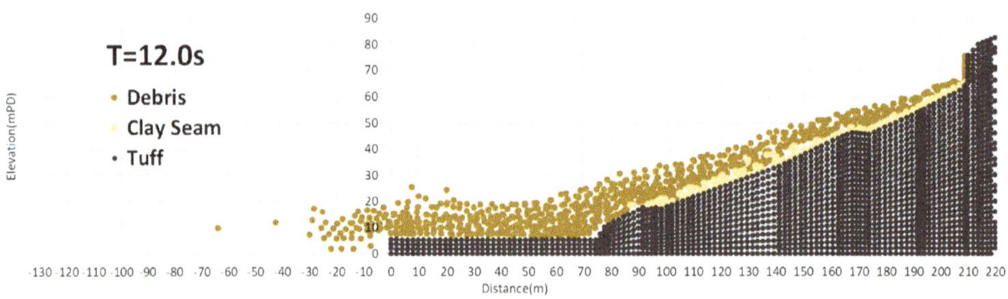

Figure 8.30 Slope failure profile in Case 3-3 at T = 12.0 s.

type methods and general stochastic optimization problem for non-smooth discontinuous functions. The optimization analysis for slope stability and pile-driving backward analysis are typical problems which require optimization analysis to give the best/critical solution.

There are many types of gradient-type optimization methods which are well covered in many texts in engineering mathematics. In this section, the Powell and the DFP (Davidon-Flefchen-Powell) methods are discussed (used by Cheng), and the advantages and limitations of these two methods are discussed.

8.8.1 Powell method

The procedures for searching the extreme value of a function by Powell's method are:

1. For a trial vector with n variables, start with an initial trial z_0, and set the trial vector $S_i = (0,0,\ldots 1,.0)^T$, and 1 start at position i.
2. For iteration i, carry out a one-dimensional search and let the search vector $p_i = S_i$, and minimize $F(z_{i-1} + \alpha_i p_i)$ and let $z_i = z_{i-1} + \alpha_i p_i$, then set $i = i + 1$ and repeat the above process until $i = n$.
3. Take $p_i = p_{i+1}$, $i = 1,2,\ldots n - 1$ and let $p_n = z_n - z_0$.
4. Along direction vector $z_n - z_0$, minimize $F(z_n + \alpha_n p_n)$ and let $z_0 = z_n + \alpha_n p_n$, then go to step 2.
5. Repeat the above process n times so that n conjugate directions have undergone a one-dimensional search. The following criterion has then to be checked:

$$|\alpha_n p_n| \le \xi \tag{8.119}$$

where ξ is the precision for search.

6. If the above termination criterion is not satisfied, then repeat step 1 again.

8.8.2 DFP method

Powell's method performs well but is relatively inefficient when the number of control variables is large. For a large-scale optimization problem, the extreme point (local minimum) can be obtained by differentiation of the objective function $F(z_1,\ldots z_n)$, with the vector G given by Eq. (8.120) being a zero vector.

$$G = \left(\frac{\partial F}{\partial z_1}, \frac{\partial F}{\partial z_2}, \ldots, \frac{\partial F}{\partial z_n} \right) \tag{8.120}$$

The Hessian matrix H composed of the second derivatives of function F is required to be positive definite for the extreme point. The Hessian matrix is given by

$$H = \begin{bmatrix} \frac{\partial^2 F}{\partial z_1^2} & \frac{\partial^2 F}{\partial z_1 \partial z_2} & \cdots & \frac{\partial^2 F}{\partial z_1 \partial z_n} \\ & \frac{\partial^2 F}{\partial z_2^2} & \cdots & \frac{\partial^2 F}{\partial z_2 \partial z_n} \\ & symmetric & \ddots & \vdots \\ & & & \frac{\partial^2 F}{\partial z_n^2} \end{bmatrix} \tag{8.121}$$

The DFP method determines the true value of the Hessian matrix by a series of iteration steps. The direction search vector S based on the gradient of F can be approximately determined as

$$S_n = -A_n G_n \tag{8.122}$$

In step $n + 1$, A_{n+1} is given by

$$A_n = A_n + C_n - D_n, \text{ in which } n = 0,1,2,\ldots, \tag{8.123}$$

$$C_n = \frac{\Delta z_n \Delta z_n^T}{\Delta z_n^T Y_n}; \; D_n = \frac{A_n Y_n \left(A_n Y_n\right)^T}{Y_n^T A_n Y_n}$$

$$Y_n = G_{n+1} - G_n; \quad \Delta z_n = z_{n+1} - z_n \tag{8.124}$$

In general, the DFP method is more efficient as compared with Powell's method for a large-scale problem. The main idea of the DFP method is to make $A_0, A_1, \ldots, A_n$ approach the inverse of the Hessian matrix through iteration steps close to the extreme point. Generally A_0 is usually taken as the unit matrix for simplicity. For the Powell's and DFP methods, they are available in many open or commercial mathematics libraries and can be used with ease. The Powell or DFP functions are also available in many Fortran or C libraries. A Fortran code for GRG optimization analysis is given by Cheng et al. (2020). The advantage of this code is that the differential of the objective function is formed by finite difference instead of explicit expression, which is useful for many geotechnical problems.

For non-smooth discontinuous function optimization, the most popular method at present is the heuristic optimization method. Cheng has adopted seven types of heuristic optimization methods to develop Slope2000, and very good results have been obtained through these methods, which have been illustrated in Chapter 7.

Seven types of heuristic optimization methods can be chosen by the uses for Slope2000:

1. Simulated annealing;
2. Harmony search (original and improved);
3. Particle swarm search (original and improved);

4. Genetic search;
5. Ant-colony search;
6. Coupled search (harmony and particle swarm);
7. Tabu search.

Currently, there are hundreds of heuristic optimization methods available to engineers. Based on Cheng's experience, no single method can outperform other methods in all cases. Of these optimization methods, only limited methods require initial solutions, while most of the other methods do not require an initial solution. Actually, the availability of an initial solution has the advantage that an approximate solution can be obtained with limited slices for refined optimization analysis with more slices. Slope2000 has taken this into the design of the program in some of the optimization options. The details of different types of heuristic optimization methods can be referred to in the works by Cheng et al. (2020) and will not be repeated here.

References

Adami S (2014), *Modeling and simulation of multiphase phenomena with smoothed particle hydrodynamics*, Doktor-Ingenieurs, Technischen Universität München.

Aluru N (2000), A point collocation method based on reproducing kernel approximation, *International Journal Numerical Methods in Engineering*, 47(6), 1083–1121.

Atluri SN and Shen SP (2002) The Meshless Local Petrov-Galerkin (MLPG) Method: A simple & less-costly alternative to the finite element and boundary element methods. *CMES*, 3, 11–51.

Babuska I and Melenk JM (1997), The partition of unity method, *International Journal Numerical Methods in Engineering*, 40, 727–758.

Belinha J, Dinis LMJS, and Jorge RMN (2009), The natural neighbour radial point interpolation method: dynamic applications. *Engineering Computations*, 26(8), 911–949, https://doi.org/10.1108/02644400910996835

Belytschko T, Lu YY and Gu L (1994), Element-free Galerkin methods, *International Journal Numerical Methods in Engineering*, 37 (1994), 229–256.

Bui HH, Fukagawa R, Sako K and Ohno S (2008), Lagrangian meshfree particles method (SPH) for large deformation and failure flows of geomaterial using elastic–plastic soil constitutive model, *International Journal for Numerical and Analytical Methods in Geomechanics*, 32, 1537–1570.

Chen W and Qiu T (2011), Numerical simulations of granular materials using smoothed particle hydrodynamics method, *International Journal of Geomechanics*, 12(2), 157–164

Cheng YM (1998), Advancement and improvements in discontinuous deformation analysis, *Computers and Geotechnics*, 22(2), 153–163.

Cheng YM, Zhang YH and Chen WS (2002), Wilson non-conforming element in numerical manifold method, *Communications in Numerical Methods in Engineering*, 18, 877–884.

Cheng YM and Zhang YH (2003), Application of NMM in underground excavation, Proceedings of GeoEng2003, Beijing, China, 1159–1164.

Cheng YM and Zhang YH (2004), Application and discussion of three dimensional numerical manifold method based on hexahedron element, *Chinese Journal of Rock Mechanics and Engineering*, 1745–1754.

Cheng YM and Zhang YH (2008), Three-dimensional numerical Manifold Method–Tetrahedron and Hexahedron Mesh, *International for Rock Mechanics and Rock Engineering*, 41(4), 601–628.

Cheng YM, Li N, Fung WH and Li L (2019), Laboratory and Field Test and Distinct Element Analysis of Debris Flow, *Natural Hazards and Earth System Sciences*, 19, 181–199.

Cheng YM, Wang JH, Li L and Fung WH (2020), *Frontier in civil engineering vol.4 Numerical methods in geotechnical engineering Part I*, Benjamin Press, 2020.

Cheng YM, Wang JH, Li L and Fung WH (2020), *Frontier in civil engineering vol.4 Numerical methods in geotechnical engineering Part II*, Benjamin Press, 2020.

Chiou YJ, Tsay RJ and Chuang WL (1997), Crack propagation using manifold method. *Proceedings of the Second International Conference on Analysis of Discontinuous Deformation*, Kyoto, Japan, pp. 298–308.

Chiou YJ and Tsay RJ (2000), Nonlinear analysis of low-rise reinforced concrete shear walls by manifold method, *Journal of the Chinese Institute of Engineers*, 23(6), 721–729.

Chiou YJ, Lee YM and Tsay RJ (2002), Mixed mode fracture propagation by manifold method. *International Journal of Fracture*, 114(4), 327–347.

Crouch SL and Starfield AM (1983), *Boundary element methods in solid mechanics*, In S. L. Crouch and A. M. Starfield, George Allen & Unwin, London.

Cundall PA (1971), A computer model for simulating progressive, large-scale movements in blocky rock systems. In: *Proceedings of the international symposium on rock mechanics*. Nancy, France: [s.n.], 1971, pp. 129–136.

Cundall PA (1978), *Ball–A computer program to model granular medium using the distinct element method, Technical note TN-LN-13*, Advanced Technology Group, Dames and Moore, London, pp. 129–163.

Cundall PA (1980) *UDEC-A generalized distinct element program for modelling jointed rock*, Final Tech. Rep. Eur. Res. Office (US Army Contract DAJA37-79-C-0548).

Cundall PA and Hart RD (1985) *Development of generalized 2-D and 3-D distinct element programs for modeling jointed rock*, Misc. Paper SL-85-1, US Army Corps of Engineers.

Cundall PA and Strack ODL (1979), A discrete model for granular assemblies, *Geotechnique*, 29(1), 47–65.

Duarte CA and Oden JT (1996), h-p clouds–an h-p meshless method, Numerical Methods, *Partial Differential Equations*, 12(6), 673–705.

Eberhardt E (2003), *Rock slope stability analysis, utilization of advanced numerical techniques, Earth and Ocean Science at UBC*, University of British Columbia.

Geotechnical Engineering Office (1996), *Report on the Shum Wan road landslide of 13 August 1995*, HKSAR Government, Hong Kong.

Gingold R. and Monaghan JJ (1977), Smoothed particle hydrodynamics: Theory and application to non-spherical stars, *Monthly Notices of the Royal Astronomical Society* 181(1977), 375–389.

Goodman RE (1976), *Methods of geological engineering in discontinuous rocks*. West Publishing Company, San Francisco, CA, USA.

Hazzard JF, Maxwell SC and Young RP (1998), Micromechanical modelling of acoustic emissions, In *Proceedings of ISRM/SPE Rock Mechanics in Petroleum Engineering*, Eurock 98, Trondheim, Norway, pp. 519–526, SPE 47320.

Itasca Consulting Group Inc. (1995), PFC3D 1.0, User Guide, US.

Itasca Consulting Group Inc. (2004), PFC3D 3.1, User Guide, US.

Koo CY and Chern JC (1996), The development of DDA with third displacement function. *DDA and simulations of discontinuous media*, TSI press, pp. 342–349.

Libersky LD, Petschek AG, Carney TC, Hipp JR, Allahdadi FA (1993), High strain lagrangian hydrodynamics: A three-dimensional SPH code for dynamic material response, *Journal of Computational Physics*, 109(1), 67–75.

Lin DZ and Mo HH (1994) *Manifold method of material analysis*. The University of Oklahoma, US.

Liu GR and Liu MB (2003), Smoothed particle hydrodynamics: A meshfree particle method. World Scientific Publishing, Singapore.

Liu WK, Jun S and Zhang Y (1995), Reproducing kernel particle methods, *International Journal for Numerical Methods in Engineering*, 20, 1081–1106.
Liu WK, Li S and Belytschko T (1997), Moving least square reproducing kernel methods part i: Methodology and convergence, *Computer Methods in Applied Mechanical Engineering*, 143(1–2), 113–154.
Lucy LB (1977), A numerical approach to the testing of the fission hypothesis, *Astronomical Journal*, 82(12), 1013–1024.
Ma S, Zhang X and Qiu, XM (2009), Comparison study of MPM and SPH in modeling hypervelocity impact problems, *International Journal of Impact Engineering*, 36(2), 272–282.
Monaghan, JJ, and Lattanzio JC (1985), A refined particle method for astrophysical problems, *Astronomy and Astrophysics*, 149, 135–143.
Nayroles B, Touzot G and Villon P (1992), Generalizing the finite element method: diffuse approximation and diffuse elements, *Computational Mechanics*, 10, 307–318.
Ohtsubo H, Suzuki K, Terada K and Nakanishi K (1997), Utilization of finite covers in the manifold method for accuracy control. *Proceedings of the Second International Conference on Analysis of Discontinuous Deformation*, Kyoto, Japan, pp. 317–322.
Oñate E and Idelsohn S (1998), A mesh-free finite point method for advective-diffusive transport and fluid flow problems. *Computational Mechanics*, 21(4–5), 283–292.
Poulos HG and Davis EH (1974), *Elastic Solutions For Soil and Rock Mechanics*, John Wiley.
Qiu XJ (1996), Manifold method without use of penalty springs. *Proceedings of the First International Forum on Discontinuous Analysis (DDA) and Simulations of Discontinuous Media*, TSI Press, Berkeley, California, USA, pp. 205–249.
Shi GH (1988), *DDA-a new numerical model for the static and dynamics of block system*. Ph.D. thesis, University of California, 1988.
Shi, GH (1993), Block system modelling by discontinuous deformation analysis. *Computational Mechanics*, 11.
Shi GH and Goodman RE (1985), Two dimensional discontinuous deformation analysis. *International Journal for Numerical and Analytical Methods in Geomechanics*, 39, 541–556.
Shyu K and Salami MR (1995) *Manifold with four-node isoparametric finite element method*. Work Forum on Manifold Method of Material, California, USA. V1, pp. 165–182.
Strack ODL and Cundall PA (1984), *Fundamental studies of fabrics in granular materials, Interim report to National Science Foundation, NSF CEE-8310729*, Department of Civil and Mineral Engineering, University of Minnesota.
Sulsky D, Chen Z and Schreyer H (1994), A particle method for history-dependent materials, *Computer Methods in Applied Mechanics and Engineering*, 118(1–2), 179–196.
Thornton C (2015), *Granular dynamics, contact mechanics and particle system simulations*, Springer, Switzerland.
Tsay RJ, Chiou YJ and Chuang WL (1999), Crack growth prediction by manifold method. *Journal of Engineering Mechanics*, 125, 884–890.
Wang CY, Sheng J, Chen MH and Chuang CC (1995), *Approximation theories for the manifold method*. Vol. 1, Work Forum on Manifold Method of Material, California, USA, pp. 61–86.
Wang SL, Ge XR and Zhang G (1999), *Manifold method with four physical covers forming an element and its application*. The Third International Conference on Analysis of Discontinuous Deformation from Theory to Practice, Vail, Colorado, USA, pp. 193–201.
Yim SC (2007), *3-D Wave-Structure interaction with coastal sediments-a multi-physics/multi-solution-techniques approach*. Defense Technical Information Center, Corvallis.
Zhang GX, Sigiura Y and Hasegawa H (1997), Crack propagation and thermal fracture analysis by manifold method. *Proceedings of the Second International Conference on Analysis of Discontinuous Deformation*, Kyoto, Japan, pp. 282–297.

Zhang GX, Sugiura Y and Saito K (1999), Application of manifold method to jointed dam foundation. *The Third International Conference on Analysis of Discontinuous Deformation from Theory to Practice*, Vail, Colorado, USA, p. 211–220.

Zhang X, Chen Z and Liu Y (2017), *The material point method: A continuum-based particle method for extreme loading cases*, Tsinghua University Press Computational Mechanics Series, China.

Zhou WY, Yang Q, and Kou XD (1997), Development of high order manifold method. *Proceedings of the Second International Conference on Analysis of Discontinuous Deformation*, Kyoto, Japan, pp. 274–281.

Zhou WY, Kou XD and Yang RQ (1999), Crack propagation by manifold method coupled with element free method. *The Third International Conference on Analysis of Discontinuous Deformation from Theory to Practice*, Vail, Colorado, USA, pp. 283–289.

Zhu T and Atluri SN (1998), A modified collocation method and a penalty formulation for enforcing the essential boundary conditions in the element free Galerkin method, *Computational Mechanics*, 21(3), 211–222.

Zienkiewicz OC, Taylor RL, Zhu JZ (2011), *The finite element method: Its basis and fundamentals*, 6th edition, Butterworth Heinemann, London.

Index